FORMULAS FROM ALGEBRA

Exponents

$$a^m a^n = a^{m+n}$$

$$(a^m)^n = a^{mn}$$

$$\frac{a^m}{a^n} = a^{m-n}$$

$$(ab)^n = a^n b^n$$

$$\left(\frac{a}{b}\right)^n = \frac{a^n}{b^n}$$

Radicals

$$(\sqrt[n]{a})^n = a$$

$$\sqrt[n]{a^n} = a, \text{ if } a \geq 0$$

$$\sqrt[n]{ab} = \sqrt[n]{a}\,\sqrt[n]{b}$$

$$\sqrt[n]{\frac{a}{b}} = \frac{\sqrt[n]{a}}{\sqrt[n]{b}}$$

Logarithms

$$\log_a MN = \log_a M + \log_a N$$

$$\log_a (M/N) = \log_a M - \log_a N$$

$$\log_a (N^p) = p \log_a N$$

Factoring Formulas

$$x^2 - y^2 = (x-y)(x+y)$$

$$x^3 - y^3 = (x-y)(x^2 + xy + y^2)$$

$$x^3 + y^3 = (x+y)(x^2 - xy + y^2)$$

$$x^2 + 2xy + y^2 = (x+y)^2$$

$$x^2 - 2xy + y^2 = (x-y)^2$$

$$x^3 + 3x^2 y + 3xy^2 + y^3 = (x+y)^3$$

Binomial Formula

$$(x+y)^n = {}_nC_0 x^n y^0 + {}_nC_1 x^{n-1} y^1 + \cdots + {}_nC_{n-1} x^1 y^{n-1} + {}_nC_n x^0 y^n$$

Quadratic Formula

The solutions to $ax^2 + bx + c = 0$ are $x = \dfrac{-b \pm \sqrt{b^2 - 4ac}}{2a}$

Complex Numbers

Multiplication: $(a+bi)(c+di) = (ac - bd) + (ad + bc)i$

Polar form: $a + bi = r(\cos\theta + i\sin\theta)$ where $r = \sqrt{a^2 + b^2}$

Powers: $[r(\cos\theta + i\sin\theta)]^n = r^n(\cos n\theta + i\sin n\theta)$

Roots: $\sqrt[n]{r}\left[\cos\left(\dfrac{\theta + k\cdot 360°}{n}\right) + i\sin\left(\dfrac{\theta + k\cdot 360°}{n}\right)\right]$ $k = 0, 1, 2, \cdots, n-1$

THIRD EDITION

Algebra and Trigonometry

A PROBLEM-SOLVING APPROACH

Walter Fleming

Hamline University

Dale Varberg

Hamline University

Prentice Hall
Englewood Cliffs, New Jersey 07632

Library of Congress Cataloging-in-Publication Data

Fleming, Walter.
   Algebra and trigonometry: a problem-solving approach / Walter
Fleming, Dale Varberg. —3rd ed.
     p. cm.
   Includes index.
   ISBN 0-13-021338-1 : $26.00 (est.)
     1. Algebra.   2. Trigonometry.   I. Varberg, Dale.   II. Title.
QA154.2.F52 1988
512′.13—dc19

                                            87-31017
                                               CIP

Editorial/production supervision: Zita de Schauensee
Interior design: Christine Gehring-Wolf
Cover design: Maureen Eide
Manufacturing buyer: Paula Massenaro
Computer art for cover created by GENIGRAPHICS

Printed in the United States of America
10 9 8 7 6 5 4 3 2

Credits for quotations in text: Page 1: Philip J. Davis, "Number," in *Scientific American* (Sept. 1964),
p. 51.   Page 2: Ronald W. Clark, *Einstein: The Life and Times* (New York: World Publishing Co.,
1971), p. 12.   Page 17: G. H. Hardy: *A Mathematician's Apology* (New York: Cambridge University
Press, 1941).   Page 52: Harry M. Davis, "Mathematical Machines," in *Scientific American* (April
1949).   Pages 104 and 334: George Polya, *Mathematical Discovery* (New York: John Wiley & Sons,
1962), pp. 59 and 6-7, respectively.   Page 110: Howard Eves, *An Introduction to the History of
Mathematics* (New York: Holt, Rinehart & Winston, 1953), p. 32.   Page 137: Morris Kline, *Mathematical Thought from Ancient Times* (New York: Oxford University Press, 1972), p. 302.   Page 167: I.
Bernard Cohen: "Isaac Newton," in *Scientific American* (Dec. 1955), p. 78.   Page 188: Morris Kline,
"Geometry," in *Scientific American* (Sept. 1964), p. 69.

ISBN   0-13-021338-1   01

Prentice-Hall International (UK) Limited, *London*
Prentice-Hall of Australia Pty. Limited, *Sydney*
Prentice-Hall Canada Inc., *Toronto*
Prentice-Hall Hispanoamericana, S.A., *Mexico*
Prentice-Hall of India Private Limited, *New Delhi*
Prentice-Hall of Japan, Inc., *Tokyo*
Prentice-Hall of Southeast Asia Pte. Ltd., *Singapore*
Editora Prentice-Hall do Brasil, Ltds., *Rio de Janeiro*

Contents

George Polya

A great discovery solves a great problem but there is a grain of discovery in the solution of any problem. Your problem may be modest; but if it challenges your curiosities and brings into play your inventive faculties, and if you solve it by your own means, you may experience the tension and enjoy the triumph of discovery. Such experiences at a susceptible age may create a taste for mental work and leave their imprint on the mind and character for a life time.

—How to Solve It (p. v)

Solving a problem is similar to building a house. We must collect the right material, but collecting the material is not enough; a heap of stones is not yet a house. To construct the house or the solution, we must put together the parts and organize them into a purposeful whole.

—Mathematical Discovery (vol. 1, p. 66)

You turn the problem over and over in your mind; try to turn it so it appears simpler. The aspect of the problem you are facing at this moment may not be the most favorable. Is the problem as simply, as clearly, as suggestively expressed as possible? Could you restate the problem?

—Mathematical Discovery (vol. 2, p. 80)

We can scarcely imagine a problem absolutely new, unlike and unrelated to any formerly solved problem; but, if such a problem could exist, it would be insoluable. In fact, when solving a problem, we should always profit from previously solved problems, using their result, or their method, or the experience we acquired solving them. . . . Have you seen it before? Or have you seen the same problem in slightly different form?

—How to Solve It (p. 98)

An insect tries to escape through the windowpane, tries the same hopeless thing again and again, and does not try the next window which is open and through which it came into the room. A mouse may act more intelligently; caught in a trap, he tries to squeeze between two bars, then between the next two bars, then between other bars; he varies his trials, he explores various possibilities. A man is able, or should be able, to vary his trials more intelligently, to explore the various possibilities with more understanding, to learn by his errors and shortcomings. "Try, try again" is popular advice. It is good advice. The insect, the mouse, and the man follow it; but if one follows it with more success than the others it is because he varies his problem more intelligently.

—How to Solve It (p. 209)

These quotations are taken from George Polya, *How to Solve It*, Second Edition (Garden City, NY: Doubleday & Company, Inc., 1957) and George Polya, *Mathematical Discovery*, vols. 1 and 2 (New York: John Wiley & Sons, Inc., 1962).

Preface

Since publication of the second edition of this book, the mathematical community has lost one of its most gifted members. George Polya (1887–1985) was a researcher of first rank with some 250 published papers and several books on important mathematics. In addition, he was a great teacher, one who thought deeply about the essence of mathematics and how we learn it. In his books on mathematical pedogogy, he taught us that mathematics is preeminently the art of solving problems. And, better than anyone else, he described the strategies, illustrated the techniques, and explored the byways of this subject. The authors of this book are happy to acknowledge the influence of George Polya on their teaching and writing. In recognition of him, we cite on the previous page a number of quotations from his books. They capture something of the spirit of the man; they also say something about our philosophy in writing this book.

PROBLEM-SOLVING EMPHASIS

If emphasis on problem solving characterized earlier editions of this book, this emphasis is even more evident in the present edition. While we have rewritten parts of many sections to achieve greater clarity of exposition, most of our effort has gone into improving the problem sets. Every section of the book ends with an extensive problem set; and each of these is in two parts. First there is a collection of basic problems which have been carried over from the earlier edition with minor changes. These problems are quite straightforward and are meant to develop the skills and reinforce the main ideas of the section. Many of these problems are clustered around examples that appear within the problem set.

Following the basic problems, there is a set of miscellaneous problems. *These have been completely reworked for this edition.* Our aim for the miscellaneous problems is to give the student an exciting and challenging tour through the applications of the ideas of the section. We begin with a few easy review type problems, move in a carefully graded manner to harder and more interesting applications, and conclude with a *teaser* problem (about which we will say more shortly). We think both students and teachers will find our miscellaneous problems to be an outstanding feature of this edition.

THE TEASERS

Over our combined seventy years of teaching, we have collected a large number of intriguing problems. Many of them are our own creation; some are part of current mathematical lore; others come from mathematical history. As a special attraction for this edition, we have inserted one of these problems (identified as a *teaser*) at the end of each section. These teasers may appear difficult at first glance but in most cases become surprisingly easy when looked at the right way. In each case, the teaser relates to the ideas of the section in which it appears. As a group, the teasers form a collection of problems that we think would please even George Polya. We hope that you will page through the book with particular attention to the range and quality of the teasers.

How should the teasers be used in class? We suggest that teachers might select some of their favorite teasers and offer a prize to the student offering the best set of solutions at the end of the term. Or teachers might use these problems as the basis for a weekly problem-solving session (perhaps as an addition to the regular class sessions). Or a teacher might select from them a problem of the week offering a small prize for the best solution. Or they can simply be treated as extra stimulation for the very best students.

WRITING STYLE

We write as in our earlier books with the aim of being interesting and informative. This means that we address ourselves more to explaining than to proving, though we are careful to avoid logical sloppiness. Definitions are stated correctly but with a minimum of jargon. We do not prove the obvious and occasionally we omit the proof of something that is quite difficult. But we do help the student to see that mathematics is a logically coherent subject.

To capture student interest, we begin each section with a boxed display. This box may contain a historical anecdote, a famous quotation, an appropriate cartoon, a problem to solve, or a diagram illustrating a key idea of the section.

Calculators are now so common that we use them freely throughout the text though it is still true that most problems can be done without using these devices. Problems for which calculators are an important aid are labeled with the symbol $\boxed{\text{C}}$.

Complex numbers are introduced early (Section 1-6) but do not play a significant role until the end of Chapter 9. For teachers who wish to deemphasize their use, we have marked those problems that use complex numbers with the symbol ⊡.

FLEXIBILITY

This book can be used in either a one- or a two-semester course. The following dependence chart will help an instructor in designing a syllabus. Keep in mind that the first three chapters are a review of basic algebra. In some classes, they can be omitted or covered rapidly. Each problem set is designed so that all students can profit from the basic problems and the first few miscellaneous problems. The later miscellaneous problems should be assigned somewhat judiciously.

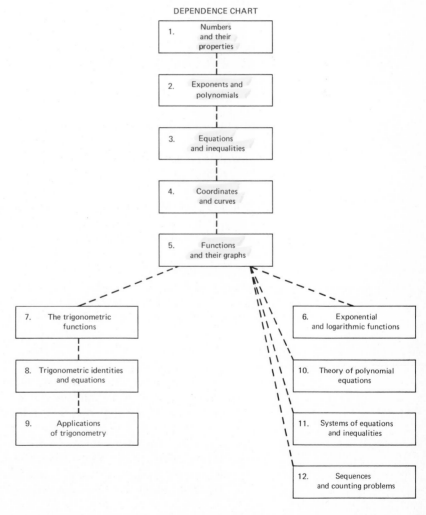

DEPENDENCE CHART

1. Numbers and their properties

2. Exponents and polynomials

3. Equations and inequalities

4. Coordinates and curves

5. Functions and their graphs

7. The trigonometric functions

8. Trigonometric identities and equations

9. Applications of trigonometry

6. Exponential and logarithmic functions

10. Theory of polynomial equations

11. Systems of equations and inequalities

12. Sequences and counting problems

SUPPLEMENTARY MATERIALS

An extensive variety of instructional aids is available from Prentice Hall.

Instructor's Manual   The instructor's manual was prepared by the authors of the textbook. It contains the following items.
- (a) Answers to all the even numbered problems (answers to the odd problems appear at the end of the textbook).
- (b) Complete solutions to the last four problems in each problem set. This includes the teaser problem.
- (c) Six versions of a chapter test for each chapter together with an answer key for these tests.
- (d) A test bank of more than 1300 problems with answers designed to aid an instructor in constructing examinations.
- (e) A set of transparencies that illustrate key ideas.

Prentice-Hall Test Generator   The test bank of more than 1300 problems is available on floppy disk for the IBM PC. This allows the instructor to generate examinations by choosing individual problems, editing them, and if desired by creating completely new problems.

Videotapes   Approximately five hours of videotaped lectures covering selected topics in college algebra are available with a qualified adoption. Contact your local Prentice Hall representative for details.

Student Solutions Manual   This manual has worked-out solutions to every third problem (not including teaser problems).

Function Plotter Software   A one-variable function plotter for the IBM PC is available with a qualified adoption. Contact your local Prentice Hall representative for details.

"How to Study Math"   Designed to help your students overcome math anxiety and to offer helpful hints regarding study habits, this useful booklet is available free with each copy sold. To request copies for your students in quantity, contact your local Prentice Hall representative.

ACKNOWLEDGMENTS

This and previous editions have profited from the warm praise and constructive criticism of many reviewers. We offer our thanks to the following people who gave helpful suggestions.

Dean Alders, *Mankato State University*
Wayne Andrepont, *The University of Southwestern Louisiana*
Karen E. Barker, *Indiana University, South Bend*
Sandra Beken, *Horry-Georgetown Technical College*

Alfred Borm, *Southwest Texas State University*
Helen Burrier, *Kirkwood Community College*
James Calhoun, *Western Illinois University*
Donald Coram, *Oklahoma State University*
Duane E. Deal, *Ball State University*
Leonard Deaton, *California State University, Los Angeles*
Phillip M. Eastman, *Boise State University*
Gary Fowler, *Louisiana State University*
Margaret Gessaman, *The University of Nebraska at Omaha*
Herbert A. Gindler, *San Diego State University*
Gerald A. Goldin, *Northern Illinois University, DeKalb*
William J. Gordon, *State University of New York, Buffalo*
Judith J. Grasso, *University of New Mexico, Albuquerque*
Mark Hale, Jr., *University of Florida*
D. W. Hall. *Michigan State University*
James E. Hall, *University of Wisconsin, Madison*
Allen Hesse, *Bowling Green State University*
James Jerkofsky, *Benedictine College*
Anne F. Landry, *Dutchess Community College, State University of New York*
Robert McMillan, *Oklahoma State University*
Jon Rahn Manon, *University of Delaware*
Eldon Miller, *University of Mississippi*
John Minnick, *DeAnza College*
James P. Muhich, *University of Wisconsin, Oshkosh*
Richard Nadel, *Florida International University*
Wallace E. Parr, *University of Maryland, Baltimore County*
H. D. Perry, *Texas A & M University*
Stephen Peterson, *University of Notre Dame*
Ricardo A. Salinas, *San Antonio College*
Jean Smith, *Middlesex Community College*
Shirley Sorensen, *University of Maryland, College Park*
Monty Strauss, *Texas Tech University*
Stuart Thomas, *Oregon State University, Corvallis*
Paul D. Trembeth, *Delaware Valley College*
Robert F. Wall, *Bryant College*
Carroll Wells, *Western Kentucky University*
Karl M. Zilm, *Lewis & Clark Community College*

The staff at Prentice Hall is to be congratulated on another fine production job. The authors wish to express appreciation especially to Priscilla McGeehon (mathematics editor), Zita de Schauensee (project editor), and Christine Gehring-Wolf (designer) for their exceptional contributions.

Walter Fleming
Dale Varberg

Numbers are an indispensable tool of civilization, serving to whip its activities into some sort of order . . . The complexity of a civilization is mirrored in the complexity of its numbers.

—Philip J. Davis

CHAPTER 1

Numbers and Their Properties

"Algebra is a merry science," Uncle Jakob would say. "We go hunting for a little animal whose name we don't know, so we call it x. When we bag our game we pounce on it and give it its right name."

Albert Einstein

1-1 What Is Algebra?

Sometimes the simplest questions seem the hardest to answer. One frustrated ninth grader responded, "Algebra is all about x and y, but nobody knows what they are." Albert Einstein was fond of his Uncle Jakob's definition, which is quoted above. A contemporary mathematician, Morris Kline, refers to algebra as generalized arithmetic. There is some truth in all of these statements, but perhaps Kline's statement is closest to the heart of the matter. What does he mean?

In arithmetic we are concerned with numbers and the four operations of addition, subtraction, multiplication, and division. We learn to understand and manipulate expressions like

$$16 - 11 \qquad \frac{3}{24} \qquad (13)(29)$$

In algebra we do the same thing, but we are more likely to write

$$a - b \qquad \frac{x}{y} \qquad mn$$

without specifying precisely what numbers these letters represent. This determination to stay uncommitted (not to know what x and y are) offers some tremendous advantages. Here are two of them.

GENERALITY AND CONCISENESS

All of us know that $3 + 4$ is the same as $4 + 3$ and that $7 + 9$ equals $9 + 7$. We could fill pages and books, even libraries, with the corresponding facts about other numbers. All of them would be correct and all would be important.

But we can achieve the same effect much more economically by writing

$$a + b = b + a$$

The simple formula says all there is to be said about adding two numbers in opposite order. It states a general law and does it on one-fourth of a line.

Or take the well-known facts that if I drive 30 miles per hour for 2 hours, I will travel 60 miles, and that if I fly 120 miles per hour for 3 hours, I will cover 360 miles. These and all other similar facts are summarized in the general formula

$$D = RT$$

which is an abbreviation for

$$\text{Distance} = \text{rate} \times \text{time}$$

FORMULAS

The formula $D = RT$ is just one of many that scientists use almost without thinking. Among these formulas are those for area and volume, which have been known since the time of the Greeks. As a premier example, we mention the formula for the area of a triangle (Figure 1), namely,

$$A = \frac{1}{2}bh$$

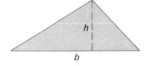

Figure 1

a formula that we will have occasion to use innumerable times in this book. Here b denotes the length of the base and h stands for the height (or altitude) of the triangle. Thus, a triangle with base $b = 24$ and height $h = 10$ has area

$$A = \frac{1}{2}\,bh = \frac{1}{2}(24)(10) = 120$$

Of course, we must be careful about units. If the base and height are given in inches, then the area is in square inches.

A more interesting formula is the familiar one

$$A = \pi r^2$$

Figure 2

for the area of a circle of radius r (Figure 2). It is interesting because of the appearance of the number π. Perhaps you have learned to approximate π by the fraction 22/7, actually, a rather poor approximation. In this course, we suggest that you use the decimal approximation 3.14159 or the even better approximation that your calculator gives (it should have a π button). Thus a circle of radius 10 centimeters has area

$$A = \pi r^2 = (3.14159)(10)(10) = 314.159$$

The answer is in square centimeters.

We are confident that you once learned all the important area and volume formulas but, because your memory may need jogging, we have listed those we will need most often in Figures 3 and 4, which accompany the first problem set.

PROBLEM SOLVING

Uncle Jakob's definition of algebra hinted at something that is very important; we call it problem solving. Algebra, like most of mathematics, is full of problems. Often these problems involve finding a number that is initially unknown but that must satisfy certain conditions. If these conditions can be translated into the symbols of algebra, it may take only a simple manipulation to find the answer or, as Uncle Jakob put it, to bag our game. Here is an illustration.

> Roger Longbottom has rented a motorboat for 5 hours from a river resort. He was told that the boat will travel 6 miles per hour upstream and 12 miles per hour downstream. How far upstream can he go and still return the boat to the resort within the allotted 5-hour time period?

We recognize immediately that this is a distance-rate-time problem; the formula $D = RT$ is certain to be important. Now what is it that we want to know? We want to find a distance, namely, how far upstream Roger dares to go. Let us call that distance x miles. Next we summarize the information that is given, keeping in mind that, since $D = RT$, it is also true that $T = D/R$.

	GOING	*RETURNING*
Distance (miles)	x	x
Rate (miles per hour)	6	12
Time (hours)	$x/6$	$x/12$

There is one piece of information we have not used; it is the key to the whole problem. The total time allowed is 5 hours, which is the sum of the time going and the time returning. Thus

$$\frac{x}{6} + \frac{x}{12} = 5$$

After multiplying both sides by 12, we have

$$2x + x = 60$$

$$3x = 60$$

$$x = 20$$

Roger can travel 20 miles upstream and still return within 5 hours.

We intend to emphasize problem solving in this book. To be able to read a mathematics book with understanding is important. To learn to calculate accurately and to manipulate symbols with ease is a worthy goal. But to be able to solve problems, easy problems and hard ones, practical problems and abstract ones, is a supreme achievement.

It is time for you to try your hand at some problems. If some of them seem difficult, do not become alarmed. All of the ideas of this section will be treated in more detail later. As the title "What Is Algebra?" suggests, we wanted to give you a preview of what lies ahead.

Problem Set 1-1

EXAMPLE A (Writing Phrases in Algebraic Notation) Use the symbols x and y to express the following phrases in algebraic notation.

(a) A number divided by the sum of twice that number and another number.

(b) The sum of 32 and $\frac{9}{5}$ of a Celsius temperature reading.

Solution.

(a) If x is the first number and y is the second, then the given phrase can be expressed by

$$\frac{x}{2x + y}$$

(b) If x represents the Celsius reading, then we can write the given phrase as

$$32 + \frac{9}{5}x$$

Perhaps you recognize this as the Fahrenheit reading corresponding to a Celsius reading of x.

Express each of the phrases in Problems 1–14 in algebraic notation using the symbols x and y.

1. One number plus one-third of another number.
2. The average of two numbers.
3. Twice one number divided by three times another.
4. The sum of a number and its square.
5. Ten percent of a number added to that number.
6. Twenty percent of the amount by which a number exceeds 50.
7. The sum of the squares of two sides of a triangle.
8. One-half the product of the base and the height of a triangle.
9. The distance in miles that a car travels in x hours at y miles per hour.
10. The time in hours it takes to go x miles at y miles per hour.
11. The rate in miles per hour of a boat that traveled y miles in x hours.
12. The total distance a car traveled in 8 hours if its rate was x miles per hour for 3 hours and y miles per hour for 5 hours.
13. The time in hours it took a boat to travel 30 miles upstream and back if its rate upstream was x miles per hour and its rate downstream was y miles per hour.
14. The time in hours it took a boat to travel 30 miles upstream and back if its rate in still water was x miles per hour and the rate of the stream was y miles per hour. Assume that x is greater than y.

In Figures 3 and 4, we have displayed the most important area and volume formulas. Use them to express the quantities of Problems 15–28 in algebraic symbols. In each

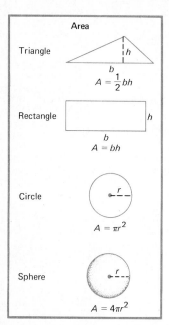

Area

Triangle

$$A = \frac{1}{2}bh$$

Rectangle

$$A = bh$$

Circle

$$A = \pi r^2$$

Sphere

$$A = 4\pi r^2$$

Figure 3

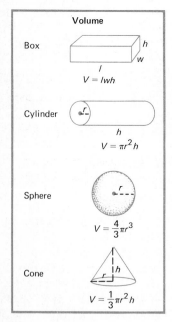

Volume

Box

$$V = lwh$$

Cylinder

$$V = \pi r^2 h$$

Sphere

$$V = \frac{4}{3}\pi r^3$$

Cone

$$V = \frac{1}{3}\pi r^2 h$$

Figure 4

problem, assume that all dimensions are given in terms of the same unit of length (such as a centimeter).

15. The area of a square of side x.
16. The area of a triangle whose height is $\frac{1}{3}$ the length of its base.
17. The surface area of a cube of side x.
18. The area of the surface of a rectangular box whose dimensions are x, $2x$, and $3x$.
19. The surface area of a sphere whose diameter is x.
20. The area of a Norman window whose shape is that of a square of side x topped with a semicircle.
21. The volume of a box with square base of length x and height 10.
22. The volume of a cylinder whose radius and height are both x.
23. The volume of a sphere of diameter x.
24. The volume of what is left of a cylinder of radius 10 and altitude 12 when a hole of radius x is drilled along the center axis of the cylinder. (Assume x is less than 10.)
25. The volume of what is left when a round hole of radius 2 is drilled through a cube of side x. (Assume that the hole is drilled perpendicular to a face and that x is greater than 4.)
26. The volume of what is left of a cylinder when a square hole of width x is drilled along the center axis of the cylinder of radius $3x$ and height y.
27. The area and perimeter of a running track in the shape of a square with semi-circular ends if the square has side x. (Recall that the circumference of a circle satisfies $C = 2\pi r$, where r is the radius.)
28. The area of what is left of a circle of radius r after removing an isosceles triangle which has a diameter as a base and the opposite vertex on the circumference of the circle.

EXAMPLE B (Stating Sentences as Algebraic Equations) Express each of the following sentences as an equation using the symbol x. Then solve for x.
(a) The sum of a number and three-fourths of that number is 21.
(b) A rectangular field which is 125 meters longer than it is wide has a perimeter of 650 meters.

Solution.

(a) Let x be the number. The sentence can be written as

$$x + \frac{3}{4}x = 21$$

If we multiply both sides by 4, we have

$$4x + 3x = 84$$

$$7x = 84$$

$$x = 12$$

It is always wise to make at least a mental check of the solution. The sum of 12 and three-fourths of 12—that is, the sum of 12 and 9—does equal 21.

Now imagine that both n and m are written as products of primes. As we saw above, both n^2 and m^2 must then have an even number of 2's as factors. Thus in the above equation, the prime 2 appears on the left an odd number of times but on the right an even number of times. This is clearly impossible. What can be wrong? The only thing that can be wrong is our supposition that $\sqrt{2}$ was a rational number.

To let one number, $\sqrt{2}$, through the dike was bad enough. But, as Euclid realized, a host of others came pouring through with it. Exactly the same proof shows that $\sqrt{3}$, $\sqrt{5}$, $\sqrt{7}$, and, in fact, the square roots of all primes are irrational. The Greeks, who steadfastly insisted that all measurements must be based on whole numbers and their ratios, could not find a satisfactory way out of this dilemma. Today we recognize that the only adequate solution is to enlarge the number system.

THE REAL NUMBERS

Let us take a bold step and simply declare that every line segment shall have a number that measures its length. The set of all numbers that can measure lengths, together with their negatives and zero, constitute the **real numbers.** Thus the rational numbers are automatically real numbers; the positive rational numbers certainly measure lengths.

Consider again the thermometer on its side, the calibrated line (Figure 10). We may have thought we had labeled every point. Not so; there were many holes corresponding to what we now call $\sqrt{2}$, $\sqrt{5}$, π, and so on. But with the introduction of the real numbers, all the holes are filled in; every point has a number label. Because of this, we often call this calibrated line the **real line.**

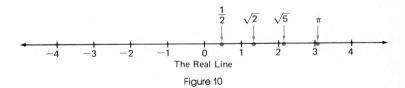

The Real Line

Figure 10

DECIMALS

There is another important way to describe the real numbers. It calls for review of a basic idea. Recall that

$$.4 = \frac{4}{10} \qquad .7 = \frac{7}{10}$$

Similarly,

$$.41 = \frac{4}{10} + \frac{1}{100} = \frac{40}{100} + \frac{1}{100} = \frac{41}{100}$$

$$.731 = \frac{7}{10} + \frac{3}{100} + \frac{1}{1000} = \frac{700}{1000} + \frac{30}{1000} + \frac{1}{1000} = \frac{731}{1000}$$

It is a simple matter to locate a decimal on the number line. For example, to locate 1.4, we divide the interval from 1 to 2 into 10 equal parts and pick the fourth point of division (Figure 11).

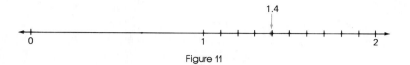

Figure 11

If the interval from 1.4 to 1.5 is divided into 10 equal parts, the second point of division corresponds to 1.42 (Figure 12).

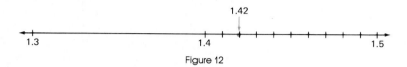

Figure 12

$$
\begin{array}{r}
.875 \\
8\overline{)7.000} \\
64 \\
\hline
60 \\
56 \\
\hline
40 \\
40 \\
\hline
\end{array}
$$

$$
\begin{array}{r}
.33333 \\
3\overline{)1.00000} \\
9 \\
\hline
10 \\
9 \\
\hline
10 \\
9 \\
\hline
10 \\
9 \\
\hline
10 \\
9 \\
\hline
1
\end{array}
$$

Figure 13

We can find the decimal corresponding to a rational number by long division. For example, the first division in Figure 13 shows that $\frac{7}{8}$ = .875. If we try the same procedure on $\frac{1}{3}$, something different happens. The decimal just keeps on going; it is an **unending decimal.** Actually, the terminating decimal .875 can be thought of as unending if we annex zeros. Thus

$$\frac{7}{8} = .875 = .8750000 \ldots$$

$$\frac{1}{3} = .3333333 \ldots$$

$$\frac{1}{6} = .16666 \ldots$$

Let us take an example that is a bit more complicated, $\frac{2}{7}$.

$$
\begin{array}{r}
.285714 \\
7\overline{)2.0000000} \\
14 \\
\hline
60 \\
56 \\
\hline
40 \\
35 \\
\hline
50 \\
49 \\
\hline
10 \\
7 \\
\hline
30 \\
28 \\
\hline
20
\end{array}
$$

If we continue the division, the pattern must repeat (note the circled 20's). Thus

$$\frac{2}{7} = .285714285714285714 \ldots$$

which can also be written

$$\frac{2}{7} = .\overline{285714}$$

The bar indicates that the block of digits 285714 repeats indefinitely.

In fact, the decimal expansion of any rational number must inevitably start repeating, because there are only finitely many different possible remainders in the division process (at most as many as the divisor). It is a remarkable coincidence that the converse statement is also true. A repeating decimal must inevitably represent a rational number (see Problems 25–31). Thus the rational numbers are precisely those numbers that can be represented by repeating decimals.

What about the nonrepeating, unending decimals like

$$.12112111211112 \ldots$$

They represent the **irrational numbers.** And they, together with the rational numbers, constitute the real numbers (Figure 14).

We showed that $\sqrt{2}$ is not rational (that is, it is irrational). It too has a decimal expansion.

$$\sqrt{2} = 1.414213 \ldots$$

Actually, the decimal expansion of $\sqrt{2}$ is known to several thousand places. It does not repeat. It cannot. It is a fact of mathematics.

We have supposed that you are familiar with the symbols $\sqrt{2}, \sqrt{3}, \ldots$ from earlier courses. For completeness, we now make the meaning of these symbols precise.

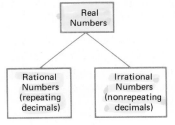

Figure 14

SQUARE ROOTS AND CUBE ROOTS

Every positive number has two square roots. For example, 3 and -3 are the two square roots of 9; this is so because $3^2 = 9$ and $(-3)^2 = 9$. If a is positive, the symbol $\sqrt{a}$ always denotes the *positive* square root of a. Thus $\sqrt{9} = 3$, and the two square roots of 7 are $\sqrt{7}$ and $-\sqrt{7}$. For the present, $\sqrt{-9}$ is a meaningless symbol since there is no real number whose square is -9. To make sense of the square root of a negative number requires a further extension of the number system, a topic we take up in Section 1-6.

In contrast, $\sqrt[3]{a}$, called the cube root of a, makes sense for any real number a. It is the unique real number whose cube is a. Thus $\sqrt[3]{8} = 2$ since $2^3 = 8$, and $\sqrt[3]{-64} = -4$ since $(-4)^3 = -64$.

We postpone the discussion of general nth roots until Section 6-1. In the meantime, we shall use square roots and cube roots in examples and problems.

CAUTION

$\sqrt{64} = \pm 8$

$\sqrt{64} = 8$

Problem Set 1-3

EXAMPLE A (Prime Factorization) Find the prime factorizations of 168 and 420.

Solution. Factor out as many 2's as possible, then 3's, then 5's, and so on.

$$168 = 2 \cdot 84 = 2 \cdot 2 \cdot 42 = 2 \cdot 2 \cdot 2 \cdot 21 = 2 \cdot 2 \cdot 2 \cdot 3 \cdot 7$$
$$420 = 2 \cdot 210 = 2 \cdot 2 \cdot 105 = 2 \cdot 2 \cdot 3 \cdot 35 = 2 \cdot 2 \cdot 3 \cdot 5 \cdot 7$$

Write the prime factorization of each number.

1. 250   2. 504   3. 200   4. 2079   5. 2100   6. 1650

EXAMPLE B (Least Common Multiple) The *least common multiple* (lcm) of several positive integers is the smallest positive integer that is a multiple of all of them. Find the least common multiple of 168 and 420.

Solution. We found the prime factorizations of these two numbers in Example A. To find the least common multiple of 168 and 420, write down the product of all factors that occur in either number, repeating a factor according to the greatest number of times it occurs in either number.

$$\text{lcm}(168, 420) = 2 \cdot 2 \cdot 2 \cdot 3 \cdot 5 \cdot 7 = 840$$

Find each of the following using the information you obtained in Problems 1–6.

7. lcm(250, 200)   8. lcm(504, 2079)   9. lcm(250, 2100)
10. lcm(504, 1650)   11. lcm(250, 200, 2100)   12. lcm(504, 2079, 1650)

EXAMPLE C (Least Common Denominator) Calculate $\frac{5}{168} + \frac{13}{420}$.

Solution. Write both fractions with a common denominator. The best choice of common denominator is the least common multiple of 168 and 420—namely, 840—obtained in Example B. We call it the *least common denominator*.

$$\frac{5}{168} + \frac{13}{420} = \frac{5 \cdot 5}{840} + \frac{13 \cdot 2}{840} = \frac{25 + 26}{840} = \frac{51}{840} = \frac{17}{280}$$

Calculate each of the following, using the results of Problems 7–12.

13. $\dfrac{3}{250} + \dfrac{17}{200}$ 14. $\dfrac{5}{504} - \dfrac{1}{2079}$ 15. $\dfrac{7}{250} - \dfrac{1}{2100}$

16. $\dfrac{13}{504} + \dfrac{13}{1650}$ 17. $\dfrac{3}{250} - \dfrac{17}{200} + \dfrac{11}{2100}$ 18. $\dfrac{13}{504} + \dfrac{13}{1650} - \dfrac{17}{2079}$

EXAMPLE D (Rational Numbers as Repeating Decimals) Write $\frac{68}{165}$ as a repeating decimal.

Solution.

$$165\overline{)68.0000}^{\,.412}$$
$$\underline{660}$$
$$200$$
$$\underline{165}$$
$$350$$
$$\underline{330}$$
$$200$$

Answer: $\dfrac{68}{165} = .41212\ldots = .41\overline{2}$

Find the repeating decimal expansion for each number. Use the bar notation for your answer.

19. $\frac{2}{3}$ 20. $\frac{3}{5}$ 21. $\frac{5}{8}$ 22. $\frac{13}{11}$ 23. $\frac{6}{13}$ 24. $\frac{4}{13}$

EXAMPLE E (Repeating Decimals as Rational Numbers)   Write $.\overline{24}$ as the ratio of two integers.

Solution.   Let $x = .\overline{24} = .242424\ldots.$ Then $100x = 24.242424\ldots.$ Subtract x from $100x$ and simplify.

$$100x = 24.2424\ldots$$
$$x = .2424\ldots$$
$$99x = 24$$
$$x = \frac{24}{99} = \frac{8}{33}$$

We multiplied x by 100 because x is a decimal that repeats in a two-digit group. If the decimal had repeated in a three-digit group, we would have multiplied by 1000.

Write each of the following as a ratio of two integers.

25. $.\overline{7}$ 26. $.\overline{123}$ 27. $.\overline{235}$ 28. $.875$

29. $.3\overline{25}$ 30. $.5\overline{21}$ 31. $.3\overline{21}$

MISCELLANEOUS PROBLEMS

32. Write the prime factorizations of 420 and 630.

33. Calculate $\frac{11}{420} \cdot \frac{11}{3} - \left(\frac{13}{420} + \frac{11}{630}\right)$ and write your answer in reduced form.

34. Find the decimal expansion of $\frac{1}{7} + \frac{1}{9}$.

35. Write $.\overline{27} + .\overline{23}$ as the ratio of two integers.

36. Show that $.2\overline{9}$ and $.3\overline{0}$ are decimal expansions of the same rational number. What numbers have two decimal expansions?

37. Show that $\sqrt{3}$ is irrational by mimicking the proof given in the text for $\sqrt{2}$.

38. Show that $\sqrt{p}$ is irrational if p is a prime.

39. Show that both the sum and product of two rational numbers is rational.

40. Show by example that the sum and product of two irrational numbers can be rational.

41. Show that $\sqrt{2} + \frac{2}{3}$ is irrational. *Hint*: Let $\sqrt{2} + \frac{2}{3} = r$. By adding $-\frac{2}{3}$ to both sides, show that it is impossible for r to be rational.

42. Show that the sum of a rational number and an irrational number is irrational.

43. Show that $\frac{2}{3}\sqrt{2}$ is irrational.

44. Show that the product of a nonzero rational number and an irrational number is irrational.

45. Which of the following numbers are rational?

    (a) $\sqrt{5} + .\overline{12}$ (b) $\sqrt{3}(\sqrt{3} + 2)$ (c) $\sqrt{\frac{6}{18}}$

    (d) $\sqrt{2}\sqrt{8}$ (e) $(1 + \sqrt{2})(1 - \sqrt{2})$ (f) $\sqrt{\frac{27}{75}}$

    (g) $.12\sqrt{2}$ (h) $.\overline{12}\sqrt{2}$ (i) $(.12)(.\overline{12})$

46. Write a positive rational number smaller than .000001 and a positive irrational number smaller than $(.000001)\sqrt{2}$. Is there a smallest positive rational number? A smallest positive irrational number?

47. Show that the sum of the squares of the diagonals of a parallelogram is equal to the sum of the squares of the four sides.

48. Consider the rectangular box shown in Figure 15 with sides of length a, b, and c. Find a formula for d, the distance between the diagonally opposite corners A and B.

49. For the box of Figure 15, find a formula for d in terms of the lengths e, f, and g of the face diagonals that meet at A.

50. **TEASER** Consider the box of Figure 15 to be closed. Find (in terms of a, b, and c) the length of the shortest path that a spider could take in crawling from A to B on the surface of the box.

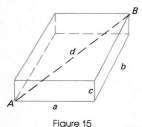

Figure 15

Playing by the Rules

  Chess has rules; so does mathematics. You can't move a pawn backwards; you must not divide by zero. Chess without rules is no game at all. Numbers without definite properties are worthless curiosities. A good chess player and a good mathematician play by the rules.

1-4   Fundamental Properties of the Real Numbers

Now we face an important question. Are the real numbers adequate to handle all the applications we are likely to encounter? Or will some problems arise that will force us to enlarge our number system again? The answer is that the real numbers are sufficient for most purposes. Except for one type of problem, to be described in Section 1-6, we will do our work within the context of the real numbers. From now on, when we say number with no qualifying adjective, we mean real number. You can count on it.

All of this suggests that we ought to pay some attention to the fundamental properties of the real numbers. By a fundamental property, we mean something so basic that we must understand it, and to understand means more than to memorize. Understanding a property means to see the purpose the property serves, to recognize its implications, and to be able to derive other things from it.

ASSOCIATIVE PROPERTY

Addition and multiplication are the fundamental operations; subtraction and division are offshoots of them. These operations are binary operations, that is, they work on two numbers at a time. Thus $3 + 4 + 5$ is technically meaningless. We ought to write either $3 + (4 + 5)$ or $(3 + 4) + 5$. But, luckily, it really does not matter; we get the same answer either way. Addition is **associative.**

$$a + (b + c) = (a + b) + c$$

Thus we can write $3 + 4 + 5$ or even $3 + 4 + 5 + 6 + 7$ without ambiguity. The answer will be the same regardless of the order in which the additions are done.

Addition and multiplication are like Siamese twins: What is true for one is quite likely to be true for the other. Thus multiplication, too, is associative.

$$a \cdot (b \cdot c) = (a \cdot b) \cdot c$$

If we wish, we can write $a \cdot b \cdot c$ with no parentheses at all.

COMMUTATIVE PROPERTY

It makes some difference whether you first put on your slippers and then take a bath or vice versa—the difference between wet and dry slippers. But for addition or multiplication, the order of the two numbers does not matter. Both operations are **commutative.**

$$a + b = b + a$$

$$a \cdot b = b \cdot a$$

Thus

$$3 + 4 = 4 + 3$$

$$3 + 4 + 6 = 4 + 3 + 6 = 6 + 4 + 3$$

and

$$7 \cdot 8 \cdot 9 = 8 \cdot 7 \cdot 9 = 9 \cdot 8 \cdot 7$$

NEUTRAL ELEMENTS

To be neutral is to sit on the sidelines and refuse to do battle. A neutral party can be ignored; the outcome is not affected by its presence. Thus we call 0 the **neutral element** for addition; its presence can be ignored in an addition.

$$a + 0 = 0 + a = a$$

Similarly, 1 is the neutral element for multiplication since

$$a \cdot 1 = 1 \cdot a = a$$

The numbers 0 and 1 are also called the **identity elements** for addition and multiplication, respectively.

While 0 is inactive in addition, its effect in multiplication is overwhelming; any number multiplied by 0 is completely wiped out.

$$a \cdot 0 = 0 \cdot a = 0$$

We have not put this property in a box; it is not quite as basic as the others since it can be derived from them (see Problem 37). However, you should still know it.

INVERSES

The numbers 3 and -3 are like an acid and a base: When you add them together, they neutralize each other. We refer to them as inverses of each other. In fact, every number a has a unique **additive inverse** $-a$ (also called the negative of a). It satisfies

$$a + (-a) = (-a) + a = 0$$

Similarly, every number a different from 0 has a unique **multiplicative inverse** a^{-1} satisfying

$$a \cdot a^{-1} = a^{-1} \cdot a = 1$$

Thus, $3^{-1} = \frac{1}{3}$ since $3 \cdot \frac{1}{3} = 1$. In fact, for any $a \neq 0$, $a^{-1} = 1/a$, and for this reason we often say the "reciprocal of a" rather than the "multiplicative inverse of a."

DISTRIBUTIVE PROPERTY

When we indicate an addition and a multiplication in the same expression, we face a problem. For example, what do we mean by $3 + 4 \cdot 2$? If we mean $(3 + 4) \cdot 2$, the answer is 14; if we mean $3 + (4 \cdot 2)$, the answer is 11. Most of us would not give the first answer. We are so familiar with a convention that we use it without thinking, but now is a good time to emphasize it. In any expression involving additions and multiplications which has no parentheses, we agree to do all the multiplications first. Thus

$$4 \cdot 5 + 3 = 20 + 3 = 23$$

and

$$4 \cdot 5 + 6 \cdot 2 = 20 + 12 = 32$$

We can always overrule this agreement by inserting parentheses. For example,

$$4 \cdot (5 + 3) = 4 \cdot 8 = 32$$

The agreement just described is a matter of convenience; no law forces it upon us. However, the **distributive property,**

$$a \cdot (b + c) = a \cdot b + a \cdot c$$
$$(b + c) \cdot a = b \cdot a + c \cdot a$$

$3 \cdot (4 + 2)$

$3 \cdot 4 + 3 \cdot 2$

Figure 16

is not a matter of choice or convenience. Rather, it is another of the fundamental properties of numbers. That it must hold for the positive integers is almost obvious, as may be seen by examining Figure 16 in the margin. But we assert that it is just as true that

$$\frac{1}{2} \cdot (\sqrt{2} + \pi) = \frac{1}{2} \cdot \sqrt{2} + \frac{1}{2} \cdot \pi$$

Actually, we use the distributive property all the time, often without realizing it. The familiar calculation in Figure 17 is really a shorthand version of

$$65 \cdot 34 = (5 + 60)34 = 5 \cdot 34 + 60 \cdot 34$$
$$= 170 + 2040 = 2210$$

$$
\begin{array}{r}
34 \\
\times\, 65 \\
\hline
170 \\
+\, 204 \\
\hline
2210
\end{array}
$$

Figure 17

SUBTRACTION AND DIVISION

Addition and multiplication are the basic operations; subtraction and division are dependent on them. Subtraction is the addition of an additive inverse and division is multiplication by a reciprocal. Thus

$$a - b = a + (-b)$$

and

$$\frac{a}{b} = a \div b = a \cdot b^{-1} = a \cdot \frac{1}{b}$$

Clearly $3 - 5 \neq 5 - 3$ and $3 \div 5 \neq 5 \div 3$, which tell us that subtraction and division are not commutative. Nor are they associative. In spite of these drawbacks, these operations are important to us.

There is one restriction: we never divide by 0. Why do we exclude $\frac{4}{0}$, $\frac{6}{0}$, $\frac{10}{0}$, and similar expressions from our consideration? We exclude them to maintain consistency. If $\frac{4}{0}$ were a number q, that is, if $\frac{4}{0} = q$, then $4 = 0 \cdot q = 0$, which is nonsense. The symbol $\frac{0}{0}$ is meaningless for a different reason: If $\frac{0}{0} = p$, then $0 = 0 \cdot p$, which is true for any number p. We choose to avoid such ambiguities by excluding division by zero.

CAUTION

$$\frac{10}{0} = 0 \qquad \frac{10}{0} = \infty$$

$\frac{10}{0}$ is meaningless

Problem Set 1-4

EXAMPLE A (Simple Calculations) Calculate $31.9 + 45 + 68.1 + 155 + 43.2$.

Solution. Intelligent use of the associative and commutative properties makes this a breeze.

$$(31.9 + 68.1) + (45 + 155) + 43.2 = 100 + 200 + 43.2 = 343.2$$

Find the following sums or products using the basic properties we have introduced. If you can do it all in your head, that will be fine.

1. $420 + 431 + 580$
2. $99{,}985 + 67 + 15$
3. $983 + 400 + 300 + 17$
4. $8.75 + 14 + 36 + 1.25$
5. $\frac{11}{13} + 43 + \frac{2}{13} + 17$
6. $\frac{15}{8} + \frac{5}{6} + \frac{1}{6} + \frac{9}{8}$
7. $6 \cdot \frac{3}{4} \cdot \frac{1}{6} \cdot 4$
8. $99 + 98 + 97 + 3 + 2 + 1$
9. $5 \cdot \frac{1}{3} \cdot \frac{2}{5} \cdot 6 \cdot \frac{1}{2}$
10. $(.25)(363)(400)(\frac{1}{3})$

EXAMPLE B (Number Properties) What properties justify each of the following?
  (a) $6 + [5 + (-6)] = [6 + (-6)] + 5$
  (b) $-\frac{1}{3}(\frac{6}{7} - \frac{9}{11}) = -\frac{2}{7} + \frac{3}{11}$

Solution.

  (a) Commutative and associative properties for addition:

$$6 + [5 + (-6)] = 6 + [(-6) + 5] = [6 + (-6)] + 5$$

  (b) The definition of subtraction and the distributive property:

$$-\frac{1}{3}(\frac{6}{7} - \frac{9}{11}) = -\frac{1}{3}(\frac{6}{7} + (-\frac{9}{11})) = -\frac{1}{3} \cdot \frac{6}{7} + (-\frac{1}{3})(-\frac{9}{11}) = -\frac{2}{7} + \frac{3}{11}$$

Name the properties that justify each of the following.

11. $\frac{5}{6}(\frac{3}{4} \cdot 12) = (12 \cdot \frac{5}{6})\frac{3}{4}$

12. $4 + (.52 - 2) = (4 - 2) + .52$

13. $(2 + 3) + 4 = 4 + (2 + 3)$

14. $9(48) = 9(40) + 9(8)$

15. $6 + [-6 + 5]$
    $= [6 + (-6)] + 5 = 5$

16. $6[\frac{1}{6} \cdot 4] = [6 \cdot \frac{1}{6}]4 = 4$

17. $-(\sqrt{5} + \sqrt{3} - 5)$
    $= -\sqrt{5} - \sqrt{3} + 5$

18. $\left(\frac{b}{c}\right)\left(\frac{b}{c}\right)^{-1} = 1$

19. $(x + 4)(x + 2)$
    $= (x + 4)x + (x + 4)2$

20. $(a + b)a^{-1} = 1 + \frac{b}{a}$

EXAMPLE C (More on Properties) Which of the following are true for all real numbers a, b, and c?

(a) $(a - b) - c = a - (b - c)$

(b) $(a + b) \div c = (a \div c) + (b \div c)$

Solution.

(a) Not true; subtraction is not associative. For example, if $a = 12$, $b = 9$, and $c = 5$, then

$$(a - b) - c = (12 - 9) - 5 = 3 - 5 = -2$$

$$a - (b - c) = 12 - (9 - 5) = 12 - 4 = 8$$

(b) True, provided $c \neq 0$:

$$(a + b) \div c = (a + b)\left(\frac{1}{c}\right) = a\left(\frac{1}{c}\right) + b\left(\frac{1}{c}\right)$$

$$= (a \div c) + (b \div c)$$

Which of the equalities in Problems 21–32 are true for all choices of a, b, and c? If false, provide an example. If true, provide a demonstration similar to that in part (b) above. Assume divisions by 0 are excluded.

21. $a - (b - c) = a - b + c$

22. $a + bc = ac + bc$

23. $a \div (b + c)$
    $= (a \div b) + (a \div c)$

24. $(a + b)^{-1} = a^{-1} + b^{-1}$

25. $ab(a^{-1} + b^{-1}) = b + a$

26. $a(a + b + c) = a^2 + ab + ac$

27. $(a + b)(a^{-1} + b^{-1}) = 1$

28. $(a + b)(a + b)^{-1} = 1$

29. $(a + b)(a + b) = a^2 + b^2$

30. $(a + b)(a - b) = a^2 - b^2$

31. $a \div (b \div c) = (a \div b) \div c$

32. $a \div (b \div c) = (a \cdot c) \div b$

MISCELLANEOUS PROBLEMS

33. Let # denote the exponentiation operation, that is, $a \# b = a^b$.
    (a) Calculate $4 \# 3$, $2(3 \# 2)$, and $2 \# (3 \# 2)$.
    (b) Is # commutative?
    (c) Is # associative?

34. Which of the following operations commute with each other, that is, which give the same answer when performed on a number in either order?
   (a) Doubling and tripling.
   (b) Doubling and cubing.
   (c) Squaring and cubing.
   (d) Squaring and taking the negative.
   (e) Squaring and taking the reciprocal.
   (f) Adding 3 and taking the negative.
   (g) Multiplying by 3 and dividing by 2.

35. Show that if a and b are nonzero, then $(ab)^{-1} = b^{-1}a^{-1}$, $(a^{-1})^{-1} = a$, and $(a/b)^{-1} = b/a$.

36. Show that if b, c, and d are nonzero, then $a/b \div c/d = (a \div c)/(b \div d)$.

37. We claimed in the text that we could show that $a \cdot 0 = 0$ for all a using only the properties displayed in boxes. Justify each of the following equalities by using one of these properties.

$$
\begin{aligned}
0 &= -(a \cdot 0) + a \cdot 0 \\
&= -(a \cdot 0) + a \cdot (0 + 0) \\
&= -(a \cdot 0) + (a \cdot 0 + a \cdot 0) \\
&= [-(a \cdot 0) + a \cdot 0] + a \cdot 0 \\
&= 0 + a \cdot 0 \\
&= a \cdot 0
\end{aligned}
$$

38. Demonstrate $(-a) \cdot b = -(a \cdot b)$ using only the properties displayed in boxes and Problem 37.

39. Demonstrate $(-a) \cdot (-b) = a \cdot b$ using only the properties displayed in boxes and Problems 37 and 38.

40. TEASER Let {Ann, Betty, Connie, Debra, and Eva} be five girls arranged according to height so Ann is shorter than Betty, Betty is shorter than Connie, and so on. Define $X + Y$ to be the taller of X and Y and $X \cdot Y$ to be the shorter of X and Y. It is to be understood that $X + X = X$ and $X \cdot X = X$.
   (a) Calculate   Connie $\cdot$ (Betty $\cdot$ Connie),   Eva $+$ (Debra $\cdot$ Betty),   and Ann $\cdot$ (Debra $+$ Connie).
   (b) Which of the properties displayed in boxes hold?
   (c) The distributive law $X + (Y \cdot Z) = (X + Y) \cdot (X + Z)$ of addition over multiplication fails for numbers. Does it hold in the five-girl system?

"I wonder, sir, if you would be willing to move to the end?"

1-5 Order and Absolute Value

We may question the bartender's tact, but not his mathematical taste. He has a deep feeling for an important notion that we call order; it is intimately tied up with the real number system. To describe this notion, we introduce a special symbol $<$; it stands for the phrase "is less than."

We begin by recalling that every real number (except 0) falls into one of two classes. Either it is positive or it is negative. Then, given two real numbers a and b, we say that

$$a < b \text{ if } b - a \text{ is positive}$$

Thus $-3 < -2$ since $-2 - (-3)$ is positive. Similarly, $3 < \pi$ since

$$\pi - 3 = 3.14159 \ldots - 3 = .14159 \ldots$$

which is a positive number.

Another and more intuitive way to think about the symbol $<$ is to relate it to the real number line. To say that $a < b$ means that a is to the left of b on the real line (Figure 18).

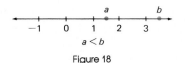

a < b

Figure 18

The symbol $<$ has a twin, denoted by $>$, which is read "is greater than." If you know how $<$ behaves, you automatically know how $>$ behaves. Thus there is no need to say much about $>$. It is enough to note that $b > a$ means exactly the same thing as $a < b$. In particular, $b > 0$ and $0 < b$ mean the same thing; both say that b is a positive number. Relations like $a < b$ and $b > a$ are called **inequalities.**

PROPERTIES OF $<$

If Homer is shorter than Ichabod and Ichabod is shorter than Jehu, then of course Homer is shorter than Jehu (Figure 19). This and other properties of the "less than" relation seem almost obvious. The following is a formal statement of the three most important properties.

Homer   Ichabod   Jehu

Figure 19

PROPERTIES OF INEQUALITIES

1. (Transitivity).   If $a < b$ and $b < c$, then $a < c$.
2. (Addition).   If $a < b$, then $a + c < b + c$.
3. (Multiplication).   If $a < b$ and $c > 0$, then $a \cdot c < b \cdot c$
     If $a < b$ and $c < 0$, then $a \cdot c > b \cdot c$

Property 2 says that you can add the same number to both sides of an inequality. It also says that you can subtract the same number from both sides, since c can be negative. Property 3 has a catch. Notice that if we multiply both sides by a positive number, we preserve the direction of the inequality; however, if we multiply by a negative number, we reverse the direction of the inequality. Thus

$$2 < 3$$

and

$$2 \cdot 4 < 3 \cdot 4$$

but

$$2(-4) > 3(-4)$$

These facts are shown on the number line in Figure 20.

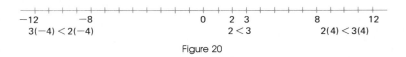

Figure 20

Division, which is equivalent to multiplication by the reciprocal, also satisfies Property 3.

THE $\leq$ RELATION

In addition to its twin $>$, the symbol $<$ has another relative. It is denoted by $\leq$ and is read "is less than or equal to." We say

---

$a \leq b$ if $b - a$ is either positive or zero

For example, it is correct to say $2 \leq 3$; it is also correct to say $2 \leq 2$. This new relation behaves very much like $<$. In fact, if we put a bar under every $<$ and

$>$ in the properties of inequalities displayed above, the resulting statements will be correct. Naturally, $b \geq a$ means the same thing as $a \leq b$.

INTERVALS

We can use the order symbols $<$ and $\leq$ to describe intervals on the real line. When we write $-1.5 < x \leq 3.2$, we mean that x is simultaneously greater than -1.5 and less than or equal to 3.2. The set of all such numbers is the interval shown in Figure 21.

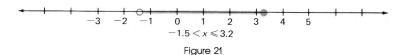

$$-1.5 < x \leq 3.2$$

Figure 21

The small circle at the left indicates that -1.5 is left out; the heavy dot at the right indicates that 3.2 is included. Figure 22 illustrates other possibilities.

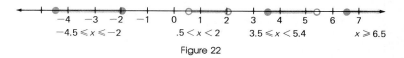

Figure 22

Sometimes we use set notation to describe an interval. For example, to denote the interval at the far left above, we could write $\{x: -4.5 \leq x \leq -2\}$, which is read "the set of all x such that x is greater than or equal to -4.5 and less than or equal to -2."

We should not write nonsense such as

$$3 < x < 2 \quad \text{or} \quad 2 > x < 3$$

The first is simply a contradiction. There is no number both greater than 3 and less than 2. The second says that x is both less than 2 and less than 3. This should be written simply as $x < 2$.

ABSOLUTE VALUE

Often we want to describe the size of a number, not caring whether it is positive or negative. To do this, we introduce the concept of absolute value, symbolized by two vertical bars $|\ \ |$. It is defined by

$$|a| = \begin{cases} a & \text{if } a \geq 0 \\ -a & \text{if } a < 0 \end{cases}$$

This two-pronged definition can cause confusion. It says that if a number is positive or zero, its absolute value is itself. But if a number is negative, then its absolute value is its additive inverse (which is a positive number). For example,

$$|7| = 7$$

since $a = 7$ is positive. On the other hand,

$$|-7| = -(-7) = 7$$

since $a = -7$ is negative. Note that $|0| = 0$.

We may also think of $|a|$ geometrically. It represents the distance between a and 0 on the number line. More generally, $|a - b|$ is the distance between a and b (Figure 23). The number $|b - a|$ represents this same distance.

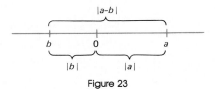

Figure 23

You should satisfy yourself that this is true for arbitrary choices of a and b, for example, that $|7 - (-3)|$ really is the distance between 7 and -3.

The properties of absolute values are straightforward and easy to remember.

PROPERTIES OF ABSOLUTE VALUES

1. $|a \cdot b| = |a| \cdot |b|$
2. $\left|\dfrac{a}{b}\right| = \dfrac{|a|}{|b|}$
3. $|a + b| \le |a| + |b|$
4. $|-a| = |a|$
5. $|a|^2 = a^2$

Actually, Properties 4 and 5 follow immediately from Property 1, since

$$|-a| = |(-1) \cdot a| = |-1| \cdot |a| = 1 \cdot |a| = |a|$$

and

$$|a|^2 = |a| \cdot |a| = |a \cdot a| = |a^2| = a^2$$

There is an important connection between absolute values and square roots, namely,

$$\boxed{\sqrt{a^2} = |a|}$$

For example, $\sqrt{6^2} = |6| = 6$ and $\sqrt{(-6)^2} = |-6| = 6$.

Problem Set 1-5

In Problems 1–12, replace the symbol # by the appropriate symbol: $<$, $>$, or $=$.

1. $1.5 \# -1.6$
2. $-2 \# -3$
3. $\sqrt{2} \# 1.4$
4. $\pi \# 3.15$
5. $\frac{1}{5} \# \frac{1}{6}$
6. $-\frac{1}{5} \# -\frac{1}{6}$
7. $5 - \sqrt{2} \# 5 - \sqrt{3}$
8. $\sqrt{2} - 5 \# \sqrt{3} - 5$
9. $-\frac{3}{16}\pi \# -\frac{3}{17}\pi$
10. $(\frac{16}{17})^2 \# (\frac{17}{18})^2$
11. $|-\pi + (-2)| \# |-\pi| + |-2|$
12. $|\pi - 2| \# \pi - 2$

13. Order the following numbers from least to greatest.

$$\frac{3}{4}, \, -2, \, \sqrt{2}, \, \frac{-\pi}{2}, \, -\frac{3}{2}\sqrt{2}, \, \frac{43}{24}$$

.75   1.41   -1.57   -2.12   1.79

14. Order the following numbers from least to greatest.

$$5 - 5, \, .37, \, -\sqrt{3}, \, \frac{14}{33}, \, -\frac{7}{4}, \, -\frac{49}{35}, \, \frac{3}{8}$$

∅   .37 -1.73 .42 -1.75 -1.4 .375

Use a real number line to show the set of numbers that satisfy each given inequality.

15. $x < -4$
16. $x < 3$
17. $x \geq -2$
18. $x \leq 3$
19. $-1 < x < 3$
20. $2 < x < 5$
21. $0 < x \leq 3$
22. $-3 \leq x < 2$
23. $-\frac{1}{2} \leq x \leq \frac{3}{2}$
24. $-\frac{7}{4} \leq x \leq -\frac{3}{4}$

Write an inequality for each interval.

25.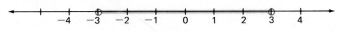
26.
27.
28.
29.
30.
31.
32.

EXAMPLE (Removing Absolute Values) Write each of the following without the absolute value symbol. Then show the corresponding interval(s) on the real number line.

(a) $|x| < 3$ (b) $|x - 2| < 3$ (c) $|x - 4| \geq 2$

Solution.

(a) Since $|x|$, the distance between x and 0 on the number line, is less than 3, x must be between -3 and 3. We write $-3 < x < 3$; this interval is displayed in Figure 24.

Figure 24

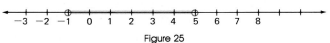

(b) Here $x - 2$, instead of x, must be between -3 and 3, that is,

$$-3 < x - 2 < 3$$

Adding 2 to each quantity gives $-1 < x < 5$, the interval shown in Figure 25.

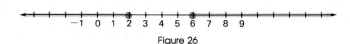

Figure 25

(c) The distance between x and 4 is greater than or equal to 2. This means that $x - 4 \leq -2$ or $x - 4 \geq 2$. That is,

$$x \leq 2 \quad \text{or} \quad x \geq 6$$

The corresponding two-part set is graphed in Figure 26.

Figure 26

In set notation, we may write the solution set for the inequality in part (c) as $\{x : x \leq 2 \text{ or } x \geq 6\}$ or equivalently as $\{x : x \leq 2\} \cup \{x : x \geq 6\}$.

Write each of the following without the absolute value symbol. Show the corresponding interval(s) on the real number line.

33. $|x| \leq 4$ 34. $|x| \geq 2$ 35. $|x - 3| < 2$
36. $|x - 5| < 1$ 37. $|x + 1| \leq 3$ 38. $|x + \frac{3}{2}| \leq \frac{1}{2}$
39. $|x - 5| > 5$ 40. $|x + 3| < 3$

MISCELLANEOUS PROBLEMS

41. Restate each of the following in mathematical symbols.
   (a) x is not greater than 12.
   (b) x is less than 3 and greater than or equal to -11.
   (c) x is not more than 4 units away from y.
   (d) x is at least 3 units away from 7.
   (e) x is closer to 5 than it is to y.
42. Show that if $a < b$, then $a < (a + b)/2 < b$.
43. Given that $\sqrt{2} = 1.414213\ldots$, arrange the following numbers in increasing order.

$$1.414, \ 1.41\overline{4}, \ 1.4\overline{14}, \ 1.\overline{414}, \ \sqrt{2}, \ \frac{\sqrt{2} + 1.414}{2}$$

44. Show that if a, b, c, and d are positive numbers, then $a/b < c/d$ if and only if $ad < bc$.
45. Indicate which number is the larger.
   (a) $\frac{11}{46}$ or $\frac{6}{25}$ (b) $\frac{4}{17}$ or $\frac{7}{29.8}$ (c) $\frac{17.1}{85}$ or $\frac{33}{165}$ (d) $\frac{11}{13}$ or $.\overline{846153}$

46. Given that $0 < a < b$, indicate how each of the following are related and justify your answer.
    (a) a^2 and b^2.
    (b) $\sqrt{a}$ and $\sqrt{b}$.
    (c) $1/a$ and $1/b$.
47. Let a, b, and A denote the altitude, base, and area, respectively, of a triangle. If $10 < b < 12$ and $50 < A < 60$, what can you say about a?
48. If the radius r and height h of a cylinder satisfy $4.9 < r < 5.1$ and $10.2 < h < 10.4$, what can you say about the volume V of the cylinder?
49. If $y = 1/x$ and $|x - 4| < 2$, what can you say about y?
50. Show that $|a| < 1$ and $a \neq 0$ imply $a^2 < |a|$, but $|a| > 1$ implies $a^2 > |a|$.
51. Under what conditions is $|a + b| = |a| + |b|$?
52. Show that if $|a| \leq 1$, then $|2a + 3a^2 + 4a^3| \leq 9$.
53. Show that $\sqrt{a^2 + b^2} \leq |a| + |b|$.
54. TEASER Sam Slugger had a better batting average than Wes Weakbat during 1985 and again in 1986. If the hits and at bats are lumped together for the two years, does it follow that Sam has a better combined batting average than Wes? Be sure to justify your conclusion.

$$i = \sqrt{-1}$$

"The Divine Spirit found a sublime outlet in that wonder of analysis, that portent of the ideal world, that amphibian between being and not-being, which we call the imaginary root of negative unity."

Gottfried Wilhelm Leibniz
1646–1716

1-6   The Complex Numbers

As early as 1550, the Italian mathematician Raffael Bombelli had introduced numbers like $\sqrt{-1}$, $\sqrt{-2}$, and so on, to solve certain equations. But mathematicians had a hard time deciding whether they should be considered legitimate numbers. Even Leibniz, who ranks with Newton as the greatest mathematician of the late seventeenth century, called them amphibians between being and not being. He wrote them down, he used them in calculations, but he carefully covered his tracks by calling them imaginary numbers. Unfortunately that name (which was actually first used by Descartes) has stuck, though these numbers are now well accepted by all mathematicians and have numerous applications in science. Let us see what these new numbers are and why they are needed.

Go back to the whole numbers 0, 1, 2, 3, ... We can easily solve the equation $x + 3 = 7$ within this system ($x = 4$). On the other hand, the equation $x + 7 = 3$ has no whole number solution. To solve it, we need the negative integer -4. Similarly, we cannot solve $3x = 2$ in the integers. We can say that the solution is $\frac{2}{3}$ only after the rational numbers have been introduced. To their dismay, the Greeks discovered that $x^2 - 2 = 0$ had no rational solution. We conquered that problem by enlarging our family of numbers to the real numbers. But there are still simple equations without solutions. Consider $x^2 + 1 = 0$. Try as you will, you will never solve it within the real number system.

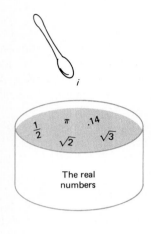

The real numbers

By now, our procedure is well established. When we need new numbers, we invent them. This time we invent a number denoted by i (or by $\sqrt{-1}$) which satisfies $i^2 = -1$. However, we cannot get by with just one number. For after we have adjoined it to the real numbers, we still must be able to multiply and add. Thus with i, we also need numbers such as

$$2i \qquad -4i \qquad (\tfrac{3}{2})i$$

which are called pure imaginary numbers. We also need

$$3 + 2i \qquad 11 + (-4i) \qquad \tfrac{3}{4} + \tfrac{3}{2}i$$

and it appears that we need even more complicated things such as

$$(3 + 8i + 2i^2 + 6i^3)(5 + 2i)$$

Actually, this last number can be simplified to $1 + 12i$, as we shall see later. In fact, no matter how many additions and multiplications we do, after the expressions are simplified we shall never have anything more complicated than a number of the form $a + bi$ (a fact that Figure 27 is meant to illustrate). Such numbers, that is, numbers of the form $a + bi$, where a and b are real, are called **complex numbers**. We refer to a as the **real part** and b as the **imaginary part** of $a + bi$. Since we shall agree that $0 \cdot i = 0$, it follows that $a + 0i = a$, and so every real number is automatically a complex number. If $b \neq 0$, then $a + bi$ is nonreal, and in this case $a + bi$ is said to be an imaginary number.

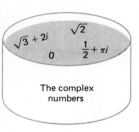

The complex numbers

Figure 27

ADDITION AND MULTIPLICATION

We cannot say anything sensible about operations for complex numbers until we agree on the meaning of equality. The definition that seems most natural is this:

$$a + bi = c + di \quad \text{means} \quad a = c \text{ and } b = d$$

That is, two complex numbers are equal if and only if their real parts and their imaginary parts are equal. As an example, suppose $x^2 + yi = 4 - 3i$. Then we know that $x = \pm 2$ and $y = -3$.

Now we can consider addition. Actually we have already used the plus sign in $a + bi$. That was like trying to add apples and bananas. The addition can be indicated but no further simplification is possible. We do not get apples and we do not get bananas; we get fruit salad.

When we have two numbers of the form $a + bi$, we can actually perform an addition. We just add the real parts and the imaginary parts separately—that is, we add the apples and we add the bananas. Thus

$$(3 + 2i) + (6 + 5i) = 9 + 7i$$

and more generally,

$$(a + bi) + (c + di) = (a + c) + (b + d)i$$

When we consider multiplication, our desire to maintain the properties of Section 1.4 leads to a definition that looks complicated. Thus let us first look at some examples.

(i)	$2(3 + 4i) = 6 + 8i$	(distributive property)
(ii)	$(3i)(-4i) = -12i^2 = 12$	(commutative and associative properties and $i^2 = -1$)
(iii)	$(3 + 2i)(6 + 5i) = (3 + 2i)6 + (3 + 2i)5i$	(distributive property)
	$= 18 + 12i + 15i + 10i^2$	(distributive, commutative, and associative properties)
	$= 18 + (12 + 15)i - 10$	(distributive property)
	$= 8 + 27i$	(commutative property)

The same kind of reasoning applied to the general case leads to

$$(a + bi)(c + di) = (ac - bd) + (ad + bc)i$$

which we take as the definition of multiplication for complex numbers.

Actually there is no need to memorize the formula for multiplication. Just do what comes naturally (that is, use familiar properties) and then replace i^2 by -1 wherever it arises, as in the following example.

$$(2 - 3i)(5 + 4i) = (10 - 12i^2) + (8i - 15i)$$

$$= (10 + 12) + (-7i)$$

$$= 22 - 7i$$

Consider the more complicated expression mentioned earlier. After noting that $i^3 = i^2 i = -i$, we have

$$(3 + 8i + 2i^2 + 6i^3)(5 + 2i) = (3 + 8i - 2 - 6i)(5 + 2i)$$

$$= (1 + 2i)(5 + 2i)$$

$$= (5 + 4i^2) + (2i + 10i)$$

$$= 1 + 12i$$

SUBTRACTION AND DIVISION

Subtraction is easy; we simply subtract corresponding real and imaginary parts. For example,

$$(3 + 6i) - (5 + 2i) = (3 - 5) + (6i - 2i)$$

$$= -2 + 4i$$

and

$$(5 + 2i) - (3 + 7i) = (5 - 3) + (2i - 7i)$$

$$= 2 + (-5i)$$

$$= 2 - 5i$$

Division is somewhat more difficult. We first note that $a - bi$ is called the **conjugate** of $a + bi$. Thus $2 - 3i$ is the conjugate of $2 + 3i$ and $-2 + 5i$ is the conjugate of $-2 - 5i$. Next, we observe that a complex number times its conjugate is a real number. For example,

$$(3 + 4i)(3 - 4i) = 9 + 16 = 25$$

and in general

$$(a + bi)(a - bi) = a^2 + b^2$$

To simplify the quotient of two complex numbers, multiply the numerator and denominator by the conjugate of the denominator, as illustrated below.

$$\frac{2 + 3i}{3 + 4i} = \frac{(2 + 3i)(3 - 4i)}{(3 + 4i)(3 - 4i)} = \frac{18 + i}{9 + 16} = \frac{18}{25} + \frac{1}{25} i$$

The effect of this multiplication is to replace a complex denominator by a real one. Notice that the result of the division is a number in the form $a + bi$.

A GENUINE EXTENSION

We assert that the complex numbers constitute a genuine enlargement of the real numbers. This means first of all that they include the real numbers, since any real number a can be written as $a + 0i$. Second, the complex numbers satisfy all the properties we discussed in Section 1-4 for the real numbers. The order properties of Section 1-5, however, do not apply to the complex numbers.

Figure 28 summarizes our development of the number systems. Is there any need to enlarge the number system again? The answer is no, and for a good reason. With the complex numbers, we can solve any equation that arises in algebra. Right now, we expect you to take this statement on faith.

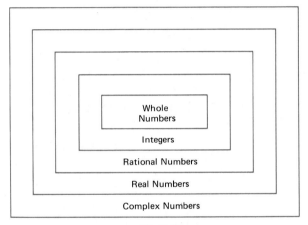

Figure 28

Problem Set 1-6

Carry out the indicated operations and write the answer in the form $a + bi$.

1. $(2 + 3i) + (-4 + 5i)$
2. $(3 - 4i) + (-5 - 6i)$
3. $5i - (4 + 6i)$
4. $(-3i + 4) + 8i$
5. $(3i - 6) + (3i + 6)$
6. $(6 + 3i) + (6 - 3i)$
7. $4i^2 + 7i$
8. $i^3 + 2i$
9. $i(4 - 11i)$
10. $(3 + 5i)i$
11. $(3i + 5)(2i + 4)$
12. $(2i + 3)(3i - 7)$
13. $(3i + 5)^2$
14. $(-3i - 5)^2$
15. $(5 + 6i)(5 - 6i)$
16. $(\sqrt{3} + \sqrt{2}i)(\sqrt{3} - \sqrt{2}i)$
17. $\dfrac{5 + 2i}{1 - i}$
18. $\dfrac{4 + 9i}{2 + 3i}$
19. $\dfrac{5 + 2i}{i}$
20. $\dfrac{-4 + 11i}{-i}$
21. $\dfrac{(2 + i)(3 + 2i)}{1 + i}$
22. $\dfrac{(4 - i)(5 - 2i)}{4 + i}$

EXAMPLE A (Multiplicative Inverses) Find $(5 + 4i)^{-1}$, the multiplicative inverse of $5 + 4i$.

Solution. The multiplicative inverse is just the reciprocal. Thus

$$(5 + 4i)^{-1} = \frac{1}{5 + 4i} = \frac{1(5 - 4i)}{(5 + 4i)(5 - 4i)} = \frac{5 - 4i}{25 - 16i^2}$$

$$= \frac{5 - 4i}{25 + 16} = \frac{5 - 4i}{41} = \frac{5}{41} - \frac{4}{41}i$$

Find each value.

23. $(2 - i)^{-1}$ 24. $(3 + 2i)^{-1}$ 25. $(\sqrt{3} + i)^{-1}$ 26. $(\sqrt{2} - \sqrt{2}i)^{-1}$

Use the results above to perform the following divisions. For example, division by $2 - i$ is the same as multiplication by $(2 - i)^{-1}$.

27. $\dfrac{2 + 3i}{2 - i}$ 28. $\dfrac{1 + 2i}{3 + 2i}$ 29. $\dfrac{4 - i}{\sqrt{3} + i}$ 30. $\dfrac{\sqrt{2} + \sqrt{2}i}{\sqrt{2} - \sqrt{2}i}$

EXAMPLE B (High Powers of i) Simplify (a) i^{51}; (b) $(2i)^6 i^{19}$.

Solution. Keep in mind that $i^2 = -1$, $i^3 = i^2 i = -i$, and $i^4 = i^2 i^2 = (-1)(-1) = 1$. Then use the usual rules of exponents.
(a) $i^{51} = i^{48} i^3 = (i^4)^{12} i^3 = (1)^{12}(-i) = -i$
(b) $(2i)^6 i^{19} = 2^6 i^6 i^{19} = 64 i^{25} = 64 i^{24} i = 64i$

Simplify each of the following.

31. i^{94} 32. i^{39} 33. $(-i)^{17}$ 34. $(2i)^3 i^{12}$

35. $\dfrac{(3i)^{16}}{(9i)^5}$ 36. $\dfrac{2^8 i^{19}}{(-2i)^{11}}$ 37. $(1 + i)^3$ 38. $(2 - i)^4$

EXAMPLE C (Complex Roots) We say that a is a 4th root of b if $a \cdot a \cdot a \cdot a = b$ (more briefly, $a^4 = b$). Thus 3 is a 4th root of 81 because $3 \cdot 3 \cdot 3 \cdot 3 = 81$. Show that $1 + i$ is a 4th root of -4.

Solution.

$$(1 + i)(1 + i)(1 + i)(1 + i) = (1 + i)^2(1 + i)^2$$

$$= (1 + 2i - 1)(1 + 2i - 1)$$

$$= (2i)(2i)$$

$$= 4i^2$$

$$= -4$$

39. Show that i is a 4th root of 1. Can you find three other 4th roots of 1?
40. Show that $-1 - i$ is a 4th root of -4.
41. Show that $1 - i$ is a 4th root of -4.
42. Show that $-1 + i$ is a 4th root of -4.

43. Perform the indicated operations and simplify.
    (a) $2 + 3i - i(4 - 3i)$ (b) $(i^7 + 2i^2)/i^3$
    (c) $3 + 4i + (3 + 4i)(-1 + 2i)$ (d) $i^{14} + (5 + 2i)/(5 - 2i)$
    (e) $\dfrac{5 - 2i}{3 + 4i} + \dfrac{5 + 2i}{3 - 4i}$ (f) $\dfrac{(3 + 2i)(3 - 2i)}{2\sqrt{3} + i}$

44. Simplify.
    (a) $1 + i + i^2 + i^3 + \cdots + i^{16}$

    (b) $1 + \dfrac{1}{i} + \dfrac{1}{i^2} + \dfrac{1}{i^3} + \cdots + \dfrac{1}{i^{16}}$

    (c) $(1 - i)^{16}$

45. Find a and b so that each is true.
    (a) $(2 + i)(2 - i)(a + bi) = 10 - 4i$
    (b) $(2 + i)(a - bi) = 8 - i$

46. Show that -2, $1 + \sqrt{3}i$, and $1 - \sqrt{3}i$ are each cube roots of -8.

47. Let $x = -1/2 + (\sqrt{3}/2)i$. Find and simplify.
    (a) x^3
    (b) $1 + x + x^2$
    (c) $(1 - x)(1 - x^2)$

48. Show that $2 + i$ and $-2 - i$ are both solutions to $x^2 - 3 - 4i = 0$.

49. Find the two square roots of i, writing your answers in the form $a + bi$.

50. Is $1/(a + bi)$ ever equal to $i/a + 1/bi$?

51. A standard notation for the conjugate of the complex number x is $\bar{x}$. Thus, $\overline{a + bi} = a - bi$. Show that each of the following is true in general.
    (a) $\overline{x + y} = \bar{x} + \bar{y}$ (b) $\overline{xy} = \bar{x}\,\bar{y}$
    (c) $\overline{x^{-1}} = (\bar{x})^{-1}$ (d) $\overline{x/y} = \bar{x}/\bar{y}$

52. **TEASER**   The absolute value of a complex number x is defined by $|x| = \sqrt{x\bar{x}}$. Note that this is consistent with the meaning of absolute value when x is a real number. Prove the triangle inequality for complex numbers, that is, show $|x + y| \le |x| + |y|$.

Chapter Summary

Algebra is generalized arithmetic. In it, we use letters to stand for numbers and then manipulate these letters according to definite rules.

The number systems used in algebra are the **whole numbers,** the **integers,** the **rational numbers,** the **real numbers,** and the **complex numbers.** Except for the complex numbers, we can visualize all these numbers as labels for points along a line, called the **real line.** Rational numbers can be expressed as ratios of integers. Real numbers (and therefore rational numbers) can be expressed as **decimals.** In fact, the rational numbers can be expressed as **repeating decimals** and the **irrational** (not rational) numbers as **nonrepeating decimals.**

The fundamental properties of numbers include the **commutative, asso-ciative,** and **distributive** laws. Because of their special properties, 0 and 1 are called the **neutral elements** for addition and multiplication. Within the rational number system, the real number system, and the complex number system, all numbers have **additive inverses** and all but 0 have **multiplicative inverses.**

The real numbers are **ordered** by the relation $<$ (is less than); its properties should be well understood as should those of its relatives $\leq$, $>$, and $\geq$. The symbol $|\ |$, read **absolute value,** denotes the magnitude of a number regardless of whether it is positive or negative.

Chapter Review Problem Set

1. An open box with a square base of length x centimeters has height y centimeters. Express its volume V and surface area S in terms of x and y.

2. Suppose that an airplane can fly x miles per hour in still air and that the wind is blowing from east at y miles per hour. Express in terms of x and y the time it will take for the plane to fly 100 miles due east and then return.

3. Perform the indicated operations and simplify.

   (a) $\dfrac{13}{24} - \dfrac{7}{12} + \dfrac{2}{3}$

   (b) $\dfrac{9}{24} \cdot \dfrac{15}{18} \cdot \dfrac{8}{5}$

   (c) $\dfrac{1}{3}\left(\dfrac{1}{2} - \left(\dfrac{5}{6} - \dfrac{2}{3}\right)\right)$

   (d) $\dfrac{\frac{11}{30}}{\frac{33}{18}}$

   (e) $\dfrac{\frac{3}{4} - \frac{1}{12} + \frac{3}{8}}{\frac{3}{4} + \frac{5}{12} - \frac{7}{8}}$

   (f) $3 + \dfrac{\frac{3}{4} - \frac{7}{8}}{\frac{5}{12}}$

4. Find the prime factorizations of 1000 and 180.

5. Find the least common multiple of 1000 and 180.

6. Express $\frac{5}{13}$ and $\frac{11}{7}$ as repeating decimals.

7. Express $.\overline{257}$ and $1.2\overline{3}$ as ratios of two integers.

8. Is $(a \div b) \times c = (a \times c) \div b$? Is $(a + b)^{-1} = a^{-1} + b^{-1}$?

9. State the associative law of multiplication and the commutative law of addition.

10. Order the following numbers from least to greatest: $\frac{29}{20}$, 1.4, $1.\overline{4}$, $\sqrt{2}$, $\frac{13}{8}$.

11. Write the inequalities $|x + 2| < 6$ and $|2x - 1| \leq 3$ without using the absolute value symbol. Then show the corresponding intervals on the real line.

12. Is $|-x| = x$? Explain.

⊡ 13. Write each of the following in the form $a + bi$.
   (a) $(3 - 2i) + (-7 + 6i) - (2 - 3i)$
   (b) $(3 - 2i)(3 + 2i) - 4i^2 + (2i)^3$
   (c) $2 + 3i + (3 - i)/(2 + 3i)$
   (d) $(2 + i)^3$
   (e) $(5 - 3i)^{-1}$

14. Assuming that $\sqrt{5}$ is irrational, show that $5 + \sqrt{5}$ and $5\sqrt{5}$ are irrational.

15. A rectangle with a perimeter of 72 centimeters is 4 centimeters longer than it is wide. Find its area.

Thus under the Descartes-Fermat scheme points became pairs of numbers, and curves became collections of pairs of numbers subsumed in equations. The properties of curves could be deduced by algebraic processes applied to the equations. With this development, the relation between number and geometry had come full circle. The classical Greeks had buried algebra in geometry, but now geometry was eclipsed by algebra. As the mathematicians put it, geometry was arithmetized.

—Morris Kline

CHAPTER 2

Exponents and Polynomials

The sermon was long; the sanctuary was hot. It was even hard to stay awake. For awhile, young Franklin Figit amused himself by doodling on the Sunday bulletin. Growing tired of that, he tried folding the bulletin in more and more complicated ways. That's when a bright idea hit him. "I wonder what would happen," he said to himself, "if I took a mammoth bulletin, folded it in half, then in half again, and so on until I had folded it 40 times. How high a stack of paper would that make?"

2-1   Integral Exponents

Young Franklin has posed a very interesting question. What do you think the answer is? 10 inches? 3 feet? 500 feet? Make a guess and write it in the margin. When we finally work out the solution late in this section, you are likely to be surprised at the answer.

Let us make a start on the problem right away. If the bulletin were c units thick ($c = .01$ inches would be a reasonable value), then after folding once, it would be $2c$ units thick. After two folds, it would measure $2 \cdot 2c$ units thick, and after 40 folds it would have a thickness of

$$2 \cdot 2 \cdot 2 \cdot 2 \cdot 2 \cdot 2 \cdot 2 \cdot 2 \cdot 2 \cdot 2 \cdot 2 \cdot 2 \cdot 2 \cdot 2 \cdot 2 \cdot 2 \cdot 2 \cdot 2 \cdot 2 \cdot 2$$
$$\cdot 2 \cdot c$$

Nobody with a sense of economy and elegance would write a product of forty 2's in this manner. To indicate such a product, most ordinary people and all mathematicians prefer to write 2^{40}. The number 40 is called an **exponent;** it tells you how many 2's to multiply together. The number 2^{40} is called a **power** of 2 and we read it "2 to the 40th power."

In the general case, if b is any number ($\frac{3}{4}$, π, $\sqrt{5}$, i, . . .) and n is a positive integer, then

$$b^n = \underbrace{b \cdot b \cdot b \cdots b}_{n \text{ factors}}$$

Thus

$$b^3 = b \cdot b \cdot b \qquad b^5 = b \cdot b \cdot b \cdot b \cdot b$$

How do we write the product of 1000 b's? (Honey is not the answer we have in mind.) The product of 1000 b's is written as b^{1000}.

RULES FOR EXPONENTS

The behavior of exponents is excellent, being governed by a few simple rules that are easy to remember. Consider multiplication first. If we multiply 2^5 by 2^8, we have

$$2^5 \cdot 2^8 = \underbrace{(2 \cdot 2 \cdot 2 \cdot 2 \cdot 2)}_{5}\underbrace{(2 \cdot 2 \cdot 2 \cdot 2 \cdot 2 \cdot 2 \cdot 2 \cdot 2)}_{8}$$

$$= \underbrace{2 \cdot 2 \cdot 2 \cdot 2 \cdot 2 \cdot 2 \cdot 2 \cdot 2 \cdot 2 \cdot 2 \cdot 2 \cdot 2 \cdot 2}_{13}$$

$$= 2^{13}$$

$$= 2^{5+8}$$

This suggests that to find the product of powers of 2, you should add the exponents. There is nothing special about 2; it could just as well be 5, $\frac{2}{3}$, or π. We can put the general rule in a nutshell by using the symbols of algebra.

$$b^m \cdot b^n = b^{m+n}$$

Here b can stand for any number, but (for now) think of m and n as positive integers. Be careful with this rule. If you write

$$3^4 \cdot 3^5 = 3^9$$

or

$$\pi^9 \cdot \pi^{12} \cdot \pi^2 = \pi^{23}$$

that is fine. But do not try to use the rule on $2^4 \cdot 3^5$ or $a^2 \cdot b^3$; it just does not apply.

Next consider the problem of raising a power to a power. By definition, $(2^{10})^3$ is $2^{10} \cdot 2^{10} \cdot 2^{10}$, which allows us to apply the rule above. Thus

$$(2^{10})^3 = 2^{10} \cdot 2^{10} \cdot 2^{10} = 2^{10+10+10} = 2^{10 \cdot 3}$$

It appears that to raise a power to a power we should multiply the exponents; in symbols

$$(b^m)^n = b^{m \cdot n}$$

Try to convince yourself that this rule is true for any number b and for any positive integer exponents m and n.

Sometimes we need to simplify quotients like

$$\frac{8^6}{8^6} \qquad \frac{2^9}{2^5} \qquad \frac{10^4}{10^6}$$

The first one is easy enough; it equals 1. Furthermore,

$$\frac{2^9}{2^5} = \frac{2^5 \cdot 2^4}{2^5} = 2^4 = 2^{9-5}$$

and

$$\frac{10^4}{10^6} = \frac{10^4}{10^4 \cdot 10^2} = \frac{1}{10^2} = \frac{1}{10^{6-4}}$$

These illustrate the general rules.

$$\frac{b^m}{b^n} = 1 \qquad \text{if } m = n$$

$$\frac{b^m}{b^n} = b^{m-n} \qquad \text{if } m > n$$

$$\frac{b^m}{b^n} = \frac{1}{b^{n-m}} \qquad \text{if } n > m$$

In each case, we assume $b \neq 0$.

We did not put a box around these rules simply because we are not happy with them. It took three lines to describe what happens when you divide powers of the same number. Surely we can do better than that, but first we shall have to extend the notion of exponents to numbers other than positive integers.

ZERO AND NEGATIVE EXPONENTS

So far, symbols like 4^0 and 10^{-3} have not been used. We want to give them meaning and do it in a way that is consistent with what we have already learned. For example, 4^0 must behave so that

$$4^0 \cdot 4^7 = 4^{0+7} = 4^7$$

This can happen only if $4^0 = 1$. More generally, we require that

$$b^0 = 1$$

Here b can be any number except 0 (0^0 will be left undefined).

What about 10^{-3}? If it is to be admitted to the family of powers, it too must abide by the rules. Thus we insist that

$$10^{-3} \cdot 10^3 = 10^{-3+3} = 10^0 = 1$$

This means that 10^{-3} has to be the reciprocal of 10^3. Consequently, we are led to make the definition

$$b^{-n} = \frac{1}{b^n} \qquad b \neq 0$$

$$\begin{aligned}
2^3 &= 2 \cdot 2 \cdot 2 \\
2^2 &= 2 \cdot 2 \\
2^1 &= 2 \\
2^0 &= 1 \\
2^{-1} &= \frac{1}{2} \\
2^{-2} &= \frac{1}{2} \cdot \frac{1}{2} \\
2^{-3} &= \frac{1}{2} \cdot \frac{1}{2} \cdot \frac{1}{2}
\end{aligned}$$

Figure 1

This definition results in the nice pattern illustrated in Figure 1. But what is more significant, it allows us to state the law for division of powers in a very simple form. To lead up to that law, consider the following manipulations.

$$\frac{b^0}{b^{-6}} = \frac{1}{1/b^6} = b^6 = b^{0-(-6)}$$

$$\frac{b^4}{b^9} = \frac{b^4}{b^4 \cdot b^5} = \frac{1}{b^5} = b^{-5} = b^{4-9}$$

$$\frac{b^5}{b^5} = 1 = b^0 = b^{5-5}$$

$$\frac{b^{-3}}{b^{-9}} = \frac{1/b^3}{1/b^9} = \frac{b^9}{b^3} = b^6 = b^{-3-(-9)}$$

In fact, for any choice of integers m and n, we find that

$$\frac{b^m}{b^n} = b^{m-n} \qquad b \neq 0$$

What about the two rules we learned earlier? Are they still valid when m and n are arbitrary (possibly negative) integers? The answer is yes. A few illustrations may help convince you.

$$b^{-3} \cdot b^7 = \frac{1}{b^3} \cdot b^7 = \frac{b^7}{b^3} = b^4 = b^{-3+7}$$

$$(b^{-5})^2 = \left(\frac{1}{b^5}\right)^2 = \frac{1}{b^5} \cdot \frac{1}{b^5} = \frac{1}{b^{10}} = b^{-10} = b^{(-5)(2)}$$

POWERS OF PRODUCTS AND QUOTIENTS

Expressions like $(ab)^n$ and $(a/b)^n$ often arise; we need rules for handling them. Notice that

$$(ab)^n = \underbrace{(ab)(ab) \ldots (ab)}_{n \text{ factors}} = \underbrace{a \cdot a \cdots a}_{n \text{ factors}} \cdot \underbrace{b \cdot b \cdots b}_{n \text{ factors}} = a^n b^n$$

$$\left(\frac{a}{b}\right)^n = \underbrace{\left(\frac{a}{b}\right)\left(\frac{a}{b}\right) \cdots \left(\frac{a}{b}\right)}_{n \text{ factors}} = \frac{a \cdot a \cdots a}{b \cdot b \cdots b} = \frac{a^n}{b^n}$$

Our demonstrations are valid for any positive integer n, but the results are correct even if n is negative or zero. Thus for any integer n,

$$(ab)^n = a^n b^n$$

$$\left(\frac{a}{b}\right)^n = \frac{a^n}{b^n}$$

Thus

$$(3x^2y)^4 = 3^4(x^2)^4y^4 = 81x^8y^4$$

and

$$\left(\frac{2x^{-1}}{y}\right)^3 = \frac{2^3(x^{-1})^3}{y^3} = \frac{2^3x^{-3}}{y^3} = \frac{8}{x^3y^3}$$

We summarize our discussion of exponents by stating the five main rules together. In using these rules, it is always understood that division by zero is to be avoided.

RULES FOR EXPONENTS

In each case, a and b are (real or complex) numbers and m and n are any integers.

1. $b^m b^n = b^{m+n}$
2. $(b^m)^n = b^{mn}$
3. $\dfrac{b^m}{b^n} = b^{m-n}$
4. $(ab)^n = a^n b^n$
5. $\left(\dfrac{a}{b}\right)^n = \dfrac{a^n}{b^n}$

THE PAPER-FOLDING PROBLEM AGAIN

It is now a simple matter to solve Franklin Figit's paper-folding problem especially if we are satisfied with a reasonable approximation to the answer. To be specific, let us approximate 1 foot by 10 inches, 1 mile by 5000 feet, and 2^{10} (which is really 1024) by 1000. Then a bulletin of thickness .01 inch will make a stack of the following height when folded 40 times ($\approx$ means "is approximately equal to").

$$(.01)2^{40} = (.01) \cdot 2^{10} \cdot 2^{10} \cdot 2^{10} \cdot 2^{10} \text{ inches}$$

$$\approx \frac{1}{10^2} \cdot 10^3 \cdot 10^3 \cdot 10^3 \cdot 10^3 \text{ inches}$$

$$= 10^{10} \text{ inches}$$

$$\approx 10^9 \text{ feet}$$

$$\approx \frac{10 \cdot 10^8}{5 \cdot 10^3} \text{ miles}$$

$$= 2 \cdot 10^5 \text{ miles}$$

$$= 200{,}000 \text{ miles}$$

That is a stack of paper that would reach almost to the moon.

Problem Set 2-1

Use the rules for exponents to simplify each of the following. Then calculate the result.

1. $\dfrac{3^2 3^5}{3^4}$ 2. $\dfrac{2^6 2^7}{2^{10}}$ 3. $\dfrac{(2^2)^4}{2^6}$ 4. $\dfrac{(5^4)^3}{5^9}$ 5. $\dfrac{(3^2 2^3)^3}{6^6}$ 6. $6^6 \left(\dfrac{2}{3^2}\right)^3$

EXAMPLE A (Removing Negative Exponents)   Rewrite without negative exponents and simplify.
   (a) -4^{-2} (b) $(-4)^{-2}$ (c) $[(\tfrac{3}{4})^{-1}]^2$ (d) $2^5 3^{-2} 2^{-3}$

Solution.
   (a) The exponent -2 applies just to 4.

$$-4^{-2} = -\frac{1}{4^2} = -\frac{1}{4 \cdot 4} = -\frac{1}{16}$$

   (b) The exponent -2 now applies to -4.

$$(-4)^{-2} = \frac{1}{(-4)^2} = \frac{1}{16}$$

   (c) First apply the rule for a power of a power.

$$\left[\left(\frac{3}{4}\right)^{-1}\right]^2 = \left(\frac{3}{4}\right)^{-2} = \frac{1}{(\frac{3}{4})^2} = \frac{1}{\frac{9}{16}} = \frac{16}{9}$$

   (d) Note that the two powers of 2 can be combined.

$$2^5 3^{-2} 2^{-3} = 2^5 2^{-3} 3^{-2} = 2^2 \cdot \frac{1}{3^2} = \frac{4}{9}$$

Write without negative exponents and simplify.

7. 5^{-2} 8. $(-5)^{-2}$ 9. -5^{-2} 10. 2^{-5}

11. $(-2)^{-5}$ 12. $\left(\dfrac{1}{5}\right)^{-2}$ 13. $\left(\dfrac{-2}{3}\right)^{-3}$ 14. $-\dfrac{2^{-3}}{3}$

15. $\dfrac{2^{-2}}{3^{-3}}$ 16. $\left[\left(\dfrac{2}{3}\right)^{-2}\right]^2$ 17. $\left[\left(\dfrac{3}{2}\right)^{-2}\right]^{-2}$ 18. $\dfrac{4^0 + 0^4}{4^{-1}}$

19. $\dfrac{2^{-2} - 4^{-3}}{(-2)^2 + (-4)^0}$ 20. $\dfrac{3^{-1} + 2^{-3}}{(-1)^3 + (-3)^2}$

21. $3^3 \cdot 2^{-3} \cdot 3^{-5}$ 22. $4^2 \cdot 4^{-4} \cdot 3^0$

EXAMPLE B (Products and Quotients)   Use Rules 4 and 5 to simplify

   (a) $(2x)^6$;   (b) $\left(\dfrac{2x}{3}\right)^4$;   (c) $(x^{-1}y^2)^{-3}$.

Solution.
   (a) $(2x)^6 = 2^6 x^6 = 64x^6$ (b) $\left(\dfrac{2x}{3}\right)^4 = \dfrac{(2x)^4}{3^4} = \dfrac{2^4 x^4}{3^4} = \dfrac{16x^4}{81}$

   (c) $(x^{-1}y^2)^{-3} = (x^{-1})^{-3}(y^2)^{-3} = x^3 y^{-6} = x^3 \cdot \dfrac{1}{y^6} = \dfrac{x^3}{y^6}$

Simplify, writing your answer without negative exponents.

23. $(3x)^4$

24. $\left(\dfrac{2}{y}\right)^5$

25. $(xy^2)^6$

26. $\left(\dfrac{y^2}{3z}\right)^4$

27. $\left(\dfrac{2x^2y}{w^3}\right)^4$

28. $\left(\dfrac{\sqrt{2x}}{3}\right)^4$

29. $\left(\dfrac{3x^{-1}y^2}{z^2}\right)^3$

30. $\left(\dfrac{2x^{-2}y}{z^{-1}}\right)^2$

31. $\left(\dfrac{\sqrt{5}i}{x^{-2}}\right)^4$

32. $(i\sqrt{3}x^{-2})^6$

33. $(4y^3)^{-2}$

34. $(x^3z^{-2})^{-1}$

35. $\left(\dfrac{5x^2}{ab^{-2}}\right)^{-1}$

36. $\left(\dfrac{3x^2y^{-2}}{2x^{-1}y^4}\right)^{-3}$

$$\left(\dfrac{a}{b}\right)^{-1} = \dfrac{b}{a}$$

EXAMPLE C (Simplifying Complicated Expressions) Simplify

(a) $\dfrac{4ab^{-2}c^3}{a^{-3}b^3c^{-1}}$;

(b) $\left[\dfrac{(2xz^{-2})^3(x^{-2}z)}{2xz^2}\right]^4$;

(c) $(a^{-1} + b^{-2})^{-1}$.

Solution.

(a) $\dfrac{4ab^{-2}c^3}{a^{-3}b^3c^{-1}} = \dfrac{4a(1/b^2)\cdot c^3}{(1/a^3)b^3(1/c)} = \dfrac{4ac^3/b^2}{b^3/(a^3c)} = \dfrac{4ac^3}{b^2}\cdot\dfrac{a^3c}{b^3} = \dfrac{4a^4c^4}{b^5}$

In simplifying expressions like the one above, *a factor can be moved from numerator to denominator or vice versa by changing the sign of its exponent.* That is important enough to remember. Let us do part (a) again using this fact.

$$\dfrac{4ab^{-2}c^3}{a^{-3}b^3c^{-1}} = \dfrac{4aa^3c^3c}{b^2b^3} = \dfrac{4a^4c^4}{b^5}$$

(b) $\left[\dfrac{(2xz^{-2})^3(x^{-2}z)}{2xz^2}\right]^4 = \left[\dfrac{8x^3z^{-6}x^{-2}z}{2xz^2}\right]^4$

$$= \left[\dfrac{8xz^{-5}}{2xz^2}\right]^4$$

$$= \left[\dfrac{4}{z^2z^5}\right]^4 = \dfrac{256}{z^{28}}$$

(c) $(a^{-1} + b^{-2})^{-1} = \left(\dfrac{1}{a} + \dfrac{1}{b^2}\right)^{-1} = \left(\dfrac{b^2 + a}{ab^2}\right)^{-1} = \dfrac{ab^2}{b^2 + a}$

Note the difference between a product and a sum.

$$(a^{-1}\cdot b^{-2})^{-1} = a\cdot b^2$$

but

$$(a^{-1} + b^{-2})^{-1} \neq a + b^2$$

CAUTION

$(a^{-1} + b^{-1})^{-1} = a + b$

$(a^{-1} + b^{-1})^{-1} = \left(\dfrac{1}{a} + \dfrac{1}{b}\right)^{-1}$

$= \left(\dfrac{b + a}{ab}\right)^{-1} = \dfrac{ab}{b + a}$

Simplify, leaving your answer free of negative exponents.

37. $\dfrac{2x^{-3}y^2z}{x^3y^4z^{-2}}$

38. $\dfrac{3x^{-5}y^{-3}z^4}{9x^2yz^{-1}}$

39. $\left(\dfrac{-2xy}{z^2}\right)^{-1}(x^2y^{-3})^2$

40. $(4ab^2)^3\left(\dfrac{-a^3}{2b}\right)^2$

41. $\dfrac{ab^{-1}}{(ab)^{-1}} \cdot \dfrac{a^2b}{b^{-2}}$

42. $\dfrac{3(b^{-2}d)^4(2bd^3)^2}{(2b^2d^3)(b^{-1}d^2)^5}$

43. $\left[\dfrac{(3b^{-2}d)(2)(bd^3)^2}{12b^3d^{-1}}\right]^5$

44. $\left[\dfrac{(ab^2)^{-1}}{(ba^2)^{-2}}\right]^{-1}$

45. $(a^{-2} + a^{-3})^{-1}$

46. $a^{-2} + a^{-3}$

47. $\dfrac{x^{-1}}{y^{-1}} - \left(\dfrac{x}{y}\right)^{-1}$

48. $(x^{-1} - y^{-1})^{-1}$

MISCELLANEOUS PROBLEMS

Simplify the expressions in Problems 49–58, leaving your answer free of negative exponents.

49. $(\tfrac{1}{2}x^{-1}y^2)^{-3}$

50. $(x + x^{-1})^2$

51. $\dfrac{2^{-2}}{1 + \dfrac{3^{-1}}{1 + 3^{-1}}}$

52. $[(\tfrac{1}{2} + \tfrac{2}{3})^{-1} + (\tfrac{1}{4} + \tfrac{1}{3})^{-1}]^{-1}$

53. $\dfrac{(2x^{-1}y^2)^2}{2xy} \cdot \dfrac{x^{-3}}{y^3}$

54. $\left(\dfrac{\sqrt{2}x^2y}{xy^{-2}z^2}\right)^4$

55. $[(\tfrac{1}{2}x^{-2})^3(4xy^{-1})^2]^2$

56. $\left[\dfrac{4y^2z^{-3}}{x^3(2x^{-1}z^2)^3}\right]^{-2}$

57. $(x^{-1} + y^{-1})^{-1}(x + y)$

58. $[1 - (1 + x^{-1})^{-1}]^{-1}$

59. Express each of the following as a single power of 2.

(a) $\tfrac{1}{2} \cdot \tfrac{1}{4} \cdot \tfrac{1}{8} \cdot \tfrac{1}{16} \cdot \tfrac{1}{32}$ (b) $\tfrac{1}{2} + \tfrac{1}{4} + \tfrac{1}{8} + \tfrac{1}{16} + \tfrac{1}{32} + \tfrac{1}{32}$

60. Express $8(\tfrac{2}{3})^4 - 4(\tfrac{2}{3})^5 + 2(\tfrac{2}{3})^6 + 6(\tfrac{2}{3})^7$ in the form $2^m/3^n$.

61. Which is larger, 2^{1000} or $(10)^{300}$?

62. Consider Franklin Figit's paper-folding problem, with which we began this section. If the pile of paper stands on 1 square inch after 40 folds, about how much area did it cover at the beginning?

63. G. P. Jetty has agreed to pay his new secretary according to the following plan: 1¢ the first day, 2¢ the second day, 4¢ the third day, and so on, doubling each day.

(a) How much will the secretary make during the first 4 days? 5 days? 6 days?

(b) From part (a), you should see a pattern. How much will the secretary make during the first n days?

(c) Assume Jetty is worth $2 billion and that his secretary started work on January 1. About when will Jetty go broke?

64. **TEASER** By a^{b^c}, mathematicians mean $a^{(b^c)}$; that is, in a tower of exponents we start at the top and work down. For example $2^{2^{2^2}} = 2^{2^4} = 2^{16} = 65{,}536$. Arrange the following numbers (all with four 2s) from smallest to largest. You should be able to do it without making use of a calculator.

$$2222, \quad 222^2, \quad 22^{22}, \quad 2^{222}, \quad 22^{2^2}, \quad 2^{22^2}, \quad 2^{2^{2^2}}$$

A new revolution is taking place in technology today. It both parallels and completes the Industrial Revolution that started a century ago. The first phase of the Industrial Revolution meant the mechanization, then the electrification of brawn. The new revolution means the mechanization and electrification of brains.

Harry M. Davis

2-2   Calculators and Scientific Notation

Most people can do arithmetic if they have to. But few do it with either enthusiasm or accuracy. Frankly, it is a rather dull subject. The spectacular sales of pocket electronic calculators demonstrate both our need to do arithmetic and our distaste for it.

Pocket calculators vary greatly in what they can do. The simplest perform only the four arithmetic operations and sell for under $10. The most sophisticated are programmable and can automatically perform dozens of operations in sequence. For this course, a standard scientific calculator is ideal. In addition to the four arithmetic operations, it will calculate values for the exponential and logarithmic functions and the trigonometric and inverse trigonometric functions.

Two kinds of logic are commonly used in pocket calculators, **reverse Polish** logic and **algebraic** logic. The former avoids the use of parentheses and is highly efficient once you learn its rules. Algebraic logic uses parentheses and mimics the procedures of ordinary algebra. For this reason, we have chosen to illustrate calculator operations using a typical algebraic calculator. However, we warn our readers that not all that we say may be valid for your calculator. We cannot emphasize this too strongly; you must learn the operating rules for your own calculator. If you do, you will find that a calculator will become a powerful tool in helping you solve problems. From here on in the book, we have marked with the symbol ⓒ those problems for which a calculator is particularly helpful.

Since calculators can display only a fixed number of digits (usually 8 or 10), we face an immediate problem. How shall we handle very large or very small numbers on a calculator? The answer depends on a notational device that was invented long before pocket calculators.

SCIENTIFIC NOTATION

It is in science that we are most likely to meet very large or very small numbers. For example, the speed of light is 29,979,000,000 centimeters per second. At the other extreme, the mass of the proton is .00000000000000000000000167 grams. These numbers are unwieldy primarily because of the large number of zeros, zeros which serve only to place the decimal point. The significant digits are 29979 in the first case and 167 in the second. Now note that

$$29,979,000,000 = 2.9979 \times 10^{10}$$

$$.00000000000000000000000167 = 1.67 \times 10^{-24}$$

Both of these numbers have been rewritten in scientific notation.

A positive number N is in **scientific notation** when it is written in the form

$$N = c \times 10^n$$

where n is an integer and c is a real number such that $1 \leq c < 10$. To put N in scientific notation, place the decimal point after the first nonzero digit of N, the so-called standard position. Then count the number of places from there to the original position of the decimal point. This number, taken as positive if counted to the right and negative if counted to the left, is the exponent n to be used as the power of 10. Here is the process illustrated for the number 3,651,000.

$$3651000 = 3.651 \times 10^6$$
$$\underbrace{\qquad\qquad}_{6 \text{ places}}$$

Calculations with large or small numbers are easily accomplished in scientific notation. Suppose we wish to calculate

$$p = \frac{(3,200,000,000)(.0000000284)}{.00000000128}$$

First, we write

$$3,200,000,000 = 3.2 \times 10^9$$

$$.0000000284 = 2.84 \times 10^{-8}$$

$$.00000000128 = 1.28 \times 10^{-9}$$

Then

$$p = \frac{(3.2 \times 10^9)(2.84 \times 10^{-8})}{1.28 \times 10^{-9}}$$

$$= \frac{(3.2)(2.84)}{1.28} \times 10^{9-8-(-9)}$$

$$= 7.1 \times 10^{10}$$

With scientific notation in hand, we can explain how to get a number into a typical calculator. To enter 238.75, simply press in order the keys 2 3 8 . 7 5.

If the negative of this number is desired, press the same keys and then press the change sign key $\boxed{+/-}$. Numbers larger than 10^8 or smaller than 10^{-8} must be entered in scientific notation. For example, 2.3875×10^{19} would be entered as follows.

$$2.3875 \; \boxed{\text{EE}} \; 19$$

The key $\boxed{\text{EE}}$ (which stands for *enter exponent*) controls the two places at the extreme right of the display. They are reserved for the exponent on 10. After pressing the indicated keys, the display will read

$$\boxed{2.3875 \qquad 19}$$

If you press

$$2.3875 \; \boxed{+/-} \; \boxed{\text{EE}} \; 19 \; \boxed{+/-}$$

the display will read

$$\boxed{-2.3875 \quad -19}$$

which stands for the number -2.3875×10^{-19}.

In making a complicated calculation, you may enter some numbers in standard notation and others in scientific notation. The calculator understands either form and makes the proper translations. If any of the entered data is in scientific notation, it will display the answer in this format. Also, if the answer is too large or too small for standard format, the calculator will automatically convert the result of a calculation to scientific notation.

DOING ARITHMETIC

The five keys $\boxed{+}$, $\boxed{-}$, $\boxed{\times}$, $\boxed{\div}$, and $\boxed{=}$ are the workhorses for arithmetic in any calculator using algebraic logic. To perform the calculation

$$175 + 34 - 18$$

simply press the keys indicated below.

$$175 \; \boxed{+} \; 34 \; \boxed{-} \; 18 \; \boxed{=}$$

The answer 191 will appear in the display.

Or consider $(175)(14)/18$. Press

$$175 \; \boxed{\times} \; 14 \; \boxed{\div} \; 18 \; \boxed{=}$$

and the calculator will display 136.11111.

An expression involving additions (or subtractions) and multiplications (or divisions) may be ambiguous. For example, $2 \times 3 + 4 \times 5$ could have several meanings depending on which operations are performed first.

(i) $\qquad 2 \times (3 + 4) \times 5 = 70$

(ii) $\qquad (2 \times 3) + (4 \times 5) = 26$

(iii) $\qquad [(2 \times 3) + 4] \times 5 = 50$

(iv) $\qquad 2 \times [3 + (4 \times 5)] = 46$

Parentheses are used in mathematics and in calculators to indicate the order in which operations are to be performed. To do calculation (i), press

$$2 \; \boxed{\times} \; \boxed{(} \; 3 \; \boxed{+} \; 4 \; \boxed{)} \; \boxed{\times} \; 5 \; \boxed{=}$$

Similarly to do calculation (iii), press

$$\boxed{(} \; \boxed{(} \; 2 \; \boxed{\times} \; 3 \; \boxed{)} \; \boxed{+} \; 4 \; \boxed{)} \; \boxed{\times} \; 5 \; \boxed{=}$$

Recall that, in arithmetic, we have an agreement that when no parentheses are used, multiplications and divisions are done before additions and subtractions. Thus

$$2 \times 3 + 4 \times 5$$

is interpreted as $(2 \times 3) + (4 \times 5)$. The same convention is used in most calculators. Pressing

$$2 \; \boxed{\times} \; 3 \; \boxed{+} \; 4 \; \boxed{\times} \; 5 \; \boxed{=}$$

will yield the answer 26. Similarly, for calculation (iii), pressing

$$\boxed{(} \; 2 \; \boxed{\times} \; 3 \; \boxed{+} \; 4 \; \boxed{)} \; \boxed{\times} \; 5 \; \boxed{=}$$

will yield 50, since within the parentheses, the calculator will do the multiplication first. However, when in doubt use parentheses, since without them it is easy to make errors.

SPECIAL FUNCTIONS

Most scientific calculators have keys for finding powers and roots of a number. On our sample calculator, the $\boxed{y^x}$ key is used to raise a number y to the xth power. For example, to calculate $2.75^{-.34}$, press

$$2.75 \; \boxed{y^x} \; .34 \; \boxed{+/-} \; \boxed{=}$$

and the correct result, .70896841, will appear in the display.

Finding a root is the inverse of raising to a power. For example, taking a cube root is the inverse operation of cubing. Thus, to calculate $\sqrt[3]{17}$, press $17 \; \boxed{\text{INV}} \; \boxed{y^x} \; 3 \; \boxed{=}$ and you will get 2.5712816. In using the $\boxed{y^x}$ key, the calculator insists that y be positive. However, x may be either positive or negative.

Square roots occur so often that there is a special key for them on some calculators. To calculate $\sqrt{17}$, simply press 17 $\boxed{\sqrt{}}$ and you will immediately get 4.1231056.

Our introduction to pocket calculators has been very brief. The problem set will give you practice in using your particular model. To clear up any difficulties and to learn other features of your calculator, consult your instruction book.

THE METRIC SYSTEM

Several of the examples of this section have involved metric units; it is time we said a word about the metric system of measurement. It has long been recognized by most countries that the metric system, with its emphasis on 10 and powers of 10, offers an attractive way to measure length, volume and weight. Only in the United States do we hang on to our hodgepodge of inches, feet, miles, pounds, and quarts. Even here, however, it appears that the metric system will gradually win acceptance.

Table 1 summarizes the metric system. It highlights the relationship to powers of 10 within the metric system and gives some of the conversion factors that relate the metric system to our English system. Using the table, you should be able to answer questions such as the following.

1. How many kilometers per hour are equivalent to 65 miles per hour?
2. How many grams are equivalent to 200 pounds?

The answers to these two questions are

1. 65 miles per hour $\approx 65/.62$, or 105 kilometers per hour;
2. 200 pounds $\approx 200(453.6)$ grams $= 9.07 \times 10^4$ grams.

TABLE 1   The Metric System

Length	Volume	Weight
kilometer = 10^3 meter	kiloliter = 10^3 liter	kilogram = 10^3 gram
hectometer = 10^2 meter	hectoliter = 10^2 liter	hectogram = 10^2 gram
dekameter = 10 meter	dekaliter = 10 liter	dekagram = 10 gram
meter = 1 meter	liter = 1 liter	gram = 1 gram
decimeter = 10^{-1} meter	deciliter = 10^{-1} liter	decigram = 10^{-1} gram
centimeter = 10^{-2} meter	centiliter = 10^{-2} liter	centigram = 10^{-2} gram
millimeter = 10^{-3} meter	milliliter = 10^{-3} liter	milligram = 10^{-3} gram
1 kilometer $\approx$ .62 miles	1 liter $\approx$ 1.057 quarts	1 kilogram $\approx$ 2.20 pounds
1 inch $\approx$ 2.54 centimeters	1 liter = 10^3 cubic centimeters	1 pound $\approx$ 453.6 grams

Problem Set 2-2

Write each of the following numbers in scientific notation.

1. 341,000,000
2. 25 billion
3. .0000000513
4. .00000000012
5. .0000000001245
6. .0000000000012578

Calculate, leaving your answers in scientific notation.

7. $(1.2 \times 10^5)(7 \times 10^{-9})$
8. $(2.4 \times 10^{-11})(1.2 \times 10^{16})$
9. $\dfrac{(0.000021)(240000)}{7000}$
10. $\dfrac{(36,000,000)(.000011)}{.0000033}$
11. $(54)(.00005)(2,000,000)^2$
12. $\dfrac{(3400)^2(400,000)^3}{(.017)^2}$

EXAMPLE A (Conversions Between Units)

   (a) Convert 2.56×10^4 kilometers to centimeters.
   (b) Convert 3.42×10^2 kilograms to ounces.
   (c) Convert 43.8 cubic meters to liters.

Solution.

   (a) 2.56×10^4 kilometers $= (2.56 \times 10^4)(10^3)$ meters

$$= (2.56 \times 10^4)(10^3)(10^2) \text{ centimeters}$$

$$= 2.56 \times 10^9 \text{ centimeters}$$

   (b) 3.42×10^2 kilograms $\approx (3.42 \times 10^2)(2.20)$ pounds

$$= (3.42 \times 10^2)(2.20)(16) \text{ ounces}$$

$$\approx 1.20 \times 10^4 \text{ ounces}$$

   (c) 43.8 cubic meters $= (43.8)(100)^3$ cubic centimeters

$$= \frac{(43.8)(100)^3}{10^3} \text{ liters}$$

$$= 4.38 \times 10^4 \text{ liters}$$

Express each of the following in scientific notation.

13. The number of centimeters in 413.2 meters.
14. The number of millimeters in 1.32×10^4 kilometers.
15. The number of kilometers in 4×10^{15} millimeters.
16. The number of meters in 1.92×10^8 centimeters.
17. The number of millimeters in one yard (36 inches).
18. The number of grams in 4.1×10^3 pounds.

EXAMPLE B (Hierarchy of Operations in a Calculator)   Most scientific calculators (with algebraic logic) perform operations in the following order.

1. Unary operations, such as taking square roots.
2. Multiplications and divisions from left to right.
3. Additions and subtractions from left to right.

Parentheses are used just as in ordinary algebra. Pressing the $\boxed{=}$ key will cause all pending operations to be performed. Calculate

$$\frac{3.12 + (4.15)(5.79)}{5.13 - 3.76}$$

Solution.   This can be done in more than one way, and calculators may vary. On the author's model, either of the following sequences of keys will give the correct result.

3.12 $\boxed{+}$ 4.15 $\boxed{\times}$ 5.79 $\boxed{=}$ $\boxed{\div}$ $\boxed{(}$ 5.13 $\boxed{-}$ 3.76 $\boxed{)}$ $\boxed{=}$

5.13 $\boxed{-}$ 3.76 $\boxed{=}$ $\boxed{1/x}$ $\boxed{\times}$ $\boxed{(}$ 3.12 $\boxed{+}$ 4.15 $\boxed{\times}$ 5.79 $\boxed{)}$ $\boxed{=}$

The result is 19.816423, or 19.82 rounded to two decimal places.

$\boxed{c}$ *Use your pocket calculator to perform the calculations in Problems 19–38. We suggest that you begin by making a mental estimate of the answer. For example, the answer to Problem 21 might be estimated as $(3 - 6)(14 \times 50) = -2100$. Similarly, the answer to Problem 35 might be estimated as $(1.5)(10)/(20) = .75$. This will help you catch errors caused by pressing the wrong keys or failing to use parentheses properly.*

19. $34.1 - 49.95 + 64.2$

20. $7.465 + 3.12 - .0156$

21. $(3.42 - 6.71)(14.3 \times 51.9)$

22. $(21.34 + 2.37)(74.13 - 26.3)$

23. $\dfrac{514 + 31.9}{52.6 - 50.8}$

24. $\dfrac{547.3 - 832.7}{.0567 + .0416}$

25. $\dfrac{(6.34 \times 10^7)(537.8)}{1.23 \times 10^{-5}}$

26. $\dfrac{(5.23 \times 10^{16})(.0012)}{1.34 \times 10^{11}}$

27. $\dfrac{6.34 \times 10^7}{.00152 + .00341}$

28. $\dfrac{3.134 \times 10^{-8}}{5.123 + 6.1457}$

29. $\dfrac{532 + 1.346}{34.91}(1.75 - 2.61)$

30. $\dfrac{39.95 - 42.34}{15.76 - 16.71}(5.31 \times 10^4)$

31. $(1.214)^3$

32. $(3.617)^{-2}$

33. $\sqrt[3]{1.215}$

34. $\sqrt[3]{1.5789}$

35. $\dfrac{(1.34)(2.345)^3}{\sqrt{364}}$

36. $\dfrac{(14.72)^{12}(59.3)^{11}}{\sqrt{17.1}}$

37. $\dfrac{\sqrt{130} - \sqrt{5}}{15^6 - 4^8}$

38. $\dfrac{\sqrt{143.2} + \sqrt{36.1}}{(234.1)^4 - (11.2)^2}$

EXAMPLE C (A Speed of Light Problem) How long will it take a light ray to reach the earth from the sun? Assume that light travels 2.9979×10^{10} centimeters per second, that 1 mile is equivalent to 1.609 kilometers, and that it is 9.30×10^7 miles from the sun to the earth.

Solution. The speed of light in kilometers per second is 2.9979×10^5, which we shall round to 2.998×10^5. The time required is

$$\frac{(9.30 \times 10^7)(1.609)}{2.998 \times 10^5} = 4.9912 \times 10^2 \approx 499$$

seconds.

[c] 39. How long will it take a light ray to travel from the moon to the earth? (The distance to the moon is 2.39×10^5 miles.)

[c] 40. How long would it take a rocket traveling 4500 miles per hour to reach the sun from the earth?

[c] 41. A light year is the distance light travels in one year (365.24 days). Our nearest star is 4.300 light years away. How many meters is that?

[c] 42. How long would it take a rocket going 4500 miles per hour to get from the earth to our nearest star (see Problem 41)?

[c] 43. What is the area in square meters of a rectangular field 2.3 light years by 4.5 light years?

[c] 44. What is the volume in cubic meters of a cube 4.3 light years on a side?

MISCELLANEOUS PROBLEMS

[c] 45. Calculate, writing your answer in scientific notation.

(a) $\dfrac{(3.151 \times 10^2)^4(32{,}400)}{(21{,}300)^2}$

(b) $\dfrac{(0.433)^3 - (2.31)^{-4} + \sqrt{0.0932}}{5.23 \times 10^3}$

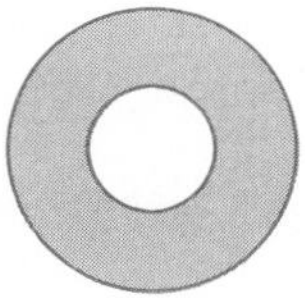

[c] 46. If, on a flat plane, I walk 24.51 meters due east and then 57.24 meters due north, how far will I be from my starting point?

[c] 47. Find the area of the ring (annulus) in Figure 2 if the outer circle has radius 26.25 centimeters and the inner circle has radius 14.42 centimeters.

[c] 48. Figure 3 shows a trapezoid with unequal sides of lengths 9.2 and 14.3 and altitude 10.4, all in inches. Find its area.

[c] 49. Let $x_1, x_2, x_3, \ldots, x_n$ denote n numbers. We define the mean $\bar{x}$ and the standard deviation s by

$$\bar{x} = \frac{1}{n}(x_1 + x_2 + \cdots + x_n) \qquad s = \sqrt{\left[\frac{1}{n}(x_1^2 + x_2^2 + \cdots + x_n^2) - \bar{x}^2\right]}$$

For the six numbers 121, 132, 155, 161, 133, and 175, calculate $\bar{x}$ and s.

Figure 2

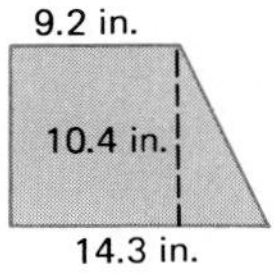

9.2 in.

10.4 in.

14.3 in.

Figure 3

© 50. A book has 516 pages (that is, 258 sheets) each $7\frac{1}{2}$ inches by $9\frac{1}{4}$ inches. If the paper to make this book could be laid out as a giant square, what would the length of a side be?

© 51. Mary Cartwright is 86 years old today. If a year has 365.24 days and a heart beats 75 times a minute, how old is Mary measured in heartbeats?

© 52. The earth has a radius of 3960 miles, the sun a radius of 400,000 miles. In terms of volume, how many times as large as the earth is the sun?

© 53. Sound waves travel at 1100 feet per second, and radio waves travel at 186,000 miles per second. Hilda sat at the back of an auditorium 200 feet from the speaker. Her husband, Hans, listened on the radio in a city 2000 miles away. Who heard the speaker first and how much sooner?

© 54. Assuming the universe is a sphere of radius 10^9 light years, what is the volume of the universe in cubic miles?

© 55. One mole of any substance is an amount equal to its molecular weight in grams. Avogadro's number (6.02×10^{23}) is the number of molecules in a mole. The atomic weights of hydrogen, carbon, and oxygen are 1, 12, and 16, respectively. Thus, the molecular weights of water (H_2O) and of carbon dioxide (CO_2) are $2 \cdot 1 + 1 \cdot 16 = 18$ and $1 \cdot 12 + 2 \cdot 16 = 44$, respectively.
(a) How many molecules are there in 20 grams of water?
(b) How many molecules are there in 30 grams of carbon dioxide?

© 56. How much will 10^{24} molecules of cholesterol ($C_{27}H_{46}O$) weigh (see Problem 55)?

© 57. Given a seed value x_1, we can generate a sequence of new values by repeated use of a *recursion formula* which relates a new value to an old value. A typical example of a recursion formula stated in words is "new x equals the square of old x." In mathematical notation, this is written as $x_{n+1} = (x_n)^2$. With a seed value of $x_1 = 3$, we see that

$$x_2 = (x_1)^2 = 3^2 = 9$$
$$x_3 = (x_2)^2 = 9^2 = 81$$

and so on.

(a) Find x_8.
(b) If we choose any seed value greater than 1, what happens to x_n as n gets very large?
(c) If we choose any seed value between 0 and 1, what happens to x_n as n gets very large?

© 58. **TEASER** This problem will require considerable experimentation with your calculator, using various positive seed values. In each case, describe what happens to x_n as the indicated recursion formula is applied a very large number of times, that is, as n becomes very large. Refer to Problem 57 for the ideas involved.

(a) $x_{n+1} = \sqrt{x_n}$
(b) $x_{n+1} = 2\sqrt{x_n}$
(c) $x_{n+1} = 3\sqrt{x_n}$
(d) $x_{n+1} = k\sqrt{x_n}$

$$3x^2 + 11x + 9$$

$$\frac{3}{2}x - 5$$

$$\sqrt{3}x^5 + \pi x^3 + 2x^2 + \frac{5}{4}$$

$$-3x^4 + 2x^3 - 3x$$

$$5x^{12}$$

$$39$$

Poly–what?

polygamist   A person who has several spouses.
polyglot   A person who speaks several languages.
polygon   A plane figure which has several sides.
polymath   A person who knows much about many subjects.
polynomial   An expression which is the sum of several terms.
polyphony   A musical combination of several harmonizing parts.

2-3   Polynomials

The dictionary definition of polynomial given above is suggestive, but it is not nearly precise enough for mathematicians. The expression $2x^{-2} + \sqrt{x} + 1/x$ is a sum of several terms, but it is not a polynomial. On the other hand, $3x^4$ is just one term; yet it is a perfectly good polynomial. Fundamentally, a **real polynomial in x** is any expression that can be obtained from the real numbers and x using only the operations of addition, subtraction, and multiplication. For example, we can get $3 \cdot x \cdot x \cdot x \cdot x$ or $3x^4$ by multiplication. We can get $2x^3$ by the same process and then have

$$3x^4 + 2x^3$$

by addition. We could never get $2x^{-2} = 2/x^2$ or $\sqrt{x}$; the first involves a division and the second involves taking a root. Thus $2/x^2$ and $\sqrt{x}$ are not polynomials. Try to convince yourself that all of the expressions in the box above are polynomials.

There is another way to define a polynomial in fewer words. A **real polynomial in x** is any expression of the form

$$a_n x^n + a_{n-1}x^{n-1} + a_{n-2}x^{n-2} + \cdots + a_1 x + a_0$$

where the a's are real numbers and n is a nonnegative integer. We shall not have much to do with complex polynomials in x, but they are defined in exactly the same way, except that the a's are allowed to be complex numbers. In either case, we refer to the a's as **coefficients.** The **degree** of a polynomial is the largest exponent that occurs in the polynomial in a term with a nonzero coefficient. Here are several more examples.

1. $\frac{3}{2}x - 5$ is a first degree (or **linear**) polynomial in x.
2. $3y^2 - 2y + 16$ is a second degree (or **quadratic**) polynomial in y.
3. $\sqrt{3}t^5 - \pi t^2 - 17$ is a fifth degree polynomial in t.
4. $5x^3$ is a third degree polynomial in x. It is also called a **monomial**, since it has only one term.
5. 13 is a polynomial of degree zero. Does that seem strange to you? If it helps, think of it as $13x^0$. In general, any nonzero constant polynomial has degree zero. We do not define the degree of the zero polynomial.

The whole subject of polynomials would be pretty dull and almost useless if it stopped with a definition. Fortunately, polynomials—like numbers—can be manipulated. In fact, they behave something like the integers. Just as the sum, difference, and product of two integers are integers, so the sum, difference, and product of two polynomials are polynomials. Notice that we did not mention division; that subject is discussed in Section 2-5 and in Chapter 10.

ADDITION

Adding two polynomials is a snap. Treat x like a number and use the commutative, associative, and distributive properties freely. When you are all done, you will discover that you have just grouped like terms (that is, terms of the same degree) and added their coefficients. Here, for example, is how we add $x^3 + 2x^2 + 7x + 5$ and $x^2 - 3x - 4$.

$$(x^3 + 2x^2 + 7x + 5) + (x^2 - 3x - 4)$$
$$= x^3 + (2x^2 + x^2) + (7x - 3x)$$
$$+ (5 - 4) \qquad \text{(associative and commutative properties)}$$
$$= x^3 + (2 + 1)x^2 + (7 - 3)x + 1 \qquad \text{(distributive property)}$$
$$= x^3 + 3x^2 + 4x + 1$$

How important are the parentheses in this example? Actually, they are indispensable only in the third line, where we used the distributive property. Why did we use them in the first and second lines? Because they emphasize what is happening. In the first line, they show which polynomials are added; in the second line, they draw attention to the terms being grouped. To shed additional light, notice that

$$x^3 + 3x^2 + 4x + 1$$

is the correct answer not only for

$$(x^3 + 2x^2 + 7x + 5) + (x^2 - 3x - 4)$$

but also for

$$(x^3 + 2x^2) + (7x + 5 + x^2) + (-3x - 4)$$

and even for

$$(x^3) + (2x^2) + (7x) + (5) + (x^2) + (-3x) + (-4)$$

SUBTRACTION

How do we subtract two polynomials? We replace the subtracted polynomial by its negative and add. For example, we rewrite

$$(3x^2 - 5x + 2) - (5x^2 + 4x - 4)$$

as

$$(3x^2 - 5x + 2) + (-5x^2 - 4x + 4)$$

Then, after grouping like terms, we obtain

$$(3x^2 - 5x^2) + (-5x - 4x) + (2 + 4)$$
$$= (3 - 5)x^2 + (-5 - 4)x + (2 + 4)$$
$$= -2x^2 - 9x + 6$$

If you can go directly from the original problem to the answer, do so; we do not want to make simple things complicated. But be sure to note that

$$(3x^2 - 5x + 2) - (5x^2 + 4x - 4) \neq 3x^2 - 5x + 2 - 5x^2 + 4x - 4$$

The minus sign in front of $(5x^2 + 4x - 4)$ changes the sign of all three terms.

MULTIPLICATION

The distributive property is the basic tool in multiplication. Here is a simple example using $a(b + c) = ab + ac$.

$$(3x^2)(2x^3 + 7) = (3x^2)(2x^3) + (3x^2)(7)$$
$$= 6x^5 + 21x^2$$

Here is a more complicated example, which uses the distributive property $(a + b)c = ac + bc$ at the first step.

$$(3x - 4)(2x^3 - 7x + 8) = (3x)(2x^3 - 7x + 8) + (-4)(2x^3 - 7x + 8)$$
$$= (3x)(2x^3) + (3x)(-7x) + (3x)(8) + (-4)(2x^3)$$
$$+ (-4)(-7x) + (-4)(8)$$
$$= 6x^4 - 21x^2 + 24x - 8x^3 + 28x - 32$$
$$= 6x^4 - 8x^3 - 21x^2 + 52x - 32$$

Notice that each term of $3x - 4$ multiplies each term of $2x^3 - 7x + 8$.

If the process just illustrated seems unwieldy, you may find the format below helpful.

$$
\begin{array}{r}
2x^3 - 7x + 8 \\
3x - 4 \\
\hline
6x^4 \qquad -21x^2 + 24x \\
- 8x^3 \qquad + 28x - 32 \\
\hline
6x^4 - 8x^3 - 21x^2 + 52x - 32
\end{array}
$$

When both polynomials are linear (that is, of the form $ax + b$), there is a handy shortcut. For example, just one look at $(x + 4)(x + 5)$ convinces us that the product has the form $x^2 + (\quad) + 20$. It is the middle term that may cause a little trouble. Think of it this way.

$$\overbrace{(x + \underbrace{4)(x}_{4x} + 5)}^{5x} = x^2 + (\;\;5x + 4x\;\;) + 20$$
$$= x^2 + 9x + 20$$

Similarly,

$$\overbrace{(2x - \underbrace{3)(x}_{-3x} + 5)}^{10x} = 2x^2 + (\;\;10x - 3x\;\;) - 15$$
$$= 2x^2 + 7x - 15$$

Finally,

$$\overbrace{(3x + \underbrace{2)(5x}_{10x} - 7)}^{-21x} = 15x^2 + (\;\;-21x + 10x\;\;) - 14$$
$$= 15x^2 - 11x - 14$$

The FOIL Method

$(x + 4)(x + 5) =$

F O I L

$x^2 + 5x + 4x + 20$

The four terms are the products of the Firsts, Outers, Inners, and Lasts.

Figure 4

Soon you should be able to find such simple products in your head. Some people find the FOIL method helpful (Figure 4).

THREE SPECIAL PRODUCTS

CAUTION

~~$(x + 5)^2 = x^2 + 25$~~

$(x + 5)^2 = x^2 + 10x + 25$

Some products occur so often that they deserve to be highlighted.

$$(x + a)(x - a) = x^2 - a^2$$
$$(x + a)^2 = x^2 + 2ax + a^2$$
$$(x - a)^2 = x^2 - 2ax + a^2$$

Here are some illustrations.

$$(x + 7)(x - 7) = x^2 - 7^2 = x^2 - 49$$
$$(x + 3)^2 = x^2 + 2 \cdot 3x + 3^2 = x^2 + 6x + 9$$
$$(x - 4)^2 = x^2 - 2 \cdot 4x + 4^2 = x^2 - 8x + 16$$

POLYNOMIALS IN SEVERAL VARIABLES

In the box above, we assumed that you would think of x as a variable and a as a constant. However, that is not necessary. If we consider both x and a to be variables, then expressions like $x^2 - a^2$ and $x^2 + 2ax + a^2$ are polynomials in two variables. Other examples are

$$x^2y + 3xy + y \qquad u^3 + 3u^2v + 3uv^2 + v^3$$

Polynomials in several variables present little in the way of new ideas; we defer further consideration of them to Example C.

Problem Set 2-3

Decide whether the given expression is a polynomial. If it is, give its degree.

1. $3x^2 - x + 2$
2. $4x^5 - x$
3. $\pi s^5 - \sqrt{2}$
4. $3\sqrt{2}t$
5. $16\sqrt{2}$
6. $511/\sqrt{2}$
7. $3t^2 + \sqrt{t} + 1$
8. $t^2 + 3t + 1/t$
9. $3t^{-2} + 2t^{-1} + 5$
10. $5 + 4t + 6t^{10}$

Perform the indicated operations and simplify. Write your answer as a polynomial in descending powers of the variable.

11. $(2x - 7) + (-4x + 8)$
12. $(\frac{3}{2}x - \frac{1}{4}) + \frac{5}{6}x$
13. $(2x^2 - 5x + 6) + (2x^2 + 5x - 6)$
14. $(\sqrt{3}t + 5) + (6 - 4 - 2\sqrt{3}t)$
15. $(5 - 11x^2 + 4x) + (x - 4 + 9x^2)$
16. $(x^2 - 5x + 4) + (3x^2 + 8x - 7)$
17. $(2x - 7) - (-4x + 8)$
18. $(\frac{3}{2}x - \frac{1}{4}) - \frac{5}{6}x$
19. $(2x^2 - 5x + 6) - (2x^2 + 5x - 6)$
20. $y^3 - 4y + 6 - (3y^2 + 6y - 3)$
21. $5x(7x - 11) + 19$
22. $-x^2(7x^3 - 5x + 1)$
23. $(t + 5)(t + 11)$
24. $(t - 5)(t + 13)$
25. $(x + 9)(x - 10)$
26. $(x - 13)(x - 7)$
27. $(2t - 1)(t + 7)$
28. $(3t - 5)(4t - 2)$
29. $(4 + y)(y - 2)$
30. $1 + y(y - 2)$
31. $(3.41x - 2.53)(2.34x + 1.77)$ $\boxed{c}$
32. $(.66x + .87)(.41x - 3.12)$ $\boxed{c}$

EXAMPLE A (More on Special Products) Find each of the following products.

(a) $(2t + 9)(2t - 9)$
(b) $(2x^2 - 3x)^2$
(c) $[(x^2 + 2) + x][(x^2 + 2) - x]$

Solution. We appeal to the boxed results on page 64.

(a) $(2t + 9)(2t - 9) = (2t)^2 - 9^2 = 4t^2 - 81$
(b) $(2x^2 - 3x)^2 = (2x^2)^2 - 2(2x^2)(3x) + (3x)^2 = 4x^4 - 12x^3 + 9x^2$

(c) We apply the first formula in the box with $x^2 + 2$ and x playing the roles of x and a, respectively.

$$[(x^2 + 2) + x][(x^2 + 2) - x] = (x^2 + 2)^2 - x^2$$
$$= x^4 + 4x^2 + 4 - x^2$$
$$= x^4 + 3x^2 + 4$$

Use the special product formulas to perform the following multiplications. Write your answer as a polynomial in descending powers of the variable.

33. $(x + 10)^2$ 34. $(y + 12)^2$

35. $(x + 8)(x - 8)$ 36. $(t - 5)(t + 5)$

37. $(2t - 5)^2$ 38. $(3s + 11)^2$

39. $(2x^4 + 5x)(2x^4 - 5x)$ 40. $(u^3 + 2u^2)(u^3 - 2u^2)$

41. $[(t + 2) + t^3]^2$ 42. $[(1 - x) + x^2]^2$

43. $[(t + 2) + t^3][(t + 2) - t^3]$ 44. $[(1 - x) + x^2][(1 - x) - x^2]$

ⓒ 45. $(2.3x - 1.4)^2$ ⓒ 46. $(2.43x - 1.79)(2.43x + 1.79)$

EXAMPLE B (Expanding Cubes) It is easy to show that

$$(x + a)^3 = x^3 + 3ax^2 + 3a^2x + a^3$$
$$(x - a)^3 = x^3 - 3ax^2 + 3a^2x - a^3$$

Use these results to find (a) $(x - 5)^3$; (b) $(2x^2 + 3x)^3$.

Solution.

(a) $(x - 5)^3 = x^3 - 3 \cdot 5x^2 + 3 \cdot 5^2x - 5^3$
             $= x^3 - 15x^2 + 75x - 125$

(b) $(2x^2 + 3x)^3 = (2x^2)^3 + 3(2x^2)^2(3x) + 3(2x^2)(3x)^2 + (3x)^3$
                  $= 8x^6 + 36x^5 + 54x^4 + 27x^3$

Expand the following cubes.

47. $(x + 2)^3$ 48. $(x - 4)^3$ 49. $(2t - 3)^2$

50. $(3u + 1)^3$ 51. $(2t + t^2)^3$ 52. $(4 - 3t^2)^3$

53. $[(2t + 1) + t^2]^3$ 54. $[u^2 - (u + 1)]^3$

EXAMPLE C (Multiplying Polynomials in Several Variables) Find the following products.

(a) $(3x + 2y)(x - 5y)$ (b) $(2xy - 3z)^2$

Solution.

(a) $(3x + 2y)(x - 5y) = 3x^2 - 13xy - 10y^2$

(b) $(2xy - 3z)^2 = (2xy)^2 - 2(2xy)(3z) + (3z)^2$
                $= 4x^2y^2 - 12xyz + 9z^2$

Perform the indicated operations and simplify.

55. $(x - 3y)^2$

56. $(2x + y)^2$

57. $(3x - 2y)(3x + 2y)$

58. $(u^2 - 2v)(u^2 + 2v)$

59. $(3x - y)(4x + 5y)$

60. $(2x + y)(6x + 5y)$

61. $(2x^2y + z)(x^2y - z)$

62. $(3x + 5y^2z)(x - y^2z)$

63. $(t + 1 + s)(t + 1 - s)$

64. $(3u + 2v + 1)^2$

65. $(2t - 3s)^3$

66. $(u + 4v)^3$

MISCELLANEOUS PROBLEMS

In Problems 67–86, perform the indicated operations and simplify.

67. $(2x^2 - 3y)(2x^2 + 3y)$

68. $(3x - y^3)^2$

69. $(2s^3 + 3t)(s^3 - 4t)$

70. $2u(3u - 1)(3u + 1)$

71. $(2u - v^2)^3$

72. $(x^2 + 3y)^3$

73. $2x(3x^2 - 6x + 4) - 3x[2x^2 - 4(x - 1)]$

74. $(y + 3)(2y - 5) - 2(3y - 2)(y + 2)$

75. $(2s + 3)^2 - (2s + 3)(2s - 3)$

76. $(x - \sqrt{2}y)(x + \sqrt{2}y) - (x - \sqrt{2}y)^2$

77. $(x^2 + 2x - 3)(x^2 + 2x + 3)$

78. $(y^2 - 3y + 2)^2$

79. $(2x^2 + x - 1)(x + 2)$

80. $(y^2 + 2y - 3)(2y + 1)$

81. $(x^2 - 2xy)(x^2 + 2xy)(x + y)$

82. $xy(2x^2 - 3y^2)(3x^2 + 2y^2)$

83. $(x^2 + 2xy + 4y^2)(x - 2y)$

84. $(2x - y)^3 + 12x^2y - 6xy^2$

85. $(x^2 + xy + y^2)(x^2 - xy + y^2)$

86. $(2x + y)^4$

87. Find the coefficient of x^3 in the expansion of $(x^2 + 2x + 3)(x^3 - 3x^2 + 2x + 1)$.   TRY WITHOUT WRITING ANYTHING DOWN

88. Find the coefficient of x^5 in the expansion of $(x^4 + 2x^3 + 3x^2)^2$.

89. A triple (a, b, c) of positive integers is called a **Pythagorean triple** if $a^2 + b^2 = c^2$. For example, $(3, 4, 5)$ and $(5, 12, 13)$ are Pythagorean triples. Show that $(2m, m^2 - 1, m^2 + 1)$ is a Pythagorean triple, provided m is an integer greater than 1.

90. Show that $(m^2 - n^2, 2mn, m^2 + n^2)$ is a Pythagorean triple, provided m and n are positive integers with $m > n$.

91. If $r > 0$ and $(r + r^{-1})^2 = 5$, find $r^3 + r^{-3}$.

92. If $a + b = 1$ and $a^2 + b^2 = 2$, find $a^3 + b^3$.

93. If $x + y = \sqrt{11}$ and $x^2 + y^2 = 16$, find $x^4 + y^4$. *Hint:* It will be helpful to know that $(x + y)^4 = x^4 + 4x^3y + 6x^2y^2 + 4xy^3 + y^4$.

94. TEASER   Show that 1 plus the product of four consecutive integers is always a perfect square.

Polynomial to be Factored	Johnny's Answer	Teacher's Comments
1. $x^6 + 2x^2$	$x^2(x^3 + 2)$	Wrong. Have you forgotten that $x^2 x^3 = x^5$? Right answer: $x^2(x^4 + 2)$
2. $x^2 + 5x + 6$	$(x + 6)(x + 1)$	Wrong. You didn't check the middle term. Right answer: $(x + 2)(x + 3)$
3. $x^2 - 4y^2$	$(x + 2y)(x - 2y)$	Right.
4. $x^2 + 4y^2$	$(x + 2y)^2$	Wrong. $(x + 2y)^2 = x^2 + 4xy + 4y^2$ Right answer: $x^2 + 4y^2$ doesn't factor using real coefficients.
5. $x^2 y^2 + 6xy + 9$	Impossible	Wrong. $x^2 y^2$ and 9 are squares. You should have suspected a perfect square. Right answer: $(xy + 3)^2$

2-4   Factoring Polynomials

To factor 90 means to write it as a product of smaller numbers; to factor it completely means to write it as a product of primes, that is, numbers that cannot be factored further. Thus we have factored 90 when we write $90 = 9 \cdot 10$, but it is not factored completely until we write

$$90 = 2 \cdot 3 \cdot 3 \cdot 5$$

Similarly to **factor** a polynomial means to write it as a product of simpler polynomials; to **factor** a polynomial **completely** is to write it as a product of polynomials that cannot be factored further. Thus when we write

$$x^3 - 9x = x(x^2 - 9)$$

we have factored $x^3 - 9x$, but not until we write

$$x^3 - 9x = x(x + 3)(x - 3)$$

have we factored $x^3 - 9x$ completely.

Now why can't Johnny factor? He can't factor because he can't multiply. If he doesn't know that

$$(x + 2)(x + 3) = x^2 + 5x + 6$$

he certainly is not going to know how to factor $x^2 + 5x + 6$. That is why we urge you to memorize the special product formulas on page 69. Of course, a product formula is also a factoring formula when read from right to left.

To urge memorization may be a bit old-fashioned, but we suggest that a fact, once memorized, becomes a permanent friend. It is best to memorize in words. For example, read formula 3 as follows: the square of a sum of two terms is the first squared plus twice their product plus the second squared.

$$\boxed{\begin{aligned}
&\text{PRODUCT FORMULAS} \longrightarrow \\
&\longleftarrow \text{FACTORING FORMULAS} \\[4pt]
&1.\ a(x + y + z) = ax + ay + az \\
&2.\ (x + a)(x + b) = x^2 + (a + b)x + ab \\
&3.\ (x + y)^2 = x^2 + 2xy + y^2 \\
&4.\ (x - y)^2 = x^2 - 2xy + y^2 \\
&5.\ (x + y)(x - y) = x^2 - y^2 \\
&6.\ (x + y)^3 = x^3 + 3x^2y + 3xy^2 + y^3 \\
&7.\ (x - y)^3 = x^3 - 3x^2y + 3xy^2 - y^3 \\
&8.\ (x + y)(x^2 - xy + y^2) = x^3 + y^3 \\
&9.\ (x - y)(x^2 + xy + y^2) = x^3 - y^3
\end{aligned}}$$

TAKING OUT A COMMON FACTOR

This, the simplest factoring procedure, is based on formula 1 above. You should always try this process first. Take Johnny's first problem as an example. Both terms of $x^6 + 2x^2$ have x^2 as a factor, so we take it out.

$$x^6 + 2x^2 = x^2(x^4 + 2)$$

Always factor out as much as you can. Taking 2 out of $4xy^2 - 6x^3y^4 + 8x^4y^2$ is not nearly enough, though it is a common factor; taking out $2xy$ is not enough either. You should take out $2xy^2$. Then

$$4xy^2 - 6x^3y^4 + 8x^4y^2 = 2xy^2(2 - 3x^2y^2 + 4x^3)$$

FACTORING BY TRIAL AND ERROR

In factoring, as in life, success often results from trying and trying again. What does not work is systematically eliminated; eventually, effort is rewarded. Let us see how this process works on $x^2 - 5x - 14$. We need to find numbers a and b such that

$$x^2 - 5x - 14 = (x + a)(x + b)$$

Since ab must equal -14, two possibilities immediately occur to us: $a = 7$ and $b = -2$ or $a = -7$ and $b = 2$. Try them both to see if one works.

$$(x + 7)(x - 2) = x^2 + 5x - 14$$

$$(x - 7)(x + 2) = x^2 - 5x - 14 \qquad \text{Success!}$$

The brackets help us calculate the middle term, the crucial step in this kind of factoring.

Here is a tougher factoring problem: Factor $2x^2 + 13x - 15$. It is a safe bet that if $2x^2 + 13x - 15$ factors at all, then

$$2x^2 + 13x - 15 = (2x + a)(x + b)$$

Since $ab = -15$, we are likely to try combinations of 3 and 5 first.

$$(2x + 5)(x - 3) = 2x^2 - x - 15$$

$$(2x - 5)(x + 3) = 2x^2 + x - 15$$

$$(2x + 3)(x - 5) = 2x^2 - 7x - 15$$

$$(2x - 3)(x + 5) = 2x^2 + 7x - 15$$

Discouraging, isn't it? But that is a poor reason to give up. Maybe we have missed some possibilities. We have, since combinations of 15 and 1 might work.

$$(2x - 15)(x + 1) = 2x^2 - 13x - 15$$

$$(2x + 15)(x - 1) = 2x^2 + 13x - 15 \qquad \text{Success!}$$

When you have had a lot of practice, you will be able to speed up the process. You will simply write

$$2x^2 + 13x - 15 = (2x + ?)(x + ?)$$

and mentally try the various possibilities until you find the right one. Of course, it may happen, as in the case of $2x^2 - 4x + 5$, that you cannot find a factorization.

PERFECT SQUARES

Certain second degree (quadratic) polynomials are a breeze to factor.

$$x^2 + 10x + 25 = (x + 5)(x + 5) = (x + 5)^2$$

$$x^2 - 12x + 36 = (x - 6)^2$$

$$4x^2 + 12x + 9 = (2x + 3)^2$$

These are modeled after the special product formulas 3 and 4, which we now write with a and b replacing x and y.

$$a^2 + 2ab + b^2 = (a + b)^2$$
$$a^2 - 2ab + b^2 = (a - b)^2$$

We look for first and last terms that are squares, say of a and b. Then we ask if the middle term is twice their product.

But we need to be very flexible; a and b might be quite complicated. Consider $x^4 + 2x^2y^3 + y^6$. The first term is the square of x^2 and the last is the square of y^3; the middle term is twice their product.

$$x^4 + 2x^2y^3 + y^6 = (x^2)^2 + 2(x^2)(y^3) + (y^3)^2$$
$$= (x^2 + y^3)^2$$

Similarly,

$$y^2z^2 - 6ayz + 9a^2 = (yz - 3a)^2.$$

However,

$$a^4b^2 + 6a^2bc + 4c^2 \neq (a^2b + 2c)^2$$

since the middle term does not check.

DIFFERENCE OF SQUARES

Do you see a common feature in the following polynomials?

$$x^2 - 16 \qquad y^2 - 100 \qquad 4y^2 - 9b^2$$

Each is the difference of two squares. From one of our special product formulas (formula 5), we know that

$$\boxed{a^2 - b^2 = (a + b)(a - b)}$$

Thus

$$x^2 - 16 = (x + 4)(x - 4)$$
$$y^2 - 100 = (y + 10)(y - 10)$$
$$4y^2 - 9b^2 = (2y + 3b)(2y - 3b)$$

SUM AND DIFFERENCE OF CUBES

Now we are ready for some high-class factoring. Consider $8x^3 + 27$ and $x^3z^3 - 1000$. The first is a sum of cubes and the second is a difference of cubes. The secrets to success are the two special product formulas for cubes, restated here.

$$\boxed{\begin{aligned} a^3 + b^3 &= (a + b)(a^2 - ab + b^2) \\ a^3 - b^3 &= (a - b)(a^2 + ab + b^2) \end{aligned}}$$

To factor $8x^3 + 27$, replace a by $2x$ and b by 3 in the first formula.

$$8x^3 + 27 = (2x)^3 + 3^3 = (2x + 3)[(2x)^2 - (2x)(3) + 3^2]$$
$$= (2x + 3)(4x^2 - 6x + 9)$$

Similarly, to factor $x^3 z^3 - 1000$, let $a = xz$ and $b = 10$ in the second formula.

$$x^3 z^3 - 1000 = (xz)^3 - 10^3 = (xz - 10)(x^2 z^2 + 10xz + 100)$$

Someone is sure to make a terrible mistake and write

$$x^3 + y^3 = (x + y)^3 \qquad \text{Wrong!!}$$

Remember that

$$(x + y)^3 = x^3 + 3x^2 y + 3xy^2 + y^3$$

TO FACTOR OR NOT TO FACTOR

Which of the following can be factored?

(i) $x^2 - 4$

(ii) $x^2 - 6$

(iii) $x^2 + 16$

Did you say only the first one? You are correct if we insist on integer coefficients, or as we say, if we **factor over the integers.** But if we factor over the real numbers (that is, insist only that the coefficients be real), then the second polynomial can be factored.

$$x^2 - 6 = (x + \sqrt{6})(x - \sqrt{6})$$

If we factor over the complex numbers, even the third polynomial can be factored.

$$x^2 + 16 = (x + 4i)(x - 4i)$$

For this reason, we should always spell out what kind of coefficients we permit in the answer. We give specific directions in the following problem set. Incidently, before trying the problems, review the opening panel of this section. It should make good sense now.

Problem Set 2-4

Factor completely over the integers (that is, allow only integer coefficients in your answers).

1. $x^2 + 5x$ 2. $y^3 + 4y^2$ 3. $x^2 + 5x - 6$

4. $x^2 + 5x + 4$ 5. $y^4 - 6y^3$ 6. $t^4 + t^2$

7. $y^2 + 4y - 12$ 8. $z^2 - 3z - 40$ 9. $y^2 + 8y + 16$
10. $9x^2 + 24x + 16$ 11. $4x^2 - 12xy + 9y^2$ 12. $9x^2 - 6x + 1$
13. $y^2 - 64$ 14. $x^2 - 4y^2$ 15. $1 - 25b^2$
16. $9x^2 - 64y^2$ 17. $4z^2 - 4z - 3$ 18. $7x^2 - 19x - 6$
19. $20x^2 + 3xy - 2y^2$ 20. $4x^2 + 13xy - 12y^2$ 21. $x^3 + 27$
22. $y^3 - 27$ 23. $a^3 - 8b^3$ 24. $8a^3 - 27b^3$
25. $x^3 - x^3y^3$ 26. $x^6 + x^3y^3$ 27. $x^2 - 3$
28. $y^2 - 5$ 29. $3x^2 - 4$

Factor completely over the real numbers.

30. $x^2 - 3$ 31. $y^2 - 5$ 32. $3x^2 - 4$
33. $5z^2 - 4$ 34. $t^4 - t^2$ 35. $t^4 - 2t^2$
36. $x^2 + 2\sqrt{2}x + 2$ 37. $y^2 - 2\sqrt{3}y + 3$ 38. $x^2 + 4y^2$
39. $x^2 + 9$ 40. $4x^2 + 1$

□ *Factor completely over the complex numbers.*

41. $x^2 + 9$ 42. $4x^2 + 1$

EXAMPLE A (Factoring by Substitution)   Factor completely over the integers.
(a) $3x^4 + 10x^2 - 8$
(b) $(x + 2y)^2 - 3(x + 2y) - 10$

Solution.
(a) Replace x^2 by u (or some other favorite letter of yours). Then
$$3x^4 + 10x^2 - 8 = 3u^2 + 10u - 8$$
But we know how to factor the latter:
$$3u^2 + 10u - 8 = (3u - 2)(u + 4)$$
Thus, when we go back to x, we get
$$3x^4 + 10x^2 - 8 = (3x^2 - 2)(x^2 + 4)$$
Neither of these quadratic polynomials factors further (using integer coefficients), so we are done.
(b) Here we could let $u = x + 2y$ and then factor the resulting quadratic polynomial. But this time, let us do that step mentally and write
$$(x + 2y)^2 - 3(x + 2y) - 10 = [(x + 2y) + ?][(x + 2y) - ?]$$
$$= (x + 2y + 2)(x + 2y - 5)$$

Factor completely over the integers. If you can factor without actually making a substitution, that is fine.

43. $x^6 + 9x^3 + 14$ 44. $x^4 - x^2 - 6$
45. $4x^4 - 37x^2 + 9$ 46. $6y^4 + 13y^2 - 5$
47. $(x + 4y)^2 + 6(x + 4y) + 9$ 48. $(m - n)^2 + 5(m - n) + 4$
49. $x^4 - x^2y^2 - 6y^4$ 50. $x^4y^4 + 5x^4y^2 + 6x^4$

EXAMPLE B (Factoring in Stages)   Factor $x^6 - y^6$ over the integers.

Solution.   First think of this expression as a difference of squares. Then factor again.

$$x^6 - y^6 = (x^3 - y^3)(x^3 + y^3)$$
$$= (x - y)(x^2 + xy + y^2)(x + y)(x^2 - xy + y^2)$$

Factor completely over the integers.

51. $x^6 - 64$

52. $x^4 - y^4$

53. $x^8 - x^4y^4$

54. $x^9 - 64x^3$

55. $x^6 + y^6$ (sum of cubes)

56. $a^6 + 64$

EXAMPLE C (Factoring by Grouping)   Factor.
  (a) $am - an - bm + bn$
  (b) $a^2 - 4ab + 4b^2 - c^2$

Solution.   To factor expressions involving more than three terms will usually require grouping of some of the terms together.
  (a) $am - an - bm + bn = (am - an) - (bm - bn)$
$$= a(m - n) - b(m - n)$$
$$= (a - b)(m - n)$$

Note that $m - n$ was a common factor of both terms. Only after removing that factor did we have a factored form.
  (b) $a^2 - 4ab + 4b^2 - c^2 = (a^2 - 4ab + 4b^2) - c^2$
$$= (a - 2b)^2 - c^2 \quad \text{(difference of squares)}$$
$$= [(a - 2b) + c][(a - 2b) - c]$$
$$= (a - 2b + c)(a - 2b - c)$$

Factor completely over the integers.

57. $x^3 - 4x^2 + x - 4$

58. $y^3 + 3y^2 - 2y - 6$

59. $4x^2 - 4x + 1 - y^2$

60. $9a^2 - 4b^2 - 12b - 9$

61. $3x + 3y - x^2 - xy$

62. $y^2 - 3y + xy - 3x$

63. $x^2 + 6xy + 9y^2 + 2x + 6y$

64. $y^2 + 4xy + 4x^2 - 3y - 6x$

65. $x^2 + 2xy + y^2 + 3x + 3y + 2$

66. $a^2 - 2ab + b^2 - c^2 + 4cd - 4d^2$

EXAMPLE D (Factoring by Adding and Subtracting the Same Thing)   Factor $x^4 + 4$ over the integers.

Solution.   Most people would bet money that this cannot be factored. We have to admit that it is tricky. But see what happens when we add and subtract $4x^2$.

$$x^4 + 4 = x^4 + 4x^2 + 4 - 4x^2$$
$$= (x^4 + 4x^2 + 4) - 4x^2$$
$$= (x^2 + 2)^2 - (2x)^2 \qquad \text{(difference of squares)}$$
$$= (x^2 + 2 + 2x)(x^2 + 2 - 2x)$$
$$= (x^2 + 2x + 2)(x^2 - 2x + 2)$$

Factor completely over the integers.

67. $x^4 + 64$

68. $y^8 + 4$

69. $x^4 + x^2 + 1$

70. $x^8 + 3x^4 + 4$

MISCELLANEOUS PROBLEMS

In Problems 71–94, factor completely over the integers.

71. $4 - 9m^2$

72. $9m^2 + 6m + 1$

73. $6x^2 - 5x + 1$

74. $6x^2 - 5x - 6$

75. $5x^3 - 20x$

76. $3x^2 - 18x + 27$

77. $6x^3 - 5x^2 + x$

78. $4x^5 - 32x^2$

79. $2u^4 - 7u^2 + 5$

80. $x^3y^2 + x^2y^3 + 2xy^4$

81. $(a + 2b)^2 - 3(a + 2b) - 28$

82. $(2a + b)^3 - 8$

83. $(a + 3b)^4 - 1$

84. $x^2 + 4xy + 4y^2 - x^2y^2$

85. $x^2 - 6xy + 9y^2 + 4x - 12y$

86. $x^2 - 2xy + y^2 + 3x - 3y + 2$

87. $9x^4 - 24x^2y^2 + 16y^4 - y^2$

88. $9x^4 + 15x^2y^2 + 16y^4$

89. $x^4 - 3x^2y^2 + y^4$

90. $8x^2(x - 2) + 8x(x - 2) + 2(x - 2)$

91. $(x + 3)^2(x + 2)^3 - 20(x + 3)(x + 2)^2$

92. $(x - 3y)(x + 5y)^4 - 4(x - 3y)(x + 5y)^2$

93. $x^{2n} + 3x^n + 2$

94. $x^{n+3} + 5x^n + x^3 + 5$

95. Calculate each of the following the easy way.

(a) $(547)^2 - (453)^2$

(b) $\dfrac{2^{20} - 2^{17} + 7}{2^{17} + 1}$

(c) $\left(1 - \dfrac{1}{2^2}\right)\left(1 - \dfrac{1}{3^2}\right)\left(1 - \dfrac{1}{4^2}\right) \cdots \left(1 - \dfrac{1}{29^2}\right)$

96. If n is an integer greater than 1, then $a^n - b^n = (a - b)P$, where P is a polynomial in a and b. Find the general form for P.

97. Note that $2^3 = 3^2 - 1^2$ and $3^3 = 6^2 - 3^2$. Show that the cube of any integer n can be written as the difference of the squares of two integers. *Hint:* One of these integers can be chosen as $n(n + 1)/2$.

98. TEASER If $a + b + c = 1$, $a^2 + b^2 + c^2 = 2$, and $a^3 + b^3 + c^3 = 3$, find $a^4 + b^4 + c^4$. *Hint:* This generalizes Problem 92 of Section 2-3.

Thomas
Jefferson

Riddle: What do the following have in common?

(i) The Declaration of Independence

(ii) $\dfrac{x^2y^5 \;+\; 13xy^7 \;+\; 25x^4y^2 \;+\; 17}{3xy^3 \;+\; 105y^2 \;+\; 14x \;-\; 2x^5y^7}$

Answer: They are both rational expressions.

"I'd say the second expression was a
mite more rational!"

2-5   Rational Expressions

We admit that the riddle above is a bit silly, but it allowed us to insert
Jefferson's picture and to make a point that is not well known. Thomas
Jefferson was trained in mathematics and wrote that he had been profoundly
influenced by his college mathematics teacher.

Recall that the quotient (ratio) of two integers is a rational number. A
quotient (ratio) of two polynomials is called a **rational expression.** Here are
some examples involving one variable.

$$\frac{x}{x+1} \qquad \frac{3x^2+6}{x^2+2x} \qquad \frac{x^{15}+3x}{6x^2+11}$$

The example in the riddle is a rational expression in two variables.

We add, subtract, multiply, and divide rational expressions by the same
rules we used for rational numbers (see Section 1-2). The result is always a
rational expression. That should not surprise us, since we had the same kind
of experience with rational numbers.

Our immediate task is to define the reduced form for a rational expression. This, too, generalizes a notion we had for rational numbers.

CAUTION

$$\frac{x^3+3x^2}{x^3} = 1 + 3x^2$$

$$\frac{x^3+3x^2}{x^3} = \frac{x^2(x+3)}{x^3}$$

$$= \frac{x+3}{x}$$

REDUCED FORM

A rational expression is in **reduced form** if its numerator and denominator
have no (nontrivial) common factor. For example, $x/(x+2)$ is in reduced
form, but $x^2/(x^2+2x)$ is not, since

$$\frac{x^2}{x^2+2x} = \frac{\cancel{x} \cdot x}{\cancel{x}(x+2)} = \frac{x}{x+2}$$

To reduce a rational expression, we factor numerator and denominator and divide out, or cancel, common factors. Here are two examples.

$$\frac{x^2 + 7x + 10}{x^2 - 25} = \frac{(x + 5)(x + 2)}{(x + 5)(x - 5)} = \frac{x + 2}{x - 5}$$

$$\frac{2x^2 + 5xy - 3y^2}{2x^2 + xy - y^2} = \frac{(2x - y)(x + 3y)}{(2x - y)(x + y)} = \frac{x + 3y}{x + y}$$

ADDITION AND SUBTRACTION

We add (or subtract) rational expressions by rewriting them so that they have the same denominator and then adding (or subtracting) the new numerators. Suppose we want to add

$$\frac{3}{x} + \frac{2}{x + 1}$$

The appropriate common denominator is $x(x + 1)$. Remember that we can multiply numerator and denominator of a fraction by the same thing. Accordingly,

$$\frac{3}{x} + \frac{2}{x + 1} = \frac{3(x + 1)}{x(x + 1)} + \frac{x \cdot 2}{x(x + 1)} = \frac{3x + 3 + 2x}{x(x + 1)} = \frac{5x + 3}{x(x + 1)}$$

The same procedure is used in subtraction.

$$\frac{2}{x + 1} - \frac{3}{x} = \frac{x \cdot 2}{x(x + 1)} - \frac{3(x + 1)}{x(x + 1)} = \frac{2x - (3x + 3)}{x(x + 1)}$$

$$= \frac{2x - 3x - 3}{x(x + 1)} = \frac{-x - 3}{x(x + 1)}$$

Here is a more complicated example. Study each step carefully.

$(x - 1)^2 (x + 1)$ is the lowest common denominator

Forgetting the parentheses around $2x^2 + 3x + 1$ would be a serious blunder

No cancellation is possible since $x^2 - 6x - 1$ doesn't factor over the integers

$$\frac{3x}{x^2 - 1} - \frac{2x + 1}{x^2 - 2x + 1} = \frac{3x}{(x - 1)(x + 1)} - \frac{2x + 1}{(x - 1)^2}$$

$$= \frac{3x(x - 1)}{(x - 1)^2(x + 1)} - \frac{(2x + 1)(x + 1)}{(x - 1)^2(x + 1)}$$

$$= \frac{3x^2 - 3x - (2x^2 + 3x + 1)}{(x - 1)^2(x + 1)}$$

$$= \frac{3x^2 - 3x - 2x^2 - 3x - 1}{(x - 1)^2(x + 1)}$$

$$= \frac{x^2 - 6x - 1}{(x - 1)^2(x + 1)}$$

MULTIPLICATION

We multiply rational expressions in the same manner as we do rational numbers; that is, we multiply numerators and multiply denominators. For example,

$$\frac{3}{x + 5} \cdot \frac{x + 2}{x - 4} = \frac{3(x + 2)}{(x + 5)(x - 4)} = \frac{3x + 6}{x^2 + x - 20}$$

Sometimes we need to reduce the product, if we want the simplest possible answer. Here is an illustration.

$$\frac{2x - 3}{x + 5} \cdot \frac{x^2 - 25}{6xy - 9y} = \frac{(2x - 3)(x^2 - 25)}{(x + 5)(6xy - 9y)}$$

$$= \frac{(2x - 3)(x - 5)(x + 5)}{(x + 5)(3y)(2x - 3)}$$

$$= \frac{x - 5}{3y}$$

This example shows that it is a good idea to do as much factoring as possible at the outset. That is what set up the cancellation.

DIVISION

There are no real surprises with division, as we simply invert the divisor and multiply. Here is a nontrivial example.

$$\frac{x^2 - 5x + 4}{2x + 6} \div \frac{2x^2 - x - 1}{x^2 + 5x + 6} = \frac{x^2 - 5x + 4}{2x + 6} \cdot \frac{x^2 + 5x + 6}{2x^2 - x - 1}$$

$$= \frac{(x^2 - 5x + 4)(x^2 + 5x + 6)}{(2x + 6)(2x^2 - x - 1)}$$

$$= \frac{(x - 4)(x - 1)(x + 2)(x + 3)}{2(x + 3)(x - 1)(2x + 1)}$$

$$= \frac{(x - 4)(x + 2)}{2(2x + 1)}$$

Problem Set 2-5

Reduce each of the following.

1. $\dfrac{x + 6}{x^2 - 36}$

2. $\dfrac{x^2 - 1}{4x - 4}$

3. $\dfrac{y^2 + y}{5y + 5}$

4. $\dfrac{x^2 - 7x + 6}{x^2 - 4x - 12}$

5. $\dfrac{(x + 2)^3}{x^2 - 4}$

6. $\dfrac{x^3 + a^3}{(x + a)^2}$

7. $\dfrac{zx^2 + 4xyz + 4y^2z}{x^2 + 3xy + 2y^2}$

8. $\dfrac{x^3 - 27}{3x^2 + 9x + 27}$

Perform the indicated operations and simplify.

9. $\dfrac{5}{x - 2} + \dfrac{4}{x + 2}$

10. $\dfrac{5}{x - 2} - \dfrac{4}{x + 2}$

11. $\dfrac{5x}{x^2 - 4} + \dfrac{3}{x + 2}$

12. $\dfrac{3}{x} - \dfrac{2}{x + 3} + \dfrac{1}{x^2 + 3x}$

13. $\dfrac{2}{xy} + \dfrac{3}{xy^2} - \dfrac{1}{x^2y^2}$

14. $\dfrac{x + y}{xy^3} - \dfrac{x - y}{y^4}$

15. $\dfrac{x + 1}{x^2 - 4x + 4} + \dfrac{4}{x^2 + 3x - 10}$

16. $\dfrac{x^2}{x^2 - x + 1} - \dfrac{x + 1}{x}$

EXAMPLE A (The Three Signs of a Fraction) Simplify

$$\frac{x}{3x - 6} - \frac{2}{2 - x}$$

Solution.

$$\frac{x}{3x - 6} - \frac{2}{2 - x} = \frac{x}{3(x - 2)} - \frac{2}{2 - x}$$

Now we make a crucial observation. Notice that

$$-(2 - x) = -2 + x = x - 2$$

CAUTION

$$\dfrac{5}{x - 2} - \dfrac{x + 2}{x - 2} = \dfrac{5 - x + 2}{x - 2}$$

$$= \dfrac{7 - x}{x - 2}$$

$$\dfrac{5}{x - 2} - \dfrac{x + 2}{x - 2} = \dfrac{5 - x - 2}{x - 2}$$

$$= \dfrac{3 - x}{x - 2}$$

That is, $2 - x$ and $x - 2$ are negatives of each other. Thus the expression above may be rewritten as

$$\frac{x}{3(x - 2)} - \frac{2}{2 - x} = \frac{x}{3(x - 2)} - \frac{2}{-(x - 2)}$$

$$= \frac{x}{3(x - 2)} + \frac{2}{x - 2}$$

$$= \frac{x}{3(x - 2)} + \frac{6}{3(x - 2)}$$

$$= \frac{x + 6}{3(x - 2)}$$

When we replaced $-\dfrac{2}{-(x - 2)}$ by $\dfrac{2}{x - 2}$, we used the fact that

$$-\frac{a}{-b} = \frac{a}{b}$$

Keep in mind that a fraction has three sign positions: numerator, denominator, and total fraction. You may change any two of them without changing the value of the fraction. Thus

$$\frac{a}{b} = -\frac{a}{-b} = -\frac{-a}{b} = \frac{-a}{-b}$$

Simplify.

17. $\dfrac{4}{2x - 1} + \dfrac{x}{1 - 2x}$

18. $\dfrac{x}{6x - 2} - \dfrac{3}{1 - 3x}$

19. $\dfrac{2}{6y - 2} + \dfrac{y}{9y^2 - 1} - \dfrac{2y + 1}{1 - 3y}$

20. $\dfrac{x}{4x^2 - 1} + \dfrac{2}{4x - 2} - \dfrac{3x + 1}{1 - 2x}$

21. $\dfrac{m^2}{m^2 - 2m + 1} - \dfrac{1}{3 - 3m}$

22. $\dfrac{2x}{x^2 - y^2} + \dfrac{1}{x + y} + \dfrac{1}{y - x}$

In Problems 23–30, multiply and express in simplest form, as illustrated in the text.

23. $\dfrac{5}{2x - 1} \cdot \dfrac{x}{x + 1}$

24. $\dfrac{3}{x^2 - 2x} \cdot \dfrac{x - 2}{x}$

25. $\dfrac{x + 2}{x^2 - 9} \cdot \dfrac{x + 3}{x^2 - 4}$

26. $\left(1 + \dfrac{1}{x + 2}\right)\left(\dfrac{4}{3x + 9}\right)$

27. $x^2 y^4\left(\dfrac{x}{y^2} - \dfrac{y}{x^2}\right)$

28. $\dfrac{5x^2}{x^3 + y^3}\left(\dfrac{1}{xy^2} - \dfrac{1}{x^2 y^2} + \dfrac{1}{x^3 y}\right)$

29. $\left(\dfrac{x^2 + 5x}{x^2 - 16}\right)\left(\dfrac{x^2 - 2x - 24}{x^2 - x - 30}\right)$

30. $\left(\dfrac{x^3 - 125}{2x^3 - 10x^2}\right)\left(\dfrac{7x}{x^3 + 5x^2 + 25x}\right)$

Express the quotients in Problems 31–37 in simplest form, as illustrated in the text.

31. $\dfrac{\dfrac{5}{2x - 1}}{\dfrac{x}{x + 1}}$

32. $\dfrac{\dfrac{5}{2x - 1}}{\dfrac{x}{4x^2 - 1}}$

33. $\dfrac{\dfrac{x + 2}{x^2 - 4}}{x}$

34. $\dfrac{\dfrac{x + 2}{x^2 - 3x}}{\dfrac{x^2 - 4}{x}}$

35. $\dfrac{\dfrac{x^2 + a^3}{x^3 - a^3}}{\dfrac{x + 2a}{(x - a)^2}}$

36. $\dfrac{1 + \dfrac{2}{b}}{1 - \dfrac{4}{b^2}}$

37. $\dfrac{\dfrac{y^2 + y - 2}{y^2 + 4y}}{\dfrac{2y^2 - 8}{y^2 + 2y - 8}}$

EXAMPLE B (A Quotient Arising in Calculus) Simplify

$$\frac{\dfrac{2}{x+h}-\dfrac{2}{x}}{h}$$

Solution.   This expression may look artificial, but it is one you are apt to find in calculus. It represents the average rate of change in $2/x$ as x changes to $x+h$. We begin by doing the subtraction in the numerator.

$$\frac{\dfrac{2}{x+h}-\dfrac{2}{x}}{h}=\frac{\dfrac{2x-2(x+h)}{(x+h)x}}{h}=\frac{\dfrac{2x-2x-2h}{(x+h)x}}{\dfrac{h}{1}}$$

$$=\frac{-2h}{(x+h)x}\cdot\frac{1}{h}=\frac{-2}{(x+h)x}$$

Simplify each of the following.

38. $\dfrac{\dfrac{4}{x+h}-\dfrac{4}{x}}{h}$

39. $\dfrac{\dfrac{1}{2x+2h+3}-\dfrac{1}{2x+3}}{h}$

40. $\dfrac{\dfrac{x+h}{x+h+4}-\dfrac{x}{x+4}}{h}$

41. $\dfrac{\dfrac{1}{(x+h)^2}-\dfrac{1}{x^2}}{h}$

EXAMPLE C (Four-Story Fractions) Simplify

$$\frac{\dfrac{x}{x-4}-\dfrac{3}{x+3}}{\dfrac{1}{x}+\dfrac{2}{x-4}}$$

Solution.   *Method 1*   (Simplify the numerator and denominator separately and then divide.)

$$\frac{\dfrac{x}{x-4}-\dfrac{3}{x+3}}{\dfrac{1}{x}+\dfrac{2}{x-4}}=\frac{\dfrac{x(x+3)-3(x-4)}{(x-4)(x+3)}}{\dfrac{x-4+2x}{x(x-4)}}$$

$$=\frac{\dfrac{x^2+3x-3x+12}{(x-4)(x+3)}}{\dfrac{3x-4}{x(x-4)}}$$

$$=\frac{x^2+12}{(x-4)(x+3)}\cdot\frac{x(x-4)}{3x-4}$$

$$=\frac{x(x^2+12)}{(x+3)(3x-4)}$$

Method 2 (Multiply the fractions in the numerator and denominator by a common denominator, in this case, $(x-4)(x+3)x$.)

$$\frac{\dfrac{x}{x-4}-\dfrac{3}{x+3}}{\dfrac{1}{x}+\dfrac{2}{x-4}} = \frac{\left(\dfrac{x}{x-4}-\dfrac{3}{x+3}\right)(x-4)(x+3)x}{\left(\dfrac{1}{x}+\dfrac{2}{x-4}\right)(x-4)(x+3)x}$$

$$= \frac{x^2(x+3)-3x(x-4)}{(x-4)(x+3)+2x(x+3)}$$

$$= \frac{x^3+3x^2-3x^2+12x}{(x+3)(x-4+2x)}$$

$$= \frac{x(x^2+12)}{(x+3)(3x-4)}$$

Simplify, using either of the above methods.

42. $$\dfrac{\dfrac{1}{x+2}-\dfrac{3}{x^2-4}}{\dfrac{3}{x-2}}$$

43. $$\dfrac{\dfrac{y}{y+4}-\dfrac{2}{y^2+5y+4}}{\dfrac{4}{y+1}+\dfrac{3}{y+4}}$$

44. $$\dfrac{\dfrac{1}{x}-\dfrac{1}{x-2}+\dfrac{3}{x^2-2x}}{\dfrac{x}{x-2}+\dfrac{3}{x}}$$

45. $$\dfrac{\dfrac{a^2}{b^2}-\dfrac{b^2}{a^2}}{\dfrac{a}{b}-\dfrac{b}{a}}$$

46. $$\dfrac{n-\dfrac{n^2}{n-m}}{1+\dfrac{m^2}{n^2-m^2}}$$

47. $$\dfrac{\dfrac{x^2}{x-y}-x}{\dfrac{y^2}{x-y}+y}$$

48. $1 - \dfrac{x-(1/x)}{1-(1/x)}$

49. $$\dfrac{y-\dfrac{1}{1+(1/y)}}{y+\dfrac{1}{y-(1/y)}}$$

MISCELLANEOUS PROBLEMS

In Problems 50–59, perform the indicated operations and simplify.

50. $\dfrac{1-2x}{x-4}+\dfrac{6x+2}{3x-12}$

51. $x+y+\dfrac{y^2}{x-y}$

52. $x+y+\dfrac{y^2}{x-y}+\dfrac{x^2}{x+y}$

53. $\left(\dfrac{1}{x}+\dfrac{1}{y}\right)\left(x+y-\dfrac{x^2+y^2}{x+y}\right)$

54. $\dfrac{x}{x^2-5x+6}+\dfrac{3}{x^2-7x+12}$

55. $\dfrac{x}{x^2+11x+30}-\dfrac{5}{x^2+9x+20}$

56. $\dfrac{\dfrac{a-b}{a+b} - \dfrac{a+b}{a-b}}{\dfrac{ab}{a-b}}$

57. $\dfrac{\dfrac{a^2+4a+3}{a} - \dfrac{2a+2}{a-1}}{\dfrac{a+1}{a^2-a}}$

58. $\dfrac{x^3-8y^3}{x^2-4y^2} + \dfrac{2xy}{x+2y}$

59. $\dfrac{18x^2y - 27x^2 - 8y + 12}{6xy - 4y - 9x + 6}$

60. Let x, y, and z be positive numbers. How does $(x+z)/(y+z)$ compare in size with x/y?

61. Arnold Thinkhard simplified

$$\frac{x^3+y^3}{x^3+(x-y)^3} \quad \text{to} \quad \frac{x+y}{x+(x-y)}$$

by canceling all the 3s. Even though his method was entirely wrong, show that he got the right answer.

62. A number of the form

$$a_0 + \cfrac{1}{a_1 + \cfrac{1}{a_2 + \cfrac{1}{a_3 + \cdots}}}$$

where a_k is an integer with $a_k \geq 0$ if $k \geq 1$, is called a **continued fraction**. Here the dots indicate a pattern that may either continue indefinitely or terminate. Simplify each of the following terminating continued fractions.

(a) $2 + \cfrac{1}{2 + \cfrac{1}{2 + \frac{1}{2}}}$

(b) $1 + \cfrac{1}{2 + \cfrac{1}{3 + \frac{1}{4}}}$

(c) $x + \cfrac{1}{x + \cfrac{1}{x + \frac{1}{x}}}$

63. A terminating continued fraction clearly represents a rational number (Why?). Conversely, every rational number can be represented as a terminating continued fraction. For example,

$$\frac{7}{4} = 1 + \frac{3}{4} = 1 + \frac{1}{\frac{4}{3}} = 1 + \cfrac{1}{1 + \frac{1}{3}}$$

Find the continued fraction expansion of each of the following.

(a) $\dfrac{13}{5}$

(b) $\dfrac{29}{11}$

(c) $-\dfrac{5}{4}$

64. **TEASER** Evaluate the nonterminating continued fraction

$$\phi = 1 + \cfrac{1}{1 + \cfrac{1}{1 + \cfrac{1}{1 + \cdots}}}$$

It would spoil the problem to give a real hint but Problem 63 implies that ϕ is irrational and we will tell you that ϕ satisfies a simple quadratic equation. Our choice of the Greek letter ϕ to denote this number is deliberate. The ancient Greeks loved this number and gave it a special name.

Chapter Summary

An **exponent** is a numerical superscript placed on a number to indicate that a certain operation is to be performed on that number. In particular

$$b^4 = b \cdot b \cdot b \cdot b \qquad b^0 = 1 \qquad b^{-3} = \frac{1}{b^3} = \frac{1}{b \cdot b \cdot b}$$

Exponents mix together according to five laws called the **rules of exponents.**

A number is in **scientific notation** when it appears in the form $c \times 10^n$, where $1 \le |c| < 10$ and n is an integer. Very small and very large numbers are commonly written this way. Pocket calculators often use scientific notation in their displays. These electronic marvels are designed to take the drudgery out of arithmetic calculations and are a valuable tool in an algebra-trigonometry course.

An expression of the form

$$a_n x^n + a_{n-1} x^{n-1} + \cdots + a_1 x + a_0$$

is called a **polynomial** in x. The exponent n is its **degree** (provided $a_n \ne 0$); the a's are its **coefficients.** Polynomials can be added, subtracted, and multiplied, the result in each case being another polynomial.

To **factor** a polynomial is to write it as a product of simpler polynomials; to **factor over the integers** is to write a polynomial as a product of polynomials with integer coefficients. Here are five examples.

$x^2 - 2ax = x(x - 2a)$	Common factor
$4x^2 - 25 = (2x + 5)(2x - 5)$	Difference of squares
$6x^2 + x - 15 = (2x - 3)(3x + 5)$	Trial and error
$x^2 + 14x + 49 = (x + 7)^2$	Perfect square
$x^3 + 1000 = (x + 10)(x^2 - 10x + 100)$	Sum of cubes

A quotient (ratio) of two polynomials is called a **rational expression.** The expression is in **reduced form** if its numerator and denominator have no nontrivial common factors. We add, subtract, multiply, and divide rational expressions in much the same way as we do rational numbers.

Chapter Review Problem Set

Simplify, leaving your answer free of negative exponents.

1. $\left(\dfrac{3}{4}\right)^{-2}$

2. $\left(\dfrac{5}{6}\right)^2 \left(\dfrac{5}{6}\right)^{-4}$

3. $\left(\dfrac{5}{6} + \dfrac{1}{3}\right)^{-2}$

4. $\dfrac{2x^{-2}y^2}{4xy^{-3}}$

5. $(3x^{-2}y^3)^{-3}$

6. $\dfrac{(a^{-2}b)^2(2ab^{-3})^{-1}}{(ab^{-2})^3}$

Express in scientific notation.

7. 1,382,000

8. $(3.1)10^4(2.2)10^{-7}$

9. $\dfrac{(6.5)10^4}{(1.3)10^{-3}}$

Decide which of the following are polynomials. Give the degree of each polynomial.

10. $\frac{3}{2}x^2 - \sqrt{2}x + \pi$

11. $10x^3 + 5\sqrt{x} + 6$

12. $4t^{-2} + 5t^{-1} + 6$

13. $\dfrac{4}{x^3 + 11}$

Perform the indicated operations and simplify.

14. $(3x - 5) + (2x^2 - 2x + 11)$

15. $5 - x^3 - (x^2 - 2x^3 + x - 2)$

16. $(2x - 3)(x + 5)$

17. $(2x - 1)^2 - 4x^2$

18. $(z^3 + 4)(z^3 - 4)$

19. $(2x^2 - 3w)(x^2 + 4w)$

20. $(x + 2a)(x^2 - 2ax + 4a^2)$

21. $(2y - 3)(3y^2 - 5y + 6)$

22. $(3t^2 - t + 1)^2$

23. $(a + bcd)(a - bcd)$

In Problems 24–33, factor completely over the integers.

24. $2x^4 - x^3 + 11x^2$

25. $y^2 - 7y + 12$

26. $6z^2 + z - 1$

27. $49a^2 - 25$

28. $9c^2 - 24cd + 16d^2$

29. $a^3 - 27$

30. $8a^3b^3 + 1$

31. $x^8 - x^4y^4$

32. $x^2 + 2xy + y^2 - z^4$

33. $4c^2 - d^2 - 6c - 3d$

34. Factor $9x^2 - 11$ over the real numbers.

⊡ 35. Factor $x^2 + 16$ over the complex numbers.

Reduce each of the following.

36. $\dfrac{x^3 - 8}{2x - 4}$

37. $\dfrac{2x - 2x^2}{x^3 - 2x^2 + x}$

Perform the indicated operations and simplify.

38. $\dfrac{18}{x^2 + 3x} - \dfrac{4}{x} + \dfrac{6}{x + 3}$

39. $\dfrac{x^2 + x - 6}{x^2 - 1} \cdot \dfrac{x^2 + x - 2}{x^2 + 5x + 6}$

40. $\dfrac{\dfrac{x}{x - 3} - \dfrac{2}{x^2 - 4x + 3}}{\dfrac{5}{x - 1} + \dfrac{5}{x - 3}}$

*As the sun eclipses the stars
by its brilliancy, so the man
of knowledge will eclipse the
fame of others in the
assemblies of the people if he
proposes algebraic problems,
and still more if he solves
them.*

—Brahmagupta

CHAPTER 3

Equations
and Inequalities

Euclid's *Elements*, composed about 300 B.C., has been reprinted in several hundred editions. It is, next to the Bible, the most influential book ever written. The common notions at the left are four of the ten basic axioms with which Euclid begins his book.

3-1   Equations

It was part of Euclid's genius to recognize that our usage of the word *equals* is fundamental to all that we do in mathematics. But to describe that usage may not be as simple as Euclid thought. When we write

$$.25 + \frac{3}{4} + \frac{1}{3} - (.3333\ldots) = 1$$

Equality?

"We hold these truths to be self-evident, that all men are created equal ... "

we certainly do not mean that the symbol on the left coincides with the one on the right. Instead, we mean that both symbols, the complicated one and the simple one, stand for (or name) the same number. That is the basic meaning of *equals* as used in this book.

But having said that, we must make another distinction. When we write

$$x^2 - 25 = (x - 5)(x + 5)$$

and

$$x^2 - 25 = 0$$

we have two quite different things in mind. In the first case, we are making an assertion. We claim that no matter what number x represents, the expressions on the left and right of the equality stand for the same number. This certainly cannot be our meaning in the second case. There we are asking a question: What numbers can x symbolize so that both sides of the equality $x^2 - 25 = 0$ stand for the same number?

An equality that is true for all values of the variable is called an **identity.** One that is true only for some values is called a conditional **equation.** And here are the corresponding jobs for us to do. We **prove** identities, but we **solve** (or find the solutions of) equations. Both jobs will be very important in this book; however, it is the second that interests us most right now.

SOLVING EQUATIONS

Sometimes we can solve an equation by inspection. It takes no mathematical apparatus and little imagination to see that

$$x + 4 = 6$$

has $x = 2$ as a solution. On the other hand, to solve

$$2x^2 + 8x = 8x + 18$$

is quite a different matter. For this kind of equation, we need some machinery. Our general strategy is to modify an equation one step at a time until it is in a form where the solution is obvious. Of course, we must be careful that the modifications we make do not change the solutions. Here, too, Euclid pointed the way.

RULES FOR MODIFYING EQUATIONS

1. Adding the same quantity to (or subtracting the same quantity from) both sides of an equation does not change its solutions.
2. Multiplying (or dividing) both sides of an equation by the same nonzero quantity does not change its solutions.

Consider $2x^2 + 8x = 8x + 18$ again. One way to solve this equation is to use the following steps.

Given equation:	$2x^2 + 8x = 8x + 18$
Subtract $8x$:	$2x^2 = 18$
Divide by 2:	$x^2 = 9$
Take square roots:	$x = 3$ or -3

Thus the solutions of $2x^2 + 8x = 8x + 18$ are 3 and -3. In Section 3-4, we will solve equations like this one by other methods, but the rules stated above will continue to play a fundamental role.

LINEAR EQUATIONS

The simplest kind of equation to solve is one in which the *variable* (also called the *unknown*) occurs only to the first power. Consider

$$12x - 9 = 5x + 5$$

Our procedure is to use the rules for modifying equations to bring all the terms in x to one side and the constant terms to the other and then to divide by the coefficient of x. The result is that we have x all alone on one side of the equation and a number (the solution) on the other.

$$\begin{array}{ll}
\text{Given equation:} & 12x - 9 = 5x + 5 \\
\text{Add 9:} & 12x = 5x + 14 \\
\text{Subtract } 5x: & 7x = 14 \\
\text{Divide by 7:} & x = 2
\end{array}$$

It is always a good idea to check your answer. In the original equation, replace x by the value that you found, to see if a true statement results.

$$12(2) - 9 \stackrel{?}{=} 5(2) + 5$$

$$15 = 15$$

An equation of the form $ax + b = 0$ $(a \neq 0)$ is called a **linear equation.** It has one solution, $x = -b/a$. Many equations not initially in this form can be transformed to it using the rules we have learned.

EQUATIONS THAT CAN BE CHANGED TO LINEAR FORM

Consider

$$\frac{2}{x + 1} = \frac{3}{2x - 2}$$

If we agree to exclude $x = -1$ and $x = 1$ from consideration, then $(x + 1)(2x - 2)$ is not zero, and we may multiply both sides by that expression. We get

$$\frac{2}{x + 1}(x + 1)(2x - 2) = \frac{3}{2x - 2}(x + 1)(2x - 2)$$

$$2(2x - 2) = 3(x + 1)$$

$$4x - 4 = 3x + 3$$

$$x = 7$$

As usual, we check our solution in the original equation.

$$\frac{2}{7 + 1} \stackrel{?}{=} \frac{3}{14 - 2}$$

$$\frac{2}{8} = \frac{3}{12}$$

So $x = 7$ is a solution.

The importance of checking is illustrated by our next example.

$$\frac{3x}{x - 3} = 1 + \frac{9}{x - 3}$$

To solve, we multiply both sides by $x - 3$ and then simplify.

$$\frac{3x}{x-3}(x-3) = \left(1 + \frac{9}{x-3}\right)(x-3)$$

$$3x = x - 3 + 9$$

$$2x = 6$$

$$x = 3$$

When we check in the original equation, we get

$$\frac{3 \cdot 3}{3 - 3} \overset{?}{=} 1 + \frac{9}{3 - 3}$$

This is nonsense, since it involves division by zero. What went wrong? If $x = 3$, then in our very first step we actually multiplied both sides of the equation by zero, a forbidden operation. Thus the given equation has no solution.

The strategy of multiplying both sides by $x - 3$ in this example was appropriate, even though it initially led us to an incorrect answer. We did not worry, because we knew that in the end we were going to run a check. We should always check answers, especially in any situation in which we have multiplied by an expression involving the unknown. Such a multiplication may introduce an *extraneous* solution (but never results in the loss of a solution).

Here is another example of a similar type of equation.

$$\frac{x+4}{(x+1)(x-2)} - \frac{3}{x+1} - \frac{2}{x-2} = \frac{-8}{(x+1)(x-2)}$$

We have no choice but to multiply both sides of the equation by (x + 1) (x − 2). This gives

$$x + 4 - 3(x - 2) - 2(x + 1) = -8$$

$$x + 4 - 3x + 6 - 2x - 2 = -8$$

$$-4x + 8 = -8$$

$$-4x = -16$$

$$x = 4$$

At this point, $x = 4$ is an apparent solution; however, we are not sure until we check it in the original equation.

$$\frac{4+4}{(4+1)(4-2)} - \frac{3}{4+1} - \frac{2}{4-2} \overset{?}{=} \frac{-8}{(4+1)(4-2)}$$

$$\frac{8}{5 \cdot 2} - \frac{3}{5} - \frac{2}{2} \overset{?}{=} \frac{-8}{5 \cdot 2}$$

$$\frac{8}{10} - \frac{6}{10} - \frac{10}{10} = \frac{-8}{10}$$

It works, so $x = 4$ is a solution.

Problem Set 3-1

Determine which of the following are identities and which are conditional equations.

1. $2(x + 4) = 8$
2. $2(x + 4) = 2x + 8$
3. $3(2x - \frac{2}{3}) = 6x - 2$
4. $2x - 4 - \frac{2}{3}x = \frac{4}{3}x - 4$
5. $\frac{2}{3}x + 4 = \frac{1}{2}x - 1$
6. $3(x - 2) = 2(x - 3) + x$
7. $(x + 2)^2 = x^2 + 4$
8. $x(x + 2) = x^2 + 2x$
9. $x^2 - 9 = (x + 3)(x - 3)$
10. $x^2 - 5x + 6 = (x - 1)(x - 6)$

Solve each of the following equations.

11. $4x - 3 = 3x - 1$
12. $2x + 5 = 5x + 14$
13. $2t + \frac{1}{2} = 4t - \frac{7}{2} + 8t$
14. $y + \frac{1}{3} = 2y - \frac{2}{3} - 6y$
15. $3(x - 2) = 5(x - 3)$
16. $4(x + 1) = 2(x - 3)$
17. $\sqrt{3}z + 4 = -\sqrt{3}z + 8$
18. $\sqrt{2}x + 1 = x + \sqrt{2}$

© 19. $3.23x - 6.15 = 1.41x + 7.63$

$$\left(\text{First obtain } x = \frac{7.63 + 6.15}{3.23 - 1.41} \text{ and then use the calculator.}\right)$$

© 20. $42.1x + 11.9 = 1.03x - 4.32$

© 21. $(6.13 \times 10^{-8})x + (5.34 \times 10^{-6}) = 0$

© 22. $(5.11 \times 10^{11})x - (6.12 \times 10^{12}) = 0$

EXAMPLE A (Equations Involving Fractions) Solve $\frac{2}{3}x - \frac{3}{4} = \frac{7}{6}x + \frac{1}{2}$.

Solution. When an equation is cluttered up with many fractions, the best first step may be to get rid of them. To do this, multiply both sides by the lowest common denominator (in this case, 12). Then proceed as usual.

$$12\left(\frac{2}{3}x - \frac{3}{4}\right) = 12\left(\frac{7}{6}x + \frac{1}{2}\right)$$

$$12\left(\frac{2}{3}x\right) - 12\left(\frac{3}{4}\right) = 12\left(\frac{7}{6}x\right) + 12\left(\frac{1}{2}\right) \quad \text{(distributive property)}$$

$$8x - 9 = 14x + 6$$

$$8x = 14x + 15 \quad \text{(add 9)}$$

$$-6x = 15 \quad \text{(subtract } 14x\text{)}$$

$$x = \frac{15}{-6} = -\frac{5}{2} \quad \text{(divide by } -6\text{)}$$

The solution should now be checked in the original equation. We leave the check to the student.

CAUTION

$\frac{3}{4}x - \frac{1}{2} = 4$
$3x - 2 = 4$
$3x = 6$
$x = 2$

$\frac{3}{4}x - \frac{1}{2} = 4$
$3x - 2 = 16$
$3x = 18$
$x = 6$

Solve by first clearing the fractions.

23. $\frac{2}{3}x + 4 = \frac{1}{2}x$

24. $\frac{2}{3}x - 4 = \frac{1}{2}x + 4$

25. $\frac{9}{10}x + \frac{5}{8} = \frac{1}{5}x + \frac{9}{20}$

26. $\frac{1}{3}x + \frac{1}{4} = \frac{1}{5}x + \frac{1}{6}$

27. $\frac{3}{4}(x - 2) = \frac{9}{5}$

28. $\frac{x}{8} = \frac{2}{3}(2 - x)$

The following equations are nonlinear equations that become linear when cleared of fractions. Solve each equation and check your solutions as some might be extraneous (see page 91).

29. $\frac{5}{x + 2} = \frac{2}{x - 1}$

30. $\frac{10}{2x - 1} = \frac{14}{x + 4}$

31. $\frac{2}{x - 3} + \frac{3}{x - 7} = \frac{7}{(x - 3)(x - 7)}$

32. $\frac{2}{x - 1} + \frac{3}{x + 1} = \frac{19}{x^2 - 1}$

33. $\frac{x}{x - 2} = 2 + \frac{2}{x - 2}$

34. $\frac{x}{2x - 4} - \frac{2}{3} = \frac{7 - 2x}{3x - 6}$

Sometimes an equation that appears to be quadratic is actually equivalent to a linear equation. For example, if we subtract x^2 from both sides of the equation $x^2 + 3x = x^2 + 5$, we see that it is equivalent to $3x = 5$. Use this idea to solve each of the following.

35. $x^2 + 4x = x^2 - 3$

36. $x^2 - 2x = x^2 + 3x + 20$

37. $(x - 4)(x + 5) = (x + 2)(x + 3)$

38. $(2x - 1)(2x + 3) = 4x^2 + 6$

Sometimes an equation involving a radical becomes linear when both sides are raised to the same (appropriate) power. Solve each of the following equations. Check your answers, since raising to powers may introduce extraneous solutions.

39. $\sqrt{5 - 2x} = 5$

40. $\sqrt{3x + 7} = 4$

41. $\sqrt[3]{1 - 3x} = 4$

42. $\sqrt[3]{4x - 3} = -3$

EXAMPLE B (Solving for One Variable in Terms of Others)
   Solve $I = nE/(R + nr)$ for n.

Solution.   This is a typical problem in science in which an equality relates several variables and we want to solve for one of them in terms of the others. To do this, we proceed as if the other variables were simply numbers, which, after all, is what every variable represents.

$$I = \frac{nE}{R + nr} \qquad \text{(original equality)}$$

$$(R + nr)I = nE \qquad \text{(multiply by } R + nr)$$

$$RI + nrI = nE \qquad \text{(distributive property)}$$

$$nrI - nE = -RI \qquad \text{(subtract } nE \text{ and } RI)$$

$$n(rI - E) = -RI \qquad \text{(factor)}$$

$$n = \frac{-RI}{rI - E} \qquad \text{(divide by } rI - E)$$

Solve for the indicated variable in terms of the remaining variables.

43. $A = P + Prt$ for P

44. $R = \dfrac{E}{L - 5}$ for L

45. $I = \dfrac{nE}{R + nr}$ for r

46. $mv = Ft + mv_0$ for m

47. $A = 2\pi r^2 + 2\pi rh$ for h

48. $F = \frac{9}{5}C + 32$ for C

49. $R = \dfrac{R_1 R_2}{R_1 + R_2}$ for R_1

50. $\dfrac{1}{R} = \dfrac{1}{R_1} + \dfrac{1}{R_2} + \dfrac{1}{R_3}$ for R_2

MISCELLANEOUS PROBLEMS

In Problems 51–60, solve for x.

51. $\frac{3}{4}x - \frac{4}{3} = \frac{1}{3}x + \frac{5}{6}$

52. $0.3(0.3x - 0.1) = 0.1x - 0.3$

53. $4(3x - \frac{1}{2}) = 5x + \frac{1}{2}$

54. $(x - 2)(x + 5) = (x - 3)(x + 1)$

55. $(x - 2)(3x + 1) = (x - 2)(3x + 5)$

56. $\dfrac{2}{x - 1} = \dfrac{9}{2x - 3}$

57. $\dfrac{2x}{4x + 2} = \dfrac{x + 1}{2x - 1}$

58. $\dfrac{2}{(x + 2)(x - 1)} = \dfrac{1}{x + 2} + \dfrac{3}{x - 1}$

59. $1 + \dfrac{x}{x + 3} = \dfrac{-3}{x + 3}$

60. $\dfrac{x^2}{x - 3} = \dfrac{9}{x - 3}$

61. Solve for x in terms of a.

$$1 + \cfrac{1}{1 + \cfrac{1}{a + \cfrac{1}{x}}} = \frac{1}{a}$$

62. Solve for x in terms of a and b.

   (a) $\dfrac{ax - b}{bx - a} = \dfrac{a + b}{b}$

   (b) $\dfrac{x - a}{x + b} = \dfrac{a - ab}{a + b^2}$

63. Celsius and Fahrenheit temperatures are related by the formula $C = \frac{5}{9}(F - 32)$.
   (a) What Fahrenheit temperature corresponds to 30 degrees Celsius?
   (b) How warm is it (in degrees Fahrenheit) when a Celsius thermometer and a Fahrenheit thermometer give the same reading?
   (c) How warm is it (in degrees Celsius) when the Celsius reading is one-half the Fahrenheit reading?

64. When a principal P is invested at the simple interest rate r (written as a decimal), then the accumulated amount A after t years is given by $A = P + Prt$. For example, $2000 invested at 8 percent simple interest for 10 years will accumulate to $2000 + 2000(.08)(10)$, or $3600.
   (a) How long will it take $2000 to grow to $4000 if invested at 9 percent simple interest?

(b) A principal of $2000 grew to $6000 in 15 years when invested at the simple interest rate r. Find r in percent.

65. As a result of financial difficulties, the ABC company plans to reduce the salaries of all employees by a certain percent this year but hopes to bring them all back to their original level next year.

(a) A salary reduction of 10 percent this year will require what percent increase next year?

(b) A salary reduction of p percent this year will require what percent increase next year?

66. TEASER   When Karen inherited a small amount of money, she decided to divide it among her four children. To Alice, she gave $2 plus one-third of the remainder; to Brent, she gave $2 plus one-third of the remainder. Next, to Curtis, she gave $2 plus one-third of the remainder, and, finally, to Debra, she gave $2 plus one-third of the remainder. What was left, she divided equally among the four children. If the girls together got $35 more than the boys together, what was the size of Karen's inheritance and how much did each child get?

Rules for the Direction of the Mind

1. Reduce any kind of problem to a mathematical problem.
2. Reduce any kind of mathematical problem to a problem of algebra.
3. Reduce any problem of algebra to the solution of a single equation.

René Descartes
1596–1650

3-2   Applications Using One Unknown

Besides being a mathematician, Descartes was a first-rate philosopher. His name appears prominently in every philosophy text. We admit that his rules for the direction of the mind are overstated. Not every problem in life can be solved this way. But they do suggest a style of thinking that has been very fruitful, especially in the sciences. We intend to exploit it.

A TYPICAL WORD PROBLEM

Sometimes a little story can illustrate some big ideas.

> John plans to take his wife Helen out to dinner. Concerned about their financial situation, Helen asks him point-blank, "How much money do you have?" Never one to give a simple answer when a complicated one will do, John replies, "If I had $12 more than I have and then doubled that amount, I'd be $60 richer than I am." Helen's response is best left unrecorded.

The problem, of course, is to find out exactly how much money John has. Our task is to take a complicated word description, translate it into mathematical symbols, and then let the machinery of algebra grind out the answer. First we introduce a symbol x. It usually will stand for the principal unknown in the problem. But we need to be very precise. It is not enough to let x be John's money, or even to let x be the amount of John's money, though that is better. The symbol x must represent a number. What we should say is

Let x be the number of dollars John has.

Our story puts restrictions on x; x must satisfy a specified condition. That condition must be translated into an equation. We shall do it by bits and pieces.

WORD PHRASE	*ALGEBRAIC TRANSLATION*
How much John has	x
$12 more than he has	$x + 12$
Double that amount	$2(x + 12)$
$60 richer than he is	$x + 60$

Read John's answer again. It says that the expressions $2(x + 12)$ and $x + 60$ are equal. Thus

$$2(x + 12) = x + 60$$

$$2x + 24 = x + 60$$

$$x = 36$$

John has $36, enough for a pretty good dinner for two—even at today's prices.

A DISTANCE-RATE PROBLEM

Problems involving rates and distances occur very frequently in physics. Usually their solution involves use of the formula $D = RT$, which stands for "distance equals rate multiplied by time."

At 2:00 P.M., Slowpoke left Kansas City traveling due east at 45 miles per hour. An hour later, Speedy started after him going 60 miles per hour. When will Speedy catch up with Slowpoke?

Most of us can grasp the essential features in a picture more readily than in a mass of words. That is why all good mathematicians make sketches that summarize what is given. One such picture is shown as Figure 1.

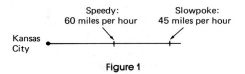

Figure 1

Next assign the unknown. Be precise.

Poor: Let t be time.
Better: Let t be the time when Speedy catches up.
Good: Let t be the number of hours after 2:00 P.M. when Speedy catches up with Slowpoke.

Notice two things:

1. Slowpoke drove t hours: Speedy, starting an hour later, drove only $t - 1$ hours.
2. Both drove the same distance.

Now use the formula $D = RT$ to conclude that Slowpoke drove a distance of $45t$ miles and Speedy drove a distance of $60(t - 1)$ miles. By statement 2, these are equal.

$$45t = 60(t - 1)$$
$$45t = 60t - 60$$
$$60 = 15t$$
$$4 = t$$

Speedy will catch up with Slowpoke 4 hours after 2:00 P.M., or at 6:00 P.M.

A MIXTURE PROBLEM

Here is a problem from chemistry.

How many liters of a 60 percent solution of nitric acid should be added to 10 liters of a 30 percent solution to obtain a 50 percent solution?

We are certain that Figure 2 will help most students with this problem.

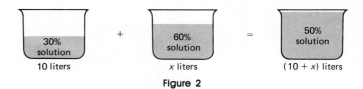

Figure 2

We have indicated on the figure what x represents, but let us be specific.

Let x be the number of liters of 60 percent solution to be added.

Now we make a crucial observation, one that will seem obvious once you think about it.

$$\begin{pmatrix} \text{The amount of} \\ \text{pure acid we} \\ \text{start with} \end{pmatrix} + \begin{pmatrix} \text{The amount of} \\ \text{pure acid we} \\ \text{add} \end{pmatrix} = \begin{pmatrix} \text{The amount of} \\ \text{pure acid we} \\ \text{end with} \end{pmatrix}$$

In symbols, this becomes

$$(.30)(10) + (.60)x = (.50)(10 + x)$$

The big job has been accomplished; we have the equation. After multiplying both sides by 10 to clear the equation of decimal fractions, we can easily solve for x.

$$(3)(10) + 6x = 5(10 + x)$$
$$30 + 6x = 50 + 5x$$
$$x = 20$$

We should add 20 liters of 60 percent solution.

SUMMARY

One hears students say that they have a mental block when it comes to "word problems." Yet most of the problems of the real world are initially stated in words. We want to destroy those mental blocks and give you one of the most satisfying experiences in mathematics. We believe you can learn to do word problems. Here are a few simple suggestions.

1. Read the problem very carefully so that you know exactly what it says and what it asks.
2. Draw a picture or make a chart that summarizes the given information.
3. Identify the unknown quantity and assign a letter to it. Be sure it represents a number (for example, of dollars, of miles, or of liters).
4. Note the condition or restriction that the problem puts on the unknown. Translate it into an equation.
5. Solve the equation. Your result is a specific number. The unknown has become known.

6. State your conclusion in words (for example, Speedy will catch up to Slowpoke at 4 hours after 2:00 P.M., or at 6:00 P.M.).

7. Note whether your answer is reasonable. If, for example, you find that Speedy will not catch up with Slowpoke until 4000 hours after 2:00 P.M., you should suspect that you have made a mistake.

Problem Set 3-2

1. The sum of 15 and twice a certain number is 33. Find that number.

2. Tom says to Jerry: I am thinking of a number. When I subtract 5 from that number and then multiply the result by 3, I get 42. What was my number? Jerry figured it out. Can you?

3. Find the number for which twice the number is 12 less than 3 times the number.

4. The result of adding 28 to 4 times a certain number is the same as subtracting 5 from 7 times that number. Find the number.

5. The sum of three consecutive positive integers is 72. Find the smallest one.

6. The perimeter (distance around) of the rectangle shown in Figure 3 is 31 inches. Find the width x.

7. A wire 130 centimeters long is bent into the shape of a rectangle which is 3 centimeters longer than it is wide. Find the width of the rectangle.

8. A rancher wants to put 2850 pounds of feed into two empty bins. If she wants the larger bin to contain 750 pounds more of the feed than the smaller one, how much must she put into each?

9. Mary scored 61 on her first math test. What must she score on a second test to bring her average up to 75?

10. Henry has scores of 61, 73, and 82 on his first three tests. How well must he do on the fourth and final test to wind up with an average of 75 for the course?

11. A change box contains 21 dimes. How many quarters must be put in to bring the total amount of change to $3.85?

12. A change box contains $3.00 in dimes and nothing else. A certain number of dimes are taken out and replaced by an equal number of quarters, with the result that the box now contains $4.20. How many dimes are taken out?

13. A woman has $4.45 in dimes and quarters in her purse. If there are 25 coins in all, how many dimes are there?

14. Young Amy has saved $8.05 in nickels and quarters. She has 29 more nickels than quarters. How many quarters does she have?

15. Read the example (in this section) about Slowpoke and Speedy again. When will Speedy be 100 miles ahead of Slowpoke?

16. Two long-distance runners start out from the same spot on an oval track which is $\frac{1}{2}$ mile around. If one runs at 6 miles per hour and the other at 7 miles an hour, when will the faster runner be one lap ahead of the slower runner?

17. The City of Harmony is 455 miles from the city of Dissension. At 12:00 noon Paul Haymaker leaves Harmony traveling at 60 miles per hour toward Dissension. Simultaneously, Nick Ploughman starts from Dissension heading toward Harmony, managing only 45 miles per hour. When will they meet?

x inches

12 inches

Figure 3

18. Suppose in Problem 17, Mr. Ploughman starts at 3:00 P.M. At what time will the two drivers meet?

19. Luella can row 1 mile upstream in the same amount of time that it takes her to row two miles downstream. If the rate of the current is 3 miles per hour, how fast can she row in still water?

20. An airplane flew with the wind for 1 hour and returned the same distance against the wind in $1\frac{1}{2}$ hours. If the speed of the plane in still air is 300 miles per hour, find the speed of the wind.

21. A father is three times as old as his son, but 15 years from now he will be only twice as old as his son. How old is his son now?

22. Jim Warmath was in charge of ticket sales at a football game. The price for general admission was $3.50, while reserved seat tickets sold for $5.00. He lost track of the ticket count, but he knew that 110 more general admission tickets had been sold than reserve seat tickets and the total gate receipts were $980. See if you can find out how many general admission tickets were sold.

23. How many cubic centimeters of a 40 percent solution of hydrochloric acid should be added to 2000 cubic centimeters of a 20 percent solution to obtain a 35 percent solution?

24. In Problem 23, how much of the 40 percent solution would have to be added in order to have a 39 percent solution?

25. A tank contains 1000 liters of 30 percent brine solution. Boiling off water from the solution will increase the percentage of salt. How much water should be boiled off to achieve a 35 percent solution?

26. Sheila Carlson invested $10,000 in a savings and loan association, some at 7 percent (simple interest) per year and the rest at $8\frac{1}{2}$ percent. How much did she invest at 7 percent if the total amount of interest for one year was $796?

27. At the Style King shop, a man's suit was marked down 15 percent and sold at $123.25. What was the original price?

28. Mr. Titus Canby bickered with a furrier over the price of a fur coat he intended to buy for his wife. The furrier offered to reduce the price by 10 percent. Titus was still not satisfied; he said he would buy the coat if the furrier would come down an additional $200 on the price. The furrier agreed and sold the coat for $1960. What was the original price?

29. The Conkwrights plan to put in a concrete drive from the street to the garage. The drive is 36 feet long and they plan to make it 4 inches thick. Since there is a delivery charge for less than 4 cubic yards of ready mixed concrete, the Conkwrights have decided to make the drive just wide enough to use 4 cubic yards. How wide should they make it?

30. Tom can do a certain job in 3 days, Dick can do it in 4 days, and Harry can do it in 5 days. How long will it take them working together? *Hint:* In one day Tom can do $\frac{1}{3}$ of the job; in x days, he can do $x/3$ of the job.

31. It takes Jack 5 days to hoe his vegetable garden. Jack and Jill together can do it in 3 days. How long would it take Jill to hoe the garden by herself?

Figure 4

EXAMPLE (Balancing Weight Problems) Susan, who weighs 80 pounds, wants to ride on a seesaw with her father, who weighs 200 pounds (Figure 4). The plank is 20 feet long and the fulcrum is at the center. If Susan sits at the very end of her side, how far from the fulcrum should her father sit to achieve balance?

Solution. Let x be the number of feet from the fulcrum to the point where Susan's dad sits. A law of physics demands that *the weight times the distance from the fulcrum* must be the same for both sides in order to have balance. For Susan, weight times distance is $80 \cdot 10$; for her dad, it is $200x$. This gives us the equation

$$200x = 800$$

from which we get $x = 4$. Susan's father should sit 4 feet from the fulcrum.

32. Where should Susan's father sit if Susan moves 2 feet closer to the fulcrum?

33. If Susan sits at one end with Roscoe, a 12-pound puppy, in her arms, where should her father sit?

Find x in each of the following. Assume in each case that the plank is 20 feet long, that the fulcrum is at the center, and that the weights balance.

34.

35.

36.

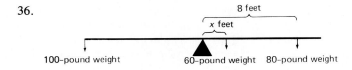

Hint: On the right side, you add the two products.

37.

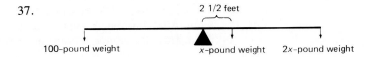

38.

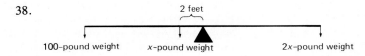

39. The sidewalk around a square garden is 3 feet wide and has area 249 square feet. What is the width of the garden?

40. What number must be added to both numerator and denominator of $\frac{3}{37}$ to bring this fraction up to $\frac{7}{24}$?

41. Professor Witquick is 4 years older than his wife, Matilda. On their 25th wedding anniversary, Witquick noted that their combined ages were double their combined ages on their wedding day. How old will Matilda be on their 50th anniversary?

42. Sylvia Swindalittle, a lawyer, charges $150 per hour for her own time and $30 per hour for that of her clerical assistant, Donald Duckwork. On a recent case, where Donald worked half again as long as Sylvia, the total charge was $2262. How long did Donald work on the case?

43. In a certain math class, the average weight for the girls was 128 pounds, for the boys, 160 pounds, and for the class as a whole, 146 pounds. How many people are in the class, given that there are 14 girls?

44. Sam Slugger got only 50 hits in his first 200 times at bat of the baseball season. During the rest of the season, he claims he will hit .300 and finish with a season average of .280. How many times at bat does he expect to have during the whole season?

45. When asked to make a contribution to the building fund, Amos Hadwiggle pledged to give $1000 more than the average of all givers. The other 88 givers contributed $44,000. How much did Amos have to give?

46. Opening drain A will empty the swimming pool in 5 hours; opening both A and B will empty it in 3 hours. How long will it take to empty the pool if only drain B is opened?

47. Here is an old riddle said to be on the tombstone of Diophantus. Diophantus lived one-sixth of his life as a child, one-twelfth of his life as a youth, and one-seventh of his life as an unmarried adult. A son was born 5 years after Diophantus was married, but this son died 4 years before his father. Diophantus lived twice as long as his son. You figure out how old Diophantus was when he died.

48. Anderson and Benson had been partners for years with Anderson owning three-fifths of the business and Benson the rest. Recently, Christenson offered to pay $100,000 to join the partnership but only on the condition that all three would then be equal partners. Anderson and Benson accepted the offer. How much of the $100,000 did each get, assuming fair division?

49. A column of soldiers 2 miles long is marching at a constant rate of 5 miles per hour. At noon, the general at the rear of the column sent forward two men on horseback traveling at 20 miles per hour.
    (a) The first, A, was told to scout the road ahead of the column and return in 2 hours. After how long should A turn around?
    (b) The second, B, was told to carry a message to the head of the column and return. How long will this take?

50. **TEASER** Here is a problem attributed to Isaac Newton, coinventor of calculus. Three pastures with grass of identical height, density, and growth rate are of size $3\frac{1}{3}$ acres, 10 acres, and 24 acres, respectively. If 12 oxen can be fed on the first pasture for 4 weeks and 21 oxen can be fed on the second for 9 weeks, how many oxen can be fed on the third pasture for 18 weeks?

The Farmer's Riddle

I have a collection of hens and rabbits. These animals have 50 heads and 140 feet. How many hens and how many rabbits do I have?

3-3 Two Equations in Two Unknowns

No one has thought more deeply or written more wisely about problem solving than George Polya. In his book *Mathematical Discovery* (Volume 1, Wiley, 1962) Polya uses the farmer's riddle as the starting point for a brilliant essay on setting up equations to solve problems. He suggests three different approaches that we might take to untangle the riddle.

Hens	Rabbits	Feet
50	0	100
0	50	200
25	25	150
28	22	144
30	20	140

Figure 5

TRIAL AND ERROR

There are 50 animals altogether. They cannot all be hens as that would give only 100 feet. Nor can all be rabbits; that would give 200 feet. Surely the right answer is somewhere between these extremes. Let us try 25 of each. That gives 50 hen-feet and 100 rabbit-feet, or a total of 150, which is too many. We need more hens and fewer rabbits. Well, try 28 hens and 22 rabbits. It does not work. Try 30 hens and 20 rabbits. There it is! That gives 60 feet plus 80 feet, just what we wanted (Figure 5).

Figure 6

BRIGHT IDEA

Let us use a little imagination. Imagine that we catch the hens and rabbits engaged in a weird new game. The hens are all standing on one foot and the rabbits are hopping around on their two hind feet (Figure 6). In this remarkable situation, only half of the feet, that is 70 feet, are in use. We can think of 70 as counting each hen once and each rabbit twice. If we subtract the total number of animals, namely, 50, we will have the number of rabbits. There it is! There have to be 70 − 50 = 20 rabbits; and that leaves 30 hens.

ALGEBRA

Trial and error is time consuming and inefficient, especially in problems with many possibilities. And we cannot expect a brilliant idea to come along for every problem. We need a systematic method that depends neither on guess work nor on sudden visions. Algebra provides such a method. To use it, we must translate the problem into algebraic symbols and set up equations.

ENGLISH	*ALGEBRAIC SYMBOLS*
The farmer has a certain number of hens	x
and	
a certain number of rabbits.	y
These animals have 50 heads	$x + y = 50$
and 140 feet.	$2x + 4y = 140$

Now we have two unknowns, x and y, but we also have two equations relating them. We want to find the values for x and y that satisfy both equations at the same time. There are two standard methods.

METHOD OF ADDITION OR SUBTRACTION

We learned two rules for modifying equations in Section 3-1. Here is another rule, especially useful in solving a **system of equations,** that is, a set of several equations in several unknowns.

> **RULE 3**
>
> You may add one equation to another (or subtract one equation from another) without changing the simultaneous solutions of a system of equations.

Here is how this rule is used to solve the farmer's riddle.

Given equations:
$$\begin{cases} x + y = 50 \\ 2x + 4y = 140 \end{cases}$$

Multiply the first equation by (-2): $\qquad -2x - 2y = -100$
Write the second equation: $\qquad \underline{2x + 4y = 140}$

Add the two equations: $\qquad 2y = 40$

Multiply by $\frac{1}{2}$: $\qquad \boxed{y = 20}$

Substitute $y = 20$ into one of the original equations (in this case, we shall use the first one): $\qquad x + 20 = 50$

Add -20: $\qquad \boxed{x = 30}$

The key idea is this: Multiplying the first equation by -2 makes the coefficients of x in the two equations negatives of each other. Addition of the two equations eliminates x, leaving one equation in the single unknown y. The resulting equation can be solved by methods learned earlier (see Section 3-1).

METHOD OF SUBSTITUTION

Consider the same pair of equations again.

$$\begin{cases} x + y = 50 \\ 2x + 4y = 140 \end{cases}$$

We may solve the first equation for y in terms of x and substitute the result in the second equation

$$y = 50 - x$$
$$2x + 4(50 - x) = 140$$
$$2x + 200 - 4x = 140$$
$$-2x = -60$$
$$\boxed{x = 30}$$

Then substitute the value obtained for x in the expression for y.

$$\boxed{y = 50 - 30 = 20}$$

Naturally, our results agree with those obtained earlier.

Whichever method we use, it is a good idea to check our answer against the original problem. Thirty chickens and 20 rabbits do have a total of 50 heads and they do have $(30)(2) + (20)(4) = 140$ feet in all.

A DISTANCE-RATE PROBLEM

An airplane, flying with the help of a strong tail wind, covered 1200 miles in 2 hours. However, the return trip flying into the wind took $2\frac{1}{2}$ hours. How fast would the plane have flown in still air and what was the speed of the wind, assuming both rates to be constant?

Here is the solution. Let

$$x = \text{speed of the plane in still air in miles per hour}$$

$$y = \text{speed of the wind in miles per hour}$$

Then

$$x + y = \text{speed of the plane with the wind}$$

$$x - y = \text{speed of the plane against the wind}$$

Next, we recall the familiar formula $D = RT$, or distance equals rate multiplied by time. Applying it in the form $TR = D$ to the two trips yields

$$2(x + y) = 1200 \qquad \text{(with wind)}$$
$$\tfrac{5}{2}(x - y) = 1200 \qquad \text{(against wind)}$$

or equivalently

$$2x + 2y = 1200$$
$$5x - 5y = 2400$$

To eliminate y, we multiply the first equation by 5 and the second by 2, and then add the two equations.

$$10x + 10y = 6000$$
$$\underline{10x - 10y = 4800}$$
$$20x = 10{,}800$$
$$x = 540$$

Finally, substituting $x = 540$ in the first of the original equations gives

$$2 \cdot 540 + 2y = 1200$$
$$1080 + 2y = 1200$$
$$2y = 120$$
$$y = 60$$

We conclude that the plane's speed in still air was 540 miles per hour and that the wind speed was 60 miles per hour.

A check against the original statement of the problem shows that our answers are correct. With the wind, the plane will travel at 600 miles per hour and will cover 1200 miles in 2 hours. Against the wind, the plane will fly at 480 miles per hour and take $2\tfrac{1}{2}$ hours to cover 1200 miles.

Problem Set 3-3

In each of the following, find the values for the two unknowns that satisfy both equations. Use whichever method you prefer.

1. $2x + 3y = 13$
   $y = 13$

2. $2x - 3y = 7$
   $x = -4$

3. $2u - 5v = 23$
   $2u = 3$

4. $5s + 6t = 2$
   $3t = -4$

5. $7x + 2y = -1$
   $y = 4x + 7$

6. $7x + 2y = -1$
   $x = -5y + 14$

7. $y = -2x + 11$
   $y = 3x - 9$

8. $x = 5y$
   $x = -3y - 24$

9. $x - y = 14$
   $x + y = -2$

10. $2x - 3y = 8$
    $4x + 3y = 16$

11. $2s - 3t = -10$
    $5s + 6t = 29$

12. $2w - 3z = -23$
    $8w + 2z = -22$

13. $5x - 4y = 19$
    $7x + 3y = 18$

14. $4x - 2y = 16$
    $6x + 5y = 24$

Hint: In Problem 13, multiply top equation by 3 and the bottom one by 4; then add.

15. $7x - 4y = 0$
    $2x + 7y = 57$

16. $2a + 3b = 0$
    $3a - 2b = \frac{13}{2}$

17. $\frac{2}{3}x + y = 4$
    $x + 2y = 5$

18. $\frac{3}{4}x - \frac{1}{2}y = 12$
    $x + y = -8$

19. $.125x - .2y = 3$
    $.75x + .3y = 10.5$

20. $.13x - .24y = 1$
    $2.6x + 4y = -2.4$

21. $\dfrac{4}{x} + \dfrac{3}{y} = 17$

    $\dfrac{1}{x} - \dfrac{3}{y} = -7$

22. $\dfrac{4}{x} - \dfrac{2}{y} = 12$

    $\dfrac{5}{x} + \dfrac{1}{y} = 8$

Hint: Let $u = 1/x$ and $v = 1/y$. Solve for u and v and then find x and y.

23. $\dfrac{2}{\sqrt{x}} - \dfrac{1}{\sqrt{y}} = \dfrac{2}{3}$

    $\dfrac{1}{\sqrt{x}} + \dfrac{2}{\sqrt{y}} = \dfrac{7}{6}$

24. $\dfrac{1}{x-2} - \dfrac{2}{y+3} = 3$

    $\dfrac{5}{x-2} + \dfrac{2}{y+3} = 3$

EXAMPLE (Three Unknowns) Find values for x, y, and z that satisfy all three of the following equations.

$$2x + y - 3z = -9$$
$$x - 2y + 4z = 17$$
$$3x - y - z = 2$$

Solution. The idea is to eliminate one of the unknowns from two different pairs of equations. This gives us just two equations in two unknowns, which we solve as before. Let us eliminate y from the first two equations. To do this, we multiply the first equation by 2 and add to the second.

$$4x + 2y - 6z = -18$$
$$\underline{x - 2y + 4z = 17}$$
$$5x \qquad - 2z = -1$$

Next, we eliminate y from the first and third equations by simply adding them.

$$2x + y - 3z = -9$$
$$3x - y - z = 2$$
$$5x \quad - 4z = -7$$

Our problem thus reduces to solving the following system of two equations in two unknowns.

$$5x - 2z = -1$$
$$5x - 4z = -7$$

We leave this for you to do. You should get $z = 3$ and $x = 1$. If you substitute these values for x and z in any of the original equations, you will find that $y = -2$. Thus

$$x = 1 \qquad y = -2 \qquad z = 3$$

Solve each of these systems for x, y, and z.

25. $4x - y + 2z = 2$
    $-3x + y - 4z = -1$
    $x \quad + 5z = 1$

26. $x + 4y - 8z = -10$
    $3x - y + 5z = 12$
    $-4x + 2y + z = -9$

27. $2x + 3y + 4z = -6$
    $-x + 4y - 6z = 6$
    $3x - 2y + 2z = 2$

28. $3x \quad + z = 0$
    $3x + 2y + z = 4$
    $9x + 5y + 10z = 3$

29. $3x - 2y + z = -2$
    $4x + 3y - 5z = 5$
    $5x - 5y + 3z = -4$

30. $2x - 4y - 2z = -3$
    $2x + 6y + 3z = 7$
    $3x + 3y + 4z = 12$

MISCELLANEOUS PROBLEMS

31. Solve for r and s if $\frac{1}{3}r + \frac{2}{9}s = 24$ and $\frac{2}{9}r + \frac{1}{3}s = 26$.

32. Solve for x and y if $3/(x - 1) + 2/(y - 2) = 10$ and $2/(x - 1) + 3/(y - 2) = 10$.

33. Rodney Roller sold two cars for which he received a total of $13,000. If he received $1400 more for one car than the other, what was the selling price of each car?

34. Janice Stockton's estate is valued at $5000 more than three times her husband's estate. The combined value of their estates is $185,000. Find the value of each estate.

35. Find two numbers whose sum is $\frac{1}{3}$ but whose difference is 3.

36. The sum of the digits of a two-digit number is 12, and reversing the digits increases the value of the number by 54. Find the number.

37. If the numerator and denominator of a fraction are each increased by 1, the result is $\frac{2}{3}$; but if the numerator and denominator are each decreased by 1, the result is $\frac{5}{9}$. Find the fraction.

38. Ella Goldthorpe needed to borrow $80,000 for a business venture. For part of the needed funds, she was able to get a 10 percent simple interest loan from her credit union; for the rest, she paid 12 percent simple interest at a bank. If her total interest for a year was $9360, how much did she borrow from each source?

39. The attendance at a professional football game was 45,000 and the gate receipts were $495,000. If each person bought either a $10 ticket or a $15 ticket, how many tickets of each kind were sold?

40. Flying with a wind of 60 miles per hour, it took a plane 2 hours to get from A to B; returning against the wind took 2 hours and 30 minutes. What was the plane's airspeed (speed in still air) and how far is it from A to B?

41. It took a motorboat 5 hours to go downstream from A to B and 7 hours to return. How long would it have taken Huckleberry Finn to float on his raft from A to B?

42. A certain commuter travels by bus and then by train to work each day. It takes him 45 minutes and costs him $3.05. He estimates that the train averages 50 miles per hour and the bus, only 25 miles per hour. If the bus fare is 12 cents per mile and the train fare is 10 cents per mile, how far does he go by each mode of transportation? Assume that no time is lost in transferring from the bus to the train.

43. A grocer has some coffee worth $3.60 per pound and some worth only $2.90 per pound. How much of each kind should she mix together to get 100 pounds of a blend worth $3.30 per pound?

44. A solution that is 40 percent alcohol is to be mixed with one that is 90 percent alcohol to obtain 50 liters of 50 percent solution. How many liters of each should be used?

45. In stocking her apparel shop, Susan Sharp paid $4800 for some dresses and coats, paying $40 for each dress and $100 for each coat. She was able to sell the dresses at a profit of 20 percent of the selling price and the coats at a profit of 50 percent of the selling price, giving her a total profit of $1800. How many coats and how many dresses did she buy?

46. Workers in a certain factory are classified into two groups depending on the skills required for two jobs. Group 1 workers are paid $12.00 per hour and group 2 workers are paid $7.50 per hour. In negotiations for a new contract, the union is demanding that workers in the second group have their hourly wages brought up to $\frac{2}{3}$ of that for workers in the first group. The factory has 55 workers in group 1 and 40 workers in group 2, all of whom work a 40-hour week. If the management is prepared to increase the weekly payroll by $8640, what hourly wages should it propose for each class of workers?

47. Solve the following system of equations for x and y in terms of z.

$$3x - 4y + 2z = -5$$
$$4x - 5y + 5z = 2$$

48. **TEASER** Here is a slight modification of a problem posed by the famous Swiss mathematician Leonhard Euler. For 100 crowns, a certain farmer bought some horses, goats, and sheep. For a horse, he paid $3\frac{1}{2}$ crowns, for a goat, $1\frac{1}{3}$ crowns, and for a sheep, $\frac{1}{2}$ crown. Assuming he bought at least 7 of each and that he bought 100 animals in all, how many horses, goats, and sheep did he buy? Note that the answers must be integers.

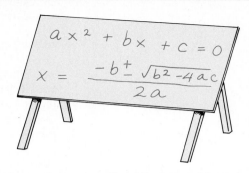

3-4   Quadratic Equations

A linear (first degree) equation may be put in the form $ax + b = 0$. We have seen that such an equation has exactly one solution, $x = -b/a$. That is simple and straightforward; no one is likely to stumble over it. But even the ancient Babylonians knew that equation solving goes far beyond this simple case. In fact, a good part of mathematical history revolves around attempts to solve more and more complicated equations.

The next case to consider is the second degree, or **quadratic,** equation— that is, an equation of the form

$$ax^2 + bx + c = 0 \qquad (a \neq 0)$$

Here are some examples.

(i) $\qquad\qquad\qquad x^2 - 4 = 0$

(ii) $\qquad\qquad\qquad x^2 - x - 6 = 0$

(iii) $\qquad\qquad\qquad 8x^2 - 2x = 1$

(iv) $\qquad\qquad\qquad x^2 = 6x - 2$

Although equations (iii) and (iv) do not quite fit the pattern, we accept them because they readily transform to equations in standard form.

(iii) $\qquad\qquad\qquad 8x^2 - 2x - 1 = 0$

(iv) $\qquad\qquad\qquad x^2 - 6x + 2 = 0$

SOLUTION BY FACTORING

All of us remember that 0 times any number is 0. Just as important but sometimes forgotten is the fact that if the product of two numbers is 0, then one or both of the factors must be 0.

$$\boxed{\begin{array}{l} \text{If } u = 0 \text{ or } v = 0, \text{ then } u \cdot v = 0 \\ \text{If } u \cdot v = 0, \text{ then either } u = 0, \text{ or } v = 0, \text{ or both.} \end{array}}$$

This fact allows us to solve any quadratic equation that has 0 on one side provided we can factor its other side. Simply factor, set each factor equal to 0, and solve the resulting linear equations. We illustrate.

(i)
$$x^2 - 4 = 0$$
$$(x - 2)(x + 2) = 0$$
$$x - 2 = 0 \qquad x + 2 = 0$$
$$x = 2 \qquad\quad x = -2$$

(ii)
$$x^2 - x - 6 = 0$$
$$(x - 3)(x + 2) = 0$$
$$x - 3 = 0 \qquad x + 2 = 0$$
$$x = 3 \qquad\quad x = -2$$

(iii)
$$8x^2 - 2x - 1 = 0$$
$$(4x + 1)(2x - 1) = 0$$
$$4x + 1 = 0 \qquad 2x - 1 = 0$$
$$x = -\frac{1}{4} \qquad x = \frac{1}{2}$$

Equation (iv) remains unsolved; we do not know how to factor its left side. For this equation, we need a more powerful method. First, however, we need a brief discussion of square roots.

SQUARE ROOTS

The number 9 has two square roots, 3 and -3. In fact, every positive number has two square roots, one positive and the other negative. If a is positive, its **positive square root** is denoted by $\sqrt{a}$. Thus $\sqrt{9} = 3$. Do not write $\sqrt{9} = -3$ or $\sqrt{9} = \pm 3$; both are wrong. But you can say that the two square roots of 9 are $\pm\sqrt{9}$ (or ± 3) and that the two square roots of 7 are $\pm\sqrt{7}$.

Here are two important properties of square roots, valid for any positive numbers a and b.

$$\boxed{\begin{array}{l} \sqrt{ab} = \sqrt{a}\sqrt{b} \\ \sqrt{\dfrac{a}{b}} = \dfrac{\sqrt{a}}{\sqrt{b}} \end{array}}$$

For example,

$$\sqrt{4 \cdot 16} = \sqrt{4}\sqrt{16} = 2 \cdot 4 = 8$$

$$\sqrt{28} = \sqrt{4 \cdot 7} = \sqrt{4}\sqrt{7} = 2\sqrt{7}$$

$$\sqrt{\frac{4}{9}} = \frac{\sqrt{4}}{\sqrt{9}} = \frac{2}{3}$$

The square roots of a negative number are imaginary. For example, the two square roots of -9 are $3i$ and $-3i$, since

$$(3i)^2 = 3^2 i^2 = 9(-1) = -9$$

$$(-3i)^2 = (-3)^2 i^2 = 9(-1) = -9$$

In fact, if a is positive, the two square roots of $-a$ are $\pm\sqrt{a}i$. And in this case, the symbol $\sqrt{-a}$ will denote $\sqrt{a}i$. Thus $\sqrt{-7} = \sqrt{7}i$.

COMPLETING THE SQUARE

Consider equation (iv) again.

$$x^2 - 6x + 2 = 0$$

We may write it as

$$x^2 - 6x = -2$$

Now add 9 to both sides, making the left side a perfect square, and factor.

$$x^2 - 6x + 9 = -2 + 9$$

$$(x - 3)^2 = 7$$

This means that $x - 3$ must be one of the two square roots of 7—that is,

$$x - 3 = \pm\sqrt{7}$$

Hence

$$x = 3 + \sqrt{7} \quad \text{or} \quad x = 3 - \sqrt{7}$$

You may ask how we knew that we should add 9. Any expression of the form $x^2 + px$ becomes a perfect square when $(p/2)^2$ is added, since

$$x^2 + px + \left(\frac{p}{2}\right)^2 = \left(x + \frac{p}{2}\right)^2$$

For example, $x^2 + 10x$ becomes a perfect square when we add $(10/2)^2$ or 25.

$$x^2 + 10x + 25 = (x + 5)^2$$

The rule for completing the square (namely, add $(p/2)^2$) works only when the coefficient of x^2 is 1. However, that fact causes no difficulty for quadratic equations. If the leading coefficient is not 1, we simply divide both sides by this coefficient and then complete the square. We illustrate with the equation

$$2x^2 - x - 3 = 0$$

We divide both sides by 2 and proceed as before.

$$x^2 - \frac{1}{2}x - \frac{3}{2} = 0$$

$$x^2 - \frac{1}{2}x = \frac{3}{2}$$

$$x^2 - \frac{1}{2}x + \left(\frac{1}{4}\right)^2 = \frac{3}{2} + \left(\frac{1}{4}\right)^2$$

$$\left(x - \frac{1}{4}\right)^2 = \frac{25}{16}$$

$$x - \frac{1}{4} = \pm\frac{5}{4}$$

$$x = \frac{1}{4} + \frac{5}{4} = \frac{3}{2} \quad \text{or} \quad x = \frac{1}{4} - \frac{5}{4} = -1$$

THE QUADRATIC FORMULA

The method of completing the square works on any quadratic equation. But there is a way of doing this process once and for all. Consider the general quadratic equation

$$ax^2 + bx + c = 0$$

with real coefficients $a \neq 0$, b, and c. First add $-c$ to both sides and then divide by a to obtain

$$x^2 + \frac{b}{a}x = -\frac{c}{a}$$

Next complete the square by adding $(b/2a)^2$ to both sides and then simplify.

$$x^2 + \frac{b}{a}x + \left(\frac{b}{2a}\right)^2 = -\frac{c}{a} + \left(\frac{b}{2a}\right)^2$$

$$\left(x + \frac{b}{2a}\right)^2 = -\frac{c}{a} + \frac{b^2}{4a^2}$$

$$\left(x + \frac{b}{2a}\right)^2 = \frac{b^2 - 4ac}{4a^2}$$

Finally take the square root of both sides.

$$x + \frac{b}{2a} = \pm\frac{\sqrt{b^2 - 4ac}}{2a}$$

or

$$x = \frac{-b}{2a} \pm \frac{\sqrt{b^2 - 4ac}}{2a}$$

We call this result the **quadratic formula** and normally write it as follows.

$$x = \frac{-b \pm \sqrt{b^2 - 4ac}}{2a}$$

Let us see how it works on example (iv).

$$x^2 - 6x + 2 = 0$$

Here $a = 1$, $b = -6$, and $c = 2$. Thus

$$x = \frac{-(-6) \pm \sqrt{36 - 4 \cdot 2}}{2}$$

$$= \frac{6 \pm \sqrt{28}}{2}$$

$$= \frac{6 \pm \sqrt{4 \cdot 7}}{2}$$

$$= \frac{6 \pm 2\sqrt{7}}{2}$$

$$= \frac{2(3 \pm \sqrt{7})}{2}$$

$$= 3 \pm \sqrt{7}$$

As a second example, consider $2x^2 - 4x + \frac{25}{8} = 0$. Here $a = 2$, $b = -4$, and $c = \frac{25}{8}$. The quadratic formula gives

$$x = \frac{4 \pm \sqrt{16 - 25}}{4} = \frac{4 \pm \sqrt{-9}}{4} = \frac{4 \pm 3i}{4}$$

The expression $b^2 - 4ac$ that appears under the square root sign in the quadratic formula is called the **discriminant.** It determines the character of the solutions.

1. If $b^2 - 4ac > 0$, there are two real solutions.
2. If $b^2 - 4ac = 0$, there is one real solution.
3. If $b^2 - 4ac < 0$, there are two nonreal (imaginary) solutions.

Problem Set 3-4

EXAMPLE A (Simplifying Square Roots)   Simplify

(a) $\sqrt{54}$;   (b) $\dfrac{2 + \sqrt{48}}{4}$;   (c) $\dfrac{\sqrt{6}}{\sqrt{150}}$.

Solution.

(a) $\sqrt{54} = \sqrt{9 \cdot 6} = \sqrt{9}\sqrt{6} = 3\sqrt{6}$

Here we factored out the largest square in 54, namely, 9, and then used the first property of square roots.

(b) $\dfrac{2 + \sqrt{48}}{4} = \dfrac{2 + \sqrt{16 \cdot 3}}{4} = \dfrac{2 + 4\sqrt{3}}{4} = \dfrac{2(1 + 2\sqrt{3})}{4}$

$= \dfrac{1 + 2\sqrt{3}}{2}$

If you are tempted to continue as follows,

$$\dfrac{1 + \cancel{2}\sqrt{3}}{\cancel{2}} = 1 + \sqrt{3} \qquad \text{Wrong!}$$

resist the temptation. That cancellation is wrong, because 2 is not a factor of the entire numerator but only of $2\sqrt{3}$.

(c) $\dfrac{\sqrt{6}}{\sqrt{150}} = \sqrt{\dfrac{6}{150}} = \sqrt{\dfrac{1}{25}} = \dfrac{1}{5}$

Sometimes, as in this case, it is best to write a quotient of two square roots as a single square root and then simplify.

Simplify each of the following.

1. $\sqrt{50}$
2. $\sqrt{300}$
3. $\sqrt{\tfrac{1}{4}}$
4. $\sqrt{\dfrac{3}{27}}$
5. $\dfrac{\sqrt{45}}{\sqrt{20}}$
6. $\sqrt{.04}$
7. $\sqrt{11^2 \cdot 4}$
8. $\dfrac{\sqrt{2^3 \cdot 5}}{\sqrt{2 \cdot 5^3}}$
9. $\dfrac{5 + \sqrt{72}}{5}$
10. $\dfrac{4 - \sqrt{12}}{2}$
11. $\boxed{\text{i}}$ $\dfrac{18 + \sqrt{-9}}{6}$
12. $\boxed{\text{i}}$ $\dfrac{3 + \sqrt{-8}}{3}$

EXAMPLE B (Quadratics That Are Already Perfect Squares)   Solve

(a) $x^2 = 9$; (b) $(x + 3)^2 = 17$; (c) $(2y - 5)^2 = 16$.

Solution.   The easiest way to solve these equations is to take square roots of both sides.

(a) $x = 3$ or $x = -3$

(b) $x + 3 = \pm\sqrt{17}$

$x = -3 + \sqrt{17}$ or $x = -3 - \sqrt{17}$

(c) $2y - 5 = \pm 4$

$2y = 5 + 4$ or $2y = 5 - 4$

$y = \frac{9}{2}$ or $y = \frac{1}{2}$

Solve by the method above.

13. $x^2 = 25$ 14. $x^2 = 14$ 15. $(x - 3)^2 = 16$

16. $(x + 4)^2 = 49$ 17. $(2x + 5)^2 = 100$ 18. $(3y - \frac{1}{3})^2 = 25$

☐ 19. $m^2 = -9$ ☐ 20. $(m - 6)^2 = -36$

Solve by factoring.

21. $x^2 = 3r$ 22. $2x^2 - 5x = 0$

23. $x^2 - 9 = 0$ 24. $x^2 - \frac{9}{4} = 0$

25. $m^2 - .0144 = 0$ 26. $x^2 - x - 2 = 0$

27. $x^2 - 3x - 10 = 0$ 28. $x^2 + 13x + 22 = 0$

29. $3x^2 + 5x - 2 = 0$ 30. $3x^2 + x - 2 = 0$

31. $6x^2 - 13x - 28 = 0$ 32. $10x^2 + 19x - 15 = 0$

Solve by completing the square.

33. $x^2 + 8x = 9$ 34. $x^2 - 12x = 45$ 35. $z^2 - z = \frac{3}{4}$

36. $x^2 + 5x = 2\frac{3}{4}$ ☐ 37. $x^2 + 4x = -9$ ☐ 38. $x^2 - 14x = -65$

Solve by using the quadratic formula.

39. $x^2 + 8x + 12 = 0$ 40. $x^2 - 2x - 15 = 0$

41. $x^2 + 5x + 3 = 0$ 42. $z^2 - 3z - 8 = 0$

43. $3x^2 - 6x - 11 = 0$ 44. $4t^2 - t - 3 = 0$

45. $x^2 + 5x + 5 = 0$ 46. $y^2 + 8y + 10 = 0$

☐ 47. $2z^2 - 6z + 11 = 0$ ☐ 48. $x^2 + x + 1 = 0$

© *Solve using the quadratic formula. Write your answers rounded to four decimal places.*

49. $2x^2 - \pi x - 1 = 0$ 50. $3x^2 - \sqrt{2}x - 3\pi = 0$

51. $x^2 + .8235x - 1.3728 = 0$ 52. $5x^2 - \sqrt{3}x - 4.3213 = 0$

EXAMPLE C (Solving for One Variable in Terms of the Other)   Solve for y in terms of x.

(a) $y^2 - 2xy - x^2 - 2 = 0$ (b) $(y - 3x)^2 + 3(y - 3x) - 4 = 0$

Solution.

(a) We use the quadratic formula with $a = 1$, $b = -2x$, and $c = -x^2 - 2$.

$$y = \frac{2x \pm \sqrt{4x^2 + 4(x^2 + 2)}}{2} = \frac{2x \pm \sqrt{8x^2 + 8}}{2}$$

$$= \frac{2x \pm 2\sqrt{2x^2 + 2}}{2}$$

$$= x \pm \sqrt{2x^2 + 2}$$

So $y = x + \sqrt{2x^2 + 2}$ or $y = x - \sqrt{2x^2 + 2}$.

(b) If we substitute z for $y - 3x$, the equation becomes

$$z^2 + 3z - 4 = 0$$

We solve this equation for z.

$$(z + 4)(z - 1) = 0$$
$$z = -4 \quad \text{or} \quad z = 1$$

Thus

$$y - 3x = -4 \quad \text{or} \quad y - 3x = 1$$
$$y = 3x - 4 \quad \text{or} \quad y = 3x + 1$$

Solve for y in terms of x.

53. $(y - 2)^2 = 4x^2$

54. $(y + 3x)^2 = 9$

55. $(y + 3x)^2 = 9x^2$

56. $4y^2 + 4xy - 5 + x^2 = 0$

57. $(y + 2x)^2 - 8(y + 2x) + 15 = 0$

58. $(x - 2y + 3)^2 - 3(x - 2y + 3) + 2 = 0$

MISCELLANEOUS PROBLEMS

Solve the equations in Problems 59–76.

59. $(2x - 1)^2 = \frac{9}{4}$

60. $3x^2 = 1 + 2x$

61. $2x^2 = 4x - 2$

62. $2x^2 + 3x = x^2 + 2x + 12$

63. $y^2 + 2y - 4 = 0$

64. $9y^2 - 3y - 1 = 0$

☐ 65. $2m^2 + 2m + 1 = 0$

☐ 66. $\sqrt{3}m^2 - 2m + 3\sqrt{3} = 0$

☐ 67. $x^4 - 5x^2 - 6 = 0$

68. $x - 4\sqrt{x} + 3 = 0$

69. $\left(x - \frac{4}{x}\right)^2 - 7\left(x - \frac{4}{x}\right) + 12 = 0$

70. $(x^2 + x)^2 - 8(x^2 + x) = -12$

71. $\dfrac{x^2 + 2}{x^2 - 1} = \dfrac{x + 4}{x + 1}$

72. $\dfrac{x}{x^2 + 1} = \dfrac{x - 3}{x^2 - 7}$

73. $\dfrac{1}{x} + \dfrac{1}{x - 1} = \dfrac{8}{3}$

74. $\dfrac{1}{x + 2} + \dfrac{1}{x - 2} = 1$

75. $\sqrt{2x + 1} = \sqrt{x} + 1$

76. $\dfrac{7}{x - 1} - \dfrac{2}{\sqrt{x - 1}} + \dfrac{1}{7} = 0$

Solve the systems of equations in Problems 77–80.

77. $xy = 20$
    $-3 = 2x - y$

78. $x^2 + y^2 = 25$
    $3x + y = 15$

79. $2x^2 - xy + y^2 = 14$
    $x - 2y = 0$

80. $x^2 + 4y^2 = 13$
    $2x^2 - y^2 = 17$

81. A rectangle has a perimeter of 26 and an area of 30. Find its dimensions.

82. The sum of the squares of three consecutive positive odd integers is 683. Find these integers.

83. A square piece of cardboard was used to construct a tray by cutting 2-inch squares out of each of the four corners and turning up the flaps (Figure 7). Find the size of the original square if the resulting tray has a volume of 128 square inches.

Figure 7

84. To harvest a rectangular field of wheat, which is 720 meters by 960 meters, a farmer cut swaths around the outside, thus forming a steadily growing border of cut wheat and leaving a steadily shrinking rectangle of uncut wheat in the middle. How wide was the border when the farmer was half through?

85. At noon, Tom left point A walking due north; an hour later, Dick left point A walking due east. Both boys walked at 4 miles per hour and both carried walkie-talkies with a range of 8 miles. At what time did they lose contact with each other?

86. The diameter AB of the circle in Figure 8 has length 12, and CD, which is perpendicular to the diameter, has length 5. Find the length of AD.

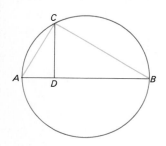

Figure 8

87. (The golden ratio)   As Euclid suggested, divide a line segment into two unequal parts so that the ratio of the whole to the longer part is equal to the ratio of the longer part to the shorter. Find the ratio ϕ of the longer to the shorter.

88. **TEASER**   Here is a version of a nifty problem posed by the American puzzlist Sam Lloyd. A square formation of soldiers, 50 feet on a side, is marching forward at a constant rate.
    (a) Running at constant speed, the company mascot ran from the left rear to the left front of the formation and back again in the time it took the company to march forward 50 feet. How many feet did the dog travel?
    (b) Running at a faster rate, the mascot ran completely around the square formation in the time the company marched 50 feet forward. How many feet did the dog travel this time?

    Suggestion: Because this problem is tricky, we follow Martin Gardner in suggesting a method of attack. Measure distances in units of 50 feet and time in units corresponding to the time for the company to march 50 feet. In these units, let x be the rate of the dog and note that the rate of the company is 1. Now find and simplify the equation x must satisfy. In (a), this leads to a quadratic equation that is easy to solve; in (b), this leads to a fourth-degree equation that is solvable only by advanced methods. However, you can get an approximate answer by experimenting with your calculator. Of course, once you know x in the two parts, you need only multiply by 50 to get the answers to the two questions.

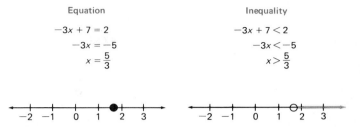

	Equation	Inequality
	$-3x + 7 = 2$	$-3x + 7 < 2$
	$-3x = -5$	$-3x < -5$
	$x = \dfrac{5}{3}$	$x > \dfrac{5}{3}$

Each problem that I solved became a rule which served afterwards to solve other problems.

Descartes

3-5 Inequalities

Solving an inequality is very much like solving an equation, as the example above demonstrates. However, there are dangers in proceeding too mechanically. It will be important to think at every step.

Recall the distinction we made between identities and equations in Section 3-1. A similar distinction applies to inequalities. An inequality which is true for all values of the variables is called an **unconditional inequality.** Examples are

$$(x - 3)^2 + 1 > 0$$

and

$$|x| \le |x| + |y|$$

Most inequalities (for example, $-3x + 7 < 2$) are true only for some values of the variables; we call them **conditional inequalities.** Our primary task in this section is to solve conditional inequalities, that is, to find all those numbers which make a conditional inequality true.

LINEAR INEQUALITIES

To solve the linear inequality $Ax + B < C$, we try to rewrite it in successive steps until the variable x stands by itself on one side of the inequality (see opening display). This depends primarily on the properties stated in Section 1-5 and repeated here.

> **PROPERTIES OF INEQUALITIES**
>
> 1. (Transitivity). If $a < b$ and $b < c$, then $a < c$.
> 2. (Addition). If $a < b$, then $a + c < b + c$.
> 3. (Multiplication). If $a < b$ and $c > 0$, then $a \cdot c < b \cdot c$.
>   If $a < b$ and $c < 0$, then $a \cdot c > b \cdot c$.

We illustrate the use of these properties, applied to $\leq$ rather than $<$, by solving the following inequality.

$$-2x + 6 \leq 18 + 4x$$

Add $-4x$: $\qquad -6x + 6 \leq 18$

Add -6: $\qquad\qquad -6x \leq 12$

Multiply by $-\frac{1}{6}$: $\qquad\qquad x \geq -2$

By rights, we should check this solution. All we know so far is that any value of x that satisfies the original inequality satisfies $x \geq -2$. Can we go in the opposite direction? Yes, because every step is reversible. For example, starting with $x \geq -2$, we can multiply by -6 to get $-6x \leq 12$. In practice, we do not actually carry out this check as we recognize that Property 2 can be restated:

$$a < b \text{ is equivalent to } a + c < b + c.$$

There are similar restatements of Property 3.

QUADRATIC INEQUALITIES

To solve

$$x^2 - 2x - 3 > 0$$

we first factor, obtaining

$$(x + 1)(x - 3) > 0$$

Next we ask ourselves when the product of two numbers is positive. There are two cases; either both factors are negative or both factors are positive.

Case 1 (Both negative) We want to know when both factors are negative, that is, we seek to solve $x + 1 < 0$ and $x - 3 < 0$ simultaneously. The first gives $x < -1$ and the second gives $x < 3$. Together they give $x < -1$.

Case 2 (Both positive) Both factors are positive when $x + 1 > 0$ and $x - 3 > 0$, that is, when $x > -1$ and $x > 3$. These give $x > 3$.

The solution set for the original inequality is the union of the solution sets for the two cases. In set notation, it may be written either as

$$\{x : x < -1 \text{ or } x > 3\}$$

or as

$$\{x : x < -1\} \cup \{x : x > 3\}$$

Figure 9 summarizes what we have learned.

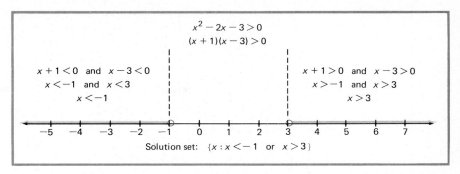

Figure 9

SPLIT-POINT METHOD

The preceding example could have been approached in a slightly different way using the notion of split-points. The solutions of $(x + 1)(x - 3) = 0$, which are -1 and 3, serve as split-points that divide the real line into the three intervals: $x < -1$, $-1 < x < 3$, and $3 < x$. Since $(x + 1)(x - 3)$ can change sign only at a split-point, it must be of one sign (that is, be either always positive or always negative) on each of these intervals. To determine which of them make up the solution set of the inequality $(x + 1)(x - 3) > 0$, all we need to do is pick a single (arbitrary) point from each interval and test it for inclusion in the solution set. If it passes the test, the entire interval from which it was drawn belongs to the solution set.

To show how this method works, let us consider the third degree inequality

$$(x + 2)(x - 1)(x - 4) < 0$$

The solutions of the corresponding equation

$$(x + 2)(x - 1)(x - 4) = 0$$

are -2, 1, and 4. They break the real line into the four intervals $x < -2$, $-2 < x < 1$, $1 < x < 4$, and $4 < x$. Suppose we pick -3 as the test point for the interval $x < -2$. We see that -3 makes each of the three factors $x + 2$, $x - 1$, and $x - 4$ negative, and so it makes $(x + 2)(x - 1)(x - 4)$ negative. You should pick test points from each of the other three intervals to verify the results shown in Figure 10 at the top of the next page.

INEQUALITIES WITH ABSOLUTE VALUES

In solving inequalities involving absolute values, it will be helpful to recall two basic facts we learned in Section 1-5.

Let $a > 0$. Then,
1. $|x| < a$ is equivalent to $-a < x < a$;
2. $|x| > a$ is equivalent to $x < -a$ or $x > a$.

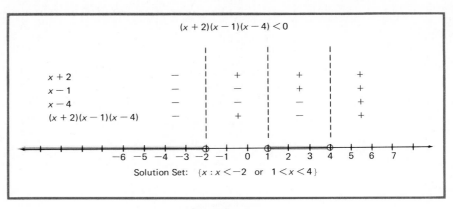

$$(x + 2)(x - 1)(x - 4) < 0$$

$x + 2$	$-$	$+$	$+$	$+$
$x - 1$	$-$	$-$	$+$	$+$
$x - 4$	$-$	$-$	$-$	$+$
$(x + 2)(x - 1)(x - 4)$	$-$	$+$	$-$	$+$

Solution Set: $\{x : x < -2 \ \text{or} \ 1 < x < 4\}$

Figure 10

$$|x - 5| < 2$$
$$x - 5 < 2$$
$$x < 7$$

$$|x - 5| < 2$$
$$-2 < x - 5 < 2$$
$$3 < x < 7$$

$$|x - 5| > 4$$
$$x - 5 > 4$$
$$x > 9$$

$$|x - 5| > 4$$
$$x - 5 > 4 \quad \text{or} \quad x - 5 < -4$$
$$x > 9 \quad \text{or} \quad x < 1$$

To solve the inequality $|3x - 2| < 4$, proceed as follows.

Given inequality:	$\lvert 3x - 2 \rvert < 4$
Remove absolute value:	$-4 < 3x - 2 < 4$
Add 2:	$-2 < 3x \qquad < 6$
Multiply by $\frac{1}{3}$:	$-\frac{2}{3} < x \qquad < 2$

Here, as in the preceding examples, we could use the split-point method. The solutions of the equation $|3x - 2| = 4$, which are $-\frac{2}{3}$ and 2, would be the split-points. The solution would again be $-\frac{2}{3} < x < 2$.

AN APPLICATION

A student wishes to get a grade of B in Mathematics 16. On the first four tests, he got 82 percent, 63 percent, 78 percent, and 90 percent, respectively. A grade of B requires an average between 80 percent and 90 percent, inclusive. What grade on the fifth test would qualify this student for a B?

Let x represent the grade (in percent) on the fifth test. The inequality to be satisfied is

$$80 \leq \frac{82 + 63 + 78 + 90 + x}{5} \leq 90$$

This can be rewritten successively as

$$80 \leq \frac{313 + x}{5} \leq 90$$

$$400 \leq 313 + x \leq 450$$
$$87 \leq \qquad x \leq 137$$

A score greater than 100 is impossible, so the actual solution to this problem is $87 \leq x \leq 100$.

Problem Set 3-5

Which of the following inequalities are unconditional and which are conditional?

1. $x \geq 0$
2. $x^2 \geq 0$
3. $x^2 + 1 > 0$
4. $x^2 > 1$
5. $x - 2 < -5$
6. $2x + 3 > -1$
7. $x(x + 4) \leq 0$
8. $(x - 1)(x + 2) > 0$
9. $(x + 1)^2 > x^2$
10. $(x - 2)^2 \leq x^2$
11. $(x + 1)^2 > x^2 + 2x$
12. $(x - 2)^2 > x(x - 4)$

Solve each of the following inequalities and show the solution set on the real number line.

13. $3x + 7 < x - 5$
14. $-2x + 11 > x - 4$
15. $\frac{2}{3}x + 1 > \frac{1}{2}x - 3$
    Hint: First get rid of the fractions.
16. $3x - \frac{1}{2} < \frac{1}{2}x + 4$
17. $\frac{3}{4}x - \frac{1}{2} < \frac{1}{6}x + 2$
18. $\frac{2}{7}x + \frac{1}{3} \leq -\frac{2}{3}x + \frac{15}{14}$
19. $(x - 2)(x + 5) \leq 0$
20. $(x + 1)(x + 4) \geq 0$
21. $(2x - 1)(x + 3) > 0$
22. $(3x + 2)(x - 2) < 0$
23. $x^2 - 5x + 4 \geq 0$
24. $x^2 + 4x + 3 \leq 0$
    Hint: Factor the left side.
25. $2x^2 - 7x + 3 < 0$
26. $3x^2 - 5x - 2 > 0$
27. $|2x + 3| < 2$
28. $|2x - 4| \leq 3$
29. $|-2x - 1| \leq 1$
30. $|-2x + 3| > 2$
31. $(x + 4)x(x - 3) \geq 0$
32. $(x + 3)x(x - 3) \geq 0$
33. $(x - 2)^2(x - 5) < 0$
34. $(x + 1)^2(x - 1) > 0$

EXAMPLE A (Inequalities Involving Quotients) Solve the following inequality.

$$\frac{3}{x - 2} > \frac{2}{x}$$

Solution. Our natural inclination would be to multiply both sides by $x(x - 2)$ to obtain $3x > 2(x - 2)$. But in doing this, we would be assuming that $x(x - 2)$ is positive, something that is clearly illegal. Rather, we rewrite the given inequality as follows.

$$\frac{3}{x - 2} - \frac{2}{x} > 0 \qquad \left(\text{add } -\frac{2}{x} \text{ to both sides}\right)$$

$$\frac{3x - 2(x - 2)}{(x - 2)x} > 0 \qquad \text{(combine fractions)}$$

$$\frac{x + 4}{(x - 2)x} > 0 \qquad \text{(simplify numerator)}$$

The factors $x + 4$, x, and $x - 2$ in the numerator and denominator determine the three split-points -4, 0, and 2. The chart in Figure 11

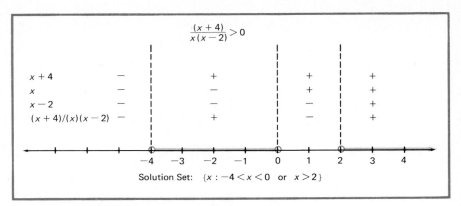

Figure 11

shows the signs of $x + 4, x, x - 2$, and $(x + 4)/(x - 2)x$ on each of the intervals determined by the split-points, as well as the solution set.

Solve each of the following inequalities.

35. $\dfrac{x - 5}{x + 2} \le 0$ 36. $\dfrac{x + 3}{x - 2} > 0$ 37. $\dfrac{x(x + 2)}{x - 5} > 0$

38. $\dfrac{x - 1}{(x - 3)(x + 3)} \ge 0$ 39. $\dfrac{5}{x - 3} > \dfrac{4}{x - 2}$ 40. $\dfrac{-3}{x + 1} < \dfrac{2}{x - 4}$

EXAMPLE B (Rewriting with Absolute Values)   Write $-2 < x < 8$ as an inequality involving absolute values.

Solution.   Look at this interval on the number line (Figure 12). It is 10 units long and its midpoint is at 3. A number x is in this interval provided that it is within a radius of 5 of this midpoint, that is, if

$$|x - 3| < 5$$

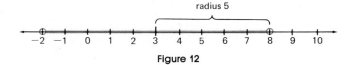

Figure 12

We can check that $|x - 3| < 5$ is equivalent to the original inequality by writing it as

$$-5 < x - 3 < 5$$

and then adding 3 to each member.

Write each of the following as an inequality involving absolute values.

41. $0 < x < 6$ 42. $0 < x < 12$ 43. $-1 \le x \le 7$

44. $-3 \le x \le 7$ 45. $2 < x < 11$ 46. $-10 < x < -3$

EXAMPLE C (Quadratic Inequalities That Cannot Be Factored by Inspection)
Solve the inequality

$$x^2 - 5x + 3 \geq 0$$

Solution. Even though $x^2 - 5x + 3$ cannot be factored by inspection, we can solve the quadratic equation $x^2 - 5x + 3 = 0$ by use of the quadratic formula. We obtain the two solutions $(5 - \sqrt{13})/2 \approx .7$ and $(5 + \sqrt{13})/2 \approx 4.3$. These two numbers split the real line into three intervals from which the numbers 0, 1, and 5 could be picked as test points. Notice that $x = 0$ makes $x^2 - 5x + 3$ positive, $x = 1$ makes it negative, and $x = 5$ makes it positive. This gives us the solution set

$$\left\{ x : x \leq \frac{5 - \sqrt{13}}{2} \text{ or } x \geq \frac{5 + \sqrt{13}}{2} \right\}$$

which can be pictured as shown in Figure 13.

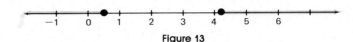

Figure 13

The split-points are included in the solution set because our inequality $x^2 - 5x + 3 \geq 0$ includes equality.

Solve each of the following inequalities and display the solution set on the number line.

47. $x^2 - 7 < 0$ 48. $x^2 - 12 > 0$
49. $x^2 - 4x + 2 \geq 0$ 50. $x^2 - 4x - 2 \leq 0$
ⓒ 51. $x^2 + 6.32x + 3.49 > 0$ ⓒ 52. $x^2 + 4.23x - 2.79 < 0$

EXAMPLE D (Finding Least Values) Find the least value that $x^2 - 4x + 9$ can take on.

Solution. We use the method of completing squares to write $x^2 - 4x + 9$ as the sum of a perfect square and a constant.

$$x^2 - 4x + 9 = (x^2 - 4x + 4) + 5 = (x - 2)^2 + 5$$

Since the smallest value $(x - 2)^2$ can take on is zero, the smallest value $(x - 2)^2 + 5$ can assume is 5.

Find the least value each of the following expressions can take on.

53. $x^2 + 8x + 20$ 54. $x^2 + 10x + 40$
55. $x^2 - 2x + 101$ 56. $x^2 - 4x + 104$

MISCELLANEOUS PROBLEMS

Solve the inequalities in Problems 57–70.

57. $\frac{1}{2}x + \frac{3}{4} > \frac{2}{3}x - \frac{4}{3}$ 58. $3(x - \frac{2}{5}) \leq 5x - \frac{1}{3}$
59. $2x^2 + 5x - 3 < 0$ 60. $x^2 + x - 1 \leq 0$

61. $(x + 1)^2(x - 1)(x - 4)(x - 8) < 0$

62. $(x - 3)(x^3 - x^2 - 6x) \geq 0$

63. $\dfrac{1}{x - 2} + 1 < \dfrac{2}{x + 2}$

64. $\dfrac{2}{x - 2} < \dfrac{3}{x - 3}$

65. $|4x - 3| \geq 2$

66. $|2x - 1| > 3$

67. $|3 - 4x| < 7$

68. $|x| < x + 3$

69. $|x - 2| < |x + 3|$

70. $||x + 2| - x| < 5$

71. $|x^2 - 2x - 4| > 4$

72. $|x^2 - 14x + 44| < 4$

73. For what values of k will the following equations have real solutions?
    (a) $x^2 + 4x + k = 0$
    (b) $x^2 - kx + 9 = 0$
    (c) $x^2 + kx + k = 0$
    (d) $x^2 + kx + k^2 = 0$

74. Amy scored 73, 82, 69, and 94 on four 100-point tests in Mathematics 16. Suppose that a grade of B requires an average between 75 percent and 85 percent.
    (a) What score on a 100-point final exam would qualify Amy for a B in the course?
    (b) What score on a 200-point final exam would qualify Amy for a B in the course?

75. The mathematics department at Podunk University has a staff of 6 people with an average salary of $32,000 per year. An additional professor is to be hired. In what range can a salary S be offered if the department average must be between $31,000 and $35,000?

76. Company A will loan out a car for $35 per day plus 10¢ per mile, whereas company B charges $30 per day plus 12¢ per mile. I need a car for 5 days. For what range of mileage will I be ahead financially if I rent a car from company B?

77. A ball thrown upward with a velocity of 64 feet per second from the top edge of a building 80 feet high will be at height $(-16t^2 + 64t + 80)$ feet above the ground after t seconds.
    (a) What is the greatest height attained by the ball (see Example D)?
    (b) During what time period is the ball higher than 96 feet?
    (c) At what time did the ball hit the ground, assuming it missed the building on the way down?

78. For what numbers is it true that the sum of the number and its reciprocal is greater than 2?

79. Let a, b, and c denote the lengths of the two legs and hypotenuse of a right triangle so that $a^2 + b^2 = c^2$. Show that if n is an integer greater than 2, then $a^n + b^n < c^n$.

80. **TEASER** Sophus Slybones, the proprieter of a specialty coffee shop, weighs coffee for customers by using the two-pan balance shown in Figure 14. Long ago, he dropped the balance on the concrete floor and now the left arm is slightly longer than the right one, that is, $a > b$. This poses no problem for Sophus. When a customer orders two kilograms of coffee, he first places a one-kilogram weight in the left pan balancing it with some coffee in the right pan, which he then pours in a sack. Next, he places the one-kilogram weight in the right pan balancing it with coffee in the left pan, which he also pours in the sack. That makes exactly 2 kilograms says Sophus, a claim you must analyze. Assume the balance arms are weightless but that the pans weigh 100 grams each.
    (a) Show that Sophus has been cheating himself all these years.

Figure 14

(b) If the amount Sophus gives a customer is actually 2.01 kilograms, determine a/b.

If you want to make the problem a little harder, show that the conclusion in (a) is valid even if you don't assume the balance arms are weightless.

Johnny's Dilemma

Although he had been warned against it a thousand times, Johnny still walked across the railroad bridge when he was in a hurry—which was most of the time. It was the day of the biggest football game of the season and Johnny was late. He was already one fourth of the way across the bridge when he heard something. Turning to look, he saw the train coming; it was exactly one bridge-length away from him. Which way should he run?

3-6 More Applications (Optional)

We would not suggest that the problem above is a typical application of algebra. Johnny would be well advised to forget about algebra, make an instant decision, and hope for the best. Nevertheless, Johnny's problem is intriguing. Moreover, it can serve to reemphasize the important principles of real life problem solving that we mentioned briefly at the end of Section 3-2.

First, **clarify the question that is asked**. "Which way should he run?" must mean "Which way should Johnny run to have the best chance of surviving?" That is still too vaguely stated for mathematical analysis. We think the question really means, "Which of the two directions allows Johnny to run at the slowest rate and still avoid the train?" Most questions that come to us from the real world are loosely stated. Our first job is always to pin down precisely (at least in our own minds) what the real question is. It is foolish to try to answer a question that we do not understand. That should be obvious, but it is often overlooked.

Johnny's problem appears to be difficult for another reason; there does not seem to be enough information. We do not know how long the bridge is, we do not know how fast the train is traveling, and we do not know how fast Johnny can run. We could easily despair of making any progress on the problem!

This leads us to state our second principle. **Organize the information**

you have. The best way to do this is to draw a diagram or picture that somehow captures the essence of the problem. It should be an abstract or idealized picture. We shall represent people and trains by points and train tracks and bridges by line segments. These are only approximations to the real situation, but they are necessary if progress is to be made. After the picture is drawn, **label the key quantities of the problem with letters and write down what they represent.** One way to represent Johnny's situation is in Figure 15.

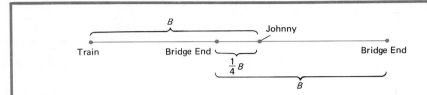

B = length of bridge in feet
S = speed of train in feet per second
Y = speed in feet per second that Johnny must run toward train to escape
Z = speed in feet per second that Johnny must run away from train to escape

Figure 15

Our next principle is this. **Write down the algebraic relationships that exist between the symbols you have introduced.** In Johnny's case, we need to remember the formula $D = RT$ (or $T = D/R$), which relates distance, rate, and time. Now, if Johnny runs toward the train just fast enough to escape, then his time to run the distance $B/4$ must equal the time for the train to go $3B/4$. That gives

(i)
$$\frac{B/4}{Y} = \frac{3B/4}{S}$$

On the other hand, if Johnny runs away from the train, he will have to cover $3B/4$ in the same time that the train goes $7B/4$. Thus,

(ii)
$$\frac{3B/4}{Z} = \frac{7B/4}{S}$$

Now we have done what is often the hardest part of the problem, translating it into algebraic equations. Our job now is to **manipulate the equations until a solution appears.** In the example we are considering, we must solve the equations for Y and Z in terms of S. When we do this, we get

(i)
$$Y = \frac{1}{3}S$$

(ii)
$$Z = \frac{3}{7}S$$

There remains a crucial step: **Interpret the result in the language of the original problem.** In Johnny's case, this step is easy. If he runs toward the train, he will need to run at $\frac{1}{3}$ the speed of the train to escape. If he runs away from the train, he must run at $\frac{3}{7}$ of the speed of the train. Clearly, he will have a better chance of making it if he runs toward the train.

Perhaps by now you need to be reassured. We do not expect you to go through a long-winded analysis like ours for every problem you meet. But we do think that you will have to apply the principles we have stated if you are to be successful at problem solving. The problem set that follows is designed to let you apply these principles to problems from several areas of applied mathematics.

Problem Set 3-6

The problems below are arranged according to type. Only the first set (rate-time problems) relates directly to the example of this section. However, all of them should require some use of the principles we have enunciated. Do not expect to solve all of the problems. But do accept them as challenges worthy of genuine effort.

Rate-Time Problems

1. Sound travels at 1100 feet per second in air and at 5000 feet per second in water. An explosion on a distant ship was recorded by the underwater recording device at a monitoring station. Thirteen seconds later the operator heard the explosion. How far was the ship from the station?

2. The primary wave and secondary wave of an earthquake travel away from the epicenter at rates of 8 and 4.8 kilometers per second, respectively. If the primary wave arrives at a seismic station 15 seconds ahead of the secondary wave, how far is the station from the epicenter?

3. Ricardo often flies from Clear Lake to Sun City and returns on the same day. On a windless day, he can average 120 miles per hour for the round trip, while on a windy day, he averaged 140 miles per hour one way and 100 miles per hour the other. It took him 15 minutes longer on the windy day. How far apart are Clear Lake and Sun City?

4. A classic puzzle problem goes like this: If a column of men 3 miles long is marching at 5 miles per hour, how long will it take a courier on a horse traveling at 25 miles per hour to deliver a message from the end of the column to the front and then return?

5. How long after 4:00 P.M. will the minute hand on a clock overtake the hour hand?

6. At what time between 4:00 and 5:00 do the hands of a clock form a straight line?

7. An old machine is able to do a certain job in 8 hours. Recently a new machine was installed. Working together, the two machines did the same job in 3 hours. How long would it take the new machine to do the job by itself?

8. Center City uses fire trucks to fill the village swimming pool. It takes one truck 3 hours to do the job and another 2 hours. How long would it take them working together?

9. An airplane takes off from a carrier at sea and flies west for 2 hour at 600 miles per hour. It then returns at 500 miles per hour. In the meantime, the carrier has traveled west at 30 miles per hour. When will the two meet?

10. A passenger train 480 feet long traveling at 75 miles per hour meets a freight train 1856 feet long traveling on parallel tracks at 45 miles per hour. How long does it take the trains to pass each other? *Note:* Sixty miles per hour is equivalent to 88 feet per second.

11. A medical company makes two types of heart valves, standard and deluxe. It takes 5 minutes on the lathe and 10 minutes on the drill press to make a standard valve, but 9 minutes on the lathe and 15 minutes on the drill press for a deluxe valve. On a certain day the lathe will be available for 4 hours and the drill press for 7 hours. How many valves of each kind should the company make that day if both machines are to be fully utilized?

[c] 12. Two boats travel at right angles to each other after leaving the dock at 1:00 P.M. At 3:00 P.M., they are 16 miles apart. If the first boat travels 6 miles per hour faster than the second, what are their rates?

13. A car is traveling at an unknown rate. If it traveled 15 miles per hour faster, it would take 90 minutes less to go 450 miles. How fast is the car going?

14. A boy walked along a level road for awhile, then up a hill. At the top of the hill he turned around and walked back to his starting point. He walked 4 miles per hour on level ground, 3 miles per hour uphill, and 6 miles per hour downhill, with the total trip taking 5 hours. How far did he walk altogether?

15. Jack and Jill live at opposite ends of the same street. Jack wanted to deliver a box at Jill's house and Jill wanted to leave some flowers at Jack's house. They started at the same moment, each walking at a constant speed. They met the first time 300 meters from Jack's house. On their return trip, they met 400 meters from Jill's house. How long is the street? (Assume that neither loitered at the other house nor when they met.)

[c] 16. Two cars are traveling toward each other on a straight road at the same constant speed. A plane flying at 350 miles per hour passes over the second car 2 hours after passing over the first car. The plane continues to fly in the same direction and is 2400 miles from the cars when they pass. Find the speed of the cars.

Science Problems

17. The illumination I in foot-candles on a surface d feet from a light source of c candlepower is given by the formula $I = c/d^2$. How far should an 80-candlepower light be placed from a surface to give the same illumination as a 20-candlepower light at 10 feet?

[c] 18. By experiment, it has been found that a car traveling v miles per hour will require d feet to stop, where $d = 0.044v^2 + 1.1v$. Find the velocity of a car if it took 176 feet to stop.

[c] 19. A bridge 200 feet long was built in the winter with no provision for expansion. In the summer, the supporting beams expanded in length by 8 inches, forcing the center of the bridge to drop. Assuming, for simplicity, that the bridge took the shape of a V, how far did the center drop? First guess at the answer and then work it out.

20. The distance s (in feet) traveled by an object in t seconds when it has an initial velocity v_0 and a constant acceleration a is given by $s = v_0 t + \frac{1}{2}at^2$. An object was observed to travel 32 feet in 4 seconds and 72 feet in 6 seconds. Find the initial velocity and the acceleration.

21. A chemist has 5 kiloliters of 20 percent sulfuric acid solution. She wishes to increase its strength to 30 percent by draining off some and replacing it with 80 percent soluton. How much should she drain off?

22. How many liters each of a 35 percent alcohol solution and a 95 percent alcohol solution must be mixed to obtain 12 liters of an 80 percent alcohol solution?

23. One atom of carbon combines with 2 atoms of oxygen to form one molecule of carbon dioxide. The atomic weights of carbon and oxygen are 12.0 and 16.0, respectively. How many milligrams of oxygen are required to produce 4.52 milligrams of carbon dioxide?

24. Four atoms of iron (atomic weight 55.85) combine with 6 atoms of oxygen (atomic weight 16.00) to form 2 molecules of rust. How many grams of iron would there be in 79.85 grams of rust?

ⓒ 25. A sample weighing .5000 grams contained only sodium chloride and sodium bromide. The chlorine and bromine from this sample were precipitated together as silver chloride and silver bromide. This precipitate weighed 1.100 grams. Sodium chloride is 60.6 percent chlorine, sodium bromide is 77.6 percent bromine, silver chloride is 24.7 percent chlorine, and silver bromide is 42.5 percent bromine. Calculate the weights of sodium chloride and sodium bromide in the sample.

Business Problems

26. Sarah Tyler bought stock in the ABC Company on Monday. The stock went up 10 percent on Tuesday and then dropped 10 percent on Wednesday. If she sold the stock on Wednesday for $1000, what did she pay on Monday?

27. If Jane Witherspoon has $4182 in her savings account and wants to buy stock in the ABC Company at $59 a share, how many shares can she buy and still maintain a balance of at least $2000 in her account?

28. Six men plan to take a charter flight to Bear Lake in Canada for a fishing trip, sharing the cost equally. They discover that if they took three more men along, each share of the original six would be reduced by $150. What is the total cost of the charter?

29. Susan has a job at Jenny's Nut Shop. Jenny asks Susan to prepare 25 pounds of mixed nuts worth $1.74 per pound by using walnuts valued at $1.30 per pound and cashews valued at $2.30 per pound. How many pounds of each kind should Susan use?

30. Alec Brown plans to sell toy gizmos at the state fair. He can buy them at 40¢ apiece and will sell them for 65¢ each. It will cost him $200 to rent a booth for the 7-day fair. How many gizmos must he sell to just break even?

31. The cost of manufacturing a product is the sum of fixed plant costs (real estate taxes, utilities, and so on) and variable costs (labor, raw materials, and so on) that depend on the number of units produced. The profit P that the company makes in a year is given by

$$P = TR - (FC + VC)$$

where TR is the total revenue (total sales), FC is the total fixed cost, and VC is the total variable cost. A company that makes one product has $32,000 in total fixed costs. If the variable cost of producing one unit is $4 and if units can be sold at $6 each, find out how many units must be produced to give a profit of $15,000.

32. The ABC company has total fixed costs of $100,000 and total variable costs equal to 80 percent of total sales. What must the total sales be to yield a profit of $40,000? (See Problem 31.)

33. Do Problem 32 assuming that the company pays 30 percent income taxes on its profit and wants a profit of $40,000 after taxes.

34. The XYZ company has total fixed yearly costs of $120,000 and last year had total variable costs of $350,000 while selling 200,000 gizmos at $2.50 each. Competition will force the manager to reduce the sales price to $2.00 each next year. If total fixed costs and variable costs per unit are expected to remain the same, how many gizmos will the company have to sell to have the same profit as last year?

35. Podunk University wishes to maintain a student-faculty ratio of 1 faculty member for every 15 undergraduates and 1 faculty member for each 6 graduate students. It costs the university $600 for each undergraduate student and $900 for each graduate student over and above what is received in tuition. If the university expects $2,181,600 in gifts (beyond tuition) next year and will have 300 faculty members, how many undergraduate and how many graduate students should it admit?

36. A department store purchased a number of smoke detectors at a total cost of $2000. In unpacking them, the stock boys damaged 8 of them so badly that they could not be sold. The remaining detectors were sold at a profit of $25 each, and a total profit of $400 was realized when all of them were sold. How many smoke detectors were originally purchased?

Geometry Problems

37. A flag that is 6 feet by 8 feet has a blue cross of uniform width on a white background. The cross extends to all 4 edges of the flag. The area of the cross and the background are equal. Find the width of the cross.

38. If a right triangle has hypotenuse 2 units longer than one leg and 4 units longer than the other, find the dimensions of the triangle.

39. The area and perimeter of a right triangle are both 30. Find its dimensions.

40. Assume that the earth is a sphere of radius 4000 miles. How far is the horizon from an airplane 5 miles high?

41. Three mutually tangent circles have centers, A, B, and C and radii a, b, and c, respectively. The lengths of the segments AB, BC, and CA are 13, 15, and 18, respectively. Find the lengths of the radii. Assume that each circle is outside of the other two.

42. A rectangle is inscribed in a circle of radius 5. Find the dimensions of the rectangle if its area is 40.

[c] 43. A ladder is standing against a house with its lower end 10 feet from the house. When the lower end is pulled 2 feet farther from the house, the upper end slides 3 feet down the house. How long is the ladder?

[c] 44. A trapezoid (a quadrilateral with two sides parallel) is inscribed in a square 12 inches on a side. One of the parallel sides is the diagonal of the square. If the trapezoid has area 24 square inches, how far apart are its parallel sides?

45. A 40-inch length of wire is cut in two. One of the pieces is bent to form a square and the other is bent to form a rectangle three times as long as wide. If the combined area of the square and the rectangle is $55\frac{3}{4}$ square inches, where was the wire cut?

46. TEASER Figure 16 shows a ring-shaped sidewalk surrounding an irregular pond. We wish to determine the area A of the sidewalk but do not know the radii

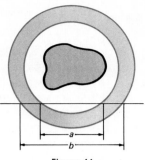

Figure 16

of the two circles. In fact, we do not even know where their common center is and we have no intention of getting our feet wet. However, we can easily make the measurements a and b shown in the figure. Express A in terms of a and b. Note that this shows that all annuli (ring-shaped regions) that intersect a line in the same two line segments have the same area.

47. TEASER   H. E. Dudeney, an English puzzlist, gave us this problem. A metal flagpole broke at a certain point with the top part tipping over like a hinge and the tip hitting the ground at a point 20 feet from the base. It was rewelded but again it broke, this time at a point 5 feet lower with the tip hitting the ground at a point 30 feet from the base. How tall was the flagpole?

48. TEASER   Audrey, Betty, and Candy, who live in different houses, agree to take a cab to the ball game and share expenses. The cab picked up Audrey first, Betty 3 miles later, and Candy 2 miles after Betty. The ballpark is 1 mile from Candy's house. Their total bill was $19.80. The cab driver wanted to charge Candy $3.30 since she went one-sixth of the way, but she rightly refused. What should each of the women pay?

49. TEASER   We do not know the origin of this old teaser, which appears in many mathematical books. Arnold Thinkhard rowed upstream with a uniform (still-water) speed of v miles per hour against a current of c miles per hour. When he had gone 1 mile, the wind blew his hat into the water just out of reach. It was an old hat, so he decided to let it go. After rowing another hour, he suddenly remembered that he had stuck his train ticket in the hatband of that old hat. Turning around immediately, he rowed back and caught up with his hat at his original starting point. Determine the rate of the current.

50. TEASER   This problem is attributed to that great eighteenth-century mathematician Leonhard Euler. A group of heirs divided the estate left by their rich uncle according to the following novel plan. The first heir took c dollars plus an nth of the remainder, the second took $2c$ dollars plus an nth of what still remained, and so on. Each succeeding heir took c dollars more than the previous heir plus one-nth of the new remainder. They were able to follow this scheme and still all heirs got the same amount of money. Express the total inheritance in terms of n and c. Then determine the number of heirs if $n = 16$.

Chapter Summary

The equalities $(x + 1)^2 = x^2 + 2x + 1$ and $x^2 = 4$ are quite different in character. The first, called an **identity**, is true for all values of x. The second, called a **conditional equation**, is true only for certain values of x, in fact, only for $x = 2$ and $x = -2$. To **solve** an equation is to find those values of the unknown which make the equality true; it is one of the major tasks of mathematics.

The equation $ax + b = 0 \, (a \neq 0)$ is called a **linear equation** and has exactly one solution, $x = -b/a$. Similarly, $ax^2 + bx + c = 0 \, (a \neq 0)$ is a **quadratic equation** and usually has two solutions. Sometimes they can be found by **factoring** the left side and setting both factors equal to zero. Another method that always works is **completing the square**, but the best general method is simply substituting in the **quadratic formula**.

$$x = \frac{-b \pm \sqrt{b^2 - 4ac}}{2a}$$

Here $b^2 - 4ac$, called the **discriminant**, plays a critical role. The equation has two real solutions, one real solution, or two nonreal solutions according as the discriminant is positive, zero, or negative.

Equations arise naturally in the study of word problems. Such problems may lead to one equation in one unknown but often lead to a **system** of several equations in several unknowns. In the latter case, our task is to find the values of the unknowns that satisfy all the equations of the system simultaneously.

Inequalities look like equations with the equal sign replaced by $<, \leq, >$, or $\geq$. The methods for solving **conditional inequalities** are very similar to those for conditional equations. One difference is that the direction of an inequality sign is reversed upon multiplication or division by a negative number. Another is that the set of solutions normally consists of one or more **intervals** of numbers, rather than a finite set. For example, the inequality $3x - 2 < 5$ has the solution set $\{x: x < \frac{7}{3}\}$.

Chapter Review Problem Set

1. Which are identities and which are conditional equations?
   (a) $3(x - 2) = 3x - 6$ (b) $3x - 2 = x - 6$
   (c) $(x + 2)^2 = x^2 + 4$ (d) $x^2 + 5x + 6 = (x + 3)(x + 2)$

2. Solve the following equations.

   (a) $3\left(x + \frac{1}{2}\right) = x - \frac{1}{3}$

   (b) $\frac{6}{x - 5} = \frac{21}{x}$

   (c) $(x - 1)(2x + 1) = (2x - 3)(x + 2)$ (d) $\frac{x}{2x + 2} - 1 = \frac{8 - 3x}{6x + 6}$

3. In $s = \frac{1}{2}at^2 + v_0 t$, solve for v_0 in terms of the other variables.

4. Recall that the Fahrenheit and Celsius temperature scales are related by the equation $F = \frac{9}{5}C + 32$.
   (a) How cold is it in degrees Celsius when the Celsius reading is twice the Fahrenheit reading?
   (b) For what Fahrenheit temperatures is the Fahrenheit reading higher than the Celsius reading?

5. Solve the systems below.

   (a) $2x - 3y = 7$

   $x + 4y = -2$

   (b) $\frac{1}{3}x - \frac{5}{6}y = 2$

   $\frac{1}{2}x + y = -\frac{3}{2}$

6. Simplify.

   (a) $\frac{6 + \sqrt{18}}{12}$ (b) $\frac{10 - \sqrt{300}}{2}$ (c) $\frac{-2 + \sqrt{8}}{6}$

Solve the quadratic equations in Problems 7–16 by any method you choose.

7. $x^2 = 49$

8. $(x - 2)(x + 5) = 0$

9. $(x + 3)^2 = 0$

10. $(x - 2)^2 = 25$

11. $(2x + 1)^2 = 81$

12. $x^2 - 9x + 20 = 0$

13. $x^2 = 4x$

14. $x^2 + 2x - 4 = 0$

15. $3x^2 + x - 1 = 0$

16. $x^2 + mx + 2n = 0$

Solve for y in terms of x.

17. $(y - 2x)^2 = 4$

18. $(2y + 3x)^2 - 4(2y + 3x) + 3 = 0$

Solve the inequalities in Problems 19–22.

19. $-5x + 3 \geq 2x - 9$

20. $x^2 + 5x - 6 \geq 0$

21. $x^2 + x - 3 < 0$

22. $\dfrac{x - 4}{x + 1} > 0$

23. Jill Garcia has $10,000 to invest. How much should she put in the bank at 6 percent interest and how much in the credit union at 8 percent interest if she hopes to have $730 interest at the end of one year?

24. A fast train left Chicago at 6:00 A.M. traveling at 60 miles per hour. At 8:00 A.M., a slower train left St. Paul traveling at a constant rate. If these two cities are 450 miles apart and the two trains crashed head on at 11:00 A.M., what was the rate of the slower train?

25. John Appleseed rowed upstream for a distance of 4 miles in 2 hours. If he had rowed twice as hard and the current had been half as strong, he could have done it in $\frac{4}{7}$ hour. What was the rate of the current?

And so Fermat and Descartes turned to the application of algebra to the study of geometry. The subject they created is called coordinate, or analytic, geometry; its central idea is the assocation of algebraic equations with curves and surfaces. This creation ranks as one of the richest and most fruitful veins of thought ever struck in mathematics.

—Morris Kline

CHAPTER 4

Coordinates and Curves

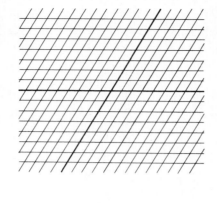

Pierre de Fermat
1601–1665

Why are Fermat and Descartes staring at each other so intently?
Because both have axes to grind.

René Descartes
1596–1650

4-1 The Cartesian Coordinate System

Two Frenchmen deserve credit for the idea of a coordinate system. Pierre de Fermat was a lawyer who made mathematics his hobby. In 1629, he wrote a paper which makes explicit use of coordinates to describe points and curves. René Descartes was a philosopher who thought mathematics could unlock the secrets of the universe. He published *La Géométrie* in 1637. It is a famous book and though it does emphasize the role of algebra in solving geometric problems, one finds only a hint of coordinates there. By virtue of having the idea first and more explicitly, Fermat ought to get the major credit. History can be a fickle friend; coordinates are known as Cartesian coordinates, named after René Descartes.

No matter who gets the credit, it was an idea whose time had come. It made possible the invention of calculus, one of the greatest inventions of the human mind. That invention was to come in 1665 at the hands of a 23-year-old genius named Isaac Newton. You will probably study calculus later on. There you will use the ideas of this chapter over and over.

REVIEW OF THE REAL LINE

Recall the real line (Figure 1), which was introduced in Section 1-3.

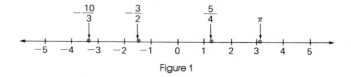

Figure 1

Every point on this line can be given a label, a real number, which specifies exactly where the point is. We call this label the **coordinate** of the point.

Consider now two points A and B with coordinates a and b, respectively. We will need a formula for the distance between A and B in terms of the coordinates a and b. The formula is

$$d(A, B) = |b - a|$$

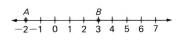
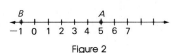

and it is correct whether A is to the right or to the left of B. Note the two examples in Figure 2. In the first case,

$$d(A, B) = |3 - (-2)| = |5| = 5$$

In the second,

$$d(A, B) = |-1 - 5| = |-6| = 6$$

Figure 2

CARTESIAN COORDINATES

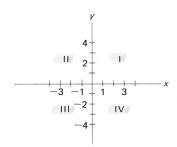

Figure 3

In the plane, produce two copies of the real line, one horizontal and the other vertical, so that they intersect at the zero points of the two lines. The two lines are called **coordinate axes;** their intersection is labeled with O and is called the **origin.** By convention, the horizontal line is called the **x-axis** and the vertical line is called the **y-axis.** The positive half of the x-axis is to the right; the positive half of the y-axis is upward. The coordinate axes divide the plane into four regions called **quadrants,** labeled I, II, III, and IV, as shown in Figure 3.

Each point P in the plane can now be assigned a pair of numbers called its **Cartesian coordinates.** If vertical and horizontal lines through P intersect the x- and y-axes at a and b, respectively, as in Figure 4, then P has coordinates (a, b). We call (a, b) an **ordered pair** of numbers because it makes a difference which number is first. The first number a is the **x-coordinate** (or abscissa); the second number b is the **y-coordinate** (or ordinate).

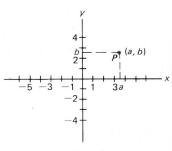

Figure 4

Conversely, take any ordered pair (a, b) of real numbers. The vertical line through a on the x-axis and the horizontal line through b on the y-axis meet in a point P whose coordinates are (a, b).

Think of it this way: The coordinates of a point are the address of that point. If you have found a house (or a point), you can read its address. Conversely, if you know the address of a house (or a point), you can always locate it. In Figure 5 at the top of the next page, we have shown the coordinates (addresses) of several points.

THE DISTANCE FORMULA

Consider the points P_1 and P_2 with coordinates $(-1, -2)$ and $(3, 1)$, respectively. The segment joining P_1 and P_2 is the hypotenuse of a right triangle with right angle at $P_3(3, -2)$ (see Figure 6). We easily calculate the lengths of the two legs.

$$d(P_1, P_3) = |3 - (-1)| = 4 \qquad d(P_2, P_3) = |1 - (-2)| = 3$$

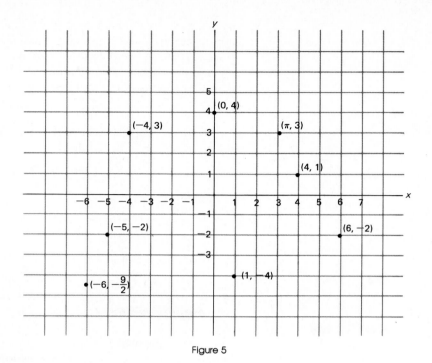

Figure 5

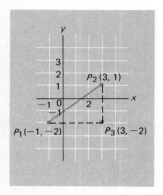

Figure 6

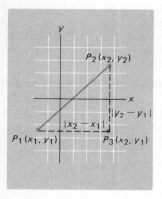

Figure 7

By the Pythagorean theorem (see Section 1-3),

$$d(P_1, P_2) = \sqrt{4^2 + 3^2} = \sqrt{25} = 5$$

Next consider two arbitrary points $P_1(x_1, y_1)$ and $P_2(x_2, y_2)$ (as in Figure 7) that are not on the same horizontal or vertical line. They determine a right triangle with legs of length $|x_2 - x_1|$ and $|y_2 - y_1|$. By the Pythagorean theorem,

$$d(P_1, P_2) = \sqrt{(x_2 - x_1)^2 + (y_2 - y_1)^2}$$

This formula is known as the **distance formula.** You should check that it is valid even if P_1 and P_2 lie on the same vertical or horizontal line.

THE MIDPOINT FORMULA

Consider two points $A(x_1, y_1)$ and $B(x_2, y_2)$ in the plane and let $P(x, y)$ be the midpoint of the segment joining them. Drop perpendiculars from $A, P,$ and B to the x-axis as shown in Figure 8. Then x is midway between x_1 and x_2, so

$$x - x_1 = x_2 - x$$

$$2x = x_1 + x_2$$

$$x = \frac{x_1 + x_2}{2}$$

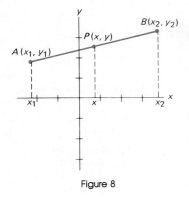

Figure 8

Similar reasoning applies to y. The result, called the **midpoint formula,** says that the coordinates (x, y) of the midpoint P are given by

$$x = \frac{x_1 + x_2}{2} \qquad y = \frac{y_1 + y_2}{2}$$

For example, the midpoint of the segment joining $A(-3, -2)$ and $B(5, 9)$ has coordinates.

$$\left(\frac{-3 + 5}{2}, \frac{-2 + 9}{2} \right) = \left(1, \frac{7}{2} \right)$$

Problem Set 4-1

Find $d(A, B)$ where A and B are points on the number line having the given coordinates a and b.

1. $a = -3, b = 2$
2. $a = 5, b = -6$
3. $a = \frac{11}{4}, b = -\frac{5}{4}$
4. $a = \frac{31}{8}, b = \frac{13}{8}$
5. $a = 3.26, b = 4.96$
6. $a = -1.45, b = -5.65$
7. $a = 2 - \pi, b = \pi - 3$
8. $a = 4 + 2\sqrt{3}, b = -6 + \sqrt{3}$

Again, A and B are points on the number line. Given a and $d(A, B)$, what are the two possible values of b?

9. $a = 5, d(A, B) = 2$
10. $a = 9, d(A, B) = 5$
11. $a = -2, d(A, B) = 4$
12. $a = 3, d(A, B) = 7$
13. $a = \frac{5}{2}, d(A, B) = \frac{3}{4}$
14. $a = 1.8, d(A, B) = 2.4$

Plot $A, B, C,$ and D on a coordinate system. Name the quadrilateral ABCD—that is, is it a square, a rectangle, or what?

15. $A(4, 3), B(4, -3), C(-4, 3), D(-4, -3)$
16. $A(-1, 6), B(0, 5), C(3, 2), D(5, 0)$
17. $A(1, 3), B(2, 6), C(4, 7), D(3, 4)$
18. $A(0, 2), B(3, 0), C(2, -1), D(-1, 1)$

In Problems 19–26, find $d(P_1, P_2)$ where P_1 and P_2 have the given coordinates. Also find the coordinates of the midpoint of the segment P_1P_2.

19. $(2, -1), (5, 3)$
20. $(2, 1), (7, 13)$

21. $(4, 2)$, $(2, 4)$ 22. $(-1, 5)$, $(6, 7)$

23. $(\sqrt{3}, 0)$, $(0, \sqrt{6})$ 24. $(\sqrt{2}, 0)$, $(0, -\sqrt{7})$

[c] 25. $(1.234, -5.132)$, $(6.714, 8.341)$

[c] 26. $(-42.1, 16.3)$, $(12.2, -5.3)$

27. The points A, B, C, and D of Problem 17 form a parallelogram.
    (a) Find $d(A, B)$, $d(B, C)$, $d(C, D)$, and $d(D, A)$.
    (b) Find the coordinates of the midpoints of the diagonals AC and BD of the parallelogram $ABCD$.
    (c) What facts about a parallelogram do your answers to parts (a) and (b) agree with?

28. The points $(3, -1)$ and $(3, 3)$ are two vertices of a square. Give two pairs of other possible vertices. Can you give a third pair?

29. Let $ABCD$ be a rectangle whose sides are parallel to the coordinate axes. Find the coordinates of B and D if the coordinates of A and C are as given.
    (a) $(-2, 0)$ and $(4, 3)$ (Draw a picture.)
    (b) $(2, -1)$ and $(8, 7)$

30. Show that the triangle whose vertices are $(5, 3)$, $(-2, 4)$ and $(10, 8)$ is isosceles.

31. Use the distance formula to show that the triangle whose vertices are $(2, -4)$, $(4, 0)$, and $(8, -2)$ is a right triangle.

32. (a) Find the point on the y-axis that is equidistant from the points $(3, 1)$ and $(6, 4)$. *Hint:* Let the unknown point be $(0, y)$.
    (b) Find the point on the x-axis that is equidistant from $(3, 1)$ and $(6, 4)$.

33. Use the distance formula to show that the three points are on a line.
    (a) $(0, 0)$, $(3, 4)$, $(-6, -8)$
    (b) $(-4, 1)$, $(-1, 5)$, $(5, 13)$

EXAMPLE (Point-of-Division Formula) Let $A(x_1, y_1)$ and $B(x_2, y_2)$ be the endpoints of a line segment and let t be a number between 0 and 1. The point $P(x, y)$ on this segment satisfying

$$d(A, P) = t \cdot d(A, B)$$

has coordinates given by

$$x = (1 - t)x_1 + tx_2$$
$$y = (1 - t)y_1 + ty_2$$

Note that $t = \frac{1}{2}$ gives the midpoint formula. Use the formula to find the point $P(x, y)$ two-thirds of the way from $A(-2, 5)$ to $B(4, -4)$.

Solution. Substitution of $t = \frac{2}{3}$ gives

$$x = \frac{1}{3}(-2) + \frac{2}{3}(4) = \frac{6}{3} = 2$$

$$y = \frac{1}{3}(5) + \frac{2}{3}(-4) = -\frac{3}{3} = -1$$

The required point is $(2, -1)$; see Figure 9.

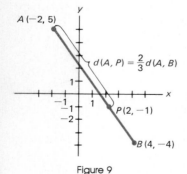

$A(-2, 5)$

$d(A, P) = \frac{2}{3} d(A, B)$

$P(2, -1)$

$B(4, -4)$

Figure 9

Use the point-of-division formula to find $P(x, y)$ for the given value of t and given points A and B. Plot the points A, P, and B.

34. $A(5, -8)$, $B(11, 4)$, $t = \frac{1}{3}$ 35. $A(5, -8)$, $B(11, 4)$, $t = \frac{2}{3}$
36. $A(5, -8)$, $B(11, 4)$, $t = \frac{5}{6}$ 37. $A(4, 9)$, $B(104, 209)$, $t = \frac{13}{100}$

MISCELLANEOUS PROBLEMS

38. Show that the triangle with vertices $(2, -2)$, $(-2, 2)$, and $(2\sqrt{3}, 2\sqrt{3})$ is equilateral.

39. Cities A, B and C are located at $(0, 0)$, $(214, 17)$, and $(230, 179)$, respectively, with distances in miles. There are straight roads from A to B and from B to C, but only an air route goes directly from A to C. It costs \$3.71 per mile to ship a certain item by truck and \$4.81 per mile by air. Find the cheaper way to ship this item from A to C and calculate the savings made by choosing this form of shipment.

40. Find x such that the distance between the points $(3, 0)$ and $(x, 3)$ is 5.

41. Find the point (x, y) in the first quadrant that together with $(0, 0)$ and $(-3, 4)$ are vertices of an equilateral triangle.

42. Consider the triangle with vertices $A(0, 0)$, $B(3, 0)$, and $C(0, 4)$. Find the length of the altitude from A to BC.

43. Show that the midpoint of the hypotenuse of a right triangle is equidistant from the three vertices. *Hint:* Place the triangle in the coordinate system so its vertices are $(0, 0)$, $(a, 0)$, and (a, b) with $a > 0$ and $b > 0$.

44. Find the point (x, y) that is equidistant from $(1, 3)$, $(4, 2)$, and $(-3, 1)$.

45. Find the point (x, y) such that $(4, 5)$ is two-thirds of the way from $(2, 1)$ to (x, y) on the segment connecting these points.

46. Show that the midpoints of the sides of a triangle are the vertices of another triangle whose area is one-fourth that of the original triangle. *Hint:* You may suppose the vertices of the original triangle to be $(0, 0)$, $(0, a)$, and (b, c) with a, b, and c positive.

47. Given $A(-4, 3)$ and $B(21, 38)$, find the coordinates of the four points that divide AB into five equal parts.

48. Find the point on the segment joining $(1, 3)$ and $(6, 7)$ that is $\frac{11}{13}$ of the way from the first point to the second.

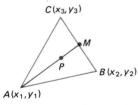

$C(x_3, y_3)$
M
P
$B(x_2, y_2)$
$A(x_1, y_1)$

Figure 10

49. Consider the general triangle shown in Figure 10. The line segment from vertex A to the midpoint M of the opposite side is called a median.
    (a) Express the coordinates of the point P two-thirds of the way from A to M in terms of the coordinates of the three vertices of the triangle.
    (b) What do you conclude from (a) about the three medians of a triangle?

50. Prove that the diagonals of a parallelogram bisect each other.

51. Let $ABCD$ be the vertices of a rectangle, labeled in cyclic order. Suppose that P is a point in the same plane as the rectangle and at distances a, b, c, and d from A, B, C, and D, respectively. Show that $a^2 + c^2 = b^2 + d^2$. *Hint:* Introduce a coordinate system in an intelligent way.

52. TEASER Arnold Thinkhard has been ordered to attach four guy wires to the top of a tall pole. These four guy wires are anchored to four stakes A, B, C, and D

which form the vertices (in cyclic order) of a rectangle on the ground. After attaching wires of length 210, 60, and 180 feet from A, B, and C, Arnold discovers that the fourth wire from D is too short to reach. A new one must be cut, but how long should it be? Even though you don't know the dimensions of the rectangle or the height of the pole, it is your job to solve Arnold's problem. It is not enough to get the answer; you must demonstrate that your answer is correct before poor Arnold climbs back up the pole.

Algebra   Geometry

"As long as algebra and geometry traveled separate paths, their advance was slow and their applications limited. But when the two sciences joined company, they drew from each other fresh vitality and thenceforward marched on at a rapid pace toward perfection."

Joseph-Louis Lagrange

4-2   Algebra and Geometry United

The Greeks were preeminent geometers but poor algebraists. Though they were able to solve a host of geometry problems, their limited algebraic skills kept others beyond their grasp. By 1600, geometry was a mature and eligible bachelor. Algebra was a young woman only recently come of age. Fermat and Descartes were the matchmakers; they brought the two together. The resulting union is called **analytic geometry**, or coordinate geometry.

THE GRAPH OF AN EQUATION

An equation is an algebraic object. By means of a coordinate system, it can be transformed into a curve, a geometric object. Here is how it is done.

Consider the equation $y = x^2 - 3$. Its set of solutions is the set of ordered pairs (x, y) that satisfy the equation. These ordered pairs are the coordinates of points in the plane. The set of all such points is called the **graph** of the equation. The graph of an equation in two variables x and y will usually be a curve.

To obtain this graph, we follow a definite procedure.

1. Obtain the coordinates of a few points.
2. Plot those points in the plane.
3. Connect the points with a smooth curve in the order of increasing x values.

The best way to do step 1 is to make a **table of values.** Assign values to one of the variables, say x, determine the corresponding values of the other, and then list the pairs of values in tabular form. The whole three-step procedure is illustrated in Figure 11 for the previously mentioned equation, $y = x^2 - 3$.

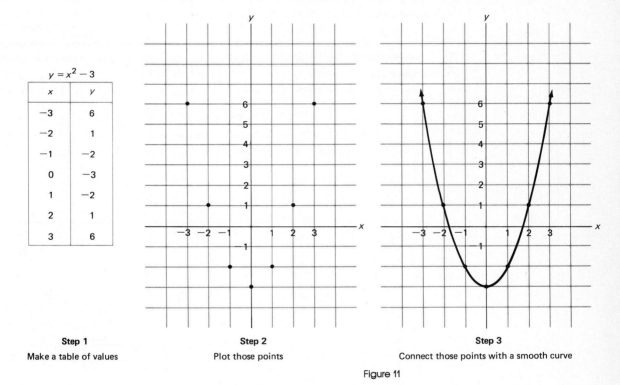

$y = x^2 - 3$

x	y
-3	6
-2	1
-1	-2
0	-3
1	-2
2	1
3	6

Step 1
Make a table of values

Step 2
Plot those points

Step 3
Connect those points with a smooth curve

Figure 11

Of course, you need to use common sense and even a little faith. When you connect the points you have plotted with a smooth curve, you are assuming that the curve behaves nicely between consecutive points; that is faith. This is why you should plot enough points so the outline of the curve seems very clear; the more points you plot, the less faith you will need. Also you should recognize that you can seldom display the whole curve. In our example, the curve has infinitely long arms opening wider and wider. But our graph does show the essential features. That is what we always aim to do—show enough of the graph so the essential features are visible.

SYMMETRY OF A GRAPH

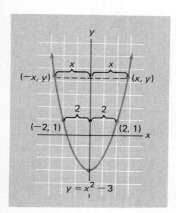

Figure 12

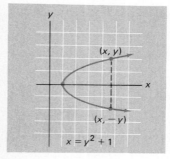

Figure 13

The graph of $y = x^2 - 3$, drawn earlier and again in Figure 12, has a nice property of symmetry. If the coordinate plane were folded along the y-axis the two branches would coincide. For example, $(3, 6)$ would coincide with $(-3, 6)$; $(2, 1)$ would coincide with $(-2, 1)$; and, more generally, (x, y) would coincide with $(-x, y)$. Algebraically, this corresponds to the fact that we may replace x by $-x$ in the equation $y = x^2 - 3$ without changing it. More precisely, $y = x^2 - 3$ and $y = (-x)^2 - 3$ are equivalent equations. *EVEN FUNCTION*

Whenever an equation is unchanged by replacing (x, y) with $(-x, y)$, the ✱ graph of the equation is **symmetric with respect to the y-axis.** Likewise, if the equation is unchanged when (x, y) is replaced by $(x, -y)$, its graph is **symmetric with respect to the x-axis.** The equation $x = 1 + y^2$ is of the latter type; its graph is shown in Figure 13. *ODD FUNCTION* ✱

A third type of symmetry is **symmetry with respect to the origin.** It occurs whenever replacing (x, y) by $(-x, -y)$ produces no change in the equation. The equation $y = x^3$ is a good example as $-y = (-x)^3$ is equivalent to $y = x^3$. The graph is shown in Figure 14. Note that the dotted line segment from $(-x, -y)$ to (x, y) is bisected by the origin.

In graphing $y = x^3$, we used a smaller scale on the y-axis than on the x-axis. This made it possible to show a larger portion of the graph. We suggest that before putting scales on the two axes, you should examine your table of

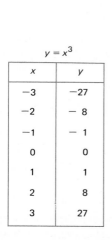

$y = x^3$

x	y
−3	−27
−2	− 8
−1	− 1
0	0
1	1
2	8
3	27

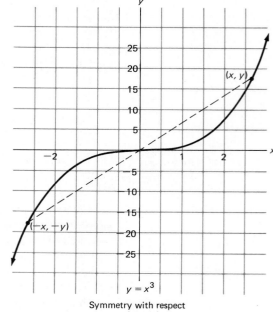

Symmetry with respect
to the origin

Figure 14

values. Choose scales so that all of your points can be plotted and still keep your graph of reasonable size.

Graphing an equation is an extremely important operation. It gives us a picture to look at. Most of us can absorb qualitative information from a picture much more easily than from symbols. But if we want precise quantitative information, then symbols are better; they are easier to manipulate. That is why we must be able to reverse the process just described, which is our next topic.

THE EQUATION OF A GRAPH

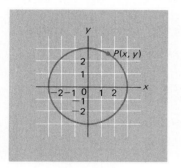

Figure 15

A graph is a geometric object, a picture. How can we turn it into an algebraic object, an equation? Sometimes it is easy, but not always. As an example, consider a circle of radius 3. It consists of all points 3 units from a fixed point, called the center. Figure 15 shows this circle with its center at the origin of a coordinate system.

Take *any* point P on the circle and label its coordinates (x, y). It must satisfy the equation

$$d(P, O) = 3$$

From the distance formula of the previous section, we have

$$\sqrt{(x - 0)^2 + (y - 0)^2} = 3$$

or equivalently (after squaring both sides)

$$x^2 + y^2 = 9$$

This is the equation we sought.

We could move to other types of curves, attempting to find their equations. In fact, we will do exactly that in later sections. Right now, we shall consider more general circles, that is, circles with arbitrary radii and arbitrary centers.

THE STANDARD EQUATION OF A CIRCLE

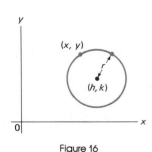

Figure 16

Consider a circle of radius r with center at (h, k), as in Figure 16. To find its equation, take an arbitrary point on the circle with coordinates (x, y). According to the distance formula, it must satisfy the equation

$$\sqrt{(x - h)^2 + (y - k)^2} = r$$

or, equivalently,

$$(x - h)^2 + (y - k)^2 = r^2$$

Every circle has an equation of this form; we call it the **standard equation of a circle.**

As an example, let us find the equation of a circle of radius 6 centered at $(2, -1)$. We use the boxed equation with $h = 2$, $k = -1$, and $r = 6$. This gives

$$(x - 2)^2 + (y + 1)^2 = 36$$

Let us go the other way. What circle has the equation

$$(x + 3)^2 + (y - 4)^2 = 49$$

Think of this as

$$[x - (-3)]^2 + (y - 4)^2 = 49$$

Then it is clear that this is the equation of a circle with center $(-3, 4)$ and radius 7.

Consider this last equation again. Notice that it could be written as

$$x^2 + 6x + 9 + y^2 - 8y + 16 = 49$$

or equivalently as

$$x^2 + y^2 + 6x - 8y = 24$$

A natural question to ask is whether every equation of the form

$$x^2 + y^2 + Dx + Ey = F$$

is the equation of a circle. Take

$$x^2 + y^2 - 6y + 16x = 8$$

as an example. Recalling a skill we learned in Section 3-4 (completing the square), we may rewrite this as

$$(x^2 + 16x + \quad) + (y^2 - 6y + \quad) = 8$$

or

$$(x^2 + 16x + 64) + (y^2 - 6y + 9) = 8 + 64 + 9$$

or

$$(x + 8)^2 + (y - 3)^2 = 81$$

We recognize this to be the equation of a circle with center at $(-8, 3)$ and radius 9. In fact, we will always have a circle unless the number on the right side in the last step is negative or zero.

Problem Set 4-2

Graph each of the following equations, showing enough of the graph to bring out its essential features. Begin by noting any of the three kinds of symmetry discussed in the text.

1. $y = 3x - 2$ 2. $y = 2x + 1$ 3. $y = -x^2 + 4$

4. $y = -x^2 - 2x$ 5. $y = x^2 - 4x$ 6. $y = x^3 + 2$

7. $y = -x^3$ 8. $y = \dfrac{12}{x^2 + 4}$ 9. $y = \dfrac{4}{x^2 + 1}$

10. $y = x^3 + x$

EXAMPLE (More Graphing)   Graph the equation $x = y^2 - 2y + 4$.

Solution.   Assign values to y and calculate the corresponding values of x. Then plot the corresponding points and draw the graph (Figure 17). Note that this graph is symmetric with respect to the line $y = 1$.

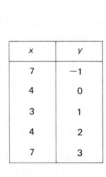

x	y
7	-1
4	0
3	1
4	2
7	3

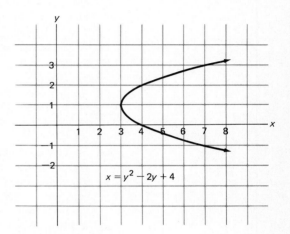

Figure 17

Graph each of the following equations.

11. $x = 2y - 1$ 12. $x = -3y + 1$ 13. $x = -2y^2$

14. $x = 2y - y^2$ 15. $x = y^3$ 16. $x = 8 - y^3$

Write the equation of the circle with the given center and radius.

17. Center $(0, 0)$, radius 6.   18. Center $(2, 3)$, radius 3.

19. Center $(4, 1)$, radius 5.   20. Center $(2, -1)$, radius $\sqrt{7}$.

21. Center $(-2, 1)$, radius $\sqrt{3}$.   22. Center $(\pi, \frac{3}{4})$, radius $\frac{1}{2}$.

Graph the following equations.

23. $(x - 2)^2 + y^2 = 16$ 24. $(x - 2)^2 + (y + 2)^2 = 25$

25. $(x + 1)^2 + (y - 3)^2 = 64$ 26. $(x + 4)^2 + (y + 6)^2 = \frac{49}{4}$

Find the center and radius of each of the following circles. Hint: Complete the squares.

27. $x^2 + y^2 + 2x - 10y + 25 = 0$ 28. $x^2 + y^2 - 6y = 16$

29. $x^2 + y^2 - 12x + 35 = 0$ 30. $x^2 + y^2 - 10x + 10y = 0$
31. $4x^2 + 4y^2 + 4x - 12y + 1 = 0$ 32. $3x^2 + 3y^2 - 2x + 4y = \frac{20}{3}$

MISCELLANEOUS PROBLEMS

Sketch the graphs of the equations in Problems 33–38. Begin by deciding if the graph has any of the three kinds of symmetry discussed in the text.

33. $y = 12/x$ 34. $y = 3 + 4/x^2$
35. $y = 2(x - 1)^2$ 36. $x = -y^2 + 8$
37. $x = 4y - y^2$ 38. $x^2 + y^2 - 2y = 8$

39. Find the equation of the circle with center $(5, -7)$ that is tangent to the x-axis.

40. Find the equation of the circle with center on the line $y = x$ that is tangent to the y-axis at $(0, 5)$.

41. Write the equation of the circle with center $(-3, 2)$ that passes through $(4, 3)$.

42. Find the equation of the circle that has AB as diameter given that $A = (1, 2)$ and $B = (5, 12)$.

43. Identify each of the following as a circle (giving its radius and center), or as a point (giving its coordinates), or as the empty set.
    (a) $x^2 + y^2 - 4x + 6y = -13$ (b) $2x^2 + 2y^2 - 2x + 6y = 3$
    (c) $4x^2 + 4y^2 - 8x - 4y = -7$ (d) $\sqrt{3}x^2 + \sqrt{3}y^2 - 6y = 2\sqrt{3}$

44. Find the equation of the circle of radius 5 with center in the first quadrant that passes through $(3, 0)$ and $(0, 1)$.

45. The four points $(1, 0)$, $(8, 1)$, $(7, 8)$, and $(0, 7)$ are vertices of a square. Find the equations of (a) the circle circumscribed about this square and (b) the circle inscribed in this square.

46. A belt fits tightly around the two circles (pulleys) with equations $x^2 - 2x + y^2 + 4y = 11$ and $x^2 - 14x + y^2 - 12y = -69$. Find the length of this belt.

47. A belt fits tightly around the two circles with equations $x^2 + y^2 = 1$ and $(x - 6)^2 + y^2 = 16$. Find the length of this belt.

48. **TEASER** Circles of radius 2 are centered at the three vertices of a triangle with sides of lengths 8, 11, and 12. Find the length of a belt that fits tightly around these three circles as shown in Figure 18. Next, solve the same problem for a quadrilateral with sides of lengths 8, 9, 10, and 12. Finally, discover the very general result of which the above examples are special cases.

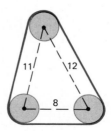

Figure 18

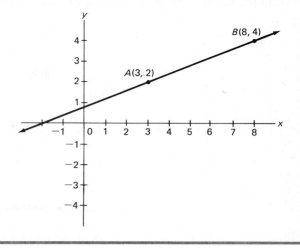

A point is that which has no part.

A line is breadthless length.

A straight line is a line which lies evenly with the points of itself.

Euclid

300 B.C.

4-3 The Straight Line

Euclid's definition of a straight line is not very helpful, but neither are most of the alternatives we have heard. Fortunately, we all know what we mean by a straight line even if we cannot seem to describe it in terms of more primitive ideas. There is one thing on which we must agree: Given two points (for example, *A* and *B* above), there is one and only one straight line that passes through them. And contrary to Euclid, let us agree that the word *line* shall always mean straight line.

A line is a geometric object. When it is placed in a coordinate system, it ought to have an equation just as a circle does. How do we find the equation of a line? To answer this question we will need the notion of slope.

THE SLOPE OF A LINE

Consider the line in our opening diagram. From point *A* to point *B*, there is a **rise** (vertical change) of 2 units and a **run** (horizontal change) of 5 units. We say that the line has a slope of $\frac{2}{5}$. In general (Figure 19), for a line through $A(x_1, y_1)$ and $B(x_2, y_2)$, where $x_1 \neq x_2$, we define the **slope** *m* of that line by

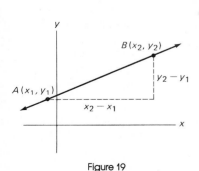

Figure 19

$$m = \frac{\text{rise}}{\text{run}} = \frac{y_2 - y_1}{x_2 - x_1}$$

You should immediately raise a question. A line has many points. Does the value we get for the slope depend on which pair of points we use for *A* and *B*? The similar triangles in Figure 20 at the top of the next page show us that

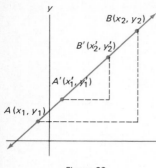

Figure 20

$$\frac{y_2' - y_1'}{x_2' - x_1'} = \frac{y_2 - y_1}{x_2 - x_1}$$

Thus, points A' and B' would do just as well as A and B. It does not even matter whether A is to the left or right of B since

$$\frac{y_1 - y_2}{x_1 - x_2} = \frac{y_2 - y_1}{x_2 - x_1}$$

All that matters is that we subtract the coordinates in the same order in numerator and denominator.

The slope m is a measure of the steepness of a line, as Figure 21 illustrates. Notice that a horizontal line has zero slope and a line that rises to the right has positive slope. The larger this positive slope is, the more steeply the line rises. A line that falls to the right has negative slope. The concept of slope for a vertical line makes no sense since it would involve division by zero. Therefore the notion of slope for a vertical line is left undefined.

CAUTION

A horizontal line has no slope.

A horizontal line has zero slope.
A vertical line has no slope slope.

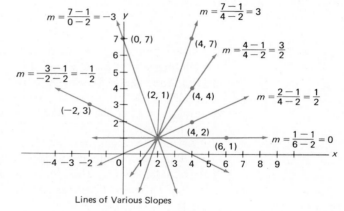

Lines of Various Slopes

Figure 21

THE POINT-SLOPE FORM

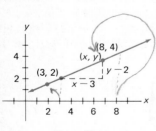

Figure 22

Consider again the line of our opening diagram; it is reproduced in Figure 22. We know:

1. it passes through $(3, 2)$;
2. it has slope $\frac{2}{5}$.

Take any other point on that line, such as one with coordinates (x, y). If we use this point together with $(3, 2)$ to measure slope, we must get $\frac{2}{5}$; that is,

$$\frac{y - 2}{x - 3} = \frac{2}{5}$$

or, after multiplying by $x - 3$,

$$y - 2 = \frac{2}{5}(x - 3)$$

Notice that this last equation is satisfied by all points on the line, even by $(3, 2)$. Moreover, no points not on the line can satisfy this equation.

What we have just done in an example can be done in general. The line passing through the (fixed) point (x_1, y_1) with slope m has equation

$$y - y_1 = m(x - x_1)$$

We call it the **point-slope** form for the equation of a line.

Consider once more the line of our example. That line passes through $(8, 4)$ as well as $(3, 2)$. If we use $(8, 4)$ as (x_1, y_1), we get the equation

$$y - 4 = \frac{2}{5}(x - 8)$$

which looks quite different from

$$y - 2 = \frac{2}{5}(x - 3)$$

However, both can be simplified to $5y - 2x = 4$; they are equivalent.

THE SLOPE-INTERCEPT FORM

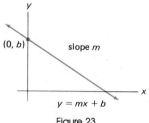

$y = mx + b$

Figure 23

The equation of a line can be expressed in various forms. Suppose we are given the slope m for a line and the y-intercept b (that is, the line intersects the y-axis at $(0, b)$ as in Figure 23). Choosing $(0, b)$ as (x_1, y_1) and applying the point-slope form, we get

$$y - b = m(x - 0)$$

which we can rewrite as

$$y = mx + b$$

The latter is called the **slope-intercept** form for the equation of a line.

Why get excited about that, you ask? Because any time we see an equation written this way, we recognize it as the equation of a line and can immediately read its slope and y-intercept. For example, consider the equation

$$3x - 2y + 4 = 0$$

If we solve for y, we get

$$y = \frac{3}{2}x + 2$$

It is the equation of a line with slope $\frac{3}{2}$ and y-intercept 2.

EQUATION OF A VERTICAL LINE

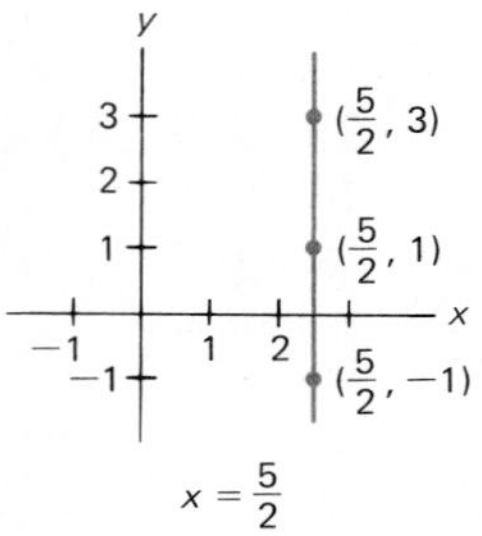

Vertical lines do not fit within the discussion above; they do not have slopes. But they do have equations, very simple ones. The line in Figure 24 has equation $x = \frac{5}{2}$, since every point on the line satisfies this equation. The equation of any vertical line can be put in the form

$$x = k$$

where k is a constant. It should be noted that the equation of a horizontal line can be written in the form $y = k$.

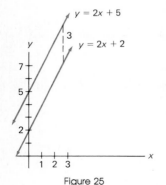

$x = \frac{5}{2}$

Figure 24

THE FORM $Ax + By + C = 0$

It would be nice to have a form that covered all lines including vertical lines. Consider for example,

(i) $y - 2 = -4(x + 2)$
(ii) $y = 5x - 3$
(iii) $x = 5$

These can be rewritten (by taking everything to the left side) as follows:

(i) $4x + y + 6 = 0$
(ii) $-5x + y + 3 = 0$
(iii) $x + 0y - 5 = 0$

All are of the form

$$Ax + By + C = 0$$

which we call the **general linear equation.** It takes only a moment's thought to see that the equation of any line can be put in this form. Conversely, the graph of $Ax + By + C = 0$ is always a line (if A and B are not both zero (see Problem 54)).

PARALLEL LINES

If two lines have the same slope, they are parallel. Thus $y = 2x + 2$ and $y = 2x + 5$ represent parallel lines; both have a slope of 2. The second line is 3 units above the first for every value of x (Figure 25).

Similarly, the lines with equations $-2x + 3y + 12 = 0$ and $4x - 6y = 5$ are parallel. To see this, solve these equations for y (that is, find the slope-intercept form); you get $y = \frac{2}{3}x - 4$ and $y = \frac{2}{3}x - \frac{5}{6}$, respectively. Both have slope $\frac{2}{3}$; they are parallel.

Figure 25

We may summarize by stating that *two nonvertical lines are parallel if and only if they have the same slope*.

PERPENDICULAR LINES

Is there a simple slope condition which characterizes perpendicular lines? Yes; *two nonvertical lines are perpendicular if and only if their slopes are negative reciprocals of each other*. We are not going to prove this, but an example will help explain why it is true. The slopes of the lines $y = \frac{3}{4}x$ and $y = -\frac{4}{3}x$ are negative reciprocals of each other. Both lines pass through the origin. The points (4, 3) and (3, −4) are on the first and second lines, respectively. The two right triangles shown in Figure 26 are congruent with $\angle \alpha = \angle \delta$ and $\angle \beta = \angle \gamma$. But

$$\angle \alpha + \angle \beta = 90°$$

and therefore

$$\angle \alpha + \angle \gamma = 90°$$

That says the two lines are perpendicular to each other.

The lines $2x - 3y = 5$ and $3x + 2y = -4$ are also perpendicular, since—after solving them for y—we see that the first has slope $\frac{2}{3}$ and the second has slope $-\frac{3}{2}$.

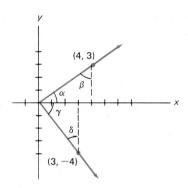

Figure 26

Problem Set 4-3

Find the slope of the line containing the given two points.

1. (2, 3) and (4, 8)
2. (4, 1) and (8, 2)
3. (−4, 2) and (3, 0)
4. (2, −4) and (0, −6)
5. (3, 0) and (0, 5)
6. (−6, 0) and (0, 6)
ⓒ 7. (−1.732, 5.014) and (4.315, 6.175)
ⓒ 8. $(\pi, \sqrt{3})$ and $(1.642, \sqrt{2})$

Find an equation for each of the following lines. Then write your answer in the form $Ax + By + C = 0$. $y = Mx + b$

9. Through (2, 3) with slope 4.
10. Through (4, 2) with slope 3.
11. Through (3, −4) with slope −2.
12. Through (−5, 2) with slope −1.
13. With y-intercept 4 and slope −2.
14. With y-intercept −3 and slope 1.
15. With y-intercept 5 and slope 0.
16. With y-intercept 1 and slope −1.

17. Through $(2, 3)$ and $(4, 8)$.

18. Through $(4, 1)$ and $(8, 2)$.

19. Through $(3, 0)$ and $(0, 5)$.

20. Through $(-6, 0)$ and $(0, 6)$.

[c] 21. Through $(\sqrt{3}, \sqrt{7})$ and $(\sqrt{2}, \pi)$.

[c] 22. Through $(\pi, \sqrt{3})$ and $(\pi + 1, 2\sqrt{3})$.

23. Through $(2, -3)$ and $(2, 5)$.

24. Through $(-5, 0)$ and $(-5, 4)$.

In Problems 25–32, find the slope and y-intercept of each line.

25. $y = 3x + 5$ 26. $y = 6x + 2$

27. $3y = 2x - 4$ 28. $2y = 5x + 2$

29. $2x + 3y = 6$ 30. $4x + 5y = -20$

31. $y + 2 = -4(x - 1)$ 32. $y - 3 = 5(x + 2)$

33. Write the equation of the line through $(3, -3)$:
   (a) parallel to the line $y = 2x + 5$;
   (b) perpendicular to the line $y = 2x + 5$;
   (c) parallel to the line $2x + 3y = 6$;
   (d) perpendicular to the line $2x + 3y = 6$;
   (e) parallel to the line through $(-1, 2)$ and $(3, -1)$;
   (f) parallel to the line $x = 8$;
   (g) perpendicular to the line $x = 8$.

34. Find the value of k for which the line $4x + ky = 5$:
   (a) passes through the point $(2, 1)$;
   (b) is parallel to the y-axis;
   (c) is parallel to the line $6x - 9y = 10$;
   (d) has equal x- and y-intercepts;
   (e) is perpendicular to the line $y - 2 = 2(x + 1)$.

35. Write the equation of the line through $(0, -4)$ that is perpendicular to the line $y + 2 = -\frac{1}{2}(x - 1)$.

36. Find the value of k such that the line $kx - 3y = 10$:
   (a) is parallel to the line $y = 2x + 4$;
   (b) is perpendicular to the line $y = 2x + 4$;
   (c) is perpendicular to the line $2x + 3y = 6$.

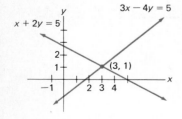

Figure 27

EXAMPLE A (Intersection of Two Lines)   Find the coordinates of the point of intersection of the lines $3x - 4y = 5$ and $x + 2y = 5$ (Figure 27).

Solution.   We simply solve the two equations simultaneously (see Section 3-3). Multiply the second equation by -3 and then add the two equations.

$$3x - 4y = 5$$
$$\underline{-3x - 6y = -15}$$
$$-10y = -10$$
$$y = 1$$
$$x = 3$$

Find the coordinates of the point of intersection in each problem below. Then write the equation of the line through that point perpendicular to the line given first.

37. $2x + 3y = 4$
$-3x + y = 5$

38. $4x - 5y = 8$
$2x + y = -10$

39. $3x - 4y = 5$
$2x + 3y = 9$

40. $5x - 2y = 5$
$2x + 3y = 6$

EXAMPLE B (Distance from a Point to a Line) It can be shown that the distance d from the point (x_1, y_1) to the line $Ax + By + C = 0$ is

$$d = \frac{|Ax_1 + By_1 + C|}{\sqrt{A^2 + B^2}}$$

Find the distance from $(1, 2)$ to $3x - 4y = 5$ (Figure 28).

Solution. First write the equation as $3x - 4y - 5 = 0$. The formula gives

$$d = \frac{|3 \cdot 1 - 4 \cdot 2 - 5|}{\sqrt{3^2 + (-4)^2}} = \frac{|-10|}{5} = 2$$

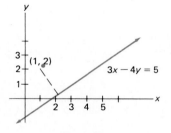

Figure 28

In each case, find the distance from the given point to the given line.

41. $(-3, 2)$, $3x + 4y = 6$

42. $(4, -1)$, $2x - 2y + 4 = 0$

43. $(-2, -1)$, $5y = 12x + 1$

44. $(3, -1)$, $y = 2x - 5$

Find the (perpendicular) distance between the given parallel lines. Hint: First find a point on one of the lines.

45. $3x + 4y = 6$, $3x + 4y = 12$

46. $5x + 12y = 2$, $5x + 12y = 7$

MISCELLANEOUS PROBLEMS

47. Which pairs of lines are parallel, which are perpendicular, and which are neither?
(a) $y = 3x - 2$, $6x - 2y = 0$
(b) $x = -2$, $y = 4$
(c) $x = 2(y - 2)$, $y = -\frac{1}{2}(x - 1)$
(d) $2x + 5y = 3$, $10x - 4y = 7$

48. Find the equation of the line through the point $(2, -1)$ that:
(a) passes through $(-3, 5)$;
(b) is parallel to $2x - 3y = 5$;
(c) is perpendicular to $x + 2y = 3$;
(d) is perpendicular to the y-axis.

49. Find the equation of the line passing through the intersection of the lines $x + 2y = 1$ and $3x + 2y = 5$ that is parallel to the line $3x - 2y = 4$.

50. A line L is perpendicular to the line $2x + 3y = 6$ and passes through $(-3, 1)$. Where does L cut the y-axis?

51. Find the equation of the perpendicular bisector of the line segment connecting $(3, -2)$ and $(7, 6)$.

52. The center of the circle that circumscribes a triangle is at the intersection of the perpendicular bisectors of the sides. Use this fact to find the center of the circle that goes through the three points $(0, 8)$, $(6, 2)$, and $(12, 14)$.

53. Show that the equation of the line with x-intercept a and y-intercept b (both a and b nonzero) can be written in the form

$$\frac{x}{a} + \frac{y}{b} = 1$$

54. Show that the graph of $Ax + By + C = 0$ is always a line (provided A and B are not both 0). *Hint:* Consider two cases: (1) $B = 0$ and (2) $B \neq 0$.

55. A line passes through $(3, 2)$ and its nonzero y-intercept is twice its x-intercept. Find the equation of the line.

56. For each k, the equation $2x - y + 4 + k(x + 3y - 6) = 0$ represents a line (why?). One value of k determines a line with slope $\frac{3}{4}$. Where does this line cut the y-axis?

57. The ABC company makes zeebos, which it sells for $20 apiece. The material and labor to make a zeebo cost $16 and the company has fixed yearly costs (utilities, real estate taxes, and so on) of $8500. Write an expression for the company's profit P in a year in which it makes and sells x zeebos. What is its profit in a year in which it makes only 2000 zeebos?

58. A piece of equipment purchased today for $80,000 will depreciate linearly to a scrap value of $2000 after 20 years. Write a formula for its value V after t years.

59. Find the distance between the parallel lines $12x - 5y = 2$ and $12x - 5y = 7$.

60. Find a formula for the distance between the lines $y = mx + b$ and $y = mx + B$ in terms of m, b, and B.

61. Show that the line through the midpoints of two sides of a triangle is parallel to the third side.

62. Let the three vertices of a triangle lie on a circle with two of them being the ends of a diameter. Show that the triangle is a right triangle. *Hint:* Place the triangle in the coordinate system so two of its vertices are $(-a, 0)$ and $(a, 0)$.

63. Find two points in the plane that are simultaneously equidistant from the lines $y = \pm x$ and the point $(5, 3)$.

64. Using the same axes, sketch the graphs of $x^2 + y^2 = 1$ and $|x| + |y| = 1$ and compute the area of the region between the two graphs.

65. A line through $(4, 4)$ is tangent to the circle $x^2 + y^2 = 4$ at a point P in the fourth quadrant. Find the coordinates of P.

66. **TEASER** Suppose that 2 million points in the plane are given. Show that there is a line with exactly 1 million of these points on each side of it. Can you always find a circle with exactly 1 million of these points inside it?

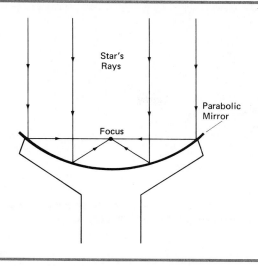

Star's
Rays

Parabolic
Mirror

Focus

Reflecting Telescopes
A reflecting telescope, such as the one at Mount Palomar in California, makes use of a parabolic mirror. The rays of light from a star (parallel lines) are focused by the mirror at a single point. This is a characteristic property of the mathematical curve that we call a parabola.

4-4 The Parabola

We have seen that the graph $y = ax + b$ is always a line. Now we want to study the graph of $y = ax^2 + bx + c (a \neq 0)$. As we shall discover, this graph is always a smooth, cup-shaped curve something like the cross section of the mirror shown in the opening diagram. We call it a **parabola.**

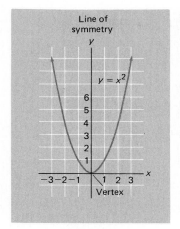

Figure 29

THE GRAPH OF $y = x^2$

The simplest case of all is $y = x^2$. The graph of this equation is shown in Figure 29. Two important features should be noted.

1. The curve is symmetric about the y-axis. This follows from the fact that the equation $y = x^2$ is not changed if we replace (x, y) by $(-x, y)$.
2. The curve reaches its lowest point at $(0, 0)$, the point where the curve intersects the line of symmetry. We call this point the **vertex** of the parabola.

Next we consider how the graph of $y = x^2$ is modified as we look successively at $y = ax^2$, $y = x^2 + k$, $y = (x - h)^2$, and $y = a(x - h)^2 + k$.

THE GRAPH OF $y = ax^2$

In Figure 30, we show the graphs of $y = x^2$, $y = 3x^2$, $y = -2x^2$, and $y = \frac{1}{2}x^2$. They suggest the following general facts. The graph of $y = ax^2$,

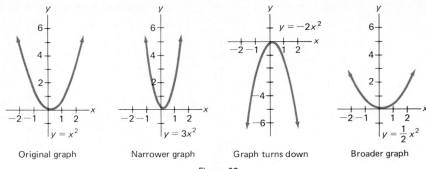

| Original graph | Narrower graph | Graph turns down | Broader graph |

Figure 30

$a \neq 0$, is a parabola with vertex at the origin, opening upward if $a > 0$ and downward if $a < 0$. Increasing $|a|$ makes the graph narrower.

THE GRAPHS OF $y = x^2 + k$ AND $y = (x - h)^2$

The graphs of $y = x^2 + 4$, $y = x^2 - 6$, $y = (x - 2)^2$, and $y = (x + 1)^2$ can all be obtained by shifting (translating) the graph of $y = x^2$, while maintaining its shape. They are shown in Figure 31. After studying these graphs carefully, you will understand the general situation we now describe.

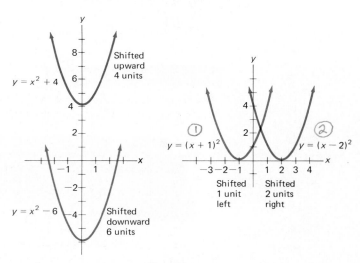

Figure 31

$$Y = (x - h)^2$$

② $Y = (x - (+h))^2$ ← h is +

① $Y = (x - (-h))^2$ ← h is −

The graph of $y = x^2 + k$ is obtained by shifting the graph of $y = x^2$ vertically $|k|$ units, upward if $k > 0$ and downward if $k < 0$. The vertex is at $(0, k)$.

The graph of $y = (x - h)^2$ is obtained by a horizontal shift of $|h|$ units, to the right if $h > 0$ and to the left if $h < 0$. The vertex is at $(h, 0)$.

THE GRAPH OF $y = a(x - h)^2 + k$ OR $y - k = a(x - h)^2$

We can get the graph of $y = 2(x - 3)^2 - 4$ by shifting the graph of $y = 2x^2$ three units to the right and four units down. This puts the vertex at $(3, -4)$. More generally, the graph of $y = a(x - h)^2 + k$ is the graph of $y = ax^2$ shifted horizontally $|h|$ units and vertically $|k|$ units, so that the vertex is at (h, k). The graphs in Figure 32 illustrate these facts.

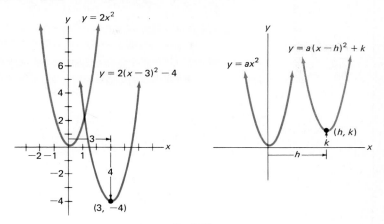

Figure 32

What about the graph of $y = -2(x - 4)^2 + 6$? It has the same shape as the graph of $y = 2x^2$, but turns down with its vertex at $(4, 6)$.

For the circle $x^2 + y^2 = r^2$, recall that replacing x by $x - h$ and y by $y - k$ to obtain $(x - h)^2 + (y - k)^2 = r^2$ had the effect of shifting the center from $(0, 0)$ to (h, k). A similar thing happens to the parabola $y = ax^2$. Replacing x by $x - h$ and y by $y - k$ changes $y = ax^2$ to $y - k = a(x - h)^2$ (which you will note is equivalent to $y = a(x - h)^2 + k$) and correspondingly shifts the vertex from $(0, 0)$ to (h, k).

THE GRAPH OF $y = ax^2 + bx + c$

The most general equation considered so far is $y = a(x - h)^2 + k$. If we expand $a(x - h)^2$ and collect terms on the right side, the equation takes the form $y = ax^2 + bx + c$. Conversely, $y = ax^2 + bx + c \, (a \neq 0)$ always represents a parabola with a vertical line of symmetry, as we shall now show. We use the method of completing the square to rewrite $y = ax^2 + bx + c$ as follows.

$$y = a\left(x^2 + \frac{b}{a}x + \quad\right) + c$$

$$= a\left[x^2 + \frac{b}{a}x + \left(\frac{b}{2a}\right)^2\right] + c - a\left(\frac{b}{2a}\right)^2$$

$$= a\left(x + \frac{b}{2a}\right)^2 + \left(c - \frac{b^2}{4a}\right)$$

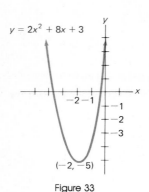

$y = 2x^2 + 8x + 3$

$(-2, -5)$

Figure 33

This is the equation of a parabola with vertex at $(-b/2a, c - b^2/4a)$ and line of symmetry $x = -b/2a$.

What we have just said leads to an important conclusion. The graph of the equation $y = ax^2 + bx + c(a \neq 0)$ is always a parabola with vertical axis of symmetry. Moreover, we have the following facts.

> **The Parabola $y = ax^2 + bx + c$**
>
> 1. The x-coordinate of the vertex is $-b/2a$; the y-coordinate is easily found by substitution in the equation.
> 2. The parabola turns upward if $a > 0$ and downward if $a < 0$. It is a fat or thin parabola, according as $|a|$ is small or large.

As an example, consider $y = 2x^2 + 8x + 3$. Its graph is a parabola with vertex at $x = -b/2a = -8/4 = -2$. The y-coordinate of the vertex, obtained by substituting $x = -2$ in the equation, is $y = -5$. The parabola turns upward and it is rather thin. A sketch is shown in Figure 33.

APPLICATIONS

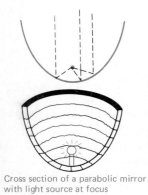

Cross section of a parabolic mirror with light source at focus

Figure 34

Our opening panel hinted at an important application of the parabola related to its optical properties. When light rays, parallel to the axis of a parabolic mirror, hit the mirror, they are reflected to a single point called the *focus*. Conversely, if a light source is placed at the focus of a parabolic mirror, the reflected rays of light are parallel to the axis, a principle used in flashlights (Figure 34).

The parabola is used also in the design of suspension bridges and load-bearing arches. If equal weights are placed along a line, equally spaced, and suspended from a thin flexible cable, the cable will assume a shape that closely approximates a parabola.

We turn to a very different kind of application. From physics, we learn that the path of a projectile is a parabola. It is known, for example, that a projectile fired at an angle of 45° from the horizontal with an initial speed of $320\sqrt{2}$ feet per second follows a curve with equation

$$y = -\frac{1}{6400}x^2 + x$$

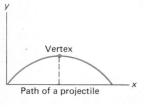

Vertex

Path of a projectile

Figure 35

where the coordinate axes are placed as shown in Figure 35. Taking this for granted, we may ask two questions.

1. What is the maximum height attained by the projectile?
2. What is the range (horizontal distance traveled) of the projectile?

To find the maximum height is simply to find the y-coordinate of the vertex. First we find the x-coordinate.

$$x = \frac{-b}{2a} = -\frac{1}{-2/6400} = 3200$$

When we substitute this value in the equation, we get

$$y = -\frac{1}{6400}(3200)^2 + 3200 = -1600 + 3200 = 1600$$

The greatest height is thus 1600 feet.

The range of the projectile is the x-coordinate of the point where it lands. By symmetry, this is simply twice the x-coordinate of the vertex; that is,

$$\text{range} = 2(3200) = 6400 \text{ feet}$$

This value could also be obtained by solving the quadratic equation

$$-\frac{1}{6400}x^2 + x = 0$$

since the x-coordinate of the landing point is the value of x when $y = 0$.

Problem Set 4-4

The equations in Problems 1–10 represent parabolas. Sketch the graph of each parabola, indicating the coordinates of the vertex.

1. $y = 3x^2$
2. $y = -2x^2$
3. $y = x^2 + 5$
4. $y = 2x^2 - 4$
5. $y = (x - 4)^2$
6. $y = -(x + 3)^2$
7. $y = 2(x - 1)^2 + 5$
8. $y = 3(x + 2)^2 - 4$
9. $y = -4(x - 2)^2 + 1$
10. $y = \frac{1}{2}(x + 3)^2 + 3$

Write each of the following in the form $y = ax^2 + bx + c$.

11. $y = 2(x - 1)^2 + 7$
12. $y = -3(x + 2)^2 + 5$
13. $-2y + 5 = (x - 5)^2$
14. $3y + 6 = (x + 3)^2$

Sketch the graph of each equation. Begin by plotting the vertex and at least one point on each side of the vertex. Recall that the x-coordinate of the vertex for $y = ax^2 + bx + c$ is $x = -b/2a$.

15. $y = x^2 + 2x$
16. $y = 3x^2 - 6x$
17. $y = -2x^2 + 8x + 1$
18. $y = -3x^2 + 6x + 4$

EXAMPLE A (Horizontal Parabolas) Sketch the parabolas (a) $x = 2y^2$; (b) $x = 2(y - 3)^2 - 4$.

Solution. Note that the roles of x and y are interchanged when compared to earlier examples. The line of symmetry will therefore be horizontal. The vertex in (a) is $(0, 0)$; in (b), it is $(-4, 3)$. The second graph can be obtained by shifting the first one 4 units to the left and 3 units upward. The results are shown in Figure 36 at the top of the next page.

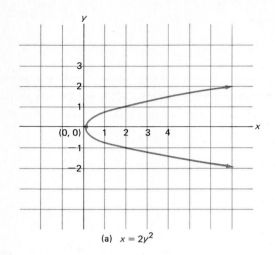

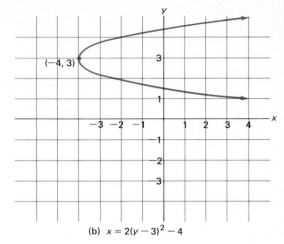

(a) $x = 2y^2$

(b) $x = 2(y - 3)^2 - 4$

Figure 36

In Problems 19–24, sketch the graph of the equation and indicate the coordinates of the vertex.

19. $x = -2y^2$
20. $x = -2y^2 + 8$
21. $x = -2(y + 2)^2 + 8$
22. $x = 3(y - 1)^2 + 6$
23. $x = y^2 + 4y + 2$. *Note:* The y-coordinate of the vertex is at $-b/2a$.
24. $x = 4y^2 - 8y + 10$

EXAMPLE B (Intersection of a Line and a Parabola)   Find the points of intersection of the line $y = -2x + 2$ and the parabola $y = 2x^2 - 4x - 2$ (see Figure 37).

Solution.   We must solve the two equations simultaneously. This is easy to do by equating the two expressions for y and then solving the resulting equation for x.

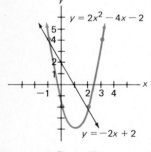

Figure 37

$$-2x + 2 = 2x^2 - 4x - 2$$
$$0 = 2x^2 - 2x - 4$$
$$0 = 2(x^2 - x - 2)$$
$$0 = 2(x - 2)(x + 1)$$
$$x = -1 \qquad x = 2$$

By substitution, we find the corresponding values of y to be 4 and -2; the intersection points are therefore $(-1, 4)$ and $(2, -2)$.

Find the points of intersection for the given line and parabola.

25. $y = -x + 1$
    $y = x^2 + 2x + 1$
26. $y = -x + 4$
    $y = -x^2 + 2x + 4$
27. $y = -2x + 1$
    $y = -x^2 - x + 3$
28. $y = -3x + 15$
    $y = 3x^2 - 3x + 12$

29. $y = 1.5x + 3.2$
    $y = x^2 - 2.9x$ ⃞c

30. $y = 2.1x - 6.4$
    $y = -1.2x^2 + 4.3$ ⃞c

EXAMPLE C (Using Side Conditions) Find the equation of the vertical parabola with vertex $(2, -3)$ and passing through the point $(9, -10)$.

Solution. The equation must have the form

$$y + 3 = a(x - 2)^2$$

To find a, we substitute $x = 9$ and $y = -10$.

$$-7 = a(7)^2$$

Thus $a = -\frac{1}{7}$, and the required equation is $y + 3 = -\frac{1}{7}(x - 2)^2$.

In Problems 31–34, find the equation of the vertical parabola that satisfies the given conditions.

31. Vertex $(0, 0)$, passing through $(-6, 3)$.
32. Vertex $(0, 0)$, passing through $(3, -6)$.
33. Vertex $(-2, 0)$, passing through $(6, -8)$.
34. Vertex $(3, -1)$, passing through $(-2, 5)$.
35. Find the equation of the parabola passing through $(1, 1)$ and $(2, 7)$ and having the y-axis as the line of symmetry. *Hint:* The equation has the form $y = ax^2 + c$.
36. Find the equation of the parabola passing through $(1, 2)$ and $(-2, -7)$ and having the y-axis as the line of symmetry.

MISCELLANEOUS PROBLEMS

37. In each case, find the coordinates of the vertex and then sketch the graph. Use the same coordinate axes for all three graphs.
    (a) $y = 2x^2 - 8x + 4$ (b) $y = 2x^2 - 8x + 8$
    (c) $y = 2x^2 - 8x + 11$
38. Sketch the following parabolas using the same coordinate axes.
    (a) $y = -3x^2 + 6x + 9$ (b) $y = -3x^2 + 6x - 3$
    (c) $y = -3x^2 + 6x - 9$
39. Sketch the graphs of the following parabolas.
    (a) $y = \frac{1}{2}x^2 - 2x$ (b) $x = \frac{1}{2}y^2 - 2y$
40. Recall that the discriminant d of $ax^2 + bx + c$ is $d = b^2 - 4ac$ (see page 114). Calculate d for each of the parabolas of Problem 37. Then show that the graph of $y = ax^2 + bx + c$ will cross the x-axis, just touch the x-axis, or not meet the x-axis according as $d > 0$, $d = 0$, or $d < 0$.
41. Find the points of intersection of the following pairs of curves.
    (a) $y = x^2 - 2x + 6$ (b) $y = x^2 - 4x + 6$
        $y = -3x + 8$ $y = 2x - 3$
    (c) $y = -x^2 + 2x + 4$ (d) $y = x^2 - 2x + 7$
        $y = -2x + 9$ $y = 11 - x^2$
42. For what values of k does the parabola $y = x^2 - kx + 4$ have two x-intercepts?

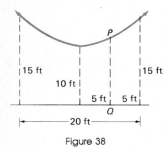

15 ft   15 ft
  10 ft
    5 ft   5 ft
       Q
⊢——— 20 ft ———⊣

Figure 38

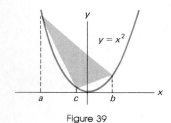

y
$y = x^2$
a c b x

Figure 39

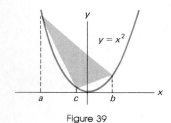

y
$x^2 = 4py$
$F(0, p)$ G
           R
                   x

Figure 40

43. The parabola $y = a(x - 2)(x - 8)$ passes through $(10, 40)$. Find a and the vertex of this parabola.

44. Find the equation of the vertical parabola that passes through $(-1, -2)$, $(0, 3)$, and $(2, 7)$.

45. If the curve shown in Figure 38 is part of a parabola, find the distance $\overline{PQ}$. *Hint:* Begin by finding the equation of the parabola, assuming its vertex is at the origin.

46. The parabolic cable for a suspension bridge is attached to the two towers at points 400 feet apart and 90 feet above the horizontal bridge deck. The cables drop to a point 10 feet above the deck. Find the xy-equation of the cable, assuming it is symmetric about the y-axis with vertex at $(0, 10)$.

47. A retailer has learned from experience that if she charges x dollars apiece for toy trucks, she can sell $300 - 100x$ of them. The trucks cost her $2 each. Write a formula for her total profit P in terms of x. Then determine what she should charge for each truck to maximize her profit.

48. A company that makes fancy golf carts has fixed overhead costs of $12,000 per year and direct costs (labor and materials) of $80 per cart. It sells its carts to a certain retailer at a nominal price of $120 each. However, the company offers a discount of 1 percent for 100 carts, 2 percent for 200 carts, and, in general, $x/100$ percent for x carts. Assume the retailer will buy as many carts as the company can produce but that its production facilities limit this to a maximum of 1800 units.
   (a) Write a formula for C, the cost of producing x carts.
   (b) Show that its total revenue R in dollars is $R = 120x - .012x^2$.
   (c) What is the smallest number of carts it can produce and still break even?
   ☐ (d) What number of carts will produce the maximum profit and what is this profit?

49. Find the area (in terms of a and b) of the triangle shown in Figure 39. The point c is midway between a and b.

50. Starting at $(0, 0)$, a ball travels along the path $y = ax^2 + bx$. Find a and b if the ball reaches a height of 75 at $x = 50$ and its maximum height at $x = 100$.

51. Let $P(x, y)$ move so that its distance from a fixed point $F(0, 3)$ is always equal to its perpendicular distance from the fixed line $y = -3$. Derive and simplify the equation of the path. *Note:* If you do this correctly, the equation should be that of a parabola with vertex at the origin.

52. Problem 51 gives a special case of the following very important result. Suppose that $P(x, y)$ moves so that its distance from the fixed point $F(0, p)$ is equal to its perpendicular distance from the line $y = -p$. Show that the equation of the path is $x^2 = 4py$, the equation of a parabola. *Note:* The fixed point is called the *focus* and the fixed line, the *directrix*.

53. In Figure 40, FG is parallel to the x-axis and RG is parallel to the y-axis. Show that the length $L = \overline{FR} + \overline{RG}$ is independent of where R is on the parabola by finding a formula for L in terms of p alone.

54. TEASER   Let $\{a_1, a_2, \ldots, a_n\}$ and $\{b_1\ b_2, \ldots, b_n\}$ be two sets of n numbers each. Show that the inequality
$$(a_1b_1 + a_2b_2 + \cdots + a_nb_n)^2$$
$$\leq (a_1^2 + a_2^2 + \cdots + a_n^2)(b_1^2 + b_2^2 + \cdots + b_n^2)$$
always holds. *Hint:* See if you can use the fact that if $Ax^2 + 2Bx + C \geq 0$ for all x; then $B^2 - AC \leq 0$ (see Problem 40).

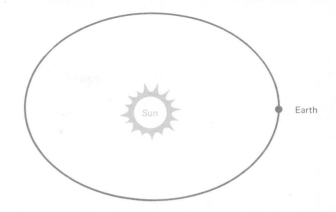

Planetary Orbits

"What," Halley asked Newton, "would be the curve described by the planets on the supposition that gravity diminished as the square of the distance?" Newton answered without hesitation: "An ellipse". How did he know that? "Why," replied Newton, "I have calculated it."

I. Bernard Cohen

4-5　Ellipses and Hyperbolas (Optional)

Roughly speaking, an ellipse is a flattened circle. In the case of the earth's orbit about the sun, there is very little flattening (less than the opening panel indicates). But all ellipses, be they nearly circular or very flat, have important properties in common. As with parabolas, we shall begin by discussing equations of ellipses.

EQUATIONS OF ELLIPSES

Consider the equation

$$\frac{x^2}{25} + \frac{y^2}{16} = 1$$

Because x and y can be replaced by $-x$ and $-y$, respectively, without changing the equation, the graph is symmetric with respect to both axes and the origin. To find the x-intercepts, we let $y = 0$ and solve for x.

$$\frac{x^2}{25} = 1$$
$$x^2 = 25$$
$$x = \pm 5$$

Thus the graph intersects the x-axis at $(\pm 5, 0)$. By a similar procedure (letting $x = 0$), we find that the graph intersects the y-axis at $(0, \pm 4)$. Plotting these points and a few others leads to the graph in Figure 41 on the next page.

This curve is an example of an ellipse. The dotted line segment PQ with endpoints on the ellipse and passing through the origin is called a **diameter**.

x	y
0	±4
±2	±3.67
±3	±3.20
±5	0

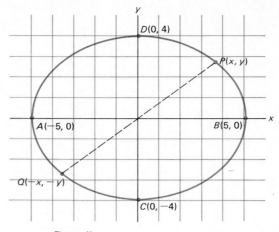

Figure 41

The longest diameter, *AB*, is the **major diameter** (sometimes called the major axis) and the shortest one, *CD*, is the **minor diameter**. The origin, which is a bisector of every diameter, is appropriately called the **center** of the ellipse. The endpoints of the major diameter are called the **vertices** of the ellipse.

More generally, if *a* and *b* are any positive numbers, the equation

$$\frac{x^2}{a^2} + \frac{y^2}{b^2} = 1$$

represents an ellipse with center at the origin and intersecting the *x*- and *y*-axes at $(\pm a, 0)$ and $(0, \pm b)$, respectively. If $a > b$, it is called a *horizontal* ellipse (because the major diameter is horizontal); if $a = b$, it is a *circle* of radius *a*; and if $a < b$, it is called a *vertical* ellipse (see Figure 42).

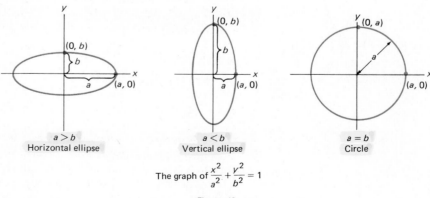

The graph of $\frac{x^2}{a^2} + \frac{y^2}{b^2} = 1$

Figure 42

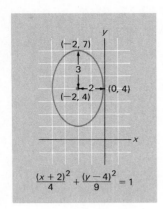

$$\frac{(x + 2)^2}{4} + \frac{(y - 4)^2}{9} = 1$$

Figure 43

TRANSLATING THE ELLIPSE

If we move the ellipse $x^2/a^2 + y^2/b^2 = 1$ (without turning it) so that its center is at (h, k) rather than the origin, its equation takes the form

$$\frac{(x - h)^2}{a^2} + \frac{(y - k)^2}{b^2} = 1$$

This is called the **standard form** for the equation of an ellipse. Again, the relative sizes of a and b determine whether it is a horizontal ellipse, a vertical ellipse, or a circle. For example, the graph of

$$\frac{(x + 2)^2}{4} + \frac{(y - 4)^2}{9} = 1$$

is a vertical ellipse centered at $(-2, 4)$ with major diameter of length $2 \cdot 3 = 6$ and minor diameter of length $2 \cdot 2 = 4$ (see Figure 43).

EQUATIONS OF HYPERBOLAS

What a difference a change in sign can make! The graphs of

$$\frac{x^2}{25} + \frac{y^2}{16} = 1$$

and

$$\frac{x^2}{25} - \frac{y^2}{16} = 1$$

are as different as night and day. The first is an ellipse; the second is a hyperbola. Let us see what we can find out about this hyperbola.

First note that it has x-intercepts ± 5 but no y-intercepts (setting $x = 0$ in the equation yields $y^2 = -16$). Since x can be replaced by $-x$ and y can be replaced by $-y$ without changing the equation, the graph is symmetric with respect to both axes and the origin. This makes it appropriate to call the origin the **center** of the hyperbola. If we solve for y in terms of x, we get

$$y = \pm \tfrac{4}{5}\sqrt{x^2 - 25}$$

This implies first that we must have $|x| \geq 5$; so the graph has no points between $x = -5$ and $x = 5$. Second, since for large $|x|$, $\tfrac{4}{5}\sqrt{x^2 - 25}$ behaves very much like $\tfrac{4}{5}x$, the hyperbola must draw closer and closer to the lines $y = \pm \tfrac{4}{5}x$. These lines are called **asymptotes** of the graph.

When we put all of this information together and plot a few points, we are led to the graph in Figure 44 at the top of the next page. You can see from it why $(-5, 0)$ and $(5, 0)$ are called **vertices** of the hyperbola.

The example analyzed above suggests the general situation. If a and b are positive numbers, the graphs of

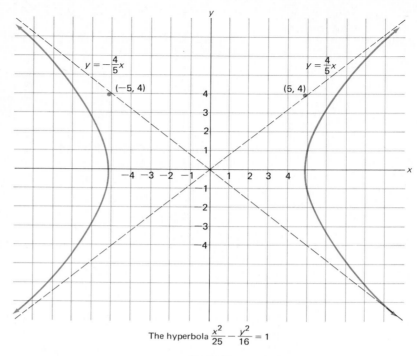

$y = -\dfrac{4}{5}x$ $y = \dfrac{4}{5}x$

$(-5, 4)$ $(5, 4)$

The hyperbola $\dfrac{x^2}{25} - \dfrac{y^2}{16} = 1$

Figure 44

CAUTION

In contrast to the case of the ellipse, whether a hyperbola is vertical or horizontal is not determined by the relative of a and b.
Rather, it is determined by whether the minus sign is associated with the x- or the y-term.

$$\frac{x^2}{a^2} - \frac{y^2}{b^2} = 1 \quad \text{or} \quad \frac{y^2}{b^2} - \frac{x^2}{a^2} = 1$$

are hyperbolas. In the first case, the hyperbola is said to be *horizontal*, since the vertices are at $(\pm a, 0)$, on a horizontal line. In the second case, the hyperbola is *vertical* with vertices $(0, \pm b)$. In both cases, the asymptotes are the lines $y = \pm(b/a)x$. Note in Figure 45 how the numbers a and b determine a rectangle with the asymptotes as diagonals.

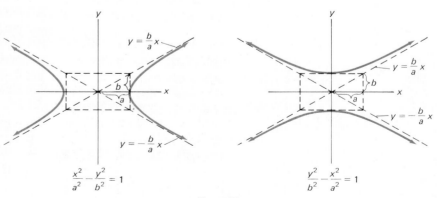

$y = \dfrac{b}{a}x$ $y = -\dfrac{b}{a}x$ $\dfrac{x^2}{a^2} - \dfrac{y^2}{b^2} = 1$

$y = \dfrac{b}{a}x$ $y = -\dfrac{b}{a}x$ $\dfrac{y^2}{b^2} - \dfrac{x^2}{a^2} = 1$

Figure 45

As an example of a vertical hyperbola, consider

$$\frac{y^2}{16} - \frac{x^2}{25} = 1$$

The vertices are at $(0, \pm 4)$ and the asymptotes have equations $y = \pm \frac{4}{5}x$.

TRANSLATING THE HYPERBOLA

If we translate the general hyperbolas discussed above so that their centers are at (h, k) rather than the origin, their equations take the form

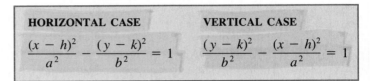

HORIZONTAL CASE	VERTICAL CASE
$\dfrac{(x - h)^2}{a^2} - \dfrac{(y - k)^2}{b^2} = 1$	$\dfrac{(y - k)^2}{b^2} - \dfrac{(x - h)^2}{a^2} = 1$

These are called the **standard forms** for the equation of a hyperbola.
As a final example, consider

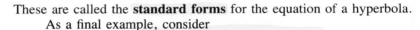

$$\frac{(y - 1)^2}{9} - \frac{(x + 2)^2}{4} = 1$$

This is the equation of a vertical hyperbola with center at $(-2, 1)$ and vertices 3 units above and below the center—that is, $(-2, 4)$ and $(-2, -2)$. The graph, including the asymptotes, is shown in Figure 46.

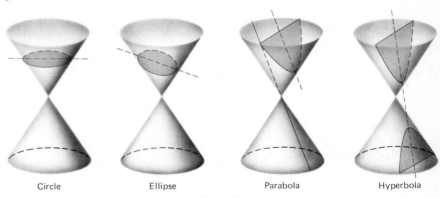

THE CONIC SECTIONS

The Greeks called the four curves—circles, parabolas, ellipses, and hyperbolas—*conic sections*. If you take a cone with two nappes and pass planes through it at various angles, the curve of intersection in each instance is one of these four curves (assuming that the intersecting plane does not pass through the apex of the cone). The diagrams in Figure 47 illustrate this important fact.

Circle   Ellipse   Parabola   Hyperbola

Figure 47

$\dfrac{(y - 1)^2}{9} - \dfrac{(x + 2)^2}{4} = 1$

Figure 46

$(-2, 4)$

$(-2, 1)$

$(-2, -2)$

Problem Set 4-5

Each of the following equations represents an ellipse. Find its center and the endpoints of the major and minor diameters and graph the ellipse.

1. $\dfrac{x^2}{25} + \dfrac{y^2}{9} = 1$

2. $\dfrac{x^2}{36} + \dfrac{y^2}{16} = 1$

3. $\dfrac{x^2}{9} + \dfrac{y^2}{25} = 1$

4. $\dfrac{x^2}{4} + \dfrac{y^2}{25} = 1$

5. $\dfrac{(x-2)^2}{25} + \dfrac{(y+1)^2}{9} = 1$

6. $\dfrac{(x-3)^2}{36} + \dfrac{(y-4)^2}{25} = 1$

7. $\dfrac{(x+3)^2}{9} + \dfrac{y^2}{16} = 1$

8. $\dfrac{x^2}{4} + \dfrac{(y-4)^2}{49} = 1$

In Problems 9–14, find the center of the ellipse with AB and CD as major and minor diameters, respectively. Then write the equation of the ellipse.

9. $A(6, 0),\ B(-6, 0),\ C(0, 3),\ D(0, -3)$

10. $A(5, 0),\ B(-5, 0),\ C(0, 1),\ D(0, -1)$

11. $A(0, 6),\ B(0, -6),\ C(4, 0),\ D(-4, 0)$

12. $A(0, 3),\ B(0, -3),\ C(2, 0),\ D(-2, 0)$

13. $A(-4, 3),\ B(8, 3),\ C(2, 1),\ D(2, 5)$
    Hint: Use the midpoint formula to find the center.

14. $A(3, -4),\ B(11, -4),\ C(7, -7),\ D(7, -1)$

Each of the following equations represents a hyperbola. Find its center and its vertices. Then graph the equation.

15. $\dfrac{x^2}{16} - \dfrac{y^2}{9} = 1$

16. $\dfrac{x^2}{36} - \dfrac{y^2}{16} = 1$

17. $\dfrac{y^2}{9} - \dfrac{x^2}{16} = 1$

18. $\dfrac{y^2}{16} - \dfrac{x^2}{36} = 1$

19. $\dfrac{(x-3)^2}{9} - \dfrac{(y+2)^2}{16} = 1$

20. $\dfrac{(x+1)^2}{16} - \dfrac{(y-4)^2}{36} = 1$

21. $\dfrac{(y+3)^2}{4} - \dfrac{x^2}{25} = 1$

22. $\dfrac{y^2}{49} - \dfrac{(x-2)^2}{9} = 1$

EXAMPLE A (Hyperbolas with Side Conditions)   Find the equation of the hyperbola satisfying the given conditions.

(a) Its vertices are $(4, 0)$ and $(-4, 0)$ and one of its asymptotes has slope $\frac{3}{2}$.

(b) The center is at $(2, 1)$, one vertex is at $(2, -4)$, and the equation of one asymptote is $5x - 7y = 3$.

Solution.

(a) This is a horizontal hyperbola with center at $(0, 0)$. Since $\frac{3}{2} = b/a = b/4$, we get $b = 6$. The required equation is

$$\frac{x^2}{16} - \frac{y^2}{36} = 1$$

(b) Since the vertex $(2, -4)$ is 5 units directly below the center, the hyperbola is vertical and $b = 5$. Solving $5x - 7y = 3$ for y in terms of x gives $y = \frac{5}{7}x - \frac{3}{7}$, so the given asymptote has slope $\frac{5}{7}$. Thus $\frac{5}{7} = b/a = 5/a$, so $a = 7$. The required equation is

$$\frac{(y - 1)^2}{25} - \frac{(x - 2)^2}{49} = 1$$

In Problems 23–28, find the equation of the hyperbola satisfying the given conditions.

23. The vertices are $(4, 0)$ and $(-4, 0)$ and one asymptote has slope $\frac{5}{4}$.
24. The vertices are $(7, 0)$ and $(-7, 0)$ and one asymptote has slope $\frac{2}{7}$.
25. The center is $(6, 3)$, one vertex is $(8, 3)$, and one asymptote has slope 1.
26. The center is $(3, -3)$, one vertex is $(-2, -3)$, and one asymptote has slope $\frac{6}{5}$.
27. The vertices are $(4, 3)$ and $(4, 15)$ and the equation of one asymptote is $3x - 2y + 6 = 0$.
28. The vertices are $(5, 0)$ and $(5, 14)$ and the equation of one asymptote is $7x - 5y = 0$.

EXAMPLE B (Changing to Standard Form)   Change each of the following equations to standard form. Decide whether the corresponding curve is an ellipse or a hyperbola and whether it is horizontal or vertical. Find the center and the vertices.
(a) $x^2 + 4y^2 - 8x + 16y = -28$
(b) $4x^2 - 9y^2 + 24x + 36y + 36 = 0$

Solution.
(a) We use the familiar process of completing the square.

$$(x^2 - 8x + \quad) + 4(y^2 + 4y + \quad) = -28$$
$$(x^2 - 8x + 16) + 4(y^2 + 4y + 4) = -28 + 16 + 16$$
$$(x - 4)^2 + 4(y + 2)^2 = 4$$
$$\frac{(x - 4)^2}{4} + \frac{(y + 2)^2}{1} = 1$$

We recognize this to be a horizontal ellipse (the larger denominator is in the x-term). Its center is at $(4, -2)$ and its vertices are at $(2, -2)$ and $(6, -2)$.
(b) Again we complete the squares.

$$4(x^2 + 6x + \quad) - 9(y^2 - 4y + \quad) = -36$$
$$4(x^2 + 6x + 9) - 9(y^2 - 4y + 4) = -36 + 36 - 36$$
$$4(x + 3)^2 - 9(y - 2)^2 = -36$$
$$\frac{(y - 2)^2}{4} - \frac{(x + 3)^2}{9} = 1$$

This is a vertical hyperbola (the y-term is positive). The center is at $(-3, 2)$ and the vertices are at $(-3, 0)$ and $(-3, 4)$, 2 units below and above the center.

For each of the following, change the equation to standard form and decide whether the corresponding curve is an ellipse or a hyperbola and whether it is horizontal or vertical. Find the center and the vertices.

29. $9x^2 + 16y^2 + 36x - 96y = -36$
30. $x^2 + 4y^2 - 2x + 32y = -61$
31. $4x^2 - 9y^2 - 16x - 18y - 29 = 0$
32. $9x^2 - y^2 + 90x + 8y + 200 = 0$
33. $25x^2 + y^2 - 4y = 96$
34. $25x^2 + 9y^2 - 100x - 125 = 0$
35. $4x^2 - y^2 - 32x - 4y + 69 = 0$
36. $x^2 - y^2 + 8x + 4y + 13 = 0$

MISCELLANEOUS PROBLEMS

In Problems 37–44, decide whether the graph of the given equation is a circle, an ellipse, or a hyperbola. Then sketch the graph.

37. $\dfrac{x^2}{64} + \dfrac{y^2}{16} = 1$

38. $\dfrac{x^2}{16} + \dfrac{y^2}{16} = 1$

39. $\dfrac{x^2}{64} - \dfrac{y^2}{16} = 1$

40. $\dfrac{x^2}{4} - \dfrac{y^2}{9} = 1$

41. $\dfrac{(x-2)^2}{25} + \dfrac{(y-1)^2}{4} = 1$

42. $\dfrac{(x-2)^2}{4} - \dfrac{(y+2)^2}{9} = 1$

43. $x^2 - 2y^2 - 6x + 8y = 1$

44. $x^2 + 2y^2 - 6x + 8y = 1$

45. Find the equation of the ellipse whose vertices are $(-4, 2)$ and $(10, 2)$ and whose minor diameter has length 10.

46. Find the equation of the ellipse that passes through the point $(3, 2)$ and has its vertices at $(\pm 5, 0)$.

47. Find the equation of the hyperbola having vertices $(\pm 6, 0)$ and one of its asymptotes passing through $(3, 2)$.

48. Find the equation of the hyperbola with vertices at $(0, \pm 2)$ and passing through the point $(2, 4)$.

ⓒ 49. Find the y-coordinates of the points for which $x = \pm 2.5$ on the ellipse

$$\frac{x^2}{24} + \frac{y^2}{19} = 1$$

ⓒ 50. Find the y-coordinates of the points for which $x = \pm 5$ on the hyperbola

$$\frac{x^2}{17} - \frac{y^2}{11} = 1$$

ⓒ 51. The area of the ellipse $x^2/a^2 + y^2/b^2 = 1$ is πab. Find the areas of the following ellipses.

(a) $\dfrac{x^2}{7} + \dfrac{y^2}{11} = 1$　　(b) $\dfrac{x^2}{111} + y^2 = 1$

Figure 48

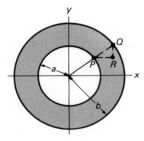

Figure 49

[c] 52. Find the radius of the circle which has the same area as the given ellipse.

(a) $\dfrac{x^2}{256} + \dfrac{y^2}{89} = 1$　　(b) $\dfrac{x^2}{50} + \dfrac{y^2}{19} = 1$

53. The cross section of the mug in Figure 48 is a hyperbola with center on the base of the mug and vertex 1 centimeter above the base. The mug is 9 centimeters high and has an opening at the top with diameter 8 centimeters. How deep is the mug at a point 2 centimeters from its central axis?

54. A doorway has the shape of an elliptical arch (a semi-ellipse) that is 6 feet wide at the base and 8 feet high at the center. A heavy rectangular box that is 4 feet wide is to be slid through this doorway. How high can the box be?

55. Arnold Thinkhard stood high on a ladder 25 feet long that was leaning against a vertical wall (the positive y-axis). When the bottom of the ladder began to slide along the ground (the positive x-axis), Arnold pressed his nose against the ladder and hung on for dear life. If his nose was 4 feet from the top end of the ladder, what was the equation of the path this important organ took in descending to the ground?

56. As the wheel in Figure 49 turns, the point $R(x, y)$ traces out a path in the xy-plane. Find the equation of this path, assuming that R is always on a vertical line through Q and a horizontal line through P.

57. Let $F_1(-4, 0)$ and $F_2(4, 0)$ be two given points and let $P(x, y)$ move so that

$$d(P, F_1) + d(P, F_2) = 10$$

By deriving the equation of the path of P, show that this path is an ellipse.

58. Suppose in Problem 57 that the two fixed points F_1 and F_2 are at $(-c, 0)$ and $(c, 0)$ and that $P(x, y)$ moves so that

$$d(P, F_1) + d(P, F_2) = 2a$$

where $a > c > 0$. Show that the equation of the path is $x^2/a^2 + y^2/b^2 = 1$, where $b^2 = a^2 - c^2$. The given distance condition is often taken as the definition of an ellipse and the two fixed points are called *foci* (plural of *focus*).

59. Let $F_1(-5, 0)$ and $F_2(5, 0)$ be two given points and let $P(x, y)$ move so that

$$\left| d(P, F_1) - d(P, F_2) \right| = 6$$

By deriving the equation of the path of P, show that this path is a hyperbola.

60. Suppose in Problem 59 that the two fixed points F_1 and F_2 are at $(-c, 0)$ and $(c, 0)$ and that $P(x, y)$ moves so that

$$\left| d(P, F_1) - d(P, F_2) \right| = 2a$$

where $c > a > 0$. Show that the equation of the path is $x^2/a^2 - y^2/b^2 = 1$, where $b^2 = c^2 - a^2$. The given distance condition is often taken as the definition of a hyperbola and, as with the ellipse, the two fixed points are called foci.

61. Stakes are driven into the ground at $(\pm 15, 0)$. A loop of rope 80 feet long is thrown over the stakes and attached to Fido's collar. Fido pulls on the rope and,

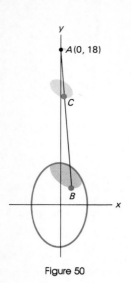

$A(0, 18)$

C

B

Figure 50

keeping it taut, races around the stakes, thus completing a closed path. Write the xy-equation of Fido's path. *Hint:* See Problem 58.

62. TEASER   In Figure 50, A is a fixed point at $(0, 18)$ in the xy-plane. As B painted the elliptical region bounded by $x^2/9 + y^2/16 = 1$, C also painted a region by staying on the line AB and exactly one-third of the way from A to B. Determine the shape of C's region and how much paint C needed if B used exactly 1 gallon.

Chapter Summary

Like a city planner, we introduce in the plane two main streets, one vertical (the **y-axis**) and the other horizontal (the **x-axis**). Relative to these axes, we can specify any point by giving its address (x, y). The numbers x and y, called **Cartesian coordinates,** measure the directed distances from the vertical and horizontal axes, respectively. And given two points A and B with addresses (x_1, y_1) and (x_2, y_2), we may calculate the distance between them from the **distance formula:**

$$d(A, B) = \sqrt{(x_2 - x_1)^2 + (y_2 - y_1)^2}$$

In **analytic geometry,** we use the notion of coordinates to combine algebra and geometry. Thus we may graph the equation $y = x^2$ (algebra), thereby turning it into a curve (geometry). Conversely, we may take a circle with radius 6 and center $(-1, 4)$ and give it the equation

$$(x + 1)^2 + (y - 4)^2 = 36$$

The simplest of all curves is a **line.** If a line passes through (x_1, y_1) and (x_2, y_2) with $x_1 \neq x_2$, then its **slope** m is given by

$$m = \frac{\text{rise}}{\text{run}} = \frac{y_2 - y_1}{x_2 - x_1}$$

There are two important forms for the equation of a nonvertical line.

Point-slope form:	$y - y_1 = m(x - x_1)$
Slope-intercept form:	$y = mx + b$

Vertical lines do not have slope; their equations take the form $x = k$. All lines (vertical and nonvertical) can be written in the form of the **general linear equation**

$$Ax + By + C = 0$$

The distance d from the point (x_1, y_1) to the line $Ax + By + C = 0$ is

$$d = \frac{|Ax_1 + By_1 + C|}{\sqrt{A^2 + B^2}}$$

Nonvertical lines are parallel if their slopes are equal, and they are perpendicular if their slopes are negative reciprocals.

Somewhat more complicated curves are the *conic sections:* **parabolas, circles, ellipses,** and **hyperbolas.** Here are typical equations for them.

Parabola: $y - k = a(x - h)^2$

Circle: $(x - h)^2 + (y - k)^2 = r^2$

Ellipse: $\dfrac{(x - h)^2}{a^2} + \dfrac{(y - k)^2}{b^2} = 1$

Hyperbola: $\dfrac{(x - h)^2}{a^2} - \dfrac{(y - k)^2}{b^2} = 1$

Chapter Review Problem Set

1. Name the graph of each of the following.
   (a) $y = -4x + 3$
   (b) $x^2 + y^2 = 25$
   (c) $2x - 3y = 0$
   (d) $y = x^2$
   (e) $(y - 2)^2 = -3(x + 1)$
   (f) $(x - 1)^2 + (y + 2)^2 = 9$
   (g) $\dfrac{x^2}{9} + \dfrac{y^2}{49} = 1$
   (h) $\dfrac{x^2}{9} - \dfrac{y^2}{49} = 1$
   (i) $x^2 - 4y = 16$
   (j) $x^2 + 4y^2 = 16$
   (k) $x^2 - 4y^2 = 16$
   (l) $x^2 - 4y^2 = -16$

2. Sketch the graph of each of the following equations. For some, you will need to make a table of values.
   (a) $y = -3x + 4$
   (b) $x^2 + y^2 = 16$
   (c) $y = x^2 + \dfrac{1}{x}$
   (d) $y = x^3 - 4x$
   (e) $y = \sqrt{x} + 2$
   (f) $\dfrac{x^2}{25} + \dfrac{y^2}{9} = 1$

3. Consider the triangle determined by the points $A(-2, 3)$, $B(3, 5)$ and $C(1, 9)$.
   (a) Sketch the triangle.
   (b) Find the lengths of the three sides.
   (c) Find the slopes of the lines which contain the three sides.
   (d) Write an equation for each of these lines.
   (e) Find the midpoints of AB, BC, and CA.
   (f) Find the equation of the line through A parallel to BC.
   (g) Find the equation of the line through A perpendicular to BC.
   (h) Find the length of the altitude from A to side BC.
   (i) Calculate the area of the triangle.
   (j) Show that the line segment joining the midpoints of AB and AC is parallel to BC and is one-half the length of BC.

4. Write the equation of the line satisfying the given conditions in the form $Ax + By + C = 0$.

(a) It is vertical and passes through $(4, -1)$.

(b) It is horizontal and passes through $(4, -1)$.

(c) It is parallel to the line $3x + 2y = 4$ and passes through $(4, -1)$.

(d) It passes through $(2, -5)$ and has x-intercept 3.

(e) It is perpendicular to the line $2y = -3x$ and passes through $(-2, 3)$.

(f) It is parallel to the lines $y = 4x + 5$ and $y = 4x - 7$ and is midway between them.

(g) It passes through $(4, 6)$ and has equal x- and y-intercepts.

(h) It is tangent to the circle $x^2 + y^2 = 169$ at the point $(12, 5)$.

5. Find the vertex of each of the following parabolas and decide which way it opens (up, down, to the right, or to the left).

(a) $x^2 = -6y$ (b) $y^2 = 16x$

(c) $y - 1 = \frac{1}{10}(x + 2)^2$ (d) $x - \frac{1}{4} = -\frac{1}{3}(y + \frac{1}{2})^2$

(e) $y = x^2 - 4x$ (f) $2y^2 + 12y = x - 1$

6. Write the equation of the parabola satisfying the following conditions.

(a) Vertex at $(0, 0)$, y-axis as line of symmetry, and passing through $(10, -5)$.

(b) Vertex at $(3, 0)$, axis of symmetry $y = 0$, and passing through $(1, 12)$.

(c) Vertex at $(-3, 5)$, horizontal line of symmetry, and passing through $(-1, 6)$.

7. Find the points of intersection of the parabola $y = x^2 + x$ and the line $6x - y - 4 = 0$.

8. By completing the squares, identify each of the following conic sections and give pertinent information about it.

(a) $x^2 + 2x + y^2 - 4y = 2$

(b) $2x^2 + y^2 + 4x - 4y = 14$

(c) $2x^2 - y^2 + 4x - 4y = 14$

(d) $2x^2 + 4x - y = 14$

9. Find the equation of the ellipse satisfying the given conditions.

(a) Vertices $(0, \pm 5)$; passing through $(2, 5\sqrt{5}/3)$.

(b) Vertices $(-2, 0)$ and $(6, 0)$; length of minor diameter 4.

10. Find the equation of the hyperbola with vertices $(-2, 5)$ and $(-2, 11)$ and with one of its asymptotes passing through $(0, 14)$.

Mathematicians do not deal in objects, but in relations between objects; thus, they are free to replace some objects by others so long as the relations remain unchanged. Content to them is irrelevant; they are interested in form only.

Henri Poincaré

CHAPTER 5

Functions and Their Graphs

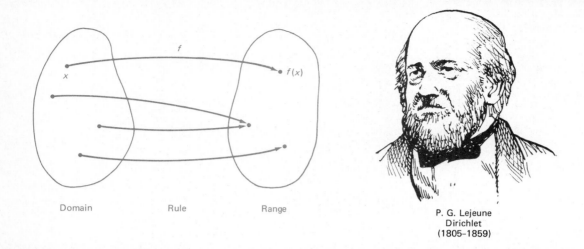

Domain   Rule   Range

P. G. Lejeune
Dirichlet
(1805–1859)

5-1   Functions

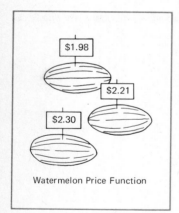

Watermelon Price Function

Figure 1

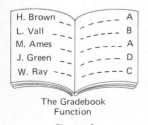

The Gradebook
Function

Figure 2

One of the most important ideas in mathematics is that of a function. For a long time, mathematicians and scientists wanted a precise way to describe the relationships that may exist between two variables. It is somewhat surprising that it took so long for the idea to crystallize into a clear, unambiguous concept. The French mathematician P. G. Lejeune Dirichlet (1805–1859) is credited with the modern definition of function.

DEFINITION

*A **function** is a rule which assigns to each element in one set (called the **domain** of the function) exactly one value from another set. The set of all assigned values is called the **range** of the function.*

Three examples will help clarify this idea. When a grocer puts a price tag on each of the watermelons for sale (Figure 1), a function is determined. Its domain is the set of watermelons, its range is the set of prices, and the rule is the procedure the grocer uses in assigning prices (perhaps a specified amount per pound.)

When a professor assigns a grade to each student in a class (Figure 2), he or she is determining a function. The domain is the set of students and the range is the set of grades, but who can say what the rule is? It varies from professor to professor; some may even prefer to keep it a secret.

A much more typical function from our point of view is the *squaring* function displayed in Figure 3. It takes a number from the domain $\{-2, -1, 0, 1, 2, 3\}$ and squares it, producing a number in the range $\{0, 1, 4, 9\}$. This function is typical for two reasons: Both the domain and range are sets of

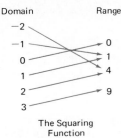

Domain   Range

The Squaring
Function

Figure 3

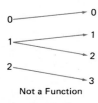

Not a Function

Figure 4

numbers and the rule can be specified by giving an algebraic formula. Most functions in this book will be of this type.

The definition says that a function assigns exactly one value to each element of the domain. Thus Figure 4 is not a diagram for a function, since two values are assigned to the element 1.

FUNCTIONAL NOTATION

Long ago, mathematicians introduced a special notation for functions. A single letter like f (or g or h) is used to name the function. Then $f(x)$, read f of x or f at x, denotes the value that f assigns to x. Thus, if f names the squaring function,

$$f(-2) = 4 \qquad f(2) = 4 \qquad f(-1) = 1$$

and, in general,

$$f(x) = x^2$$

We call this last result the *formula* for the function f. It tells us in a concise algebraic way what f does to any number. Notice that the given formula and

$$f(y) = y^2 \qquad f(z) = z^2$$

all say the same thing; the letter used for the domain variable is a matter of no significance, though it does happen that we shall usually use x. Many functions do not have simple formulas (see Problems 17 and 18), but in this book, most of them do.

For a further example, consider the function that cubes a number and then subtracts 1 from the result. If we name this function g, then

$$g(2) = 2^3 - 1 = 7$$
$$g(-1) = (-1)^3 - 1 = -2$$
$$g(.5) = (.5)^3 - 1 = -.875$$
$$g(\pi) = \pi^3 - 1 \approx 30$$

and, in general,

$$g(x) = x^3 - 1$$

Few students would have trouble using this formula when x is replaced by a specific number. However, it is important to be able to use it when x is replaced by anything whatever, even an algebraic expression. Be sure you understand the following calculations.

$$g(a) = a^3 - 1$$
$$g(y^2) = (y^2)^3 - 1 = y^6 - 1$$
$$g\left(\frac{1}{z}\right) = \left(\frac{1}{z}\right)^3 - 1 = \frac{1}{z^3} - 1$$

$$g(2 + h) = (2 + h)^3 - 1 = 8 + 12h + 6h^2 + h^3 - 1$$
$$= h^3 + 6h^2 + 12h + 7$$
$$g(x + h) = (x + h)^3 - 1 = x^3 + 3x^2h + 3xh^2 + h^3 - 1$$

DOMAIN AND RANGE

The rule of correspondence is the heart of a function, but a function is not completely determined until its domain is given. Recall that the **domain** is the set of elements to which the function assigns values. In the case of the squaring function f (reproduced in Figure 5), we gave the domain as the set $\{-2, -1, 0, 1, 2, 3\}$. We could just as well have specified the domain as the set of all real numbers; the formula $f(x) = x^2$ would still make perfectly good sense. In fact, there is a common agreement that if no domain is specified, it is understood to be the largest set of real numbers for which the rule for the function makes sense and gives real number values. We call it the **natural domain** of the function. Thus if no domain is specified for the function with formula $g(x) = x^3 - 1$, it is assumed to be the set of all real numbers. Similarly, for

$$h(x) = \frac{1}{x - 1}$$

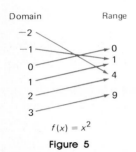

Domain   Range

-2

$-1 \quad\quad 0$

$0 \quad\quad 1$

$1 \quad\quad 4$

$2 \quad\quad 9$

3

$f(x) = x^2$

Figure 5

we would take the natural domain to consist of all real numbers except 1. Here, the number 1 is excluded to avoid division by zero.

Once the domain is understood and the rule of correspondence is given, the **range** of the function is determined. It is the set of values of the function. Here are several examples.

RULE	INDEPENDENT VARIABLE DOMAIN	DEPENDENT VARIABLE RANGE
$F(x) = 4x$	All reals	All reals
$G(x) = \sqrt{x - 3}$	$\{x : x \geq 3\}$	Nonnegative reals (INCLUDES ZERO)
$H(x) = \dfrac{1}{(x - 2)^2}$	$\{x : x \neq 2\}$	Positive reals (EXCLUDES ZERO)

(Handwritten annotations:)

✗ CAN'T EQUAL TWO,
THE X AXIS IS AN ASYMTOTE,
AS X GETS LARGER, H(X)
APPROACHES ∅.
THE LINE X=2 IS AN ASYMTOTE,
AS X APPROACHES IT FROM
EITHER SIDE, H(X)
APPROACHES ∞.

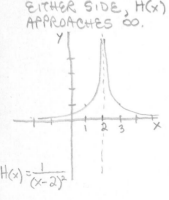

$H(x) = \dfrac{1}{(x-2)^2}$

INDEPENDENT AND DEPENDENT VARIABLES

Scientists like to use the language of variables in talking about functions. Let us illustrate by referring to an object falling under the influence of gravity near the earth's surface. If the object is very dense (so air resistance can be neglected), it will fall according to the formula

$$d = f(t) = 16t^2$$

Here d represents the distance in feet that the object falls during the first t seconds of falling. In this case, t is called the **independent variable** and d, which depends on t, is the **dependent variable.** Also, d is said to be a function of t.

FUNCTIONS OF TWO OR MORE VARIABLES

All functions illustrated so far have been functions of one variable. Suppose that a function f associates a value with each ordered pair (x, y). In this case, we say that f is a function of two variables. As an example, let

$$f(x, y) = x^2 + 3y^2$$

Then,

$$f(3, -2) = 3^2 + 3(-2)^2 = 21$$

and

$$f(0, 6) = (0)^2 + 3(6)^2 = 108$$

The natural domain for this function is the set of all ordered pairs (x, y), that is, the whole Cartesian plane. Its range is the set of nonnegative numbers.

For a second example, let $g(x, y, z) = 2x^2 - 4y + z$. We say that g is a function of three variables. Note that

$$g(4, 2, -3) = 2(4)^2 - 4(2) + (-3) = 21$$

and

$$g\left(u^3, 2v, \frac{1}{w}\right) = 2(u^3)^2 - 4(2v) + \frac{1}{w}$$

$$= 2u^6 - 8v + \frac{1}{w}$$

VARIATION

Sometimes we describe a functional relationship between a dependent variable and one or more independent variables by using the language of variation (or proportion). To say that y **varies directly** as x (or that y is proportional to x) means that $y = kx$ for some constant k. To say that y **varies inversely** as x means that $y = k/x$ for some constant k. Finally, to say that z **varies jointly** as x and y means that $z = kxy$ for some constant k.

In variation problems, we are often given not only the form of the relationship, but also a specific set of values satisfied by the variables involved. This allows us to evaluate the constant k and obtain an explicit formula for the dependent variable in terms of the independent variables. Here is an illustration.

It is known that y varies directly as x and that $y = 10$ when $x = -2$. Find an explicit formula for y in terms of x.

We substitute the given values in the equation $y = kx$ and get $10 = k(-2)$. Thus $k = -5$, and we can write the explicit formula $y = -5x$.

We offer a second example, this time using functional notation.

It is given that a function f varies jointly with x and y and that $f(4, 3) = 2$. Find the explicit formula for f and also evaluate $f(3, 12)$.

We know that $f(x, y) = kxy$. Also, $f(4, 3) = k(4)(3) = 12k = 2$, which implies that $k = \frac{1}{6}$. Thus, the explicit formula for $f(x, y)$ is

$$f(x, y) = \frac{1}{6}xy$$

and therefore

$$f(3, 12) = \frac{1}{6}(3)(12) = 6$$

Problem Set 5-1

1. If $f(x) = x^2 - 4$, evaluate each expression.
   (a) $f(-2)$ (b) $f(0)$ (c) $f(\frac{1}{2})$ (d) $f(.1)$
   (e) $f(\sqrt{2})$ (f) $f(a)$ (g) $f(1/x)$ (h) $f(x + 1)$

2. If $f(x) = (x - 4)^2$, evaluate each expression in Problem 1.

3. If $f(x) = 1/(x - 4)$, evaluate each expression.
   (a) $f(8)$ (b) $f(2)$ (c) $f(\frac{9}{2})$ (d) $f(\frac{31}{8})$
   (e) $f(4)$ (f) $f(4.01)$ (g) $f(1/x)$ (h) $f(x^2)$
   (i) $f(2 + h)$ (j) $f(2 - h)$

4. If $f(x) = x^2$ and $g(x) = 2/x$, evaluate each expression.
   (a) $f(-7)$ (b) $g(-4)$ (c) $f(\frac{1}{4})$ (d) $1/f(4)$
   (e) $g(\frac{1}{4})$ (f) $1/g(4)$ (g) $g(0)$ (h) $g(1)f(1)$
   (i) $f(g(1))$ (j) $f(1)/g(1)$

EXAMPLE A (Finding Natural Domains)   Find the natural domain of
   (a) $f(x) = 4x/[(x + 2)(x - 3)]$;   (b) $g(x) = \sqrt{x^2 - 4}$.

Solution.   We recall that the natural domain is the largest set of real numbers for which the formula makes sense and gives real number values. Thus in part (a), the domain consists of all real numbers except -2 and 3; in part (b) we must have $x^2 \geq 4$, which is equivalent to $|x| \geq 2$. Notice that if $|x| < 2$, we would be taking the square root of a negative number, so the result would not be a real number.

In Problems 5–16, find the natural domain of the given function.

5. $f(x) = x^2 - 4$ 6. $f(x) = (x - 4)^2$

7. $g(x) = \dfrac{1}{x^2 - 4}$ 8. $g(x) = \dfrac{1}{9 - x^2}$

9. $h(x) = \dfrac{2}{x^2 - x - 6}$ 10. $h(x) = \dfrac{1}{2x^2 + 3x - 2}$

   $(x - 3)(x + 2)$ $(2x - 1)(x + 2)$

11. $F(x) = \dfrac{1}{x^2 + 4}$

12. $F(x) = \dfrac{1}{9 + x^2}$

13. $G(x) = \sqrt{x - 2}$

14. $G(x) = \sqrt{x + 2}$

15. $H(x) = \dfrac{1}{5 - \sqrt{x}}$

16. $H(x) = \dfrac{1}{\sqrt{x + 1} - 2}$

17. Not all functions arising in mathematics have rules given by simple algebraic formulas. Let $f(n)$ be the nth digit in the decimal expansion of

$$\pi = 3.14159265358979323846 \ldots$$

Thus $f(1) = 3$ and $f(3) = 4$. Find (a) $f(6)$; (b) $f(9)$; (c) $f(16)$. What is the natural domain for this function?

18. Let g be the function which assigns to each positive integer the number of factors in its prime factorization. Thus

$$g(2) = 1$$

$$g(4) = g(2 \cdot 2) = 2$$

$$g(36) = g(2 \cdot 2 \cdot 3 \cdot 3) = 4$$

Find (a) $g(24)$; (b) $g(37)$; (c) $g(64)$; (d) $g(162)$. Can you find a formula for this function?

EXAMPLE B (Functions Generated on a Calculator)   Show how the function $f(x) = 2\sqrt{x} + 5$ can be generated on a calculator. Then calculate $f(\pi)$.

Solution.   To avoid confusion with the times sign, we use # rather than x for an arbitrary number. Then on a typical algebraic logic calculator,

$$f(\#) = 2 \; \boxed{\times} \; \# \; \boxed{\sqrt{x}} \; \boxed{+} \; 5 \; \boxed{=}$$

In particular,

$$f(\pi) = 2 \; \boxed{\times} \; \pi \; \boxed{\sqrt{x}} \; \boxed{+} \; 5 \; \boxed{=}$$

which yields the value 8.544908.

© *In Problems 19–26, write the sequence of keys that will generate the given function, using # for an arbitrary number. Then use the sequence to calculate $f(2.9)$.*

19. $f(x) = (x + 2)^2$

20. $f(x) = \sqrt{x + 3}$

21. $f(x) = 3(x + 2)^2 - 4$

22. $f(x) = 4\sqrt{x + 3} - 11$

23. $f(x) = \left(3x + \dfrac{2}{\sqrt{x}}\right)^3$

24. $f(x) = \left(\dfrac{x}{3} + \dfrac{3}{x}\right)^5$

25. $f(x) = \dfrac{\sqrt{x^5 - 4}}{2 + 1/x}$

26. $f(x) = \dfrac{(\sqrt{x} - 1)^3}{x^2 + 4}$

© *In Problems 27–32, write the algebraic formula for the function that is generated by the given sequence of calculator keys.*

27. $\# \; \boxed{+} \; 5 \; \boxed{\sqrt{x}} \; \boxed{=}$

28. $\boxed{(} \; \# \; \boxed{+} \; 5 \; \boxed{)} \; \boxed{\sqrt{x}} \; \boxed{=}$

29. $2 \; \boxed{\times} \; \# \; \boxed{x^2} \; \boxed{+} \; 3 \; \boxed{\times} \; \# \; \boxed{=}$

30. $3 \; \boxed{+/-} \; \boxed{\times} \; \# \; \boxed{x^2} \; \boxed{-} \; 7 \; \boxed{=}$

31. $([)$ 3 $[\times]$ $([)$ $[\#]$ $[-]$ 2 $[)]$ $[x^2]$ $[+]$ 9 $[)]$ $[\sqrt{x}]$ $[=]$

32. $([)$ $[\#]$ $[1/x]$ $[\times]$ 2 $[+]$ 4 $[)]$ $[\sqrt{x}]$ $[=]$

Problems 33–44 deal with functions of two variables. Let $g(x, y) = 3xy - 5x$ and $G(x, y) = (5x + 3y)/(2x - y)$. Find each of the following.

33. $g(2, 5)$ 34. $g(-1, 3)$ 35. $g(5, 2)$ 36. $g(3, -1)$

37. $G(1, 1)$ 38. $G(3, 3)$ 39. $G(\frac{1}{2}, 1)$ 40. $G(5, 0)$

41. $g(2x, 3y)$ 42. $G(2x, 4y)$ 43. $g(x, 1/x)$ 44. $G(x - y, y)$

EXAMPLE C (More on Variation)   Suppose that w varies jointly as x and the square root of y and that $w = 14$ when $x = 2$ and $y = 4$. Find an explicit formula for w.

Solution.   We translate the first statement into mathematical symbols as

$$w = kx\sqrt{y}$$

To evaluate k, we substitute the given values for w, x, and y.

$$14 = k \cdot 2\sqrt{4} = 4k$$

or

$$k = \frac{14}{4} = \frac{7}{2}$$

Thus the explicit formula for w is

$$w = \frac{7}{2}x\sqrt{y}$$

In Problems 45–50, find an explicit formula for the dependent variable.

45. y varies directly as x, and $y = 12$ when $x = 3$.

46. y varies directly as x^2, and $y = 4$ when $x = 0.1$.

47. y varies inversely as x (that is, $y = k/x$), and $y = 5$ when $x = \frac{1}{5}$.

48. V varies jointly as r^2 and h, and $V = 75$ when $r = 5$ and $h = 9$.

49. I varies directly as s and inversely as d^2, and $I = 9$ when $s = 4$ and $d = 12$.

50. W varies directly as x and inversely as the square root of yz, and $W = 5$ when $x = 7.5$, $y = 2$, and $z = 18$.

51. The maximum range of a projectile varies as the square of the initial velocity. If the range is 16,000 feet when the initial velocity is 600 feet per second,
    (a) write an explicit formula for R in terms of v, where R is the range in feet and v is the initial velocity in feet per second;
    (b) use this formula to find the range when the initial velocity is 800 feet per second.

52. Suppose that the amount of gasoline used by a car varies jointly as the distance traveled and the square root of the average speed. If a car used 8 gallons on a 100-mile trip going at an average speed of 64 miles per hour, how many gallons would that car use on a 160-mile trip at an average speed of 25 miles per hour?

MISCELLANEOUS PROBLEMS

53. If $f(x) = x^2 - (2/x)$, evaluate and simplify each expression.
    (a) $f(2)$
    (b) $f(-1)$
    (c) $f(\frac{1}{2})$
    (d) $f(\sqrt{2})$
    (e) $f(2 - \sqrt{2})$
    (f) $f(.01)$
    (g) $f(1/x)$
    (h) $f(a^2)$
    (i) $f(a + b)$

54. If $f(x, y) = (x^2 - xy)/(x + y)$, evaluate and simplify each expression.
    (a) $f(2, 1)$
    (b) $f(3, 0)$
    (c) $f(-2, 1)$
    (d) $f(3\sqrt{2}, \sqrt{2})$
    (e) $f(1/x, 2y)$
    (f) $f(a + b, b)$

55. Determine the natural domain of each function.
    (a) $f(t) = \dfrac{t + 2}{t^2(t + 3)}$
    (b) $g(t) = \sqrt{t^2 - 4}$
    (c) $h(t) = \dfrac{3t + 1}{1 - \sqrt{2t}}$
    (d) $k(s, t) = \dfrac{3st\sqrt{9 - s^2}}{t^2 - 1}$

56. Let $f(n)$ be the nth digit in the decimal expansion of $\frac{5}{13}$. Determine the domain and range of f.

57. A 2-mile race track has the shape of a rectangle with semicircular ends of radius x. If $A(x)$ is the area of the region inside the track, determine the domain and range of A.

58. Cut a yardstick in two pieces of length x and $3 - x$ and, together with a 1-foot stick, form a triangle of area $A(x)$. Determine the domain and range of A. *Note:* The triangle of maximum area will be isosceles.

59. Write a formula for $F(x)$ in each case.
    (a) $F(x)$ is the area of an equilateral triangle of perimeter x.
    (b) $F(x)$ is the area of a regular hexagon inscribed in a circle of radius x.
    (c) $F(x)$ is the volume of water of depth x in a conical tank with vertex downward. The tank is 8 feet high and has diameter 6 feet at the top.
    (d) $F(x)$ is the average cost per unit of producing x refrigerators in a day for a company that has daily overhead of $1300 and pays direct costs (labor and materials) of $240 to make each refrigerator.
    (e) $F(x)$ is the dollar cost of renting a car for 10 days and driving x miles if the rental company charges $18 per day and 22 cents per mile for mileage beyond the first free 100 miles.

60. Some of the following equations determine a function f with formula of the form $y = f(x)$. For those that do, find $f(x)$. *Recall:* A function must associate just one value with each x.
    (a) $x^2 + 3y^2 = 1$
    (b) $xy + 2y = 5 - 2x$
    (c) $x = \sqrt{2y - 1}$
    (d) $2x = (y + 1)/y$

10 ft   6 in.

2 in.

Load

Figure 6

61. The safe load $S(x, y, z)$ of a horizontal beam supported at both ends varies directly as its breadth x and the square of its depth y and inversely as its length z. If a 2-inch by 6-inch white pine joist 10 feet long safely supports 1000 pounds when placed on edge (as shown in Figure 6), find an explicit formula for $S(x, y, z)$. Then determine its safe load when placed flatwise.

62. Which of the following functions satisfy $f(x + y) = f(x) + f(y)$ for all real numbers x and y?
    (a) $f(t) = 2t$
    (b) $f(t) = t^2$
    (c) $f(t) = 2t + 3$
    (d) $f(t) = -3t$

63. Determine the formula for $f(t)$ if for all real numbers x and y

$$f(x)f(y) - f(xy) = x + y$$

64. **TEASER** A function f satisfying $f(x + y) = f(x) + f(y)$ for all real numbers x and y is said to be an *additive function*. Prove that if f is additive, then there is a number m such that $f(t) = mt$ for all rational numbers t.

Geometry, however, supplies sustenance and meaning to bare formulas . . . . One can still believe Plato's statement that "geometry draws the soul toward truth."

Morris Kline

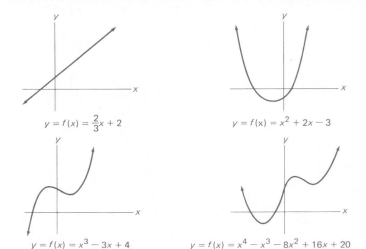

$y = f(x) = \frac{2}{3}x + 2$

$y = f(x) = x^2 + 2x - 3$

$y = f(x) = x^3 - 3x + 4$

$y = f(x) = x^4 - x^3 - 8x^2 + 16x + 20$

5-2 Graphs of Functions

We have said that functions are usually specified by giving formulas. Formulas are fine for manipulation and almost essential for exact quantitative information, but to grasp the overall qualitative features of a function, we need a picture. The best picture of a function is its graph. And the **graph of a function** f is simply the graph of the equation $y = f(x)$. We learned how to graph equations in the previous chapter.

POLYNOMIAL FUNCTIONS

We look first at polynomial functions, that is, functions of the form

$$f(x) = a_n x^n + a_{n-1} x^{n-1} + \cdots + a_1 x + a_0$$

Four typical graphs are shown above. We know from the last chapter that the graph of $f(x) = ax + b$ is always a straight line and that, if $a \neq 0$, the graph of $f(x) = ax^2 + bx + c$ is necessarily a parabola.

The graphs of higher degree polynomial functions are harder to describe, but after we have studied two examples, we can offer some general guidelines. Consider first the cubic function

$$f(x) = x^3 - 3x + 4$$

With the help of a table of values, we sketch its graph (Figure 7).

x	y
−3	−14
−2	2
−1	6
0	4
1	2
2	6
3	22

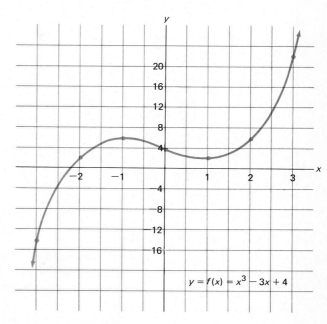

$$y = f(x) = x^3 - 3x + 4$$

Figure 7

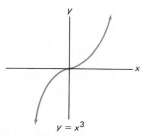

$$y = x^3$$

Figure 8

Notice that for large positive values of x, the values of y are large and positive; similarly, for large negative values of x, y is large and negative. This is due to the dominance of the leading term x^3 for large $|x|$. This dominance is responsible for a drooping left arm and a right arm held high on the graph. Notice also that the graph has one hill and one valley. This is typical of the graph of a cubic function, though it is possible for it to have no hills or valleys. The graph of $y = x^3$ illustrates this latter behavior (Figure 8).

Next consider a typical fourth degree polynomial function.

$$f(x) = -x^4 + 4x^3 + 2x^2 - 12x - 3$$

A table of values and the graph are shown in Figure 9 on the next page.

The leading term $-x^4$, which is negative for all values of x, determines that the graph has two drooping arms. Note that there are two hills and one valley.

In general, we can make the following statements about the graph of

$$f(x) = a_n x^n + a_{n-1} x^{n-1} + \cdots + a_1 x + a_0, \qquad a_n \neq 0$$

x	y
−2	−19
−1	6
0	− 3
1	−10
2	− 3
3	6
4	−19

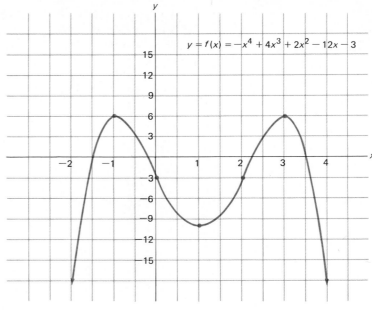

Figure 9

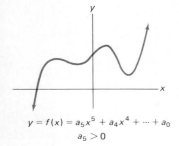

$$y = f(x) = a_5 x^5 + a_4 x^4 + \cdots + a_0$$
$$a_5 > 0$$

Figure 10

1. If n is even and $a_n < 0$, the graph will have two drooping arms; if n is even and $a_n > 0$, it will have both arms raised. This is due to the dominance of $a_n x^n$ for large values of $|x|$.

2. If n is odd, one arm droops and the other points upward. Again, this is dictated by the dominance of $a_n x^n$.

3. The combined number of hills and valleys cannot exceed $n - 1$, although it can be less.

Based on these facts, we expect the graph of a fifth degree polynomial function with positive leading coefficient to look something like the graph in Figure 10.

FACTORED POLYNOMIAL FUNCTIONS

The task of graphing can be simplified considerably if our polynomial is factored. The real solutions of $f(x) = 0$ correspond to the x-intercepts of the graph of $y = f(x)$—that is, to the x-coordinates of the points where the graph intersects the x-axis. If the polynomial is factored, these intercepts are easy to find. Consider as an example.

$$y = f(x) = x(x + 3)(x - 1)$$

The solutions of $f(x) = 0$ are 0, −3, and 1; these are the x-intercepts of the graph. Clearly, $f(x)$ cannot change signs between adjacent x-intercepts since only at these points can any of the linear factors change sign. The signs of $f(x)$ on the four intervals determined by $x = -3$, $x = 0$, and $x = 1$ are shown in Figure 11 (to check this, try substituting an x-value from each of these intervals, as in the split-point method of Section 3-5).

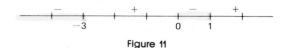

Figure 11

This information and a few plotted points lead to the graph of Figure 12.

x	y
−2	6
−1	4
.5	−.88
2	10

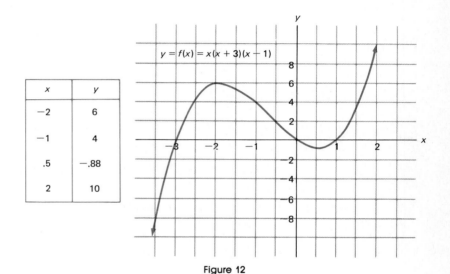

Figure 12

FUNCTIONS WITH MULTI-PART RULES

Sometimes a function has polynomial components even though it is not a polynomial function. Especially notable is the absolute value function $f(x) = |x|$, which has the two-part rule

$$f(x) = \begin{cases} -x & \text{if } x < 0 \\ x & \text{if } x \geq 0 \end{cases}$$

For $x < 0$, the graph coincides with the line $y = -x$; for $x \geq 0$, it coincides with the line $y = x$. Note the sharp corner at the origin (Figure 13).

Here is a more complicated example.

$$g(x) = \begin{cases} x + 2 & \text{if } x < 0 \\ x^2 & \text{if } 0 \leq x \leq 2 \\ 4 & \text{if } x > 2 \end{cases}$$

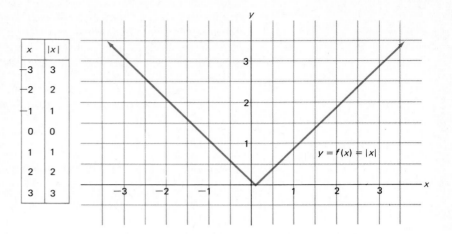

x	$\mid x \mid$
−3	3
−2	2
−1	1
0	0
1	1
2	2
3	3

Though this way of describing a function may seem strange, it is not at all unusual in more advanced courses. The graph of g consists of three pieces (Figure 14).

1. A part of the line $y = x + 2$.
2. A part of the parabola $y = x^2$.
3. A part of the horizontal line $y = 4$.

Note the use of the circle at $(0, 2)$ to indicate that this point is not part of the graph.

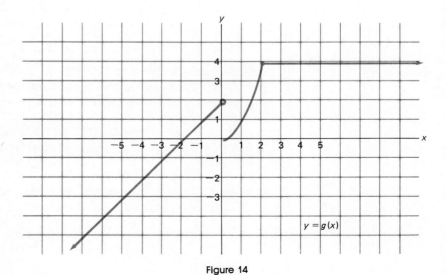

Figure 14

Problem Set 5-2

Graph each of the following polynomial functions. The first two are called constant functions.

1. $f(x) = 5$
2. $f(x) = -4$
3. $f(x) = -3x + 5$
4. $f(x) = 4x - 3$
5. $f(x) = x^2 - 5x + 4$
6. $f(x) = x^2 + 2x - 3$
7. $f(x) = x^3 - 9x$
8. $f(x) = x^3 - 16x$
9. $f(x) = 2.12x^3 - 4.13x + 2$ ©
10. $f(x) = -1.2x^3 + 2.3x^2 - 1.4x$ ©

Graph each of the following functions.

11. $f(x) = 2|x|$
12. $f(x) = |x| - 2$
13. $f(x) = |x - 2|$
14. $f(x) = |x| + 2$

15. $f(x) = \begin{cases} x & \text{if } x < 0 \\ 2 & \text{if } x \ge 0 \end{cases}$

16. $f(x) = \begin{cases} -1 & \text{if } x \le 0 \\ 2x & \text{if } x > 0 \end{cases}$

17. $f(x) = \begin{cases} -5 & \text{if } x \le -3 \\ 4 - x^2 & \text{if } -3 < x \le 3 \\ -5 & \text{if } x > 3 \end{cases}$

18. $f(x) = \begin{cases} 9 & \text{if } x < 0 \\ 9 - x^2 & \text{if } 0 \le x \le 3 \\ x^2 - 9 & \text{if } x > 3 \end{cases}$

EXAMPLE A (Symmetry Properties)   A function f is called an **even function** if $f(-x) = f(x)$ for all x in its domain. The graph of an even function is symmetric with respect to the y-axis. A function g is called an **odd function** if $g(-x) = -g(x)$ for all x in its domain; its graph is symmetric with respect to the origin (see Section 4-2 for a full discussion of symmetry). Graph the following two functions, observing their symmetries.

(a) $f(x) = x^4 + x^2 - 3$

(b) $g(x) = x^3 + 2x$

Solution.   Notice that

$$f(-x) = (-x)^4 + (-x)^2 - 3 = x^4 + x^2 - 3 = f(x)$$

$$g(-x) = (-x)^3 + 2(-x) = -x^3 - 2x = -g(x)$$

Thus f is even and g is odd. Their graphs are sketched in Figure 15 on the next page. Note that a polynomial function involving only even powers of x is even, while one involving only odd powers of x is odd.

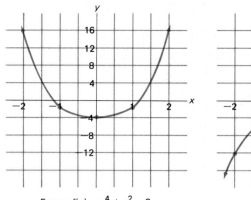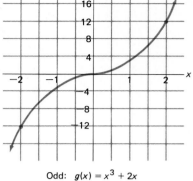

Even: $f(x) = x^4 + x^2 - 3$ Odd: $g(x) = x^3 + 2x$

Figure 15

Determine which of the following are even functions, which are odd functions, and which are neither. Then sketch the graphs of those that are even or odd, making use of the symmetry properties.

19. $f(x) = 2x^2 - 5$ EVEN
20. $f(x) = -3x^2 + 2$ EVEN
21. $f(x) = x^2 - x + 1$ NEITHER
22. $f(x) = -2x^3$ ODD
23. $f(x) = 4x^3 - x$ ODD
24. $f(x) = x^3 + x^2$ NEITHER
25. $f(x) = 2x^4 - 5x^2$ EVEN
25. $f(x) = 3x^4 + x^2$ EVEN

EXAMPLE B (More on Factored Polynomials)   Graph

$$f(x) = (x - 1)^2(x - 3)(x + 2)$$

Solution.   The x-intercepts are at 1, 3, and -2. The new feature is that $x - 1$ occurs as a square. The factor $(x - 1)^2$ never changes sign, so the graph does not cross the x-axis at $x = 1$; it merely touches the axis there (Figure 16). Note the entries in the table of values corresponding to $x = 0.9$ and $x = 1.1$.

Sketch the graph of each of the following.

27. $f(x) = (x + 1)(x - 1)(x - 3)$
28. $f(x) = x(x - 2)(x - 4)$
29. $f(x) = x^2(x - 4)$
30. $f(x) = x(x + 2)^2$
31. $f(x) = (x + 2)^2(x - 2)^2$
32. $f(x) = x(x - 1)^3$

MISCELLANEOUS PROBLEMS

33. Recall that a function assigns exactly one value to each element in its domain. What must be true about a graph for it to be the graph of a function with rule of the form $y = f(x)$?

x	y
−4	350
−3	96
−2	0
−1	− 16
0	− 6
0.9	− .06
1	0
1.1	− .06
2	−4
3	0
4	54

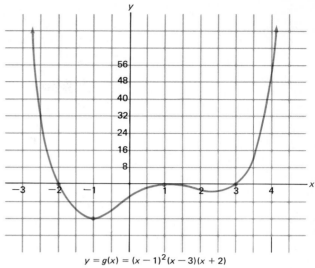

$$y = g(x) = (x - 1)^2(x - 3)(x + 2)$$

Figure 16

34. Which of the graphs in Figure 17 are graphs of functions with rules of the form $y = f(x)$?

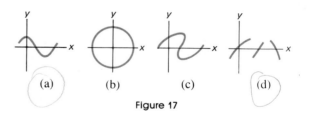

(a) (b) (c) (d)

Figure 17

In Problems 35–42, sketch the graph of the given function.

35. $f(x) = \sqrt{x}$

36. $f(x) = 2\sqrt[3]{x}$

37. $f(x) = (x + 2)(x - 1)^2(x - 3)$

38. $f(x) = x(x - 2)^3$

39. $f(x) = \begin{cases} 4 - x^2 & \text{if } -2 \le x < 2 \\ x - 1 & \text{if } \ \ 2 \le x \le 4 \end{cases}$

40. $f(x) = \begin{cases} |x| & \text{if } \ -2 \le x \le 1 \\ x^2 & \text{if } \ \ \ 1 < x \le 2 \\ 4 & \text{if } \ \ \ 2 < x \le 4 \end{cases}$

41. $f(x) = x^4 + x^2 + 2, \ -2 \le x \le 2$

42. $f(x) = x^5 + x^3 + x, \ -1 \le x \le 1$

43. The graphs of the three functions $f(x) = x^2$, $g(x) = x^4$, and $h(x) = x^6$ all pass through the points $(-1, 1)$, $(0, 0)$, and $(1, 1)$. Draw sketches of these three functions using the same axes. Be sure to show clearly how they differ for $-1 < x < 1$.

44. Sketch the graph of $f(x) = x^{50}$ for $-1 \leq x \leq 1$. Be sure to calculate $f(.5)$ and $f(.9)$. What simple figure does the graph resemble?

45. The function $f(x) = [x]$ is called the **greatest integer function.** It assigns to each real number x the largest integer that is less than or equal to x. For example, $[\frac{5}{2}] = 2$, $[5] = 5$, and $[-1.5] = -2$. Graph this function on the interval $-2 \leq x \leq 6$.

46. Graph each of the following functions on the interval $-2 \leq x \leq 2$.
    (a) $f(x) = 3[x]$ (b) $g(x) = [3x]$

47. Suppose that the cost of shipping a package is 15¢ for anything weighing less than an ounce and 25¢ for anything weighing at least 1 ounce but less than 2 ounces. Beyond that the pattern continues with the cost increased 10¢ for each additional ounce or fraction thereof. Write a formula for the cost $C(x)$ of shipping a package weighing x ounces using the symbol [ ] and graph this function.

48. It costs the XYZ company $1000 + 10\sqrt{x}$ dollars to make x dolls, which sell for $8 each. Express the total profit $T(x)$ in terms of x and then graph this function.

c 49. Calculating values for higher-degree polynomials is messy but can be simplified, as we illustrate for $f(x) = 4x^5 - 3x^4 + 2x^3 - x^2 + 7x - 3$. Write this polynomial as

$$f(x) = ((((4x - 3)x + 2)x - 1)x + 7)x - 3$$

Use this to calculate $f(3)$, $f(4.3)$, and $f(-1.6)$. You should be able to make your calculator do these calculations without using any parentheses.

50. An open box is to be made from a piece of 12-inch by 18-inch cardboard by cutting a square of side x inches from each corner and turning up the sides. Express the volume $V(x)$ of the box in terms of x and graph the resulting function. What is the domain of V? Use your graph to help you find the value of x that makes $V(x)$ a maximum. What is this maximum value?

51. The function $f(x) = \langle x \rangle$ will denote the **distance to nearest integer function.** It assigns to each real number x the distance to the integer nearest to x. For example, $\langle 1.2 \rangle = .2$, $\langle 1.7 \rangle = .3$, and $\langle 2 \rangle = 0$. Graph this function on the interval $0 \leq x \leq 4$ and then find the area of the region between this graph and the x-axis.

52. TEASER   Consider the function $f(x) = \langle x \rangle / 10^{[x]}$ on the infinite interval $x \geq 0$. Sketch the graph of this function and calculate the *total* area of the region between the graph and the x-axis. Write this area first as an unending decimal and then as a ratio of two integers.

Rational Functions

The graph of the rational function

$$f(x) = \frac{p(x)}{q(x)}$$

exhibits spectacular behavior whenever the denominator $q(x)$ nears 0. It must either blow up to plus infinity or down to minus infinity.

5-3　Graphing Rational Functions

If $f(x)$ is given by

$$f(x) = \frac{p(x)}{q(x)}$$

where $p(x)$ and $q(x)$ are polynomials, then f is called a **rational function.** For simplicity, we shall assume that $f(x)$ is in reduced form, that is, that $p(x)$ and $q(x)$ have no common nontrivial factors. Typical examples of rational functions are

$$f(x) = \frac{x + 1}{x^2 - x + 6} = \frac{x + 1}{(x - 3)(x + 2)}$$

$$g(x) = \frac{(x + 2)(x - 5)}{(x + 3)^3}$$

Graphing a rational function can be tricky, primarily because of the denominator $q(x)$. Whenever it is zero, something dramatic is sure to happen to the graph. That is the point of our opening cartoon.

THE GRAPHS OF $1/x$ AND $1/x^2$

Let us consider two simple cases.

$$f(x) = \frac{1}{x} \qquad g(x) = \frac{1}{x^2}$$

Notice that f is an odd function ($f(-x) = -f(x)$), while g is even ($g(-x) = g(x)$). These facts imply that the graph of f is symmetric with respect to the origin, and that the graph of g is symmetric with respect to the y-axis. Thus we need to use only positive values of x to calculate y-values. Each calculation yields two points on the graph. Observe particularly the behavior of each graph near $x = 0$ (Figure 18).

x	$1/x$	$1/x^2$
0	—	—
.01	100	10000
.1	10	100
1	1	1
4	.25	.06

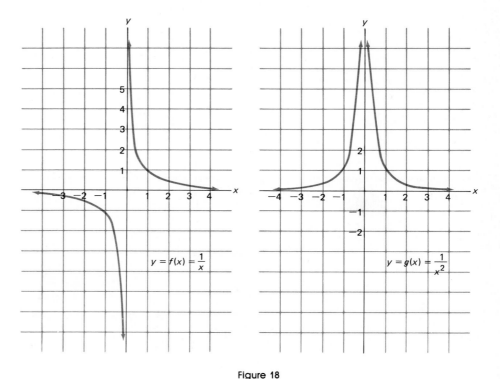

Figure 18

In both cases, the x- and y-axes play special roles; we call them *asymptotes* for the graphs. If, as a point moves away from the origin along a curve, the distance between it and a line becomes closer and closer to zero, then that line is called an **asymptote** for the curve. Clearly the line $x = 0$ is a vertical asymptote for both of our curves and the line $y = 0$ is a horizontal asymptote for both of them.

THE GRAPHS OF $1/(x - 2)$ AND $1/(x - 2)^2$

If we replace x by $x - 2$ in our two functions, we get two new functions.

$$h(x) = \frac{1}{x - 2} \qquad k(x) = \frac{1}{(x - 2)^2}$$

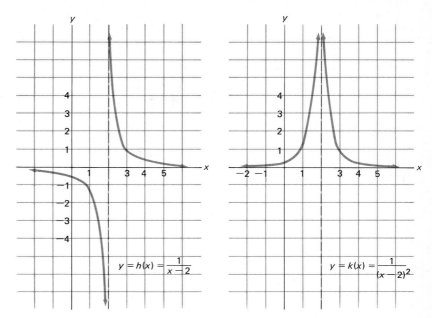

$$y = h(x) = \frac{1}{x-2}$$

$$y = k(x) = \frac{1}{(x-2)^2}$$

Figure 19

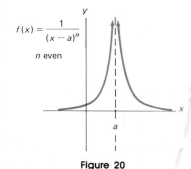

$$f(x) = \frac{1}{(x-a)^n}$$

n even

Figure 20

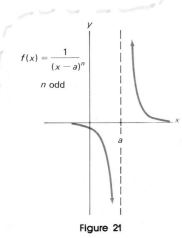

$$f(x) = \frac{1}{(x-a)^n}$$

n odd

Figure 21

Their graphs are just like those of f and g except that they are moved two units to the right, as you can see in Figure 19.

Two observations should be made. The vertical asymptote (dotted line) occurs where the denominator is zero; that is, it is the line $x = 2$. The horizontal asymptote is again the line $y = 0$.

In general, the graph of $f(x) = 1/(x - a)^n$ has the line $y = 0$ as a horizontal asymptote and the line $x = a$ as a vertical asymptote. The behavior of the graph near $x = a$ for n even and n odd is illustrated in Figures 20 and 21.

MORE COMPLICATED EXAMPLES

Consider next the rational function determined by

$$y = f(x) = \frac{x}{x^2 + x - 6} = \frac{x}{(x - 2)(x + 3)}$$

We expect its graph to have vertical asymptotes at $x = 2$ and $x = -3$. Again, the line $y = 0$ will be a horizontal asymptote since, as $|x|$ gets large, the term x^2 in the denominator will dominate, so that y will behave much like x/x^2 or $1/x$ and will thus approach zero. The graph crosses the x-axis where the numerator is zero, namely, at $x = 0$. Finally, with the help of a table of values, we sketch the graph, shown in Figure 22 at the top of the next page.

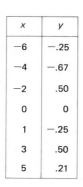

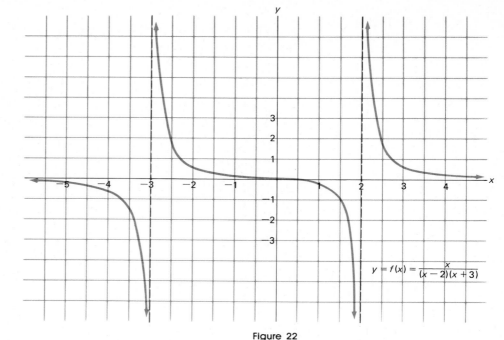

$$y = f(x) = \frac{x}{(x-2)(x+3)}$$

Figure 22

Lastly we consider

$$y = f(x) = \frac{2x^2 + 2x}{x^2 - 4x + 4} = \frac{2x(x+1)}{(x-2)^2}$$

The graph will have one vertical asymptote, at $x = 2$. To check on a horizontal asymptote, we note that for large $|x|$, the numerator behaves like $2x^2$ and the denominator behaves like x^2. It follows that $y = 2$ is a horizontal asymptote. The graph crosses the x-axis where the numerator $2x(x+1)$ is zero, namely, at $x = 0$ and $x = -1$. Figure 23 exhibits a good approximation to the graph.

A GENERAL PROCEDURE

Here is an outline of the procedure for graphing a rational function

$$y = f(x) = \frac{p(x)}{q(x)}$$

which is in reduced form.

1. Check for symmetry with respect to the y-axis and the origin.
2. Factor the numerator and denominator.
3. Determine the vertical asymptotes (if any) by checking where the denom-

x	y
−30	1.70
−10	1.25
−6	.94
−4	.67
−1	0
0	0
1	4
3	24
6	4.5
10	3.44
20	2.59

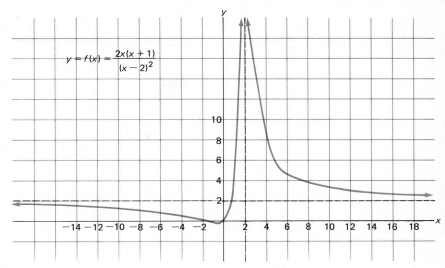

$$y = f(x) = \frac{2x(x+1)}{(x-2)^2}$$

Figure 23

inator is zero. Draw a dotted line for each asymptote. Be sure to examine the behavior of the graph near a vertical asymptote.

4. Determine the horizontal asymptote (if any) by asking what y approaches as x becomes large. This is accomplished by examining the quotient of the leading terms from numerator and denominator. Indicate any horizontal asymptote with a dotted line.

5. Determine the x-intercepts (if any). These occur where the numerator is zero.

6. Make a small table of values and plot corresponding points.

7. Sketch the graph.

Problem Set 5-3

Sketch the graph of each of the following functions.

1. $f(x) = \dfrac{2}{x+2}$

2. $f(x) = \dfrac{-1}{x+2}$

3. $f(x) = \dfrac{2}{(x+2)^2}$

4. $f(x) = \dfrac{1}{(x-3)^2}$

5. $f(x) = \dfrac{2x}{x+2}$

6. $f(x) = \dfrac{x+2}{x-3}$

7. $f(x) = \dfrac{1}{(x+2)(x-1)}$

8. $f(x) = \dfrac{3}{x^2-9}$

9. $f(x) = \dfrac{x + 1}{(x + 2)(x - 1)}$

10. $f(x) = \dfrac{3x}{x^2 - 9}$

11. $f(x) = \dfrac{2x^2}{(x + 2)(x - 1)}$

12. $f(x) = \dfrac{x^2 - 4}{x^2 - 9}$

EXAMPLE A (No Vertical Asymptotes)   Sketch the graph of

$$f(x) = \frac{x^2 - 4}{x^2 + 1} = \frac{(x - 2)(x + 2)}{x^2 + 1}$$

Solution.   Note that f is an even function, so the graph will be symmetric with respect to the y-axis. The denominator is not zero for any real x, so there are no vertical asymptotes. The line $y = 1$ is a horizontal asymptote, since for large $|x|$, $f(x)$ behaves like x^2/x^2. The x-intercepts are $x = 2$ and $x = -2$. The graph is shown in Figure 24.

x	y
0	−4
1	−1.5
2	0
4	.71
6	.86

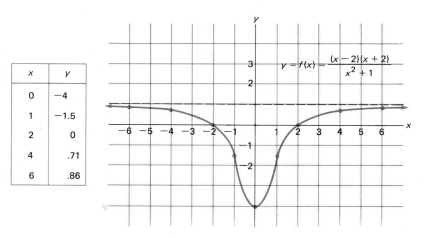

Figure 24

Sketch the graph of each of the following.

13. $f(x) = \dfrac{1}{x^2 + 2}$

14. $f(x) = \dfrac{x^2 - 2}{x^2 + 2}$

15. $f(x) = \dfrac{x}{x^2 + 2}$

16. $f(x) = \dfrac{x^3}{x^2 + 2}$

EXAMPLE B (Oblique Asymptotes)   Sketch the graph of

$$f(x) = \frac{x^2}{x + 1}$$

$$\begin{array}{r}
x - 1 \\
x + 1 \overline{\smash{\big)}\ x^2} \\
\underline{x^2 + x} \\
-x \\
\underline{-x - 1} \\
1
\end{array}$$

Figure 25

Solution. From our earlier discussion, we expect a vertical asymptote at $x = -1$. There is no horizontal asymptote. However, when we do a long division (as in Figure 25), we find that

$$f(x) = x - 1 + \frac{1}{x + 1}$$

As $|x|$ gets larger and larger, the term $1/(x + 1)$ tends to zero, and so $f(x)$ gets closer and closer to $x - 1$. This means that the line $y = x - 1$ is an asymptote, called an **oblique asymptote.** Its significance is indicated on the graph in Figure 26. In general, we can expect an oblique asymptote for the graph of a rational function whenever the degree of the numerator is exactly one more than that of the denominator.

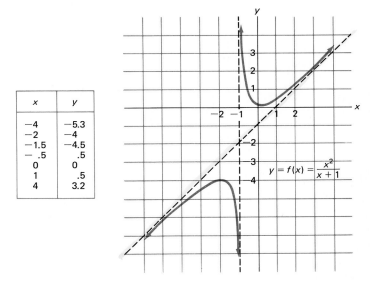

x	y
−4	−5.3
−2	−4
−1.5	−4.5
− .5	.5
0	0
1	.5
4	3.2

$$y = f(x) = \frac{x^2}{x + 1}$$

Figure 26

Sketch the graphs of the following rational functions.

17. $f(x) = \dfrac{2x^2 + 1}{2x}$

18. $f(x) = \dfrac{x^2 - 2}{x}$

19. $f(x) = \dfrac{x^2}{x - 1}$

20. $f(x) = \dfrac{x^3}{x^2 + 1}$

EXAMPLE C (Rational Functions That Are Not in Reduced Form) Sketch the graph of

$$f(x) = \frac{x^2 + x - 6}{x - 2}$$

Solution. Notice that

$$f(x) = \frac{(x + 3)(x - 2)}{x - 2}$$

You have the right to expect that we will cancel the factor $x - 2$ from numerator and denominator and graph

$$g(x) = x + 3$$

But note that 2 is in the domain of g but not in the domain of f. Thus f and g and their graphs are exactly alike except at one point, namely, at $x = 2$. Both graphs are shown in Figure 27. You will notice the hole in the graph of $y = f(x)$ at $x = 2$. This technical distinction is occasionally important.

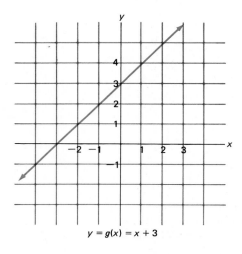

$$y = g(x) = x + 3$$

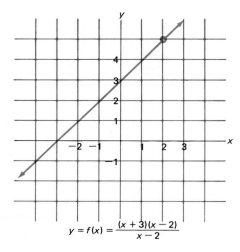

$$y = f(x) = \frac{(x + 3)(x - 2)}{x - 2}$$

Figure 27

Sketch the graph of each of the following rational functions, which, you will note, are not in reduced form.

21. $f(x) = \dfrac{(x + 2)(x - 4)}{x + 2}$

22. $f(x) = \dfrac{x^2 - 4}{x - 2}$

23. $f(x) = \dfrac{x^3 - x^2 - 12x}{x + 3}$

24. $f(x) = \dfrac{x^3 - 4x}{x^2 - 2x}$

MISCELLANEOUS PROBLEMS

Sketch the graphs of the rational functions in Problems 25–30.

25. $f(x) = \dfrac{x}{x + 5}$

26. $f(x) = \dfrac{x - 2}{x + 3}$

27. $f(x) = \dfrac{x^2 - 9}{x^2 - x - 2}$

28. $f(x) = \dfrac{x - 2}{(x + 3)^2}$

29. $f(x) = \dfrac{x^2 - 9}{x^2 - x - 6}$

30. $f(x) = \dfrac{x - 2}{x^2 + 3}$

31. Sketch the graphs of $f(x) = x^n/(x^2 + 1)$ for $n = 1, 2,$ and 3, being careful to show all asymptotes.

32. Where does the graph of $f(x) = (x^3 + x^2 - 2x + 1)/(x^3 + 2x^2 - 2)$ cross its horizontal asymptote?

33. Determine all asymptotes (vertical, horizontal, oblique) for the graph of $f(x) = x^4/(x^n - 1)$ in each case.
    (a) $n = 1$　　　　(b) $n = 2$　　　　(c) $n = 3$
    (d) $n = 4$　　　　(e) $n = 5$　　　　(f) $n = 6$

34. Consider the graph of the general rational function

$$f(x) = \frac{a_n x^n + a_{n-1} x^{n-1} + \cdots + a_1 x + a_0}{b_m x^m + b_{m-1} x^{m-1} + \cdots + b_1 x + b_0}, \quad a_n \neq 0, \, b_m \neq 0$$

Identify its horizontal asymptote in each case.
(a) $m > n$　　　　(b) $m = n$　　　　(c) $m < n$

35. A manufacturer of gizmos has overhead of $20,000 per year and direct costs (labor and material) of $50 per gizmo. Write an expression for $U(x)$, the average cost per unit, if the company makes x gizmos per year. Graph the function U and then draw some conclusions from your graph.

36. A cylindrical can is to contain 10π cubic inches. Write a formula for $S(r)$, the total surface area, in terms of the radius r. Graph the function S and use it to estimate the radius of the can that will require the least material to make.

37. Find a formula for $f(x)$ if f is a rational function whose graph goes through $(2, 5)$ and has exactly two asymptotes, namely, $y = 2x + 3$ and $x = 3$.

38. **TEASER**　Sketch the graphs of $f(x) = [1/x]$ and $g(x) = (1/x)$ for $0 < x \leq 1$. The symbols $[\ ]$ and $(\ )$ were defined in Problems 45 and 51 of Section 5-2.

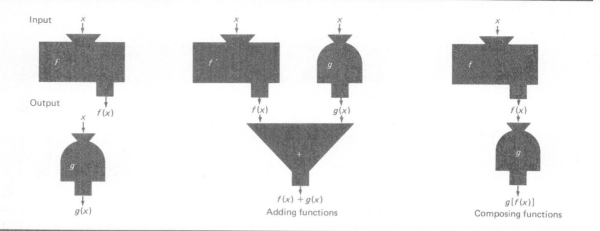

Input

Output

$f(x)$

x

$g(x)$

$f(x)$ $g(x)$

$f(x) + g(x)$

Adding functions

$f(x)$

$g[f(x)]$

Composing functions

5-4   Putting Functions Together

There is still another way to visualize a function. Think of the function named f as a machine. It accepts a number x as input, operates on it, and then presents the number $f(x)$ as output. Machines can be hooked together to make more complicated machines; similarly, functions can be combined to produce more complicated functions. That is the subject of this section.

SUMS, DIFFERENCES, PRODUCTS, AND QUOTIENTS

The simplest way to make new functions from old ones is to use the four arithmetic operations on them. Suppose, for example, that the functions f and g have the formulas

$$f(x) = \frac{x - 3}{2} \qquad g(x) = \sqrt{x}$$

We can make a new function $f + g$ by having it assign to x the value $(x - 3)/2 + \sqrt{x}$; that is,

$$(f + g)(x) = f(x) + g(x) = \frac{x - 3}{2} + \sqrt{x}$$

Of course, we must be a little careful about domains. Clearly, x must be a number on which both f and g can operate. In other words, the domain of $f + g$ is the intersection (common part) of the domains of f and g.

The functions $f - g$, $f \cdot g$, and f/g are defined in a completely analogous way. Assuming that f and g have their respective natural domains—namely, all reals and the nonnegative reals, respectively—we have the following.

$$(f + g)(x) = f(x) + g(x) = \frac{x - 3}{2} + \sqrt{x} \qquad x \geq 0$$

$$(f - g)(x) = f(x) - g(x) = \frac{x - 3}{2} - \sqrt{x} \qquad x \geq 0$$

$$(f \cdot g)(x) = f(x) \cdot g(x) = \frac{x - 3}{2} \sqrt{x} \qquad x \geq 0$$

$$(f/g)(x) = f(x)/g(x) = \frac{x - 3}{2\sqrt{x}} \qquad x > 0$$

To graph the function $f + g$, it is often best to graph f and g separately in the same coordinate plane and then add the y-coordinates together along vertical lines. We illustrate this method (called **addition of ordinates**) in Figure 28.

The graph of $f - g$ can be handled similarly. Simply graph f and g in the same coordinate plane and subtract ordinates. We can even graph $f \cdot g$ and f/g in the same manner, but that is harder.

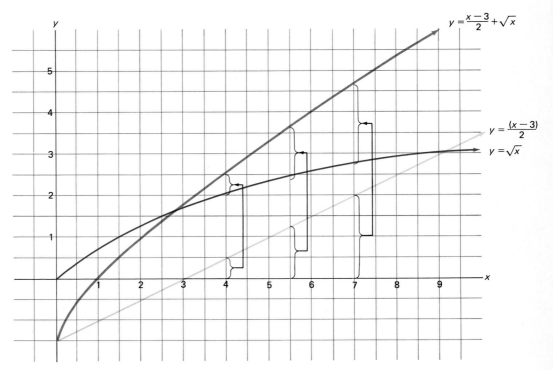

Figure 28

COMPOSITION OF FUNCTIONS

Figure 29

To compose functions is to string them together in tandem. Part of our opening display (reproduced in Figure 29) shows how this is done. If f operates on x to produce $f(x)$ and then g operates on $f(x)$ to produce $g(f(x))$, we say that we have composed g and f. The resulting function, called the **composite of g with f,** is denoted by $g \circ f$. Thus

$$(g \circ f)(x) = g(f(x))$$

Recall our earlier examples, $f(x) = (x - 3)/2$ and $g(x) = \sqrt{x}$. We may compose them in two ways.

$$(g \circ f)(x) = g(f(x)) = g\left(\frac{x - 3}{2}\right) = \sqrt{\frac{x - 3}{2}}$$

$$(f \circ g)(x) = f(g(x)) = f(\sqrt{x}) = \frac{\sqrt{x} - 3}{2}$$

We note one thing right away: Composition of functions is not commutative; $g \circ f$ and $f \circ g$ are not the same. We must also be careful in describing the domain of a composite function. The domain of $g \circ f$ is that part of the domain of f for which g can accept $f(x)$ as input. In our example, the domain of $g \circ f$ is $x \geq 3$, not all x or $x \geq 0$ as we might have thought at first glance. Figure 30 offers another view of these matters. The shaded portion of the domain of f is not in the domain of $g \circ f$; for x in this portion, $f(x)$ is outside the domain of g.

In calculus, we shall often need to take a given function and decompose it, that is, break it into composite pieces. Usually, this can be done in several ways. Take $p(x) = \sqrt{x^2 + 3}$ for example. We may think of it as

$$p(x) = g(f(x)) \quad \text{where} \quad g(x) = \sqrt{x} \quad \text{and} \quad f(x) = x^2 + 3$$

or as

$$p(x) = g(f(x)) \quad \text{where} \quad g(x) = \sqrt{x + 3} \quad \text{and} \quad f(x) = x^2$$

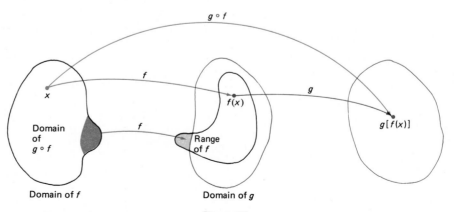

Figure 30

TRANSLATIONS

Observing how a function is built up from simpler ones using the operations of this section can be a big aid in graphing. This is especially true of *translations*, which result from the composition $f(x - h)$ and/or the simple addition of a constant k.

Consider, for example, the graphs of

$$y = f(x) \qquad y = f(x - 3) \qquad y = f(x) + 2 \qquad y = f(x - 3) + 2$$

for the case $f(x) = |x|$. The four graphs are shown in Figure 31.

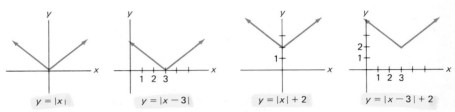

$$y = |x| \qquad\qquad y = |x - 3| \qquad\qquad y = |x| + 2 \qquad\qquad y = |x - 3| + 2$$

Figure 31

Notice that all four graphs have the same shape; the last three are just translations (rigid movements) of the first. Replacing x by $x - 3$ translates the graph 3 units to the right; adding 2 translates the graph 2 units upward.

What happened with $f(x) = |x|$ is typical. In Figure 32, we give another illustration, this time for the function $f(x) = x^3 + x^2$.

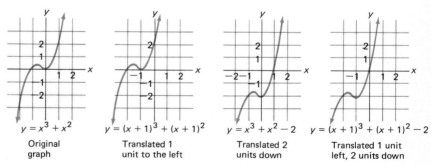

$y = x^3 + x^2$	$y = (x + 1)^3 + (x + 1)^2$	$y = x^3 + x^2 - 2$	$y = (x + 1)^3 + (x + 1)^2 - 2$
Original graph	Translated 1 unit to the left	Translated 2 units down	Translated 1 unit left, 2 units down

Figure 32

Exactly the same principles apply in the general situation. Figure 33 on the next page shows what happens when h and k are both positive. If $h < 0$, the translation is to the left; if $k < 0$, the translation is downward.

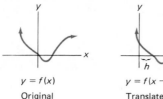

$y = f(x)$ $y = f(x - h)$ $y = f(x) + k$ $y = f(x - h) + k$

Original graph   Translated h units to the right   Translated k units up   Translated h units to the right and k units up

Figure 33

Problem Set 5-4

1. Let $f(x) = x^2 - 2x + 2$ and $g(x) = 2/x$. Calculate each of the following.
   (a) $(f + g)(2)$ (b) $(f + g)(0)$ (c) $(f - g)(1)$
   (d) $(f \cdot g)(-1)$ (e) $(f/g)(2)$ (f) $(g/f)(2)$
   (g) $(f \circ g)(-1)$ (h) $(g \circ f)(-1)$ (i) $(g \circ g)(3)$

2. Let $f(x) = 3x + 5$ and $g(x) = |x - 2|$. Perform the calculations in Problem 1 for these functions.

In each of the following, write the formulas for $(f + g)(x)$, $(f - g)(x)$, $(f \cdot g)(x)$, and $(f/g)(x)$ and give the domains of these four functions.

3. $f(x) = x^2$, $g(x) = x - 2$ 4. $f(x) = x^3 - 1$, $g(x) = x + 3$

5. $f(x) = x^2$, $g(x) = \sqrt{x}$ 6. $f(x) = 2x^2 + 5$, $g(x) = \dfrac{1}{x}$

7. $f(x) = \dfrac{1}{x - 2}$, $g(x) = \dfrac{x}{x - 3}$ 8. $f(x) = \dfrac{1}{x^2}$, $g(x) = \dfrac{1}{5 - x}$

For each of the following, write the formulas for $(g \circ f)(x)$ and $(f \circ g)(x)$ and give the domains of these composite functions.

9. $f(x) = x^2$, $g(x) = x - 2$ 10. $f(x) = x^3 - 1$, $g(x) = x + 3$

11. $f(x) = \dfrac{1}{x}$, $g(x) = x + 3$ 12. $f(x) = 2x^2 + 5$, $g(x) = \dfrac{1}{x}$

13. $f(x) = \sqrt{x - 2}$, $g(x) = x^2 - 2$ 14. $f(x) = \sqrt{2x}$, $g(x) = x^2 + 1$

15. $f(x) = 2x - 3$, $g(x) = \frac{1}{2}(x + 3)$ 16. $f(x) = x^3 + 1$, $g(x) = \sqrt[3]{x - 1}$

EXAMPLE (Decomposing Functions) In each of the following, H can be thought of as a composite function $g \circ f$. Write formulas for $f(x)$ and $g(x)$.

(a) $H(x) = (2 + 3x)^2$ (b) $H(x) = \dfrac{1}{(x^2 + 4)^3}$

Solution.

(a) Think of how you might calculate $H(x)$. You would first calculate $2 + 3x$ and then square the result. That suggests

$$f(x) = 2 + 3x \qquad g(x) = x^2$$

(b) Here there are two obvious ways to proceed. One way would be to let

$$f(x) = x^2 + 4 \qquad g(x) = \frac{1}{x^3}$$

Another selection, which is just as good, is

$$f(x) = \frac{1}{x^2 + 4} \qquad g(x) = x^3$$

We could actually think of H as the composite of four functions. Let

$$f(x) = x^2 \qquad g(x) = x + 4 \qquad h(x) = x^3 \qquad j(x) = \frac{1}{x}$$

Then

$$H = j \circ h \circ g \circ f$$

You should check this result.

In each of the following, write formulas for g(x) and f(x) so that $H = g \circ f$. The answer is not unique.

17. $H(x) = (x + 4)^3$
18. $H(x) = (2x + 1)^3$
19. $H(x) = \sqrt{x + 2}$
20. $H(x) = \sqrt[3]{2x + 1}$
21. $H(x) = \dfrac{1}{(2x + 5)^3}$
22. $H(x) = \dfrac{6}{(x + 4)^3}$
23. $H(x) = |x^3 - 4|$
24. $H(x) = |4 - x - x^2|$

Use the method of addition or subtraction of ordinates to graph each of the following. That is, graph $y = f(x)$ and $y = g(x)$ in the same coordinate plane and then obtain the graph of $f + g$ or $f - g$ by adding or subtracting ordinates.

25. $f + g$, where $f(x) = x^2$ and $g(x) = x - 2$.
26. $f + g$, where $f(x) = |x|$ and $g(x) = x$.
27. $f - g$, where $f(x) = 1/x$ and $g(x) = x$.
28. $f - g$, where $f(x) = x^3$ and $g(x) = -x + 1$.

In each of the following, graph the function f carefully and then use translations to sketch the graphs of the functions g, h, and j.

29. $f(x) = x^2$, $g(x) = (x - 2)^2$, $h(x) = x^2 - 4$, and $j(x) = (x - 2)^2 + 1$
30. $f(x) = x^3$, $g(x) = (x + 2)^3$, $h(x) = x^3 + 4$, and $j(x) = (x + 2)^3 - 2$
31. $f(x) = \sqrt{x}$, $g(x) = \sqrt{x - 3}$, $h(x) = \sqrt{x} + 2$, and $j(x) = \sqrt{x - 3} - 2$
32. $f(x) = \dfrac{1}{x}$, $g(x) = \dfrac{1}{x - 4}$, $h(x) = \dfrac{1}{x} + 3$, and $j(x) = \dfrac{1}{x - 4} - 5$

MISCELLANEOUS PROBLEMS

33. Let $f(x) = 2x + 3$ and $g(x) = x^3$. Write formulas for each of the following.
    (a) $(f + g)(x)$ (b) $(g - f)(x)$ (c) $(f \cdot g)(x)$ (d) $(f/g)(x)$
    (e) $(f \circ g)(x)$ (f) $(g \circ f)(x)$ (g) $(f \circ f)(x)$ (h) $(g \circ g \circ g)(x)$

34. If $f(x) = 1/(x - 1)$ and $g(x) = \sqrt{x + 1}$, write formulas for $(f \circ g)(x)$ and $(g \circ f)(x)$ and give the domains of these composite functions.

35. If $f(x) = x^2 - 4$, $g(x) = |x|$, and $h(x) = 1/x$, write a formula for $(h \circ g \circ f)(x)$ and indicate its domain.

36. In general, how many different functions can be obtained by composing three different functions f, g, and h in different orders?

37. In calculus, the *difference quotient*

$$\frac{f(x + h) - f(x)}{h}$$

arises repeatedly. Calculate this expression and simplify it for each of the following.
   (a) $f(x) = x^2$ (b) $f(x) = 2x + 3$
   (c) $f(x) = 1/x$ (d) $f(x) = 2/(x - 2)$

38. Calculate $[g(x - h) - g(x)]/h$ for each of the following. Simplify your answer.
   (a) $g(x) = 4x - 9$ (b) $g(x) = x^2 + 2x$
   (c) $g(x) = x + 1/x$ (d) $g(x) = x^3$

39. Find $(f \circ g)(x)$ and $(g \circ f)(x)$ in each case.
   (a) $f(x) = x^2$, $g(x) = \sqrt{x}$ (b) $f(x) = x^3$, $g(x) = \sqrt[3]{x}$
   (c) $f(x) = x^2$, $g(x) = x^3$ (d) $f(x) = x^2$, $g(x) = 1/x^3$

40. Let $f(x) = (x - 3)/(x + 1)$. Show that if $x \neq \pm 1$, then $f(f(f(x))) = x$.

41. Let $f(x) = [(1 - \sqrt{x})/(1 + \sqrt{x})]^2$. Solve for x if $f(f(x)) = x^2 + \frac{1}{4}$ and $0 < x < 1$.

42. Let $f(x) = x^2 + 5x$. Solve for x if $f(f(x)) = f(x)$.

43. Sketch the graph of $f(x) = |x + 1| - |x| + |x - 1|$ on the interval $-2 \leq x \leq 2$. Then calculate the area of the region between this graph and the x-axis.

44. The *greatest integer function* $[\ ]$ was defined in Problem 45 of Section 5-2. Graph each of the following functions on the interval $-2 \leq x \leq 6$.
   (a) $f(x) = 2[x]$ (b) $g(x) = 2 + [x]$
   (c) $h(x) = [x - 2]$ (d) $k(x) = x - [x]$

45. Let f be an even function (meaning $f(-x) = f(x)$) and let g be an odd function (meaning $g(-x) = -g(x)$), both functions having the whole real line as their domains. Which of the following are even? Odd? Neither even nor odd?
   (a) $f(x)g(x)$ (b) $f(x)/g(x)$ (c) $[g(x)]^2$
   (d) $[g(x)]^3$ (e) $f(x) + g(x)$ (f) $g(g(x))$
   (g) $f(f(x))$ (h) $3f(x) + [g(x)]^2$ (i) $g(x) + g(-x)$

46. Show that any function f having the whole real line as its domain can be represented as the sum of an even function and an odd function. *Hint:* Consider $f(x) + f(-x)$ and $f(x) - f(-x)$.

47. The *distance to the nearest integer function* $(\ )$ was defined in Problem 51 of Section 5-2.
   (a) Sketch the graphs of $f(x) = (x)$, $g(x) = (2x)/2$, $h(x) = (4x)/4$ and $F(x) = f(x) + g(x) + h(x)$ on the interval $0 \leq x \leq 4$.
   (b) Find the areas of the regions between the graph of each of these functions and the x-axis on the interval $0 \leq x \leq 4$. Note that the sum of the first three areas is the fourth.

48. **TEASER**   Generalize Problem 47 by considering the graph of

$$F_n(x) = \langle x \rangle + \frac{1}{2}\langle 2x \rangle + \frac{1}{4}\langle 4x \rangle + \cdots + \frac{1}{2^n}\langle 2^n x \rangle$$

on the interval $0 \le x \le 4$. Find a nice formula for the area A_n of the region between this graph and the x-axis. What happens to A_n as n grows without bound? *Note:* The limiting form of F_n plays an important role in advanced mathematics giving an example of a function whose graph is continuous but does not have a tangent line at any point.

A one-to-one function   has an inverse.

5-5   Inverse Functions

Some processes are reversible; most are not. If I take off my shoes, I may put them back on again. The second operation undoes the first one and brings things back to the original state. But if I throw my shoes in the fire, I will have a hard time undoing the damage I have done.

A function f operates on a number x to produce a number $y = f(x)$. It may be that we can find a function g that will operate on y and give back x. For example, if

$$y = f(x) = 2x + 1$$

then

$$g(x) = \frac{1}{2}(x - 1)$$

is such a function, since

$$g(y) = g(f(x)) = \frac{1}{2}(2x + 1 - 1) = x$$

When we can find such a function g, we call it the *inverse* of f. Not all functions have inverses. Whether they do or not has to do with a concept called one-to-oneness.

ONE-TO-ONE FUNCTIONS

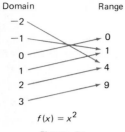

$f(x) = x^2$

Figure 34

In Figure 34, we have reproduced an example we studied earlier, the squaring function with domain $\{-2, -1, 0, 1, 2, 3\}$. It is a perfectly fine function, but it does have one troublesome feature. It may assign the same value to two different x's. In particular, $f(-2) = 4$ and $f(2) = 4$. Such a function cannot possibly have an inverse g. For what would g do with 4? It would not know whether to give back -2 or 2 as the value.

In contrast, consider $f(x) = 2x + 1$, pictured in Figure 35. Notice that this function never assigns the same value to two different values of x. Therefore there is an unambiguous way of undoing it.

We say that a function f is **one-to-one** if $x_1 \neq x_2$ implies $f(x_1) \neq f(x_2)$, that is, if different values for x always result in different values for $f(x)$. Some functions are one-to-one; some are not. It would be nice to have a graphical criterion for deciding.

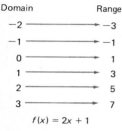

$f(x) = 2x + 1$

Figure 35

Consider the functions $f(x) = x^2$ and $f(x) = 2x + 1$ again, but now let the domains be the set of all real numbers. Their graphs are shown in Figure 36. In the first case, certain horizontal lines (those which are above the x-axis) meet the graph in two points; in the second case, every horizontal line meets the graph in exactly one point. Notice on the first graph that $f(x_1) = f(x_2)$ even though $x_1 \neq x_2$. On the second graph, this cannot happen. Thus we have the important fact that *if every horizontal line meets the graph of a function f in at most one point, then f is one-to-one.*

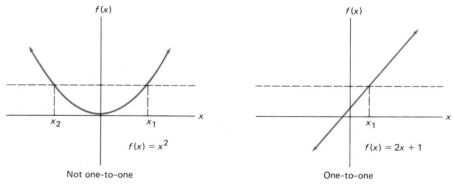

Figure 36

INVERSE FUNCTIONS

Now we are ready to give a formal definition of the main idea of this section.

DEFINITION

Let f be a one-to-one function with domain X and range Y. Then the function g with domain Y and range X which satisfies

$$g(f(x)) = x$$

*for all x in X is called the **inverse of f.***

We make several important observations. First, the boxed formula simply says that g undoes what f did. Second, if g undoes f, then f will undo g, that is,

$$f(g(y)) = y$$

for all y in Y. Third, the function g is usually denoted by the symbol f^{-1}. You are cautioned to remember that f^{-1} does *not* mean $1/f$, as you have the right to expect. Mathematicians decided long ago that f^{-1} should stand for the inverse function (the undoing function). Thus

$$(f^{-1} \circ f)(x) = x \quad \text{and} \quad (f \circ f^{-1})(y) = y$$

For example, if $f(x) = 4x$, then $f^{-1}(y) = \frac{1}{4}y$ since

$$(f^{-1} \circ f)(x) = f^{-1}(f(x)) = f^{-1}(4x) = \frac{1}{4}(4x) = x$$

and

$$(f \circ f^{-1})(y) = f(f^{-1}(y)) = f(\tfrac{1}{4}y) = 4(\tfrac{1}{4}y) = y$$

The boxed results are illustrated in Figure 37.

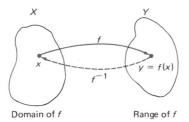
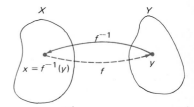

$$f^{-1}[f(x)] = x \qquad\qquad f[f^{-1}(y)] = y$$

Figure 37

FINDING A FORMULA FOR f^{-1}

If f adds 2, then f^{-1} ought to subtract 2. To say it in symbols, if $f(x) = x + 2$, then we might expect $f^{-1}(y) = y - 2$. And we are right, for

$$f^{-1}[f(x)] = f^{-1}(x + 2) = x + 2 - 2 = x$$

If f divides by 3 and then subtracts 4, we expect f^{-1} to add 4 and multiply by 3. Symbolically, if $f(x) = x/3 - 4$, then we expect $f^{-1}(y) = 3(y + 4)$. Again we are right, for

$$f^{-1}[f(x)] = f^{-1}\left(\frac{x}{3} - 4\right) = 3\left(\frac{x}{3} - 4 + 4\right) = x$$

Note that you must undo things in the reverse order in which you did them (that is, we divided by 3 and then subtracted 4, so to undo this, we first add 4 and then multiply by 3).

When we get to more complicated functions, it is not always easy to find the formula for the inverse function. Here is an important way to look at it.

$$x = f^{-1}(y) \quad \text{if and only if} \quad y = f(x)$$

That means that we can get the formula for f^{-1} by solving the equation $y = f(x)$ for x. Here is an example. Let $y = f(x) = 3/(x - 2)$. Follow the steps below.

$$y = \frac{3}{x - 2}$$

$$(x - 2)y = 3$$

$$xy - 2y = 3$$

$$xy = 3 + 2y$$

$$x = \frac{3 + 2y}{y}$$

Thus

$$f^{-1}(y) = \frac{3 + 2y}{y}$$

In the formula for f^{-1} just derived, there is no need to use y as the variable. We might use u or t or even x. The formulas

$$f^{-1}(u) = \frac{3 + 2u}{u}$$

$$f^{-1}(t) = \frac{3 + 2t}{t}$$

$$f^{-1}(x) = \frac{3 + 2x}{x}$$

all say the same thing in the sense that they give the same rule. It is conventional to give formulas for functions using x as the variable, and so we would write $f^{-1}(x) = (3 + 2x)/x$ as our answer. Let us summarize. To find the formula for $f^{-1}(x)$, use the following steps.

> **THREE-STEP PROCEDURE FOR FINDING $f^{-1}(x)$**
>
> 1. Solve $y = f(x)$ for x in terms of y.
> 2. Use $f^{-1}(y)$ to name the resulting expression in y.
> 3. Replace y by x to get the formula for $f^{-1}(x)$.

THE GRAPHS OF f AND f^{-1}

Since $y = f(x)$ and $x = f^{-1}(y)$ are equivalent, the graphs of these two equations are the same. Suppose we want to compare the graphs of $y = f(x)$ and $y = f^{-1}(x)$ (where, you will note, we have used x as the domain variable in both cases). To get $y = f^{-1}(x)$ from $x = f^{-1}(y)$, we interchange the roles of x and y. Graphically, this corresponds to folding (reflecting) the graph across the 45° line—that is, across the line $y = x$ (Figure 38). This is the same as saying that if the point (a, b) is on one graph, then (b, a) is on the other.

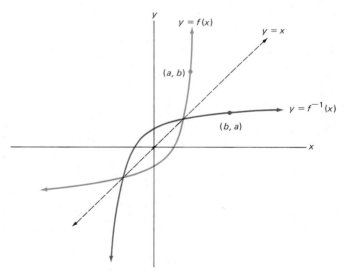

Reflecting a graph across the line $y = x$

Figure 38

Here is a simple example. Let $f(x) = x^3$; then $f^{-1}(x) = \sqrt[3]{x}$, the cube root of x. The graphs of $y = x^3$ and $y = \sqrt[3]{x}$ are shown in Figure 39 at the top of the next page, first separately and then on the same coordinate plane. Note that $f(2) = 8$ and $f^{-1}(8) = 2$.

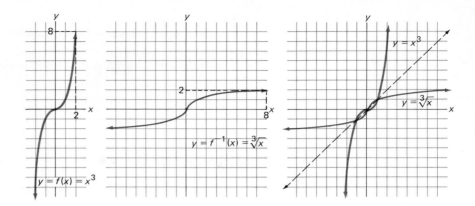

Figure 39

Problem Set 5-5

1. Examine the graphs in Figure 40.
   (a) Which of these are the graphs of functions with x as domain variable?
   (b) Which of these functions are one-to-one?
   (c) Which of them have inverses?

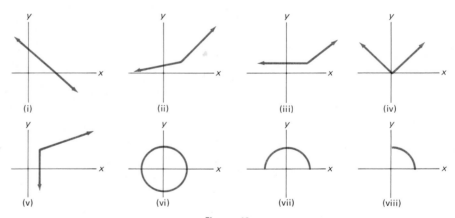

Figure 40

2. Let each of the following functions have its natural domain. Which of them are one-to-one? *Hint:* Consider their graphs.

   (a) $f(x) = x^4$ (b) $f(x) = x^3$ (c) $f(x) = \dfrac{1}{x}$

   (d) $f(x) = \dfrac{1}{x^2}$ (e) $f(x) = x^2 + 2x + 3$ (f) $f(x) = |x|$

   (g) $f(x) = \sqrt{x}$ (h) $f(x) = -3x + 2$

3. Let $f(x) = 3x - 2$. To find $f^{-1}(2)$, note that $f^{-1}(2) = a$ if $f(a) = 2$, that is, if $3a - 2 = 2$; so $f^{-1}(2) = a = \frac{4}{3}$. Find each of the following.
   (a) $f^{-1}(1)$ (b) $f^{-1}(-3)$ (c) $f^{-1}(14)$

4. Let $g(x) = 1/(x - 1)$. Find each of the following.
   (a) $g^{-1}(1)$ (b) $g^{-1}(-1)$ (c) $g^{-1}(14)$

EXAMPLE A (Finding $f^{-1}(x)$)   Use the three-step procedure to find $f^{-1}(x)$ if $f(x) = 2x^3 - 1$. Check your result by calculating $f(f^{-1}(x))$.

Solution.

     Step 1   We solve $y = 2x^3 - 1$ for x in terms of y.

$$2x^3 = y + 1$$

$$x^3 = \frac{y + 1}{2}$$

$$x = \sqrt[3]{\frac{y + 1}{2}}$$

     Step 2   Call the result $f^{-1}(y)$.

$$f^{-1}(y) = \sqrt[3]{\frac{y + 1}{2}}$$

     Step 3   Replace y by x.

$$f^{-1}(x) = \sqrt[3]{\frac{x + 1}{2}}$$

$$\text{Check:} \quad f(f^{-1}(x)) = 2\left(\sqrt[3]{\frac{x + 1}{2}}\right)^3 - 1 = 2\left(\frac{x + 1}{2}\right) - 1 = x$$

Each of the functions in Problems 5–14 has an inverse (using its natural domain). Find the formula for $f^{-1}(x)$. Then check your result by calculating $f(f^{-1}(x))$.

5. $f(x) = 5x$ 6. $f(x) = -4x$ 7. $f(x) = 2x - 7$

8. $f(x) = -3x + 2$ 9. $f(x) = \sqrt{x} + 2$ 10. $f(x) = 2\sqrt{x} - 6$

11. $f(x) = \dfrac{x}{x - 3}$ 12. $f(x) = \dfrac{x - 3}{x}$ 13. $f(x) = (x - 2)^3 + 2$

14. $f(x) = \frac{1}{3}x^5 - 2$

15. In the same coordinate plane, sketch the graphs of $y = f(x)$ and $y = f^{-1}(x)$ for $f(x) = \sqrt{x} + 2$ (see Problem 9).

16. In the same coordinate plane, sketch the graphs of $y = f(x)$ and $y = f^{-1}(x)$ for $f(x) = x/(x - 3)$ (see Problem 11).

17. Sketch the graph of $y = f^{-1}(x)$ if the graph of $y = f(x)$ is as shown in Figure 41.

18. Show that $f(x) = 2x/(x - 1)$ and $g(x) = x/(x - 2)$ are inverses of each other by calculating $f(g(x))$ and $g(f(x))$.

19. Show that $f(x) = 3x/(x + 2)$ and $g(x) = 2x/(3 - x)$ are inverses of each other.

20. Sketch the graph of $f(x) = x^3 + 1$ and note that f is one-to-one. Find a formula for $f^{-1}(x)$.

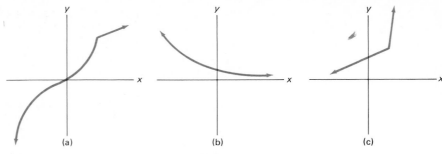

(a)   (b)   (c)

Figure 41

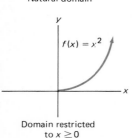

Natural domain

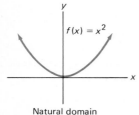

Domain restricted
to $x \geq 0$

Figure 42

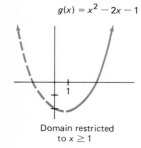

Domain restricted
to $x \geq 1$

Figure 43

EXAMPLE B (Restricting the Domain) The function $f(x) = x^2$ does not have an inverse if we use its natural domain (all real numbers). However, if we restrict its domain to $x \geq 0$ so that we are considering only its right branch (see Figure 42), then it has an inverse, $f^{-1}(x) = \sqrt{x}$. Use the same idea to show that $g(x) = x^2 - 2x - 1$ has an inverse when its domain is appropriately restricted. Find $g^{-1}(x)$.

Solution. The graph of $g(x)$ is shown in Figure 43; it is a parabola with vertex at $x = 1$. Accordingly, we can restrict the domain to $x \geq 1$. To find the formula for $g^{-1}(x)$, we first solve $y = x^2 - 2x - 1$ for x using an old trick, completing the square.

$$y + 1 = x^2 - 2x$$
$$y + 1 + 1 = x^2 - 2x + 1$$
$$y + 2 = (x - 1)^2$$
$$\pm\sqrt{y + 2} = x - 1$$
$$1 \pm \sqrt{y + 2} = x$$

Notice that there are two expressions for x; they correspond to the two halves of the parabola. We chose to make $x \geq 1$, so $x = 1 + \sqrt{y + 2}$ is the correct expression for $g^{-1}(y)$. Thus

$$g^{-1}(x) = 1 + \sqrt{x + 2}$$

If we had chosen to make $x \leq 1$, the correct answer would have been $g^{-1}(x) = 1 - \sqrt{x + 2}$.

In each of the following, restrict the domain so that f has an inverse. Describe the restricted domain and find a formula for $f^{-1}(x)$. Note: Different restrictions of the domain are possible.

21. $f(x) = (x - 1)^2$ 22. $f(x) = (x + 3)^2$

23. $f(x) = (x + 1)^2 - 4$ 24. $f(x) = (x - 2)^2 + 3$

25. $f(x) = x^2 + 6x + 7$ 26. $f(x) = x^2 - 4x + 9$

27. $f(x) = |x + 2|$

28. $f(x) = 2|x - 3|$

29. $f(x) = \dfrac{(x - 1)^2}{1 + 2x - x^2}$

30. $f(x) = \dfrac{-1}{x^2 + 4x + 3}$

MISCELLANEOUS PROBLEMS

31. Sketch the graph of $f(x) = 1/(x - 1)$. Is f one-to-one? Calculate each of the following.
    (a) $f(3)$ (b) $f^{-1}(\frac{1}{2})$ (c) $f(0)$
    (d) $f^{-1}(-1)$ (e) $f^{-1}(3)$ (f) $f^{-1}(-2)$

32. If $f(x) = x/(x - 2)$, find the formula for $f^{-1}(x)$.

33. Find the formula for $f^{-1}(x)$ if $f(x) = 1/(x - 1)$ and sketch the graph of $y = f^{-1}(x)$. Compare this graph with the graph of $y = f(x)$ that you sketched in Problem 31.

34. Find the formula for $f^{-1}(x)$ if $f(x) = (x^3 + 2)/(x^3 + 3)$.

35. Sketch the graph of $f(x) = x^2 - 2x - 3$ and observe that it is not one-to-one. Restrict its domain so it is and then find a formula for $f^{-1}(x)$.

36. Let $f(x) = (x - 3)/(x + 1)$. Show that $f^{-1}(x) = f(f(x))$.

37. Let $f(x) = (2x^2 - 4x - 1)/(x - 1)^2$. If we restrict the domain so $x > 1$, then f has an inverse. Find the formula for $f^{-1}(x)$.

38. Suppose that f and g have inverses. Show that in this case $f \circ g$ has an inverse and that $(f \circ g)^{-1} = g^{-1} \circ f^{-1}$.

39. What must be true of the graph of f if f is *self-inverse* (meaning f is its own inverse)? What does this mean for the xy-equation determining f?

40. Let $y = f(x) = x/(x - 1)$. Show that the condition found in Problem 39 is satisfied. Now check that f is self-inverse by showing that $f(f(x)) = x$.

41. Let $f(x) = (ax + b)/(cx + d)$ where $bc - ad \neq 0$.
    (a) Find the formula for $f^{-1}(x)$.
    (b) Why did we impose the condition $bc - ad \neq 0$?
    (c) What relation connecting a and d will make f self-inverse?

42. **TEASER** Let $f_1(x) = x$, $f_2(x) = 1/x$, $f_3(x) = 1 - x$, $f_4(x) = 1/(1 - x)$, $f_5(x) = (x - 1)/x$, and $f_6(x) = x/(x - 1)$. Note that

$$f_4(f_3(x)) = \frac{1}{1 - f_3(x)} = \frac{1}{1 - (1 - x)} = \frac{1}{x} = f_2(x)$$

$\circ$	f_1	f_2	f_3	f_4	f_5	f_6
f_1						
f_2						
f_3						
f_4		f_2				
f_5						
f_6						

that is, $f_4 \circ f_3 = f_2$. In fact, if we compose any two of these six functions, we will get one of the six functions. Complete the adjoining composition table and then use it to find each of the following (which will also be one of the six functions).
(a) $f_3 \circ f_3 \circ f_3 \circ f_3 \circ f_3$
(b) $f_1 \circ f_2 \circ f_3 \circ f_4 \circ f_5 \circ f_6$
(c) f_6^{-1}
(d) $(f_3 \circ f_6)^{-1}$
(e) F if $f_2 \circ f_5 \circ F = f_5$

Chapter Summary

A **function** f is a rule which assigns to each element x in one set (called the **domain**) a value $f(x)$ from another set. The set of all these values is called the **range** of the function. Numerical functions are usually specified by formulas (for example, $g(x) = (x^2 + 1)/(x + 1)$). The **natural domain** for such a function is the largest set of real numbers for which the formula makes sense and gives real values (thus, the natural domain for g consists of all real numbers except $x = -1$). Related to the notion of function is that of **variation.**

The **graph** of a function f is simply the graph of the equation $y = f(x)$. Of special interest are the graphs of **polynomial functions** and **rational functions.** In graphing them, we should show the hills, the valleys, the x-**intercepts,** and, in the case of rational functions, the vertical and horizontal **asymptotes.**

Functions can be combined in many ways. Of these, composition is perhaps the most significant. The **composite** of f with g is defined by $(f \circ g)(x) = f(g(x))$.

Some functions are **one-to-one;** some are not. Those that are one-to-one have undoing functions called inverses. The **inverse** of f, denoted by f^{-1}, satisfies $f^{-1}(f(x)) = x$. Finding a formula for $f^{-1}(x)$ can be tricky; therefore, we described a definite procedure for doing it.

Chapter Review Problem Set

1. Let $f(x) = x^2 - 1$ and $g(x) = 2/x$. Calculate if possible.
   (a) $f(4)$ (b) $g(\frac{1}{2})$ (c) $g(0)$
   (d) $f(1)/g(3)$ (e) $f(g(4))$ (f) $g(f(4))$

2. Find the natural domain of f if $f(x) = \sqrt{x + 1}/(x - 1)$.

3. If y varies directly as the cube of x and $y = 1$ when $x = 2$, find an explicit formula for y in terms of x.

4. If z varies directly as x and inversely as the square of y, and if $z = 1$ when x and y are both 3, find z when $x = 16$ and $y = 2$.

5. Graph each function.
   (a) $f(x) = (x - 2)^2$ (b) $f(x) = x^3 + 2x$

   (c) $f(x) = \dfrac{1}{x^2 - x - 2}$ (d) $f(x) = \begin{cases} 0 & \text{if } x \leq 0 \\ x^2 & \text{if } 0 < x < 1 \\ 1 & \text{if } x \geq 1 \end{cases}$

6. Suppose that g is an even function satisfying $g(x) = \sqrt{x}$ for $x \geq 0$. Sketch its graph on $-4 \leq x \leq 4$.

7. Sketch the graph of $h(x) = x^2 + 1/x$ by first graphing $y = x^2$ and $y = 1/x$ and then adding ordinates.

8. If $f(x) = x^3 + 1$ and $g(x) = x + 2$, give formulas for each of the following.
   (a) $f(g(x))$ (b) $g(f(x))$ (c) $f(f(x))$
   (d) $g(g(x))$ (e) $f^{-1}(x)$ (f) $g^{-1}(x)$
   (g) $g(x + h)$ (h) $[g(x + h) - g(x)]/h$ (i) $f(3x)$

9. How does the graph of $y = f(x - 2) + 3$ relate to the graph of $y = f(x)$?

10. Which of the following functions are even? Odd? One-to-one?

    (a) $f(x) = 1/(x^2 - 1)$ (b) $f(x) = 1/x$
    (c) $f(x) = |x|$ (d) $f(x) = 3x + 4$

11. Let $f(x) = x/(x - 2)$. Find a formula for $f^{-1}(x)$. Graph $y = f(x)$ and $y = f^{-1}(x)$ using the same coordinate axes.

12. How could we restrict the domain of $f(x) = (x + 2)^2$ so that f has an inverse?

The method of logarithms, by reducing to a few days the labor of many months, doubles as it were, the life of the astronomer, besides freeing him from the errors and disgust inseparable from long calculation.

P. S. Laplace

CHAPTER 6

Exponential and Logarithmic Functions

The discovery of the quadratic formula led to a long search for a corresponding formula for the cubic equation. Success had to wait until the sixteenth century, when the Italian school of mathematicians at Bologna found formulas for both the cubic and quartic equations. A typical result is Cardano's solution to $x^3 + ax = b$ which takes the form

$$x = \sqrt[3]{\sqrt{\frac{a^3}{27} + \frac{b^2}{4}} + \frac{b}{2}} - \sqrt[3]{\sqrt{\frac{a^3}{27} + \frac{b^2}{4}} - \frac{b}{2}}$$

Hieronimo Cardano
1501–1576

6-1   Radicals

Historically, interest in radicals has been associated with the desire to solve equations. Even the general cubic equation leads to very complicated radical expressions. Today, powerful iterative methods make results like Cardano's solution historical curiosities. Yet the need for radicals continues; it is important that we know something about them.

Raising a number to the 3rd power (or cubing it) is a process which can be undone. The inverse process—taking the 3rd root—is denoted by $\sqrt[3]{}$. We call $\sqrt[3]{a}$ a *radical* and read it "the cube root of a." Thus $\sqrt[3]{8} = 2$ and $\sqrt[3]{-125} = -5$ since $2^3 = 8$ and $(-5)^3 = -125$.

Our first goal is to give meaning to the symbol $\sqrt[n]{a}$ when n is any positive integer. Naturally, we require that $\sqrt[n]{a}$ be a number which yields a when raised to the nth power; that is

$$(\sqrt[n]{a})^n = a$$

When n is odd, that is all we need to say, since for any real number a, there is exactly one real number whose nth power is a.

When n is even, we face two serious problems, problems that are already apparent when $n = 2$. We have already discussed square roots, using the symbol $\sqrt{}$ rather than $\sqrt[2]{}$ (see Section 3-4). Recall that if $a < 0$, then $\sqrt{a}$ is not a real number (for example, $\sqrt{-4} = 2i$). Even if $a > 0$, we are in trouble since there are always two real numbers with squares equal to a. For example, both -3 and 3 have squares equal to 9. We agree that in this ambiguous case, $\sqrt{a}$ shall always denote the positive square root of a. Thus $\sqrt{9}$ is equal to 3, not -3.

We make a similar agreement about $\sqrt[n]{a}$ for n an even number greater than 2. First, we shall avoid the case $a < 0$. Second, when $a \geq 0$, $\sqrt[n]{a}$ will always denote the nonnegative number whose nth power is a. Thus $\sqrt[4]{81} = 3$, $\sqrt[4]{16} = 2$, and $\sqrt[4]{0} = 0$; however, the symbol $\sqrt[4]{-16}$ will be assigned no meaning in this book. Let us summarize.

> If n is odd, $\sqrt[n]{a}$ is the unique real number satisfying $(\sqrt[n]{a})^n = a$.
> If n is even and $a \geq 0$, $\sqrt[n]{a}$ is the unique nonnegative real number satisfying $(\sqrt[n]{a})^n = a$.

The symbol $\sqrt[n]{a}$, as we have defined it, is called the **principal nth root of a**; for brevity, we often drop the adjective *principal*.

RULES FOR RADICALS

Radicals, like exponents, obey certain rules. The most important ones are listed below, where it is assumed that all radicals name real numbers.

> **RULES FOR RADICALS**
>
> 1. $(\sqrt[n]{a})^n = a$
> 2. $\sqrt[n]{a^n} = a \qquad (a \geq 0)$
> 3. $\sqrt[n]{ab} = \sqrt[n]{a}\,\sqrt[n]{b}$
> 4. $\sqrt[n]{\dfrac{a}{b}} = \dfrac{\sqrt[n]{a}}{\sqrt[n]{b}}$

Rule 2 holds also for $a < 0$ if n is odd; for example, $\sqrt[5]{(-2)^5} = -2$.

These rules can all be proved, but we believe that the following illustrations will be more helpful to you than proofs.

$$(\sqrt[4]{7})^4 = 7$$
$$\sqrt[14]{3^{14}} = 3$$
$$\sqrt{2} \cdot \sqrt{18} = \sqrt{36} = 6$$
$$\frac{\sqrt[3]{750}}{\sqrt[3]{6}} = \sqrt[3]{\frac{750}{6}} = \sqrt[3]{125} = 5$$

SIMPLIFYING RADICALS

One use of the four rules given above is to simplify radicals. Here are two examples.

$$\sqrt[3]{54x^4y^6} \qquad \sqrt[4]{x^8 + x^4y^4}$$

We assume that x and y represent positive numbers.

In the first example, we start by factoring out the largest possible third power.

$$\sqrt[3]{54x^4y^6} = \sqrt[3]{(27x^3y^6)(2x)}$$
$$= \sqrt[3]{(3xy^2)^3(2x)}$$
$$= \sqrt[3]{(3xy^2)^3}\,\sqrt[3]{2x} \qquad \text{(Rule 3)}$$
$$= 3xy^2\,\sqrt[3]{2x} \qquad \text{(Rule 2)}$$

In the second example, it is tempting to write $\sqrt[4]{x^8 + x^4y^4} = x^2 + xy$, thereby pretending that $\sqrt[4]{a^4 + b^4} = a + b$. This is wrong, because $(a + b)^4 \neq a^4 + b^4$. Here is what we can do.

$$\sqrt[4]{x^8 + x^4y^4} = \sqrt[4]{x^4(x^4 + y^4)}$$
$$= \sqrt[4]{x^4}\,\sqrt[4]{x^4 + y^4} \qquad \text{(Rule 3)}$$
$$= x\sqrt[4]{x^4 + y^4} \qquad \text{(Rule 2)}$$

We were able to take x^4 out of the radical because it is a 4th power and a factor of $x^8 + x^4y^4$.

RATIONALIZING DENOMINATORS

For some purposes (including hand calculations), fractions with radicals in their denominators are considered to be needlessly complicated. Fortunately, we can usually rewrite a fraction so that its denominator is free of radicals. The process we go through is called **rationalizing the denominator.** Here are two examples.

$$\frac{1}{\sqrt[5]{x}} \qquad \frac{x}{\sqrt{x} + \sqrt{y}}$$

In the first case, we multiply numerator and denominator by $\sqrt[5]{x^4}$, which gives the 5th root of a 5th power in the denominator.

$$\frac{1}{\sqrt[5]{x}} = \frac{1 \cdot \sqrt[5]{x^4}}{\sqrt[5]{x} \cdot \sqrt[5]{x^4}} = \frac{\sqrt[5]{x^4}}{\sqrt[5]{x \cdot x^4}} = \frac{\sqrt[5]{x^4}}{\sqrt[5]{x^5}} = \frac{\sqrt[5]{x^4}}{x}$$

In the second case, we make use of the identity $(a + b)(a - b) = a^2 - b^2$. If we multiply numerator and denominator of the fraction by $\sqrt{x} - \sqrt{y}$, the radicals in the denominator disappear

$$\frac{x}{\sqrt{x} + \sqrt{y}} = \frac{x(\sqrt{x} - \sqrt{y})}{(\sqrt{x} + \sqrt{y})(\sqrt{x} - \sqrt{y})} = \frac{x\sqrt{x} - x\sqrt{y}}{x - y}$$

We should point out that this manipulation is valid provided $x \neq y$.

Problem Set 6-1

Simplify the following radical expressions. This will involve removing perfect powers from radicals and rationalizing denominators. Assume that all letters represent positive numbers.

1. $\sqrt{9}$
2. $\sqrt[3]{-8}$
3. $\sqrt[5]{32}$
4. $\sqrt[4]{16}$
5. $(\sqrt[3]{7})^3$
6. $(\sqrt{\pi})^2$
7. $\sqrt[3]{(\frac{3}{2})^3}$
8. $\sqrt[3]{(-2/7)^5}$
9. $(\sqrt{5})^4$
10. $(\sqrt[3]{5})^6$
11. $\sqrt{3}\sqrt{27}$
12. $\sqrt{2}\sqrt{32}$
13. $\sqrt[3]{16}/\sqrt[3]{2}$
14. $\sqrt[4]{48}/\sqrt[4]{3}$
15. $\sqrt[3]{10^{-6}}$
16. $\sqrt[4]{10^8}$
17. $1/\sqrt{2}$
18. $1/\sqrt{3}$
19. $\sqrt{10}/\sqrt{2}$
20. $\sqrt{6}/\sqrt{3}$
21. $\sqrt[3]{54x^4y^5}$
22. $\sqrt[3]{-16x^3y^8}$
23. $\sqrt[4]{(x+2)^4y^7}$
24. $\sqrt[4]{x^5(y-1)^8}$
25. $\sqrt{x^2+x^2y^2}$
26. $\sqrt{25+50y^4}$
27. $\sqrt[3]{x^6-9x^3y}$
28. $\sqrt[4]{16x^{12}+64x^8}$
29. $\sqrt[3]{x^4y^{-6}z^6}$
30. $\sqrt[4]{32x^{-4}y^9}$
31. $\dfrac{2}{\sqrt{x}+3}$
32. $\dfrac{4}{\sqrt{x}-2}$
33. $\dfrac{2}{\sqrt{x+3}}$
34. $\dfrac{4}{\sqrt{x-2}}$
35. $\dfrac{1}{\sqrt[4]{8x^3}}$
36. $\dfrac{1}{\sqrt[3]{5x^2y^4}}$
37. $\sqrt[3]{2x^{-2}y^4}\sqrt[3]{4xy^{-1}}$
38. $\sqrt[4]{125x^5y^3}\sqrt[4]{5x^{-9}y^5}$
39. $\sqrt{50}-2\sqrt{18}+\sqrt{8}$
40. $\sqrt[3]{24}+\sqrt[3]{375}$

CAUTION

$$\sqrt{a^4+a^4b^2} \neq a^2+a^2b$$

$$\sqrt{a^4+a^4b^2} = \sqrt{a^4(1+b^2)}$$
$$= a^2\sqrt{1+b^2}$$

EXAMPLE A (Equations Involving Radicals) Solve the following equations.

(a) $\sqrt[3]{x-2}=3$ (b) $x=\sqrt{2-x}$

Solution.

(a) Raise both sides to the 3rd power and solve for x.

$$(\sqrt[3]{x-2})^3 = 3^3$$
$$x-2 = 27$$
$$x = 29$$

(b) Square both sides and solve for x.

$$x^2 = 2-x$$
$$x^2+x-2 = 0$$
$$(x-1)(x+2) = 0$$
$$x=1 \qquad x=-2$$

Let us check our answers in part (b) by substituting them in the original equation. When we substitute these numbers for x in $x = \sqrt{2-x}$, we find that 1 works but -2 does not.

$$1 = \sqrt{2 - 1} \qquad -2 \neq \sqrt{2 - (-2)}$$

In squaring both sides of $x = \sqrt{2 - x}$, we introduced an extraneous solution. That happened because $a = b$ and $a^2 = b^2$ are not equivalent statements. Whenever you square both sides of an equation (or raise both sides of an equation to any even power), be sure to check your answers.

Solve each of the following equations.

41. $\sqrt{x - 1} = 5$

42. $\sqrt{x + 2} = 3$

43. $\sqrt[3]{2x - 1} = 2$

44. $\sqrt[3]{1 - 5x} = 6$

45. $\sqrt{\dfrac{x}{x + 2}} = 4$

46. $\sqrt[3]{\dfrac{x - 2}{x + 1}} = -2$

47. $\sqrt{x^2 + 4} = x + 2$

48. $\sqrt{x^2 + 9} = x - 3$

49. $\sqrt{2x + 1} = x - 1$

50. $\sqrt{x} = 12 - x$

EXAMPLE B (Combining Fractions Involving Radicals) Sums and differences of fractions involving radicals occur often in calculus. It is usually desirable to combine these fractions. Do so in

(a) $\dfrac{1}{\sqrt[3]{x + h}} - \dfrac{1}{\sqrt[3]{x}};$ (b) $\dfrac{x}{\sqrt{x^2 + 4}} - \dfrac{\sqrt{x^2 + 4}}{x}.$

Solution.

(a) $\dfrac{1}{\sqrt[3]{x + h}} - \dfrac{1}{\sqrt[3]{x}} = \dfrac{\sqrt[3]{x}}{\sqrt[3]{x}\sqrt[3]{x + h}} - \dfrac{\sqrt[3]{x + h}}{\sqrt[3]{x}\sqrt[3]{x + h}} = \dfrac{\sqrt[3]{x} - \sqrt[3]{x + h}}{\sqrt[3]{x}\sqrt[3]{x + h}}$

(b) $\dfrac{x}{\sqrt{x^2 + 4}} - \dfrac{\sqrt{x^2 + 4}}{x} = \dfrac{x^2}{x\sqrt{x^2 + 4}} - \dfrac{\sqrt{x^2 + 4}\sqrt{x^2 + 4}}{x\sqrt{x^2 + 4}}$

$\qquad\qquad = \dfrac{x^2 - (x^2 + 4)}{x\sqrt{x^2 + 4}} = \dfrac{-4}{x\sqrt{x^2 + 4}}$

Combine the fractions in each of the following. Do not bother to rationalize denominators.

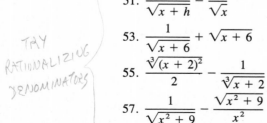

TRY
RATIONALIZING
DENOMINATORS

51. $\dfrac{2}{\sqrt{x + h}} - \dfrac{2}{\sqrt{x}}$

52. $\dfrac{\sqrt{x}}{\sqrt{x + 2}} - \dfrac{1}{\sqrt{x}}$

53. $\dfrac{1}{\sqrt{x + 6}} + \sqrt{x + 6}$

54. $\dfrac{\sqrt{x + 1}}{\sqrt{x + 3}} - \dfrac{\sqrt{x + 3}}{x + 1}$

55. $\dfrac{\sqrt[3]{(x + 2)^2}}{2} - \dfrac{1}{\sqrt[3]{x + 2}}$

56. $\dfrac{\sqrt{x + 7}}{\sqrt{x - 2}} - \dfrac{\sqrt{x - 2}}{x + 7}$

57. $\dfrac{1}{\sqrt{x^2 + 9}} - \dfrac{\sqrt{x^2 + 9}}{x^2}$

58. $\dfrac{x}{\sqrt{x^2 + 3}} + \dfrac{\sqrt{x^2 + 3}}{x}$

MISCELLANEOUS PROBLEMS

59. Simplify each expression (including rationalizing denominators). Assume all letters represent positive numbers.

(a) $\sqrt[4]{16a^4b^8}$ (b) $\sqrt{27}\sqrt{3b^3}$

(c) $\sqrt{12} + \sqrt{48} - \sqrt{27}$ (d) $\sqrt{250a^4b^6}$

(e) $\sqrt[3]{\dfrac{-32x^2y^7}{4x^5y}}$ (f) $\left(\sqrt[3]{\dfrac{y}{2x}}\right)^6$

(g) $\sqrt{8a^5} + \sqrt{18a^3}$ (h) $\sqrt[6]{512} - \sqrt{50} + \sqrt[6]{128}$

(i) $\sqrt[4]{a^4 + a^4b^4}$ (j) $\dfrac{1}{\sqrt[3]{7bc^3}}$

(k) $\dfrac{2}{\sqrt{a} - b}$ (l) $\sqrt{a}\left(\sqrt{a} + \dfrac{1}{\sqrt{a^3}}\right)$

60. If a is *any* real number and n is even, then $\sqrt[n]{a^n} = |a|$. Use this to simplify each of the following.

(a) $\sqrt{a^4 + 4a^2}$ (b) $\sqrt[4]{a^4 + a^4b^4}$ (c) $\sqrt{(a - b)^2c^4}$

61. Solve each equation for x.

(a) $\sqrt[3]{1 - 5x} = -4$ (b) $\sqrt{4x + 1} = x + 1$

(c) $\sqrt{x + 3} = 2 + \sqrt{x - 5}$ (d) $\sqrt{12 + x} = 4 + \sqrt{4 + x}$

(e) $x - \sqrt{x} - 6 = 0$ (f) $\sqrt[3]{x^2} - 2\sqrt[3]{x} - 8 = 0$

□ 62. Most scientific calculators have a key for roots (on some you must use the two keys $\boxed{\text{INV}}$ $\boxed{y^x}$). Calculate each of the following.

(a) $\sqrt[5]{31}$ (b) $\sqrt[3]{240}$

(c) $\sqrt[10]{78}$ (d) $\sqrt{282} - \sqrt{280}$

(e) $\sqrt[4]{.012}(\sqrt{30} - \sqrt{29})^2$ (f) $\dfrac{\sqrt[4]{29} + \sqrt[3]{6}}{\sqrt{14}}$

63. We know that $f(x) = x^5$ and $g(x) = \sqrt[5]{x}$ are inverse functions. Sketch their graphs using the same coordinate axes.

64. Rewrite $1/(\sqrt{2} + \sqrt{3} - \sqrt{5})$ with a rational denominator.

65. Figure 1 shows a right triangle. Determine $\overline{AC}$ so that the routes ACB and ADB from A to B have the same length.

66. In calculus, it is sometimes advantageous to rationalize the numerator. Rewrite each of the following with a rational numerator.

(a) $\dfrac{\sqrt{x} - \sqrt{y}}{\sqrt{x} + \sqrt{y}}$ (b) $\dfrac{\sqrt{x + h} - \sqrt{x}}{h}$ (c) $\dfrac{\sqrt[3]{x} - \sqrt[3]{y}}{x - y}$

67. Show each of the following to be true.

(a) $\dfrac{\sqrt{6} + \sqrt{2}}{2} = \sqrt{2 + \sqrt{3}}$ (b) $\sqrt{2 + \sqrt{3}} + \sqrt{2 - \sqrt{3}} = \sqrt{6}$

(c) $\sqrt[3]{9\sqrt{3} - 11\sqrt{2}} = \sqrt{3} - \sqrt{2}$

68. TEASER   Find the exact value of x if $x = \sqrt[3]{9 + 4\sqrt{5}} + \sqrt[3]{9 - 4\sqrt{5}}$. *Hint*: You might guess at the answer by using your calculator but you will not be done until you have given an algebraic demonstration that your guess is correct.

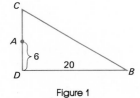

Figure 1

$$2^2 = 2 \cdot 2$$
$$2^3 = 2 \cdot 2 \cdot 2$$
$$2^4 = 2 \cdot 2 \cdot 2 \cdot 2$$
$$2^{4.6} = 2 \cdot 2 \cdot 2 \cdot 2 \cdot 2$$

One of the authors once asked a student to write the definition of $2^{4.6}$ on the blackboard. After thinking deeply for a minute, he wrote:

6-2　Exponents and Exponential Functions

After you have criticized the student mentioned above, ask yourself how you would define $2^{4.6}$. Of course, integral powers of 2 make perfectly good sense, although 2^{-3} and 2^0 became meaningful only after we had *defined* a^{-n} to be $1/a^n$ and a^0 to be 1 (see Section 2-1). Those were good definitions because they were consistent with the familiar rules of exponents. Now we ask what meaning we can give to powers like $2^{1/2}$, $2^{4.6}$, and even 2^π so that these familiar rules still hold.

RATIONAL EXPONENTS

We assume throughout this section that $a > 0$. If n is any positive integer, we want

$$(a^{1/n})^n = a^{(1/n) \cdot n} = a^1 = a$$

But we know that $(\sqrt[n]{a})^n = a$. Thus we define

$$\boxed{a^{1/n} = \sqrt[n]{a}}$$

For example, $2^{1/2} = \sqrt{2}$, $27^{1/3} = \sqrt[3]{27} = 3$, and $(16)^{1/4} = \sqrt[4]{16} = 2$.

Next, if m and n are positive integers, we want

$$(a^{1/n})^m = a^{m/n} \quad \text{and} \quad (a^m)^{1/n} = a^{m/n}$$

This forces us to define

$$\boxed{a^{m/n} = (\sqrt[n]{a})^m = \sqrt[n]{a^m}}$$

Accordingly,

$$2^{3/2} = (\sqrt{2})^3 = \sqrt{2}\,\sqrt{2}\,\sqrt{2} = 2\sqrt{2}$$

Rules for Exponents

1. $a^m a^n = a^{m+n}$

2. $\dfrac{a^m}{a^n} = a^{m-n}$

3. $(a^m)^n = a^{mn}$

---

and

$$27^{2/3} = (\sqrt[3]{27})^2 = 3^2 = 9$$

Lastly, we define

$$a^{-m/n} = \frac{1}{a^{m/n}}$$

so that

$$2^{-1/2} = \frac{1}{2^{1/2}} = \frac{1}{\sqrt{2}}$$

and

$$4^{-3/2} = \frac{1}{4^{3/2}} = \frac{1}{(\sqrt{4})^3} = \frac{1}{8}$$

We have just succeeded in defining a^x for all rational numbers x (recall that a rational number is a ratio of two integers). What is more important is that we have done it in such a way that the rules of exponents still hold. Incidentally, we can now answer the question in our opening display.

$$2^{4.6} = 2^4 2^{.6} = 2^4 2^{6/10} = 16(\sqrt[10]{2})^6$$

For simplicity, we have assumed that a is positive in our discussion of $a^{m/n}$. But we should point out that the definition of $a^{m/n}$ given above is also appropriate for the case in which a is negative and n is odd. For example,

$$(-27)^{2/3} = (\sqrt[3]{-27})^2 = (-3)^2 = 9$$

REAL EXPONENTS

Irrational powers such as 2^π and $3^{\sqrt{2}}$ are intrinsically more difficult to define than are rational powers. Rather than attempt a technical definition, we ask you to consider what 2^π might mean. The decimal expansion of π is $3.14159\ldots$. Thus we could look at the sequence of rational powers

$$2^3, \ 2^{3.1}, \ 2^{3.14}, \ 2^{3.141}, \ 2^{3.1415}, \ 2^{3.14159}, \ \ldots$$

As you should suspect, when the exponents get closer and closer to π, the corresponding powers of 2 get closer and closer to a definite number. We shall call the number 2^π.

The process of starting with integral exponents and then extending to rational exponents and finally to real exponents can be clarified by means of three graphs (Figure 2 on the next page). Note the table of values in the margin.

The first graph suggests a curve rising from left to right. The second graph makes the suggestion stronger. The third graph leaves nothing to the imagination; it is a continuous curve and it shows 2^x for all values of x, rational and irrational. As x increases in the positive direction, the values of 2^x increase

x	2^x
-3	$\frac{1}{8}$
-2	$\frac{1}{4}$
-1	$\frac{1}{2}$
0	1
$\frac{1}{2}$	$\sqrt{2} \approx 1.4$
1	2
$\frac{3}{2}$	$2\sqrt{2} \approx 2.8$
2	4
3	8

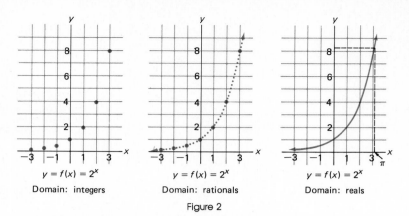

$y = f(x) = 2^x$
Domain: integers

$y = f(x) = 2^x$
Domain: rationals

$y = f(x) = 2^x$
Domain: reals

Figure 2

without bound; in the negative direction, the values of 2^x approach 0. Notice that 2^π is a little less than 9; its value correct to seven decimal places is

$$2^\pi = 8.8249778$$

See if your calculator gives this value.

EXPONENTIAL FUNCTIONS

The function $f(x) = 2^x$, graphed above, is one example of an exponential function. But what has been done with 2 can be done with any positive real number a. In general, the formula

$$f(x) = a^x$$

determines a function called an **exponential function with base a.** Its domain is the set of all real numbers and its range is the set of positive numbers.

Let us see what effect the size of a has on the graph of $f(x) = a^x$. We choose $a = 2$, $a = 3$, $a = 5$, and $a = \frac{1}{3}$, showing all four graphs in Figure 3.

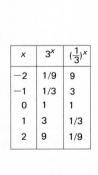

x	3^x	$(\frac{1}{3})^x$
-2	$1/9$	9
-1	$1/3$	3
0	1	1
1	3	$1/3$
2	9	$1/9$

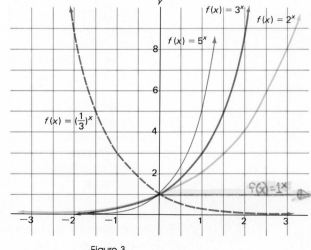

Figure 3

The graph of $f(x) = 3^x$ looks much like the graph of $f(x) = 2^x$, although it rises more rapidly. The graph of $f(x) = 5^x$ is even steeper. All three of these functions are *increasing functions,* meaning that the values of $f(x)$ increase as x increases; more formally, $x_2 > x_1$ implies $f(x_2) > f(x_1)$. The function $f(x) = (\frac{1}{3})^x$, on the other hand, is a *decreasing function.* In fact, you can get the graph of $f(x) = (\frac{1}{3})^x$ by reflecting the graph of $f(x) = 3^x$ about the y-axis. This is because $(\frac{1}{3})^x = 3^{-x}$.

We can summarize what is suggested by our discussion as follows.

If $a > 1$, $f(x) = a^x$ is an increasing function.
If $0 < a < 1$, $f(x) = a^x$ is a decreasing function.

In both of these cases, the graph of f has the x-axis as an asymptote. The case $a = 1$ is not very interesting since it yields the constant function $f(x) = 1$.

PROPERTIES OF EXPONENTIAL FUNCTIONS

It is easy to describe the main properties of exponential functions, since they obey the rules we learned in Section 2-1. Perhaps it is worth repeating them, since we do want to emphasize that they now hold for all *real* exponents x and y (at least for the case where a and b are both positive).

1. $a^x a^y = a^{x+y}$

2. $\dfrac{a^x}{a^y} = a^{x-y}$

3. $(a^x)^y = a^{xy}$

4. $(ab)^x = a^x b^x$

5. $\left(\dfrac{a}{b}\right)^x = \dfrac{a^x}{b^x}$

Here are a number of examples that are worth studying.

$$3^{1/2} 3^{3/4} = 3^{1/2+3/4} = 3^{5/4}$$

$$\frac{\pi^4}{\pi^{5/2}} = \pi^{4-5/2} = \pi^{3/2}$$

$$(2^{\sqrt{3}})^4 = 2^{4\sqrt{3}}$$

$$(5^{\sqrt{2}} 5^{1-\sqrt{2}})^{-3} = (5^1)^{-3} = 5^{-3}$$

Problem Set 6-2

Write each of the following as a power of 7.

1. $\sqrt[3]{7}$ 2. $\sqrt[5]{7}$ 3. $\sqrt[3]{7^2}$ 4. $\sqrt[5]{7^3}$ 5. $\dfrac{1}{\sqrt[3]{7}}$

6. $\dfrac{1}{\sqrt[5]{7}}$ 7. $\dfrac{1}{\sqrt[3]{7^2}}$ 8. $\dfrac{1}{\sqrt[5]{7^3}}$ 9. $7\sqrt[3]{7}$ 10. $7\sqrt[5]{7}$

Rewrite each of the following using exponents instead of radicals. For example, $\sqrt[5]{x^3} = x^{3/5}$.

11. $\sqrt[3]{x^2}$ 12. $\sqrt[4]{x^3}$ 13. $x^2\sqrt{x}$ 14. $x\sqrt[3]{x}$

15. $\sqrt{(x+y)^3}$ 16. $\sqrt[3]{(x+y)^2}$ 17. $\sqrt{x^2+y^2}$ 18. $\sqrt[3]{x^3+8}$

Rewrite each of the following using radicals instead of fractional exponents. For example, $(xy^2)^{3/7} = \sqrt[7]{x^3y^6}$

19. $4^{2/3}$ 20. $10^{3/4}$ 21. $8^{-3/2}$

22. $12^{-5/6}$ 23. $(x^4+y^4)^{1/4}$ 24. $(x^2+xy)^{1/2}$

25. $(x^2y^3)^{2/5}$ 26. $(3ab^2)^{2/3}$ 27. $(x^{1/2}+y^{1/2})^{1/2}$

28. $(x^{1/3}+y^{2/3})^{1/3}$

Simplify each of the following. Give your answer without any exponents.

29. $25^{1/2}$ 30. $27^{1/3}$

31. $8^{2/3}$ 32. $16^{3/2}$

33. $9^{-3/2}$ 34. $64^{-2/3}$

35. $(-.008)^{2/3}$ 36. $(-.027)^{5/3}$

37. $(.0025)^{3/2}$ 38. $(1.44)^{3/2}$

39. $5^{2/3}5^{-5/3}$ 40. $4^{3/4}4^{-1/4}$

41. $16^{7/6}16^{-5/6}16^{-4/3}$ 42. $9^2 9^{2/3} 9^{-7/6}$

43. $(8^2)^{-2/3}$ 44. $(4^{-3})^{3/2}$

EXAMPLE A (Simplifying Expressions Involving Exponents) Simplify and write the answer without negative exponents.

(a) $\dfrac{x^{1/3}(8x)^{-2/3}}{x^{-3/4}}$ (b) $\left(\dfrac{2x^{-1/2}}{y}\right)^4\left(\dfrac{x}{y}\right)^{-1}(3x^{10/3})$

Solution.

4/12 − 8/12 + 9/12

(a) $\dfrac{x^{1/3}(8x)^{-2/3}}{x^{-3/4}} = x^{1/3}8^{-2/3}x^{-2/3}x^{3/4} = \dfrac{x^{1/3-2/3+3/4}}{8^{2/3}} = \dfrac{x^{5/12}}{4}$

(b) $\left(\dfrac{2x^{-1/2}}{y}\right)^4\left(\dfrac{x}{y}\right)^{-1}(3x^{10/3}) = \left(\dfrac{16x^{-2}}{y^4}\right)\left(\dfrac{y}{x}\right)(3x^{10/3})$

$= \dfrac{48x^{-2-1+10/3}}{y^{4-1}} = \dfrac{48x^{1/3}}{y^3}$

Simplify, writing your answer without negative exponents.

45. $(3a^{1/2})(-2a^{3/2})$

46. $(2x^{3/4})(5x^{-3/4})$

47. $(2^{1/2}x^{-2/3})^6$

48. $(\sqrt{3}x^{-1/4}y^{3/4})^4$

49. $(xy^{-2/3})^3(x^{1/2}y)^2$

50. $(a^2b^{-1/4})^2(a^{-1/3}b^{1/2})^3$

51. $\dfrac{(2x^{-1}y^{2/3})^2}{x^2y^{-2/3}}$

52. $\left(\dfrac{a^{1/2}b^{1/3}}{c^{5/6}}\right)^{12}$

53. $\left(\dfrac{x^{-2}y^{3/4}}{x^{1/2}}\right)^{12}$

54. $\dfrac{x^{1/3}y^{-3/4}}{x^{-2/3}y^{1/2}}$

55. $y^{2/3}(2y^{4/3} - y^{-5/3})$

56. $x^{-3/4}\left(-x^{7/4} + \dfrac{2}{\sqrt[4]{x}}\right)$

57. $(x^{1/2} + y^{1/2})^2$

58. $(a^{3/2} + \pi)^2$

EXAMPLE B (Combining Fractions)   Perform the following addition.

$$\frac{(x + 1)^{2/3}}{x} + \frac{1}{(x + 1)^{1/3}}$$

Solution.

$$\frac{(x + 1)^{2/3}}{x} + \frac{1}{(x + 1)^{1/3}} = \frac{(x + 1)^{2/3}(x + 1)^{1/3}}{x(x + 1)^{1/3}} + \frac{x}{x(x + 1)^{1/3}}$$

$$= \frac{x + 1 + x}{x(x + 1)^{1/3}} = \frac{2x + 1}{x(x + 1)^{1/3}}$$

Combine the fractions in each of the following.

59. $\dfrac{(x + 2)^{4/5}}{3} + \dfrac{2x}{(x + 2)^{1/5}}$

60. $\dfrac{(x - 3)^{1/3}}{4} - \dfrac{1}{(x - 3)^{2/3}}$

61. $(x^2 + 1)^{1/3} - \dfrac{2x^2}{(x^2 + 1)^{2/3}}$

62. $(x^2 + 2)^{1/4} + \dfrac{x^2}{(x^2 + 2)^{3/4}}$

EXAMPLE C (Mixing Radicals of Different Orders)   Express $\sqrt{2}\sqrt[3]{5}$ using just one radical.

Solution.   Square roots and cube roots mix about as well as oil and water, but exponents can serve as a blender. They allow us to write both $\sqrt{2}$ and $\sqrt[3]{5}$ as sixth roots.

$$\sqrt{2}\sqrt[3]{5} = 2^{1/2} \cdot 5^{1/3}$$

$$= 2^{3/6} \cdot 5^{2/6}$$

$$= (2^3 \cdot 5^2)^{1/6}$$

$$= \sqrt[6]{200}$$

Express each of the following in terms of at most one radical in simplest form.

63. $\sqrt{2}\sqrt[3]{2}$

64. $\sqrt[3]{2}\sqrt[4]{2}$

65. $\sqrt[4]{2}\sqrt[6]{x}$

66. $\sqrt[3]{5}\sqrt{x}$

67. $\sqrt[3]{x}\sqrt{x}$

68. $\sqrt{x\sqrt[3]{x}}$

Use a calculator to find an approximate value of each of the following.

69. $2^{1.34}$ 70. $2^{-.79}$ 71. $\pi^{1.34}$ 72. π^{π}

73. $(1.46)^{\sqrt{2}}$ 74. $\pi^{\sqrt{2}}$ 75. $(.9)^{50.2}$ 76. $(1.01)^{50.2}$

Sketch the graph of each of the following functions.

77. $f(x) = 4^x$ 78. $f(x) = 4^{-x}$ 79. $f(x) = (\tfrac{2}{3})^x$

80. $f(x) = (\tfrac{2}{3})^{-x}$ 81. $f(x) = \pi^x$ 82. $f(x) = (\sqrt{2})^x$

MISCELLANEOUS PROBLEMS

83. Rewrite using exponents in place of radicals and simplify.
    (a) $\sqrt[6]{b^3}$ (b) $\sqrt[8]{x^4}$ (c) $\sqrt[3]{a^2 + 2ab + b^2}$

84. Simplify.
    (a) $(32)^{-6/5}$ (b) $(-.008)^{2/3}$ (c) $(5^{-1/2}\, 5^{3/4}\, 5^{1/8})^{16}$

85. Simplify, writing your answer without either radicals or negative exponents.
    (a) $(27)^{2/3}(.0625)^{-3/4}$ (b) $\sqrt[3]{4}\sqrt{2} + \sqrt[6]{2}$
    (c) $\sqrt[3]{a^2}\sqrt[4]{a^3}$ (d) $\sqrt{a\sqrt[3]{a^2}}$
    (e) $[a^{3/2} + a^{-3/2}]^2$ (f) $[a^{1/4}(a^{-5/4} + a^{3/4})]^{-1}$
    (g) $\left(\dfrac{\sqrt[3]{a^3 b^2}}{\sqrt[4]{a^6 b^3}}\right)^{-1}$ (h) $\left(\dfrac{a^{-2} b^{2/3}}{b^{-1/2}}\right)^{-4}$
    (i) $\left[\dfrac{(27)^{4/3} - (27)^0}{(3^2 + 4^2)^{1/2}}\right]^{3/4}$ (j) $(16 a^2 b^3)^{3/4} - 4ab^2 (a^2 b)^{1/4}$
    (k) $(\sqrt{3})^{3\sqrt{3}} - (3\sqrt{3})^{\sqrt{3}} + (\sqrt{3}^{\sqrt{3}})^{\sqrt{3}}$
    (l) $(a^{1/3} - b^{1/3})(a^{2/3} + a^{1/3}b^{1/3} + b^{2/3})$

86. Combine and simplify, writing your answer without negative exponents.
    (a) $4x^2(x^2 + 2)^{-2/3} - 3(x^2 + 2)^{1/3}$
    (b) $x^3(x^3 - 1)^{-3/4} - (x^3 - 1)^{1/4}$

87. Solve for x.
    (a) $4^{x+1} = (1/2)^{2x}$ (b) $5^{x^2-x} = 25$
    (c) $2^{4x}4^{x-3} = (64)^{x-1}$ (d) $(x^2 + x + 4)^{3/4} = 8$
    (e) $x^{2/3} - 3x^{1/3} = -2$ (f) $2^{2x} - 2^{x+1} - 8 = 0$

88. Using the same axes, sketch the graph of each of the following.
    (a) $f(x) = 2^x$ (b) $g(x) = -2^x$ (c) $h(x) = 2^{-x}$
    (d) $k(x) = 2^x + 2^{-x}$ (e) $m(x) = 2^{x-4}$

89. Sketch the graph of $f(x) = 2^{-|x|}$.

90. Using the same axes, sketch the graphs of $f(x) = x^\pi$ and $g(x) = \pi^x$ on the interval $2 \le x \le 3.5$. One solution of $x^\pi = \pi^x$ is π. Use your graphs to help you find another one (approximately).

91. Give a simple argument to show that an exponential function $f(x) = a^x (a > 0$, $a \ne 1)$ is not equivalent to any polynomial function.

92. TEASER   If a and b are irrational, does it follow that a^b is irrational? *Hint:* Consider $\sqrt{2}^{\sqrt{2}}$ and $(\sqrt{2}^{\sqrt{2}})^{\sqrt{2}}$.

A Packing Problem

World population is growing at about 2 percent per year. If this continues indefinitely, how long will it be until we are all packed together like sardines? In answering, assume that "sardine packing" for humans is one person per square foot of land area.

6-3 Exponential Growth and Decay

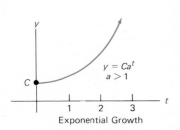

Exponential Growth

Figure 4

The phrase *exponential growth* is used repeatedly by professors, politicians, and pessimists. Population, energy use, mining of ores, pollution, and the number of books about these things are all said to be growing exponentially. Most people probably do not know what exponential growth means, except that they have heard it guarantees alarming consequences. For students of this book, it is easy to explain its meaning. For y to grow exponentially with time t means that it satisfies the relationship

$$y = Ca^t$$

for constants C and a, with $C > 0$ and $a > 1$ (see Figure 4). Why should so many ingredients of modern society behave this way? The basic cause is population growth.

POPULATION GROWTH

Simple organisms reproduce by cell division. If, for example, there is one cell today, that cell may split so that there are two cells tomorrow. Then each of those cells may divide giving four cells the following day (Figure 5). As this process continues, the numbers of cells on successive days form the sequence

$$1, 2, 4, 8, 16, 32, \ldots$$

If we start with 100 cells and let $f(t)$ denote the number present t days from now, we have the results indicated in the table of Figure 6.

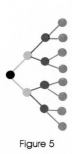

Figure 5

t	0	1	2	3	4	5
$f(t)$	100	200	400	800	1600	3200

Figure 6

It seems that

$$f(t) = (100)2^t$$

A perceptive reader will ask if this formula is really valid. Does it give the right answer when $t = 5.7$? Is not population growth a discrete process, occurring in unit amounts at distinct times, rather than a continuous process as the formula implies? The answer is that the exponential growth model provides a very good approximation to the growth of simple organisms, provided the initial population is large.

The mechanism of reproduction is different (and more interesting) for people, but the pattern of population growth is similar. World population is presently growing at about 2 percent per year. In 1975, there were about 4 billion people. Accordingly, the population in 1976 in billions was $4 + 4(.02) = 4(1.02)$, in 1977 it was $4(1.02)^2$, in 1978 it was $4(1.02)^3$, and so on. If this trend continues, there will be $4(1.02)^{20}$ billion people in the world in 1995, that is, 20 years after 1975. It appears that world population obeys the formula.

$$p(t) = 4(1.02)^t$$

where $p(t)$ represents the number of people (in billions) t years after 1975.

In general, if $A(t)$ is the amount at time t of a quantity growing exponentially at the rate of r (written as a decimal), then

$$A(t) = A(0)(1 + r)^t$$

Here $A(0)$ is the initial amount—that is, the amount present at $t = 0$.

DOUBLING TIMES

One way to get a feeling for the spectacular nature of exponential growth is via the concept of **doubling time;** this is the length of time required for an exponentially growing quantity to double in size. It is easy to show that if a quantity doubles in an initial time interval of length T, it will double in size in *any* time interval of length T. Consider the world population problem as an example. By the table of Figure 7, $(1.02)^{35} \approx 2$, so world population doubles in 35 years. Since it was 4 billion in 1975, it should be 8 billion in 2010, 16 billion in 2045, and so on. This alarming information is displayed graphically in Figure 8.

Now we can answer the question about sardine packing in our opening display. There are slightly more than 1,000,000 billion square feet of land area on the surface of the earth. Sardine packing for humans is about 1 square foot per person. Thus we are asking when $4(1.02)^t$ billion will equal 1,000,000 billion. This leads to the equation

$$(1.02)^t = 250,000$$

To solve this exponential equation, we use the following approximations

t	$(1.02)^t$
5	1.104
10	1.219
15	1.346
20	1.486
25	1.641
30	1.811
35	2.000
40	2.208
45	2.438
50	2.692
55	2.972
60	3.281
65	3.623
70	4.000
75	4.416
80	4.875
85	5.383
90	5.943

Figure 7

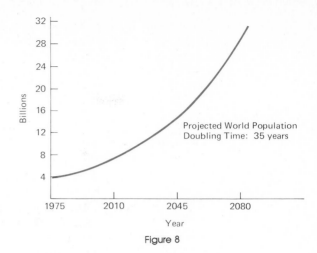

Figure 8

$$(1.02)^{35} \approx 2 \qquad 250,000 \approx 2^{18}$$

Our equation can then be rewritten as

$$[(1.02)^{35}]^{t/35} = 2^{18}$$

or

$$2^{t/35} = 2^{18}$$

We conclude that

$$\frac{t}{35} = 18$$

$$t = (18)(35) = 630$$

Thus, after about 630 years, we will be packed together like sardines. If it is any comfort, war, famine, or birth control will change population growth patterns before then.

COMPOUND INTEREST

One of the best practical illustrations of exponential growth is money earning compound interest. Suppose that Amy puts $1000 in a bank today at 8 percent interest compounded annually. Then at the end of one year the bank adds the interest of $(.08)(1000) = \$80$ to her $1000, giving her a total of $1080. But note that $1080 = 1000(1.08)$. During the second year, $1080 draws interest. At the end of that year, the bank adds $(.08)(1080)$ to the account, bringing the total to

$$1080 + (.08)(1080) = (1080)(1.08)$$
$$= 1000(1.08)(1.08)$$
$$= 1000(1.08)^2$$

Continuing in this way, we see that Amy's account will have grown to $1000(1.08)^3$ by the end of 3 years, $1000(1.08)^4$ by the end of 4 years, and so on. By the end of 15 years, it will have grown to

$$1000(1.08)^{15} \approx 1000(3.172169)$$

$$= \$3172.17$$

To calculate $(1.08)^{15}$, we used Table 2 at the end of the problem set. We could have used a calculator.

How long would it take for Amy's money to double—that is, when will

$$1000(1.08)^t = 2000$$

This will occur when $(1.08)^t = 2$. According to the table just mentioned, this happens at $t \approx 9$, or in about 9 years.

EXPONENTIAL DECAY

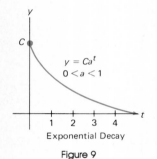

$y = Ca^t$
$0 < a < 1$

Exponential Decay

Figure 9

Fortunately, not all things grow; some decline or decay. In fact, some things— notably the radioactive elements—decay exponentially. This means that the amount y present at time t satisfies

$$y = Ca^t$$

for some constants C and a with $C > 0$ and $0 < a < 1$ (see Figure 9).

Here an important idea is that of **half-life,** the time required for half of a substance to disappear. For example, radium decays with a half-life of 1620 years. Thus if 1000 grams of radium are present now, 1620 years from now 500 grams will be present, $2(1620) = 3240$ years from now only 250 grams will be present, and so on.

The precise nature of radioactive decay is used to date old objects. If an object contains radium and lead (the product to which radium decays) in the ratio 1 to 3, then it is believed that an original amount of pure radium has decayed to $\frac{1}{4}$ its original size. The object must be two half-lives, or 3240 years, old. Two important assumptions have been made: (1) decay of radium is exactly exponential over long periods of time; and (2) no lead was originally present. Recent research raises some question about the correctness of such assumptions.

Problem Set 6-3

1. In each of the following, indicate whether y grows exponentially or decays exponentially with t.

   (a) $y = 128\left(\dfrac{1}{2}\right)^t$

   (b) $y = 5\left(\dfrac{5}{3}\right)^t$

   (c) $y = 4(10)^9(1.03)^t$

   (d) $y = 1000(.99)^t$

2. Find the values of y corresponding to $t = 0$, $t = 1$, and $t = 2$ for each case in Problem 1.

3. Use the table at the end of the problem set or a calculator with a $\boxed{y^x}$ key to find each value.
   (a) $(1.08)^{20}$ (b) $(1.12)^{25}$ (c) $1000(1.04)^{40}$ (d) $2000(1.02)^{80}$

4. Evaluate each of the following.
   (a) $(1.01)^{100}$ (b) $(1.02)^{40}$ (c) $100(1.12)^{50}$ (d) $500(1.04)^{30}$

5. Silver City's present population of 1000 is expected to grow exponentially over the next 10 years at 4 percent per year. How many people will it have at the end of that time? *Hint:* $A(t) = A(0)[1 + r]^t$.

6. The value of houses in Longview is said to be growing exponentially at 12 percent per year. What will a house valued at \$100,000 today be worth after 8 years?

7. Under the assumptions concerning world population used in this section, what will be the approximate number of people on earth in each year?
   (a) 1990 (that is, 15 years after 1975)
   (b) 2000
   (c) 2065

8. A certain radioactive substance has a half-life of 40 minutes. What fraction of an initial amount of this substance will remain after 1 hour, 20 minutes (that is, after 2 half-lives)? After 2 hours, 40 minutes?

EXAMPLE A (Compound Interest)   Roger put \$1000 in a money market fund at 15 percent interest compounded annually. How much was it worth after 4 years?

Solution.   We will have to use a calculator since $(1.15)^n$ is not in our table. The answer is

$$1000(1.15)^4 = \$1749.01$$

9. If you put \$100 in the bank for 8 years, how much will it be worth at the end of that time at
   (a) 8 percent compounded annually;
   (b) 12 percent compounded annually?

10. If you invest \$500 in the bank today, how much will it be worth after 25 years at
    (a) 8 percent compounded annually;
    (b) 4 percent compounded annually;
    (c) 12 percent compounded annually?

11. If you put \$3500 in the bank today, how much will it be worth after 40 years at
    (a) 8 percent compounded annually;
    (b) 12 percent compounded annually?

12. Approximately how long will it take for money to accumulate to twice its value if
    (a) it is invested at 8 percent compounded annually;
    (b) it is invested at 12 percent compounded annually?

13. Suppose that you invest P dollars at r percent compounded annually. Write an expression for the amount accumulated after n years.

EXAMPLE B (More on Compound Interest)   If $1000 is invested at 8 percent compounded quarterly, find the accumulated amount after 15 years.

Solution.   Interest calculated at 2 percent ($\frac{1}{4}$ of 8 percent) is converted to principal every 3 months. By the end of the first 3-month period, the account has grown to $1000(1.02) = \$1020$; by the end of the second 3-month period, it has grown to $1000(1.02)^2$; and so on. The accumulated amount after 15 years, or 60 conversion periods, is

$$1000(1.02)^{60} \approx 1000(3.28103)$$

$$= \$3281.03$$

Suppose more generally that P dollars is invested at a rate r (written as a decimal), which is compounded m times per year. Then the accumulated amount A after t years is given by

$$A = P\left(1 + \frac{r}{m}\right)^{tm}$$

In our example

$$A = 1000\left(1 + \frac{.08}{4}\right)^{15 \cdot 4} = 1000(1.02)^{60}$$

Find the accumulated amount for the indicated initial principal, compound interest rate, and total time period.

14. $2000; 8 percent compounded annually; 15 years.
15. $5000; 8 percent compounded semiannually; 5 years.
16. $5000; 12 percent compounded monthly; 5 years.
ⓒ 17. $3000; 9 percent compounded annually; 10 years.
ⓒ 18. $3000; 9 percent compounded semiannually; 10 years.
ⓒ 19. $3000; 9 percent compounded quarterly; 10 years.
ⓒ 20. $3000; 9 percent compounded monthly; 10 years.
ⓒ 21. $1000; 8 percent compounded monthly; 10 years.
ⓒ 22. $1000; 8 percent compounded daily; 10 years.   *Hint:* Assume there are 365 days in a year, so that the interest rate per day is .08/365.

MISCELLANEOUS PROBLEMS

23. Let $y = 5400(2/3)^t$. Evaluate y for $t = -1, 0, 1, 2,$ and 3.
24. For what value of t in Problem 23 is $y = 3200/3$?
25. If $(1.023)^T = 2$, find the value of $100(1.023)^{3T}$.
26. If $(.67)^H = \frac{1}{2}$, find the value of $32(.67)^{4H}$.
ⓒ 27. Suppose the population of a certain city follows the formula

$$p(t) = 4600(1.016)^t$$

where $p(t)$ is the population t years after 1980.

(a) What will the population be in 2020? In 2080?

(b) Experiment with your calculator to find the doubling time for this population.

28. The number of bacteria in a certain culture is known to triple every hour. Suppose the count at 12:00 noon is 162,000. What was the count at 11:00 A.M.? At 8:00 A.M.?

29. If $100 is invested today, how much will it be worth after 5 years at 8 percent interest if interest is:

(a) compounded annually;

(b) compounded quarterly;

ⓒ (c) compounded monthly;

ⓒ (d) compounded daily? (There are 365 days in a year.)

30. How long does it take money to double if invested at 12 percent compounded monthly? (Use the compound interest table.)

31. About how long does it take an exponentially growing population to double if its rate of growth is;

(a) 8 percent per year? (Use the compound interest table.)

ⓒ (b) 6.5 percent per year? (Experiment with your calculator.)

ⓒ 32. A manufacturer of radial tires found that the percentage P of tires still usable after being driven m miles was given by

$$P = 100(2.71)^{-.000025m}$$

What percentage of tires are still usable at 80,000 miles?

33. A certain radioactive element has a half-life of 1690 years. Starting with 30 milligrams there will be $q(t)$ milligrams left after t years, where $q(t) = 30(1/2)^{kt}$.

(a) Determine the constant k.

ⓒ (b) How much will be left after 2500 years?

34. One method of depreciation allowed by IRS is the double-declining-balance method. In this method, the original value C of an item is depreciated each year by $100(2/N)$ percent of its value at the beginning of that year, N being the useful life of the item.

(a) Write a formula for the value V of the item after n years.

ⓒ (b) If an item cost $10,000 and has a useful life of 15 years, calculate its value after 10 years. After 15 years.

(c) Does the value of an item ever become zero by this depreciation method?

ⓒ 35. (Carbon dating)   All living things contain carbon-12, which is a stable element, and carbon-14, which is radioactive. While a plant or animal is alive, the ratio of these two isotopes of carbon remains unchanged, since carbon-14 is constantly renewed; but after death, no more carbon-14 is absorbed. The half-life of carbon-14 is 5730 years. Bones from a human body were found to contain only 76 percent of the carbon-14 in living bones. How long before did the person die?

ⓒ 36. Manhattan Island is said to have been bought from the Indians by Peter Minuit in 1626 for $24. If, instead of making this purchase, Minuit had put the money in a savings account drawing interest at 6 percent compounded annually, what would that account be worth in the year 2000?

ⓒ 37. Hamline University was founded in 1854 with a gift of $25,000 from Bishop Hamline of the Methodist Church. Suppose that Hamline University had wisely put $10,000 of this gift in an endowment drawing 10 percent interest compounded annually, promising not to touch it until 1988 (exactly 134 years later). How much could it then withdraw each year and still maintain this endowment at the 1988 level?

38. **TEASER** Suppose one water lily growing exponentially at the rate of 8 percent per day is able to cover a certain pond in 50 days. How long would it take 10 of these lilies to cover the pond?

TABLE 2 Compound Interest Table

n	$(1.01)^n$	$(1.02)^n$	$(1.04)^n$	$(1.08)^n$	$(1.12)^n$
1	1.01000000	1.02000000	1.04000000	1.08000000	1.12000000
2	1.02010000	1.04040000	1.08160000	1.16640000	1.25440000
3	1.03030100	1.06120800	1.12486400	1.25971200	1.40492800
4	1.04060401	1.08243216	1.16985856	1.36048896	1.57351936
5	1.05101005	1.10408080	1.21665290	1.46932808	1.76234168
6	1.06152015	1.12616242	1.26531902	1.58687432	1.97382269
7	1.07213535	1.14868567	1.31593178	1.71382427	2.21068141
8	1.08285671	1.17165938	1.36856905	1.85093021	2.47596318
9	1.09368527	1.19509257	1.42331181	1.99900463	2.77307876
10	1.10462213	1.21899442	1.48024428	2.15892500	3.10584821
11	1.11566835	1.24337431	1.53945406	2.33163900	3.47854999
12	1.12682503	1.26824179	1.60103222	2.51817012	3.89597599
15	1.16096896	1.34586834	1.80094351	3.17216911	5.47356576
20	1.22019004	1.48594740	2.19112314	4.66095714	9.64629309
25	1.28243200	1.64060599	2.66583633	6.84847520	17.00006441
30	1.34784892	1.81136158	3.24339751	10.06265689	29.95992212
35	1.41660276	1.99988955	3.94608899	14.78534429	52.79961958
40	1.48886373	2.20803966	4.80102063	21.72452150	93.05097044
45	1.56481075	2.43785421	5.84117568	31.92044939	163.98760387
50	1.64463182	2.69158803	7.10668335	46.90161251	289.00218983
55	1.72852457	2.97173067	8.64636692	68.91385611	509.32060567
60	1.81669670	3.28103079	10.51962741	101.25706367	897.59693349
65	1.90936649	3.62252311	12.79873522	148.77984662	1581.87249060
70	2.00676337	3.99955822	15.57161835	218.60640590	2787.79982770
75	2.10912847	4.41583546	18.94525466	321.20452996	4913.05584077
80	2.21671522	4.87543916	23.04979907	471.95483426	8658.48310008
85	2.32978997	5.38287878	28.04360494	693.45648897	15259.20568055
90	2.44863267	5.94313313	34.11933334	1018.91508928	26891.93422336
95	2.57353755	6.56169920	41.51138594	1497.12054855	47392.77662369
100	2.70481383	7.24464612	50.50494818	2199.76125634	83522.26572652

An active participant in the political and religious battles of his day, the Scot. John Napier amused himself by studying mathematics and science. He was interested in reducing the work involved in the calculations of spherical trigonometry, especially as they applied to astronomy. In 1614 he published a book containing the idea that made him famous. He gave it the name *logarithm*.

John Napier
(1550–1617)

6-4   Logarithms and Logarithmic Functions

Napier's approach to logarithms is out of style, but the goal he had in mind is still worth considering. He hoped to replace multiplications by additions. He thought additions were easier to do, and he was right.

Consider the exponential function $f(x) = 2^x$ and recall that

$$2^x \cdot 2^y = 2^{x+y}$$

On the left, we have a multiplication and on the right, an addition. If we are to fulfill Napier's objective, we want logarithms to behave like exponents. That suggests a definition. The logarithm of N to the base 2 is the exponent to which 2 must be raised to yield N. That is,

$$\log_2 N = x \quad \text{if and only if} \quad 2^x = N$$

Thus

$$\log_2 4 = 2 \quad \text{since} \quad 2^2 = 4$$
$$\log_2 8 = 3 \quad \text{since} \quad 2^3 = 8$$
$$\log_2 \sqrt{2} = \tfrac{1}{2} \quad \text{since} \quad 2^{1/2} = \sqrt{2}$$

and, in general,

$$\log_2 (2^x) = x \quad \text{since} \quad 2^x = 2^x$$

Has Napier's goal been achieved? Does the logarithm turn a product into a sum? Yes, for note that

$$\log_2(2^x \cdot 2^y) = \log_2(2^{x+y}) \qquad \text{(property of exponents)}$$
$$= x + y \qquad \text{(definition of } \log_2)$$
$$= \log_2(2^x) + \log_2(2^y)$$

Thus

$$\log_2(2^x \cdot 2^y) = \log_2(2^x) + \log_2(2^y)$$

which has the form

$$\log_2(M \cdot N) = \log_2 M + \log_2 N$$

THE GENERAL DEFINITION

What has been done for 2 can be done for any base $a > 1$. The **logarithm of N to the base a** is the exponent x to which a must be raised to yield N. Thus

$$\boxed{\log_a N = x \quad \text{if and only if} \quad a^x = N}$$

Now we can calculate many kinds of logarithms.

$$\log_4 16 = 2 \quad \text{since} \quad 4^2 = 16$$
$$\log_{10} 1000 = 3 \quad \text{since} \quad 10^3 = 1000$$
$$\log_{10}(.001) = -3 \quad \text{since} \quad 10^{-3} = \frac{1}{1000} = .001$$

What is $\log_{10} 7$? We are not ready to answer that yet, except to say it is a number x satisfying $10^x = 7$ (see Section 6-6).

We point out that negative numbers and zero do not have logarithms. Suppose -4 and 0 did have logarithms, that is, suppose

$$\log_a(-4) = m \quad \text{and} \quad \log_a 0 = n$$

Then

$$a^m = -4 \quad \text{and} \quad a^n = 0$$

But that is impossible; we learned earlier that a^x is always positive.

PROPERTIES OF LOGARITHMS

There are three main properties of logarithms.

> **PROPERTIES OF LOGARITHMS**
>
> 1. $\log_a(M \cdot N) = \log_a M + \log_a N$
> 2. $\log_a(M/N) = \log_a M - \log_a N$
> 3. $\log_a(M^p) = p \log_a M$

To establish Property 1, let

$$x = \log_a M \quad \text{and} \quad y = \log_a N$$

Then, by definition,

$$M = a^x \quad \text{and} \quad N = a^y$$

so that

$$M \cdot N = a^x \cdot a^y = a^{x+y}$$

Thus $x + y$ is the exponent to which a must be raised to yield $M \cdot N$, that is,

$$\log_a (M \cdot N) = x + y = \log_a M + \log_a N$$

Properties 2 and 3 are demonstrated in a similar fashion.

THE LOGARITHMIC FUNCTION

The function determined by

$$g(x) = \log_a x$$

is called the **logarithmic function with base a.** We can get a feeling for the behavior of this function by drawing its graph for $a = 2$ (Figure 10).

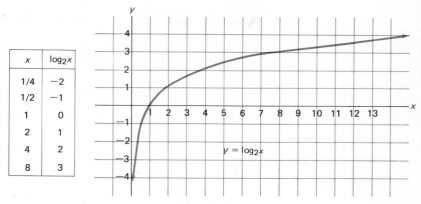

x	$\log_2 x$
1/4	−2
1/2	−1
1	0
2	1
4	2
8	3

Figure 10

Several properties of $y = \log_2 x$ are apparent from this graph. The domain consists of all positive real numbers. If $0 < x < 1$, $\log_2 x$ is negative; if $x > 1$, $\log_2 x$ is positive. The y-axis is a vertical asymptote of the graph since very small positive x values yield large negative y values. Although $\log_2 x$ continues to increase as x increases, even this small part of the complete graph indicates how slowly it grows for large x. In fact, by the time x reaches 1,000,000, $\log_2 x$ is still loafing along at about 20. In this sense, it behaves in a manner opposite to the exponential function 2^x, which grows more and more rapidly as x increases. There is a good reason for this opposite behavior; the two functions are inverses of each other.

INVERSE FUNCTIONS

We begin by emphasizing two facts that you must not forget.

$$a^{\log_a x} = x$$
$$\log_a(a^x) = x$$

For example, $2^{\log_2 7} = 7$ and $\log_2(2^{-19}) = -19$. Both of these facts are direct consequences of the definition of logarithms; the second is also a special case of Property 3, stated earlier. What these facts tell us is that the logarithmic and exponential functions undo each other.

Let us put this in the language of Section 5-5. If $f(x) = a^x$ and $g(x) = \log_a x$, then

$$f(g(x)) = f(\log_a x) = a^{\log_a x} = x$$

and

$$g(f(x)) = g(a^x) = \log_a(a^x) = x$$

Thus g is really f^{-1}. This fact also tells us something about the graphs of g and f: They are simply reflections of each other about the line $y = x$ (Figure 11).

Note finally that $f(x) = a^x$ has the set of all real numbers as its domain and the positive real numbers as its range. Thus its inverse $f^{-1}(x) = \log_a x$ has domain consisting of the positive real numbers and range consisting of all real numbers. We emphasize again a fact that is important to remember. *Negative numbers and zero do not have logarithms.*

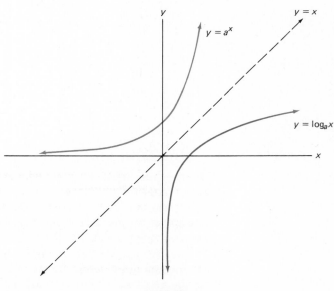

Figure 11

Problem Set 6-4

Write each of the following in logarithmic form. For example, $3^4 = 81$ can be written as $\log_3 81 = 4$.

1. $4^3 = 64$ 2. $7^3 = 343$ 3. $27^{1/3} = 3$

4. $16^{1/4} = 2$ 5. $4^0 = 1$ 6. $81^{-1/2} = \frac{1}{9}$

7. $125^{-2/3} = \frac{1}{25}$ 8. $2^{9/2} = 16\sqrt{2}$ 9. $10^{\sqrt{3}} = a$ *a ≈ 53.96*

10. $5^{\sqrt{2}} = b$ 11. $10^a = \sqrt{3}$ 12. $b^x = y$

b ≈ 9.74   *a ≈ .23856*

Write each of the following in exponential form. For example, $\log_5 125 = 3$ can be written as $5^3 = 125$.

$\log_{10}\left(\frac{1}{100}\right) = -2$

13. $\log_5 625 = 4$ 14. $\log_6 216 = 3$ 15. $\log_4 8 = \frac{3}{2}$

16. $\log_{27} 9 = \frac{2}{3}$ 17. $\log_{10}(.01) = -2$ 18. $\log_3(\frac{1}{27}) = -3$

19. $\log_c c = 1$ 20. $\log_b N = x$ 21. $\log_c Q = y$

Determine the value of each of the following logarithms.

22. $\log_4 16$ 23. $\log_5 25$ 24. $\log_7 \frac{1}{7}$

25. $\log_3 \frac{1}{3}$ 26. $\log_4 2$ 27. $\log_{27} 3$

28. $\log_{10}(10^{-6})$ 29. $\log_{10}(.0001)$ 30. $\log_8 1$

31. $\log_3 1$ 32. $\log_{100} 1000$ 33. $\log_8 16$

Find the value of c in each of the following.

34. $\log_c 25 = 2$ 35. $\log_c 8 = 3$ 36. $\log_4 c = -\frac{1}{2}$

37. $\log_9 c = -\frac{3}{2}$ 38. $\log_2(2^{5.6}) = c$ 39. $\log_3(3^{-2.9}) = c$

* 40. $8^{\log_8 11} = c$ * 41. $5^{2\log_5 7} = c$ * 42. $3^{4\log_3 2} = c$

Given $\log_{10} 2 = .301$ and $\log_{10} 3 = .477$, calculate each of the following without the use of tables. For example, in Problem 43, $\log_{10} 6 = \log_{10} 2 \cdot 3 = \log_{10} 2 + \log_{10} 3$, and in Problem 50, $\log_{10} 54 = \log_{10} 2 \cdot 3^3 = \log_{10} 2 + \log_{10} 3^3 = \log_{10} 2 + 3 \log_{10} 3$.

43. $\log_{10} 6$ 44. $\log_{10} \frac{3}{2}$ 45. $\log_{10} 16$

46. $\log_{10} 27$ 47. $\log_{10} \frac{1}{4}$ 48. $\log_{10} \frac{1}{27}$

49. $\log_{10} 24$ 50. $\log_{10} 54$ 51. $\log_{10} \frac{8}{9}$

52. $\log_{10} \frac{3}{8}$ 53. $\log_{10} 5$ 54. $\log_{10} \sqrt[3]{3}$

[c] *Your scientific calculator has a $\log_{10}$ key (which may be abbreviated log). Use it to find each of the following.*

55. $\log_{10} 34$ 56. $\log_{10} 1417$ 57. $\log_{10}(.0123)$

58. $\log_{10}(.3215)$ 59. $\log_{10} 9723$ 60. $\log_{10}(\frac{21}{312})$

EXAMPLE A (Combining Logarithms)   Write the following expression as a single logarithm.

$$2 \log_{10} x + 3 \log_{10}(x + 2) - \log_{10}(x^2 + 5)$$

IN GENERAL,
$$a^{z\log_a x} = x^z$$

$$2\log_{10} x + 3\log_{10}(x+2) - \log_{10}(x^2+5)$$

Solution. We use the properties of logarithms to rewrite this as

$$\log_{10} x^2 + \log_{10}(x+2)^3 - \log_{10}(x^2+5) \qquad \text{(Property 3)}$$

$$= \log_{10} x^2(x+2)^3 - \log_{10}(x^2+5) \qquad \text{(Property 1)}$$

$$= \log_{10}\left[\frac{x^2(x+2)^3}{x^2+5}\right] \qquad \text{(Property 2)}$$

CAUTION

$$\log_b 6 + \log_b 4 = \log_b 10$$
$$\log_b 30 - \log_b 5 = \log_b 25$$

$$\log_b 6 + \log_b 4 = \log_b 24$$
$$\log_b 30 - \log_b 5 = \log_b 6$$

Write each of the following as a single logarithm.

61. $3\log_{10}(x+1) + \log_{10}(4x+7)$
62. $\log_{10}(x^2+1) + 5\log_{10} x$
63. $3\log_2(x+2) + \log_2 8x - 2\log_2(x+8)$
64. $2\log_5 x - 3\log_5(2x+1) + \log_5(x-4)$
65. $\frac{1}{2}\log_6 x + \frac{1}{3}\log_6(x^3+3)$
66. $-\frac{2}{3}\log_3 x + \frac{5}{2}\log_3(2x^2+3)$

EXAMPLE B (Solving Logarithmic Equations)   Solve the equation

$$\log_2 x + \log_2(x+2) = 3$$

Solution.   First we note that we must have $x > 0$ so that both logarithms exist. Next we rewrite the equation using the first property of logarithms and then the definition of a logarithm.

$$\log_2 x(x+2) = 3$$
$$x(x+2) = 2^3$$
$$x^2 + 2x - 8 = 0$$
$$(x+4)(x-2) = 0$$

We reject $x = -4$ (because $-4 < 0$) and keep $x = 2$. To make sure that 2 is a solution, we substitute 2 for x in the original equation.

$$\log_2 2 + \log_2(2+2) \overset{?}{=} 3$$
$$1 + 2 = 3$$

Solve each of the following equations.

67. $\log_7(x+2) = 2$
68. $\log_5(3x+2) = 1$
69. $\log_2(x+3) = -2$
70. $\log_4(\frac{1}{64}x+1) = -3$
71. $\log_2 x - \log_2(x-2) = 3$
72. $\log_3 x - \log_3(2x+3) = -2$
73. $\log_2(x-4) + \log_2(x-3) = 1$
74. $\log_{10} x + \log_{10}(x-3) = 1$

EXAMPLE C (Change of Base)   In Problems 89 and 90, you will be asked to establish the **change-of-base formula**

$$\log_b x = \frac{\log_a x}{\log_a b}$$

Use this formula and the $\boxed{\log_{10}}$ key on a calculator to find $\log_2 13$.

Solution.

$$\log_2 13 = \frac{\log_{10} 13}{\log_{10} 2} \approx \frac{1.1139433}{.30103} \approx 3.7004$$

☐ *Use the method of Example C to find the following.*

75. $\log_2 128$ 76. $\log_3 128$ 77. $\log_3 82$

78. $\log_5 110$ 79. $\log_6 39$ 80. $\log_2(.26)$

MISCELLANEOUS PROBLEMS

81. Find the value of x in each of the following.
   (a) $x = \log_6 36$ (b) $x = \log_4 2$ (c) $\log_{25} x = \frac{3}{2}$
   (d) $\log_4 x = \frac{5}{2}$ (e) $\log_x 10\sqrt{10} = \frac{3}{2}$ (f) $\log_x \frac{1}{8} = -\frac{3}{2}$

82. Write each of the following as a single logarithm.
   (a) $3 \log_2 5 - 2 \log_2 7$
   (b) $\frac{1}{2} \log_5 64 + \frac{1}{3} \log_5 27 - \log_5(x^2 + 4)$
   (c) $\frac{2}{3} \log_{10}(x + 5) + 4 \log_{10} x - 2 \log_{10}(x - 3)$

83. Evaluate $\dfrac{(\log_{27} 3)(\log_{27} 9)(3^{2\log_3 2})}{\log_3 27 - \log_3 9 + \log_3 1}$

84. Solve for x.
   (a) $2(\log_4 x)^2 + 3 \log_4 x - 2 = 0$
   (b) $(\log_x 8)^2 - \log_x 8 - 6 = 0$
   (c) $\log_x \sqrt{3} + \log_x 3^5 + \log_x(\frac{1}{27}) = \frac{5}{4}$

85. Solve for x.
   (a) $\log_5(2x - 1) = 2$

   (b) $\log_4\left(\dfrac{x - 2}{2x + 3}\right) = 0$

   (c) $\log_4(x - 2) - \log_4(2x + 3) = 0$
   (d) $\log_{10} x + \log_{10}(x - 15) = 2$

   (e) $\dfrac{\log_2(x + 1)}{\log_2(x - 1)} = 2$

   (f) $\log_8[\log_4(\log_2 x)] = 0$

86. Solve for x.
   (a) $2^{\log_2 x} = 16$ (b) $2^{\log_x 2} = 16$ (c) $x^{\log_2 x} = 16$
   (d) $\log_2 x^2 = 2$ (e) $(\log_2 x)^2 = 1$ (f) $x = (\log_2 x)^{\log_2 x}$

87. Solve for y in terms of x.
   (a) $\log_a(x + y) = \log_a x + \log_a y$
   (b) $x = \log_a(y + \sqrt{y^2 - 1})$

88. Show that $f(x) = \log_a(x + \sqrt{1 + x^2})$ is an odd function.

89. Show that $\log_2 x = \log_{10} x \log_2 10$, where $x > 0$. *Hint:* Let $\log_{10} x = c$. Then $x = 10^c$. Next take $\log_2$ of both sides.

90. Use the technique outlined in Problem 89 to show that for $a > 0$, $b > 0$, and $x > 0$

$$\log_a x = \log_b x \log_a b$$

This is equivalent to the change-of-base formula of Example *C*.

91. Show that $\log_a b = 1/\log_b a$, where a and b are positive.

92. If $\log_b N = 2$, find $\log_{1/b} N$.

93. Graph the equations $y = 3^x$ and $y = \log_3 x$ using the same coordinate axes.

94. Find the solution set for each of the following inequalities.

   (a) $\log_2 x < 0$ (b) $\log_{10} x \geq -1$

   (c) $2 < \log_3 x < 3$ (d) $-2 \leq \log_{10} x \leq -1$

   (e) $2^x > 10$ (f) $2^x < 3^x$

95. Sketch the graph of each of the following functions using the same coordinate axes.

   (a) $f(x) = \log_2 x$ (b) $g(x) = \log_2(x + 1)$ (c) $h(x) = 3 + \log_2 x$

96. TEASER   Let log represent $\log_{10}$. Evaluate.

   (a) $\log \frac{1}{2} + \log \frac{2}{3} + \log \frac{3}{4} + \cdots + \log \frac{98}{99} + \log \frac{99}{100}$

   (b) $\log \dfrac{3}{4} + \log \dfrac{8}{9} + \log \dfrac{15}{16} + \cdots + \log \dfrac{99^2 - 1}{99^2} + \log \dfrac{100^2 - 1}{100^2}$

   (c) $\log_2 3 \cdot \log_3 4 \cdot \log_4 5 \cdot \log_5 6 \cdots \log_{63} 64$

The collected works of this brilliant Swiss mathematician will fill 74 volumes when completed. No other person has written so profusely on mathematical topics. Remarkably, 400 of his research papers were written after he was totally blind. One of his contributions was the introduction of the number $e = 2.71828 \ldots$ as the base for natural logarithms.

Leonhard Euler
1707–1783

6-5   Natural Logarithms and Applications

Napier invented logarithms to simplify arithmetic calculations. Computers and calculators have reduced that application to minor significance, though we shall discuss such a use of logarithms later in this chapter. Here we have in mind deeper applications such as solving exponential equations, defining power functions, and modeling physical phenomena.

In order to make any progress, we shall need an easy way to calculate logarithms. Fortunately, this has been done for us as tables of logarithms to several bases are available. For our purposes in this section, one base is as good as another. Base 10 would be an appropriate choice, but we would rather defer discussion of logarithms to base 10 (common logarithms) until Section 6-6. We have chosen rather to introduce you to the number e (after Euler), which is used as a base of logarithms in all advanced mathematics courses. You will see the importance of logarithms to this base (**natural logarithms**) when you study calculus (also see Figure 12). An approximate value of e is

$$e \approx 2.71828$$

e and interest

Suppose one dollar is invested at 100 percent interest compounded n times per year. Then it will be worth $(1 + \frac{1}{n})^n$ dollars at the end of the year. As n increases without bound (that is, as compounding occurs more and more often)

$$(1 + \frac{1}{n})^n \to e$$

and we have what is called continuous compounding. Under this plan, \$1 grows to \$2.72 in one year.

Figure 12

and, like π, e is an irrational number. Table 3 on page 256 shows a table of values of natural logarithms, which we shall denote by ln instead of $\log_e$.

Since ln denotes a genuine logarithm function, we have as in Section 6-4

$$\ln N = x \quad \text{if and only if} \quad e^x = N$$

and consequently

$$\ln e^x = x \quad \text{and} \quad e^{\ln N} = N$$

Moreover the three properties of logarithms hold.

1. $\ln (MN) = \ln M + \ln N$
2. $\ln(M/N) = \ln M - \ln N$
3. $\ln (N^p) = p \ln N$

SOLVING EXPONENTIAL EQUATIONS

Consider first the simple equation

$$5^x = 1.7$$

We call it an *exponential equation* because the unknown is in the exponent. To solve it, we take natural logarithms of both sides.

$$5^x = 1.7$$

$$\ln(5^x) = \ln 1.7$$

$$x \ln 5 = \ln 1.7 \qquad \text{(Property 3)}$$

$$x = \frac{\ln 1.7}{\ln 5}$$

$$x \approx \frac{.531}{1.609} \approx .330 \quad \text{(Table 3)}$$

We point out that the last step can also be done on a scientific calculator, which has a key for calculating natural logarithms.

TABLE 3 Table of Natural Logarithms

x	$\ln x$	x	$\ln x$	x	$\ln x$
		4.0	1.386	8.0	2.079
0.1	-2.303	4.1	1.411	8.1	2.092
0.2	-1.609	4.2	1.435	8.2	2.104
0.3	-1.204	4.3	1.459	8.3	2.116
0.4	-0.916	4.4	1.482	8.4	2.128
0.5	-0.693	4.5	1.504	8.5	2.140
0.6	-0.511	4.6	1.526	8.6	2.152
0.7	-0.357	4.7	1.548	8.7	2.163
0.8	-0.223	4.8	1.569	8.8	2.175
0.9	-0.105	4.9	1.589	8.9	2.186
1.0	0.000	5.0	1.609	9.0	2.197
1.1	0.095	5.1	1.629	9.1	2.208
1.2	0.182	5.2	1.649	9.2	2.219
1.3	0.262	5.3	1.668	9.3	2.230
1.4	0.336	5.4	1.686	9.4	2.241
1.5	0.405	5.5	1.705	9.5	2.251
1.6	0.470	5.6	1.723	9.6	2.262
1.7	0.531	5.7	1.740	9.7	2.272
1.8	0.588	5.8	1.758	9.8	2.282
1.9	0.642	5.9	1.775	9.9	2.293
2.0	0.693	6.0	1.792	10	2.303
2.1	0.742	6.1	1.808	20	2.996
2.2	0.788	6.2	1.825	30	3.401
2.3	0.833	6.3	1.841	40	3.689
2.4	0.875	6.4	1.856	50	3.912
2.5	0.916	6.5	1.872	60	4.094
2.6	0.956	6.6	1.887	70	4.248
2.7	0.993	6.7	1.902	80	4.382
2.8	1.030	6.8	1.917	90	4.500
2.9	1.065	6.9	1.932	100	4.605
3.0	1.099	7.0	1.946		
3.1	1.131	7.1	1.960		
3.2	1.163	7.2	1.974	e	1.000
3.3	1.194	7.3	1.988		
3.4	1.224	7.4	2.001	π	1.145
3.5	1.253	7.5	2.015		
3.6	1.281	7.6	2.028		
3.7	1.308	7.7	2.041		
3.8	1.335	7.8	2.054		
3.9	1.361	7.9	2.067		

To find the natural logarithm of a number N which is either smaller than 0.1 or larger than 10, write N in scientific notation, that is, write $N = c \times 10^k$. Then $\ln N = \ln c + k \ln 10 = \ln c + k(2.303)$. A more complete table of natural logarithms appears as Table A of the Appendix.

Here is a more complicated example.

$$5^{2x-1} = 7^{x+2}$$

Begin by taking natural logarithms of both sides and then solve for x.

$$\ln(5^{2x-1}) = \ln(7^{x+2})$$

$$(2x - 1)\ln 5 = (x + 2)\ln 7 \qquad \text{(Property 3)}$$

$$2x \ln 5 - \ln 5 = x \ln 7 + 2 \ln 7$$

$$2x \ln 5 - x \ln 7 = \ln 5 + 2 \ln 7$$

$$x(2 \ln 5 - \ln 7) = \ln 5 + 2 \ln 7$$

$$x = \frac{\ln 5 + 2 \ln 7}{2 \ln 5 - \ln 7}$$

$$\approx \frac{1.609 + 2(1.946)}{2(1.609) - 1.946} \quad \begin{array}{l}\text{(Table 3 or a} \\ \text{calculator)}\end{array}$$

$$\approx 4.325$$

You get a more accurate answer, 4.321, if you use a calculator all the way.

THE GRAPHS OF ln x AND e^x

We have already pointed out that $\ln e^x = x$ and $e^{\ln x} = x$. Thus $f(x) = \ln x$ and $g(x) = e^x$ are inverse functions, which means that their graphs are reflections of each other across the line $y = x$. They are shown in Figure 13.

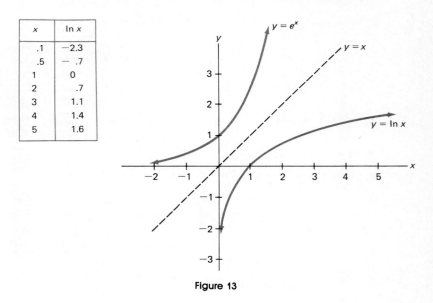

x	ln x
.1	−2.3
.5	− .7
1	0
2	.7
3	1.1
4	1.4
5	1.6

Figure 13

Since ln x is not defined for $x \leq 0$, it is of some interest to consider ln $|x|$, which is defined for all x except 0. Its graph is shown in Figure 14 at the top of the next page. Note the symmetry with respect to the y-axis.

EXPONENTIAL FUNCTIONS VERSUS POWER FUNCTIONS

Look closely at the formulas below.

$$f(x) = 2^x \qquad f(x) = x^2$$

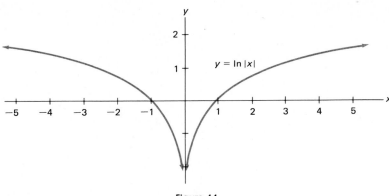

Figure 14

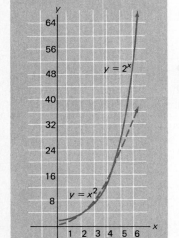

Figure 15

They are very different, yet easily confused. The first is an exponential function, while the second is called a power function. Both grow rapidly for large x, but the exponential function ultimately gets far ahead (see the graphs in Figure 15).

The situation described above is a special instance of two very general classes of functions.

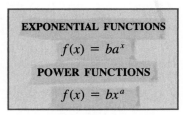

EXPONENTIAL FUNCTIONS

$$f(x) = ba^x$$

POWER FUNCTIONS

$$f(x) = bx^a$$

CURVE FITTING

A recurring theme in science is to fit a mathematical curve to a set of experimental data. Suppose that a scientist, studying the relationship between two variables x and y, obtained the data plotted in Figure 16. In searching for curves to fit these data, the scientist naturally thought of exponential curves and power curves. How did he or she decide if either was appropriate? The scientist took logarithms. Let us see why.

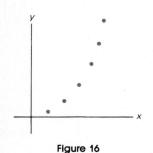

Figure 16

MODEL 1	MODEL 2
$y = ba^x$	$y = bx^a$
$\ln y = \ln b + x \ln a$	$\ln y = \ln b + a \ln x$
$Y = B + Ax$	$Y = B + aX$

Here the scientist made the substitutions $Y = \ln y$, $B = \ln b$, $A = \ln a$, and $X = \ln x$.

In both cases, the final result is a linear equation. But note the difference.

In the first case, ln y is a linear function of x, whereas in the second case, ln y is a linear function of ln x. These considerations suggest the following procedures. Make two additional plots of the data. In the first, plot ln y against x, and in the second, plot ln y against ln x. If the first plotting gives data nearly along a straight line, Model 1 is appropriate; if the second does, then Model 2 is appropriate. If neither plot approximates a straight line, our scientist should look for a different and perhaps more complicated model.

We have used natural logarithms in the discussion above; we could also have used common logarithms (logarithms to the base 10). In the latter case, special kinds of graph paper are available to simplify the curve fitting process. On semilog paper, the vertical axis has a logarithmic scale; on log-log paper, both axes have logarithmic scales. The xy-data can be plotted *directly* on this paper. If semilog paper gives an (approximately) straight line, Model 1 is indicated; if log-log paper does so, then Model 2 is appropriate. You will have ample opportunity to use these special kinds of graph paper in your science courses.

LOGARITHMS AND PHYSIOLOGY

The human body appears to have a built-in logarithmic calculator. What do we mean by this statement?

In 1834, the German physiologist E. Weber noticed an interesting fact. Two heavy objects must differ in weight by considerably more than two light objects if a person is to perceive a difference between them. Other scientists noted the same phenomenon when human subjects tried to differentiate loudness of sounds, pitches of musical tones, brightness of light, and so on. Experiments suggested that people react to stimuli on a logarithmic scale, a result formulated as the Weber-Fechner law (see Figure 17).

$$S = C \ln\left(\frac{R}{r}\right)$$

Here R is the actual intensity of the stimulus, r is the threshold value (smallest value at which the stimulus is observed), C is a constant depending on the type of stimulus, and S is the perceived intensity of the stimulus. Note that a change in R is not as perceptible for large R as for small R because as R increases, the graph of the logarithmic functions gets steadily flatter.

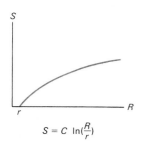

$S = C \ln\left(\frac{R}{r}\right)$

Figure 17

Problem Set 6-5

In Problems 1–8, find the value of each natural logarithm.

1. $\ln e$
2. $\ln(e^2)$
3. $\ln 1$
4. $\ln\left(\frac{1}{e}\right)$

5. $\ln\sqrt{e}$
6. $\ln(e^{1.1})$
7. $\ln\left(\frac{1}{e^3}\right)$
8. $\ln(e^n)$

For Problems 9–14, find each value. Assume that ln a = 2.5 and ln b = −.4.

9. $\ln(ae)$

10. $\ln\left(\dfrac{e}{b}\right)$

11. $\ln\sqrt{b}$

12. $\ln(a^2 b^{10})$

13. $\ln\left(\dfrac{1}{a^3}\right)$

14. $\ln(a^{4/5})$

Use the table of natural logarithms to calculate the values in Problems 15–20.

15. $\ln 120 = \ln 60 + \ln 2$ 16. $\ln 150$ 17. $\ln 690$

18. $\ln 84$ 19. $\ln \frac{6}{5}$ 20. $\ln 20{,}000$

In Problems 21–26, use the natural logarithm table to find N.

21. $\ln N = 2.208$ 22. $\ln N = 1.808$ 23. $\ln N = -.105$

24. $\ln N = -.916$ 25. $\ln N = 4.500$ 26. $\ln N = 9.000$

C *Use your calculator ($\boxed{\ln}$ or $\boxed{\log_e}$ key) to find each of the following.*

27. $\ln 4.31$ 28. $\ln 517$ 29. $\ln(.127)$

30. $\ln(.00424)$ 31. $\ln\left(\dfrac{6.71}{42.3}\right)$ 32. $\ln\sqrt{457}$

33. $\dfrac{\ln 6.71}{\ln 42.3}$ 34. $\sqrt{\ln 457}$ 35. $\ln(51.4)^3$

36. $\ln(31.2 + 43.1)$ 37. $(\ln 51.4)^3$ 38. $3 \ln 51.4$

C *Use your calculator to find N in each of the following. Hint: ln N = 5.1 if and only if $N = e^{5.1}$. On some calculators, there is an $\boxed{e^x}$ key; on others, you use the two keys* $\boxed{\text{INV}}\ \boxed{\ln}$.

39. $\ln N = 2.12$ 40. $\ln N = 5.63$ 41. $\ln N = -.125$

42. $\ln N = .00257$ 43. $\ln\sqrt{N} = 3.41$ 44. $\ln N^3 = .415$

EXAMPLE A (Exponential Equations) Solve $4^{3x-2} = 15$ for x.

Solution. Take natural logarithms of both sides and then solve for x. Complete the solution using the ln table or a calculator.

$$\ln 4^{3x-2} = \ln 15$$

$$(3x - 2)\ln 4 = \ln 15$$

$$3x \ln 4 - 2 \ln 4 = \ln 15$$

$$3x \ln 4 = 2 \ln 4 + \ln 15$$

$$x = \frac{2 \ln 4 + \ln 15}{3 \ln 4}$$

$$x \approx 1.32$$

Solve for x using the method above.

45. $3^x = 20$ 46. $5^x = 40$ 47. $2^{x-1} = .3$

48. $4^x = 3^{2x-1}$ 49. $(1.4)^{x+2} = 19.6$ 50. $5^x = \frac{1}{2}(4^x)$

EXAMPLE B (Doubling Time)   How long will it take money to double in value if it is invested at 9.5 percent interest compounded annually?

Solution.   For convenience, consider investing \$1. From Section 6-3, we know that this dollar will grow to $(1.095)^t$ dollars after t years. Thus we must solve the exponential equation $(1.095)^t = 2$. This we do by the method of Example A, using a calculator. The result is

$$t = \frac{\ln 2}{\ln 1.095} \approx 7.64$$

[c] 51. How long would it take money to double at 12 percent compounded annually?

[c] 52. How long would it take money to double at 12 percent compounded monthly? *Hint:* After t months, \$1 is worth $(1.01)^t$ dollars.

[c] 53. How long would it take money to double at 15 percent compounded quarterly?

[c] 54. A certain substance decays according to the formula $y = 100e^{-.135t}$, where t is in years. Find its half-life.

[c] 55. By finding the natural logarithm of the numbers in each pair, determine which is larger.
   (a) $10^5, 5^{10}$ (b) $10^9, 9^{10}$
   (c) $10^{20}, 20^{10}$ (d) $10^{1000}, 1000^{10}$

56. What do your answers in Problem 55 confirm about the growth of 10^x and x^{10} for large x?

57. On the same coordinate plane, graph $y = 3^x$ and $y = x^3$ for $0 \le x \le 4$.

58. By means of a change of variable(s) (as explained in the text), transform each equation below to a linear equation. Find the slope and Y-intercept of the resulting line.
   (a) $y = 3e^{2x}$ (b) $y = 2x^3$ (c) $xy = 12$
   (d) $y = x^e$ (e) $y = 5(3^x)$ (f) $y = ex^{1.1}$

t	N
0	100
1	700
2	5000
3	40,000

Figure 18

EXAMPLE C (Curve Fitting)   The table in Figure 18 shows the number N of bacteria in a certain culture found after t hours. Which is a better description of these data,

$$N = ba^t \quad \text{or} \quad N = bt^a$$

Find a and b.

Solution.   Following the discussion of curve fitting in the text, we begin by plotting $\ln N$ against t. If the resulting points lie along a line, we choose $N = ba^t$ as the appropriate model. If not, we will plot $\ln N$ against $\ln t$ to check on the second model. Since the fit to a line is quite good (Figure 19 at the top of the next page,), we accept $N = ba^t$ as our model. To find a and b, we write $N = ba^t$ in the form

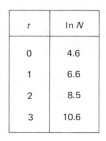

t	ln N
0	4.6
1	6.6
2	8.5
3	10.6

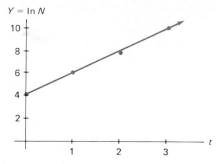

Figure 19

$$\ln N = \ln b + t \ln a \quad \text{or} \quad Y = \ln b + (\ln a)t$$

Examination of the line shows that it has a Y-intercept of 4.6 and a slope of about 2; so for its equation we write $Y = 4.6 + 2t$. Comparing this with $Y = \ln b + (\ln a)t$ gives

$$\ln b = 4.6 \qquad \ln a = 2$$

Finally, we use the natural logarithm table or a calculator to find

$$b \approx 100 \qquad a \approx 7.4$$

Thus the original data are described reasonably well by the equation

$$N = 100(7.4)^t$$

For the data sets below, decide whether $y = ba^x$ or $y = bx^a$ is the better model. Then determine a and b.

59.

x	1	2	3	4
y	96	145	216	325

60.

x	0	1	2	4
y	243	162	108	48

61.

x	1	2	3	5
y	12	190	975	7490

62.

x	1	4	9
y	16	128	432

MISCELLANEOUS PROBLEMS

63. Evaluate without using a calculator or tables.

   (a) $\ln(e^{4.2})$
   (b) $e^{2 \ln 2}$
   (c) $\dfrac{\ln 3e}{2 + \ln 9}$

[c] 64. Solve for x.
   (a) $\ln(5 + x) = 1.2$
   (b) $e^{x^2 - x} = 2$
   (c) $e^{2x} = (.6)8^x$

65. Evaluate without use of a calculator or tables.
   (a) $\ln[(e^{3.5})^2]$
   (b) $(\ln e^{3.5})^2$
   (c) $\ln(1/\sqrt{e})$
   (d) $(\ln 1)/(\ln \sqrt{e})$
   (e) $e^{3 \ln 5}$
   (f) $e^{\ln(1/2) + \ln(2/3)}$

[c] 66. Since $a^x = e^{\ln a^x} = e^{x \ln a}$, the study of exponential functions can be subsumed under the study of the function e^{kx}. Determine k so that each of the following is true. *Hint:* In (a) rewrite the equation as $3^x = (e^k)^x$ which implies that $3 = e^k$.

Now take natural logarithms.

(a) $3^x = e^{kx}$

(b) $\pi^x = e^{kx}$

(c) $(1/3)^x = e^{kx}$

© 67. Solve for x.

(a) $10^{2x+3} = 200$

(b) $10^{2x} = 8^{x-1}$

(c) $10^{x^2+3x} = 200$

(d) $e^{-.32x} = 1/2$

(e) $x^{\ln x} = 10$

(f) $(\ln x)^{\ln x} = x$

(g) $x^{\ln x} = x$

(h) $\ln x = (\ln x)^{\ln x}$

(i) $(x^2 - 5)^{\ln x} = x$

68. Show that each of the following is an identity. Assume $x > 0$.

(a) $(\sqrt{3})^{3\sqrt{3}} = (3\sqrt{3})^{\sqrt{3}}$

(b) $2.25^{3.375} = 3.375^{2.25}$

(c) $a^{\ln(x^b)} = (a^{\ln x})^b$

(d) $x^x = e^{x\ln x}$

(e) $(\ln x)^x = e^{x\ln(\ln x)}$

(f) $\dfrac{\ln\left(\dfrac{x+1}{x}\right)^x}{\ln\left(\dfrac{x+1}{x}\right)^{x+1}} = \dfrac{\left(\dfrac{x+1}{x}\right)^x}{\left(\dfrac{x+1}{x}\right)^{x+1}}$

© 69. A certain substance decays according to the formula $y = 100e^{-3t}$, where y is the amount present after t years. Find its half-life.

© 70. Suppose the number of bacteria in a certain culture t hours from now will be $200e^{.468t}$. When will the count reach 10,000?

© 71. A radioactive substance decays exponentially with a half-life of 240 years. Determine k in the formula $A = A_0e^{-kt}$, where A is the amount present after t years and A_0 is the initial amount.

72. By means of a change of variables using natural logarithms, transform each of the following equations to a linear equation. Then find the slope and the Y-intercept of the resulting line.

(a) $xy^2 = 40$

(b) $y = 9e^{2x}$

© 73. Sketch the graph of

$$y = \frac{1}{\sqrt{2\pi}}e^{-(1/2)x^2}$$

This is the famous *normal curve*, so important in statistics.

© 74. In calculus it is shown that

$$e^x \approx 1 + x + \frac{x^2}{2} + \frac{x^3}{6} + \frac{x^4}{24} + \frac{x^5}{120}$$

Use this to approximate e and $e^{-1/2}$.

75. From Problem 74 or from looking at the graph of $y = e^x$, you might guess the true result that $e^x > 1 + x$ for all $x > 0$. Use this and the obvious fact that $(\pi/e) - 1 > 0$ to demonstrate algebraically that $e^\pi > \pi^e$.

© 76. It is important to have a feeling for how various functions grow for large x. Let $\ll$ symbolize the phrase *grows slower than*. Use a calculator to convince yourself that

$$\ln x \ll \sqrt{x} \ll x \ll x^2 \ll e^x \ll x^x$$

© 77. In calculus, it is shown that $(1 + r/m)^m$ gets closer and closer to e^r as m gets larger and larger. Now if P dollars is invested at rate r (written as a decimal) compounded m times per year, it will grow to $P(1 + r/m)^{mt}$ dollars at the end of t years (Example B of Section 6-3). If interest is compounded continuously (see the box *e* **and interest** on page 255), P dollars will grow to Pe^{rt} dollars at the end of t years. Use these facts to calculate the value of $100 after 10 years if interest is at 12 percent (that is, $r = .12$) and is compounded (a) monthly; (b) daily; (c) hourly; (d) continuously.

78. **TEASER** The *harmonic* sum $S_n = 1 + \frac{1}{2} + \frac{1}{3} + \frac{1}{4} + \cdots + 1/n$ has intrigued both amateur and professional mathematicians since the time of the Greeks. In calculus, it is shown that for $n > 1$

$$\ln n < 1 + \frac{1}{2} + \frac{1}{3} + \frac{1}{4} + \cdots + \frac{1}{n} < 1 + \ln n$$

This shows that S_n grows arbitrarily large but that it grows very very slowly.
(a) About how large would n need to be for $S_n > 100$?
(b) Show how you could stack a pile of identical bricks each of length 1 foot (one brick to a tier as shown in Figure 20) to achieve an overhang of 50 feet. Could you make the overhang 50 million feet? Yes, it does have something to do with part (a).

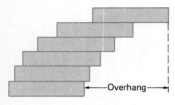

Figure 20

Henry Briggs 1561–1631

When 10 is used as the base for logarithms, some very nice things happen as the display at the right shows. Realizing this, John Napier and his friend Henry Briggs decided to produce a table of these logarithms. The job was accomplished by Briggs after Napier's death and published in 1624 in the famous book *Arithmetica Logarithmica*.

$\log .0853 = .9309 - 2$

$\log .853 \ = .9309 - 1$

$\log 8.53 \ = .9309$

$\log 85.3 \ = .9309 + 1$

$\log 853 \ = .9309 + 2$

$\log 8530 = .9309 + 3$

6-6   Common Logarithms (Optional)

In Section 6-4, we defined the logarithm of a positive number N to the base a as follows.

$$\log_a N = x \quad \text{if and only if} \quad a^x = N$$

Then in Section 6-5, we introduced base e, calling the result natural logarithms. These are the logarithms that are most important in advanced branches of mathematics such as calculus.

In this section, we shall consider logarithms to the base 10, often called **common logarithms,** or Briggsian logarithms. They have been studied by high school and college students for centuries as an aid to computation, a subject we take up in the next section. We shall write $\log N$ instead of $\log_{10} N$ (just as we write $\ln N$ instead of $\log_e N$). Note that

$$\log N = x \quad \text{if and only if} \quad 10^x = N$$

In other words, the common logarithm of any power of 10 is simply its exponent. Thus

$$\log 100 = \log 10^2 = 2$$

$$\log 1 = \log 10^0 = 0$$

$$\log .001 = \log 10^{-3} = -3$$

But how do we find common logarithms of numbers that are not integral powers of 10, such as 8.53 or 14,600? That is the next topic.

FINDING COMMON LOGARITHMS

We know that $\log 1 = 0$ and $\log 10 = 1$. If $1 < N < 10$, we correctly expect $\log N$ to be between 0 and 1. Table B (in the appendix) gives us four-place approximations of the common logarithms of all three-digit numbers between 1 and 10. For example,

$$\log 8.53 = .9309$$

We find this value by locating 8.5 in the left column and then moving right to the entry with 3 as heading. Similarly,

$$\log 1.08 = .0334$$

and

$$\log 9.69 = .9863$$

You should check these values.

You may have noticed in the opening box that log 8.53, log 8530, and log .0853 all have the same positive fractional part, .9309, called the **mantissa** of the logarithm. They differ only in the integer part, called the **characteristic** of the logarithm. To see why this is so, recall that

$$\log(M \cdot N) = \log M + \log N$$

Thus

$$\log 8530 = \log(8.53 \times 10^3) = \log 8.53 + \log 10^3$$

$$= .9309 + 3$$

$$\log .0853 = \log(8.53 \times 10^{-2}) = \log 8.53 + \log 10^{-2}$$

$$= .9309 - 2$$

Clearly the mantissa .9309 is determined by the sequence of digits 8, 5, 3, while the characteristic is determined by the position of the decimal point. Let us say that the decimal point is in **standard position** when it occurs immediately after the first nonzero digit. The characteristic for the logarithm of a number with the decimal point in standard position is 0. The characteristic is k if the decimal point is k places to the right of standard position; it is $-k$ if it is k places to the left of standard position.

8530000.   log 8,530,000 = .9309 + 6

6 places right

.0000853   log .0000853 = .9309 − 5

5 places left

Here is another way of describing the mantissa and characteristic. If $c \times 10^n$ is the scientific notation for N, then log c is the mantissa and n is the characteristic of log N.

FINDING ANTILOGARITHMS

If we are to make significant use of common logarithms, we must know how to find a number when its logarithm is given. This process is called finding the inverse logarithm, or the **antilogarithm.** The process is simple: Use the mantissa to find the sequence of digits and then let the characteristic tell you where to put the decimal point.

Suppose, for example, that you are given

$$\log N = .4031 - 4$$

Locate .4031 in the body of Table B. You will find it across from 2.5 and below 3. Thus the number N must have 2, 5, 3 as its sequence of digits. Since the characteristic is -4, put the decimal point 4 places to the left of standard position. The result is

$$N = .000253$$

As a second example, let us find antilog 5.9547. The mantissa .9547 gives us the digits 9, 0, 1. Since the characteristic is 5,

$$\text{antilog } 5.9547 = 901,000$$

LINEAR INTERPOLATION

Suppose that for some function f we know $f(a)$ and $f(b)$, but we want $f(c)$, where c is between a and b (see the diagrams in Figure 21). As a reasonable approximation, we may pretend that the graph of f is a straight line between a and b. Then

$$f(c) \approx f(a) + d$$

where, by similarity of triangles (Figure 22),

$$\frac{d}{f(b) - f(a)} = \frac{c - a}{b - a}$$

That is,

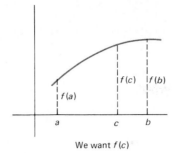

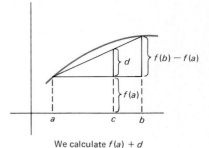

We want $f(c)$ We calculate $f(a) + d$

Figure 21

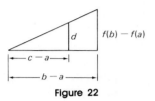

Figure 22

$$d = \frac{f(b) - f(a)}{b - a}(c - a)$$

The process just described is called **linear interpolation.** The process of linear interpolation for the logarithm function is explained in Examples A and B of the problem set, as well as at the beginning of the logarithm tables at the back of the book.

Problem Set 6-6

Find the common logarithm of each of the following numbers.

1. 10,000
2. 1,000,000
3. .01
4. .0001
5. $10^4 10^{3/2}$
6. $10^3 10^{4/3}$
7. $(10^3)^{-5}$
8. $(10^5)^{-1/10}$

Find N in each case.

9. $\log N = 4$
10. $\log N = 6$
11. $\log N = -2$
12. $\log N = -5$
13. $\log N = \frac{3}{2}$
14. $\log N = \frac{1}{3}$
15. $\log N = -\frac{3}{4}$
16. $\log N = -\frac{1}{6}$

Use Table B to find the following logarithms.

17. $\log 4.32$
18. $\log 3.09$
19. $\log 158$
20. $\log 47.3$
21. $\log .0329$
22. $\log .0715$
23. $\log 563{,}000$
24. $\log 420{,}000$
25. $\log(9.23 \times 10^8)$
26. $\log(2.83 \times 10^{-11})$

Find N in each case.

27. $\log N = 1.5159$
28. $\log N = 3.9015$
29. $\log N = .0043 - 2$
30. $\log N = .8627 - 4$
31. $\log N = 8.5999$
32. $\log N = 4.7427$

Find the antilogarithm of each number.

33. 2.2201
34. 3.8639
35. $.9232 - 1$
36. $.8500 - 5$

EXAMPLE A (Linear Interpolation in Finding Logarithms) Find log 34.67.

Solution. Our table gives the logarithms of 34.6 and 34.7, so we use linear interpolation to get an intermediate value. Here is how we arrange our work.

$$.10\left[.07\left[\begin{array}{l}\log 34.60 = 1.5391\\ \log 34.67 = \quad ?\\ \log 34.70 = 1.5403\end{array}\right]d\right].0012$$

$$\frac{d}{.0012} = \frac{.07}{.10} = \frac{7}{10}$$

$$d = \frac{7}{10}(.0012) \approx .0008$$

$$\log 34.67 \approx \log 34.60 + d \approx 1.5391 + .0008 = 1.5399$$

Use linear interpolation in Table B to find each value.

37. log 5.237   38. log 9.826   39. log 7234
40. log 68.04   41. log .001234   42. log .09876

EXAMPLE B (Interpolation in Finding Antilogarithms) Find antilog 2.5285.

Solution. We find .5285 sandwiched between .5276 and .5289 in the body of Table B.

$$.0013\left[.0009\left[\begin{array}{l}\text{antilog } 2.5276 = 337.0\\ \text{antilog } 2.5285 = \quad ?\\ \text{antilog } 2.5289 = 338.0\end{array}\right]d\right]1.0$$

$$\frac{d}{1.0} = \frac{.0009}{.0013} = \frac{9}{13}$$

$$d = \frac{9}{13}(1.0) \approx .7$$

$$\text{antilog } 2.5285 \approx 337.0 + .7 = 337.7$$

Find the antilogarithm of each of the following using linear interpolation.

43. 0.8497   44. 0.8516   45. 3.9130
46. 1.9849   47. .6004 − 2   48. .4946 − 4

MISCELLANEOUS PROBLEMS

49. Without using a calculator, find the common logarithm of each of the following.
    (a) $10^{3/2}\,10^{-1/4}$ (b) $(.0001)^{1/3}$
    (c) $\sqrt[3]{10}\,\sqrt{10}$ (d) $10^{\log(.001)}$

50. Find N in each case.
    (a) $\log N = 0$ (b) $\log N = -2$
    (c) $\log N = \frac{1}{2}$ (d) $\log N = 10$

51. Use Table B with interpolation to find each of the following.
    (a) log 492.7   (b) log .04705
    (c) antilog 2.9327   (d) antilog$(.2698 - 3)$

52. Use Table B to find N. *Hint:* $-2.4473 = .5527 - 3$.
    (a) $\log N = -2.4473$ (b) $\log N = -4.0729$

c 53. Do Problem 52 using your calculator.

54. Find the characteristic of log N if:
    (a) $10^{11} < N < 10^{12}$;   (b) $.00001 < N < .0001$.

55. Use Table B to find $\log \dfrac{982 - 467}{(982)(267)}$.

56. If $b = .001\,a$ and $\log a = 5.5$, find $\log(a^3/b^4)$.

57. Evaluate antilog$\left[\log\left(\dfrac{999}{4.71 \times 328}\right) - \log 999 + \log 328\right]$.

58. Find x if $(\log x)^2 + \log(x^2) = 10^{\log 3}$

59. Let $\log N = 15.992$. In the decimal notation for N, how many digits are there before the decimal point?

60. Convince yourself that the number of digits in a positive integer N is $[\log N] + 1$. Then find the number of digits in 50^{50} when expanded out in the usual way in decimal notation.

c 61. If one can write 6 digits to the inch, about how many miles long would the number $9^{(9^9)}$ be when written in decimal notation?

62. **TEASER**   Let $N = 9^{(9^9)}$, $A =$ sum of digits in N, $B =$ sum of digits in A, and $C =$ sum of digits in B. Find C. *Hint:* If a number is divisible by 9, so is the sum of its digits (why?). Since N is divisible by 9, it follows that A, B, and C are all divisible by 9.

"The miraculous powers of modern calculation are due to three inventions: the Arabic Notation, Decimal Fractions, and Logarithms."

F. Cajori, 1897

"Electronic calculators make calculations with logarithms as obsolete as whale oil lamps."

Anonymous Reviewer, 1982

6-7 Calculations with Logarithms (Optional)

For 300 years, scientists depended on logarithms to reduce the drudgery associated with long computations. The invention of electronic computers and calculators has diminished the importance of this long-established technique. Still, we think that any student of algebra should know how products, quotients, powers, and roots can be calculated by means of common logarithms.

About all you need are the three laws stated above and Appendix Table B. A little common sense and the ability to organize your work will help.

PRODUCTS

Suppose you want to calculate $(.00872)(95,300)$. Call this number x. Then by Law 1 and Table B,

$$\log x = \log .00872 + \log 95{,}300$$
$$= (.9405 - 3) + 4.9791$$
$$= 5.9196 - 3$$
$$= 2.9196$$

Now use Table B backwards to find that antilog $.9196 = 8.31$, so $x =$ antilog $2.9196 = 831$.

Here is a good way to organize your work in a compact systematic way.

$$x = (.00872)(95{,}300)$$

$$\begin{array}{r} \log .00872 = \quad .9405 - 3 \\ (+) \ \underline{\log 95300 = 4.9791} \\ \log x = \overline{5.9196 - 3} \end{array}$$

$$x = 831$$

QUOTIENTS

Suppose we want to calculate $x = .4362/91.84$. Then by Law 2,

$$\log x = \log .4362 - \log 91.84$$
$$= (.6397 - 1) - 1.9630$$
$$= .6397 - 2.9630$$
$$= -2.3233$$

What we have done is correct; however, it is poor strategy. The result we found for $\log x$ is not in characteristic-mantissa form and therefore is not usable. Remember that the mantissa must be positive. We can bring this about by adding and subtracting 3.

$$-2.3233 = (-2.3233 + 3) - 3 = .6767 - 3$$

Actually it is better to anticipate the need for doing this and arrange the work as follows.

$$x = \frac{.4362}{91.84}$$

$$\log .4362 = .6397 - 1 = 2.6397 - 3$$
$$(-) \underline{\log 91.48 =} \qquad \underline{1.9630}$$
$$\log x = \qquad\qquad .6767 - 3$$

$$x = .00475$$

POWERS OR ROOTS

Here the main tool is Law 3. We illustrate with two examples.

$$x = (31.4)^{11}$$
$$\log x = 11 \log 31.4$$
$$\log 31.4 = 1.4969$$
$$11 \log 31.4 = 16.4659$$
$$\log x = 16.4659$$
$$x = 29{,}230{,}000{,}000{,}000{,}000$$
$$= 2.923 \times 10^{16}$$

$$x = \sqrt[4]{.427} = (.427)^{1/4}$$

$$\log x = \frac{1}{4} \log .427$$

$$\log .427 = .6304 - 1$$

$$\frac{1}{4} \log .427 = \frac{1}{4}(.6304 - 1) = \frac{1}{4}(3.6304 - 4) = .9076 - 1$$

$$\log x = .9076 - 1$$

$$x = .8084$$

Notice in the second example that we wrote $3.6304 - 4$ in place of $.6304 - 1$, so that multiplication by $\frac{1}{4}$ gave the logarithm in characteristic-mantissa form.

Problem Set 6-7

Use logarithms and Table B without interpolation to find approximate values for each of the following.

1. $(46.3)(2.76)$ 2. $(378)(9.63)$ 3. $\dfrac{46.3}{483}$

4. $\dfrac{437}{92300}$ 5. $\dfrac{.00912}{.439}$ 6. $\dfrac{.0429}{15.7}$

7. $(37.2)^5$ 8. $(113)^3$ 9. $\sqrt[3]{42.9}$

10. $\sqrt[4]{312}$ 11. $\sqrt[5]{.918}$ 12. $\sqrt[3]{.0307}$

13. $(14.9)^{2/3}$ 14. $(98.6)^{3/4}$

Use logarithms and Table B with interpolation to approximate each of the following.

15. $(31.96)(149)$ 16. $(6236)(.00108)$ 17. $\dfrac{43.98}{7.16}$

18. $\dfrac{115}{4.623}$ 19. $(.1234)^6$ 20. $(92.83)^3$

EXAMPLE A (More Complicated Calculations)   Use logarithms, without interpolation, to calculate

$$\frac{(31.4)^3(.982)}{(.0463)(824)}$$

Solution.   Let N denote the entire numerator, D the entire denominator, and x the fraction N/D. Then

$$\log x = \log N - \log D$$

where

$$\log N = 3 \log 31.4 + \log .982$$

$$\log D = \log .0463 + \log 824$$

Here is a good way to organize the work.

$$\begin{aligned}
\log 31.4 &= 1.4969 \\
3 \log 31.4 &= 4.4907 \\
(+) \quad \log .982 &= \underline{.9921 - 1} \\
\log N &= 5.4828 - 1 \\
&= 4.4828
\end{aligned}$$

$$\begin{aligned}
\log .0463 &= .6656 - 2 \\
(+) \quad \log 824 &= \underline{2.9159} \\
\log D &= 3.5815 - 2 \\
&= 1.5815
\end{aligned}$$

$$\begin{aligned}
\log N &= 4.4828 \\
(-) \quad \log D &= \underline{1.5815} \\
\log x &= 2.9013 \\
x &= 797
\end{aligned}$$

Carry out the following calculations using logarithms without interpolation.

21. $\dfrac{(.56)^2(619)}{21.8}$

22. $\dfrac{.413}{(4.9)^2(.724)}$

23. $\dfrac{(14.3)\sqrt{92.3}}{\sqrt[3]{432}}$

24. $\dfrac{(91)(41.3)^{2/3}}{42.6}$

EXAMPLE B (Solving Exponential Equations)   Solve the equation

$$2^{2x-1} = 13$$

Solution.   We begin by taking logarithms of both sides and then solving for x.

$$(2x - 1)\log 2 = \log 13$$

$$2x - 1 = \frac{\log 13}{\log 2} = \frac{1.1139}{.3010}$$

$$\log(2x - 1) = \log 1.1139 - \log .3010$$

$$= (1.0469 - 1) - (.4786 - 1)$$

$$= .5683$$

$$2x - 1 = \text{antilog } .5683 = 3.70$$

$$2x = 4.70$$

$$x = 2.35$$

Notice that we did the division of $(\log 13)/(\log 2)$ by means of logarithms. We could have done it by long division (or on a calculator) if we preferred.

Use logarithms to solve the following exponential equations. You need not interpolate.

25. $3^x = 300$ 26. $5^x = 14$ 27. $10^{2-3x} = 6240$

28. $10^{5x-1} = .00425$ 29. $2^{3x} = 3^{x+2}$ 30. $2^{x^2} = 3^x$

MISCELLANEOUS PROBLEMS

31. Use common logarithms and Table B to calculate each of the following.

(a) $\sqrt[3]{.0427}$ (b) $\dfrac{(42.9)^2(.983)}{\sqrt{323}}$ (c) $\dfrac{10^{6.42}}{8^{7.2}}$

32. Solve for x by using common logarithms.
    (a) $4^{2x} = 150$ (b) $(.975)^x = .5$

33. Solve for x.
    (a) $\log(x + 2) - \log x = 1$ (b) $\log(2x + 1) - \log(x + 3) = 0$
    (c) $\log(x + 3) - \log x + \log 2x^2 = \log 8$
    (d) $\log(x + 3) + \log(x - 1) = \log 4x$

34. Suppose that the amount Q of a radioactive substance (in grams) remaining t years from now will be $Q = (42)2^{-.017t}$. After how many years will the amount remaining be .42 grams?

35. Suppose that the bacteria count in a certain culture t hours from now is $(800)3^t$. When will the count reach 100,000?

36. Assume the population of the earth was 4.19 billion people in 1977 and that the growth rate is 2 percent per year. Then the population t years after 1977 should be $4.19(1.02)^t$ billion.
    (a) What will the population be in the year 2000?
    (b) When will the population reach 8.3 billion?

37. Answer the two questions of Problem 36 assuming the growth rate is only 1.7 percent.

38. TEASER   Show that log 2 is irrational. For what positive integers n is log n irrational?

Chapter Summary

The symbol $\sqrt[n]{a}$, the principal nth root of a, denotes one of the numbers whose nth power is a. For odd n, that is all that needs to be said. For n even and $a > 0$, we specify that $\sqrt[n]{a}$ signifies the positive nth root. Thus $\sqrt[3]{-8} = -2$ and $\sqrt{16} = \sqrt[2]{16} = 4$. (It is wrong to write $\sqrt{16} = -4$.) These symbols are also called **radicals.** These radicals obey four carefully prescribed rules (page 227). These rules allow us to simplify complicated radical expressions, in particular, to **rationalize denominators.**

The key to understanding **rational exponents** is the definition $a^{1/n} = \sqrt[n]{a}$, which implies $a^{m/n} = (\sqrt[n]{a})^m$. Thus $16^{5/4} = (\sqrt[4]{16})^5 = 2^5 = 32$. The meaning of **real exponents** is determined by considering rational approximations. For example, 2^π is the number that the sequence 2^3, $3^{3.1}$, $2^{3.14}$, ... approaches. The function $f(x) = a^x$ (and more generally $f(x) = b \cdot a^x$) is called an **exponential function.**

A variable y is **growing exponentially** or **decaying exponentially** according as $a > 1$ or $0 < a < 1$ in the equation $y = b \cdot a^x$. Typical of the former are biological populations; of the latter, radioactive elements. Corresponding key ideas are **doubling times** and **half-lives.**

Logarithms are exponents. In fact, $\log_a N = x$ means $a^x = N$, that is, $a^{\log_a N} = N$. The functions $f(x) = \log_a x$ and $g(x) = a^x$ are **inverses** of each other. Logarithms have three primary properties (page 248). **Natural logarithms** correspond to the choice of base $a = e = 2.71828 \ldots$ and play a fundamental role in advanced courses. **Common logarithms** correspond to base 10 and have historically been used to simplify arithmetic calculations.

Chapter Review Problem Set

1. Simplify, rationalizing all denominators. Assume letters represent positive numbers.

   (a) $\sqrt[3]{\dfrac{-8y^6}{z^{14}}}$

   (b) $\sqrt[4]{32x^5y^8}$

   (c) $\sqrt{4\sqrt[3]{5}}$

   (d) $\dfrac{2}{\sqrt{x} - \sqrt{y}}$

   (e) $\sqrt{50 + 25x^2}$

   (f) $\sqrt{32} + \sqrt{8}$

2. Solve the equations.

   (a) $\sqrt{x - 3} = 3$

   (b) $\sqrt{x} = 6 - x$

3. Simplify, writing your answer in exponential form with all positive exponents.

   (a) $(25a^2)^{3/2}$

   (b) $(a^{-1/2}aa^{-3/4})^2$

   (c) $\dfrac{1}{5\sqrt[4]{5^3}}$

   (d) $\dfrac{(3x^{-2}y^{3/4})^2}{3x^2y^{-2/3}}$

   (e) $(x^{1/2} - y^{1/2})^2$

   (f) $\sqrt[3]{4}\,\sqrt[4]{4}$

4. Sketch the graph of $y = (\tfrac{3}{2})^x$ and use your graph to estimate the value of $(\tfrac{3}{2})^{\pi}$.

5. A certain radioactive substance has a half-life of 3 days. What fraction of an initial amount will be left after 243 days?

6. A population grows so that its doubling time is 40 years. If this population is 1 million today, what will it be after 160 years?

7. If $100 is put in the bank at 8 percent interest compounded quarterly, what will it be worth at the end of 10 years?

8. Find x in each of the following.

   (a) $\log_4 64 = x$

   (b) $\log_2 x = -3$

   (c) $\log_x 49 = 2$

   (d) $\log_4 x = 0$

   (e) $\log_9 27 = x$

   (f) $\log_{10} x + \log_{10}(x - 3) = 1$

   (g) $a^{\log_a 10} = x$

   (h) $x = \log_a(a^{1.14})$

9. Write as a single logarithm.

$$2\log_4(3x + 1) - \frac{1}{2}\log_4 x + \log_4(x - 1)$$

10. Evaluate.

   (a) $\log_{10}\sqrt{1000}$

   (b) $\log_{27} 81$

   (c) $\ln\sqrt{e}$

   (d) $\ln(1/2^4)$

11. Use the table of natural logarithms to determine each of the following.

   (a) $\ln(\tfrac{7}{4})^3$

   (b) N if $\ln N = 2.230$

   (c) $\ln[(3.4)(9.9)]$

   (d) N if $\ln N = -0.105$

12. By taking ln of both sides, solve $2^{x+1} = 7$.

13. A certain substance decays according to the formula $y = y_0e^{-.05t}$, where t is measured in years. Find its half-life.

14. A substance initially weighing 100 grams decays exponentially according to the formula $y = 100e^{-kt}$. If 30 grams are left after 10 days, determine k.

15. Sketch the graphs of $y = \log_3 x$ and $y = 3^x$ using the same coordinate axes.

16. Use common logarithms to calculate

$$\frac{(13.2)^4\sqrt{15.2}}{29.6}$$

The great book of Nature lies open before our eyes and true philosophy is written in it. . . . But we cannot read it unless we have first learned the language and characters in which it is written. . . . It is written in mathematical language and the characters are triangles, circles, and other geometrical figures.

—Galileo

CHAPTER 7

The Trigonometric Functions

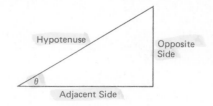

Sometime before 100 B.C., the Greeks invented trigonometry to solve problems in astronomy, navigation, and geography. The word "trigonometry" comes from Greek and means "triangle measurement." In its most basic form, trigonometry is the study of relationships between the angles and sides of a right triangle.

7-1 Right-Triangle Trigonometry

A triangle is called a *right triangle* if one of its angles is a right angle, that is, a 90° angle. The other two angles are necessarily acute angles (less than 90°) since the sum of all three angles in a triangle is 180°. Let θ (the Greek letter theta) denote one of these acute angles. We may label the three sides relative to θ: adjacent side, opposite side, and hypotenuse, as shown in the diagram above. In terms of these sides, we introduce the three fundamental ratios of trigonometry, sine θ, cosine θ, and tangent θ. Using obvious abbreviations, we give the following definitions.

$$\sin \theta = \frac{\text{opp}}{\text{hyp}}$$

$$\cos \theta = \frac{\text{adj}}{\text{hyp}}$$

$$\tan \theta = \frac{\text{opp}}{\text{adj}}$$

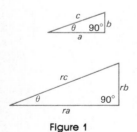

Figure 1

Thus with every acute angle θ, we associate three numbers, sin θ, cos θ, and tan θ. A careful reader might wonder whether these numbers depend only on the size of θ, or if they also depend on the lengths of the sides of the right triangle with which we started. Consider two different right triangles, each with the same angle θ (as in Figure 1). You may think of the lower triangle as a magnification of the upper one. Each of its sides has length r times that of the corresponding side in the upper triangle. If we calculate sin θ from the lower triangle, we get

$$\sin \theta = \frac{\text{opp}}{\text{hyp}} = \frac{rb}{rc} = \frac{b}{c}$$

which is the same result we get using the upper triangle. We conclude that for a given θ, sin θ has the same value no matter which right triangle is used to compute it. So do cos θ and tan θ.

SPECIAL ANGLES

We can use the Pythagorean theorem $(a^2 + b^2 = c^2)$ to find the values of sine, cosine, and tangent for the special angles $30°$, $45°$, and $60°$. Consider the two right triangles of Figure 2, which involve these angles.

Figure 2

To see that the indicated values of a are correct, note that in the first triangle, $a^2 + a^2 = 2^2$, which gives $a = \sqrt{2}$. In the second, which is half of an equilateral triangle, $a^2 + 1^2 = 2^2$, or $a = \sqrt{3}$.

From these triangles, we obtain the following important facts.

$$\sin 45° = \frac{\sqrt{2}}{2} \qquad \cos 45° = \frac{\sqrt{2}}{2} \qquad \tan 45° = 1$$

$$\sin 30° = \frac{1}{2} \qquad \cos 30° = \frac{\sqrt{3}}{2} \qquad \tan 30° = \frac{1}{\sqrt{3}} = \frac{\sqrt{3}}{3}$$

$$\sin 60° = \frac{\sqrt{3}}{2} \qquad \cos 60° = \frac{1}{2} \qquad \tan 60° = \sqrt{3}$$

OTHER ANGLES

When you need the sine, cosine, or tangent of an angle other than the special ones just considered, you may do one of two things. If you have a scientific calculator, you may simply push two or three keys and have your answer correct to eight or more significant digits. Otherwise, you will need to use Table C in the appendix.

Several facts about Table C should be noted. First, it gives answers usually to four decimal places. Second, angles are measured in degrees and tenths of degrees. By interpolation (see page 266), it is possible to consider angles measured to the nearest hundredth of a degree. Finally, notice that the left column of the table lists angles from $0°$ to $45°$. For angles from $45°$ to $90°$, use the right column; you must then also use the bottom captions. To make sure that you are reading the table (or your calculator) correctly, check that you get each of the following answers.

$$\tan 33.1° = .6519 \qquad \sin 26.9° = .4524$$

$$\cos 54.3° = .5835 \qquad \tan 82° = 7.115$$

APPLICATIONS

Suppose that you wish to measure the distance across a stream but do not want to get your feet wet. Here is how you might proceed.

Pick out a tree at C on the opposite shore and set a stone at B directly across from it on your shore (Figure 3). Set another stone at A, 100 feet up the shore from B. With an angle measuring device (for example, a protractor or a transit), measure angle θ between AB and AC. Then x, the length of BC, satisfies the following equation.

$$\tan \theta = \frac{\text{opp}}{\text{adj}} = \frac{x}{100}$$

or

$$x = 100 \tan \theta$$

For example, if θ measures 29°, you find from your scientific calculator or Table C that tan 29° = .5543. Then $x = 100(.5543) = 55.43$ feet. Since you used stones and trees for points, this suggests that you should not give your answer with such accuracy. It would be better to say that the distance x is approximately 55 feet.

As a more difficult example, consider a church with a steeple, as shown in Figure 4. The problem is to calculate the height of the steeple while standing on the ground. To find the height, mark a point B on the ground directly below the steeple and another point A 200 feet away on the ground. At A, measure the *angles of elevation* α and β to the top and bottom of the steeple. This is all the information you will need, provided you know your trigonometry.

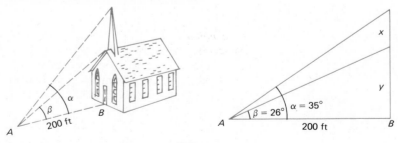

Figure 4

Let x be the height of the steeple and y be the distance from the ground to the bottom of the steeple. Suppose that $\alpha = 35°$ and $\beta = 26°$. Then

$$\tan 35° = \frac{x + y}{200}$$

$$\tan 26° = \frac{y}{200}$$

If you solve for y in the second equation and substitute the value in the first, you will get the following sequence of equations.

$$\tan 35° = \frac{x + 200 \tan 26°}{200}$$

$$200 \tan 35° = x + 200 \tan 26°$$

$$x = 200 \tan 35° - 200 \tan 26°$$

$$= 200(.7002 - .4877)$$

$$x = 42.5 \text{ feet}$$

Problem Set 7-1

In Problems 1–6, use Table C to evaluate each expression. If you have a scientific calculator, use it as a check.

1. $\sin 41.3°$ 2. $\tan 54.4°$ 3. $\cos 49.2°$

4. $\sin 89.3°$ 5. $\tan 72.3°$ 6. $\cos 38.7°$

In Problems 7–12, use Table C to find θ. We suggest you also do these problems on a calculator. For example, to do Problem 7 on many calculators, press .2164 INV sin . This will give the inverse sine of .2164, that is, the angle whose sine is .2164.

7. $\sin \theta = .2164$ 8. $\tan \theta = .3096$ 9. $\tan \theta = 2.311$

10. $\cos \theta = .9354$ 11. $\cos \theta = .3535$ 12. $\sin \theta = .7302$

Each of the remaining problems in this problem set involves a considerable amount of arithmetic that you can do by hand (using tables) or by using a calculator. (If you use a calculator to find values for the trigonometric functions, be sure that it is in the degree mode.) In Problems 13–18, find x.

13.
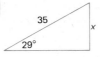

14.

15.

16.

17.
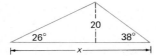

18.
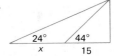

EXAMPLE A (Solving a Right Triangle Given an Angle and a Side) To solve a triangle means to determine all its unknown parts. Solve the right triangle which has hypotenuse of length 14.6 and an angle measuring 33.2°.

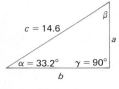

Figure 5

Solution.   First, we draw the triangle labeling the known parts and assigning letters to the unknown parts. Our convention is to use the first three Greek letters, α, β, and γ (alpha, beta, and gamma) for the angles and *a*, *b*, and *c* for the lengths of the respective sides opposite these angles (see Figure 5). We need to find β, *a*, and *b*.

(i) $\beta = 90° - 33.2° = 56.8°$

(ii) $\sin 33.2° = a/14.6$, so

$$a = 14.6 \sin 33.2° = (14.6)(.5476) \approx 7.99$$

(iii) $\cos 33.2° = b/14.6$, so

$$b = 14.6 \cos 33.2° = (14.6)(.8368) \approx 12.2$$

Notice that we gave the answers to three significant digits since the given data have three significant digits.

Solve each of the following triangles. First draw the triangle, labeling it as in the example with γ = 90°.

19. $\alpha = 42°, c = 35$ 20. $\beta = 29°, c = 50$

21. $\beta = 56.2°, c = 91.3$ 22. $\alpha = 69.9°, c = 10.6$

23. $\alpha = 39.4°, a = 120$ 24. $\alpha = 40.6°, b = 163$

EXAMPLE B (Solving a Right Triangle Given Two Sides) Solve the right triangle which has legs $a = 42.8$ and $b = 94.1$.

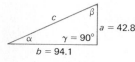

Figure 6

Solution.   First, we draw the triangle and label its parts (Figure 6). We must find α, β, and *c*.

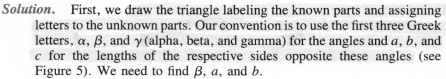

(i) $\tan \alpha = \dfrac{42.8}{94.1} \approx .4548$

Now we can find α by using Table C backwards, that is, by looking under tangent in the body of the table for .4548 and determining the corresponding angle. Or better, we can use the ⟨INV⟩⟨tan⟩ keys on a scientific calculator. On many calculators, press

$$\boxed{(}\ 42.8\ \boxed{\div}\ 94.1\ \boxed{)}\ \boxed{INV}\ \boxed{tan}$$

In either case, the result is $\alpha \approx 24.5°$.

(ii) $\beta = 90° - \alpha \approx 90° - 24.5° = 65.5°$

(iii) We could find *c* by using $c^2 = a^2 + b^2$. Instead, we use $\sin \alpha$.

$$\sin \alpha = \sin 24.5° = \frac{42.8}{c}$$

$$c = \frac{42.8}{\sin 24.5°} = \frac{42.8}{.4147} \approx 103$$

Solve the right triangles satisfying the given information in Problems 25–32, assuming that c is the hypotenuse. You can do them either with tables or a calculator.

25. $a = 9$, $b = 12$
26. $a = 24$, $b = 10$
27. $a = 40$, $c = 50$
28. $c = 41$, $a = 40$
29. $a = 14.6$, $c = 32.5$
30. $a = 243$, $c = 419$
31. $a = 9.52$, $b = 14.7$
32. $a = .123$, $b = .456$

33. A straight path leading up a hill rises 26 feet per 100 horizontal feet. What angle does it make with the horizontal?

34. A 20-foot ladder leans against a wall, making an angle of 76° with the level ground. How high up the wall is the upper end of the ladder?

35. Find the angle of elevation of the sun if a woman 5 feet 9 inches tall casts a shadow 46.8 feet long. (The *angle of elevation* is the upward angle made with the horizontal.)

36. A guy wire to a pole makes an angle of 69° with the level ground and is 14 feet from the pole at the ground. How high above the ground is the wire attached to the pole?

37. Suppose that the woman in Problem 35 is walking with her daughter Sue, who is 3 feet 10 inches tall. How long is Sue's shadow?

38. Find the length of the supporting wire in Problem 36.

MISCELLANEOUS PROBLEMS

The following problems can be solved using either a calculator or tables. We recommend using a calculator.

39. Calculate each value.
    (a) tan 14.5°
    (b) 24.6 cos 74.3°
    (c) $15.6 (\sin 14°)^2/\cos 87°$

40. Find θ in each case.
    (a) $\sin \theta = .6691$
    (b) $\cos \theta = .5519$
    (c) $\tan \theta = 5.396$

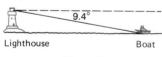

Lighthouse   Boat

Figure 7

41. From the top of a lighthouse 120 feet above sea level, the *angle of depression* (the downward angle from the horizontal) to a boat adrift on the sea is 9.4° (Figure 7). How far from the foot of the lighthouse is the boat?

42. Solve the right triangle in which $b = 67.3$ and $c = 82.9$.

43. Find x in Figure 8.

44. When the *angle of elevation* (the upward angle from the horizontal) of the sun is 28.4°, the Eiffel Tower in Paris casts a horizontal shadow 1822 feet long. How high is the tower?

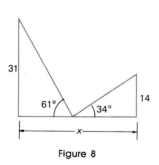

Figure 8

45. With her hands 5 feet above the ground, Sally is pulling on a kite. If the kite is 200 feet above the ground and the kite string makes an angle of 32.4° with the horizontal, how many feet of string are out?

46. A plane is flying directly away from a ground observer at a constant rate, maintaining an elevation of 15,000 feet. At a certain instant, the observer measures the angle of elevation as 44° and 15 seconds later as 31°. How fast is the plane flying in miles per hour?

47. From a window in an office building, I am looking at a television tower that is 600 meters away (horizontally). The angle of elevation of the top of the tower

48. The vertical distance from first to second floor of a certain department store is 28 feet. The escalator, which has a horizontal reach of 96 feet, takes 25 seconds to carry a person between floors. How fast does the escalator travel?

49. The Great Pyramid is about 480 feet high and its square base measures 760 feet on a side. Find the angle of elevation of one of its edges, that is, find β in Figure 9.

50. Find the angle between a principal diagonal and a face diagonal of a cube.

51. A regular hexagon (6 equal sides) is inscribed in a circle of radius 4. Find the perimeter P and area A of this hexagon.

52. A regular decagon (10 equal sides) is inscribed in a circle of radius 12. What percent of the area of the circle is the area of the decagon?

53. Find the area of the regular 6-pointed Star of David that is inscribed in a circle of radius 1 (Figure 10).

54. **TEASER** Find the area of the regular 5-pointed star (the pentagram) that is inscribed in a circle of radius 1 (Figure 11).

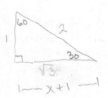

Figure 9

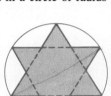

Figure 10

Figure 11

$$TAN\ 30° = \frac{1\ (x+1)}{\sqrt{3}\,(x+1)}$$

$$.57735 = \frac{x+1}{1.732x+1.732}$$

$$(1.732x+1.732)(.57735) = x+1$$

$$17.32x + 17.32 = x+1$$

$$17.32x - x = 1 - 17.32$$

$$x(17.32-1) = 1-17.32$$

$$x = \frac{-16.32}{16.32}$$

The Dynamic View of Angles

In geometry, we take a static view of angles. An angle is simply the union of two rays with a common endpoint (the vertex). In trigonometry, angles are thought of in a dynamic way. An angle is determined by rotating a ray about its endpoint from an initial position to a terminal position.

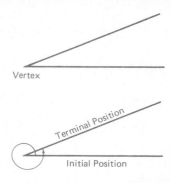

7-2 Angles and Arcs

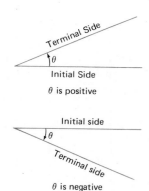

θ is positive

θ is negative

Figure 12

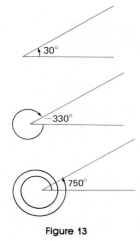

Figure 13

For the solution of right triangles (which involve acute angles), we required only the familiar and simple notion of angle from high-school geometry. But for the broader development of trigonometry, we need the new perspective on angles suggested by our opening display. Not only do we allow arbitrarily large angles, but we also distinguish between positive and negative angles. If an angle is generated by a counterclockwise rotation, it is positive; if generated by a clockwise rotation, it is negative (Figure 12). To know an angle, in trigonometry, is to know how the angle came into being. It is to know the initial side, the terminal side, and the kind of rotation that produced the angle.

DEGREE MEASUREMENT

Take a circle and divide its circumference into 360 equal parts. The angle with vertex at the center determined by one of these parts has measure one **degree** (written 1°). This way of measuring angles is due to the ancient Babylonians and is so familiar that we used it in Section 7-1 without comment. There is a refinement, however, that we avoid. The Babylonians divided each degree into 60 minutes and each minute into 60 seconds; some people still follow this cumbersome practice. If we need to measure angles to finer accuracy than a degree, we will use decimal parts. Thus we write 40.5° rather than 40°30′.

It is important that we be familiar with measuring both positive and negative angles, as well as angles resulting from large rotations. Three angles are shown in Figure 13. Note that all three have the same initial and terminal sides.

RADIAN MEASUREMENT

The best way to measure angles is in radians. Take a circle of radius r. The familiar formula $C = 2\pi r$ tells us that the circumference has 2π (about 6.28) arcs of length r around it. The angle with vertex at the center of a circle

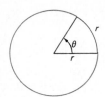

θ measures
one radian
(about 57.3°)

Figure 14

determined by an arc of length equal to its radius measures one **radian** (Figure 14). Thus an angle of size 360° measures 2π radians and an angle of size 180° measures π radians. We abbreviate the latter by writing

$$180° = \pi \text{ radians}$$

To convert from degrees to radians, all one needs to remember is the result in the box. By dividing by 2, 3, 4, and 6, respectively, we get the conversions for several special angles.

$$90° = \frac{\pi}{2} \text{ radians}$$

$$60° = \frac{\pi}{3} \text{ radians}$$

$$45° = \frac{\pi}{4} \text{ radians}$$

$$30° = \frac{\pi}{6} \text{ radians}$$

If we divide the boxed formula by 180, we get

$$1° = \frac{\pi}{180} \text{ radians}$$

and if we divide by π, we get

$$\frac{180°}{\pi} = 1 \text{ radian}$$

The following rules thus hold.

To convert from degrees to radians, multiply by $\pi/180$.
To convert from radians to degrees, multiply by $180/\pi$.

For example,

$$22° = 22\left(\frac{\pi}{180}\right) \text{ radians} \approx .38397 \text{ radians}$$

and

$$2.3 \text{ radians} = 2.3\left(\frac{180}{\pi}\right)° \approx 131.78°$$

Some scientific calculators have a key that makes these conversions automatically.

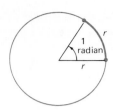

$s = 2r$

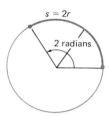

$s = tr$

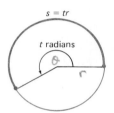

Figure 15

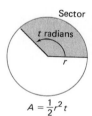

$A = \frac{1}{2}r^2 t$

Figure 16

ARC LENGTH AND AREA

Radian measure is almost invariably used in calculus because it is an intrinsic measure. The division of a circle into 360 parts was quite arbitrary; its division into parts of radius length (2π parts) is more natural. Because of this, formulas using radian measure tend to be simple, while those using degree measure are often complicated. As an example, consider arc length. Let t be the radian measure of an angle θ with vertex at the center of a circle of radius r. This angle cuts off an arc of length s which satisfies the simple formula

$$s = rt$$

This follows directly from the fact that an angle of one radian ($t = 1$) cuts off an arc of length r (see Figure 15).

A second nice formula is that for the area of the sector cut off from a circle by a central angle of t radians (Figure 16). Note that the area A of this sector is to the area of the whole circle as t is to 2π, that is, $A/\pi r^2 = t/2\pi$. Thus

$$A = \frac{1}{2}r^2 t$$

$$A = \frac{t \times r^2}{2\pi} = \frac{tr^2}{2}$$
$$= \frac{1}{2}tr^2$$

THE UNIT CIRCLE

The formula for arc length takes a particularly simple form when $r = 1$, namely, $s = t$. We emphasize its meaning. *On the unit circle, the length of an arc is the same as the radian measure of the angle it determines.*

Someone is sure to point out a difficulty in what we have just said. What happens when t is greater than 2π or when t is negative? To understand our meaning, imagine an infinitely long string on which the real number scale has been marked. Think of wrapping this string around the unit circle as shown in Figure 17.

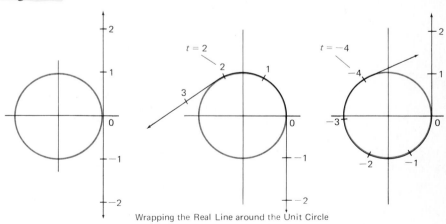

Wrapping the Real Line around the Unit Circle

Figure 17

Now if we think of the directed length (that is, the signed length) of a piece of the string, the formula $s = t$ holds no matter what t is. For example, the length of string corresponding to an angle of 8π radians is 8π. That piece of string wraps counterclockwise around the unit circle exactly 4 times. A piece of string corresponding to an angle of -3π radians would wrap clockwise around the unit circle one and a half times, its directed length being -3π.

Problem Set 7-2

Convert each of the following to radians. You may leave π in your answer.

1. 120°
2. 225°
3. 240°
4. 150°
5. 210°
6. 330°
7. 315°
8. 300°
9. 540°
10. 450°
11. −420°
12. −660°
13. 160°
14. 200°
15. $(20/\pi)°$
16. $(150/\pi)°$

Convert each of the following to degrees. Give your answer correct to the nearest tenth of a degree.

17. $\frac{4}{3}\pi$ radians *240°*
18. $\frac{5}{6}\pi$ radians
19. $-\frac{2\pi}{3}$ radians *−120°*

20. $-\frac{7\pi}{4}$ radians
21. 3π radians *540°*
22. 3 radians

c 23. 4.52 radians *258.9°*
c 24. $\frac{11}{4}$ radians
c 25. $\frac{1}{\pi}$ radians *18.2°*

c 26. $\frac{4}{3\pi}$ radians

27. Find the radian measure of the angle at the center of a circle of radius 6 inches which cuts off an arc of length

    (a) 12 inches;
    (b) 18.84 inches.

28. Find the length of the arc cut off on a circle of radius 3 feet by an angle at the center of

    (a) 2 radians;
    (b) 5.5 radians;

    (c) $\frac{\pi}{4}$ radians;
    (d) $\frac{5\pi}{6}$ radians.

29. Find the radius r for each of the following.

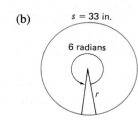

(a) s = 8.4 cm

2.8 radians

r

(b) s = 33 in.

6 radians

r

30. Through how many radians does the minute hand of a clock turn in 1 hour? The hour hand in 1 hour? The minute hand in 5 hours?

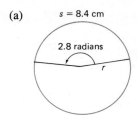

Figure 18

EXAMPLE A (Locating a Point on the Unit Circle)   Figure 18 shows a unit circle with center at the origin. Suppose that a point P moves in a counterclockwise direction around the circle starting at $(1, 0)$. In which quadrant is P when it has traveled a distance of 4 units? Of 40 units?

Solution.   Keep in mind that the distance P travels equals the radian measure of the angle through which OP turns. A distance of 4 units puts P in quadrant III since $\pi < 4 < 3\pi/2$. Once around the circle is $2\pi \approx 6.28$ units. If you divide 40 by 6.28, you get

$$40 = 6(6.28) + 2.32$$

Since 2.32 is between $\pi/2$ and π, traveling 40 units around the unit circle will put P in quadrant II.

Find the quadrant in which the point P in the example above lies when it has traveled each of the following distances.

31. 3 units

32. 3.2 units

33. 4.7 units

34. 4.8 units

35. $\left(\dfrac{5\pi}{2} + 1\right)$ units

36. $\left(\dfrac{9\pi}{2} - 1\right)$ units

37. 100 units

38. 200 units

EXAMPLE B (Angular Velocity)   A formula closely related to the arc length formula $s = rt$ is the formula

$$\boxed{v = r\omega}$$

which connects the speed (velocity) of a point on the rim of a wheel of radius r with the angular velocity ω at which the wheel is turning. Here ω is measured in radians per unit of time. Use this formula to determine the angular velocity in radians per second of a bicycle wheel of radius 16 inches if the bicycle is being ridden down the road at 30 miles per hour.

Solution.   We must use consistent units. You can check that the speed of a point on the rim of the wheel (30 miles per hour) translates to 44 feet per second and that the radius of the wheel is $\frac{4}{3}$ feet. Thus

$$44 = \frac{4}{3}\omega$$

or

$$\omega = \frac{3}{4}(44) = 33 \text{ radians per second}$$

39. Sally is pedaling her tricycle so the front wheel (radius 8 inches) turns at 4 revolutions per second. How fast is she moving down the sidewalk in feet per second? *Hint:* Four revolutions per second is 8π radians per second.

40. Suppose that the tire on a car has an outer diameter 2.5 feet. How many revolutions per minute does the tire make when the car is traveling 60 miles per hour?

41. A dead fly is stuck to a belt that passes over two pulleys 6 inches and 8 inches in radius, as shown in Figure 19. Assuming no slippage, how fast is the fly moving when the larger pulley turns at 20 revolutions per minute?

42. How fast (in revolutions per minute) is the smaller wheel in Problem 41 turning?

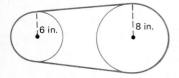

Figure 19

MISCELLANEOUS PROBLEMS

43. Convert to radians.
    (a) $-1440°$ (b) $2\frac{1}{2}$ revolutions   (c) $(60/\pi)°$

44. Convert to degrees.
    (a) $23\pi/36$ radians   (b) -4.63 radians   (c) $3/(2\pi)$ radians

45. Find the length of arc cut off on a circle of radius 4.25 centimeters by each central angle.
    (a) 6 radians   (b) $(18/13\pi)°$ (c) $17\pi/6$ radians

46. The front wheel of Tony's tricycle has a diameter of 20 inches. How far did he travel in pedaling through 60 revolutions?

47. The pedal sprocket of Maria's bicycle has radius 12 centimeters, the rear wheel sprocket has radius 3 centimeters, and the wheels have radius 40 centimeters. How far did Maria travel if she pedaled continuously for 30 revolutions of the pedal sprocket?

48. A belt traveling at the rate of 60 feet per second drives a pulley (a wheel) at the rate of 900 revolutions per minute. Find the radius of the pulley.

49. Assume that the earth is a sphere of radius 3960 miles. How fast (in miles per hour) is a point on the equator moving as a result of the earth's rotation about its axis?

$\boxed{c}$ 50. The orbit of the earth about the sun is an ellipse that is nearly circular with radius 93 million miles. Approximately, what is the earth's speed (in miles per hour) in its path around the sun? You will need the fact that a complete orbit takes 365.25 days.

51. The angle subtended by the sun at the earth (93 million miles away) is .0093 radians. Find the diameter of the sun.

$\boxed{c}$ 52. A nautical mile is the length of 1 minute ($\frac{1}{60}$ of a degree) of arc on the equator of the earth. How many miles are there in a nautical mile?

53. One of the authors (Dale Varberg) lives at exactly 45° latitude north (see Figure 20). How long would it take him to fly to the North Pole at 600 miles per hour (assuming the earth is a sphere of radius 3960 miles)?

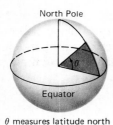

θ measures latitude north

Figure 20

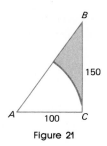

Figure 21

Figure 22

54. New York City is located at 40.5° latitude north. How far is it from there to the equator?

55. Oslo, Norway, and Leningrad, Russia, are both located at 60° latitude north. Oslo is at longitude 6° east (of the prime meridian) whereas Leningrad is at 30° east. How far apart are these two cities along the 60° parallel?

56. Find the area of the shaded region of the right triangle ABC shown in Figure 21.

57. The minute hand and hour hand of a clock are both 6 inches long and reach to the edge of the dial. Find the area of the pie-shaped region between the two hands at 5:40.

58. A cone has radius of base R and slant height L. Find the formula for its lateral surface area. *Hint:* Imagine the cone to be made of paper, slit it up the side, and lay it flat in the plane.

59. Find the area of the polar rectangle shown in Figure 22. The two curves are arcs of concentric circles.

60. TEASER Consider two circles both of radius r and with the center of each lying on the rim of the other. Find the area of the common part of the two circles.

Definitions of Sine and Cosine

Place an angle θ, whose radian measure is t, in **standard position**, that is, put θ in the coordinate plane so that its vertex is at the origin and its initial side is along the positive x-axis. Let (x,y) be the coordinates of the point of intersection of the terminal side with the unit circle. We define both sin θ (sine of θ) and sin t by

$$\sin \theta = \sin t = y.$$

Similarly,

$$\cos \theta = \cos t = x.$$

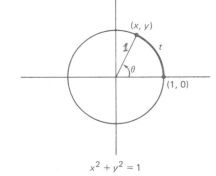

7-3 The Sine and Cosine Functions

In Section 7-1, we defined the sine and cosine for positive acute angles. The definitions in our opening display are more general and hence more widely applicable. They should be studied carefully. Notice that we have defined the sine and cosine for any angle θ and also for the corresponding number t. Both concepts are important. In geometric situations, angles play a central role; thus we are likely to need sines and cosines of angles. But in most of pure mathematics and in many scientific applications, it is the trigonometric functions of numbers that are important. In this connection, we emphasize that the number *t* may be positive or negative, large or small. And we may think of it as the radian measure of an angle, as the directed length of an arc on the unit circle, or simply as a number.

CONSISTENCY WITH EARLIER DEFINITIONS

Do the definitions given in Section 7-1 for the sine and cosine of an acute angle harmonize with those given here? Yes. Take a right triangle *ABC* with an acute angle θ. Place θ in standard position, thus determining a point $B'(x, y)$ on the unit circle and a point $C'(x, 0)$ directly below it on the *x*-axis (see Figure 23).

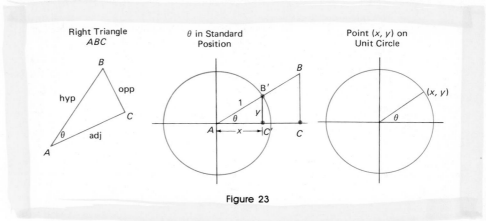

Figure 23

Notice that triangles *ABC* and *AB'C'* are similar. It follows that

$$\frac{\text{opp}}{\text{hyp}} = \frac{BC}{AB} = \frac{B'C'}{AB'} = \frac{y}{1} = y$$

$$\frac{\text{adj}}{\text{hyp}} = \frac{AC}{AB} = \frac{AC'}{AB'} = \frac{x}{1} = x$$

On the left are the old definitions of sin θ and cos θ; on the right are the new ones. They are consistent.

SPECIAL ANGLES

In Section 7-1, we learned that

$$\cos 45° = \frac{\sqrt{2}}{2} \qquad \sin 45° = \frac{\sqrt{2}}{2}$$

$$\cos 30° = \frac{\sqrt{3}}{2} \qquad \sin 30° = \frac{1}{2}$$

$$\cos 60° = \frac{1}{2} \qquad \sin 60° = \frac{\sqrt{3}}{2}$$

Making use of the consistency of the old and new definitions of sine and cosine, we conclude that the point on the unit circle corresponding to $\theta = 45° = \pi/4$ radians must have coordinates $(\sqrt{2}/2, \sqrt{2}/2)$. Similarly, the point corresponding to $\theta = 30° = \pi/6$ radians has coordinates $(\sqrt{3}/2, 1/2)$ and the point corresponding to $\theta = 60° = \pi/3$ radians has coordinates $(1/2, \sqrt{3}/2)$.

Now we can make use of obvious symmetries to find the coordinates of many other points on the unit circle. In the two diagrams of Figure 24, we show a number of these points, noting first the radian measure of the angle and then the coordinates of the corresponding point on the unit circle.

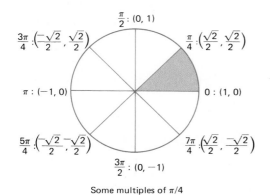

Some multiples of π/4   Some multiples of π/6

Figure 24

Notice, for example, how the coordinates of the points corresponding to $t = 5\pi/6$, $7\pi/6$, and $11\pi/6$ are related to the point corresponding to $t = \pi/6$. You should have no trouble seeing other relationships.

Once we know the coordinates of a point on the unit circle, we can state the sine and cosine of the corresponding angle. In particular, we get the values in the table in Figure 25. They are used so often that you should memorize them.

t	0	$\dfrac{\pi}{6}$	$\dfrac{\pi}{4}$	$\dfrac{\pi}{3}$	$\dfrac{\pi}{2}$	π	$\dfrac{3\pi}{2}$
$\cos t$	1	$\dfrac{\sqrt{3}}{2}$	$\dfrac{\sqrt{2}}{2}$	$\dfrac{1}{2}$	0	-1	0
$\sin t$	0	$\dfrac{1}{2}$	$\dfrac{\sqrt{2}}{2}$	$\dfrac{\sqrt{3}}{2}$	1	0	-1

Figure 25

PROPERTIES OF SINES AND COSINES

Think of what happens to x and y as t increases from 0 to 2π in Figure 26, that is, as P travels all the way around on the unit circle. For example, x steadily decreases until it reaches its smallest value of -1 at $t = \pi$; then it starts to

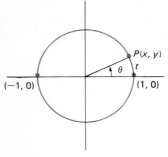

Figure 26

increase until it is back to 1 at $t = 2\pi$. We have just described the behavior of $\cos t$ (or $\cos \theta$) as t increases from 0 to 2π. You should trace the behavior of $\sin t$ in the same way. Notice that both x and y are always between -1 and 1 (inclusive). It follows that

$$-1 \le \sin t \le 1$$
$$-1 \le \cos t \le 1$$

Since P is on the unit circle, $x^2 + y^2 = 1$, and $x = \cos t$ and $y = \sin t$, it follows that

$$(\sin t)^2 + (\cos t)^2 = 1$$

It is conventional to write $\sin^2 t$ instead of $(\sin t)^2$ and $\cos^2 t$ instead of $(\cos t)^2$. Thus we have

$$\sin^2 t + \cos^2 t = 1$$

This is an identity; it is true for all t. Of course we can just as well write

$$\sin^2 \theta + \cos^2 \theta = 1$$

We have established one basic relationship between the sine and the cosine; here are two others, valid for all t.

$$\sin\left(\frac{\pi}{2} - t\right) = \cos t$$
$$\cos\left(\frac{\pi}{2} - t\right) = \sin t$$

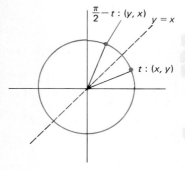

Figure 27

These relationships are easy to see when $0 < t < \pi/2$. Notice that t and $\pi/2 - t$ are measures of complementary angles (two angles with measures totaling $90°$ or $\pi/2$). That means that t and $\pi/2 - t$ determine points on the unit circle which are reflections of each other about the line $y = x$ (see Figure 27). Thus if one point has coordinates (x, y), the other has coordinates (y, x). The result given above follows from this fact.

Finally, we point out that t, $t \pm 2\pi$, $t \pm 4\pi$, ... all determine the same point on the unit circle and thus have the same sine and cosine. This repetitive behavior puts the sine and cosine into a special class of functions, for which we give the following definition. A function f is **periodic** if there is a positive number p such that

$$f(t + p) = f(t)$$

for every t in the domain of f. The smallest such p is called the **period** of f. Thus we say that sine and cosine are periodic functions with period 2π and write

$$\sin(t + 2\pi) = \sin t$$
$$\cos(t + 2\pi) = \cos t$$

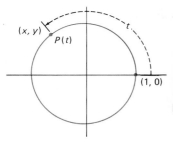

Figure 28

THE TRIGONOMETRIC POINT $P(t)$

We have introduced $\cos t$ and $\sin t$ as the x- and y-coordinates of the point on the unit circle whose directed distance from $(1, 0)$ along the unit circle is t. This point is called a **trigonometric point** and will be denoted by $P(t)$ (Figure 28). We may regard $P(t)$ to be a function of t, since for each t there is a unique point $P(t)$. This function, moreover, is periodic with period 2π—that is,

$$P(t + 2\pi) = P(t)$$

It follows that

$$P(t + k2\pi) = P(t)$$

for any integer k, a fact that allows us to find the coordinates of $P(t)$ for any t, no matter how large t is. Suppose, for example, that we wish to find the coordinates of $P(16\pi/3)$, shown in Figure 29. Since

$$\frac{16\pi}{3} = \frac{4\pi}{3} + 4\pi$$

it follows that

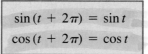

$$P\left(\frac{16\pi}{3}\right) = P\left(\frac{4\pi}{3}\right)$$

From the special angle diagrams on page 293, $P(4\pi/3)$ has coordinates $(-1/2, -\sqrt{3}/2)$. We conclude that $P(16\pi/3)$ also has these coordinates. This means that

$$\cos\left(\frac{16\pi}{3}\right) = -\frac{1}{2} \qquad \sin\left(\frac{16\pi}{3}\right) = -\frac{\sqrt{3}}{2}$$

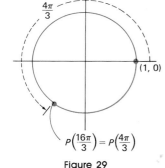

$$P\left(\frac{16\pi}{3}\right) = P\left(\frac{4\pi}{3}\right)$$

Figure 29

Problem Set 7-3

In Problems 1–8, find the coordinates of the trigonometric point P(t) for the indicated value of t. Hint: Begin by drawing a unit circle and locating P(t) on it; then relate P(t) to the diagrams on page 293.

1. $t = \dfrac{13\pi}{6}$

2. $t = \dfrac{19\pi}{6}$

3. $t = \dfrac{19\pi}{4}$

4. $t = \dfrac{15\pi}{4}$

5. $t = 24\pi + \dfrac{5\pi}{4}$

6. $t = 16\pi + \dfrac{5\pi}{6}$

7. $t = -\dfrac{7\pi}{6}$

8. $t = \dfrac{13\pi}{4}$

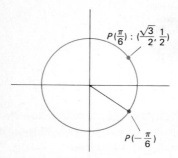

Figure 30

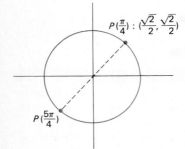

Figure 31

EXAMPLE A (Using $P(t)$ to Find Sine and Cosine Values) Find
(a) $\sin(-\pi/6)$; (b) $\cos(29\pi/4)$.

Solution.

(a) We locate $P(-\pi/6)$ on the unit circle (Figure 30) and note that its y-coordinate is $-\frac{1}{2}$ because of its position relative to $P(\pi/6)$. Thus $\sin(-\pi/6) = -\frac{1}{2}$.

(b) We simplify the problem by removing a large multiple of 2π—that is, by noting that

$$\frac{29\pi}{4} = 6\pi + \frac{5\pi}{4}$$

from which we conclude $P(29\pi/4) = P(5\pi/4)$. Then we refer to the diagrams on page 293, or better yet, we simply observe that $P(5\pi/4)$, being diametrically opposite from $P(\pi/4)$ on the unit circle, has coordinates $(-\sqrt{2}/2, -\sqrt{2}/2)$ (see Figure 31). We conclude that

$$\cos\left(\frac{29\pi}{4}\right) = -\frac{\sqrt{2}}{2}$$

Using the method of Example A, find the value of each of the following.

9. $\sin(-\pi/4)$ 10. $\sin(-5\pi/4)$ 11. $\sin(9\pi/4)$

12. $\sin(15\pi/4)$ 13. $\cos(13\pi/4)$ 14. $\cos(-7\pi/4)$

15. $\cos(10\pi/3)$ 16. $\cos(25\pi/6)$ 17. $\sin(5\pi/2)$

18. $\cos 7\pi$ 19. $\sin(-4\pi)$ 20. $\cos(7\pi/2)$

21. $\cos(19\pi/6)$ 22. $\sin(14\pi/3)$ 23. $\cos(-\pi/3)$

24. $\sin(-5\pi/6)$ 25. $\cos(125\pi/4)$ 26. $\cos(-13\pi/6)$

27. $\sin 510°$ 28. $\sin(-390°)$ 29. $\cos 840°$

30. $\cos(-720°)$ 31. $\cos(-210°)$ 32. $\sin 900°$

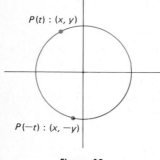

Figure 32

EXAMPLE B (Sine and Cosine of $-t$) Show that for all t

$$\sin(-t) = -\sin t$$
$$\cos(-t) = \cos t$$

that is, sine is an odd function and cosine is an even function.

Solution. The points $P(-t)$ and $P(t)$ are symmetric with respect to the x-axis (Figure 32). Thus if $P(t)$ has coordinates (x, y), $P(-t)$ has coordinates $(x, -y)$ and so

$$\sin(-t) = -y = -\sin t$$

$$\cos(-t) = x = \cos t$$

33. If $\sin 1.87 = .95557$ and $\cos 1.87 = -0.29476$, find $\sin(-1.87)$ and $\cos(-1.87)$.

34. If $\sin 15.2° = 0.2622$ and $\cos 15.2° = 0.9650$, find $\sin(-15.2°)$ and $\cos(-15.2°)$.

35. Given $P(t)$ with coordinates $(1/\sqrt{5}, -2/\sqrt{5})$.
    (a) What are the coordinates of $P(-t)$?
    (b) What are the values of $\sin(-t)$ and $\cos(-t)$?

36. If t is the radian measure of an angle in quadrant III and $\sin t = -\frac{3}{5}$, evaluate each expression.
    (a) $\sin(-t)$
    (b) $\cos t$ *Hint:* Use the fact that $\sin^2 t + \cos^2 t = 1$.
    (c) $\cos(-t)$

37. Note that $P(t)$ and $P(\pi + t)$ are symmetric with respect to the origin (Figure 33). Use this to show that
    (a) $\sin(\pi + t) = -\sin t$;   (b) $\cos(\pi + t) = -\cos t$.

38. Note that $P(t)$ and $P(\pi - t)$ are symmetric with respect to the y-axis. Use this fact to find identities analogous to those in Problem 37 for $\sin(\pi - t)$ and $\cos(\pi - t)$.

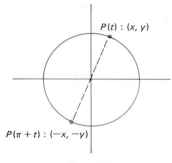

$P(t) : (x, y)$

$P(\pi + t) : (-x, -y)$

Figure 33

MISCELLANEOUS PROBLEMS

39. Use the unit circle to determine the sign (plus or minus) of each of the following.
    (a) $\cos 2$ (b) $\sin(-3)$ (c) $\cos 428°$
    (d) $\sin 21.4$ (e) $\sin(23\pi/32)$ (f) $\sin(-820°)$

40. Find the coordinates of $P(t)$ for the indicated values of t.
    (a) $t = \pi$ (b) $t = -3\pi/2$ (c) $t = -3\pi/4$
    (d) $t = 5\pi/6$ (e) $t = 44\pi/3$ (f) $t = -93.5\pi$

41. Let $P(t)$ have coordinates $(x, -1/2)$.
    (a) Find the two possible values of x.
    (b) Find the corresponding values of t.

42. With initial point $(0, -1)$, a string of length $4\pi/3$ is wound clockwise around the unit circle. What are the coordinates of the terminal point?

43. For what values of t satisfying $0 \le t < 2\pi$ are the following true?
    (a) $\sin t = \cos t$ (b) $\frac{1}{2} < \sin t < \frac{\sqrt{3}}{2}$
    (c) $\cos^2 t \ge .25$ (d) $\cos^2 t > \sin^2 t$

44. Find the four smallest positive solutions to the following equations.
    (a) $\sin t = 1$ (b) $|\cos t| = \frac{1}{2}$
    (c) $\cos t = -\frac{\sqrt{3}}{2}$ (d) $\sin t = -\frac{\sqrt{2}}{2}$

45. In each case, assume that θ is an angle in standard position with terminal side in the fourth quadrant. Use $\sin^2 \theta + \cos^2 \theta = 1$ to determine the indicated value.
    (a) $\cos \theta$ if $\sin \theta = -\frac{4}{5}$
    (b) $\sin \theta$ if $\cos \theta = \frac{24}{25}$

46. Use the unit circle to find identities for $\sin(2\pi - t)$ and $\cos(2\pi - t)$.

47. If $P(t)$ has coordinates $(\frac{4}{5}, -\frac{3}{5})$, evaluate each of the following. *Hint:* For (e) and (f), see Problems 37 and 38.
    (a) $\sin(-t)$ (b) $\sin(\frac{\pi}{2} - t)$
    (c) $\cos(2\pi + t)$ (d) $\cos(2\pi - t)$
    (e) $\sin(\pi + t)$ (f) $\cos(\pi - t)$

sin t	cos t	sin($t + \pi$)	cos($t + \pi$)	sin($\pi - t$)	sin($2\pi - t$)	Least positive value of t
$\sqrt{3}/2$	$-\frac{1}{2}$					
	$\sqrt{2}/2$	$\sqrt{2}/2$				
$-\frac{1}{2}$			$-\sqrt{3}/2$			
-1						
				$\sqrt{3}/2$	$\frac{1}{2}$	
	0				-1	

48. Fill in all the blanks in the chart above.

49. Recall that [ ] and ( ) denote "the greatest integer in" and "the distance to the nearest integer," respectively. Determine which of the following functions are periodic and, if so, specify the period.
(a) $f(x) = (x)$ (b) $f(x) = (3x)$
(c) $f(x) = [x]$ (d) $f(x) = x - [x]$

50. Suppose that $f(x)$ is periodic with period 2 and that $f(x) = 4 - x^2$ for $0 \le x < 2$. Evaluate each of the following.
(a) $f(2)$ (b) $f(4.5)$
(c) $f(-.5)$ (d) $f(8.8)$

51. Evaluate
$$\sin 1° + \sin 2° + \sin 3° + \cdots + \sin 357° + \sin 358° + \sin 359°$$

52. **TEASER**   Evaluate
$$\sin^2 1° + \sin^2 2° + \sin^2 3° + \cdots + \sin^2 357° + \sin^2 358° + \sin^2 359°$$

"Strange as it may sound, the power of mathematics rests on its evasion of all unnecessary thought and on its wonderful saving of mental operations."

Ernst Mach

New Functions from Old Ones

tangent:	$\tan t = \dfrac{\sin t}{\cos t}$
cotangent:	$\cot t = \dfrac{\cos t}{\sin t}$
secant:	$\sec t = \dfrac{1}{\cos t}$
cosecant:	$\csc t = \dfrac{1}{\sin t}$

7-4   Four More Trigonometric Functions

Without question, the sine and cosine are the most important of the six trigonometric functions. Not only do they occur most frequently in applications, but the other four functions can be defined in terms of them, as our opening box shows. This means that if you learn all you can about sines and cosines, you will automatically know a great deal about tangents, cotangents, secants, and cosecants. Ernst Mach would say that it is a way to evade unnecessary thought.

Look at the definitions in the opening box again. Naturally, we must rule out any values of t for which a denominator is zero. For example, $\tan t$ is not defined for $t = \pm\pi/2, \pm 3\pi/2, \pm 5\pi/2$, and so on. Similarly, $\csc t$ is not defined for such values as $t = 0, \pm\pi$, and $\pm 2\pi$.

PROPERTIES OF THE NEW FUNCTIONS

The wisdom of the opening paragraph will now be demonstrated. Recall the identity $\sin^2 t + \cos^2 t = 1$. Out of it come two new identities.

$$1 + \tan^2 t = \sec^2 t$$
$$1 + \cot^2 t = \csc^2 t$$

To show that the first identity is correct, we take its left side, express it in terms of sines and cosines, and do a little algebra.

$$1 + \tan^2 t = 1 + \left(\frac{\sin t}{\cos t}\right)^2$$

$$= 1 + \frac{\sin^2 t}{\cos^2 t}$$

$$= \frac{\cos^2 t + \sin^2 t}{\cos^2 t}$$

$$= \frac{1}{\cos^2 t}$$

$$= \left(\frac{1}{\cos t}\right)^2$$

$$= \sec^2 t$$

The second identity is verified in a similar fashion.

Suppose we wanted to know whether cotangent is an even or an odd function (or neither). We simply recall that $\sin(-t) = -\sin t$ and $\cos(-t) = \cos t$ and write

$$\cot(-t) = \frac{\cos(-t)}{\sin(-t)} = \frac{\cos t}{-\sin t} = -\frac{\cos t}{\sin t} = -\cot t$$

Thus cotangent is an odd function.

In a similar vein, recall the identities

(i)
$$\sin\left(\frac{\pi}{2} - t\right) = \cos t$$

(ii)
$$\cos\left(\frac{\pi}{2} - t\right) = \sin t$$

From them, we obtain

$$(iii) \qquad \tan\left(\frac{\pi}{2} - t\right) = \frac{\sin(\pi/2 - t)}{\cos(\pi/2 - t)} = \frac{\cos t}{\sin t} = \cot t$$

These three identities are examples of what are called **cofunction identities.** Sine and cosine are confunctions; so are tangent and cotangent; as are secant and cosecant. Notice that identities (i), (ii), and (iii) all have the form

$$\text{function}\left(\frac{\pi}{2} - t\right) = \text{cofunction}(t)$$

With cosecant as the function, we have

$$\csc\left(\frac{\pi}{2} - t\right) = \sec t$$

ALTERNATIVE DEFINITIONS OF THE TRIGONOMETRIC FUNCTIONS

There is another approach to trigonometry favored by some authors. Let θ be an angle in standard position and suppose that (a, b) is any point on its terminal side at a distance r from the origin (Figure 34). Then

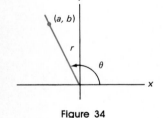

Figure 34

$$\sin\theta = \frac{b}{r} \qquad \cos\theta = \frac{a}{r}$$

$$\tan\theta = \frac{b}{a} \qquad \cot\theta = \frac{a}{b}$$

$$\sec\theta = \frac{r}{a} \qquad \csc\theta = \frac{r}{b}$$

To see that these definitions are equivalent to those we gave earlier, consider first an angle θ with terminal side in quadrant I (see Figure 35).

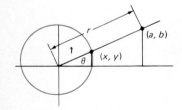

Figure 35

By similar triangles,

$$\frac{b}{r} = \frac{y}{1} \quad \text{and} \quad \frac{a}{r} = \frac{x}{1}$$

Actually these ratios are equal no matter in which quadrant the terminal side of θ is, since b and y always have the same sign, as do a and x. The first two formulas in the box now follow from our original definitions, which say that

$$\sin\theta = y \quad \text{and} \quad \cos\theta = x$$

The others are a consequence of the fact that the remaining four functions can be expressed in terms of sines and cosines.

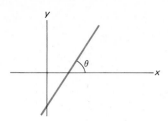

Figure 36

THE TANGENT FUNCTION AND SLOPE

Recall that the slope m of a line is the ratio of rise to run. In particular, if the line goes through the point (a, b) and also the origin, its slope is b/a. But this number b/a is also the tangent of the nonnegative angle θ that the line makes with the positive x-axis (see Figure 34).

In general, the smallest nonnegative angle θ that a line makes with the positive x-axis is called the **angle of inclination** of the line (Figure 36). It follows that for any nonvertical line, the slope m of the line satisfies

$$m = \tan \theta$$

As an example, suppose that a line has angle of inclination $120°$ and goes through the point $(1, 2)$. Then its slope is $m = \tan 120° = -\sqrt{3}$ and the line has equation

$$y - 2 = -\sqrt{3}(x - 1)$$

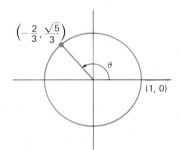

Figure 37

Problem Set 7-4

1. If $\sin t = \frac{4}{5}$ and $\cos t = -\frac{3}{5}$, evaluate each function.
   (a) $\tan t$ (b) $\cot t$ (c) $\sec t$ (d) $\csc t$
2. If $\sin t = -1/\sqrt{5}$ and $\cos t = 2/\sqrt{5}$, evaluate each function.
   (a) $\tan t$ (b) $\cot t$ (c) $\sec t$ (d) $\csc t$
3. Find the values of $\tan \theta$ and $\csc \theta$ for the angle θ of Figure 37.
4. Find $\cot \alpha$ and $\sec \alpha$ for α as shown in Figure 38.

Keeping in mind what you know about the sines and cosines of special angles, find each of the values in Problems 5–22.

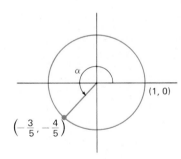

Figure 38

5. $\tan(\pi/6)$ 6. $\cot(\pi/6)$ 7. $\sec(\pi/6)$
8. $\csc(\pi/6,$ 9. $\cot(\pi/4)$ 10. $\sec(\pi/4)$
11. $\csc(\pi/3)$ 12. $\sec(\pi/3)$ 13. $\sin(4\pi/3)$
14. $\cos(4\pi/3)$ 15. $\tan(4\pi/3)$ 16. $\sec(4\pi/3)$
17. $\tan \pi$ 18. $\sec \pi$ 19. $\tan 330°$
20. $\cot 120°$ 21. $\sec 600°$ 22. $\csc(-150°)$

23. For what values of t on $0 \le t \le 4\pi$ is each of the following undefined?
    (a) $\sec t$ (b) $\tan t$ (c) $\csc t$ (d) $\cot t$
24. For which values of t on $0 \le t \le 4\pi$ is each of the following equal to 1?
    (a) $\sec t$ (b) $\tan t$ (c) $\csc t$ (d) $\cot t$

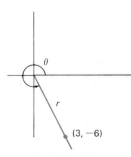

Figure 39

EXAMPLE (Using the a, b, r Definitions)   Suppose that the point $(3, -6)$ is on the terminal side of an angle in standard position (Figure 39). Find $\sin \theta$, $\tan \theta$, and $\sec \theta$.

Solution. First we find r.

$$r = \sqrt{3^2 + (-6)^2} = \sqrt{45} = 3\sqrt{5}$$

Then

$$\sin \theta = \frac{b}{r} = \frac{-6}{3\sqrt{5}} = -\frac{2}{\sqrt{5}}$$

$$\tan \theta = \frac{b}{a} = \frac{-6}{3} = -2$$

$$\sec \theta = \frac{r}{a} = \frac{3\sqrt{5}}{3} = \sqrt{5}$$

In Problems 25–28, find sin θ, tan θ, and sec θ, assuming that the given point is on the terminal side of θ.

25. $(5, -12)$ 26. $(7, 24)$ 27. $(-1, -2)$

28. $(-3, 2)$

29. If $\tan \theta = \frac{3}{4}$ and θ is an angle in the first quadrant, find $\sin \theta$ and $\sec \theta$. *Hint:* The point $(4, 3)$ is on the terminal side of θ.

30. If $\tan \theta = \frac{3}{4}$ and θ is an angle in the third quadrant, find $\cos \theta$ and $\csc \theta$. *Hint:* The point $(-4, -3)$ is on the terminal side of θ.

31. If $\sin \theta = \frac{5}{13}$ and θ is an angle in the second quadrant, find $\cos \theta$ and $\cot \theta$. *Hint:* A point with y-coordinate 5 and $r = 13$ is on the terminal side of θ. Thus the x-coordinate must be -12.

32. If $\cos \theta = \frac{4}{5}$ and $\sin \theta < 0$, find $\tan \theta$.

33. Where does the line from the origin to $(5, -12)$ intersect the unit circle?

34. Where does the line from the origin to $(-6, 8)$ intersect the unit circle?

35. Find the angle of inclination of the line $5x + 2y = 6$.

36. Find the equation of the line with angle of inclination $75°$ that passes through $(-2, 4)$.

MISCELLANEOUS PROBLEMS

37. Evaluate without use of a calculator.
   (a) $\sec(7\pi/6)$ (b) $\tan(-2\pi/3)$ (c) $\csc(3\pi/4)$
   (d) $\cot(11\pi/4)$ (e) $\csc(570°)$ (f) $\tan(180.045°)$

38. Calculate.
   (a) $\tan(\sin 2.4)$ (b) $\cot(\tan 1.49)$ (c) $\csc(\sin 11.8°)$
   (d) $\sec^2(\tan 91.2°)$ (e) $\csc(\tan \pi)$ (f) $\tan[\tan(\tan 1.5)]$

39. If $\csc t = 25/24$ and $\cos t < 0$, find each of the following.
   (a) $\sin t$ (b) $\cos t$ (c) $\tan t$
   (d) $\sec(\frac{\pi}{2} - t)$ (e) $\cot(\frac{\pi}{2} - t)$ (f) $\csc(\frac{\pi}{2} - t)$

40. Show that each of the following are identities.
   (a) $\tan(-t) = -\tan t$ (b) $\sec(-t) = \sec t$ (c) $\csc(-t) = -\csc t$

41. Find the two smallest positive values of t that satisfy each of the following.
   (a) $\tan t = -1$ (b) $\sec t = \sqrt{2}$ (c) $|\csc t| = 1$

42. Find the angle of inclination of the line that is perpendicular to the line $4x + 3y = 9$.

43. Write each of the following in terms of sines and cosines and simplify.

   (a) $\dfrac{\sec \theta \csc \theta}{\tan \theta + \cot \theta}$

   (b) $(\tan \theta)(\cos \theta - \csc \theta)$

   (c) $\dfrac{(1 + \tan \theta)^2}{\sec^2 \theta}$

   (d) $\dfrac{\sec \theta \cot \theta}{\sec^2 \theta - \tan^2 \theta}$

   (e) $\dfrac{\cot \theta - \tan \theta}{\csc \theta - \sec \theta}$

   (f) $\tan^4 \theta - \sec^4 \theta$

44. Let θ be a first quadrant angle. Express each of the other five trigonometric functions in terms of $\sin \theta$ alone.

45. Use the identities of Problem 37 in Section 7–3, namely,

$$\sin(t + \pi) = -\sin t \quad \text{and} \quad \cos(t + \pi) = -\cos t$$

to establish each of the following identities.

   (a) $\tan(t + \pi) = \tan t$

   (b) $\cot(t + \pi) = \cot t$

   (c) $\sec(t + \pi) = -\sec t$

   (d) $\csc(t + \pi) = -\csc t$

   Note: From (a) and (b), we conclude that tangent and cotangent are periodic with period π.

46. Show that $|\sec t| \geq 1$ and $|\csc t| \geq 1$ for all t for which these functions are defined.

47. If $\tan \theta = \frac{5}{12}$ and $\sin \theta < 0$, evaluate $\cos^2 \theta - \sin^2 \theta$.

48. A wheel of radius 5, centered at the origin, is rotating counterclockwise at a rate of 1 radian per second. At $t = 0$, a speck of dirt on the rim is at $(5, 0)$. What are the coordinates of the speck at time t?

49. At $t = 2\pi/3$, the speck in Problem 44 came loose and flew off along the tangent line. Where did it hit the x-axis?

50. Find the coordinates of P in Figure 40.

[c] 51. The face of a clock is in the xy-plane with center at the origin and 12 on the positive y-axis. Both hands of the clock are 5 units long.

   (a) Find the slope of the minute hand at 2:24.

   (b) Find the slope of the line through the tips of both hands at 12:50.

[c] 52. From an airplane h miles above the surface of the earth (a sphere of radius 3960 miles), I can just see a bright light on the horizon d miles away. If I measure the angle of depression of the light as $2.1°$, help me determine d and h.

53. A wheel of radius 20 centimeters is used to drive a wheel of radius 50 centimeters by means of a belt that fits around the wheels. How long is the belt if the centers of the two wheels are 100 centimeters apart?

54. TEASER   Express the length L of the crossed belt that intersects in angle 2α and fits around wheels of radius r and R (Figure 41) in terms of r, R, and α.

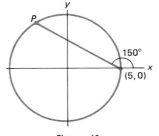

Figure 40

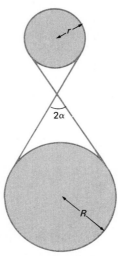

Figure 41

t (rad.)	Sin t	Tan t	Cot t	Cos t
.40	.38942	.42279	2.3652	.92106
.41	.39861	.43463	2.3008	.91712
.42	.40776	.44657	2.2393	.91309
.43	.41687	.45862	2.1804	.90897
.44	.42594	.47078	2.1241	.90475
.45	.43497	.48306	2.0702	.90045
.46	.44395	.49545	2.0184	.89605
.47	.45289	.50797	1.9686	.89157
.48	.46178	.52061	1.9208	.88699
.49	.47063	.53339	1.8748	.88233
.50	.47943	.54630	1.8305	.87758

7-5 Finding Values of the Trigonometric Functions

In order to make significant use of the trigonometric functions, we will have to be able to calculate their values for angles other than the special angles we have considered. The simplest procedure is to press the right key on a calculator and read the answer. About the only thing to remember is to make sure the calculator is in the right mode, degree or radian, depending on what we want.

Even though calculators are becoming standard equipment for most mathematics and science students, we think you should also know how to use tables. That is the subject we take up now. We might call it "what to do when your battery goes dead."

The opening display gives a small portion of a five-place table of values for sin t, tan t, cot t, and cos t. (The complete table appears as Table D at the back of the book.) From it we read the following:

$$\sin .44 = .42594 \qquad \tan .44 = .47078$$
$$\cot .44 = 2.1241 \qquad \cos .44 = .90475$$

These results are not exact; they have been rounded off to five significant digits. Keep in mind that you can think of sin .44 in two ways, as the sine of the number .44 or, if you like, as the sine of an angle of radian measure .44.

Table D appears to have two defects. First, t is given only to 2 decimal places. If we need sin .44736, we have to round or perhaps to interpolate (see page 266).

$$\sin .44736 \approx \sin .45 = .43497$$

A more serious defect appears to be the fact that values of t go only to 2.00.

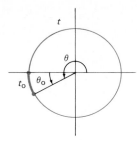

Figure 42

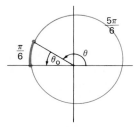

Figure 43

This limitation evaporates once we learn about reference angles and reference numbers, our next topic.

REFERENCE ANGLES AND REFERENCE NUMBERS

Let θ be any angle in standard position and let t be its radian measure. Associated with θ is an acute angle θ_0, called the **reference angle** and defined to be the smallest positive angle between the terminal side of θ and the x-axis (Figure 42). The radian measure t_0 of θ_0 is called the **reference number** corresponding to t. For example, the reference number for $t = 5\pi/6$ is $t_0 = \pi/6$ (Figure 43). Once we know t_0, we can find $\sin t$, $\cos t$, and so on, no matter what t is. Here is how we do it.

Examine the four diagrams in Figure 44.

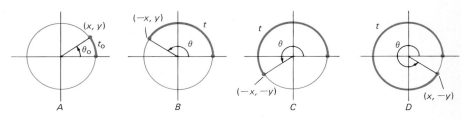

Figure 44

Each angle θ in B, C, and D has θ_0 as its reference angle and, of course, each t has t_0 as its reference number. Now we make a crucial observation. In each case, the point on the unit circle corresponding to t has the same coordinates, except for sign, as the point corresponding to t_0. It follows from this that

$$\sin t = \pm\sin t_0 \qquad \cos t = \pm\cos t_0$$

with the $+$ or $-$ sign being determined by the quadrant in which the terminal side of the angle falls. For example,

$$\sin \frac{5\pi}{6} = \sin \frac{\pi}{6} \qquad \cos \frac{5\pi}{6} = -\cos \frac{\pi}{6}$$

or, in degree notation,

$$\sin 150° = \sin 30° \qquad \cos 150° = -\cos 30°$$

We chose the plus sign for the sine and the minus sign for the cosine because in the second quadrant the sine function is positive, whereas the cosine function is negative.

What we have just said applies to all six trigonometric functions. If T stands for any one of them, then

$$T(t) = \pm T(t_0) \quad \text{and} \quad T(\theta) = \pm T(\theta_0)$$

with the plus or minus sign being determined by the quadrant in which the terminal side of θ lies. Of course $T(t_0)$ itself is always nonnegative since $0 \le t_0 \le \pi/2$.

EXAMPLES

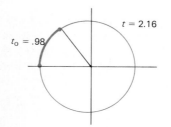

Figure 45

If we wish to calculate cos 2.16 using tables, we must first find the reference number for 2.16. Approximating π by 3.14, we find that (see Figure 45)

$$t_0 = 3.14 - 2.16 = .98$$

and thus, using Table D,

$$\cos 2.16 = -\cos .98 = -.55702$$

Notice we chose the minus sign because the cosine is negative in quadrant II.

To calculate tan 24.95 is slightly more work. First we remove as large a multiple of 2π as possible from 24.95. Using 6.28 for 2π, we get

$$24.95 = 3(6.28) + 6.11$$

The reference number for 6.11 is (see Figure 46)

$$t_0 = 6.28 - 6.11 = .17$$

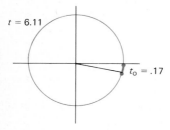

Figure 46

Thus

$$\tan 24.95 = \tan 6.11 = -\tan .17 = -.17166$$

We choose the minus sign because the tangent is negative in quadrant IV.

Now use your pocket calculator to find tan 24.95 the easy way. Be sure you put it in radian mode. You will get $-.18480$ instead of $-.17166$, a rather large discrepancy. Whom should you believe? We suggest that you trust your calculator. The reason we were so far off is that 6.28 is a rather poor approximation for 2π, and multiplying it by 3 made matters worse. Had we used 6.2832 for 2π, we would have obtained $t_0 = .1828$ and tan 24.95 $= -.18486$.

Problem Set 7-5

Find the value of each of the following using Table C or Table D.

1. sin 1.38	2. cos .67	3. cos 42.8°	4. tan 18.0°
5. cot .82	6. tan 1.11	7. sin 68.3°	8. cot 49.6°

EXAMPLE A (Finding Reference Numbers)   Find the reference number t_0 for each of the following values of t.

(a) $t = 20.59$ (b) $t = \dfrac{5\pi}{2} - .92$

$$\begin{array}{r} 3 \\ 6.28\overline{)20.59} \\ \underline{18.84} \\ 1.75 \end{array}$$

Figure 47

Solution.

(a) To get rid of the irrelevant multiples of 2π, we divide 20.59 by 6.28 ($2\pi \approx 6.28$), obtaining 1.75 as remainder (Figure 47). Since 1.75 is between $\pi/2$ and π, we subtract it from π. Thus

$$t_0 \approx \pi - 1.75 \approx 3.14 - 1.75 = 1.39$$

(b) Since $5\pi/2 - .92 = 2\pi + \pi/2 - .92$, it follows that

$$t_0 = \frac{\pi}{2} - .92 \approx 1.57 - .92 = .65$$

Find the reference number t_0 if t has the given value. Use 3.14 for π.

9. 1.84	10. 2.14	11. 3.54	12. 3.74
13. 5.18	14. 6.08	15. 10.48	16. 8.38
17. -1.12	18. -1.86	19. -2.64	20. -4.24

Find the reference number for each of the following. You may leave your answer in terms of π.

21. $13\pi/8$	22. $37\pi/36$	23. $40\pi/3$	24. $-11\pi/5$
25. $3\pi + .24$	26. $3\pi/2 + .17$	27. $3\pi - .24$	28. $3\pi/2 - .17$
29. $11\pi/2$	30. 26π		

Find the value of each of the following using Table D and $\pi = 3.14$. Calculators will give slightly different results because of this crude approximation to π.

31. $\cos 1.42$	32. $\sin .97$	33. $\tan 1.39$	34. $\cot .08$
35. $\sin 2.14$	36. $\cos 3.08$	37. $\cot 5.62$	38. $\tan 4.11$
39. $\cos(-2.54)$	40. $\sin(-4.18)$		

EXAMPLE B (Finding t When $\sin t$ or $\cos t$ Is Given)   Find 2 values of t between 0 and 2π for which (a) $\sin t = .90863$;   (b) $\cos t = -.95824$.

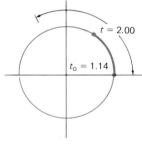

Figure 48

Solution.

(a) We get $t = 1.14$ directly from Table D (or using a calculator). Since the sine is also positive in quadrant II, we seek a value of t between $\pi/2$ and π for which 1.14 is the reference number (Figure 48). Only one number fits the bill:

$$\pi - 1.14 \approx 3.14 - 1.14 = 2.00$$

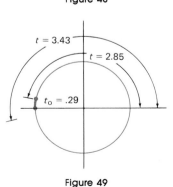

Figure 49

(b) We know that $\cos t_0 = .95824$ and so $t_0 = .29$. Now the cosine is negative in quadrants II and III. Thus we are looking for two numbers between $\pi/2$ and $3\pi/2$ with .29 as reference number (Figure 49). One is $\pi - .29 \approx 3.14 - .29 = 2.85$, and the other $\pi + .29 \approx 3.14 + .29 = 3.43$.

Find two values of t between 0 and 2π for which the given equality holds.

41. $\sin t = .94898$ 42. $\cos t = .72484$ 43. $\cos t = -.08071$

44. $\sin t = -.48818$ 45. $\tan t = 4.9131$ 46. $\cot t = 1.4007$

47. $\tan t = -3.6021$ 48. $\cot t = -.47175$

Find the reference angle (in degrees) for each of the following angles. For example, the reference angle for $\theta = 124.1°$ is $\theta_0 = 180° - 124.1° = 55.9°$.

49. $139.6°$ 50. $218.1°$ 51. $348.7°$

52. $375.4°$ 53. $-99.8°$ 54. $-224.4°$

EXAMPLE C (Finding sin θ, cos θ, and so on, When θ Is Any Angle Given in Degrees) Find the value of each of the following.

(a) $\cos 214.6°$ (b) $\cot 658°$

Solution. So far, we have used Table C to find the sine, cosine, and so on, of positive angles measuring less than 90°. Here we do this for angles of arbitrary (degree) measure.

(a) The reference angle is

$$214.6° - 180° = 34.6°$$

$$\cos 214.6° = -\cos 34.6° = -.8231$$

We used the minus sign since cosine is negative in quadrant III.

(b) First we reduce our angle by 360°

$$658° = 360° + 298°$$

CAUTION

$\cos 99° = \cos 81°$
$= .1564$

$\cos 99° = -\cos 81°$
$= -.1564$
Be sure to assign the correct sign.

The reference angle for 298° is $360° - 298°$, or 62°. In the column with cot at the bottom and 62° at the right, we find .5317. Therefore $\cot 658° = -.5317$.

Find the value of each of the following.

55. $\sin 156.1°$ 56. $\cos 138.7°$ 57. $\tan 348.9°$ 58. $\cot 224.9°$

59. $\cos(-66.1°)$ 60. $\sin 487°$ 61. $\cos 441.3°$ 62. $\sin 180.2°$

63. $\cot(-134°)$ 64. $\tan 311.6°$

Find two different degree values of θ between 0° and 360° for which the given equality holds.

65. $\sin \theta = .3633$ 66. $\cos \theta = .9907$ 67. $\tan \theta = .4942$

68. $\cot \theta = 1.2799$ 69. $\cos \theta = -.9085$ 70. $\sin \theta = -.2045$

MISCELLANEOUS PROBLEMS

71. Use Table C or D to find each of the following. You may approximate π by 3.14.
    (a) $\cos 5.63$ (b) $\sin 10.34$ (c) $\tan 8.42$
    (d) $\sin 311.3°$ (e) $\tan(-411°)$ (f) $\cos 1989°$

72. Use Tables C *and* D to calculate.
    (a) $\sin(\cos 134°)$ (b) $\sin[(\tan 1.5)°]$ (c) $\tan(-5.4°) + \tan(-5.4)$

[c] 73. Calculate.
    (a) $\cos(\sin 2.42°)$ (b) $\cos^3(\sin^2 2.42)$ (c) $\sqrt{\tan 4.21} + \ln(\sin 7.12)$

74. Use Table D and $\pi = 3.14$ to find two values of t between 0 and 2π for which each of the following is true.
    (a) $\sin t = .62879$ (b) $\cos t = -.90045$ (c) $\tan t = -4.4552$

[c] 75. Find two values of t between 0 and 2π for which each statement is true, giving your answers correct to 6 decimal places.
    (a) $\sin t = .62879$ (b) $\cos t = .34176$ (c) $\tan t = -3.14159$
    Note: On many calculators, you would press .62879 $\boxed{\text{INV}}$ $\boxed{\text{sin}}$ to get one answer to (a).

76. If $\pi/2 < t < \pi$, then $t_0 = \pi - t$. In a similar manner, express t_0 in terms of t in each case.
    (a) $3\pi/2 < t < 2\pi$ (b) $5\pi < t < 11\pi/2$ (c) $-2\pi < t < -3\pi/2$

77. If $0° < \phi < 90°$, express the reference angle θ_0 in terms of ϕ in each case.
    (a) $\theta = 180° + \phi$ (b) $\theta = 270° - \phi$ (c) $\theta = \phi - 90°$

78. Without using tables or a calculator, round to the nearest degree the smallest positive angle θ satisfying $\tan \theta = -40,000$.

79. If θ is a fourth quadrant angle whose terminal side coincides with the line $3x + 5y = 0$, find $\sin \theta$.

[c] 80. In calculus, you will learn that

$$\sin t = t - \frac{t^3}{3!} + \frac{t^5}{5!} - \frac{t^7}{7!} + \cdots$$

and

$$\cos t = 1 - \frac{t^2}{2!} + \frac{t^4}{4!} - \frac{t^6}{6!} + \cdots$$

Here, $n! = 1 \cdot 2 \cdot 3 \cdots n$ (for example, $2! = 1 \cdot 2 = 2$ and $3! = 1 \cdot 2 \cdot 3 = 6$). These series are used to construct Tables C and D. If we use just the first three terms in the sine series, we obtain

$$\sin t \approx t - \frac{t^3}{6} + \frac{t^5}{120} = \left[\left(\frac{t^2}{120} - \frac{1}{6} \right) t^2 + 1 \right] t$$

Use the first three terms of these series to approximate each of the following and compare with the corresponding value in Table D.
    (a) $\sin(.1)$ (b) $\sin(.4)$ (c) $\cos(.2)$

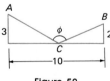

Figure 50

81. Determine ϕ in Figure 50 so that the path ACB has minimum length.

82. TEASER   Let α, β, and γ be acute angles such that $\tan \alpha = 1$, $\tan \beta = 2$, and $\tan \gamma = 3$.
    (a) Use your calculator to approximate $\alpha + \beta + \gamma$.
    (b) Make a conjecture about the exact value of $\alpha + \beta + \gamma$.
    (c) Construct a clever geometric diagram to prove your conjecture.

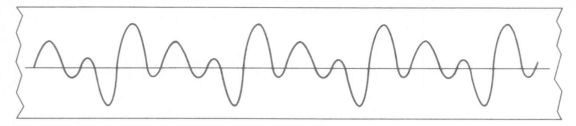

When heart beats, brain activity, or sound waves from a musical instrument are changed into visual images by means of an oscilloscope, they give a regular repetitive pattern which may look something like the diagram above. This repetitive behavior is a characteristic feature of the graphs of the trigonometric functions. In fact, almost any repetitive pattern can be approximated by appropriate combinations of the trigonometric functions.

7-6   Graphs of the Trigonometric Functions

Recall that to graph $y = f(x)$, we first construct a table of values of ordered pairs (x, y), then plot the corresponding points, and finally connect those points with a smooth curve. Here we want to graph $y = \sin t$, $y = \cos t$, and so on, and we will follow a similar procedure. Notice that we use t rather than x as the independent variable because we used t as the variable (radian measure of an angle) in our definition of the trigonometric functions.

We begin with the graphs of the sine and cosine functions. You should become so well acquainted with these two graphs that you can sketch them quickly whenever you need them. This will aid you in two ways. First, these graphs will help you remember many of the important properties of the sine and cosine functions. Second, knowing them will help you graph other more complicated trigonometric functions.

THE GRAPH OF $y = \sin t$

We begin with a table of values (Figure 51).

t	0	$\frac{\pi}{6}$	$\frac{\pi}{4}$	$\frac{\pi}{3}$	$\frac{\pi}{2}$	$\frac{3\pi}{4}$	π	$\frac{5\pi}{4}$	$\frac{3\pi}{2}$	$\frac{7\pi}{4}$	2π
$y = \sin t$	0	$\frac{1}{2}$	$\frac{\sqrt{2}}{2}$	$\frac{\sqrt{3}}{2}$	1	$\frac{\sqrt{2}}{2}$	0	$-\frac{\sqrt{2}}{2}$	-1	$-\frac{\sqrt{2}}{2}$	0

Figure 51

We have listed values of t between 0 and 2π. That is sufficient to graph one period (shown in Figure 52 as a heavy curve). From there on we can continue the curve indefinitely in either direction in a repetitive fashion, for we learned earlier that $\sin(t + 2\pi) = \sin t$.

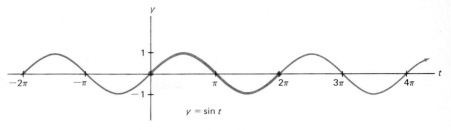

Figure 52

THE GRAPH OF $y = \cos t$

The cosine function is a copycat; its graph is just like that of the sine function but pushed $\pi/2$ units to the left (Figure 53).

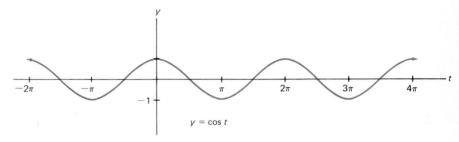

Figure 53

To see that the graph of the cosine function is correct, we might make a table of values and proceed as we did for the sine function. In fact, we ask you to do just that in Problem 1. Alternatively, we can show that

$$\cos t = \sin\!\left(t + \frac{\pi}{2}\right)$$

This follows directly from identities we have observed earlier.

$$\sin\!\left(t + \frac{\pi}{2}\right) = \sin\!\left(\frac{\pi}{2} - (-t)\right)$$

$$= \cos(-t) \qquad \text{(cofunction identity)}$$

$$= \cos t \qquad \text{(cosine is even)}$$

PROPERTIES EASILY OBSERVED FROM THESE GRAPHS

1. Both sine and cosine are periodic with 2π as period.
2. $-1 \le \sin t \le 1$ and $-1 \le \cos t \le 1$.
3. $\sin t = 0$ if $t = -\pi$, 0, π, 2π, and so on.
   $\cos t = 0$ if $t = -\pi/2$, $\pi/2$, $3\pi/2$, and so on.

4. $\sin t > 0$ in quadrants I and II.
   $\cos t > 0$ in quadrants I and IV.

5. $\sin(-t) = -\sin t$ and $\cos(-t) = \cos t$.
   The sine is an odd function; its graph is symmetric with respect to the origin. The cosine is an even function; its graph is symmetric with respect to the y-axis.

6. We can see immediately where the sine and cosine functions are increasing and where they are decreasing. For example, the sine function decreases for $\pi/2 \le t \le 3\pi/2$.

THE GRAPH OF $y = \tan t$

Since the tangent function is defined by

$$\tan t = \frac{\sin t}{\cos t}$$

we need to beware of values of t for which $\cos t = 0$: $-\pi/2, \pi/2, 3\pi/2$, and so forth. In fact, from Section 5-3, we know that we should expect vertical asymptotes at these places. Notice also that

$$\tan(-t) = \frac{\sin(-t)}{\cos(-t)} = \frac{-\sin t}{\cos t} = -\tan t$$

which means that the graph of the tangent will be symmetric with respect to the origin. Using these two pieces of information, a small table of values, and the fact that the tangent is periodic, we obtain the graph in Figure 54.

To confirm that the graph is correct near $t = \pi/2$, we suggest looking at Table D. Notice that the $\tan t$ steadily increases until at $t = 1.57$, we read $\tan t = 1255.8$. But as t takes the short step to 1.58, $\tan t$ takes a tremendous plunge to -108.65. In that short space, t has passed through $\pi/2 \approx 1.5708$ and $\tan t$ has shot up to celestial heights only to fall to a bottomless pit, from which, however, it manages to escape as t moves to the right.

While we knew the tangent would have to repeat itself every 2π units since the sine and cosine do this, we now notice that it actually repeats itself on intervals of length π. Since the word *period* denotes the length of the shortest interval after which a function repeats itself, the tangent function has period π. For an algebraic demonstration, see Problem 45 of Section 7-4.

THE GRAPH OF $y = \sec t$

Since $\sec t = 1/\cos t$, one way of getting the graph of the secant is by graphing the cosine and then taking reciprocals of the y-coordinates (Figure 55). Note that since $\cos t = 0$ at $t = -\pi/2, \pi/2, 3\pi/2$, and so on, the graph of $\sec t$ must have vertical asymptotes at these points.

Just like the cosine, the secant is an even function; that is, $\sec(-t) = \sec t$. And, like the cosine, secant has period 2π. However, notice that if $\cos t$ increases or decreases throughout an interval, $\sec t$ does just the opposite. For example, $\cos t$ decreases for $0 < t < \pi/2$, whereas $\sec t$ increases there.

t	0	$\frac{\pi}{4}$	$\frac{\pi}{3}$	$\frac{\pi}{2}$	$\frac{2\pi}{3}$	$\frac{3\pi}{4}$	π	$\frac{5\pi}{4}$	$\frac{3\pi}{2}$	$\frac{7\pi}{4}$	2π
$y = \tan t$	0	1	$\sqrt{3}$	undefined	$-\sqrt{3}$	-1	0	1	undefined	-1	0

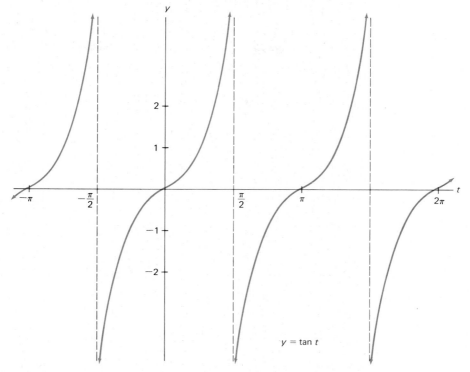

Figure 54

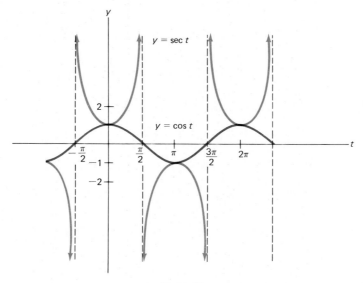

Figure 55

Be sure to study Example A below. It introduces the graph of $y = A \sin Bt$. This important topic is explored more fully in Section 9-3 in connection with simple harmonic motion.

Problem Set 7-6

1. Make a table of values and then sketch the graph of $y = \cos t$.
2. What real numbers constitute the domain of the cosine? The range?
3. Sketch the graph of $y = \cot t$ for $-2\pi \le t \le 2\pi$, being sure to show the asymptotes.
4. What real numbers constitute the entire domain of the cotangent? The range?
5. Using the corresponding fact about the cosine, demonstrate algebraically that $\sec(t + 2\pi) = \sec t$.
6. Sketch the graph of $y = \csc t$.
7. What is the domain of the secant? The range?
8. What is the domain of the cosecant? The range?
9. What is the period of the cotangent? The secant?
10. On the interval $-2\pi \le t \le 2\pi$, where is the cotangent increasing?
11. Which is true: $\cot(-t) = \cot t$ or $\cot(-t) = -\cot t$?
12. Which is true: $\csc(-t) = \csc t$ or $\csc(-t) = -\csc t$?

EXAMPLE A (Some Sine-Related Graphs) Sketch the graph of each of the following for $-2\pi \le t \le 4\pi$.
(a) $y = 2 \sin t$ (b) $y = \sin 2t$ (c) $y = 3 \sin 4t$

Solution.

(a) We could graph $y = 2 \sin t$ from a table of values. It is easier, though, to graph $\sin t$ (dotted graph below) and then multiply the ordinates by 2 (Figure 56). Since the graph bobs up and down between $y = -2$ and $y = 2$, we say that it has an **amplitude** of 2. The period is 2π, the same as for $\sin t$.

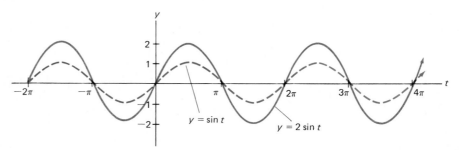

Figure 56

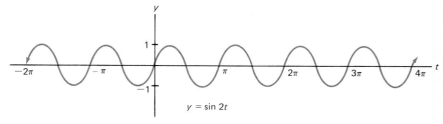

t	$-\pi$	$-\dfrac{3\pi}{4}$	$-\dfrac{\pi}{2}$	$-\dfrac{\pi}{4}$	$-\dfrac{\pi}{12}$	0	$\dfrac{\pi}{12}$	$\dfrac{\pi}{4}$	$\dfrac{\pi}{2}$	$\dfrac{3\pi}{4}$	π
$2t$	-2π	$-\dfrac{3\pi}{2}$	$-\pi$	$-\dfrac{\pi}{2}$	$-\dfrac{\pi}{6}$	0	$\dfrac{\pi}{6}$	$\dfrac{\pi}{2}$	π	$\dfrac{3\pi}{2}$	2π
$\sin 2t$	0	1	0	-1	$-\dfrac{1}{2}$	0	$\dfrac{1}{2}$	1	0	-1	0

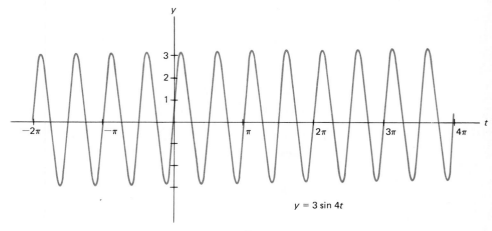

$y = \sin 2t$

Figure 57

(b) Here a table of values is advisable, since this is our first example of this type (Figure 57). This graph goes through a complete cycle as t increases from 0 to π; that is, the period of $\sin 2t$ is π instead of 2π as it was for $\sin t$. The amplitude is 1, just as for $\sin t$.

(c) We can save a lot of work once we recognize how the character of the graph of $A \sin Bt$ (and $A \cos Bt$) is determined by the numbers A and B $(B > 0)$. The amplitude (which tells how far the graph rises and falls from its median position) is given by $|A|$. The period is given by $2\pi/B$. Thus for $y = 3 \sin 4t$, the amplitude is 3 and the period is $2\pi/4 = \pi/2$. For a quick sketch, we use these two numbers to determine the high and low points and the t-intercepts, connecting these points with a smooth, wavelike curve (Figure 58).

$y = 3 \sin 4t$

Figure 58

In Problems 13–22, determine the amplitude and the period. Then sketch the graph on the indicated interval.

13. $y = 3 \cos t,\ -\pi \le t \le \pi$

14. $y = \frac{1}{2} \cos t,\ -\pi \le t \le \pi$

15. $y = -\sin t,\ -\pi \le t \le \pi$

16. $y = -2 \cos t,\ -\pi \le t \le \pi$

17. $y = \cos 4t,\ -\pi \le t \le \pi$

18. $y = \cos 3t,\ -\pi/2 \le t \le \pi/2$

19. $y = 2 \sin \frac{1}{2}t,\ -2\pi \le t \le 2\pi$

20. $y = 3 \sin \frac{1}{3}t,\ -3\pi \le t \le 3\pi$

21. $y = 2 \cos 3t,\ -\pi \le t \le \pi$

22. $y = 4 \sin 3t,\ -\pi \le t \le \pi$

EXAMPLE B (Graphing Sums of Trigonometric Functions)   Sketch the graph of the equation $y = 2 \sin t + \cos 2t$.

Solution.   We graph $y = 2 \sin t$ and $y = \cos 2t$ on the same coordinate plane (these appear as dotted-line curves in Figure 59) and then add ordinates. Notice that for any t, the ordinates (y-values) of the dotted curves are added to obtain the desired ordinate. The graph of $y = 2 \sin t + \cos 2t$ is quite different from the separate (dotted) graphs but it does repeat itself; it has period 2π.

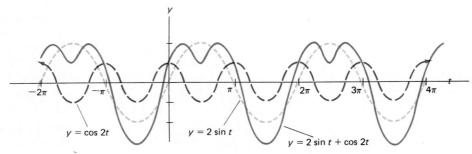

Figure 59

Sketch each graph by the method of adding ordinates. Show at least one complete period.

23. $y = 2 \sin t + \cos t$

24. $y = \sin t + 2 \cos t$

25. $y = \sin 2t + \cos t$

26. $y = \sin t + \cos 2t$

27. $y = \sin \frac{1}{2}t + \frac{1}{2} \sin t$

28. $y = \cos \frac{1}{2}t + \frac{1}{2} \cos t$

MISCELLANEOUS PROBLEMS

In Problems 29–32, sketch each graph on the indicated interval.

29. $y = -\cos t,\ -\pi \le t \le \pi$

30. $y = 3 \sin t,\ -\pi \le t \le \pi$

31. $y = \sin 4t,\ 0 \le t \le \pi$

32. $y = 3 \cos \frac{1}{2}t,\ -2\pi \le t \le 2\pi$

33. What are the amplitudes and periods for the graphs in Problems 29 and 31?

34. What are the amplitudes and periods for the graphs in Problems 30 and 32?

35. Determine the period and sketch the graph of each of the following, showing at least three periods.
    (a) $y = \tan 2t$
    (b) $y = 3 \tan(t/2)$

36. Follow the directions of Problem 35.
    (a) $y = 2 \cot 2t$ (b) $y = \sec 3t$

37. Sketch, using the same axes, the graphs of
    (a) $f(t) = \sin t$;   (b) $g(t) = 3 + \sin t$;   (c) $h(t) = \sin(t - \pi/4)$.

38. Sketch, using the same axes, the graphs of
    (a) $f(t) = \cos t$;   (b) $g(t) = -2 + \cos t$;   (c) $h(t) = \cos(t + \pi/3)$.

39. Sketch the graph of $y = \cos 3t + 2 \sin t$ for $-\pi \le t \le \pi$. Use the method of adding ordinates.

40. Sketch the graph of $y = t + \sin t$ on $-4\pi \le t \le 4\pi$. Use the method of adding ordinates.

[c] 41. **Sketch the graph** of $y = t - \cos t$ for $0 \le t \le 6$ by actually calculating the y values corresponding to $t = 0, .5, 1, 1.5, 2, 2.5, \ldots, 6$.

42. By sketching the graphs of $y = t$ and $y = 3 \sin t$ on the same coordinate axes, determine approximately all solutions of $t = 3 \sin t$.

43. The strength I of current (in amperes) in a wire of an alternating current circuit might satisfy

$$I = 30 \sin(120\pi t)$$

where time t is measured in seconds.
    (a) What is the period?
    (b) How many cycles (periods) are there in one second?
    (c) What is the maximum strength of the current?

[c] 44. Sketch the graph of $y = (\sin t)/t$ on $-3\pi \le t \le 3\pi$. Be sure to plot several points for t near 0 (for example, $t = -.5, -.2, -.1, .1, .2, .5$). What value does y seem to approach as t approaches 0?

45. Consider $y = \sin(1/t)$ on the interval $0 < t \le 1$.
    (a) Where does its graph cross the t-axis?
    (b) Evaluate y for $t = 2/\pi, 2/3\pi, 2/5\pi, 2/7\pi, \ldots$.
    (c) Sketch the graph as best you can, using a large unit on the t-axis.

46. **TEASER** How many solutions does each equation have on the indicated interval for t?
    (a) $\sin t = t/60$, $t \ge 0$
    (b) $\sin(1/t) = t/60$, $t \ge .06$
    (c) $\sin(1/t) = t/60$, $t > 0$

Chapter Summary

The word **trigonometry** means triangle measurement. In its elementary historical form, it is the study of how to **solve** right triangles when appropriate information is given. The main tools are the three trigonometric ratios sin θ, cos θ, and tan θ, which were first defined only for acute angles θ.

In order to give the subject its modern general form, we first generalized the notion of an angle θ, allowing θ to have arbitrary size and measuring it either in **degrees** or **radians.** Such an angle θ can be placed in **standard position** in a coordinate system, where it will cut off an arc of directed length t (the radian measure of θ) stretching from $(1, 0)$ to (x, y) on the unit circle. This allowed us to make the key definitions

$$\sin \theta = \sin t = y \qquad \cos \theta = \cos t = x$$

on which all of modern trigonometry rests.

From the above definitions, we derived several identities, of which the most important is

$$\sin^2 t + \cos^2 t = 1$$

We also defined four additional functions

$$\tan t = \frac{\sin t}{\cos t} \qquad \cot t = \frac{\cos t}{\sin t}$$

$$\sec t = \frac{1}{\cos t} \qquad \csc t = \frac{1}{\sin t}$$

To evaluate the trigonometric functions, we may use either a scientific calculator or Tables C and D in the Appendix. If the tables are used, the notions of **reference angle** and **reference number** become important. Finally we graphed several of the trigonometric functions, noting especially their **periodic** behavior.

Chapter Review Problem Set

1. Solve the following right triangles ($\gamma = 90°$).
   (a) $\alpha = 47.1°$, $c = 36.9$ (b) $a = 417$, $c = 573$
2. At a distance of 10 feet from a wall, the angle of elevation of the top of a mural with respect to eye level is $18°$ and the corresponding angle of depression of the bottom is $10°$. How high is the mural?
3. Change $33°$ to radians. Change $9\pi/4$ radians to degrees.
4. How far does a wheel of radius 30 centimeters roll along level ground in making 100 revolutions?
5. Calculate each of the following without use of tables or a calculator.
   (a) $\sin(7\pi/6)$ (b) $\cos(11\pi/6)$
   (c) $\tan(13\pi/4)$ (d) $\sin(41\pi/6)$
6 Evaluate.
   (a) $\sin 411°$ (b) $\cos 1312°$
   (c) $\tan 5.77$ (d) $\sin 13.12$
7. Write in terms of $\sin t$.
   (a) $\sin(-t)$ (b) $\sin(t + 4\pi)$

   (c) $\sin(\pi + t)$ (d) $\cos\left(\dfrac{\pi}{2} - t\right)$

8. For what values of t between 0 and 2π is
   (a) $\cos t > 0$ (b) $\cos 2t > 0$?

9. If $(-5, -12)$ is on the terminal side of an angle θ in standard position, find
   (a) $\cot \theta$;   (b) $\sec \theta$.

10. If $\sin \theta = \frac{2}{3}$ and θ is a second quadrant angle, find $\tan \theta$.

11. Sketch the graph of $y = 3 \cos 2t$ for $-\pi \le t \le 2\pi$.

12. Sketch the graph of $y = \sin t + \sin 2t$ using the method of adding ordinates.

13. What is the range of the sine function? Of the cosecant function?

14. Using the facts that the sine function is odd and the cosine function is even, show that cotangent is an odd function.

15. Give the general definition of $\cos t$ based on the unit circle.

For just as in nature itself there is no middle ground between truth and falsehood, so in rigorous proofs one must either establish his point beyond doubt, or else beg the question inexcusably. There is no chance of keeping one's feet by invoking limitations, distinctions, verbal distortions, or other mental acrobatics. One must with a few words and at the first assault become Caesar or nothing at all.

—Galileo

CHAPTER 8

Trigonometric Identities and Equations

$$\left(\sec\theta + \tan\theta\right)\left(1 - \sin\theta\right) = \left(\frac{1}{\cos\theta} + \frac{\sin\theta}{\cos\theta}\right)\left(1 - \sin\theta\right) = \left(\frac{1+\sin\theta}{\cos\theta}\right)\left(1 - \sin\theta\right) = \frac{1 - \sin^2\theta}{\cos\theta} = \frac{\cos^2\theta}{\cos\theta} = \cos\theta$$

8-1   Identities

Complicated combinations of the six trigonometric functions occur often in mathematics. It is important that we, like the professor above, be able to write a complicated trigonometric expression in a simpler or more convenient form. To do this requires two things. We must be good at algebra and we must know the fundamental identities of trigonometry.

THE FUNDAMENTAL IDENTITIES

We list eleven fundamental identities, which should be memorized.

1. $\tan t = \dfrac{\sin t}{\cos t}$

2. $\cot t = \dfrac{\cos t}{\sin t} = \dfrac{1}{\tan t}$

3. $\sec t = \dfrac{1}{\cos t}$

4. $\csc t = \dfrac{1}{\sin t}$

5. $\sin^2 t + \cos^2 t = 1$

6. $1 + \tan^2 t = \sec^2 t$

7. $1 + \cot^2 t = \csc^2 t$

8. $\sin\left(\dfrac{\pi}{2} - t\right) = \cos t$

9. $\cos\left(\dfrac{\pi}{2} - t\right) = \sin t$

10. $\sin(-t) = -\sin t$

11. $\cos(-t) = \cos t$

We have seen all these identities before. The first four are actually definitions; the others were established either in the text or the problem sets of Sections 7-3 and 7-4.

PROVING NEW IDENTITIES

The professor's work in our opening cartoon can be viewed in two ways. The more likely way of looking at it is that she wanted to simplify the complicated expression

$$(\sec t + \tan t)(1 - \sin t)$$

But it could be that someone had conjectured that

$$(\sec t + \tan t)(1 - \sin t) = \cos t$$

is an identity and that the professor was trying to prove it. It is this second concept we want to discuss now.

Suppose someone claims that a certain equation is an identity—that is, true for all values of the variable for which both sides make sense. How can you check on such a claim? The procedure used by the professor is one we urge you to follow. Start with the more complicated looking side and try to use a chain of equalities to produce the other side.

Suppose we wish to prove that

$$\sin t + \cos t \cot t = \csc t$$

is an identity. We begin with the left side and rewrite it step by step, using algebra and the fundamental identities, until we get the right side.

$$\sin t + \cos t \cot t = \sin t + \cos t \left(\frac{\cos t}{\sin t} \right)$$

$$= \frac{\sin^2 t + \cos^2 t}{\sin t}$$

$$= \frac{1}{\sin t}$$

$$= \csc t$$

When proving that an equation is an identity, it pays to look before you leap. Changing the more complicated side to sines and cosines, as in the above example, is often the best thing to do. But not always. Sometimes the simpler side gives us a clue as to how we should reshape the other side. For example, the left side of

$$\tan t = \frac{(\sec t - 1)(\sec t + 1)}{\tan t}$$

suggests that we try to rewrite the right side in terms of $\tan t$. This can be done by multiplying out the numerator and making use of the fundamental identity $\sec^2 t = 1 + \tan^2 t$.

$$\frac{(\sec t - 1)(\sec t + 1)}{\tan t} = \frac{\sec^2 t - 1}{\tan t} = \frac{\tan^2 t}{\tan t} = \tan t$$

Proving an identity is something like a game in that it requires a strategy. If one strategy does not work, try another, and still another, until you succeed.

A POINT OF LOGIC

Why all the fuss about working with just one side of a conjectured identity? First of all, it offers good practice in manipulating trigonometric expressions. But there is also a point of logic. If you operate on both sides simultaneously, you are in effect assuming that you already have an identity. That is bad logic and it can be corrected only by carefully checking that each step is reversible. To make this point clear, consider the equation

$$1 - x = x - 1$$

which is certainly not an identity. Yet when we square both sides we get

$$1 - 2x + x^2 = x^2 - 2x + 1$$

which is an identity. The trouble here is that squaring both sides is not a reversible operation.

The situation contrasts sharply with our procedure for solving conditional equations, in which we often perform an operation on both sides. For example, in the case of the equation

$$\sqrt{2x + 1} = 1 - x$$

we even square both sides. We are protected from error here by checking our solutions in the original equation.

Problem Set 8-1

1. Express entirely in terms of $\sin t$.
   (a) $\cos^2 t$ (b) $\tan t \cos t$
   (c) $\dfrac{3}{\csc^2 t} + 2\cos^2 t - 2$ (d) $\cot^2 t$

2. Express entirely in terms of $\cos t$.
   (a) $\sin^2 t$ (b) $\tan^2 t$
   (c) $\csc^2 t$ (d) $(1 + \sin t)^2 - 2\sin t$

3. Express entirely in terms of $\tan t$.
   (a) $\cot^2 t$ (b) $\sec^2 t$
   (c) $\sin t \sec t$ (d) $2\sec^2 t - 2\tan^2 t + 1$

4. Express entirely in terms of $\sec t$.
   (a) $\cos^4 t$ (b) $\tan^2 t$
   (c) $\tan t \csc t$ (d) $\tan^2 t - 2\sec^2 t + 5$

EXAMPLE A (Proving Identities)   Prove that the following is an identity.

$$\csc \theta - \sin \theta = \cot \theta \cos \theta$$

Solution.   The left side looks inviting, as $\csc \theta = 1/\sin \theta$. We rewrite it a step at a time.

$$\csc \theta - \sin \theta = \frac{1}{\sin \theta} - \sin \theta$$

$$= \frac{1 - \sin^2 \theta}{\sin \theta}$$

$$= \frac{\cos^2 \theta}{\sin \theta}$$

$$= \frac{\cos \theta}{\sin \theta} \cdot \cos \theta$$

$$= \cot \theta \cos \theta$$

Prove that each of the following is an identity.

5. $\cos t \sec t = 1$

6. $\sin t \csc t = 1$

7. $\tan x \cot x = 1$

8. $\sin x \sec x = \tan x$

9. $\cos y \csc y = \cot y$

10. $\tan y \cos y = \sin y$

11. $\cot \theta \sin \theta = \cos \theta$

12. $\dfrac{\sec \theta}{\csc \theta} = \tan \theta$

13. $\dfrac{\tan u}{\sin u} = \dfrac{1}{\cos u}$

14. $\dfrac{\sin u}{\csc u} + \dfrac{\cos u}{\sec u} = 1$

15. $(1 + \sin z)(1 - \sin z) = \dfrac{1}{\sec^2 z}$

16. $(\sec z - 1)(\sec z + 1) = \tan^2 z$

17. $(1 - \sin^2 x)(1 + \tan^2 x) = 1$

18. $(1 - \cos^2 x)(1 + \cot^2 x) = 1$

19. $\sec t - \sin t \tan t = \cos t$

20. $\sin t (\csc t - \sin t) = \cos^2 t$

21. $\dfrac{\sec^2 t - 1}{\sec^2 t} = \sin^2 t$

22. $\dfrac{1 - \csc^2 t}{\csc^2 t} = \dfrac{-1}{\sec^2 t}$

23. $\cos t (\tan t + \cot t) = \csc t$

24. $\dfrac{1}{\sin t \cos t} - \dfrac{\cos t}{\sin t} = \tan t$

EXAMPLE B (Expressing All Trigonometric Functions in Terms of One of Them)   If $\pi/2 < t < \pi$, express $\cos t$, $\tan t$, $\cot t$, $\sec t$, and $\csc t$ in terms of $\sin t$.

Solution.   Since $\cos^2 t = 1 - \sin^2 t$ and cosine is negative in quadrant II,

$$\cos t = -\sqrt{1 - \sin^2 t}$$

Also

$$\tan t = \frac{\sin t}{\cos t} = -\frac{\sin t}{\sqrt{1 - \sin^2 t}}$$

$$\cot t = \frac{1}{\tan t} = -\frac{\sqrt{1 - \sin^2 t}}{\sin t}$$

$$\sec t = \frac{1}{\cos t} = -\frac{1}{\sqrt{1 - \sin^2 t}}$$

$$\csc t = \frac{1}{\sin t}$$

25. If $\pi/2 < t < \pi$, express $\sin t$, $\tan t$, $\cot t$, $\sec t$, and $\csc t$ in terms of $\cos t$.
26. If $\pi < t < 3\pi/2$, express $\sin t$, $\cos t$, $\cot t$, $\sec t$, and $\csc t$ in terms of $\tan t$.
27. If $\pi/2 < t < \pi$ and $\sin t = \frac{4}{5}$, find the values of the other five functions for the same value of t. *Hint:* Use the results of Example B.
28. If $\pi < t < 3\pi/2$ and $\tan t = 2$, find $\sin t$, $\cos t$, $\cot t$, $\sec t$, and $\csc t$.

EXAMPLE C (How to Proceed When Neither Side Is Simple) Prove that

$$\frac{\sin t}{1 - \cos t} = \frac{1 + \cos t}{\sin t}$$

is an identity.

Solution. Since both sides are equally complicated, it would seem to make no difference which side we choose to manipulate. We will try to transform the left side into the right side. Seeing $1 + \cos t$ in the numerator of the right side suggests multiplying the left side by $(1 + \cos t)/(1 + \cos t)$.

$$\frac{\sin t}{1 - \cos t} = \frac{\sin t}{1 - \cos t} \cdot \frac{1 + \cos t}{1 + \cos t} = \frac{\sin t(1 + \cos t)}{1 - \cos^2 t}$$

$$= \frac{\sin t(1 + \cos t)}{\sin^2 t}$$

$$= \frac{1 + \cos t}{\sin t}$$

Prove that each of the following is an identity.

29. $\dfrac{\sec t - 1}{\tan t} = \dfrac{\tan t}{\sec t + 1}$

30. $\dfrac{1 - \tan \theta}{1 + \tan \theta} = \dfrac{\cot \theta - 1}{\cot \theta + 1}$

Hint: In Problem 30, multiply numerator and denominator of the left side by $\cot \theta$.

31. $\dfrac{\tan^2 x}{\sec x + 1} = \dfrac{1 - \cos x}{\cos x}$

32. $\dfrac{\cot x}{\csc x + 1} = \dfrac{\csc x - 1}{\cot x}$

33. $\dfrac{\sin t + \cos t}{\tan^2 t - 1} = \dfrac{\cos^2 t}{\sin t - \cos t}$

34. $\dfrac{\sec t - \cos t}{1 + \cos t} = \sec t - 1$

MISCELLANEOUS PROBLEMS

35. Express $[(\sin x + \cos x)^2 - 1]\sec x \csc^3 x$ as follows.
    (a) Entirely in terms of $\sin x$.
    (b) Entirely in terms of $\tan x$.

36. If $\sec t = 8$, find the values of (a) $\cos t$; (b) $\cot^2 t$; (c) $\csc^2 t$.

In Problems 37–56, prove that each equation is an identity. Do this by taking one side and showing by a chain of equalities that it is equal to the other side.

37. $(1 + \tan^2 t)(\cos t + \sin t) = (1 + \tan t)\sec t$

38. $1 - (\cos t + \sin t)(\cos t - \sin t) = 2 \sin^2 t$

39. $2 \sec^2 y - 1 = \dfrac{1 + \sin^2 y}{\cos^2 y}$

40. $(\sin x + \cos x)(\sec x + \csc x) = 2 + \tan x + \cot x$

41. $\dfrac{\cos z}{1 + \cos z} = \dfrac{\sin z}{\sin z + \tan z}$

42. $2 \sin^2 t + 3 \cos^2 t + \sec^2 t = (\sec t + \cos t)^2$

43. $(\csc t + \cot t)^2 = \dfrac{1 + \cos t}{1 - \cos t}$

44. $\sec^4 y - \tan^4 y = \dfrac{1 + \sin^2 y}{\cos^2 y}$

45. $\dfrac{\cos x + \sin x}{\cos x - \sin x} = \dfrac{1 + \tan x}{1 - \tan x}$

46. $\dfrac{1 + \cos x}{1 - \cos x} - \dfrac{1 - \cos x}{1 + \cos x} = 4 \cot x \csc x$

47. $(\sec t + \tan t)(\csc t - 1) = \cot t$

48. $\sec t + \cos t = \sin t \tan t + 2 \cos t$

49. $\dfrac{\cos^3 t + \sin^3 t}{\cos t + \sin t} = 1 - \sin t \cos t$

50. $\dfrac{\tan x}{1 + \tan x} + \dfrac{\cot x}{1 - \cot x} = \dfrac{\tan x + \cot x}{\tan x - \cot x}$

51. $\dfrac{1 - \cos \theta}{1 + \cos \theta} = \left(\dfrac{1 - \cos \theta}{\sin \theta}\right)^2$

52. $\dfrac{(\sec^2 \theta + \tan^2 \theta)^2}{\sec^4 \theta - \tan^4 \theta} = \sec^2 \theta + \tan^2 \theta$

53. $(\csc t - \cot t)^4(\csc t + \cot t)^4 = 1$

54. $(\sec t + \tan t)^5(\sec t - \tan t)^6 = \dfrac{1 - \sin t}{\cos t}$

55. $\sin^6 u + \cos^6 u = 1 - 3 \sin^2 u \cos^2 u$

56. $\dfrac{\cos^2 x - \cos^2 y}{\cot^2 x - \cot^2 y} = \sin^2 x \sin^2 y$

57. In a later section, we will learn that

$$\tan 3x = \frac{3 \tan x - \tan^3 x}{1 - 3 \tan^2 x}$$

Taking this for granted, show that

$$\cot 3x = \frac{3 \cot x - \cot^3 x}{1 - 3 \cot^2 x}$$

Note the similarity in form of these two identities.

58. **TEASER** Generalize Problem 57 by showing that if $\tan kx = f(\tan x)$ and if k is an odd number, then $\cot kx = f(\cot x)$. *Hint*: Let $x = \pi/2 - y$.

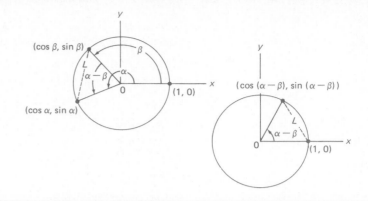

Equal Chords

In the two unit circles at the right, the two dotted chords have the same length, being chords for angles of the same size, namely, $\alpha - \beta$. Out of this simple observation, we can extract several of the most important identities of trigonometry.

8-2   Addition Laws

When you study calculus, you will meet expressions like $\cos(\alpha + \beta)$ and $\sin(\alpha - \beta)$. It will be very important to rewrite these expressions directly in terms of $\sin \alpha$, $\cos \alpha$, $\sin \beta$, and $\cos \beta$. It might be tempting to replace $\cos(\alpha + \beta)$ by $\cos \alpha + \cos \beta$ and $\sin(\alpha - \beta)$ by $\sin \alpha - \sin \beta$, but that would be terribly wrong. To see this, let's try $\alpha = \pi/3$ and $\beta = \pi/6$.

$$\cos\left(\frac{\pi}{3} + \frac{\pi}{6}\right) = \cos\frac{\pi}{2} = 0 \qquad \cos\frac{\pi}{3} + \cos\frac{\pi}{6} = \frac{1}{2} + \frac{\sqrt{3}}{2} \approx 1.4$$

$$\sin\left(\frac{\pi}{3} - \frac{\pi}{6}\right) = \sin\frac{\pi}{6} = .5 \qquad \sin\frac{\pi}{3} - \sin\frac{\pi}{6} = \frac{\sqrt{3}}{2} - \frac{1}{2} \approx .4$$

To obtain correct expressions is the goal of this section.

A KEY IDENTITY

The opening display shows two chords of equal length L. Using the formula for the distance between two points (Figure 1) and the identity $\sin^2 \theta + \cos^2 \theta = 1$, we have the following expression for the square of the chord on the right.

$$L^2 = [\cos(\alpha - \beta) - 1]^2 + \sin^2(\alpha - \beta)$$
$$= \cos^2(\alpha - \beta) - 2\cos(\alpha - \beta) + 1 + \sin^2(\alpha - \beta)$$
$$= [\cos^2(\alpha - \beta) + \sin^2(\alpha - \beta)] + 1 - 2\cos(\alpha - \beta)$$
$$= 2 - 2\cos(\alpha - \beta)$$

A similar calculation for the square of the chord on the left gives

$$L^2 = (\cos \alpha - \cos \beta)^2 + (\sin \alpha - \sin \beta)^2$$

Distance Formula

The distance between (x_1, y_1) and (x_2, y_2) is

$$\sqrt{(x_2 - x_1)^2 + (y_2 - y_1)^2}$$

Figure 1

$$= \cos^2 \alpha - 2 \cos \alpha \cos \beta + \cos^2 \beta + \sin^2 \alpha - 2 \sin \alpha \sin \beta + \sin^2 \beta$$

$$= 1 - 2 \cos \alpha \cos \beta - 2 \sin \alpha \sin \beta + 1$$

$$= 2 - 2(\cos \alpha \cos \beta + \sin \alpha \sin \beta)$$

When we equate these two expressions for L^2, we get our key identity

$$\cos(\alpha - \beta) = \cos \alpha \cos \beta + \sin \alpha \sin \beta$$

Our derivation is based on a picture in which α and β are positive angles with $\alpha > \beta$. Minor modifications would establish the identity for arbitrary angles α and β and hence also for their radian measures s and t. Thus for all real numbers s and t,

$$\boxed{\cos(s - t) = \cos s \cos t + \sin s \sin t}$$

We can use this identity to calculate $\cos(\pi/12)$ by thinking of $\pi/12$ as $\pi/3 - \pi/4$.

$$\cos \frac{\pi}{12} = \cos\left(\frac{\pi}{3} - \frac{\pi}{4}\right) = \cos \frac{\pi}{3} \cos \frac{\pi}{4} + \sin \frac{\pi}{3} \sin \frac{\pi}{4}$$

$$= \frac{1}{2} \cdot \frac{\sqrt{2}}{2} + \frac{\sqrt{3}}{2} \cdot \frac{\sqrt{2}}{2} = \frac{\sqrt{2} + \sqrt{6}}{4} \approx .9659$$

In words, this identity says: *The cosine of a difference is the cosine of the first times the cosine of the second plus the sine of the first times the sine of the second.* It is important to memorize this identity in words so you can easily apply it to $\cos(3u - v)$, $\cos[s - (-t)]$, or even $\cos[(\pi/2 - s) - t]$, as we shall have to do soon.

RELATED IDENTITIES

In the boxed identity above, we replace t by $-t$ and use the fundamental identities $\cos(-t) = \cos t$ and $\sin(-t) = -\sin t$ to get

$$\cos[s - (-t)] = \cos s \cos(-t) + \sin s \sin(-t)$$

$$= \cos s \cos t + (\sin s)(-\sin t)$$

This gives us the **addition law for cosines.**

$$\boxed{\cos(s + t) = \cos s \cos t - \sin s \sin t}$$

We illustrate this law by calculating $\cos(13\pi/12)$.

$$\cos \frac{13\pi}{12} = \cos\left(\frac{3\pi}{4} + \frac{\pi}{3}\right) = \cos \frac{3\pi}{4} \cos \frac{\pi}{3} - \sin \frac{3\pi}{4} \sin \frac{\pi}{3}$$

$$= \frac{-\sqrt{2}}{2} \cdot \frac{1}{2} - \frac{\sqrt{2}}{2} \cdot \frac{\sqrt{3}}{2} = -\frac{\sqrt{2} + \sqrt{6}}{4} \approx -.9659$$

There is also an identity involving $\sin(s + t)$. To derive this identity, we use the cofunction identity $\sin u = \cos(\pi/2 - u)$ to write

$$\sin(s + t) = \cos\left[\frac{\pi}{2} - (s + t)\right] = \cos\left[\left(\frac{\pi}{2} - s\right) - t\right]$$

Then we use our key identity for the cosine of a difference to obtain

$$\cos\left(\frac{\pi}{2} - s\right)\cos t + \sin\left(\frac{\pi}{2} - s\right)\sin t$$

Two applications of cofunction identities give us the result we want, the **addition law for sines.**

$$\sin(s + t) = \sin s \cos t + \cos s \sin t$$

Finally, replacing t by $-t$ in this last result leads to

$$\sin(s - t) = \sin s \cos t - \cos s \sin t$$

If we let $s = \pi/2$ in the latter, we get another important identity—but it is one we already know.

$$\sin\left(\frac{\pi}{2} - t\right) = \sin\frac{\pi}{2}\cos t - \cos\frac{\pi}{2}\sin t = 1 \cdot \cos t - 0 \cdot \sin t = \cos t$$

CAUTION

$$\sin\left(\theta + \frac{\pi}{2}\right) = \sin\theta + \sin\frac{\pi}{2}$$
$$= \sin\theta + 1$$

$$\sin\left(\theta + \frac{\pi}{2}\right)$$
$$= \sin\theta\cos\frac{\pi}{2} + \cos\theta\sin\frac{\pi}{2}$$
$$= \cos\theta$$

Problem Set 8-2

Find the value of each expression. Note that in each case, the answers to parts (a) and (b) are different.

1. (a) $\sin\dfrac{\pi}{4} + \sin\dfrac{\pi}{6}$ (b) $\sin\left(\dfrac{\pi}{4} + \dfrac{\pi}{6}\right)$

2. (a) $\cos\dfrac{\pi}{4} + \cos\dfrac{\pi}{6}$ (b) $\cos\left(\dfrac{\pi}{4} + \dfrac{\pi}{6}\right)$

3. (a) $\cos\dfrac{\pi}{4} - \cos\dfrac{\pi}{6}$ (b) $\cos\left(\dfrac{\pi}{4} - \dfrac{\pi}{6}\right)$

4. (a) $\sin\dfrac{\pi}{4} - \sin\dfrac{\pi}{6}$ (b) $\sin\left(\dfrac{\pi}{4} - \dfrac{\pi}{6}\right)$

Use the identities derived in this section to show that the equalities in Problems 5–12 are identities.

5. $\sin(t + \pi) = -\sin t$ 6. $\cos(t + \pi) = -\cos t$

7. $\sin\left(t + \dfrac{3\pi}{2}\right) = -\cos t$ 8. $\cos\left(t + \dfrac{3\pi}{2}\right) = \sin t$

9. $\sin\left(t - \dfrac{\pi}{2}\right) = -\cos t$ 10. $\cos\left(t - \dfrac{\pi}{2}\right) = \sin t$

11. $\cos\left(t + \dfrac{\pi}{3}\right) = \dfrac{1}{2}\cos t - \dfrac{\sqrt{3}}{2}\sin t$

12. $\sin\left(t + \dfrac{\pi}{3}\right) = \dfrac{1}{2}\sin t + \dfrac{\sqrt{3}}{2}\cos t$

EXAMPLE A (Recognizing Expressions as Single Sines or Cosines) Write as a single sine or cosine.
(a) $\sin \frac{7}{6} \cos \frac{1}{6} + \cos \frac{7}{6} \sin \frac{1}{6}$
(b) $\cos(x + h)\cos h + \sin(x + h)\sin h$

Solution.
(a) By the addition law for sines,
$$\sin \tfrac{7}{6} \cos \tfrac{1}{6} + \cos \tfrac{7}{6} \sin \tfrac{1}{6} = \sin(\tfrac{7}{6} + \tfrac{1}{6}) = \sin \tfrac{4}{3}$$
(b) This we recognize as the cosine of a difference
$$\cos(x + h)\cos h + \sin(x + h)\sin h = \cos[(x + h) - h] = \cos x$$

Write each of the following as a single sine or cosine.

13. $\cos \frac{1}{2} \cos \frac{3}{2} - \sin \frac{1}{2} \sin \frac{3}{2}$

14. $\cos 2 \cos 3 + \sin 2 \sin 3$

15. $\sin \dfrac{7\pi}{8} \cos \dfrac{\pi}{8} + \cos \dfrac{7\pi}{8} \sin \dfrac{\pi}{8}$

16. $\sin \dfrac{5\pi}{16} \cos \dfrac{\pi}{16} - \cos \dfrac{5\pi}{16} \sin \dfrac{\pi}{16}$

17. $\cos 33° \cos 27° - \sin 33° \sin 27°$

18. $\sin 49° \cos 41° + \cos 49° \sin 41°$

19. $\sin(\alpha + \beta) \cos \beta - \cos(\alpha + \beta) \sin \beta$

20. $\cos(\alpha + \beta) \cos(\alpha - \beta) - \sin(\alpha + \beta) \sin(\alpha - \beta)$

EXAMPLE B (Using the Addition Laws) Suppose that α is a first quadrant angle with $\cos \alpha = \frac{4}{5}$ and β is a second quadrant angle with $\sin \beta = \frac{12}{13}$. Evaluate $\sin(\alpha + \beta)$ and $\cos(\alpha + \beta)$ and then determine the quadrant for $\alpha + \beta$.

Solution. We are going to need $\sin \alpha$ and $\cos \beta$. We can find them by using the identity $\sin^2 \theta + \cos^2 \theta = 1$, but we have to be careful about signs.
$$\sin \alpha = \sqrt{1 - \cos^2 \alpha} = \sqrt{1 - \tfrac{16}{25}} = \tfrac{3}{5}$$
$$\cos \beta = -\sqrt{1 - \sin^2 \beta} = -\sqrt{1 - \tfrac{144}{169}} = -\tfrac{5}{13}$$
We chose the plus sign in the first case because α is a first quadrant angle

and the minus sign in the second because β is a second quadrant angle, where the cosine is negative. Then

$$\sin(\alpha + \beta) = \sin\alpha\cos\beta + \cos\alpha\sin\beta$$

$$= \left(\frac{3}{5}\right)\left(\frac{-5}{13}\right) + \left(\frac{4}{5}\right)\left(\frac{12}{13}\right) = \frac{33}{65}$$

$$\cos(\alpha + \beta) = \cos\alpha\cos\beta - \sin\alpha\sin\beta$$

$$= \left(\frac{4}{5}\right)\left(\frac{-5}{13}\right) - \left(\frac{3}{5}\right)\left(\frac{12}{13}\right) = \frac{-56}{65}$$

Since $\sin(\alpha + \beta)$ is positive and $\cos(\alpha + \beta)$ is negative, $\alpha + \beta$ is a second quadrant angle.

21. If α and β are third quadrant angles with $\sin\alpha = -\frac{4}{5}$ and $\cos\beta = -\frac{5}{13}$, find $\sin(\alpha + \beta)$ and $\cos(\alpha + \beta)$. In what quadrant does the terminal side of $\alpha + \beta$ lie?

22. Let α and β be second quadrant angles with $\sin\alpha = \frac{2}{3}$ and $\sin\beta = \frac{3}{4}$. Find $\sin(\alpha + \beta)$ and $\cos(\alpha + \beta)$ and determine the quadrant for $\alpha + \beta$.

23. Let α be a first quadrant angle with $\sin\alpha = 1/\sqrt{10}$ and β be a second quadrant angle with $\cos\beta = -\frac{1}{2}$. Find $\sin(\alpha - \beta)$ and $\cos(\alpha - \beta)$ and determine the quadrant for $\alpha - \beta$.

24. Let α and β be second and third quadrant angles, respectively, with $\cos\alpha = \cos\beta = -\frac{3}{7}$. Find $\sin(\alpha - \beta)$ and $\cos(\alpha - \beta)$ and determine the quadrant for $\alpha - \beta$.

EXAMPLE C (Tangent Identities) Verify the **addition law for tangents.**

$$\boxed{\tan(s + t) = \frac{\tan s + \tan t}{1 - \tan s \tan t}}$$

Solution.

$$\tan(s + t) = \frac{\sin(s + t)}{\cos(s + t)}$$

$$= \frac{\sin s \cos t + \cos s \sin t}{\cos s \cos t - \sin s \sin t}$$

$$= \frac{\dfrac{\sin s \cos t}{\cos s \cos t} + \dfrac{\cos s \sin t}{\cos s \cos t}}{\dfrac{\cos s \cos t}{\cos s \cos t} - \dfrac{\sin s \sin t}{\cos s \cos t}}$$

$$= \frac{\tan s + \tan t}{1 - \tan s \tan t}$$

The key step was the third one, in which we divided both the numerator and the denominator by $\cos s \cos t$.

Establish that each equation in Problems 25–28 is an identity.

25. $\tan(s - t) = \dfrac{\tan s - \tan t}{1 + \tan s \tan t}$

26. $\tan(s + \pi) = \tan s$

27. $\tan\left(t + \dfrac{\pi}{4}\right) = \dfrac{1 + \tan t}{1 - \tan t}$

28. $\tan\left(t - \dfrac{\pi}{3}\right) = \dfrac{\tan t - \sqrt{3}}{1 + \sqrt{3}\,\tan t}$

MISCELLANEOUS PROBLEMS

29. Express in terms of $\sin t$ and $\cos t$.
    (a) $\sin(t - \frac{5}{6}\pi)$ (b) $\cos(\frac{\pi}{6} - t)$

30. Express $\tan(\theta + \frac{3}{4}\pi)$ in terms of $\tan \theta$.

31. Let α and β be first and third quadrant angles, respectively, with $\sin \alpha = \frac{2}{3}$ and $\cos \beta = -\frac{1}{3}$. Evaluate each of the following exactly.
    (a) $\cos \alpha$ (b) $\sin \beta$ (c) $\cos(\alpha + \beta)$
    (d) $\sin(\alpha - \beta)$ (e) $\tan(\alpha + \beta)$ (f) $\sin(2\beta)$

32. If $0 \le t \le \pi/2$ and $\cos(t + \pi/6) = .8$, find the exact value of $\sin t$ and $\cos t$. *Hint:* $t = (t + \pi/6) - \pi/6$.

33. Evaluate each of the following (the easy way).
    (a) $\sin(t + \pi/3) \cos t - \cos(t + \pi/3) \sin t$
    (b) $\cos 175° \cos 25° + \sin 175° \sin 25°$
    (c) $\sin t \cos(1 - t) + \cos t \sin(1 - t)$

34. Find the exact value of $\cos 85° \cos 40° + \cos 5° \cos 50°$.

35. Show that each of the following is an identity.
    (a) $\sin(x + y) \sin(x - y) = \sin^2 x - \sin^2 y$
    (b) $\dfrac{\sin(x + y)}{\cos(x - y)} = \dfrac{\tan x + \tan y}{1 + \tan x \tan y}$
    (c) $\dfrac{\cos 5t}{\sin t} - \dfrac{\sin 5t}{\cos t} = \dfrac{\cos 6t}{\sin t \cos t}$

36. Show that the following are identities.
    (a) $\cot(u + v) = \dfrac{\cot u \cot v - 1}{\cot u + \cot v}$
    (b) $\dfrac{\sin(u + v)}{\sin(u - v)} = \dfrac{\tan u + \tan v}{\tan u - \tan v}$
    (c) $\dfrac{\cos 2t}{\sin t} + \dfrac{\sin 2t}{\cos t} = \csc t$

37. Let θ be the smallest counterclockwise angle from the line $y = m_1 x + b_1$ to the line $y = m_2 x + b_2$, where $m_1 m_2 \ne -1$. Show that

$$\tan \theta = \dfrac{m_2 - m_1}{1 + m_1 m_2}$$

38. Find the counterclockwise angle θ from the line $3x - 4y = 1$ to the line $2x + 6y = 3$. (See Problem 37.)

39. Use the addition and subtraction laws (the four boxed formulas of this section) to prove the following **product identities**.
    (a) $\cos s \cos t = \frac{1}{2}[\cos(s + t) + \cos(s - t)]$
    (b) $\sin s \sin t = -\frac{1}{2}[\cos(s + t) - \cos(s - t)]$
    (c) $\sin s \cos t = \frac{1}{2}[\sin(s + t) + \sin(s - t)]$
    (d) $\cos s \sin t = \frac{1}{2}[\sin(s + t) - \sin(s - t)]$

40. Use the identities of Problem 39 to prove the following **factoring identities**. *Hint:* Let $u = s + t$ and $v = s - t$.

(a) $\cos u + \cos v = 2 \cos \dfrac{u+v}{2} \cos \dfrac{u-v}{2}$

(b) $\cos u - \cos v = -2 \sin \dfrac{u+v}{2} \sin \dfrac{u-v}{2}$

(c) $\sin u + \sin v = 2 \sin \dfrac{u+v}{2} \cos \dfrac{u-v}{2}$

(d) $\sin u - \sin v = 2 \cos \dfrac{u+v}{2} \sin \dfrac{u-v}{2}$

41. Evaluate each of the following exactly.
    (a) $\cos 105° \cos 45°$ (b) $\sin 15° - \sin 75°$
    (c) $\cos 15° + \cos 30° + \cos 45° + \cos 60° + \cos 75°$

42. Show that each of the following is an identity.
    (a) $\dfrac{\cos 9t + \cos 3t}{\sin 9t - \sin 3t} = \cot 3t$ (b) $\dfrac{\sin 3u + \sin 7u}{\cos 3u + \cos 7u} = \tan 5u$
    (c) $\cos 10\beta + \cos 2\beta + 2 \cos 8\beta \cos 6\beta = 4 \cos^2 6\beta \cos 2\beta$

43. Stack three identical squares and consider angles α, β, and γ as shown in Figure 2. Prove that $\alpha + \beta = \gamma$.

44. **TEASER** Consider an oblique triangle (no right angles) with angles α, β, and γ. Prove that
    $$\tan \alpha + \tan \beta + \tan \gamma = \tan \alpha \tan \beta \tan \gamma$$

Figure 2

I could do this problem if $\sin(2t) = 2 \sin t$

Wishful Thinking

"Wishful thinking is imagining good things you don't have . . . . [It] may be bad as too much salt is bad in the soup and even a little garlic is bad in the chocolate pudding. I mean, wishful thinking may be bad if there is too much of it or in the wrong place, but it is good in itself and may be a great help in life and in problem solving."

George Polya
in *Mathematical Discovery*

8-3 Double-Angle and Half-Angle Formulas

George Polya would agree that the student in our opening panel is wishing for too much. And there is a better way than wishing to get formulas for $\sin 2t$ and $\cos 2t$. All we have to do is to think of $2t$ as $t + t$ and apply the addition laws of the previous section.

$$\sin(t + t) = \sin t \cos t + \cos t \sin t = 2 \sin t \cos t$$
$$\cos(t + t) = \cos t \cos t - \sin t \sin t = \cos^2 t - \sin^2 t$$

DOUBLE-ANGLE FORMULAS

We have just derived two very important results. They are called *double-angle formulas*, though double-number formulas would perhaps be more appropriate.

$$\sin 2t = 2 \sin t \cos t$$
$$\cos 2t = \cos^2 t - \sin^2 t$$

Suppose $\sin t = \frac{2}{5}$ and $\pi/2 < t < \pi$. Then we can calculate both $\sin 2t$ and $\cos 2t$, but we must first find $\cos t$. Since $\pi/2 < t < \pi$, the cosine is negative, and therefore

$$\cos t = -\sqrt{1 - \sin^2 t} = -\sqrt{1 - \frac{4}{25}} = -\frac{\sqrt{21}}{5}$$

The double-angle formulas now give

$$\sin 2t = 2 \sin t \cos t = 2\left(\frac{2}{5}\right)\left(\frac{-\sqrt{21}}{5}\right) = \frac{-4\sqrt{21}}{25} \approx -.73$$

$$\cos 2t = \cos^2 t - \sin^2 t = \left(\frac{-\sqrt{21}}{5}\right)^2 - \left(\frac{2}{5}\right)^2 = \frac{17}{25} \approx .68$$

CAUTION

$\cos 6\theta = 6 \cos \theta$

$\cos 6\theta = 2 \cos^2 3\theta - 1$

There are two other forms of the cosine double-angle formula that are often useful. If, in the expression $\cos^2 t - \sin^2 t$, we replace $\cos^2 t$ by $1 - \sin^2 t$, we obtain

$$\cos 2t = 1 - 2 \sin^2 t$$

and, alternatively, if we replace $\sin^2 t$ by $1 - \cos^2 t$, we have

$$\cos 2t = 2 \cos^2 t - 1$$

Of course, in all that we have done, we may replace the number t by the angle θ; hence the name double-angle formulas.

Once we grasp the generality of the four boxed formulas, we can write numerous others that follow from them. For example,

$$\sin 6\theta = 2 \sin 3\theta \cos 3\theta$$

$$\cos 4u = \cos^2 2u - \sin^2 2u$$

$$\cos t = 1 - 2 \sin^2\left(\frac{t}{2}\right)$$

$$\cos t = 2 \cos^2\left(\frac{t}{2}\right) - 1$$

The last two of these identities lead us directly to the half-angle formulas.

HALF-ANGLE FORMULAS

In the identity $\cos t = 1 - 2 \sin^2(t/2)$, we solve for $\sin(t/2)$.

$$2 \sin^2\left(\frac{t}{2}\right) = 1 - \cos t$$

$$\sin^2\left(\frac{t}{2}\right) = \frac{1 - \cos t}{2}$$

$$\sin\left(\frac{t}{2}\right) = \pm\sqrt{\frac{1 - \cos t}{2}}$$

Similarly, if we solve $\cos t = 2 \cos^2(t/2) - 1$ for $\cos(t/2)$, the result is

$$\cos\left(\frac{t}{2}\right) = \pm\sqrt{\frac{1 + \cos t}{2}}$$

In both of these formulas, the choice of the plus or minus sign is determined by the interval on which $t/2$ lies. For example,

$$\cos\left(\frac{5\pi}{8}\right) = \cos\left(\frac{5\pi/4}{2}\right) = -\sqrt{\frac{1 + \cos(5\pi/4)}{2}}$$

$$= -\sqrt{\frac{1 - \sqrt{2}/2}{2}} = -\frac{\sqrt{2 - \sqrt{2}}}{2}$$

We chose the minus sign because $5\pi/8$ corresponds to an angle in quadrant II, where the cosine is negative.

As a second example, suppose that $\cos \theta = .4$, where θ is a fourth quadrant angle. Then we can calculate $\sin(\theta/2)$ and $\cos(\theta/2)$, observing first that $\theta/2$ is necessarily a second quadrant angle.

$$\sin\left(\frac{\theta}{2}\right) = \sqrt{\frac{1 - \cos\theta}{2}} = \sqrt{\frac{1 - .4}{2}} \approx .548$$

$$\cos\left(\frac{\theta}{2}\right) = -\sqrt{\frac{1 + \cos\theta}{2}} = -\sqrt{\frac{1 + .4}{2}} \approx -.837$$

Problem Set 8-3

1. If $\cos t = \frac{4}{5}$ with $0 < t < \pi/2$, show that $\sin t = \frac{3}{5}$. Then use formulas from this section to calculate
   (a) $\sin 2t$;
   (b) $\cos 2t$;
   (c) $\cos(t/2)$;
   (d) $\sin(t/2)$.

2. If $\sin t = -\frac{2}{3}$ with $3\pi/2 < t < 2\pi$, show that $\cos t = \sqrt{5}/3$. Then calculate
   (a) $\sin 2t$;
   (b) $\cos 2t$;
   (c) $\cos(t/2)$;
   (d) $\sin(t/2)$.

Use formulas from this section to simplify the expressions in Problems 3–16. For example, $2 \sin(.5)\cos(.5) = \sin 1$.

3. $2 \sin 5t \cos 5t$

4. $2 \sin 3\theta \cos 3\theta$

5. $\cos^2(3t/2) - \sin^2(3t/2)$

6. $\cos^2(7\pi/8) - \sin^2(7\pi/8)$

7. $2 \cos^2(y/4) - 1$

8. $2 \cos^2(\alpha/3) - 1$

9. $1 - 2 \sin^2(.6t)$

10. $2 \sin^2(\pi/8) - 1$

11. $\sin^2(\pi/8) - \cos^2(\pi/8)$

12. $2 \sin(.3) \cos(.3)$

13. $\dfrac{1 + \cos x}{2}$

14. $\dfrac{1 - \cos y}{2}$

15. $\dfrac{1 - \cos 4\theta}{2}$

16. $\dfrac{1 + \cos 8u}{2}$

17. Use the half-angle formulas to calculate
    (a) $\sin(\pi/8)$;
    (b) $\cos(112.5°)$.

18. Calculate, using half-angle formulas,
    (a) $\cos 67.5°$;
    (b) $\sin(\pi/12)$.

19. Use the addition law for tangents (Example C of Section 8-2) to show that
$$\tan 2t = \frac{2 \tan t}{1 - \tan^2 t}$$

20. Use the identity of Problem 19 to evaluate $\tan 2t$ given that
    (a) $\tan t = 3$;
    (b) $\cos t = \frac{4}{5}$ and $0 < t < \pi/2$.

21. Use the half-angle formulas for sine and cosine to show that
$$\tan\left(\frac{t}{2}\right) = \pm\sqrt{\frac{1 - \cos t}{1 + \cos t}}$$

22. Use the identity of Problem 21 to evaluate
    (a) $\tan(\pi/8)$;
    (b) $\tan 112.5°$.

EXAMPLE (Using Double-Angle and Half-Angle Formulas to Prove New Identities) Prove that the following are identities.

(a) $\sin 3t = 3 \sin t - 4 \sin^3 t$

(b) $\tan \dfrac{t}{2} = \dfrac{\sin t}{1 + \cos t}$

Solution.

(a) We think of $3t$ as $2t + t$ and use the addition law for sines and then double-angle formulas.

$$
\begin{aligned}
\sin 3t &= \sin(2t + t) \\
&= \sin 2t \cos t + \cos 2t \sin t \\
&= (2 \sin t \cos t) \cos t + (1 - 2 \sin^2 t) \sin t \\
&= 2 \sin t (1 - \sin^2 t) + \sin t - 2 \sin^3 t \\
&= 2 \sin t - 2 \sin^3 t + \sin t - 2 \sin^3 t \\
&= 3 \sin t - 4 \sin^3 t
\end{aligned}
$$

(b) This is the unambiguous form for $\tan(t/2)$ (see Problem 21). To prove it, think of t as $2(t/2)$ and apply double-angle formulas to the right side.

$$
\begin{aligned}
\frac{\sin t}{1 + \cos t} &= \frac{\sin(2(t/2))}{1 + \cos(2(t/2))} \\
&= \frac{2 \sin(t/2) \cos(t/2)}{1 + 2 \cos^2(t/2) - 1} \\
&= \frac{\sin(t/2)}{\cos(t/2)} \\
&= \tan \frac{t}{2}
\end{aligned}
$$

Now prove that each of the following is an identity.

23. $\cos 3t = 4 \cos^3 t - 3 \cos t$

24. $(\sin t + \cos t)^2 = 1 + \sin 2t$

25. $\csc 2t + \cot 2t = \cot t$

26. $\sin^2 t \cos^2 t = \frac{1}{8}(1 - \cos 4t)$

27. $\dfrac{\sin \theta}{1 - \cos \theta} = \cot \dfrac{\theta}{2}$

28. $1 - 2 \sin^2 \theta = 2 \cot 2\theta \sin \theta \cos \theta$

29. $\dfrac{2 \tan \alpha}{1 + \tan^2 \alpha} = \sin 2\alpha$

30. $\dfrac{1 - \tan^2 \alpha}{1 + \tan^2 \alpha} = \cos 2\alpha$

31. $\sin 4\theta = 4 \sin \theta \, (2 \cos^3 \theta - \cos \theta)$

   Hint: $4\theta = 2(2\theta)$.

32. $\cos 4\theta = 8 \cos^4 \theta - 8 \cos^2 \theta + 1$

MISCELLANEOUS PROBLEMS

33. Write a simple expression for each of the following.
    (a) $2 \sin(x/2) \cos(x/2)$ (b) $\cos^2 3t - \sin^2 3t$
    (c) $2 \sin^2(y/4) - 1$ (d) $(\cos 4t - 1)/2$
    (e) $(1 - \cos 4t)/(1 + \cos 4t)$ (f) $(\sin 6y)/(1 + \cos 6y)$

34. Find the exact value of each of the following.
    (a) $\sin 15° \cos 15°$ (b) $\cos^2 105° - \sin^2 105°$
    (c) $\sin 15°$ (d) $\cos 105°$

35. If $\pi < t < 3\pi/2$ and $\cos t = -5/13$, find each value.
    (a) $\sin 2t$ (b) $\cos(t/2)$ (c) $\tan(t/2)$

36. If the trigonometric point $P(t)$ on the unit circle has coordinates $(-\frac{3}{5}, \frac{4}{5})$, find the coordinates for each of the following points.
    (a) $P(2t)$ (b) $P(t/2)$ (c) $P(4t)$

In Problems 37–52, prove that each equation is an identity.

37. $\cos^4 z - \sin^4 z = \cos 2z$
38. $(1 - \cos 4x)/\tan^2 2x = 2 \cos^2 2x$
39. $1 + (1 - \cos 8t)/(1 + \cos 8t) = \sec^2 4t$
40. $\sec 2t = (\sec^2 t)/(2 - \sec^2 t)$
41. $\tan(\theta/2) - \sin \theta = (-\sin \theta)/(1 + \sec \theta)$
42. $(2 - \sec^2 2\theta) \tan 4\theta = 2 \tan 2\theta$
43. $3 \cos 2t + 4 \sin 2t = (3 \cos t - \sin t)(\cos t + 3 \sin t)$
44. $\csc 2x - \cot 2x = \tan x$
45. $2(\cos 3x \cos x + \sin 3x \sin x)^2 = 1 + \cos 4x$
46. $\dfrac{1 + \sin 2x + \cos 2x}{1 + \sin 2x - \cos 2x} = \cot x$
47. $\tan 3t = \dfrac{3 \tan t - \tan^3 t}{1 - 3 \tan^2 t}$
48. $\cos^4 u = \frac{3}{8} + \frac{1}{2} \cos 2u + \frac{1}{8} \cos 4u$
49. $\sin^4 u + \cos^4 u = \frac{3}{4} + \frac{1}{4} \cos 4u$
50. $\cos^6 u - \sin^6 u = \cos 2u - \frac{1}{4} \sin^2 2u \cos 2u$
51. Prove that $\cos^2 x + \cos^2 2x + \cos^2 3x = 1 + 2 \cos x \cos 2x \cos 3x$ is an identity. *Hint:* Use half-angle formulas and factoring identities.
52. Calculate $(\sin 2t)[3 + (16 \sin^2 t - 16)\sin^2 t] - \sin 6t$ for $t = 1$, 2, and 3. Guess at an identity and then prove it.
53. If α, β, and γ are the three angles of a triangle, prove that
$$\sin 2\alpha + \sin 2\beta + \sin 2\gamma = 4 \sin \alpha \sin \beta \sin \gamma$$
54. Show that $\cos x \cos 2x \cos 4x \cos 8x \cos 16x = (\sin 32x)/(32 \sin x)$
55. Figure 3 shows two abutting circles of radius 1, one centered at the origin, the other at $(2, 0)$. Find the exact coordinates of P, the point where the line through $(-1, 0)$ and tangent to the second circle meets the first circle.
56. TEASER   Determine the exact value of
$$\sin 1° \sin 3° \sin 5° \sin 7° \cdots \sin 175° \sin 177° \sin 179°$$

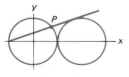

Figure 3

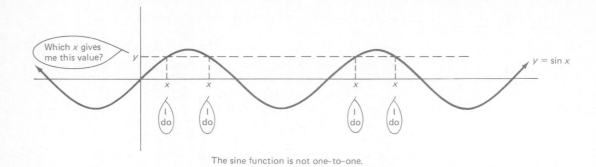

The sine function is not one–to–one.

8-4 Inverse Trigonometric Functions

The diagram above shows why the ordinary sine function does not have an inverse. We learned in Section 5-5 (a section worth reviewing now) that only one-to-one functions have inverses. To be one-to-one means that for each y there is at most one x that corresponds to it. The sine function is about as far from being one-to-one as possible. For each y between -1 and 1, there are infinitely many x's giving that y-value. To make the sine function have an inverse, we will have to restrict its domain drastically.

THE INVERSE SINE

Consider the graph of the sine function again (Figure 4). We want to restrict its domain in such a way that the sine assumes its full range of values but takes on each value only once. There are many possible choices, but the one commonly used is $-\pi/2 \le x \le \pi/2$. Notice the corresponding part of the sine graph below. From now on, whenever we need an inverse sine function, we always assume the domain of the sine has been restricted to $-\pi/2 \le x \le \pi/2$.

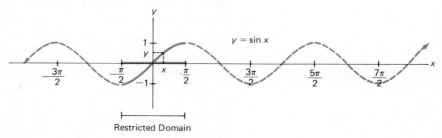

Restricted Domain

Figure 4

Having done this, we see that each y corresponds to exactly one x. We write $x = \sin^{-1} y$ (x is the inverse sine of y). Thus

$$\sin^{-1}\left(\frac{1}{2}\right) = \frac{\pi}{6}$$

$$\sin^{-1}(1) = \frac{\pi}{2}$$

$$\sin^{-1}(-1) = -\frac{\pi}{2}$$

$$\sin^{-1}\left(\frac{-\sqrt{2}}{2}\right) = -\frac{\pi}{4}$$

Please note that $\sin^{-1} y$ does not mean $1/(\sin y)$; you should not think of -1 as an exponent when used as a superscript on a function.

An alternate notation for $x = \sin^{-1} y$ is $x = \arcsin y$ (x is the arcsine of y). This is appropriate notation, since $\pi/6 = \arcsin \frac{1}{2}$ could be interpreted as saying that $\pi/6$ is the arc (on the unit circle) whose sine is $\frac{1}{2}$.

Recall from Section 5-5 that if f is a one-to-one function, then

$$x = f^{-1}(y) \quad \text{if and only if} \quad y = f(x)$$

Here the corresponding statement is

$$x = \sin^{-1} y \quad \text{if and only if} \quad y = \sin x \quad \text{and} \quad -\frac{\pi}{2} \le x \le \frac{\pi}{2}$$

Moreover

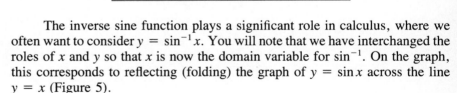

$$\sin(\sin^{-1} y) = y \quad \text{for} \quad -1 \le y \le 1$$

$$\sin^{-1}(\sin x) = x \quad \text{for} \quad -\frac{\pi}{2} \le x \le \frac{\pi}{2}$$

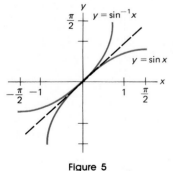

Figure 5

The inverse sine function plays a significant role in calculus, where we often want to consider $y = \sin^{-1} x$. You will note that we have interchanged the roles of x and y so that x is now the domain variable for $\sin^{-1}$. On the graph, this corresponds to reflecting (folding) the graph of $y = \sin x$ across the line $y = x$ (Figure 5).

THE INVERSE COSINE

One look at the graph of $y = \cos x$ should convince you that we cannot restrict the domain of the cosine to the same interval as that for the sine (Figure 6). We choose rather to use the interval $0 \le x \le \pi$, in which the cosine is one-to-one.

Having made the needed restriction, we may reasonably talk about $\cos^{-1}$. Moreover,

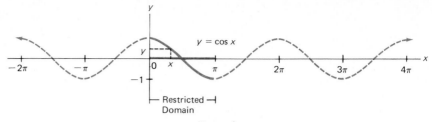

Figure 6

$$x = \cos^{-1}y \quad \text{if and only if} \quad y = \cos x \quad \text{and} \quad 0 \le x \le \pi$$

In particular,

$$\cos^{-1}1 = 0$$

$$\cos^{-1}\frac{\sqrt{3}}{2} = \frac{\pi}{6}$$

$$\cos^{-1}0 = \frac{\pi}{2}$$

$$\cos^{-1}(-1) = \pi$$

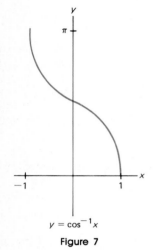

$y = \cos^{-1}x$

Figure 7

The graph of $y = \cos^{-1}x$ is shown in Figure 7. It is the graph of $y = \cos x$ reflected across the line $y = x$.

THE INVERSE TANGENT

To make $y = \tan x$ have an inverse, we restrict x to $-\pi/2 < x < \pi/2$. Thus

$$x = \tan^{-1}y \quad \text{if and only if} \quad y = \tan x \quad \text{and} \quad -\frac{\pi}{2} < x < \frac{\pi}{2}$$

The graphs of the tangent function and its inverse are shown in Figure 8.
Notice that the graph of $y = \tan^{-1}x$ has horizontal asymptotes at $y = \pi/2$ and $y = -\pi/2$.

THE INVERSE SECANT

The secant function has an inverse, provided we restrict its domain to $0 \le x \le \pi$, excluding $\pi/2$. Thus

$$x = \sec^{-1}y \quad \text{if and only if} \quad y = \sec x \quad \text{and} \quad 0 \le x \le \pi, x \ne \frac{\pi}{2}$$

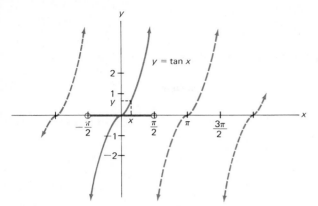

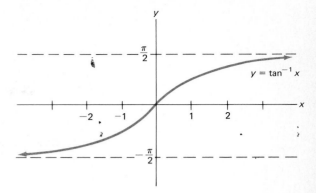

Figure 8

CAUTION

$$\sec^{-1} 2 \neq \frac{1}{\cos 2}$$

$$\sec^{-1} 2 = \cos^{-1} \frac{1}{2} = \frac{\pi}{3}$$

(Some authors choose to restrict the domain of the secant to $\{x : 0 \le x < \pi/2$ or $\pi \le x < 3\pi/2\}$. For this reason, check an author's definition before using any stated fact about the inverse secant.) The graphs of $y = \sec x$ and $y = \sec^{-1} x$ are shown in Figure 9.

Since the secant and cosine are reciprocals of each other, it is not surprising that $\sec^{-1} x$ is related to $\cos^{-1} x$. In fact, for every x in the domain of $\sec^{-1} x$, we have

$$\sec^{-1} x = \cos^{-1} \left(\frac{1}{x} \right)$$

This follows from the fact that each side is limited in value to the interval 0 to π and has the same cosine.

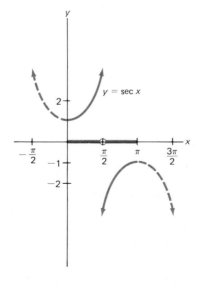

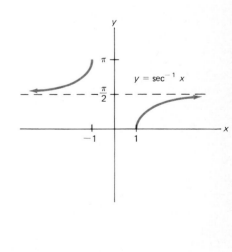

Figure 9

$$\cos(\sec^{-1} x) = \frac{1}{\sec(\sec^{-1} x)} = \frac{1}{x}$$

The two other inverse trigonometric functions, $\cot^{-1} x$ and $\csc^{-1} x$, are of less importance. They are introduced in Problem 66.

INVERSE TRIGONOMETRIC FUNCTIONS AND CALCULATORS

Most scientific calculators have been programmed to give values of $\sin^{-1}$, $\cos^{-1}$, and $\tan^{-1}$ that are consistent with the definitions we have given. For example, to obtain $\sin^{-1}(.32)$ on many calculators, press the buttons .32 $\boxed{\text{INV}}$ $\boxed{\sin}$; on other calculators, there is a button marked $\sin^{-1}$ and you simply press .32 $\boxed{\sin^{-1}}$. Normally, you cannot get $\sec^{-1}$ directly; instead you must use the identity $\sec^{-1} x = \cos^{-1}(1/x)$. Of course, in every case you must put your calculator in the appropriate mode, depending on whether you want the answer in degrees or radians.

Here is a problem that requires thinking in addition to use of a calculator. Find all values of t between 0 and 2π for which $\tan t = 2.12345$. A calculator (in radian mode) immediately gives the solution

$$t_0 = \tan^{-1}(2.12345) = 1.13067$$

However, there is another solution having t_0 as reference number, namely,

$$t = \pi + t_0 = 4.27227$$

THREE IDENTITIES

Here are three identities connecting sines, cosines, and their inverses.

(i) $\qquad\qquad\qquad \cos(\sin^{-1} x) = \sqrt{1 - x^2}$

(ii) $\qquad\qquad\qquad \sin(\cos^{-1} x) = \sqrt{1 - x^2}$

(iii) $\qquad\qquad \sin^{-1} x + \cos^{-1} x = \frac{\pi}{2}$

To prove the first identity, we let $\theta = \sin^{-1} x$. Remember that this means that $x = \sin \theta$, with $-\pi/2 \le \theta \le \pi/2$. Then

$$\cos(\sin^{-1} x) = \cos \theta = \pm\sqrt{1 - \sin^2 \theta} = \pm\sqrt{1 - x^2}$$

Finally, choose the plus sign because $\cos \theta$ is positive for $-\pi/2 \le \theta \le \pi/2$. The second identity is proved in a similar fashion.

To prove the third identity, let

$$\alpha = \sin^{-1} x \qquad \beta = \cos^{-1} x$$

and note that we must show that $\alpha + \beta = \pi/2$. Now

$$\sin(\alpha + \beta) = \sin \alpha \cos \beta + \cos \alpha \sin \beta$$
$$= \sin(\sin^{-1} x) \cos(\cos^{-1} x) + \cos(\sin^{-1} x) \sin(\cos^{-1} x)$$

$$= x \cdot x + \sqrt{1 - x^2} \cdot \sqrt{1 - x^2}$$
$$= x^2 + 1 - x^2$$
$$= 1$$

From this, we conclude that $\alpha + \beta$ is either $\pi/2$ or some number that differs from $\pi/2$ by a multiple of 2π. But since $-\pi/2 \le \alpha \le \pi/2$ and $0 \le \beta \le \pi$, it follows that

$$-\frac{\pi}{2} \le \alpha + \beta \le \frac{3\pi}{2}$$

The only possibility on this interval is $\alpha + \beta = \pi/2$.

Problem Set 8-4

Find the exact value of each of the following (without using a calculator).

1. $\sin^{-1}(\sqrt{3}/2)$
2. $\cos^{-1} \frac{1}{2}$
3. $\arcsin(\sqrt{2}/2)$
4. $\arccos(\sqrt{2}/2)$
5. $\tan^{-1} 0$
6. $\tan^{-1} 1$
7. $\tan^{-1} \sqrt{3}$
8. $\tan^{-1}(\sqrt{3}/3)$
9. $\arccos(-\frac{1}{2})$
10. $\arcsin(-\frac{1}{2})$
11. $\sec^{-1} \sqrt{2}$
12. $\sec^{-1}(-2/\sqrt{3})$

Use a calculator or Table D to find each value (in radians) in Problems 13–18.

13. $\sin^{-1} .21823$
14. $\cos^{-1} .30582$
15. $\sin^{-1}(-0.21823)$
16. $\cos^{-1}(-0.30582)$
17. $\tan^{-1} .20660$
18. $\tan^{-1}(1.2602)$

19. Calculate, using $\sec^{-1} x = \cos^{-1}(1/x)$.
    (a) $\sec^{-1} 1.4263$ (b) $\sec^{-1}(-2.6715)$
20. Calculate.
    (a) $\sec^{-1}(\pi + 1)$ (b) $\sec^{-1}(-\sqrt{5}/2)$

Solve for t, where $0 \le t < 2\pi$. Use a calculator if you have one.

21. $\sin t = .3416$
22. $\cos t = .9812$
23. $\tan t = 3.345$
24. $\sec t = 1.342$

Find the following without the use of tables or a calculator.

25. $\sin(\sin^{-1} \frac{2}{3})$
26. $\cos(\cos^{-1}(-\frac{1}{4}))$
27. $\tan(\tan^{-1} 10)$
28. $\cos^{-1}(\cos(\pi/2))$
29. $\sin^{-1}(\sin(\pi/3))$
30. $\tan^{-1}(\tan(\pi/4))$
31. $\sin^{-1}(\cos(\pi/4))$
32. $\cos^{-1}(\sin(-\pi/6))$
33. $\cos(\sin^{-1} \frac{4}{5})$
34. $\sin(\cos^{-1} \frac{3}{5})$
    Hint: Use the identities established on page 344.
35. $\cos(\tan^{-1} \frac{1}{2})$
36. $\cos(\tan^{-1}(-\frac{3}{4}))$
37. $\cos(\sec^{-1} 3)$
38. $\sec(\cos^{-1}(-.4))$
39. $\sec^{-1}(\sec(2\pi/3))$
40. $\sec(\sec^{-1} 2.56)$

41. $\cos(\sin^{-1}(-.2564))$

42. $\tan^{-1}(\sin 14.1)$

43. $\sin^{-1}(\cos 1.12)$

44. $\cos^{-1}(\cos^{-1} .91)$

45. $\tan(\sec^{-1} 2.5)$

46. $\sec^{-1}(\sin 1.67)$

EXAMPLE A (Complicated Evaluations Involving Inverses) Evaluate
(a) $\sin(2 \cos^{-1} \frac{2}{3})$;   (b) $\tan(\tan^{-1} 2 + \sin^{-1} \frac{4}{5})$.

Solution.

(a) Let $\theta = \cos^{-1}(\frac{2}{3})$ so that $\cos \theta = \frac{2}{3}$ and

$$\sin \theta = \sqrt{1 - \cos^2 \theta} = \sqrt{1 - \frac{4}{9}} = \frac{\sqrt{5}}{3}$$

Then apply the double-angle formula for $\sin 2\theta$ as indicated below.

$$\sin\left(2 \cos^{-1} \frac{2}{3}\right) = \sin 2\theta$$

$$= 2 \sin \theta \cos \theta$$

$$= 2 \frac{\sqrt{5}}{3} \cdot \frac{2}{3}$$

$$= \frac{4}{9}\sqrt{5}$$

(b) Let $\alpha = \tan^{-1} 2$ and $\beta = \sin^{-1}(\frac{4}{5})$ and apply the identity

$$\tan(\alpha + \beta) = \frac{\tan \alpha + \tan \beta}{1 - \tan \alpha \tan \beta}$$

Now $\tan \alpha = 2$ and

$$\tan \beta = \frac{\sin \beta}{\cos \beta} = \frac{\dfrac{4}{5}}{\sqrt{1 - \left(\dfrac{4}{5}\right)^2}} = \frac{\dfrac{4}{5}}{\dfrac{3}{5}} = \frac{4}{3}$$

Therefore

$$\tan(\alpha + \beta) = \frac{2 + \dfrac{4}{3}}{1 - 2 \cdot \dfrac{4}{3}} = -2$$

Evaluate by using the method of Example A, not by using a calculator.

47. $\sin(2 \cos^{-1} \frac{3}{5})$

48. $\sin(2 \cos^{-1} \frac{1}{2})$

49. $\cos(2 \sin^{-1}(-\frac{3}{5}))$

50. $\tan(2 \tan^{-1} \frac{1}{3})$

51. $\sin(\cos^{-1} \frac{3}{5} + \cos^{-1} \frac{5}{13})$

52. $\tan(\tan^{-1} \frac{1}{2} + \tan^{-1}(-3))$

53. $\cos(\sec^{-1} \frac{3}{2} - \sec^{-1} \frac{4}{3})$

54. $\sin(\sin^{-1} \frac{4}{5} + \sec^{-1} 3)$

EXAMPLE B (More Identities) Show that

$$\cos(2 \tan^{-1} x) = \frac{1 - x^2}{1 + x^2}$$

Solution. We will apply the double-angle formula

$$\cos 2\theta = 2 \cos^2 \theta - 1$$

Here $\theta = \tan^{-1} x$, so that $x = \tan \theta$. Then

$$\cos(2 \tan^{-1} x) = \cos(2\theta)$$

$$= 2 \cos^2 \theta - 1$$

$$= \frac{2}{\sec^2 \theta} - 1$$

$$= \frac{2}{1 + \tan^2 \theta} - 1$$

$$= \frac{2}{1 + x^2} - 1$$

$$= \frac{1 - x^2}{1 + x^2}$$

Show that each of the following is an identity.

55. $\tan(\sin^{-1} x) = \dfrac{x}{\sqrt{1 - x^2}}$

56. $\sin(\tan^{-1} x) = \dfrac{x}{\sqrt{1 + x^2}}$

57. $\tan(2 \tan^{-1} x) = \dfrac{2x}{1 - x^2}$

58. $\cos(2 \sin^{-1} x) = 1 - 2x^2$

59. $\cos(2 \sec^{-1} x) = \dfrac{2}{x^2} - 1$

60. $\sec(2 \tan^{-1} x) = \dfrac{1 + x^2}{1 - x^2}$

MISCELLANEOUS PROBLEMS

61. Without using tables or a calculator, find each value (in radians).
    (a) $\arcsin(-\sqrt{3}/2)$ (b) $\tan^{-1}(-\sqrt{3})$ (c) $\sec^{-1}(-2)$

62. Calculate each of the following (radian mode).
    (a) $\dfrac{2 \arccos(.956)}{3 \arcsin(-.846)}$ (b) $.3624 \sec^{-1}(4.193)$
    (c) $\cos^{-1}(2 \sin .1234)$ (d) $\sin[\arctan(4.62) - \arccos(-.48)]$

63. Without using tables or a calculator, find each value. Then check using your calculator.
    (a) $\tan[\tan^{-1}(43)]$ (b) $\cos[\sin^{-1}(\frac{5}{13})]$
    (c) $\sin[\frac{\pi}{4} + \sin^{-1}(.8)]$ (d) $\cos[\sin^{-1}(.6) + \sec^{-1}(3)]$

64. Try to calculate each of the following and then explain why your calculator gives you an error message.
    (a) $\cos[\sin^{-1}(2)]$ (b) $\cos^{-1}(\tan 2)$ (c) $\tan[\arctan 3 + \arctan(\frac{1}{3})]$

65. Solve for x.
   (a) $\cos(\sin^{-1}x) = \frac{3}{4}$
   (b) $\sin(\cos^{-1}x) = \sqrt{.19}$
   (c) $\sin^{-1}(3x - 5) = \frac{\pi}{6}$
   (d) $\tan^{-1}(x^2 - 3x + 3) = \frac{\pi}{4}$

66. To determine inverses for cotangent and cosecant, we restrict their domains to $0 < x < \pi$ and $-\pi/2 \leq x \leq \pi/2$, $x \neq 0$, respectively. With these restrictions understood, find each value.
   (a) $\cot^{-1}(\sqrt{3})$
   (b) $\cot^{-1}(-1/\sqrt{3})$
   (c) $\cot^{-1}(0)$
   (d) $\csc^{-1}(2)$
   (e) $\csc^{-1}(-1)$
   (f) $\csc^{-1}(-2/\sqrt{3})$

67. It is always true that $\sin(\sin^{-1}x) = x$, but it is not always true that $\sin^{-1}(\sin x) = x$. For example, $\sin^{-1}(\sin \pi) \neq \pi$. Instead,

$$\sin^{-1}(\sin \pi) = \sin^{-1}(0) = 0$$

   Find each value.
   (a) $\sin^{-1}[\sin(\pi/2)]$
   (b) $\sin^{-1}[\sin(3\pi/4)]$
   (c) $\sin^{-1}[\sin(5\pi/4)]$
   (d) $\sin^{-1}[\sin(3\pi/2)]$
   (e) $\cos^{-1}[\cos(3\pi)]$
   (f) $\tan^{-1}[\tan(13\pi/4)]$

68. Sketch the graph of $y = \sin^{-1}(\sin x)$ for $-2\pi \leq x \leq 4\pi$. *Hint:* See Problem 67.

69. For each of the following right triangles, write θ explicitly in terms of x.

   (a)
   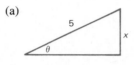

   (b)
   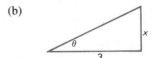

   (c)

   (d)
   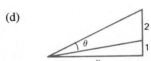

70. In some computer languages (for example, BASIC and FORTRAN), the only built-in inverse trigonometric function is $\tan^{-1}$. Establish the following identities, which show why this is sufficient.
   (a) $\sin^{-1} x = \tan^{-1}\left(\dfrac{x}{\sqrt{1 - x^2}}\right)$

   (b) $\cos^{-1} x = \dfrac{\pi}{2} - \sin^{-1} x = \dfrac{\pi}{2} - \tan^{-1}\left(\dfrac{x}{\sqrt{1 - x^2}}\right)$

71. Assume that your calculator's $\tan^{-1}$ button is working but not the $\sin^{-1}$ or $\cos^{-1}$ buttons. Use the results in Problem 70 to calculate each of the following.
   (a) $\sin^{-1}(.6)$
   (b) $\sin^{-1}(-.3)$
   (c) $\cos^{-1}(.8)$
   (d) $\cos^{-1}(-.9)$

72. Show that $\arctan(\frac{1}{4}) + \arctan(\frac{3}{5}) = \pi/4$. *Hint:* Show that both sides have the same tangent, using the formula for $\tan(\alpha + \beta)$.

73. In 1706, John Machin used the following formula to calculate π to 100 decimal places, a tremendous feat for its day. Establish this formula. *Hint:* Apply the addition formula for the tangent to the left side. Think of the right side as $4\theta = 2(2\theta)$ and use the tangent double angle formula twice.

$$\frac{\pi}{4} + \arctan\left(\frac{1}{239}\right) = 4 \arctan\left(\frac{1}{5}\right)$$

74. Show that

$$\arctan\left(\frac{1}{3}\right) + \arctan\left(\frac{1}{5}\right) + \arctan\left(\frac{1}{7}\right) + \arctan\left(\frac{1}{8}\right) = \frac{\pi}{4}$$

75. A picture 4 feet high is hung on a museum wall so that its bottom is 7 feet above the floor. A viewer whose eye level is 5 feet above the floor stands b feet from the wall.
    (a) Express θ, the vertical angle subtended by the picture at her eye, explicitly in terms of b.
    (b) Calculate θ when $b = 8$.
    (c) Determine b so $\theta = 30°$.

76. **TEASER**   A goat is tethered to a stake at the edge of a circular pond of radius r by means of a rope of length kr, $0 < k \le 2$. Find an explicit formula for its grazing area in terms of r and k.

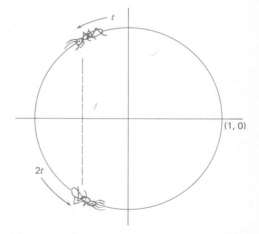

Two Bugs on a Circle

Two bugs crawl around the unit circle starting together at $(1, 0)$, one moving at one unit per second, the other moving twice as fast. When will one bug be directly above the other bug?

8-5   Trigonometric Equations

What does the bug problem have to do with trigonometric equations? Well, you should agree that after t seconds the slow bug, having traveled t units along the unit circle, is at $(\cos t, \sin t)$. The fast bug is at $(\cos 2t, \sin 2t)$. One bug will be directly above the other bug when their two x-coordinates are equal. This means we must solve the equation

$$\cos 2t = \cos t$$

Specifically, we must find the first $t > 0$ that makes this equality true. We shall solve this equation in due time, but first we ought to solve some simpler trigonometric equations.

SIMPLE EQUATIONS

Suppose we are asked to solve the equation

$$\sin t = \frac{1}{2}$$

for t. The number $t = \pi/6$ occurs to us right away. But that is not the only answer. All numbers that measure angles in the first or second quadrant and have $\pi/6$ as their reference number are solutions. Thus,

$$\ldots, -\frac{11\pi}{6}, -\frac{7\pi}{6}, \frac{\pi}{6}, \frac{5\pi}{6}, \frac{13\pi}{6}, \ldots$$

all work. In fact, one characteristic of trigonometric equations is that, if they have one solution, they have infinitely many solutions.

Let us alter the problem. Suppose we wish to solve $\sin t = \frac{1}{2}$ for $0 \le t < 2\pi$. Then the answers are $\pi/6$ and $5\pi/6$. In the following pages, we shall assume that, unless otherwise specified, we are to find only those solutions on the interval $0 \le t < 2\pi$.

For a second example, let us solve $\sin t = -.5234$ for $0 \le t < 2\pi$. Our unthinking action is to use a calculator to find $\sin^{-1}(-.5234)$, but note that this gives us the value $-.55084$, which is not on the required interval. The better way to proceed is to recognize that the given equation has two solutions but that they both have t_0 as their reference number, where

$$t_0 = \sin^{-1}(.5234) = .55084$$

The two solutions we seek are $\pi + t_0$ and $2\pi - t_0$, that is, 3.69243 and 5.73235.

Do you remember how we solved the equation $x^2 = 4x$? We rewrote it with 0 on one side, factored the other side, and then set each factor equal to 0 (see Figure 10). We follow exactly the same procedure with our next trigonometric equation. To solve

$$\cos t \cot t = -\cos t$$

use the following steps:

$$\cos t \cot t + \cos t = 0$$

$$\cos t (\cot t + 1) = 0$$

$$\cos t = 0, \cot t + 1 = 0$$

$$\cos t = 0, \qquad \cot t = -1$$

Thus our problem is reduced to solving two simple equations. The first has the two solutions, $\pi/2$ and $3\pi/2$; the second has solutions $3\pi/4$ and $7\pi/4$. Thus the set of all solutions of

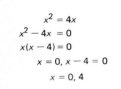

$$x^2 = 4x$$
$$x^2 - 4x = 0$$
$$x(x - 4) = 0$$
$$x = 0, x - 4 = 0$$
$$x = 0, 4$$

Figure 10

$$\cos t \cot t = -\cos t$$

on the interval $0 \le t < 2\pi$ is

$$\left\{\frac{\pi}{2}, \frac{3\pi}{4}, \frac{3\pi}{2}, \frac{7\pi}{4}\right\}$$

EQUATIONS OF QUADRATIC FORM

In Section 3-4, we solved quadratic equations by a number of techniques (factoring, taking square roots, and using the quadratic formula). We use the same techniques here. For example,

$$\cos^2 t = \frac{3}{4}$$

is analogous to $x^2 = \frac{3}{4}$. We solve such an equation by taking square roots.

$$\cos t = \pm \frac{\sqrt{3}}{2}$$

The set of solutions on $0 \le t < 2\pi$ is

$$\left\{\frac{\pi}{6}, \frac{5\pi}{6}, \frac{7\pi}{6}, \frac{11\pi}{6}\right\}$$

As a second example, consider the equation

$$2 \sin^2 t - \sin t - 1 = 0$$

Think of it as being like

$$2x^2 - x - 1 = 0$$

Now

$$2x^2 - x - 1 = (2x + 1)(x - 1)$$

and so

$$2 \sin^2 t - \sin t - 1 = (2 \sin t + 1)(\sin t - 1)$$

When we set each factor equal to zero and solve, we get

$$2 \sin t + 1 = 0 \qquad\qquad \sin t - 1 = 0$$
$$\sin t = -\tfrac{1}{2} \qquad\qquad \sin t = 1$$
$$t = \frac{7\pi}{6}, \frac{11\pi}{6} \qquad\qquad t = \frac{\pi}{2}$$

The set of all solutions on $0 \le t < 2\pi$ is

$$\left\{ \frac{\pi}{2}, \frac{7\pi}{6}, \frac{11\pi}{6} \right\}$$

USING IDENTITIES TO SOLVE EQUATIONS

Consider the equation

$$\tan^2 x = \sec x + 1$$

The identity $\sec^2 x = \tan^2 x + 1$ suggests writing everything in terms of $\sec x$.

$$\sec^2 x - 1 = \sec x + 1$$

$$\sec^2 x - \sec x - 2 = 0$$

$$(\sec x + 1)(\sec x - 2) = 0$$

$$\sec x + 1 = 0 \qquad \sec x - 2 = 0$$

$$\sec x = -1 \qquad \sec x = 2$$

$$x = \pi \qquad x = \frac{\pi}{3}, \frac{5\pi}{3}$$

Thus, the set of solutions on $0 \le t < 2\pi$ is $\{\pi/3, \pi, 5\pi/3\}$. Unfamiliarity with the secant may hinder you at the last step. If so, use $\sec x = 1/\cos x$ to write the equations in terms of cosines and solve the equations

$$\cos x = -1 \qquad \cos x = \frac{1}{2}$$

SOLUTION TO THE TWO-BUG PROBLEM

Our opening display asked when one bug would first be directly above the other. We reduced that problem to solving

$$\cos 2t = \cos t$$

for t. Using a double-angle formula, we may write

$$2 \cos^2 t - 1 = \cos t$$

$$2 \cos^2 t - \cos t - 1 = 0$$

$$(2 \cos t + 1)(\cos t - 1) = 0$$

$$\cos t = -\frac{1}{2} \qquad \cos t = 1$$

$$t = \frac{2\pi}{3}, \frac{4\pi}{3} \qquad t = 0$$

The smallest positive solution is $t = 2\pi/3$. After a little over 2 seconds, the slow bug will be directly above the fast bug.

Problem Set 8-5

Solve each of the following, finding all solutions on the interval 0 to 2π, excluding 2π.

1. $\sin t = 0$
2. $\cos t = 1$
3. $\sin t = -1$
4. $\tan t = -\sqrt{3}$
5. $\sin t = 2$
6. $\sec t = \frac{1}{2}$
7. $2 \cos x + \sqrt{3} = 0$
8. $2 \sin x + 1 = 0$
9. $\tan^2 x = 1$
10. $4 \sin^2 \theta - 3 = 0$
11. $(2 \cos \theta + 1)(2 \sin \theta - \sqrt{2}) = 0$
12. $(\sin \theta - 1)(\tan \theta + 1) = 0$
13. $\sin^2 x + \sin x = 0$
14. $2 \cos^2 x - \cos x = 0$
15. $\tan^2 \theta = \sqrt{3} \tan \theta$
16. $\cot^2 \theta = -\cot \theta$
17. $2 \sin^2 x = 1 + \cos x$
18. $\sec^2 x = 1 + \tan x$
19. $\tan^2 x - 3 \tan x + 1 = 0$ [c]
20. $\cos 2t = 3 \sin t$ [c]

CAUTION

$$\tan^2 \theta = \sqrt{3} \tan \theta$$
$$\tan \theta = \sqrt{3}$$

$$\tan^2 \theta = \sqrt{3} \tan \theta$$
$$\tan \theta (\tan \theta - \sqrt{3}) = 0$$
$$\tan \theta = 0 \qquad \tan \theta = \sqrt{3}$$

EXAMPLE A (Solving by Squaring Both Sides)   Solve

$$1 - \cos t = \sqrt{3} \sin t$$

Solution.   Since the identity relating sines and cosines involves their squares, we begin by squaring both sides. Then we express everything in terms of $\cos t$ and solve.

$$(1 - \cos t)^2 = 3 \sin^2 t$$

$$1 - 2 \cos t + \cos^2 t = 3(1 - \cos^2 t)$$

$$\cos^2 t - 2 \cos t + 1 = 3 - 3 \cos^2 t$$

$$4 \cos^2 t - 2 \cos t - 2 = 0$$

$$2 \cos^2 t - \cos t - 1 = 0$$

$$(2 \cos t + 1)(\cos t - 1) = 0$$

$$\cos t = -\frac{1}{2} \qquad \cos t = 1$$

$$t = \frac{2\pi}{3}, \frac{4\pi}{3} \qquad t = 0$$

Since squaring may introduce extraneous solutions, it is important to check our answers. We find that $4\pi/3$ is extraneous, since substituting $4\pi/3$ for t in the original equation gives us $1 + \frac{1}{2} = -\frac{3}{2}$. However, 0 and $2\pi/3$ are solutions, as you should verify.

Solve each of the following equations on the interval $0 \le t < 2\pi$; check your answers.

21. $\sin t + \cos t = 1$
22. $\sin t - \cos t = 1$
23. $\sqrt{3}(1 - \sin t) = \cos t$
24. $1 + \sin t = \sqrt{3} \cos t$
25. $\sec t + \tan t = 1$
26. $\tan t - \sec t = 1$

EXAMPLE B (Finding All of the Solutions)   Find the entire set of solutions of the equation $\cos 2t = \cos t$.

Solution.   In the text, we found 0, $2\pi/3$, and $4\pi/3$ to be the solutions for $0 \le t < 2\pi$. Clearly we get new solutions by adding 2π again and again to any of these numbers. The same holds true for subtracting 2π. In fact, the entire solution set consists of all those numbers of the form $2\pi k$, $2\pi/3 + 2\pi k$, or $4\pi/3 + 2\pi k$, where k is any integer.

Find the entire solution set of each of the following equations.

27. $\sin t = \frac{1}{2}$ 　　　　　 28. $\cos t = -\frac{1}{2}$ 　　　　　 29. $\tan t = 0$

30. $\tan t = -\sqrt{3}$ 　　　　 31. $\sin^2 t = \frac{1}{4}$ 　　　　　 32. $\cos^2 t = 1$

EXAMPLE C (Multiple-Angle Equations)   Find all solutions of $\cos 4t = \frac{1}{2}$ on the interval $0 \le t < 2\pi$.

Solution.   There will be more answers than you think. We know that $\cos 4t$ equals $\frac{1}{2}$ when

$$4t = \frac{\pi}{3}, \frac{5\pi}{3}, \frac{7\pi}{3}, \frac{11\pi}{3}, \frac{13\pi}{3}, \frac{17\pi}{3}, \frac{19\pi}{3}, \frac{23\pi}{3}$$

that is, when

$$t = \frac{\pi}{12}, \frac{5\pi}{12}, \frac{7\pi}{12}, \frac{11\pi}{12}, \frac{13\pi}{12}, \frac{17\pi}{12}, \frac{19\pi}{12}, \frac{23\pi}{12}$$

The reason that there are 8 solutions instead of 2 is that $\cos 4t$ completes 4 periods on the interval $0 \le t < 2\pi$.

Solve each of the following equations, finding all solutions on the interval $0 \le t < 2\pi$.

33. $\sin 2t = 0$ 　　　　　 34. $\cos 2t = 0$ 　　　　　 35. $\sin 4t = 1$

36. $\cos 4t = 1$ 　　　　　 37. $\tan 2t = -1$ 　　　　　 38. $\tan 3t = 0$

MISCELLANEOUS PROBLEMS

In Problems 39–56, find all solutions to the given equation on $0 \le x < 2\pi$.

39. $2 \sin^2 x = \sin x$ 　　　　　　　　　 40. $2 \cos x \sin x + \cos x = 0$

$\boxed{\text{c}}$ 41. $\cos^2 x = \frac{1}{3}$ 　　　　　　　　 $\boxed{\text{c}}$ 42. $\tan^2 x + 2 \tan x = 0$

43. $2 \tan x - \sec^2 x = 0$ 　　　　　 44. $\tan^2 x = 1 + \sec x$

45. $\tan 2x = 3 \tan x$ 　　　　　　　 46. $\cos(x/2) - \cos x = 1$

$\boxed{\text{c}}$ 47. $\sin^2 x + 3 \sin x - 1 = 0$ 　　 $\boxed{\text{c}}$ 48. $\tan^2 x - 2 \tan x - 10 = 0$

49. $\sin 2x + \sin x + 4 \cos x = -2$

50. $\cos x + \sin x = \sec x + \sec x \tan x$

51. $\sin x \cos x = -\sqrt{3}/4$ 　　　 52. $4 \sin x - 4 \sin^3 x + \cos x = 0$

$\boxed{\text{c}}$ 53. $\cos x - 2 \sin x = 2$ 　　　 $\boxed{\text{c}}$ 54. $\sin x + \cos x = \frac{1}{3}$

55. $\cos^8 x - \sin^8 x = 0$ 　　　　 56. $\cos^6 x + \sin^6 x = \frac{13}{16}$

57. A ray of light from the lamp L in Figure 11 reflects off a mirror to the object O.

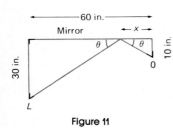

Figure 11

(a) Find the distance x.

(b) Write an equation for θ.

(c) Solve this equation.

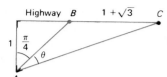

Figure 12

58. Tom and John are lost in a desert 1 mile from a highway, at point A in Figure 12. Each strikes out in a different direction to get to the highway. Tom gets to the highway at point B and John arrives at point C, $1 + \sqrt{3}$ miles farther down the road. Write an equation for θ and solve it.

59. Mr. Quincy built a slide with a 10-foot rise and 20-foot base (Figure 13). (a) Find the angle α in degrees. (b) By how much (θ in Figure 13) would the angle of the slide increase if he made the rise 15 feet, keeping the base at 20 feet?

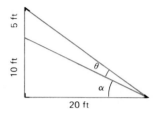

Figure 13

60. Find the angles θ_1, θ_2, and θ_3 shown in Figure 14. Your answers should convince you that the angle ABC is not trisected.

61. Solve the equation

$$\sin 4t + \sin 3t + \sin 2t = 0$$

Hint: Use the identity $\sin u + \sin v = 2 \sin((u + v)/2) \cos((u - v)/2)$.

62. Solve the equation

$$\cos 5t + \cos 3t - 2 \cos t = 0$$

Hint: Use the identity $\cos u + \cos v = 2 \cos((u + v)/2) \cos((u - v)/2)$.

63. Solve $\cos^8 u + \sin^8 u = \frac{41}{128}$ for $0 \le u \le \pi$. *Hint:* Begin by using half-angle formulas.

64. **TEASER** Show that $t = \pi/4$ is the only solution on $0 \le t \le \pi$ to the equation

$$\frac{a + b \cos t}{b + a \sin t} = \frac{a + b \sin t}{b + a \cos t}$$

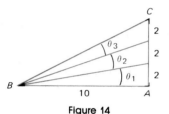

Figure 14

Chapter Summary

An **identity** is an equality that is true for all values of the unknown for which both sides of the equality make sense. Our first task was to establish the fundamental identities of trigonometry, here arranged by category.

Basic Identities

1. $\tan t = \dfrac{\sin t}{\cos t}$

2. $\cot t = \dfrac{\cos t}{\sin t} = \dfrac{1}{\tan t}$

3. $\sec t = \dfrac{1}{\cos t}$

4. $\csc t = \dfrac{1}{\sin t}$

5. $\sin^2 t + \cos^2 t = 1$

6. $1 + \tan^2 t = \sec^2 t$

7. $1 + \cot^2 t = \csc^2 t$

Cofunction Identities

8. $\sin\left(\dfrac{\pi}{2} - t\right) = \cos t$

9. $\cos\left(\dfrac{\pi}{2} - t\right) = \sin t$

Odd-Even Identities

10. $\sin(-t) = -\sin t$ 11. $\cos(-t) = \cos t$

Addition Formulas

12. $\sin(s + t) = \sin s \cos t + \cos s \sin t$

13. $\sin(s - t) = \sin s \cos t - \cos s \sin t$

14. $\cos(s + t) = \cos s \cos t - \sin s \sin t$

15. $\cos(s - t) = \cos s \cos t + \sin s \sin t$

Double-Angle Formulas

16. $\sin 2t = 2 \sin t \cos t$

17. $\cos 2t = \cos^2 t - \sin^2 t = 1 - 2 \sin^2 t = 2 \cos^2 t - 1$

Half-Angle Formulas

18. $\sin \dfrac{t}{2} = \pm \sqrt{\dfrac{1 - \cos t}{2}}$ 19. $\cos \dfrac{t}{2} = \pm \sqrt{\dfrac{1 + \cos t}{2}}$

Once we have memorized the fundamental identities, we can use them to prove thousands of other identities. The suggested technique is to take one side of a proposed identity and show by a chain of equalities that it is equal to the other.

A **trigonometric equation** is an equality involving trigonometric functions that is true only for some values of the unknown (for example, $\sin 2t = \frac{1}{2}$). Here our job is to solve the equation, that is, to find the values of the unknown that make it true.

With their natural domains, the trigonometric functions are not one-to-one and therefore do not have inverses. However, there are standard ways to restrict the domains so that inverses exist. Here are the results.

$$x = \sin^{-1} y \quad \text{means} \quad y = \sin x \quad \text{and} \quad \frac{-\pi}{2} \le x \le \frac{\pi}{2}$$

$$x = \cos^{-1} y \quad \text{means} \quad y = \cos x \quad \text{and} \quad 0 \le x \le \pi$$

$$x = \tan^{-1} y \quad \text{means} \quad y = \tan x \quad \text{and} \quad \frac{-\pi}{2} < x < \frac{\pi}{2}$$

$$x = \sec^{-1} y \quad \text{means} \quad y = \sec x \quad \text{and} \quad 0 \le x \le \pi, x \ne \frac{\pi}{2}$$

Chapter Review Problem Set

1. Prove that the following are identities.
   (a) $\cot \theta \cos \theta = \csc \theta - \sin \theta$
   (b) $\dfrac{\cos x \tan^2 x}{\sec x + 1} = 1 - \cos x$

2. Express each of the following in terms of $\sin x$ and simplify.

(a) $\dfrac{(\cos^2 x - 1)(1 + \tan^2 x)}{\csc x}$ (b) $\dfrac{\cos^2 x \csc x}{1 + \csc x}$

3. Use appropriate identities to simplify and then calculate each of the following.

(a) $2 \cos^2 22.5° - 1$

(b) $\sin 37° \cos 53° + \cos 37° \sin 53°$

(c) $\cos 108° \cos 63° + \sin 108° \sin 63°$

4. If $\cos t = -\frac{4}{5}$ and $\pi < t < 3\pi/2$, calculate

(a) $\sin 2t$;   (b) $\sin(t/2)$.

5. Prove that the following are identities.

(a) $\sin 2t \cos t - \cos 2t \sin t = \sin t$

(b) $\sec 2t + \tan 2t = \dfrac{\cos t + \sin t}{\cos t - \sin t}$

(c) $\dfrac{\cos(\alpha + \beta)}{\cos \alpha \cos \beta} = \tan \alpha(\cot \alpha - \tan \beta)$

6. Solve the following trigonometric equations for t, $0 \le t < 2\pi$.

(a) $\cos t = -\sqrt{3}/2$

(b) $(2 \sin t + 1) \tan t = 0$

(c) $\cos^2 t + 2 \cos t - 3 = 0$

(d) $\sin t - \cos t = 1$

(e) $\sin 3t = 1$

7. What is the standard way to restrict the domain of sine, cosine, and tangent so that they have inverses?

8. Calculate each of the following without the help of a calculator.

(a) $\sin^{-1}(-\sqrt{3}/2)$ (b) $\cos^{-1}(-\sqrt{3}/2)$

(c) $\tan^{-1}(-\sqrt{3})$ (d) $\tan(\tan^{-1} 6)$

(e) $\cos^{-1}(\cos 3\pi)$ (f) $\sin(\cos^{-1} \frac{2}{3})$

(g) $\cos(2 \cos^{-1} .7)$ (h) $\sin(2 \cos^{-1} \frac{5}{13})$

9. Sketch the graph of $y = \tan^{-1} x$.

10. Find an approximate value for $\tan^{-1}(-1000)$.

11. Show that

$$\frac{\pi}{4} = \arctan \frac{1}{2} + \arctan \frac{1}{3}$$

*Thus one sees in the sciences
many brilliant theories which
have remained unapplied for
a long time suddenly
becoming the foundation of
most important applications,
and likewise applications very
simple in appearance giving
birth to ideas of the most
abstract theories.*

—*Marquis de Condorcet*

CHAPTER 9

Applications of Trigonometry

The Law of Sines

Consider an arbitrary triangle with angles α, β, γ, and corresponding opposite sides a, b, c, respectively. Then

$$\frac{\sin \alpha}{a} = \frac{\sin \beta}{b} = \frac{\sin \gamma}{c}$$

Equivalently,

$$\frac{a}{\sin \alpha} = \frac{b}{\sin \beta} = \frac{c}{\sin \gamma}$$

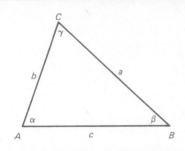

9-1 Oblique Triangles: Law of Sines

We learned in Section 7-1 how to solve a right triangle. But can we solve an oblique triangle—that is, one without a 90° angle? One valuable tool is the **law of sines,** stated above. It is valid for any triangle whatever, but we initially establish it for the case where all angles are acute.

PROOF OF THE LAW OF SINES

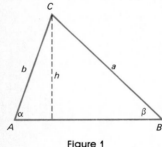

Figure 1

Consider a triangle with all acute angles, labeled as in Figure 1. Drop a perpendicular of length h from vertex C to the opposite side. Then by right-triangle trigonometry,

$$\sin \alpha = \frac{h}{b} \qquad \sin \beta = \frac{h}{a}$$

If we solve for h in these two equations and equate the results, we obtain

$$b \sin \alpha = a \sin \beta$$

Finally, dividing both sides by ab yields

$$\frac{\sin \alpha}{a} = \frac{\sin \beta}{b}$$

Since the roles of β and γ can be interchanged, the same reasoning gives

$$\frac{\sin \alpha}{a} = \frac{\sin \gamma}{c}$$

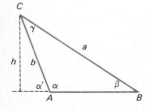

Figure 2

Next consider a triangle with an obtuse angle α ($90° < \alpha < 180°$). Drop a perpendicular of length h from vertex C to the extension of AB (see Figure 2). Notice that angle α' is the reference angle for α and so $\sin \alpha = \sin \alpha'$. It follows from right-triangle trigonometry that

$$\sin \alpha = \sin \alpha' = \frac{h}{b} \qquad \sin \beta = \frac{h}{a}$$

just as in the acute case. The rest of the argument is identical with that case.

SOLVING A TRIANGLE (AAS)

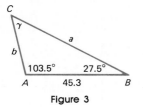

Figure 3

Suppose that we know two angles and any side of a triangle. For example, suppose that in triangle ABC, $\alpha = 103.5°$, $\beta = 27.5°$, and $c = 45.3$ (Figure 3). Our task is to find γ, a, and b.

1. Since $\alpha + \beta + \gamma = 180°$, $\gamma = 180° - (103.5° + 27.5°) = 49°$.
2. By the law of sines,

$$\frac{a}{\sin 103.5°} = \frac{45.3}{\sin 49°}$$

$$a = \frac{(45.3)(\sin 103.5°)}{\sin 49°}$$

$$= \frac{(45.3)(\sin 76.5°)}{\sin 49°}$$

$$= \frac{(45.3)(.9724)}{.7547}$$

$$\approx 58.4$$

3. Also by the law of sines,

$$\frac{b}{\sin 27.5°} = \frac{45.3}{\sin 49°}$$

$$b = \frac{(45.3)(\sin 27.5°)}{\sin 49°}$$

$$= \frac{(45.3)(.4617)}{.7547}$$

$$\approx 27.7$$

SOLVING A TRIANGLE (SSA)

Suppose that two sides and the angle opposite one of them are given. This is called the **ambiguous case** because the given information may not determine a unique triangle.

If α, a, and b are given, we consider trying to construct a triangle fitting these data by first drawing angle α, then marking off b on one of its sides thus determining vertex C. Finally, we attempt to locate vertex B by striking off a circular arc of radius a with center at C. If $a \geq b$, this can always be done in a unique way. Figure 4 on the next page illustrates this both for α acute and α obtuse. If $a < b$, there are several possibilities (Figure 5).

Fortunately, we are able to decide which of these possibilities is the case if we draw an approximate picture and then attempt to apply the law of sines. First, note that if $a \geq b$ there is one triangle corresponding to the data and for it β is an acute angle. Application of the law of sines will give sin β, which allows determination of β.

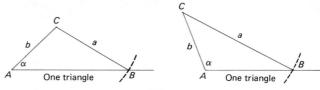

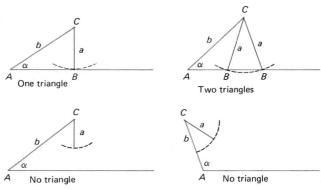

Figure 4

Figure 5

If $a < b$, we may attempt to apply the law of sines. If it yields $\sin \beta = 1$, we have a unique right triangle. If it yields $\sin \beta < 1$, we have two triangles corresponding to the two angles β_1 and β_2 (one acute, the other obtuse) with this sine. If it yields $\sin \beta > 1$, we have an inconsistency in the data; no triangle satisfying the data exists.

Suppose, for example, that we are given $\alpha = 36°$, $a = 9.4$, and $b = 13.1$ (Figure 6). We must find β, γ, and c. Since $a < b$, there may be zero, one, or two triangles. We proceed to compute $\sin \beta$.

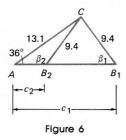

Figure 6

1. $\dfrac{\sin \beta}{13.1} = \dfrac{\sin 36°}{9.4}$

$\sin \beta = \dfrac{(13.1) \sin 36°}{9.4} = \dfrac{(13.1)(.5878)}{9.4} \approx .8191$

Since $\sin \beta < 1$, there are two triangles.

$$\beta_1 = 55° \qquad \beta_2 = 125°$$

2. $\gamma_1 = 180° - (36° + 55°) = 89°$
   $\gamma_2 = 180° - (36° + 125°) = 19°$

3. $\dfrac{c_1}{\sin 89°} = \dfrac{9.4}{\sin 36°}$

$$c_1 = \dfrac{9.4}{\sin 36°}(\sin 89°)$$

$$= \frac{9.4}{.5878}(.9998) \approx 16.0$$

$$\frac{c_2}{\sin 19°} = \frac{9.4}{\sin 36°}$$

$$c_2 = \frac{9.4}{\sin 36°}(\sin 19°)$$

$$= \frac{9.4}{.5878}(.3256) \approx 5.2$$

Problem Set 9-1

Solve the triangles of Problems 1–10 using either Table C or a calculator.

1. $\alpha = 42.6°$, $\beta = 81.9°$, $a = 14.3$
2. $\beta = 123°$, $\gamma = 14.2°$, $a = 295$
3. $\alpha = \gamma = 62°$, $b = 50$
4. $\alpha = \beta = 14°$, $c = 30$
5. $\alpha = 115°$, $a = 46$, $b = 34$
6. $\beta = 143°$, $a = 46$, $b = 84$
7. $\alpha = 30°$, $a = 8$, $b = 5$
8. $\beta = 60°$, $a = 11$, $b = 12$
9. $\alpha = 30°$, $a = 5$, $b = 8$
10. $\beta = 60°$, $a = 12$, $b = 11$

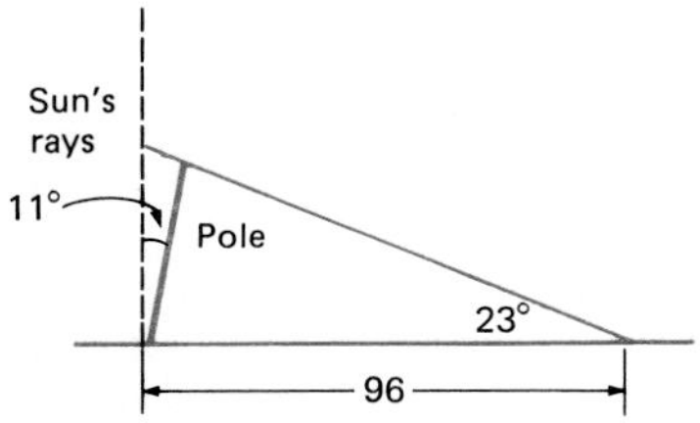

11. Two observers stationed 110 meters apart at A and B on the bank of a river are looking at a tower situated at a point C on the opposite bank. They measure angles CAB and CBA to be 43° and 57°, respectively. How far is the first observer from the tower?

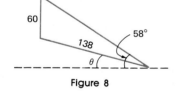

Figure 7

12. A telegraph pole leans away from the sun at an angle of 11° to the vertical. The pole casts a shadow 96 feet long on horizontal ground when the angle of elevation of the sun is 23°. Find the length of the pole (see Figure 7).

13. A vertical pole 60 feet long is standing by the side of an inclined road. It casts a shadow 138 feet long directly downhill along the road when the angle of elevation of the sun is 58°. Find the angle of inclination θ of the road (see Figure 8).

Figure 8

14. Two forest rangers 15 miles apart at points A and B observe a fire at a point C. The ranger at A measures angle CAB as 43.6° and the one at B measures angle CBA as 79.3°. How far is the fire from each ranger? How far is the fire from a straight road that goes from A to B?

EXAMPLE (An Important Area Formula) Consider a triangle with two sides b and c and included angle α. Show that the area A of the triangle is

$$A = \frac{1}{2}bc \sin \alpha$$

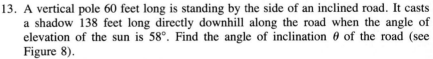

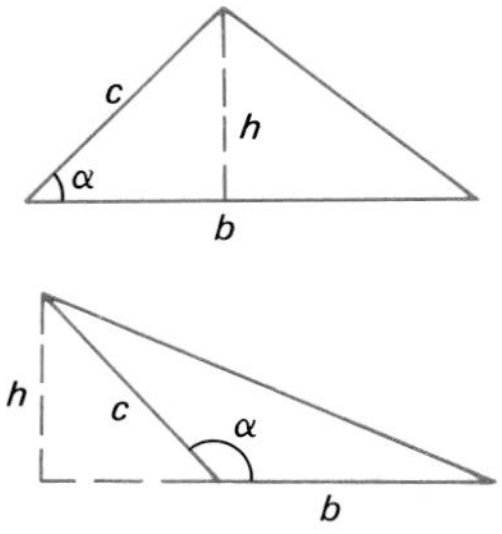

Figure 9

Solution. Let h denote the altitude of the triangle as shown in the diagrams of Figure 9. Whether α is acute or obtuse, we have $\sin \alpha = h/c$, that is,

$h = c \sin \alpha$. We conclude that

$$A = \frac{1}{2}bh = \frac{1}{2}bc \sin \alpha$$

15. Find the area of the triangle with sides $b = 20$, $c = 30$, and included angle $\alpha = 40°$.
16. Find the area of the triangle with $a = 14.6$, $b = 31.7$, and $\gamma = 130.2°$.
17. Find the area of the triangle with $c = 30.1$, $\alpha = 25.3°$, and $\beta = 112.2°$.
18. Find the area of the triangle with $a = 20$, $\alpha = 29°$, and $\gamma = 46°$.

MISCELLANEOUS PROBLEMS

19. The children's slide at the park is 30 feet long and inclines 36° from the horizontal. The ladder to the top is 18 feet long. How steep is the ladder, that is, what angle does it make with the horizontal? Assume the slide is straight and that the bottom end of the slide is at the same level as the bottom end of the ladder.

20. Prevailing winds have caused an old tree to incline 11° eastward from the vertical. The sun in the west is 32° above the horizontal. How long a shadow is cast by the tree if the tree measures 114 feet from top to bottom?

21. A rectangular room, 16 feet by 30 feet, has an open beam ceiling. The two parts of the ceiling make angles of 65° and 32° with the horizontal (an end view is shown in Figure 10). Find the total area of the ceiling.

22. Sheila Sather, traveling north on a straight road at a constant rate of 60 miles per hour, sighted flames shooting up into the air at a point 20° west of north. Exactly one hour later, the fire was 59° west of south. Determine the shortest distance from the road to the fire.

23. A lighthouse stands at a certain distance out from a straight shoreline. It throws a beam of light that revolves at a constant rate of one revolution per minute. A short time after shining on the nearest point on the shore, the beam reaches a point on the shore that is 2640 feet from the lighthouse, and 3 seconds later it reaches a point 2000 feet farther along the shore. How far is the lighthouse from the shore?

24. In Figure 11, AC is 10 meters longer than CB. Determine the length of CD.

25. Four line segments of lengths 3, 4, 5, and 6 radiate like spokes from a common point. Their outer ends are the vertices of a quadrilateral Q. Determine the maximum possible area of Q.

26. Figure 12 illustrates the Pythagorean theorem ($a^2 + b^2 = c^2$). A rubber band is stretched around this figure. Show that the area of the region enclosed by the rubber band is $2(ab + c^2)$.

27. Let 2ϕ denote the angle at a point of the regular 6-pointed star shown in Figure 13. Express the area A of this star in terms of ϕ and the edge length r.

28. TEASER   Figure 14 shows two mirrors intersecting at an angle of 15°. A light ray from S is reflected at P and again at Q and then is absorbed at R. It is given that $ST = RU = 5$, $OT = 50$, and $OU = 20$. Find the length $x + y + z$ of the path of the light ray. As indicated, the angle of incidence equals the angle of reflection.

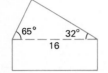

Figure 10

Figure 11

Figure 12

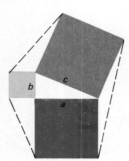

Figure 13

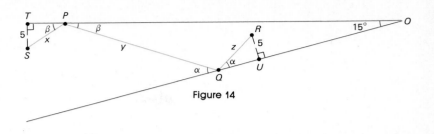

Figure 14

The Law of Cosines

Consider an arbitrary triangle with angles α, β, γ and corresponding opposite sides a, b, c, respectively. Then

$$a^2 = b^2 + c^2 - 2bc \cos \alpha$$

$$b^2 = a^2 + c^2 - 2ac \cos \beta$$

$$c^2 = a^2 + b^2 - 2ab \cos \gamma$$

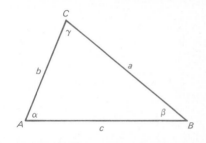

9-2   Oblique Triangles: Law of Cosines

When two sides and the included angle (SAS) or three sides (SSS) of a triangle are given, we cannot apply the law of sines to solve the triangle. Rather, we need the law of cosines, stated above in symbols. Actually it is wise to learn the law in words.

The square of any side is equal to the sum of the squares of the other two sides minus twice the product of those sides times the cosine of the angle between them.

Notice what happens when $\gamma = 90°$ so that $\cos \gamma = 0$. The law of cosines

$$c^2 = a^2 + b^2 - 2ab \cos \gamma$$

becomes

$$c^2 = a^2 + b^2$$

which is just the Pythagorean theorem. In fact, you should think of the law of cosines as a generalization of the Pythagorean theorem, with the term $-2ab \cos \gamma$ acting as a correction term when γ is not $90°$.

PROOF OF THE LAW OF COSINES

Assume first that angle α is acute. Drop a perpendicular CD from vertex C to side AB as shown in Figure 15 at the top of the next page. Label the lengths of CD, AD, and DB by h, x, and $c - x$, respectively.

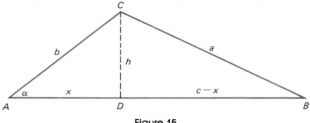

Figure 15

Consider the two right triangles ADC and BDC. By the Pythagorean theorem

$$h^2 = b^2 - x^2 \quad \text{and} \quad h^2 = a^2 - (c - x)^2$$

Equating these two expressions for h^2 gives

$$a^2 - (c - x)^2 = b^2 - x^2$$
$$a^2 = b^2 - x^2 + (c - x)^2$$
$$a^2 = b^2 - x^2 + c^2 - 2cx + x^2$$
$$a^2 = b^2 + c^2 - 2cx$$

Now $\cos \alpha = x/b$, and so $x = b \cos \alpha$. Thus

$$a^2 = b^2 + c^2 - 2cb \cos \alpha$$

which is the result we wanted.

Next we give the proof of the law of cosines for the obtuse angle case. Again drop a perpendicular from vertex C to side AB extended and label the resulting diagram as shown in Figure 16.

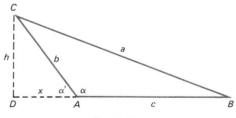

Figure 16

From consideration of triangles ADC and BDC and the Pythagorean theorem, we obtain

$$h^2 = b^2 - x^2 \quad \text{and} \quad h^2 = a^2 - (c + x)^2$$

Algebra analogous to that used in the acute angle case yields

$$a^2 = b^2 + c^2 + 2cx$$

Now α' is the reference angle for α, and so $\cos \alpha = -\cos \alpha'$. Also $\cos \alpha' = x/b$. Therefore,

$$x = b \cos \alpha' = -b \cos \alpha$$

When we substitute this expression for x in the preceding equation, we get

$$a^2 = b^2 + c^2 - 2cb \cos \alpha$$

SOLVING A TRIANGLE (SAS)

Consider a triangle with $b = 18.1$, $c = 12.3$, and $\alpha = 115°$ (Figure 17). We want to determine a, β, and γ.

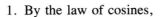

1. By the law of cosines,

$$a^2 = (18.1)^2 + (12.3)^2 - 2(18.1)(12.3) \cos 115°$$
$$= 327.61 + 151.29 - (445.26)(-\cos 65°)$$
$$= 327.61 + 151.29 + (445.26)(.4226)$$
$$= 667.08$$
$$a \approx 25.8$$

2. Now we can use the law of sines.

$$\frac{\sin \beta}{18.1} = \frac{\sin 115°}{25.8}$$

$$\sin \beta = \frac{(18.1) \sin 115°}{25.8} = \frac{(1.81)(\sin 65°)}{25.8}$$

$$= \frac{(18.1)(.9063)}{25.8} = .6358$$

$$\beta \approx 39.5°$$

3. $\gamma \approx 180° - (115° + 39.5°) = 25.5°$.

SOLVING A TRIANGLE (SSS)

If $a = 13.1$, $b = 15.5$, and $c = 17.2$, then we must determine the three angles.

1. By the law of cosines,

$$a^2 = b^2 + c^2 - 2bc \cos \alpha$$

Thus

$$\cos \alpha = \frac{b^2 + c^2 - a^2}{2bc}$$

$$= \frac{(15.5)^2 + (17.2)^2 - (13.1)^2}{2(15.5)(17.2)} = .6836$$

$$\alpha \approx 46.9°$$

2. By the law of sines,

$$\frac{\sin \beta}{15.5} = \frac{\sin 46.9°}{13.1}$$

$$\sin \beta = \frac{(15.5)(\sin 46.9°)}{13.1} = \frac{(15.5)(.7302)}{13.1} = .8640$$

$$\beta \approx 59.8°$$

3. $\gamma = 180° - (46.9° + 59.8°) = 73.3°$.

Problem Set 9-2

In Problems 1–8, solve the triangles satisfying the given data. Use either Table C or a calculator.

1. $\alpha = 60°$, $b = 14$, $c = 10$
2. $\beta = 60°$, $a = c = 8$
3. $\gamma = 120°$, $a = 8$, $b = 10$
4. $\alpha = 150°$, $b = 35$, $c = 40$
5. $a = 5$, $b = 6$, $c = 7$
6. $a = 10$, $b = 20$, $c = 25$
7. $a = 12.2$, $b = 19.1$, $c = 23.8$
8. $a = .11$, $b = .21$, $c = .31$

9. At one corner of a triangular field, the angle measures 52.4°. The sides that meet at this corner are 100 meters and 120 meters long. How long is the third side?

10. To approximate the distance between two points A and B on opposite sides of a swamp, a surveyor selects a point C and measures it to be 140 meters from A and 260 meters from B. Then she measures the angle ACB, which turns out to be 49°. What is the calculated distance from A to B?

11. Two runners start from the same point at 12:00 noon, one of them heading north at 6 miles per hour and the other heading 68° east of north at 8 miles per hour (Figure 18). What is the distance between them at 3:00 that afternoon?

12. A 50-foot pole stands on top of a hill which slants 20° from the horizontal. How long must a rope be to reach from the top of the pole to a point 88 feet directly downhill (that is, on the slant) from the base of the pole?

13. A triangular garden plot has sides of length 35 meters, 40 meters, and 60 meters. Find the largest angle of the triangle.

14. A piece of wire 60 inches long is bent into the shape of a triangle. Find the angles of the triangle if two of the sides have lengths 24 inches and 20 inches.

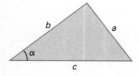

Figure 18

EXAMPLE (Heron's Area Formula) Show that a triangle with sides a, b, and c (Figure 19) and semiperimeter $s = (a + b + c)/2$ has area A given by

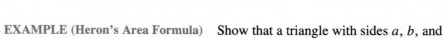

$$A = \sqrt{s(s - a)(s - b)(s - c)}$$

Figure 19

Solution. The proof is subtle, depending on the clever matching of the area formula from the last section with the law of cosines. Begin by writing the law of cosines in the form

$$2bc \cos \alpha = b^2 + c^2 - a^2$$

a formula we will use shortly. Next, take the area formula $A = \frac{1}{2} bc \sin \alpha$ in its squared form and manipulate it very carefully.

$$A^2 = \frac{1}{4}b^2c^2 \sin^2 \alpha = \frac{1}{4}b^2c^2(1 - \cos^2 \alpha)$$

$$= \frac{1}{16}(2bc)(1 + \cos \alpha)(2bc)(1 - \cos \alpha)$$

$$= \frac{1}{16}(2bc + 2bc \cos \alpha)(2bc - 2bc \cos \alpha)$$

$$= \frac{1}{16}(2bc + b^2 + c^2 - a^2)(2bc - b^2 - c^2 + a^2)$$

$$= \frac{1}{16}[(b + c)^2 - a^2][a^2 - (b - c)^2]$$

$$= \frac{(b + c + a)(b + c - a)(a - b + c)(a + b - c)}{2 \quad\quad 2 \quad\quad 2 \quad\quad 2}$$

$$= \left[\frac{a + b + c}{2}\right]\left[\frac{a + b + c}{2} - a\right]\left[\frac{a + b + c}{2} - b\right]\left[\frac{a + b + c}{2} - c\right]$$

$$= s(s - a)(s - b)(s - c)$$

15. The area of the right triangle with sides 3, 4, and 5 is 6. Confirm that Heron's formula gives the same answer.

16. Find the area of the triangle with sides 31, 42, and 53.

17. Find the area of the triangle with sides 5.9, 6.7, and 10.3.

18. Use the answer you got to Problem 16 to find the length h of the shortest altitude of the triangle with sides 31, 42, and 53.

MISCELLANEOUS PROBLEMS

19. A triangular garden plot has sides measuring 42 meters, 50 meters, and 63 meters. Find the measure of the smallest angle.

20. A diagonal and a side of a parallelogram measure 80 centimeters and 25 centimeters, respectively, and the angle between them measures 47°. Find the length of the other diagonal. Recall that the diagonals of a parallelogram bisect each other.

21. Two cars, starting from the intersection of two straight highways, travel along the highways at speeds of 55 miles per hour and 65 miles per hour, respectively. If the angle of intersection of the highways measures 72°, how far apart are the cars after 36 minutes?

22. Buoys A, B, and C mark the vertices of a triangular racing course on a lake. Buoys A and B are 4200 feet apart, buoys A and C are 3800 feet apart, and angle CAB measures 100°. If the winning boat in a race covered the course in 6.4 minutes, what was its average speed in miles per hour?

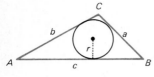

Figure 20

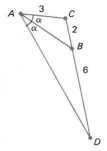

Figure 21

23. A quadrilateral Q has sides of length 1, 2, 3, and 4, respectively. The angle between the first pair of sides is 120°. Find the angle between the other pair of sides and also the exact area of Q.

24. For the triangle ABC in Figure 20, let r be the radius of the inscribed circle and let $s = (a + b + c)/2$ be its semiperimeter.
    (a) Show that the area of the triangle is rs.
    (b) Show that $r = \sqrt{(s - a)(s - b)(s - c)/s}$.
    (c) Find r for a triangle with sides 5, 6, and 7.

25. Consider a triangle with sides of length 4, 5, and 6. Show that one of its angles is twice another. *Hint:* Show that the cosine of twice one angle is equal to the cosine of another.

26. In the triangle with sides of length a, b, and c, let a_1, b_1, and c_1 denote the lengths of the corresponding medians from these sides to the opposite vertices. Show that

$$a_1^2 + b_1^2 + c_1^2 = \frac{3}{4}(a^2 + b^2 + c^2)$$

27. Determine the length of AB in Figure 21.

28. TEASER   The two hands of a clock are 4 and 5 inches long, respectively. At some time between 1:45 and 2, the tips of the hands are 8 inches apart. What time is it then?

A Piston Problem

One end of an 8-foot shaft is attached to a piston that moves up and down. The other end is attached to a wheel by means of a horizontal slotted arm which fits over a peg P on the rim. Starting at an initial position of $\theta = \pi/4$, the wheel of radius 2 feet rotates at a rate of 3 radians per second. Find a formula for d, the vertical distance from the piston to the wheel center, after t seconds.

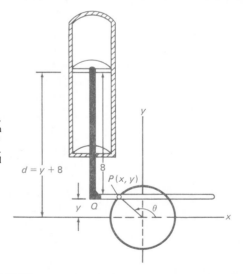

9-3   Simple Harmonic Motion

The up-and-down motion of the piston above is an example of what is called simple harmonic motion. Notice right away that the motion of the piston is essentially the same as that of the point Q. That means we want to find y; and y is just the y-coordinate of the peg P.

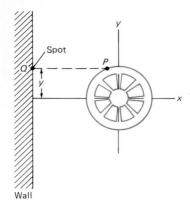

Figure 22

Perhaps the piston-wheel device seems complicated, so let's consider another version of the same problem. Imagine the wheel shown in Figure 22 to be turning at a uniform rate in the counterclockwise direction. Emanating from P, a point attached to the rim, is a horizontal beam of light, which projects a bright spot at Q on a nearby vertical wall. As the wheel turns, the spot at Q moves up and down. Our task is to express the y-coordinate of Q (which is also the y-coordinate of P) in terms of the elapsed time t.

The solution to this problem depends on a number of factors (the rate at which the wheel turns, the radius of the wheel, and the location of P at $t = 0$). We think it wise to begin with a simple case and gradually extend to more general situations.

Case 1 Suppose the wheel has radius 1, that it turns at 1 radian per second, and that it starts at $\theta = 0$. Then at time t, θ will measure t radians and P will have y-coordinate

$$y = \sin t$$

(see Figure 23). Keep in mind that this equation describes the up-and-down motion of Q.

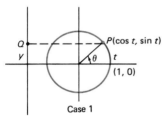

Figure 23

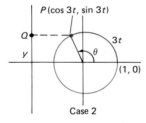

Figure 24

Case 2 Let everything be as in the first case, but now let the wheel turn at 3 radians per second (Figure 24). Then at time t, θ will measure $3t$ radians and both P and Q will have y-coordinate.

$$y = \sin 3t$$

Case 3 Next increase the radius of the wheel to 2 feet, but leave the other information as in Case 2 (Figure 25 on the next page). Now the coordinates of P are $(2 \cos 3t, 2 \sin 3t)$ and

$$y = 2 \sin 3t$$

Case 4 Finally, let the wheel start at $\theta = \pi/4$ rather than $\theta = 0$. With the help of Figure 26, we see that

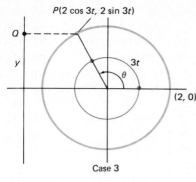

Case 3

Figure 25

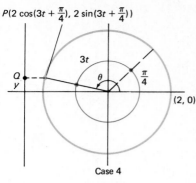

Case 4

Figure 26

$$y = 2 \sin\left(3t + \frac{\pi}{4}\right)$$

Case 4 describes the wheel of the original piston-wheel problem. The number y measures the distance between Q and the x-axis, and $d = y + 8$ is the distance from the piston to the x-axis. Thus the answer to the question first posed is

$$d = 8 + 2 \sin\left(3t + \frac{\pi}{4}\right)$$

The number 8 does not interest us; it is the sine expression that is significant. As a matter of fact, equations of the form

$$y = A \sin(Bt + C) \quad \text{and} \quad y = A \cos(Bt + C)$$

with $B > 0$ arise often in physics. Any straight-line motion which can be described by one of these formulas is called **simple harmonic motion.** Cases 1–4 are examples of this motion. Other examples from physics occur in connection with the motion of a weight attached to a vibrating spring (Figure 27) and the motion of a water molecule in an ocean wave. Voltage in an alternating current, although it does not involve motion, is given by the same kind of sine (or cosine) equation.

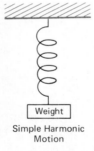

Simple Harmonic Motion

Figure 27

GRAPHS

The graphs of the four boxed equations are worthy of study. These graphs are shown in Figure 28. Note how the graph of $y = \sin t$ is progressively modified as we move from Case 1 to Case 4.

Under each graph are listed three important numbers, numbers that identify the critical features of the graph. The **period** is the length of the shortest interval after which the graph repeats itself. The **amplitude** is the maximum distance of the graph from its median position (the t-axis). The **phase shift** measures the distance the graph is shifted horizontally from its normal position.

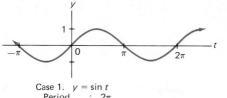

Case 1. $y = \sin t$
Period : 2π
Amplitude: 1
Phase shift: 0

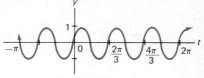

Case 2. $y = \sin 3t$
Period : $\dfrac{2\pi}{3}$
Amplitude: 1
Phase shift: 0

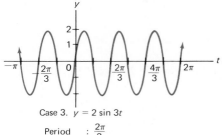

Case 3. $y = 2 \sin 3t$
Period : $\dfrac{2\pi}{3}$
Amplitude : 2
Phase shift : 0

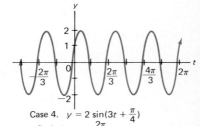

Case 4. $y = 2 \sin(3t + \frac{\pi}{4})$
Period : $\dfrac{2\pi}{3}$
Amplitude: 2
Phase shift: $-\dfrac{\pi}{12}$

Figure 28

You might have expected a phase shift of $-\pi/4$ in Case 4, since the initial angle of the wheel measured $\pi/4$ radians. But, note that factoring 3 from $3t + \pi/4$ gives

$$y = 2 \sin\left(3t + \frac{\pi}{4}\right) = 2 \sin 3\left(t + \frac{\pi}{12}\right)$$

If you recall our discussion of translations (see Section 5-4), you see why the graph is shifted $\pi/12$ units to the left. Note in particular that $y = 0$ when $t = -\pi/12$ instead of when $t = 0$.

GRAPHING IN THE GENERAL CASE

If

$$y = A \sin(Bt + C) \quad \text{or} \quad y = A \cos(Bt + C)$$

with $B > 0$, all three concepts (period, amplitude, phase shift) make good sense. We have the following formulas.

Period:	$\dfrac{2\pi}{B}$
Amplitude:	$\lvert A \rvert$
Phase shift:	$\dfrac{-C}{B}$

Knowing these three numbers is a great aid in graphing. For example, to graph

$$y = 3 \cos\left(4t - \frac{\pi}{4}\right)$$

we recall the graph of $y = \cos t$ and then modify it using the three numbers.

Period: $\quad \dfrac{2\pi}{B} = \dfrac{2\pi}{4} = \dfrac{\pi}{2}$

Amplitude: $\quad |A| = |3| = 3$

Phase shift: $\quad -\dfrac{C}{B} = \dfrac{\pi/4}{4} = \dfrac{\pi}{16}$

The result is shown in Figure 29.

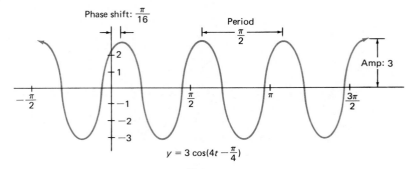

Figure 29

Problem Set 9-3

1. Sketch the graphs of the following equations in the order given. Use the interval $-2\pi \le t \le 2\pi$.
   (a) $y = \cos t$ (b) $y = \cos 2t$
   (c) $y = 4 \cos 2t$ (d) $y = 4 \cos(2t + \pi/3)$

2. Sketch the graphs of the following on $-2\pi \le t \le 4\pi$.
   (a) $y = \sin t$ (b) $y = \sin \dfrac{1}{2} t$

   (c) $y = 3 \sin \dfrac{1}{2} t$ (d) $y = 3 \sin\left(\dfrac{1}{2} t + \dfrac{\pi}{2}\right)$

EXAMPLE A (More Graphing)   Sketch the graph of $y = 3 \sin(\tfrac{1}{2}t + \pi/8)$.

Solution.   We begin by finding the three key numbers.

Period: $\quad \dfrac{2\pi}{B} = \dfrac{2\pi}{\frac{1}{2}} = 4\pi$

$$\text{Amplitude:} \quad |A| = |3| = 3$$

$$\text{Phase shift:} \quad \frac{-C}{B} = -\frac{\pi/8}{\frac{1}{2}} = -\frac{\pi}{4}$$

Then we draw the graph in Figure 30.

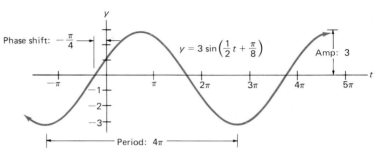

Figure 30

In Problems 3–6, find the period, amplitude, and phase shift. Then sketch the graph.

3. (a) $y = 4 \sin 2t$ (b) $y = 3 \cos\left(t + \frac{\pi}{8}\right)$

(c) $y = \sin\left(4t + \frac{\pi}{8}\right)$ (d) $y = 3 \cos\left(3t - \frac{\pi}{2}\right)$

4. (a) $y = \frac{1}{2} \cos 3t$ (b) $y = 3 \sin\left(t - \frac{\pi}{6}\right)$

(c) $y = 2 \sin\left(\frac{1}{2}t + \frac{\pi}{8}\right)$ (d) $y = \frac{1}{2} \sin(2t - 1)$

5. $y = 3 + 2 \cos\ (\frac{1}{2}t - \pi/16)$. *Hint:* The number 3 lifts the graph of $y = 2 \cos(\frac{1}{2}t - \pi/16)$ up 3 units.

6. $y = 4 + 3 \sin(2t + \pi/16)$

EXAMPLE B (Negative A) Sketch the graph of $y = -3 \cos 2t$.

Solution. We begin by asking how the graph of $y = -3 \cos 2t$ relates to that of $y = 3 \cos 2t$. Clearly, every y value has the opposite sign, which has the effect of reflecting the graph about the t-axis. Then we calculate the three crucial numbers.

$$\text{Period:} \quad \frac{2\pi}{B} = \frac{2\pi}{2} = \pi$$

$$\text{Amplitude:} \quad |A| = |-3| = 3$$

$$\text{Phase shift:} \quad \frac{-C}{B} = \frac{0}{2} = 0$$

Finally we sketch the graph (Figure 31 on the next page).

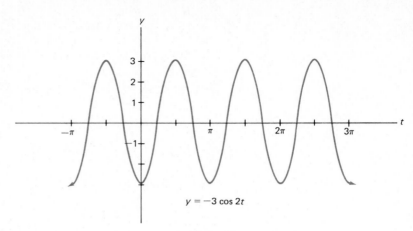

$$y = -3 \cos 2t$$

Figure 31

Now sketch the graphs of the equations in Problems 7–12.

7. $y = -2 \sin 3t$ 8. $y = -4 \cos \frac{1}{2}t$

9. $y = \sin(2t - \pi/3)$ 10. $y = -\cos(3t + \pi)$

11. $y = -2 \cos(t - \frac{1}{6})$ 12. $y = -3 \sin(3t + 3)$

13. A wheel with center at the origin is rotating counterclockwise at 4 radians per second. There is a small hole in the wheel 5 centimeters from the center. If that hole has initial coordinates (5, 0), what will its coordinates be after t seconds?

14. Answer Problem 13 if the hole is initially at (0, 5).

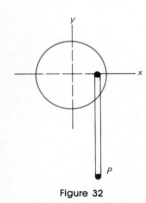

Figure 32

15. A free-hanging shaft, 8 centimeters long, is attached to the wheel of Problem 13 by putting a bolt through the hole (Figure 32). What are the coordinates of P, the bottom point of the shaft, at time t (assuming the shaft continues to hang vertically)?

16. Suppose the wheel of Problem 13 rotates at 3 revolutions per second. What are the coordinates of the hole after t seconds?

MISCELLANEOUS PROBLEMS

17. Find the period, amplitude, and phase shift for each graph.
    (a) $y = \sin 5t$ (b) $y = \frac{3}{2} \cos(\frac{1}{2}t)$
    (c) $y = 2 \cos(4t - \pi)$ (d) $y = -4 \sin(3t + 3\pi/4)$

18. Sketch the graphs of the equations in Problem 17 on the interval $-\pi \le t \le 2\pi$.

19. The weight attached to a spring (Figure 33) is bobbing up and down so that

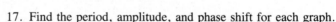

$$y = 8 + 4 \cos\left(\frac{\pi}{2}t + \frac{\pi}{4}\right)$$

where y and t are measured in feet and seconds, respectively. What is the closest the weight gets to the ceiling and when does this first happen for $t > 0$?

Figure 33

20. The equations $x = 2 + 2 \cos 4t$ and $y = 6 + 2 \sin 4t$ give the coordinates of a point moving along the circumference of a circle. Determine the center and radius of the circle. How long does it take for the point to make a complete circle?

21. Consider the wheel-piston device shown in Figure 34 (which is analogous to the

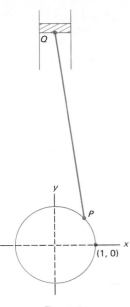

Figure 34

crankshaft and piston in an automobile engine). The wheel has a radius of 1 foot and rotates counterclockwise at 1 radian per second; the connecting rod is 5 feet long. If the point P is initially at $(1, 0)$, find the y-coordinate of Q after t seconds. Assume the x-coordinate is always zero.

22. Redo problem 21, but assume the wheel has radius 2 feet and rotates at 60 revolutions per second and that P is initially at $(2, 0)$. Is Q executing simple harmonic motion in either of these problems?

23. The voltage drop E across the terminals in a certain alternating current circuit is approximately $E = 156 \sin(110\pi t)$, where t is in seconds. What is the maximum voltage drop and what is the **frequency** (number of cycles per second) for this circuit?

24. The carrier wave for the radio wave of a certain FM station has the form $y = A \sin(2\pi \cdot 10^8 t)$, where t is measured in seconds. What is the frequency for this wave?

25. The AM radio wave for a certain station has the form

$$y = 55[1 + .02 \sin(2400\pi t)] \sin(2 \times 10^5 \pi t)$$

(a) Find y when $t = 3$.
(b) Find y when $t = .03216$.
[c] (c) Find y when $t = .0000321$.

26. In predator-prey systems, the number of predators and the number of prey tend to vary periodically. In a certain region with coyotes as predators and rabbits as prey, the rabbit population R varied according to the formula

$$R = 1000 + 150 \sin 2t$$

where t was measured in years after January 1, 1950.
(a) What was the maximum rabbit population?
(b) When was it first reached?
[c] (c) What was the population on January 1, 1953?

27. The number of coyotes C in Problem 26 satisfied

$$C = 200 + 50 \sin(2t - .7)$$

Sketch the graphs of C and R using the same coordinate system and attempt to explain the phase shift in C.

[c] 28. Sketch the graph of $y = 2^{-t} \cos 2t$ for $0 \le t \le 3\pi$. This is an example of damped harmonic motion, which is typical of harmonic motion where there is friction.

29. Use addition laws to write each of the following in the form $A_1 \sin Bt + A_2 \cos Bt$
(a) $4 \sin(2t - \frac{\pi}{4})$ (b) $3 \cos(3t + \frac{\pi}{3})$
Note: The same idea would work on any expression of the form $A \sin(Bt + C)$ or $A \cos(Bt + C)$.

30. Determine C so that

$$5 \sin 4t + 12 \cos 4t = 13 \sin(4t + C)$$

31. Suppose that A_1 and A_2 are both positive. Show that

$$A_1 \sin Bt + A_2 \cos Bt = A \sin(Bt + C)$$

where $A = \sqrt{A_1^2 + A_2^2}$ and $C = \tan^{-1}(A_2/A_1)$.

32. Generalize Problem 31 by showing that $A_1 \sin Bt + A_2 \cos Bt$ can always be written in the form $A \sin(Bt + C)$. *Hint:* Choose A as in Problem 31 and let C be the radian measure of an angle that has (A_1, A_2) on its terminal side.

33. Use the result in Problems 31 and 32 to write each of the following in the form $A \sin(Bt + C)$.
    (a) $4 \cos 2t + 3 \sin 2t$ \qquad\qquad (b) $3 \sin 4t - \sqrt{3} \cos 4t$

34. Give an argument to show that

$$A_1 \sin Bt + A_2 \cos Bt, \qquad A_1 A_2 \neq 0, B \neq 0$$

    is not a polynomial in t for any choices of A_1, A_2, and B.

35. Find the maximum and minimum values of $\cos t \pm \sin t$.

36. **TEASER** Prove that $\sin(\cos t) < \cos(\sin t)$ for all t. *Hint:* First show that

$$-\frac{\pi}{2} < \cos t < \frac{\pi}{2} - |\sin t|$$

Which Curve Has the Simpler Equation?

Parabola

Four-leaved Rose

9-4   The Polar Coordinate System

The question we have asked above makes no sense unless coordinate axes are present. Most people would then choose the parabola as having the simpler equation; but the question is still more subtle than one might think. You already know that the complexity of the equation of a curve depends on the placement of the coordinate axes. Placed just right, the equation of the parabola might be as simple as $y = x^2$. Placed less wisely, the equation might be as complicated as $x - 3 = -(y + 7)^2$, or even worse. However, the four-leaved rose has a very messy equation no matter where the x- and y-axes are placed.

But there is another aspect to the question, one that Fermat and Descartes did not think about when they gave us Cartesian coordinates. There are many different kinds of coordinate systems, that is, different ways of specifying the position of a point. One of these systems, when placed the best possible way, gives the four-leaved rose a delightfully simple equation (see Example B). This system is called the **polar coordinate system;** it simplifies many problems that arise in calculus.

POLAR COORDINATES

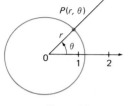

Figure 35

In place of two perpendicular axes as in Cartesian coordinates, we introduce in the plane a single horizontal ray, called the **polar axis,** emanating from a fixed point O, called the **pole.** On the polar axis, we mark off the positive half of a number scale with zero at the pole. Any point P other than the pole is the intersection of a unique circle with center O and a unique ray emanating from O (Figure 35). If r is the radius of the circle and θ is the angle the ray makes with the polar axis, then (r, θ) are the polar coordinates of P.

Points specified by polar coordinates are easiest to plot if we use polar graph paper. The grid on this paper consists of concentric circles and rays emanating from their common center. We have reproduced such a grid in Figure 36 and plotted a few points.

Of course, we can measure the angle θ in degrees as well as radians. More significantly, notice that while a pair of coordinates (r, θ) determines a unique point $P(r, \theta)$, each point has many different pairs of polar coordinates.

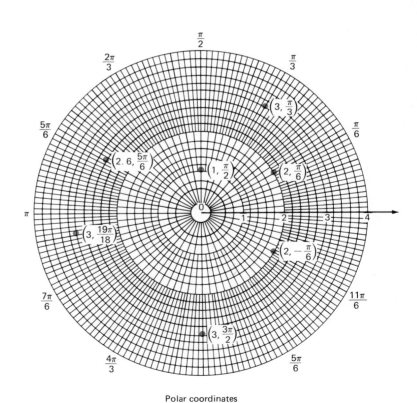

Polar coordinates

Figure 36

For example,

$$\left(2, \frac{3\pi}{2}\right) \qquad \left(2, -\frac{\pi}{2}\right) \qquad \left(2, \frac{7\pi}{2}\right)$$

are coordinates for the same point.

RELATION TO CARTESIAN COORDINATES

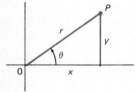

Figure 37

Let the positive x-axis of the Cartesian coordinate system serve also as the polar axis of a polar coordinate system, the origin coinciding with the pole (Figure 37). Then Cartesian coordinates and polar coordinates are related by two pairs of simple equations.

$$x = r \cos \theta \qquad\qquad r^2 = x^2 + y^2$$
$$y = r \sin \theta \qquad\qquad \tan \theta = \frac{y}{x}$$

For example, if $(4, \pi/6)$ are the polar coordinates of a point, then its Cartesian coordinates are

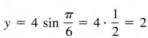

$$x = 4 \cos \frac{\pi}{6} = 4 \cdot \frac{\sqrt{3}}{2} = 2\sqrt{3}$$

$$y = 4 \sin \frac{\pi}{6} = 4 \cdot \frac{1}{2} = 2$$

On the other hand, if $(-3, \sqrt{3})$ are the Cartesian coordinates of a point (Figure 38), then

$$r = \sqrt{(-3)^2 + (\sqrt{3})^2} = \sqrt{12} = 2\sqrt{3}$$

$$\tan \theta = \frac{\sqrt{3}}{-3}$$

Figure 38

Since the point is in the second quadrant, we choose $5\pi/6$ as an appropriate value of θ. Thus one choice of polar coordinates for the point in question is $(2\sqrt{3}, 5\pi/6)$.

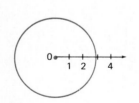

$r = 3$

Figure 39

POLAR GRAPHS

The simplest equations to graph in a polar coordinate system are of the forms

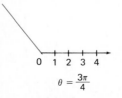

$\theta = \frac{3\pi}{4}$

Figure 40

1. $r = $ a constant
2. $\theta = $ a constant

The first is a circle with center at the pole; the second is a ray emanating from the pole. Examples are shown in Figures 39 and 40.

To graph more complicated equations such as

$$r = 2(1 + \cos \theta)$$

we follow our usual procedure of making a table of values, plotting the corresponding points, and then connecting them with a smooth curve. The graph (Figure 41) of the equation is called a *cardioid* (a heart-like curve). Notice that r approaches 0 as θ approaches π, creating a dimple in the curve.

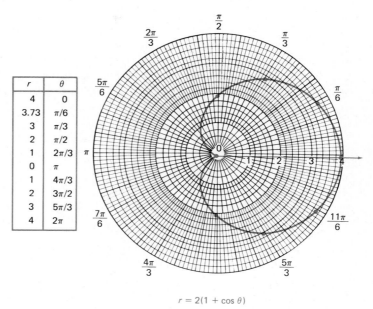

r	θ
4	0
3.73	$\pi/6$
3	$\pi/3$
2	$\pi/2$
1	$2\pi/3$
0	π
1	$4\pi/3$
2	$3\pi/2$
3	$5\pi/3$
4	2π

$r = 2(1 + \cos \theta)$

Figure 41

Problem Set 9-4

Graph each of the following points given in polar coordinates. Polar graph paper will simplify the graphing process.

1. $\left(3, \dfrac{\pi}{4}\right)$ 2. $\left(2, \dfrac{\pi}{3}\right)$ 3. $\left(\dfrac{3}{2}, \dfrac{5\pi}{6}\right)$ 4. $\left(1, \dfrac{5\pi}{3}\right)$

5. $(3, \pi)$ 6. $\left(2, \dfrac{\pi}{2}\right)$ 7. $(3, -\pi)$ 8. $\left(2, -\dfrac{3\pi}{2}\right)$

9. $(4, 70°)$ 10. $(3, 190°)$ 11. $\left(\dfrac{5}{2}, \dfrac{7\pi}{3}\right)$ 12. $\left(\dfrac{7}{2}, \dfrac{11\pi}{4}\right)$

Find the Cartesian coordinates of the point having the given polar coordinates.

13. $\left(4, \dfrac{\pi}{4}\right)$ 14. $\left(6, \dfrac{\pi}{6}\right)$ 15. $(3, \pi)$ 16. $\left(2, \dfrac{3\pi}{2}\right)$

17. $\left(10, \dfrac{4\pi}{3}\right)$ 18. $\left(8, \dfrac{11\pi}{6}\right)$ 19. $\left(2, -\dfrac{\pi}{4}\right)$ 20. $\left(3, -\dfrac{2\pi}{3}\right)$

Find polar coordinates for the point with the given Cartesian coordinates.

21. (4, 0)   22. (0, 3)   23. (−2, 0)   24. (0, −5)
25. (2, 2)   26. (2, −2)   27. (−2, 2)   28. (−2, −2)
29. $(1, -\sqrt{3})$ 30. $(-2\sqrt{3}, 2)$ 31. $(3, -\sqrt{3})$ 32. $(-\sqrt{3}, -3)$

Graph each of the following equations. Use polar graph paper if it is available.

33. $r = 2$ 34. $r = 5$
35. $\theta = \pi/3$ 36. $\theta = -2\pi/3$
37. $r = |\theta|$ (with θ in radians)   38. $r = \theta^2$
39. $r = 2(1 - \cos\theta)$ 40. $r = 3(1 + \sin\theta)$
41. $r = 2 + \cos\theta$ 42. $r = 2 - \sin\theta$

EXAMPLE A (Transforming Equations) (a) Change the Cartesian equation $(x^2 + y^2)^2 = x^2 - y^2$ to a polar equation. (b) Change $r = 2\sin 2\theta$ to a Cartesian equation.

Solution.

(a) Replacing $x^2 + y^2$ by r^2, x by $r\cos\theta$, and y by $r\sin\theta$, we get

$$(r^2)^2 = r^2\cos^2\theta - r^2\sin^2\theta$$

$$r^4 = r^2(\cos^2\theta - \sin^2\theta)$$

$$r^2 = \cos 2\theta$$

Dividing by r^2 at the last step did no harm since the graph of the last equation passes through the pole $r = 0$.

(b) $r = 2\sin 2\theta$

$$r = 2 \cdot 2\sin\theta\cos\theta$$

Multiplying both sides by r^2 gives

$$r^3 = 4(r\sin\theta)(r\cos\theta)$$

$$(x^2 + y^2)^{3/2} = 4yx$$

Transform to a polar equation.

43. $x^2 + y^2 = 4$ 44. $\sqrt{x^2 + y^2} = 6$
45. $y = x^2$ 46. $x^2 + (y - 1)^2 = 1$

Transform to a Cartesian equation.

47. $\tan\theta = 2$ 48. $r = 3\cos\theta$ 49. $r = \cos 2\theta$ 50. $r^2 = \cos\theta$

EXAMPLE B (Allowing Negative Values for r) It is sometimes useful to allow r to be negative. By the point $(-3, \pi/4)$, we shall mean the point 3 units from the pole on the ray in the opposite direction from the ray for $\theta = \pi/4$ (see Figure 42). Allowing r to be negative, graph

$$r = 2 \sin 2\theta$$

Solution. We begin with a table of values (Figure 43), plot the corresponding points, and then sketch the graph (Figure 44).

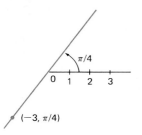

$(-3, \pi/4)$

Figure 42

θ	0	$\frac{\pi}{12}$	$\frac{\pi}{6}$	$\frac{\pi}{4}$	$\frac{\pi}{3}$	$\frac{5\pi}{12}$	$\frac{\pi}{2}$	$\frac{7\pi}{12}$	$\frac{3\pi}{4}$	$\frac{11\pi}{12}$	π	$\frac{5\pi}{4}$	$\frac{3\pi}{2}$	$\frac{7\pi}{4}$	2π
2θ	0	$\frac{\pi}{6}$	$\frac{\pi}{3}$	$\frac{\pi}{2}$	$\frac{2\pi}{3}$	$\frac{5\pi}{6}$	π	$\frac{7\pi}{6}$	$\frac{3\pi}{2}$	$\frac{11\pi}{6}$	2π	$\frac{5\pi}{2}$	3π	$\frac{7\pi}{2}$	4π
r	0	1	$\sqrt{3}$	2	$\sqrt{3}$	1	0	-1	-2	-1	0	2	0	-2	0

a   *b*   *c*   *d*

Figure 43

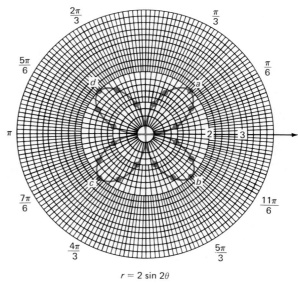

$r = 2 \sin 2\theta$

Figure 44

Note: The four leaves correspond to the four parts (a), (b), (c), and (d) of the table of values. For example, leaf (b) results from values of θ between $\pi/2$ and π where r is negative. This graph is the four-leaved rose of our opening display. Its Cartesian equation was obtained in (b) of Example A.

Graph each of the following, allowing r to be negative.

51. $r = 3 \cos 2\theta$ 52. $r = \cos 3\theta$ 53. $r = \sin 3\theta$

54. $r = 4 \cos \theta$ 55. $r = \sin 4\theta$ 56. $r = \cos 4\theta$

MISCELLANEOUS PROBLEMS

57. Transform to a Cartesian xy-equation and identify the corresponding curve.

   (a) $r = \dfrac{5}{3 \sin \theta - 2 \cos \theta}$

   (b) $r = 4 \cos \theta - 6 \sin \theta$

58. Transform to a polar equation.

   (a) $x^2 = 4y$

   (b) $(x - 5)^2 + (y + 2)^2 = 29$

Graph each of the polar equations in Problems 59–64.

59. $r = 4$

60. $\theta = -\pi/3$

61. $r = 2(1 - \sin \theta)$

62. $r = 1/\theta,\ \theta > 0$

63. $r^2 = \sin 2\theta$ *Caution:* Avoid values of θ which make r^2 negative.

64. $r = 2^\theta$ *Note:* Use both negative and positive values for θ.

65. Sketch the graphs of each pair of equations and find their points of intersection.

   (a) $r = 4 \cos \theta,\quad r \cos \theta = 1$ (b) $r = 2\sqrt{3} \sin \theta,\quad r = 2(1 + \cos \theta)$

66. Find the polar coordinates of the midpoint of the line segment joining the points with polar coordinates $(4, 2\pi/3)$ and $(8, \pi/6)$.

67. Show the distance d between the points with polar coordiantes (r_1, θ_1) and (r_2, θ_2) is given by

$$d = \sqrt{r_1^2 + r_2^2 - 2r_1 r_2 \cos(\theta_2 - \theta_1)}$$

and use this result to find the distance between $(4, 2\pi/3)$ and $(8, \pi/6)$.

68. Show that a circle of radius a and center (a, α) has polar equation $r = 2a \cos(\theta - \alpha)$. *Hint:* Law of cosines.

69. Find a formula for the area of the polar rectangle $0 < a \le r \le b$, $\alpha \le \theta \le \beta$, $\beta - \alpha < \pi$.

70. A point P moves so that its distance from the pole is always equal to its distance from the horizontal line $r \sin \theta = 4$. Show that the equation of the resulting curve (a parabola) is $r = 4/(1 + \sin \theta)$

71. A line segment L of length 4 has its two endpoints on the x- and y-axes, respectively. The point P is on L and is such that the line OP from the pole to P is perpendicular to L. Show that the set of points P satisfying this condition is a four-leaved rose by finding its polar equation.

72. **TEASER** Let F and F' be fixed points with polar coordinates $(a, 0)$ and $(-a, 0)$, respectively. A point P moves so that the product of its distances from F and F' is equal to the constant a^2 (that is, $\overline{PF} \cdot \overline{PF'} = a^2$). Find a simple polar equation (of the form $r^2 = f(\theta)$) for the resulting curve and sketch its graph.

Jean–Robert Argand (1768–1822)

Though several mathematicians (for example, De Moivre, Euler, Gauss) had thought of complex numbers as points in the plane before Argand, this obscure Swiss bookkeeper gets credit for the idea. In 1806 he wrote a small book on the geometric representation of complex numbers. It was his only contribution to mathematics.

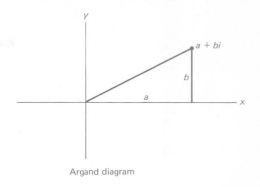

Argand diagram

9-5 Polar Representation of Complex Numbers

Throughout this book we have used the fact that a real number can be thought of as a point on a line. Now we are going to learn that a complex number can be represented as a point in the plane. This representation can be accomplished using either Cartesian or polar coordinates. The latter leads to the polar form of a complex number, which aids in multiplication and division and greatly facilitates finding powers and roots. Incidentally, if you have forgotten the basic facts about the complex numbers, you should review Section 1–6 before going on.

COMPLEX NUMBERS AS POINTS IN THE PLANE

Consider a complex number $a + bi$. It is determined by the two real numbers a and b, that is, by the ordered pair (a, b). But (a, b), in turn, determines a point in the plane. That point we now label with the complex number $a + bi$. Thus $2 + 4i$, $2 - 4i$, $-3 + 2i$, and all other complex numbers may be used as labels for points in the plane (Figure 45). The plane with points labeled this way is called the **Argand diagram** or **complex plane.** Note that $3i = 0 + 3i$ labels a point on the y-axis, which we now call the **imaginary axis,** while $4 = 4 + 0i$ corresponds to a point on the x-axis (called the **real axis**).

Recall that the absolute value of a real number a (written $|a|$) is its distance from the origin on the real line. The concept of absolute value is extended to a complex number $a + bi$ by defining

$$|a + bi| = \sqrt{a^2 + b^2}$$

which is also its distance from the origin (Figure 46). Thus while there are only two real numbers with absolute value of 5, namely, -5 and 5, there are infinitely many complex numbers with absolute value 5. They include 5, -5,

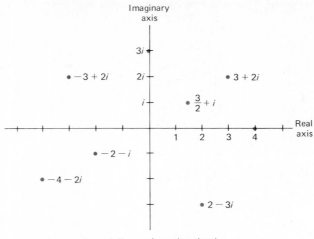

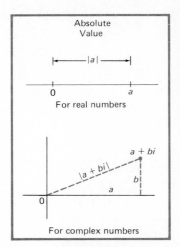

Figure 45

Figure 46

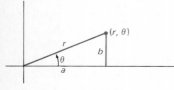

Figure 47

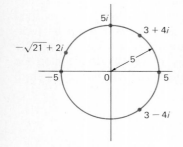

Figure 48

$5i$, $3 + 4i$, $3 - 4i$, $-\sqrt{21} + 2i$, and, in fact, all complex numbers on a circle of radius 5 centered at the origin (Figure 47).

POLAR FORM

Let (r, θ), where $r \geq 0$, be the polar coordinates of the point with Cartesian coordinates (a, b) (Figure 48). Then

$$a = r \cos \theta \qquad b = r \sin \theta$$

This means that we can write

$$a + bi = r \cos \theta + (r \sin \theta)i$$

or

$$\boxed{a + bi = r(\cos \theta + i \sin \theta)}$$

The boxed expression gives the polar form of $a + bi$. Notice that r is just the absolute value of $a + bi$; we shall refer to θ as its angle.

To put a number $a + bi$ in polar form, we use the formulas

$$r = \sqrt{a^2 + b^2} \qquad \cos \theta = \frac{a}{r}$$

For example, for $2\sqrt{3} - 2i$,

$$r = \sqrt{(2\sqrt{3})^2 + (-2)^2} = \sqrt{12 + 4} = 4$$

$$\cos \theta = \frac{2\sqrt{3}}{4} = \frac{\sqrt{3}}{2}$$

Since $2\sqrt{3} - 2i$ is in quadrant IV and $\cos \theta = \sqrt{3}/2$, θ can be chosen as an angle of $11\pi/6$ radians or $330°$. Thus

$$2\sqrt{3} - 2i = 4\left(\cos \frac{11\pi}{6} + i \sin \frac{11\pi}{6}\right)$$

$$= 4(\cos 330° + i \sin 330°)$$

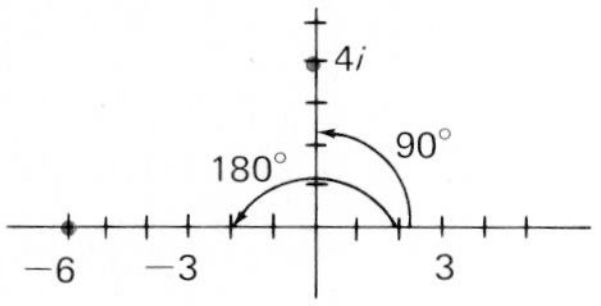

For some numbers, finding the polar form is almost trivial. Just picture in your mind (Figure 49) where -6 and $4i$ are located in the complex plane and you will know that

$$-6 = 6(\cos 180° + i \sin 180°)$$

$$4i = 4(\cos 90° + i \sin 90°)$$

Figure 49

Changing from the Cartesian form $a + bi$ to polar form is what we have just illustrated. Going in the opposite direction is much easier. For example, to change the polar form $3(\cos 240° + i \sin 240°)$ to Cartesian form, we simply calculate the sine and cosine of $240°$ and remove the parentheses.

$$3(\cos 240° + i \sin 240°) = 3\left(-\frac{1}{2} + i\frac{-\sqrt{3}}{2}\right)$$

$$= -\frac{3}{2} - \frac{3\sqrt{3}}{2}i$$

MULTIPLICATION AND DIVISION

The polar form is ideally suited for multiplying and dividing complex numbers. Let U and V be complex numbers given in polar form by

$$U = r(\cos \alpha + i \sin \alpha)$$

$$V = s(\cos \beta + i \sin \beta)$$

Then

$$U \cdot V = rs[\cos(\alpha + \beta) + i \sin(\alpha + \beta)]$$

$$\frac{U}{V} = \frac{r}{s}[\cos(\alpha - \beta) + i \sin(\alpha - \beta)]$$

In words, to multiply two complex numbers, we multiply their absolute values and add their angles. To divide two complex numbers, we divide their absolute values and subtract their angles (in the correct order). Thus if

$$U = 4(\cos 75° + i \sin 75°)$$

$$V = 3(\cos 60° + i \sin 60°)$$

then

$$U \cdot V = 12(\cos 135° + i \sin 135°)$$

$$\frac{U}{V} = \frac{4}{3}(\cos 15° + i \sin 15°)$$

To establish the multiplication formula we use a bit of trigonometry.

$$U \cdot V = r(\cos \alpha + i \sin \alpha)s(\cos \beta + i \sin \beta)$$
$$= rs(\cos \alpha \cos \beta + i \cos \alpha \sin \beta + i \sin \alpha \cos \beta + i^2 \sin \alpha \sin \beta)$$
$$= rs[(\cos \alpha \cos \beta - \sin \alpha \sin \beta) + i(\sin \alpha \cos \beta + \cos \alpha \sin \beta)]$$
$$= rs[\cos(\alpha + \beta) + i \sin(\alpha + \beta)]$$

The key step was the last one, where we used the addition laws for the cosine and the sine.

You will be asked to establish the division formula in Problem 56.

GEOMETRIC ADDITION AND MULTIPLICATION

Having learned that the complex numbers can be thought of as points in a plane, we should not be surprised that the operations of addition and multiplication have a geometric interpretation. Let U and V be any two complex numbers; that is, let

$$U = a + bi = r(\cos \alpha + i \sin \alpha)$$
$$V = c + di = s(\cos \beta + i \sin \beta)$$

Addition is accomplished algebraically by adding the real parts and imaginary parts separately.

$$U + V = (a + c) + (b + d)i$$

To accomplish the same thing geometrically, we construct the parallelogram that has O, U, and V as three of its vertices (see Figure 50). Then $U + V$ corresponds to the vertex opposite the origin, as you should be able to show by finding the coordinates of this vertex.

To multiply algebraically, we use the polar forms of U and V, adding the angles and multiplying the absolute values.

$$U \cdot V = rs[\cos(\alpha + \beta) + i \sin(\alpha + \beta)]$$

To interpret this geometrically (for the case where α and β are between $0°$ and $180°$), first draw triangle OAU, where A is the point $1 + 0i$. Then construct triangle OVW similar to triangle OAU in the manner indicated in Figure 51. We claim that $W = U \cdot V$. Certainly W has the correct angle, namely, $\alpha + \beta$. Moreover, by similarity of triangles,

$$\frac{\overline{OW}}{\overline{OV}} = \frac{\overline{OU}}{\overline{OA}} = \frac{\overline{OU}}{1}$$

(Here we are using $\overline{OW}$ for the length of the line segment from O to W.) Thus

$$|W| = \overline{OW} = \overline{OU} \cdot \overline{OV} = |U| \cdot |V|$$

so W also has the correct absolute value.

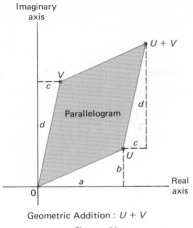

Geometric Addition : $U + V$

Figure 50

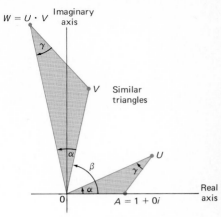

Geometric Multiplication : $U \cdot V$

Figure 51

Problem Set 9-5

In Problems 1–12, plot the given numbers in the complex plane.

1. $2 + 3i$

2. $2 - 3i$

3. $-2 - 3i$

4. $-2 + 3i$

5. 5

6. -6

7. $-4i$

8. $6i$

9. $\frac{3}{5} - \frac{4}{5}i$

10. $-\frac{5}{13} + \frac{12}{13}i$

11. $2\left(\cos \dfrac{\pi}{4} + i \sin \dfrac{\pi}{4}\right)$

12. $3\left(\cos \dfrac{7\pi}{6} + i \sin \dfrac{7\pi}{6}\right)$

13. Find the absolute values of the numbers in Problems 1, 3, 5, 7, 9, and 11.

14. Find the absolute values of the numbers in Problems 2, 4, 6, 8, and 10.

Express each of the following in the form $a + bi$.

15. $4\left(\cos \dfrac{3\pi}{2} + i \sin \dfrac{3\pi}{2}\right)$

16. $5(\cos \pi + i \sin \pi)$

17. $2(\cos 225° + i \sin 225°)$

18. $\frac{3}{2}(\cos 300° + i \sin 300°)$

Express each of the following in polar form. For example, $1 + i = \sqrt{2}\ (\cos 45° + i \sin 45°)$.

19. -4

20. 9

21. $-5i$

22. $4i$

23. $2 - 2i$

24. $-5 - 5i$

25. $2\sqrt{3} + 2i$

26. $-4\sqrt{3} + 4i$

□ 27. $5 + 4i$

□ 28. $3 + 2i$

For Problems 29–36 let $u = 2\ (\cos 140° + i \sin 140°)$, $v = 3\,(\cos 70° + i \sin 70°)$ and $w = \frac{1}{2}(\cos 55° + i \sin 55°)$. Calculate each product or quotient, leaving your answer in polar form. For example,

$$\frac{u^2}{w} = \frac{u \cdot u}{w} = \frac{4(cos\ 280° + i\ sin\ 280°)}{\frac{1}{2}(cos\ 55° + i\ sin\ 55°)} = 8(cos\ 225° + i\ sin\ 225°)$$

29. uv 30. uw 31. vw 32. uvw

33. u/v 34. uv/w 35. $1/w$ 36. $1/v$

EXAMPLE (Finding Products in Two Ways)   Find the product

$$(\sqrt{3} + i)(-4 - 4\sqrt{3}i)$$

directly and then by using the polar form.

Solution.   *Method 1*   We use the definition of multiplication given in Section 1-6 to get

$$(\sqrt{3} + i)(-4 - 4\sqrt{3}i) = (-4\sqrt{3} + 4\sqrt{3}) + (-4 - 12)i$$

$$= -16i$$

Method 2   We change both numbers to polar form, multiply by the method of this section, and finally change to $a + bi$ form.

$$\sqrt{3} + i = 2(cos\ 30° + i\ sin\ 30°)$$

$$-4 - 4\sqrt{3}i = 8(cos\ 240° + i\ sin\ 240°)$$

$$(\sqrt{3} + i)(-4 - 4\sqrt{3}i) = 16(cos\ 270° + i\ sin\ 270°)$$

$$= 16(0 - i)$$

$$= -16i$$

Find each of the following products in two ways, giving your final answer in $a + bi$ form.

37. $(4 - 4i)(2 + 2i)$ 38. $(\sqrt{3} + i)(2 - 2\sqrt{3}i)$

39. $(1 + \sqrt{3}i)(1 + \sqrt{3}i)$ 40. $(\sqrt{2} + \sqrt{2}i)(\sqrt{2} + \sqrt{2}i)$

Find the following products and quotients, giving your answers in polar form. Start by changing each of the given complex numbers to polar form.

41. $4i(2\sqrt{3} - 2i)$ 42. $(-2i)(5 + 5i)$

43. $\dfrac{4i}{2\sqrt{3} - 2i}$ 44. $\dfrac{-2i}{5 + 5i}$

45. $(2\sqrt{2} - 2\sqrt{2}i)(2\sqrt{2} - 2\sqrt{2}i)$ 46. $(1 - \sqrt{3}i)(1 - \sqrt{3}i)$

MISCELLANEOUS PROBLEMS

47. Plot the given number in the complex plane and find its absolute value.
  (a) $-5 + 12i$ (b) $-4i$ (c) $5(cos\ 60° + i\ sin\ 60°)$

48. Express in the form $a + bi$.
  (a) $5[cos(3\pi/2) + i\ sin(3\pi/2)]$ (b) $4(cos\ 180° + i\ sin\ 180°)$
  (c) $2(cos\ 315° + i\ sin\ 315°)$ (d) $3[cos(-2\pi/3) + i\ sin(-2\pi/3)]$

49. Express in polar form.
    (a) 12   (b) $-\sqrt{2} + \sqrt{2}i$
    (c) $-3i$ (d) $2 - 2\sqrt{3}i$
    (e) $4\sqrt{3} + 4i$ (f) $2(\cos 45° - i \sin 45°)$

50. Write in the form $a + bi$.
    (a) $2(\cos 37° + i \sin 37°)8(\cos 113° + i \sin 113°)$
    (b) $6(\cos 123° + i \sin 123°)/[3(\cos 33° + i \sin 33°)]$

51. Perform the indicated operations and write your answer in polar form.
    (a) $1.5(\cos 110° + i \sin 110°)4(\cos 30° + i \sin 30°)2(\cos 20° + i \sin 20°)$

    (b) $\dfrac{12(\cos 115° + i \sin 115°)}{4(\cos 55° + i \sin 55°)(\cos 20° + i \sin 20°)}$

    (c) $\dfrac{(-\sqrt{2} + \sqrt{2}\,i)(2 - 2\sqrt{3}i)}{4\sqrt{3} + 4i}$ (See Problem 49.)

52. Calculate and write your answer in the form $a + bi$

    (a) $\dfrac{(-\sqrt{2} + \sqrt{2}\,i)^2(2 - 2\sqrt{3}\,i)^3}{4\sqrt{3} + 4i}$ (See Problem 49.)

    (b) $(-1 + \sqrt{3}\,i)^5(-1 - \sqrt{3}\,i)^{-4}$

53. In each case, find two values for $z = a + bi$.
    (a) The imaginary part of z is 5 and $|z| = 13$.
    (b) The number z lies on the line $y = x$ and $|z| = 8$.

54. Find four complex numbers $z = a + bi$ that are located on the hyperbola $y^2 - x^2 = 2$ and satisfy $|z| = \sqrt{10}$.

55. Let $u = r(\cos \theta + i \sin \theta)$. Write each of the following in polar form.
    (a) u^3 (b) $\bar{u}$ ($\bar{u}$ is the conjugate of u)
    (c) $u\bar{u}$ (d) $1/u$
    (e) u^{-2} (f) $-u$

56. Prove the division formula: If $U = r(\cos \alpha + i \sin \alpha)$ and $V = s(\cos \beta + i \sin \beta)$, then

$$\frac{U}{V} = \frac{r}{s}[\cos(\alpha - \beta) + i \sin(\alpha - \beta)]$$

57. Let U and V be complex numbers. Give a geometric interpretation for (a) $|U - V|$ and (b) the angle of $U - V$.

58. By expanding $(\cos \theta + i \sin \theta)^3$ in two different ways, derive the formulas.
    (a) $\cos 3\theta = 4 \cos^3 \theta - 3 \cos \theta$
    (b) $\sin 3\theta = -4 \sin^3 \theta + 3 \sin \theta$

59. Let $z_k = 2[\cos(k\pi/4) + i \sin(k\pi/4)]$. Find the exact value of each of the following.
    (a) $z_1 z_2 z_3 \cdots z_8$
    (b) $z_1 + z_2 + z_3 + \cdots + z_8$ (Think geometrically.)

60. **TEASER**   Let $z_k = \dfrac{k}{k + 1}(\cos k° + i \sin k°)$. Find the exact value of the product

$$z_1 z_2 z_3 \cdots z_{179} z_{180}$$

Abraham De Moivre (1667–1754)

Though he was a Frenchman, De Moivre spent most of his life in London. There he became an intimate friend of the great Isaac Newton, inventor of calculus. De Moivre made many contributions to mathematics but his reputation rests most securely on the theorem that bears his name.

$$(\cos \theta + i \sin \theta)^n$$
$$= \cos n\theta + i \sin n\theta$$

De Moivre's Theorem

9-6   Powers and Roots of Complex Numbers

De Moivre's theorem tells us how to raise a complex number of absolute value 1 to an integral power. We can easily extend it to cover the case of any complex number, no matter what its absolute value. Then with a little work, we can use it to find roots of complex numbers. Here we are in for a surprise. Take the number $8i$, for example. After some fumbling around, we find that one of its cube roots is $-2i$, because $(-2i)^3 = -8i^3 = 8i$. We shall find that it has two other cube roots (both nonreal numbers). In fact, we shall see that every number has exactly three cube roots, four 4th roots, five 5th roots, and so on. To put it in a spectacular way, we claim that any number, for example $37 + 3.5i$, has 1,000,000 millionth roots.

POWERS OF COMPLEX NUMBERS

To raise the complex number $r(\cos \theta + i \sin \theta)$ to the nth power, n a positive integer, we simply find the product of n factors of $r(\cos \theta + i \sin \theta)$. But from Section 9-5, we know that we multiply complex numbers by multiplying their absolute values and adding their angles. Thus

$$[r(\cos \theta + i \sin \theta)]^n$$

$$= \underbrace{r \cdot r \cdots r}_{n \text{ factors}} [\cos (\underbrace{\theta + \theta + \cdots + \theta}_{n \text{ terms}}) + i \sin (\theta + \theta + \cdots \theta)]$$

In short,

$$[r(\cos \theta + i \sin \theta)]^n = r^n(\cos n\theta + i \sin n\theta)$$

When $r = 1$, this is De Moivre's theorem.

As a first illustration, let us find the 6th power of a complex number that is already in polar form.

$$\left[2\left(\cos\frac{\pi}{6} + i\sin\frac{\pi}{6}\right)\right]^6 = 2^6\left[\cos\left(6\cdot\frac{\pi}{6}\right) + i\sin\left(6\cdot\frac{\pi}{6}\right)\right]$$

$$= 64(\cos\pi + i\sin\pi)$$

$$= 64(-1 + i\cdot 0)$$

$$= -64$$

To find $(1 - \sqrt{3}i)^5$, we could use repeated multiplication of $1 - \sqrt{3}i$ by itself. But how much better to change $1 - \sqrt{3}i$ to polar form and use the boxed formula at the bottom of page 392.

$$1 - \sqrt{3}i = 2(\cos 300° + i\sin 300°)$$

Then

$$(1 - \sqrt{3}i)^5 = 2^5(\cos 1500° + i\sin 1500°)$$

$$= 32(\cos 60° + i\sin 60°)$$

$$= 32\left(\frac{1}{2} + i\frac{\sqrt{3}}{2}\right)$$

$$= 16 + 16\sqrt{3}i$$

THE THREE CUBE ROOTS OF $8i$

Because finding roots is tricky, we begin with an example before attempting the general case. We have already noted that $-2i$ is one cube root of $8i$, but now we claim there are two others. How shall we find them? We begin by writing $8i$ in polar form.

$$8i = 8(\cos 90° + i\sin 90°)$$

Finding cube roots is the opposite of cubing. That suggests that we take the real cube root (rather than the cube) of 8 and divide (rather than multiply) the angle 90° by 3. This would give us one cube root

$$2(\cos 30° + i\sin 30°)$$

which reduces to

$$2\left(\frac{\sqrt{3}}{2} + \frac{1}{2}i\right) = \sqrt{3} + i$$

Is this really a cube root of $8i$? For fear that you might be suspicious of the polar form, we will cube it the old-fashioned way and check.

$$(\sqrt{3} + i)^3 = (\sqrt{3} + i)(\sqrt{3} + i)(\sqrt{3} + i)$$

$$= [(3 - 1) + 2\sqrt{3}i](\sqrt{3} + i)$$

$$= 2(1 + \sqrt{3}i)(\sqrt{3} + i)$$

$$= 2(0 + 4i)$$

$$= 8i$$

Of course, the check using polar form is more direct.

$$[2(\cos 30° + i \sin 30°)]^3 = 2^3(\cos 90° + i \sin 90°)$$

$$= 8(0 + i)$$

$$= 8i$$

The process described above yielded one cube root of $8i$ (namely, $\sqrt{3} + i$); there are two others. Let us go back to our representation of $8i$ in polar form. We used the angle 90°; we could as well have used $90° + 360° = 450°$.

$$8i = 8(\cos 450° + i \sin 450°)$$

Now if we take the real cube root of 8 and divide 450° by 3 we get

$$2(\cos 150° + i \sin 150°) = 2\left(-\frac{\sqrt{3}}{2} + \frac{1}{2}i\right) = -\sqrt{3} + i$$

We could again check that this is indeed a cube root of $8i$.

What worked once might work twice. Let us write $8i$ in polar form in a third way, this time adding $2(360°)$ to its angle of 90°.

$$8i = 8(\cos 810° + i \sin 810°)$$

The corresponding cube root is

$$2(\cos 270° + i \sin 270°) = 2(0 - i) = -2i$$

This does not come as a surprise, since we knew that $-2i$ was one of the cube roots of $8i$.

If we add $3(360°)$ (that is, 1080°) to 90°, do we get still another cube root of $8i$? No, for if we write

$$8i = 8(\cos 1170° + i \sin 1170°)$$

the corresponding cube root of $8i$ would be

$$2(\cos 390° + i \sin 390°) = 2(\cos 30° + i \sin 30°)$$

But this is the same as the first cube root we found. The truth is that we have found all the cube roots of $8i$, namely, $\sqrt{3} + i$, $-\sqrt{3} + i$, and $-2i$.

Let us summarize. The number $8i$ has three cube roots given by

$$2\left[\cos\left(\frac{90°}{3}\right) + i \sin\left(\frac{90°}{3}\right)\right]$$

$$2\left[\cos\left(\frac{90° + 360°}{3}\right) + i \sin\left(\frac{90° + 360°}{3}\right)\right]$$

$$2\left[\cos\left(\frac{90° + 720°}{3}\right) + i \sin\left(\frac{90° + 720°}{3}\right)\right]$$

We can say the same thing in a shorter way by writing

$$2\left[\cos\left(\frac{90° + k \cdot 360°}{3}\right) + i \sin\left(\frac{90° + k \cdot 360°}{3}\right)\right] \qquad k = 0, 1, 2$$

ROOTS OF COMPLEX NUMBERS

We are ready to generalize. If $u \neq 0$, then

$$u = r(\cos \theta + i \sin \theta)$$

has n distinct nth roots $u_0, u_1, \ldots, u_{n-1}$ given by

$$u_k = \sqrt[n]{r}\left[\cos\left(\frac{\theta + k \cdot 360°}{n}\right) + i \sin\left(\frac{\theta + k \cdot 360°}{n}\right)\right]$$
$$k = 0, 1, 2, \ldots, n - 1$$

Recall that $\sqrt[n]{r}$ denotes the positive real nth root of $r = |u|$. In our example, it was $\sqrt[3]{|8i|} = \sqrt[3]{8} = 2$. To see that each value of u_k is an nth root, simply raise it to the nth power. In each case, you should get u.

The boxed formula assumes that θ is given in degrees. If θ is in radians, the formula takes the following form.

$$u_k = \sqrt[n]{r}\left[\cos\left(\frac{\theta + 2k\pi}{n}\right) + i \sin\left(\frac{\theta + 2k\pi}{n}\right)\right]$$
$$k = 0, 1, 2, \ldots, n - 1$$

A REAL EXAMPLE

Let us use the boxed formula to find the six 6th roots of 64. (Keep in mind that a real number is a special kind of complex number.) Changing to polar form, we write

$$64 = 64(\cos 0° + i \sin 0°)$$

Applying the formula with $r = |64| = 64$, $\theta = 0°$, and $n = 6$ gives

$$u_0 = 2(\cos 0° + i \sin 0°) = 2$$
$$u_1 = 2(\cos 60° + i \sin 60°) = 1 + \sqrt{3}i$$
$$u_2 = 2(\cos 120° + i \sin 120°) = -1 + \sqrt{3}i$$
$$u_3 = 2(\cos 180° + i \sin 180°) = -2$$
$$u_4 = 2(\cos 240° + i \sin 240°) = -1 - \sqrt{3}i$$
$$u_5 = 2(\cos 300° + i \sin 300°) = 1 - \sqrt{3}i$$

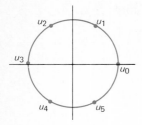

Figure 52

Notice that two of the roots, 2 and −2, are real; the other four are not real.

If you plot these six numbers (Figure 52), you will find that they lie on a circle of radius 2 centered at the origin and that they are equally spaced around the circle. This is typical of what happens in general.

Problem Set 9-6

Find each of the following, leaving your answer in polar form.

1. $\left[2\left(\cos \dfrac{\pi}{4} + i \sin \dfrac{\pi}{4}\right)\right]^3$

2. $\left[3\left(\cos \dfrac{5\pi}{6} + i \sin \dfrac{5\pi}{6}\right)\right]^2$

3. $[\sqrt{5}(\cos 11° + i \sin 11°)]^6$

4. $[\frac{1}{3}(\cos 12.5° + i \sin 12.5°)]^4$

5. $(1 + i)^8$

6. $(1 - i)^4$

Find each of the following powers. Write your answer in a + bi form.

7. $(\cos 36° + i \sin 36°)^{10}$

8. $(\cos 27° + i \sin 27°)^{10}$

9. $(\sqrt{3} + i)^5$

10. $(2 - 2\sqrt{3}i)^4$

Find the nth roots of u for the given u and n, leaving your answers in polar form. Plot these roots in the complex plane.

11. $u = 125(\cos 45° + i \sin 45°); \ n = 3$

12. $u = 81(\cos 80° + i \sin 80°); \ n = 4$

13. $u = 64\left(\cos \dfrac{\pi}{2} + i \sin \dfrac{\pi}{2}\right); \ n = 6$

14. $u = 3^8\left(\cos \dfrac{2\pi}{3} + i \sin \dfrac{2\pi}{3}\right); \ n = 8$

15. $u = 4(\cos 112° + i \sin 112°); \ n = 4$

16. $u = 7(\cos 200° + i \sin 200°); \ n = 5$

Find the nth roots of u for the given u and n. Write your answers in the a + bi form.

17. $u = 16, n = 4$

18. $u = -16, n = 4$

19. $u = 4i, n = 2$

20. $u = -27i, n = 3$

21. $u = -4 + 4\sqrt{3}i, n = 2$

22. $u = -2 - 2\sqrt{3}i, n = 4$

EXAMPLE (Roots of Unity) The *n*th roots of 1, called the **nth roots of unity,** play an important role in advanced algebra. Find the five 5th roots of unity, plot them, and show that four of the roots are powers of the 5th root.

Solution. First we represent 1 in polar form.

$$1 = 1(\cos 0° + i \sin 0°)$$

The five 5th roots are (according to the formula developed in this section)

$$u_0 = \cos 0° + i \sin 0° = 1$$

$$u_1 = \cos 72° + i \sin 72°$$

$$u_2 = \cos 144° + i \sin 144°$$

$$u_3 = \cos 216° + i \sin 216°$$

$$u_4 = \cos 288° + i \sin 288°$$

These roots are plotted in Figure 53. They lie on the unit circle and are equally spaced around it. Finally notice that

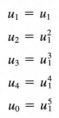

Figure 53

$$u_1 = u_1$$

$$u_2 = u_1^2$$

$$u_3 = u_1^3$$

$$u_4 = u_1^4$$

$$u_0 = u_1^5$$

Thus all the roots are powers of u_1. These powers of u_1 repeat in cycles of 5. For example, note that

$$u_1^6 = u_1^5 u_1 = u_1$$

$$u_1^7 = u_1^5 u_1^2 = u_1^2$$

In each of the following, find all the nth roots of unity for the given n and plot them in the complex plane.

23. $n = 4$ 24. $n = 6$ 25. $n = 10$ 26. $n = 12$

MISCELLANEOUS PROBLEMS

27. Calculate each of the following, leaving your answer in polar form.
 (a) $[3(\cos 20° + i \sin 20°)]^4$
 [c] (b) $[2.46(\cos 1.54 + i \sin 1.54)]^5$
 (c) $[2(\cos 50° + i \sin 50°)(\cos 30° + i \sin 30°)]^3$
 (d) $\left(\dfrac{8[\cos(2\pi/3) + i \sin(2\pi/3)]}{4[\cos(\pi/4) + i \sin(\pi/4)]} \right)^4$

28. Change to polar form, calculate, and then change back to $a + bi$ form.
 (a) $(1 - \sqrt{3}i)^5$
 (b) $[(\sqrt{3} + i)(2 - 2i)/(-1 + \sqrt{3}i)]^3$

29. Find the five 5th roots of $32(\cos 255° + i \sin 255°)$, giving your answers in polar form.

30. Find the three cube roots of $-4\sqrt{2} - 4\sqrt{2}i$, giving your answers in polar form.

31. Write the eight 8th roots of 1 in $a + bi$ form and calculate their sum and product.

32. Solve the equation $x^3 - 4 - 4\sqrt{3}i = 0$. You may give your answers in polar form.

33. Find the solution to $x^5 + \sqrt{2} - \sqrt{2}i = 0$ with the largest real part. Write your answer in the form $a + bi$

34. Solve the equation $x^3 + 27 = 0$ in two ways:
    (a) By finding the three cube roots of -27;
    (b) By writing $x^3 + 27 = (x + 3)(x^2 - 3x + 9)$ and using the quadratic formula.

35. Find the six solutions to $x^6 - 1 = 0$ by two different methods.

36. Show that $\cos(\pi/3) + i \sin(\pi/3)$ is a solution to $2x^4 + x^2 + x + 1 = 0$

37. Find all six solutions to $x^6 + x^4 + x^2 + 1 = 0$. *Hint:* The left side can be factored as $(x^2 + 1)(x^4 + 1)$.

38. Show that DeMoivre's theorem is valid when n is a negative integer.

39. If A is a nonreal number, we agree that $\sqrt{A}$ stands for the one of the two square roots with nonnegative real part. For example, the two square roots of $-4 + 4\sqrt{3}i$ are $\sqrt{2} + \sqrt{6}i$ and $-\sqrt{2} - \sqrt{6}i$, but we agree that
$$\sqrt{-4 + 4\sqrt{3}i} = \sqrt{2} + \sqrt{6}i$$
    Evaluate
    (a) $\sqrt{1 + \sqrt{3}i}$;   (b) $\sqrt{-1 + \sqrt{3}i}$.

40. The quadratic formula is valid even for quadratic equations with nonreal coefficients if we follow the agreement of Problem 39. Solve the following equations.
    (a) $x^2 - 2x + \sqrt{3}i = 0$
    (b) $x^2 - 4ix - 5 + \sqrt{3}i = 0$

41. Let n be an integer that is not divisible by 3. Simplify
$$(-1 + \sqrt{3}i)^n + (-1 - \sqrt{3}i)^n$$
    as much as possible.

42. **TEASER**   Let $1, u, u^2, u^3, \ldots, u^{15}$ be the sixteen 16th roots of unity. Calculate each of the following. Look for a simple way in each case.
    (a) $1 + u + u^2 + u^3 + \cdots + u^{15}$
    (b) $1 \cdot u \cdot u^2 \cdot u^3 \cdots u^{15}$
    (c) $(1 - u)(1 - u^2)(1 - u^3) \cdots (1 - u^{15})$
    (d) $(1 + u)(1 + u^2)(1 + u^4)(1 + u^8)(1 + u^{16})$

Chapter Summary

A triangle like the one in Figure 54 that has no right angle is called an oblique triangle. If any three of the six parts α, β, γ, a, b, and c—including at least one side—are given, we can find the remaining parts by using the **law of sines**

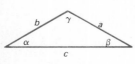

Figure 54

$$\frac{a}{\sin \alpha} = \frac{b}{\sin \beta} = \frac{c}{\sin \gamma}$$

and the **law of cosines**

$$a^2 = b^2 + c^2 - 2bc \cos \alpha \qquad \text{(one of three forms)}$$

There is, however, one case (given two sides and an angle opposite one of them) in which there might be no solution or two solutions. We call it the ambiguous case.

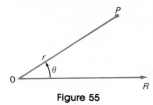

Figure 55

The equations $y = A \sin(Bt + C)$ and $y = A \cos(Bt + C)$ describe a common phenomenon known as **simple harmonic motion.** We can quickly draw the graphs of these equations by making use of three key numbers: the **amplitude,** $|A|$, the **period,** $2\pi/B$, and the **phase shift,** $-C/B$.

Some curves (for example, spirals or four-leaved roses) can be more simply described by means of **polar coordinates** than Cartesian coordinates. The polar coordinates (r, θ) of a point P measure its distance r from a fixed point O and the angle θ that OP makes with a horizontal ray OR emanating from O (Figure 55). The point O is called the **pole** and OR is the **polar axis.** The polar coordinates (r, θ) and Cartesian coordinates (x, y) are related via the equations

$$x = r \cos \theta \qquad y = r \sin \theta \qquad r^2 = x^2 + y^2$$

A complex number $a + bi$ can be represented geometrically as a point (a, b) in a plane called the **complex plane,** or **Argand diagram.** The horizontal and vertical axes are known as the **real axis** and **imaginary axis,** respectively. The distance from the origin to (a, b) is $\sqrt{a^2 + b^2}$; it is also the absolute value of $a + bi$, denoted as with real numbers by $|a + bi|$. If polar coordinates (r, θ) are used in place of Cartesian coordinates (a, b), we get the **polar form** of $a + bi$, namely,

$$r(\cos \theta + i \sin \theta)$$

This form facilitates multiplication and division and is especially helpful in finding powers and roots of a number. An important result is the formula

$$[r(\cos \theta + i \sin \theta)]^n = r^n(\cos n\theta + i \sin n\theta)$$

Chapter Review Problem Set

1. Solve each of the following triangles using Table C or a calculator.
   (a) $\alpha = 104.9°$, $\gamma = 36°$, $b = 149$
   (b) $a = 14.6$, $b = 89.2$, $c = 75.8$
   (c) $\gamma = 35°$, $a = 14$, $b = 22$
   (d) $\beta = 48.6°$, $c = 39.2$, $b = 57.6$

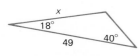

Figure 56

2. For the triangle in Figure 56, find x and the area of the triangle.

3. Find the period, amplitude, and phase shift for the graph of each equation.
   (a) $y = \cos 2t$ \qquad (b) $y = 3 \cos 4t$
   (c) $y = 2 \sin(3t - \pi/2)$ \qquad (d) $y = -2 \sin(\tfrac{1}{2}t + \pi)$

4. Sketch the graphs of the equations in Problem 3 on $-\pi \le t \le \pi$.

5. A wheel of radius 4 feet with center at the origin is rotating counterclockwise at $3\pi/4$ radians per second (Figure 57). If a paint speck P has coordinates $(-4, 0)$ initially, what will its coordinates be after t seconds?

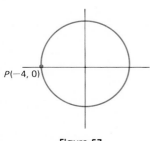

Figure 57

6. A point is moving in a straight line according to the equation $x = 3 \cos(5t + 3\pi)$, where x is in feet and t in seconds.
   (a) What is the period of the motion?
   (b) Find the initial position of the point relative to the point $x = 0$.
   (c) When is $x = 3$ for the first time?

7. Plot the points with the following polar coordinates.
   (a) $(3, 2\pi/3)$ (b) $(2, -3\pi/2)$ (c) $(6, 210°)$

8. Find the Cartesian coordinates of the points in Problem 7.

9. Find the polar coordinates of the point having the given Cartesian coordinates.
   (a) $(5, 0)$ (b) $(2\sqrt{2}, -2\sqrt{2})$ (c) $(-2\sqrt{3}, 2)$

10. Graph each of the following polar equations.
    (a) $r = 4$ (b) $r = 4 \sin \theta$ (c) $r = 4 \cos 3\theta$

11. Transform $xy = 4$ to a polar equation. Transform $r = \sin 2\theta$ to a Cartesian equation.

12. Plot the following numbers in the complex plane.
    (a) $3 - 4i$ (b) -6 (c) $5i$

    (d) $3\left(\cos\dfrac{3\pi}{4} + i \sin\dfrac{3\pi}{4}\right)$ (e) $4(\cos 300° + i \sin 300°)$

13. Find the absolute value of each number in Problem 12.

14. Express $4(\cos 150° + i \sin 150°)$ in the form $a + bi$.

15. Express in polar form.
    (a) $3i$ (b) -6 (c) $-1 - i$ (d) $2\sqrt{3} - 2i$

16. Let

$$u = 8(\cos 105° + i \sin 105°)$$

$$v = 4(\cos 40° + i \sin 40°)$$

Calculate each of the following, leaving your answer in polar form.
    (a) uv (b) u/v (c) u^3 (d) $u^2 v^3$

17. Find all the 6th roots of

$$2^6(\cos 120° + i \sin 120°)$$

leaving your answers in polar form.

18. Find the five solutions to $x^5 - 1 = 0$.

He who loves practice without theory is like the sailor who boards ship without a rudder and compass and never knows where he may cast.

—Leonardo da Vinci

CHAPTER 10

Theory of Polynomial Equations

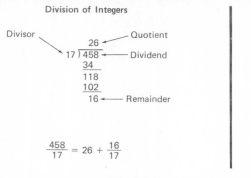 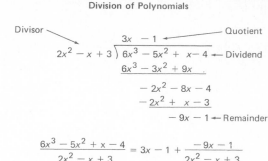

10-1   Division of Polynomials

In Section 2-4, we learned that polynomials in x can always be added, subtracted, and multiplied; the result in every case is a polynomial. For example,

$$(3x^2 + x - 2) + (3x - 2) = 3x^2 + 4x - 4$$

$$(3x^2 + x - 2) - (3x - 2) = 3x^2 - 2x$$

$$(3x^2 + x - 2) \cdot (3x - 2) = 9x^3 - 3x^2 - 8x + 4$$

Now we are going to study division of polynomials. Occasionally the division is exact.

$$(3x^2 + x - 2) \div (3x - 2) = \frac{(3x - 2)(x + 1)}{3x - 2} = x + 1$$

We say $3x - 2$ is an exact divisor, or factor, of $3x^2 + x - 2$. More often than not, the division is inexact and there is a nonzero remainder.

THE DIVISION ALGORITHM

The opening display shows that the process of division for polynomials is much the same as for integers. Both processes (we call them *algorithms*) involve subtraction. The first one shows how many times 17 can be subtracted from 458 before obtaining a remainder less than 17. The answer is 26 times, with a remainder of 16. The second algorithm shows how many times $2x^2 - x + 3$ can be subtracted from $6x^3 - 5x^2 + x - 4$ before obtaining a remainder of lower degree than $2x^2 - x + 3$. The answer is $3x - 1$ times, with a remainder of $-9x - 1$. Thus in these two examples we have

$$458 - 26(17) = 16$$

$$(6x^3 - 5x^2 + x - 4) - (3x - 1)(2x^2 - x + 3) = -9x - 1$$

Of course, we can also write these equalities as

$$458 = (17)(26) + 16$$

$$6x^3 - 5x^2 + x - 4 = (2x^2 - x + 3)(3x - 1) + (-9x - 1)$$

Notice that both of them can be summarized in the words

$$\text{Dividend} = (\text{divisor}) \cdot (\text{quotient}) + \text{remainder}$$

This statement is very important and is worth stating again for polynomials in very precise language.

THE DIVISION LAW FOR POLYNOMIALS

If $P(x)$ and $D(x)$ are any two nonconstant polynomials, then there are unique polynomials $Q(x)$ and $R(x)$ such that

$$P(x) = D(x)Q(x) + R(x)$$

where $R(x)$ is either zero or it is of lower degree than $D(x)$.

Here you should think of $P(x)$ as the dividend, $D(x)$ as the divisor, $Q(x)$ as the quotient, and $R(x)$ as the remainder.

The algorithm we use to find $Q(x)$ and $R(x)$ was illustrated in the opening display. Here is another illustration, this time for polynomials with some nonreal coefficients. Notice that we always arrange the divisor and dividend in descending powers of x before we start the division process.

$$
\begin{array}{r}
x - i \\
3x + i \overline{\smash{\big)}\ 3x^2 - 2ix + 7} \\
\underline{3x^2 + ix} \\
-3ix + 7 \\
\underline{-3ix + 1} \\
6
\end{array}
\qquad
\begin{array}{l}
Q(x) = x - i \\
R(x) = 6
\end{array}
$$

SYNTHETIC DIVISION

It is often necessary to divide by polynomials of the form $x - c$. For such a division, there is a shortcut called *synthetic division*. We illustrate how it works for

$$(2x^3 - x^2 + 5) \div (x - 2)$$

Certainly the result depends on the coefficients; the powers of x serve mainly to determine the placement of the coefficients. Below, we show the division in its usual form and then in a skeletal form with the x's omitted. Note that we leave a blank space for the missing first degree term in the long division but indicate it with a 0 in the skeletal form.

$$2x^2 + 3x + 6$$

$$x - 2 \overline{\smash{\big)}\, 2x^3 - x^2 \quad\quad + 5}$$
$$\underline{2x^3 - 4x^2}$$
$$3x^2$$
$$\underline{3x^2 - 6x}$$
$$6x + 5$$
$$\underline{6x - 12}$$
$$17$$

②   ③   ⑥
① − 2 │ 2   −1   0   5
        2   −4
        3   ⓪
        ③ − 6
            6   ⑤
            ⑥ − 12
                 17

We can condense things still more by discarding all of the circled digits. The coefficients of the quotient, 2, 3, and 6, the remainder, 17, and the numbers from which they were calculated remain. All of the important numbers appear in the diagram below, on the left. On the right, we show the final modification. There we have changed the divisor from -2 to 2 to allow us to do addition rather than subtraction at each stage.

$$-2 \,\underline{\big|\, \begin{array}{rrrr} 2 & -1 & 0 & 5 \\ & -4 & -6 & -12 \end{array}}$$
$$ \begin{array}{rrrr} 2 & 3 & 6 & 17 \end{array}$$

$$2 \,\underline{\big|\, \begin{array}{rrrr} 2 & -1 & 0 & 5 \\ & 4 & 6 & 12 \end{array}}$$
$$ \begin{array}{rrrr} 2 & 3 & 6 & 17 \end{array}$$

CAUTION

In synthetic division, be sure to insert a 0 for each missing term of the dividend.

The process shown in the final format is called **synthetic division.** We can describe it by a series of steps.

1. To divide by $x - 2$, use 2 as the synthetic divisor.
2. Write down the coefficients of the dividend. Be sure to write zeros for missing powers.
3. Bring down the first coefficient.
4. Follow the arrows, first multiplying by 2 (the divisor), then adding, multiplying the sum by 2, adding, and so on.
5. The last number in the third row is the remainder and the others are the coefficients of the quotient.

Here is another example. We use synthetic division to divide $3x^3 + x^2 - 15x - 5$ by $x + \frac{1}{3}$, that is, by $x - (-\frac{1}{3})$.

$$-\tfrac{1}{3} \,\underline{\big|\, \begin{array}{rrrr} 3 & 1 & -15 & -5 \\ & -1 & 0 & 5 \end{array}}$$
$$\phantom{-\tfrac{1}{3} \,\big|\,} \begin{array}{rrrr} 3 & 0 & -15 & 0 \end{array}$$

Since the remainder is 0, the division is exact. We conclude that

$$3x^3 + x^2 - 15x - 5 = (x + \tfrac{1}{3})(3x^2 - 15)$$

A rational expression (ratio of two polynomials) is said to be **proper** if the degree of its numerator is smaller than that of its denominator. Thus

$$\frac{x + 1}{x^2 - 3x + 2}$$

is a proper rational expression, but

$$\frac{2x^3}{x^2 - 3}$$

is improper. The division law, $P(x) = D(x)Q(x) + R(x)$, can be written as

$$\frac{P(x)}{D(x)} = Q(x) + \frac{R(x)}{D(x)}$$

It implies that any improper rational expression can be rewritten as the sum of a polynomial and a proper rational expression. For example,

$$\frac{2x^3}{x^2 - 3} = 2x + \frac{6x}{x^2 - 3}$$

a result we obtained by dividing $x^2 - 3$ into $2x^3$.

Problem Set 10-1

In Problems 1–8, find the quotient and the remainder if the first polynomial is divided by the second.

1. $x^3 - x^2 + x + 3; x^2 - 2x + 3$
2. $x^3 - 2x^2 - 7x - 4; x^2 + 2x + 1$
3. $6x^3 + 7x^2 - 18x + 15; 2x^2 + 3x - 5$
4. $10x^3 + 13x^2 + 5x + 12; 5x^2 - x + 4$
5. $4x^4 - x^2 - 6x - 9; 2x^2 - x - 3$
6. $25x^4 - 20x^3 + 4x^2 - 4; 5x^2 - 2x + 2$
7. $2x^5 - 2x^4 + 9x^3 - 12x^2 + 4x - 16; 2x^3 - 2x^2 + x - 4$
8. $3x^5 - x^4 - 8x^3 - x^2 - 3x + 12; 3x^3 - x^2 + x - 4$

In Problems 9–14, write each rational expression as the sum of a polynomial and a proper rational expression.

9. $\dfrac{x^3 + 2x^2 + 5}{x^2}$

10. $\dfrac{2x^3 - 4x^2 - 3}{x^2 + 1}$

11. $\dfrac{x^3 - 4x + 5}{x^2 + x - 2}$

12. $\dfrac{2x^3 + x - 8}{x^2 - x + 4}$

13. $\dfrac{2x^2 - 4x + 5}{x^2 + 1}$

14. $\dfrac{5x^3 - 6x + 11}{x^3 - x}$

In Problems 15–24, use synthetic division to find the quotient and remainder if the first polynomial is divided by the second.

15. $2x^3 - x^2 + x - 4;\ x - 1$ 16. $3x^3 + 2x^2 - 4x + 5;\ x - 2$

17. $3x^3 + 5x^2 + 2x - 10;\ x - 1$ 18. $2x^3 - 5x^2 + 4x - 4;\ x - 2$

19. $x^4 - 2x^2 - 1;\ x - 3$ 20. $x^4 + 3x^2 - 340;\ x - 4$

21. $x^3 + 2x^2 - 3x + 2;\ x + 1$ 22. $x^3 - x^2 + 11x - 1;\ x + 1$

23. $2x^4 + x^3 + 4x^2 + 7x + 4;\ x + \frac{1}{2}$

24. $2x^4 + x^3 + x^2 + 10x - 8;\ x - \frac{1}{2}$

EXAMPLE (Division by $x - (a + bi)$)   Find the quotient and remainder when $x^3 - 2x^2 - 6ix + 18$ is divided by $x - 2 - 3i$.

Solution.   Synthetic division works just fine even when some or all of the coefficients are nonreal. Since $x - 2 - 3i = x - (2 + 3i)$, we use $2 + 3i$ as the synthetic divisor.

$$
\begin{array}{r|rrrr}
2 + 3i & 1 & -2 & -6i & 18 \\
 & & 2 + 3i & -9 + 6i & -18 - 27i \\
\hline
 & 1 & 3i & -9 & -27i
\end{array}
$$

Quotient: $x^2 + 3ix - 9$; remainder: $-27i$.

Use synthetic division to find the quotient and remainder when the first polynomial is divided by the second.

☐ 25. $x^3 - 2x^2 + 5x + 30;\ x - 2 + 3i$

☐ 26. $2x^3 - 11x^2 + 44x + 35;\ x - 3 - 4i$

☐ 27. $x^4 - 17;\ x - 2i$

☐ 28. $x^4 + 18x^2 + 90;\ x - 3i$

MISCELLANEOUS PROBLEMS

29. Express $(2x^4 - x^3 - x^2 - 2)/(x^3 + 1)$ as a polynomial plus a proper rational expression.

30. Find the quotient and the remainder when $x^4 + 6x^3 - 2x^2 + 4x - 15$ is divided by $x^2 - 2x + 3$.

31. Find by inspection the quotient and the remainder when the first polynomial is divided by the second.
    (a) $2x^3 + 3x^2 - 11x + 9;\ x^2$
    (b) $2(x + 3)^2 + 10(x + 3) - 14;\ x + 3$
    (c) $(x - 4)^5 + x^2 + x + 1;\ (x - 4)^3$
    (d) $(x^2 + 3)^3 + 2x(x^2 + 3) + 4x - 1;\ x^2 + 3$

32. Use synthetic division to find the quotient and the remainder when the first polynomial is divided by the second.
    (a) $x^4 - 4x^3 + 29;\ x - 3$
    (b) $2x^4 - x^3 + 2x - 4;\ x + \frac{1}{2}$
    (c) $x^4 + 4x^3 + 4\sqrt{3}x^2 + 3\sqrt{3}x + 3\sqrt{3};\ x + \sqrt{3}$
    ☐ (d) $x^3 - (3 + 2i)x^2 + 10ix + 20 - 12i;\ x - 3$

33. Show that the second polynomial is a factor of the first and determine the other factor.

(a) $x^5 + x^4 - 16x - 16$; $x - 2$

(b) $x^5 + 32$; $x + 2$

(c) $x^4 - \frac{3}{2}x^3 + 3x^2 + 6x + 2$; $x + \frac{1}{2}$

☐ (d) $x^3 - 2ix^2 + x - 2i$; $x - 2i$

34. Use synthetic division to show that second polynomial is a factor of the first and determine the other factor. *Hint:* In (a), you will want to use the fact that $x^2 - 1 = (x - 1)(x + 1)$.

(a) $x^4 + x^3 - x - 1$; $x^2 - 1$

(b) $x^4 - x^3 + 2x^2 - 4x - 8$; $x^2 - x - 2$

(c) $x^4 + 2x^3 - 4x - 4$; $x^2 - 2$

☐ (d) $x^5 + x^4 - 16x - 16$; $x^2 + 4$

35. Find k so that the second polynomial is a factor of the first.

(a) $x^3 + x^2 - 10x + k$; $x - 4$

(b) $x^4 + kx + 10$; $x + 2$

(c) $k^2x^3 - 4kx + 4$; $x - 1$

36. Determine h and k so that both $x - 3$ and $x + 2$ are factors of $x^4 - x^3 + hx^2 + kx - 6$.

37. Determine a, b, and c so that $(x - 1)^3$ is a factor of $x^4 + ax^3 + bx^2 + cx - 4$.

38. **TEASER** Let a, b, c, and d be distinct integers and suppose that $x - a$, $x - b$, $x - c$, and $x - d$ are factors of the polynomial $p(x)$, which has integral coefficients. Prove that $p(n)$ is not a prime number for any integer n.

Young Scholar: And does $x^4 + 99x^3 + 21$ have a zero?

Carl Gauss: Yes.

Young Scholar: How about $\pi x^{67} - \sqrt{3}x^{19} + 4i$?

Carl Gauss: It does.

Young Scholar: How can you be sure?

Carl Gauss: When I was young, 22 I think, I proved that every non-constant polynomial, no matter how complicated, has at least one zero. Many people call it the *Fundamental Theorem of Algebra.*

Carl F. Gauss (1777–1855)
"The Prince of Mathematicians"

10-2   Factorization Theory for Polynomials

Our young scholar could have asked a harder question. How do you find the zeros of a polynomial? Even the eminent Gauss would have had trouble with that question. You see, it is one thing to know a polynomial has zeros; it is quite another thing to find them.

Even though it is a difficult task and one at which we will have only limited success, our goal for this and the next two sections is to develop methods for finding zeros of polynomials. Remember that a polynomial is an expression of the form

$$P(x) = a_n x^n + a_{n-1} x^{n-1} + \cdots + a_1 x + a_0$$

Unless otherwise specified, the coefficients (the a_i's) are allowed to be *complex* numbers. And by a **zero** of $P(x)$, we mean any complex number c (real or nonreal) such that $P(c) = 0$. The number c is also called a **solution,** or a **root,** of the equation $P(x) = 0$. Note the use of words: Polynomials have zeros, but polynomial equations have solutions.

THE REMAINDER AND FACTOR THEOREMS

Recall the division law from Section 10-1, which had as its conclusion

$$P(x) = D(x)Q(x) + R(x)$$

If $D(x)$ has the form $x - c$, this becomes

$$P(x) = (x - c)Q(x) + R$$

where R, which is of lower degree than $x - c$, must be a constant. This last equation is an identity; it is true for all values of x, including $x = c$. Thus

$$P(c) = (c - c)Q(c) + R = 0 + R$$

We have just proved an important result.

REMAINDER THEOREM

If a polynomial $P(x)$ is divided by $x - c$, then the constant remainder R is given by $R = P(c)$.

Here is a nice example. Suppose we want to know the remainder R when $P(x) = x^{1000} + x^{22} - 15$ is divided by $x - 1$. We could, of course, divide it out—but what a waste of energy, especially since we know the remainder theorem. From it we learn that

$$R = P(1) = 1^{1000} + 1^{22} - 15 = -13$$

Much more important than the mere calculation of remainders is a consequence called the *factor theorem*. Since $R = P(c)$, as we have just seen, we may rewrite the division law as

$$P(x) = (x - c)Q(x) + P(c)$$

It is plain to see that $P(c) = 0$ if and only if the division of $P(x)$ by $x - c$ is exact; that is, if and only if $x - c$ is a factor of $P(x)$.

FACTOR THEOREM

A polynomial $P(x)$ has c as a zero if and only if it has $x - c$ as a factor.

Sometimes it is easy to spot one zero of a polynomial. If so, the factor theorem may help us find the other zeros. Consider the polynomial

$$P(x) = 3x^3 - 8x^2 + 3x + 2$$

Notice that

$$P(1) = 3 - 8 + 3 + 2 = 0$$

so 1 is a zero. By the factor theorem, $x - 1$ is a factor of $P(x)$. We can use synthetic division to find the other factor.

$$
\begin{array}{r|rrrr}
1 & 3 & -8 & 3 & 2 \\
  &   & 3 & -5 & -2 \\
\hline
  & 3 & -5 & -2 & 0
\end{array}
$$

The remainder is 0 as we expected, and

$$P(x) = (x - 1)(3x^2 - 5x - 2)$$

Using the quadratic formula, we find the zeros of $3x^2 - 5x - 2$ to be $(5 \pm \sqrt{49})/6$, which simplify to 2 and $-\frac{1}{3}$. Thus $P(x)$ has 1, 2, and $-\frac{1}{3}$ as its three zeros.

COMPLETE FACTORIZATION OF POLYNOMIALS

In the example above, we did not really need the quadratic formula. If we had been clever, we would have factored $3x^2 - 5x - 2$.

$$3x^2 - 5x - 2 = (3x + 1)(x - 2)$$
$$= 3(x + \tfrac{1}{3})(x - 2)$$

Thus $P(x)$, our original polynomial, may be written as

$$P(x) = 3(x - 1)(x + \tfrac{1}{3})(x - 2)$$

from which all three of the zeros are immediately evident.

But now we make another key observation. Notice that $P(x)$ can be factored as a product of its leading coefficient and three factors of the form $(x - c)$, where the c's are the zeros of $P(x)$. This holds true in general.

COMPLETE FACTORIZATION THEOREM

If

$$P(x) = a_n x^n + a_{n-1} x^{n-1} + \cdots + a_1 x + a_0$$

is an nth degree polynomial with $n > 0$, then there are n numbers c_1, $c_2, \ldots, c_n$, not necessarily distinct, such that

$$P(x) = a_n (x - c_1)(x - c_2) \cdots (x - c_n)$$

The c's are the zeros of $P(x)$; they may or may not be real numbers.

To prove the complete factorization theorem, we must go back to Carl Gauss and our opening display. In his doctoral dissertation in 1799, Gauss gave a proof of the following important theorem, a proof that unfortunately is beyond the scope of this book.

FUNDAMENTAL THEOREM OF ALGEBRA

Every nonconstant polynomial has at least one zero.

Now let $P(x)$ be any polynomial of degree $n > 0$. By the fundamental theorem, it has a zero, which we may call c_1. By the factor theorem, $x - c_1$ is a factor of $P(x)$; that is,

$$P(x) = (x - c_1)P_1(x)$$

where $P_1(x)$ is a polynomial of degree $n - 1$ and with the same leading coefficient as $P(x)$—namely, a_n.

If $n - 1 > 0$, we may repeat the argument on $P_1(x)$. It has a zero c_2 and hence a factor $x - c_2$; that is,

$$P_1(x) = (x - c_2)P_2(x)$$

where $P_2(x)$ has degree $n - 2$. For our original polynomial $P(x)$, we may now write

$$P(x) = (x - c_1)(x - c_2)P_2(x)$$

Continuing in the pattern now established, we eventually get

$$P(x) = (x - c_1)(x - c_2) \cdots (x - c_n)P_n$$

where P_n has degree zero; that is, P_n is a constant. In fact, $P_n = a_n$, since the leading coefficient stayed the same at each step. This establishes the complete factorization theorem.

ABOUT THE NUMBER OF ZEROS

Each of the numbers c_i in

$$P(x) = a_n(x - c_1)(x - c_2) \cdots (x - c_n)$$

is a zero of $P(x)$. Are there any other zeros? No, for if d is any number different from each of the c_i's, then

$$P(d) = a_n(d - c_1)(d - c_2) \cdots (d - c_n) \neq 0$$

All of this tempts us to say that a polynomial of degree n has exactly n zeros. But hold on! The numbers $c_1, c_2, \ldots, c_n$ need not all be different. For example, the sixth degree polynomial

$$P(x) = 4(x - 2)^3(x + 1)(x - 4)^2$$

has only three distinct zeros, 2, -1, and 4. We have to settle for the following statement.

An nth degree polynomial has at most n distinct zeros.

There is a way in which we can say that there are exactly n zeros. Call c a **zero of multiplicity** k of $P(x)$ if $x - c$ appears k times in its complete factorization. For example, in

$$P(x) = 4(x - 2)^3(x + 1)(x - 4)^2$$

the zeros 2, -1, and 4 have multiplicities 3, 1, and 2, respectively. A zero of multiplicity 1 is called a **simple zero.** Notice in our example that the multiplicities add to 6, the degree of the polynomial. In general, we may say this.

> *An nth degree polynomial has exactly n zeros provided we count a zero of multiplicity k as k zeros.*

Problem Set 10-2

Use the remainder theorem to find P(c). Check your answer by substituting c for x.

1. $P(x) = 2x^3 - 5x^2 + 3x - 4$; $c = 2$
2. $P(x) = x^3 + 4x^2 - 11x - 5$; $c = 3$
3. $P(x) = 8x^4 - 3x^2 - 2$; $c = \frac{1}{2}$
4. $P(x) = 2x^4 + \frac{3}{4}x + \frac{3}{2}$; $c = -\frac{1}{2}$

Find the remainder if the first polynomial is divided by the second. Do it without actually dividing.

5. $x^{10} - 15x + 8$; $x - 1$
6. $2x^{20} + 5$; $x + 1$
7. $64x^6 + 13$; $x + \frac{1}{2}$
8. $81x^3 + 9x^2 - 2$; $x - \frac{1}{3}$

Find all of the zeros of the given polynomial and give their multiplicities.

9. $(x - 1)(x + 2)(x - 3)$
10. $(x + 2)(x + 5)(x - 7)$
11. $(2x - 1)(x - 2)^2 x^3$
12. $(3x + 1)(x + 1)^3 x^2$
13. $3(x - 1 - 2i)(x + \frac{2}{3})$
14. $5(x - 2 + \sqrt{5})(x - \frac{4}{5})$

In Problems 15–18, show that x − c is a factor of P(x).

15. $P(x) = 2x^3 - 7x^2 + 9x - 4$; $c = 1$
16. $P(x) = 3x^3 + 4x^2 - 6x - 1$; $c = 1$
17. $P(x) = x^3 - 7x^2 + 16x - 12$; $c = 3$
18. $P(x) = x^3 - 8x^2 + 13x + 10$; $c = 5$
19. In Problem 15, you know that $P(x)$ has 1 as a zero. Find the other zeros.
20. Find all of the zeros of $P(x)$ in Problem 16. Then factor $P(x)$ completely.
21. Find all of the zeros of $P(x)$ in Problem 17.
22. Find all of the zeros of $P(x)$ in Problem 18.

In Problems 23–26, factor the given polynomial into linear factors. You should be able to do it by inspection.

23. $x^2 - 5x + 6$

24. $2x^2 - 14x + 24$

25. $x^4 - 5x^2 + 4$

26. $x^4 - 13x^2 + 36$

In Problems 27–30, factor $P(x)$ into linear factors given that c is a zero of $P(x)$.

27. $P(x) = x^3 - 3x^2 - 28x + 60;\ c = 2$

28. $P(x) = x^3 - 2x^2 - 29x - 42;\ c = -2$

29. $P(x) = x^3 + 3x^2 - 10x - 12;\ c = -1$

30. $P(x) = x^3 + 11x^2 - 5x - 55;\ c = -11$

EXAMPLE A (Finding a Polynomial from Its Zeros)
(a) Find a cubic polynomial having simple zeros 3, $2i$, and $-2i$.
(b) Find a polynomial $P(x)$ with integral coefficients and having $\frac{1}{2}$ and $-\frac{2}{3}$ as simple zeros and 1 as a zero of multiplicity 2.

Solution.
(a) Let us call the required polynomial $P(x)$. Then

$$P(x) = a(x - 3)(x - 2i)(x + 2i)$$

where a can be any nonzero number. Choosing $a = 1$ and multiplying, we have

$$P(x) = (x - 3)(x^2 + 4) = x^3 - 3x^2 + 4x - 12$$

(b) $P(x) = a(x - \frac{1}{2})(x + \frac{2}{3})(x - 1)^2$
We choose $a = 6$ to eliminate fractions.

$$\begin{aligned} P(x) &= 6(x - \tfrac{1}{2})(x + \tfrac{2}{3})(x - 1)^2 \\ &= 2(x - \tfrac{1}{2})3(x + \tfrac{2}{3})(x - 1)^2 \\ &= (2x - 1)(3x + 2)(x^2 - 2x + 1) \\ &= 6x^4 - 11x^3 + 2x^2 + 5x - 2 \end{aligned}$$

CAUTION

$\cancel{(x - 3i)(x + 3i) = x^2 - 9}$

$(x - 3i)(x + 3i) = x^2 - 9i^2$
$\qquad\qquad\qquad = x^2 + 9$

In Problems 31–38, find a polynomial $P(x)$ with integral coefficients having the given zeros. Assume each zero to be simple (multiplicity 1) unless otherwise indicated.

31. 2, 1, and -4

32. 3, -2, and 5

33. $\frac{1}{2}$, $-\frac{5}{6}$

34. $\frac{3}{7}$, $\frac{3}{4}$

35. 2, $\sqrt{5}$, $-\sqrt{5}$

36. -3, $\sqrt{7}$, $-\sqrt{7}$

37. $\frac{1}{2}$ (multiplicity 2), -2 (multiplicity 3)

38. 0, -2, $\frac{3}{4}$ (multiplicity 3)

☐ *In Problems 39–42, find a polynomial $P(x)$ having only the indicated simple zeros.*

39. 2, -2, i, $-i$

40. $2i$, $-2i$, $3i$, $-3i$

41. 2, -5, $2 + 3i$, $2 - 3i$

42. -3, 2, $1 - 4i$, $1 + 4i$

EXAMPLE B (More on Zeros of Polynomials)   Show that 2 is a zero of multiplicity 3 of the polynomial

$$P(x) = 2x^5 - 17x^4 + 51x^3 - 58x^2 + 4x + 24$$

and find the remaining zeros.

Solution. We must show that $x - 2$ appears as a factor 3 times in the factored form of $P(x)$. Synthetic division can be used successively, as shown below.

The final quotient $2x^2 - 5x - 3$ factors as $(2x + 1)(x - 3)$. Therefore, the remaining two zeros are $-\frac{1}{2}$ and 3. The factored form of $P(x)$ is

$$2(x - 2)^3(x + \tfrac{1}{2})(x - 3)$$

43. Show that 1 is a zero of multiplicity 3 of the polynomial $x^5 + 2x^4 - 6x^3 - 4x^2 + 13x - 6$, find the remaining zeros, and factor completely.

44. Show that the polynomial $x^5 - 11x^4 + 46x^3 - 90x^2 + 81x - 27$ has 3 as a zero of multiplicity 3, find the remaining zeros, and factor completely.

45. Show that the polynomial

$$x^6 - 8x^5 + 7x^4 + 32x^3 + 31x^2 + 40x + 25$$

has -1 and 5 as zeros of multiplicity 2 and find the remaining zeros.

46. Show that the polynomial

$$x^6 + 3x^5 - 9x^4 - 50x^3 - 84x^2 - 72x - 32$$

has 4 as a simple zero and -2 as a zero of multiplicity 3. Find the remaining zeros.

EXAMPLE C (Factoring a Polynomial with a Given Nonreal Zero) Show that $1 + 2i$ is a zero of $P(x) = x^3 - (1 + 2i)x^2 - 4x + 4 + 8i$. Then factor $P(x)$ into linear factors.

Solution. We start by using synthetic division.

$$
\begin{array}{r|rrrr}
1 + 2i & 1 & -1 - 2i & -4 & 4 + 8i \\
 & & 1 + 2i & 0 & -4 - 8i \\
\hline
 & 1 & 0 & -4 & 0
\end{array}
$$

Therefore $1 + 2i$ is a zero and $x - 1 - 2i$ is a factor of $P(x)$, and

$$P(x) = (x - 1 - 2i)(x^2 - 4)$$
$$= (x - 1 - 2i)(x + 2)(x - 2)$$

☐ *In each of the following, factor P(x) into linear factors given that c is a zero of P(x).*

47. $P(x) = x^3 - 2ix^2 - 9x + 18i; c = 2i$

48. $P(x) = x^3 + 3ix^2 - 4x - 12i; c = -3i$

49. $P(x) = x^3 + (1 - i)x^2 - (1 + 2i)x - 1 - i; c = 1 + i$

50. $P(x) = x^3 - 3ix^2 + (3i - 3)x - 2 + 6i; c = -1 + 3i$

MISCELLANEOUS PROBLEMS

51. Find the remainder when the first polynomial is divided by the second.
    (a) $3x^{44} + 5x^{41} + 4; x + 1$
    (b) $1988x^3 - 1989x^2 + 1990x - 1991; x - 1$
    (c) $x^9 + 512; x + 2$

52. Prove that the second polynomial is a factor of the first.
    (a) $x^n - a^n; x - a$ (n a positive integer)
    (b) $x^n + a^n; x + a$ (n an odd positive integer)
    (c) $x^5 + 32a^{10}; x + 2a^2$

53. Find all zeros of the following polynomials and give their multiplicities.
    (a) $(x^2 - 4)^3$ (b) $(x^2 - 3x + 2)^2$ (c) $(x^2 + 2x - 4)^3(x + 2)^4$

54. Show that $2x^{44} + 3x^4 + 5$ has no factor of the form $x - c$, where c is real.

55. Each of the following polynomials has $\frac{1}{2}$ as a zero. Factor each polynomial into linear factors.
    (a) $12x^3 + 4x^2 - 3x - 1$
    (b) $2x^3 - x^2 - 4x + 2$
    ☐ (c) $2x^3 - x^2 + 2x - 1$

56. Find a third-degree polynomial $P(x)$ with integral coefficients that has the given numbers as zeros.
    (a) $\frac{3}{4}, 2, -\frac{2}{3}$ (b) $3, -2, -2$ (c) $3, \sqrt{2}$

57. Sketch the graph of $y = x^3 + 4x^2 - 2x$. Then find the three values of x for which $y = 8$. *Hint:* One of them is an integer.

☐ 58. Refer to the graph of Problem 57 and note that there is only one real x (namely, $x = 2$) for which $y = 20$. What does this tell you about the other two solutions of $x^3 + 4x^2 - 2x = 20$? Find these other two solutions.

59. A tray is to be constructed from a piece of sheet metal 16 inches square by cutting small identical squares of length x from the corners and then folding up the flaps (Figure 1). Determine x if the volume is to be 300 cubic inches. *Hint:* There are two possible answers, one of which is an integer.

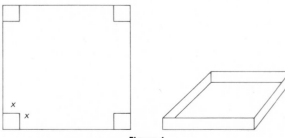

Figure 1

60. Find a polynomial $P(x) = ax^3 + bx^2 + cx + d$ that has 1, 2, and 3 as zeros and satisfies $P(0) = 36$.

61. Find $P(x) = ax^4 + bx^3 + cx^2 + dx + e$ if $P(x)$ has 1/2 as a zero of multiplicity 4 and $P(0) = 1$.

62. Write $x^5 + x^4 + x^3 + x^2 + x + 1$ as a product of linear and quadratic factors with integer coefficients.

63. Show that if $P(x) = a_nx^n + a_{n-1}x^{n-1} + a_{n-2}x^{n-2} + \cdots + a_1x + a_0$ has $n + 1$ distinct zeros, then all the coefficients must be 0.

64. Let

$$P_1(x) = x^n + a_{n-1}x^{n-1} + a_{n-2}x^{n-2} + \cdots + a_1x + a_0$$

and

$$P_2(x) = x^n + b_{n-1}x^{n-1} + b_{n-2}x^{n-2} + \cdots + b_1x + b_0$$

Show that if $P_1(x) = P_2(x)$ for n distinct values of x, then the two polynomials are equal for all x. *Hint:* Apply Problem 63 to $P(x) = P_1(x) - P_2(x)$.

65. Show that if c is zero of $x^6 - 5x^5 + 3x^4 + 7x^3 + 3x^2 - 5x + 1$, then $1/c$ is also a zero.

66. Generalize Problem 65 by showing that if c is a zero of

$$a_nx^n + a_{n-1}x^{n-1} + a_{n-2}x^{n-2} + \cdots + a_1x + a_0$$

where $a_k = a_{n-k}$ for $k = 0, 1, 2, \ldots, n$ and $a_n \neq 0$, then $1/c$ is also a zero.

67. Show that if $c_1, c_2, \ldots, c_n$ are the zeros of

$$a_nx^n + a_{n-1}x^{n-1} + a_{n-2}x^{n-2} + \cdots + a_1x + a_0$$

with $a_n \neq 0$, then
(a) $c_1 + c_2 + \cdots + c_n = -a_{n-1}/a_n$
(b) $c_1c_2 + c_1c_3 + \cdots + c_1c_n + c_2c_3 + \cdots + c_2c_n + \cdots + c_{n-1}c_n$
$= a_{n-2}/a_n$
(c) $c_1c_2 \cdots c_n = (-1)^n a_0/a_n$
Hint: Use the complete factorization theorem.

68. TEASER   Let the polynomial $x^n - kx - 1$, $n > 2$, have zeros $c_1, c_2, \ldots, c_n$. Show that
(a) If $n > 2$, then $c_1^n + c_2^n + \cdots + c_n^n = n$
(b) If $n > 3$, then $c_1^2 + c_2^2 + \cdots + c_n^2 = 0$

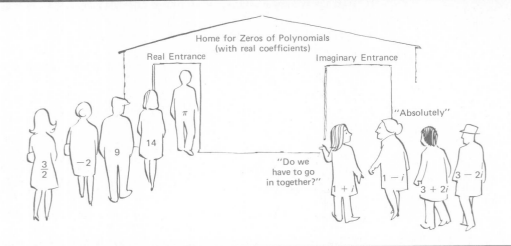

10-3 Polynomial Equations with Real Coefficients

Finding the zeros of a polynomial $P(x)$ is the same as finding the solutions of the equation $P(x) = 0$. We have solved polynomial equations before, especially in Chapter 3. In particular, recall that the two solutions of a quadratic equation with real coefficients are either both real or both nonreal. For example, the equation $x^2 - 7x + 12 = 0$ has the real solutions 3 and 4. On the other hand, $x^2 - 8x + 17 = 0$ has two nonreal solutions $4 + i$ and $4 - i$.

In this section, we shall see that if a polynomial equation has real coefficients, then its nonreal solutions (if any) must occur in pairs—conjugate pairs. In the opening display, $1 + i$ and $1 - i$ must enter together or not at all; so must $3 + 2i$ and $3 - 2i$.

PROPERTIES OF CONJUGATES

We indicate the conjugate of a complex number by putting a bar over it. If $u = a + bi$, then $\bar{u} = a - bi$. For example,

$$\overline{1 + i} = 1 - i$$
$$\overline{3 - 2i} = 3 + 2i$$
$$\overline{4} = 4$$
$$\overline{2i} = -2i$$

The operation of taking the conjugate behaves nicely in both addition and multiplication. There are two pertinent properties, stated first in words.

1. The conjugate of a sum is the sum of the conjugates.

2. The conjugate of a product is the product of the conjugates.

In symbols these properties become

1. $\overline{u_1 + u_2 + \cdots + u_n} = \overline{u}_1 + \overline{u}_2 + \overline{u}_3 + \cdots + \overline{u}_n$
2. $\overline{u_1 u_2 u_3 \cdots u_n} = \overline{u}_1 \cdot \overline{u}_2 \cdot \overline{u}_3 \cdots \overline{u}_n$

A third property follows from the second property if we set all u_i's equal to u.

3. $\overline{u^n} = (\overline{u})^n$

Rather than prove these properties, we shall illustrate them.

1. $\overline{(2 + 3i) + (1 - 4i)} = \overline{2 + 3i} + \overline{1 - 4i}$
2. $\overline{(2 + 3i)(1 - 4i)} = \overline{(2 + 3i)}\,\overline{(1 - 4i)}$
3. $\overline{(2 + 3i)^3} = \overline{(2 + 3i)}^3$

Let us check that the second statement is correct. We will do it by computing both sides independently.

$$\overline{(2 + 3i)(1 - 4i)} = \overline{2 + 12 + (-8 + 3)i} = \overline{14 - 5i} = 14 + 5i$$

$$\overline{(2 + 3i)}\,\overline{(1 - 4i)} = (2 - 3i)(1 + 4i) = 2 + 12 + (8 - 3)i = 14 + 5i$$

We can use the three properties of conjugates to demonstrate some very important results. For example, we can show that if u is a solution of the equation

$$ax^3 + bx^2 + cx + d = 0$$

where a, b, c, and d are real, then $\overline{u}$ is also a solution. To do this, we show

$$a\overline{u}^3 + b\overline{u}^2 + c\overline{u} + d = 0$$

whenever it is given that

$$au^3 + bu^2 + cu + d = 0$$

In the latter equation, take the conjugate of both sides, using the three properties of conjugates and the fact that a real number is its own conjugate.

$$\overline{au^3 + bu^2 + cu + d} = \overline{0}$$

$$\overline{au^3} + \overline{bu^2} + \overline{cu} + \overline{d} = 0$$

$$\overline{a}\,\overline{u}^3 + \overline{b}\,\overline{u}^2 + \overline{c}\,\overline{u} + \overline{d} = 0$$

$$a\overline{u}^3 + b\overline{u}^2 + c\overline{u} + d = 0$$

NONREAL SOLUTIONS OCCUR IN PAIRS

We are ready to state the main theorem of this section.

CONJUGATE PAIR THEOREM

Let

$$a_n x^n + a_{n-1} x^{n-1} + \cdots + a_1 x + a_0 = 0$$

be a polynomial equation with real coefficients. If u is a solution, its conjugate $\bar{u}$ is also a solution.

We feel confident that you will be willing to accept the truth of this theorem without further argument. The formal proof would mimic the proof given above for the cubic equation.

As an illustration of one use of this theorem, suppose that we know that $3 + 4i$ is a solution of the equation

$$x^3 - 8x^2 + 37x - 50 = 0$$

Then we know that $3 - 4i$ is also a solution. We can easily find the third solution, which incidentally must be real. (Why?) Here is how we do it.

$$
\begin{array}{r|cccc}
3 + 4i & 1 & -8 & 37 & -50 \\
& & 3 + 4i & -31 - 8i & 50 \\
\hline
3 - 4i & 1 & -5 + 4i & 6 - 8i & 0 \\
& & 3 - 4i & -6 + 8i & \\
\hline
& 1 & -2 & 0 &
\end{array}
$$

From this it follows that

$$x^3 - 8x^2 + 37x - 50 = [x - (3 + 4i)][x - (3 - 4i)][x - 2]$$

and that 2 is the third solution of our equation.

Notice something special about the product of the first two factors just displayed.

$$[x - (3 + 4i)][x - (3 - 4i)]$$
$$= x^2 - (3 + 4i)x - (3 - 4i)x + (3 + 4i)(3 - 4i)$$
$$= x^2 - 6x + 25$$

The product is a quadratic polynomial with *real* coefficients. This is not an accident. If u is any complex number, then

$$(x - u)(x - \bar{u}) = x^2 - (u + \bar{u})x + u\bar{u}$$

is a real quadratic polynomial, since both $u + \bar{u}$ and $u\bar{u}$ are real (see Problem 39). Thus the conjugate pair theorem, when combined with the complete factorization theorem of Section 10-2, has the following consequence.

REAL FACTORS THEOREM

Any polynomial with real coefficients can be factored into a product of linear and quadratic polynomials having real coefficients, where the quadratic polynomials have no real zeros.

RATIONAL SOLUTIONS

How does one get started on solving an equation of high degree? So far, all we can suggest is to guess. If you are lucky and find a solution, you can use synthetic division to reduce the degree of the equation to be solved. Eventually you may get it down to a quadratic equation, for which we have the quadratic formula.

Guessing would not be so bad if there were not so many possibilities to consider. Is there an intelligent way to guess? There is, but unfortunately it works only if the coefficients are integers, and then it only helps us find rational solutions.

Consider

$$3x^3 + 13x^2 - x - 6 = 0$$

which, as you will note, has integral coefficients. Suppose it has a rational solution c/d which is in reduced form (that is, c and d are integers without common divisors greater than 1 and $d > 0$). Then

$$3 \cdot \frac{c^3}{d^3} + 13 \cdot \frac{c^2}{d^2} - \frac{c}{d} - 6 = 0$$

or, after multiplying by d^3,

$$3c^3 + 13c^2d - cd^2 - 6d^3 = 0$$

We can rewrite this as

$$c(3c^2 + 13cd - d^2) = 6d^3$$

and also as

$$d(13c^2 - cd - 6d^2) = -3c^3$$

The first of these tells us that c divides $6d^3$ and the second that d divides $-3c^3$. But c and d have no common divisors. Therefore, c must divide 6 and d must divide 3.

The only possibilities for c are ± 1, ± 2, ± 3, and ± 6; for d, the only possibilities are 1 and 3. Thus the possible rational solutions must come from the following list.

$$\frac{c}{d}: \quad \pm 1, \pm 2, \pm 3, \pm 6, \pm \tfrac{1}{3}, \pm \tfrac{2}{3}$$

Upon checking all 12 numbers (which takes time, but a bit less time than checking *all* numbers would take!) we find that only $\tfrac{2}{3}$ works.

$$
\begin{array}{r|rrrr}
\tfrac{2}{3} & 3 & 13 & -1 & -6 \\
& & 2 & 10 & 6 \\
\hline
& 3 & 15 & 9 & 0
\end{array}
$$

We could prove the following theorem by using similar reasoning.

RATIONAL SOLUTION THEOREM (RATIONAL ROOT THEOREM)

Let

$$a_n x^n + a_{n-1} x^{n-1} + \cdots + a_1 x + a_0 = 0$$

have integral coefficients. If c/d is a rational solution in reduced form, then c divides a_0 and d divides a_n.

Problem Set 10-3

⊡ *In Problems 1–10, write the conjugate of the number.*

1. $2 + 3i$

2. $3 - 5i$

3. $4i$

4. $-6i$

5. $4 + \sqrt{6}$

6. $3 - \sqrt{5}$

7. $(2 - 3i)^8$

8. $(3 + 4i)^{12}$

9. $2(1 + 2i)^3 - 3(1 + 2i)^2 + 5$

10. $4(6 - i)^4 + 11(6 - i) - 23$

⊡ 11. If $P(x)$ is a cubic polynomial with real coefficients and has -3 and $5 - i$ as zeros, what other zero does it have?

⊡ 12. If $P(x)$ is a cubic polynomial with real coefficients and has 0 and $\sqrt{2} + 3i$ as zeros, what other zero does it have?

⊡ 13. Suppose that $P(x)$ has real coefficients and is of the fourth degree. If it has $3 - 2i$ and $5 + 4i$ as two of its zeros, what other zeros does it have?

⊡ 14. If $P(x)$ is a fourth degree polynomial with real coefficients and has $5 + 6i$ as a zero of multiplicity 2, what are its other zeros?

EXAMPLE A (Solving an Equation Given Some Solutions) Given that -1 and $1 + 2i$ are solutions of the equation

$$2x^4 - 5x^3 + 9x^2 + x - 15 = 0$$

find the other solutions.

Solution. Since the coefficients are real, $1 - 2i$ is a solution. The fourth solution is found by progressively using synthetic division.

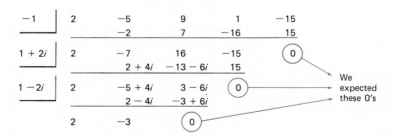

Last quotient: $2x - 3$

The 4th solution: $\dfrac{3}{2}$

□ *In Problems 15–18, one or more solutions of the specified equations are given. Find the other solutions.*

15. $2x^3 - x^2 + 2x - 1 = 0$; i
16. $x^3 - 3x^2 + 4x - 12 = 0$; $2i$
17. $x^4 + x^3 + 6x^2 + 26x + 20 = 0$; $1 + 3i$
18. $x^6 + 2x^5 + 4x^4 + 4x^3 - 4x^2 - 16x - 16 = 0$; $2i$, $-1 + i$

EXAMPLE B (Obtaining a Polynomial Given Some of Its Zeros)   Find a cubic polynomial with real coefficients that has 3 and $2 + 3i$ as zeros. Make the leading coefficient 1.

Solution.   The third zero has to be $2 - 3i$. In factored form, our polynomial is

$$(x - 3)(x - 2 - 3i)(x - 2 + 3i)$$

We multiply this out in stages.

$$(x - 3)[(x - 2)^2 + 9]$$
$$(x - 3)(x^2 - 4x + 13)$$
$$x^3 - 7x^2 + 25x - 39$$

□ *Find a polynomial with real coefficients that has the indicated degree and the given zero(s). Make the leading coefficient 1.*

19. Degree 2; zero: $2 + 5i$.
20. Degree 2; zero: $\sqrt{6}i$.
21. Degree 3; zeros: -3, $2i$.
22. Degree 3; zeros: 5, $-3i$.
23. Degree 5; zeros: 2, $3i$ (multiplicity 2).
24. Degree 5; zeros: 1, $1 - i$ (multiplicity 2).

EXAMPLE C (Finding Rational Solutions)   Find the rational solutions of the equation

$$3x^4 + 2x^3 + 2x^2 + 2x - 1 = 0$$

Then find the remaining solutions.

Solution.   The only way that c/d (in reduced form) can be a solution is for c to be 1 or -1 and for d to be 1 or 3. This means that the possibilities for c/d are ± 1 and $\pm \frac{1}{3}$. Synthetic division shows that -1 and $\frac{1}{3}$ work.

$$
\begin{array}{r|rrrrr}
-1 & 3 & 2 & 2 & 2 & -1 \\
   &   & -3 & 1 & -3 & 1 \\
\hline
\tfrac{1}{3} & 3 & -1 & 3 & -1 & 0 \\
   &   & 1 & 0 & 1 & \\
\hline
   & 3 & 0 & 3 & 0 &
\end{array}
$$

Setting the final quotient, $3x^2 + 3$, equal to zero and solving, we get

$$3x^2 = -3$$
$$x^2 = -1$$
$$x = \pm i$$

The complete solution set is $\{-1, \frac{1}{3}, i, -i\}$.

In Problems 25–30, find the rational solutions of each equation. If possible, find the other solutions.

25. $x^3 - 3x^2 - x + 3 = 0$
26. $x^3 + 3x^2 - 4x - 12 = 0$
27. $2x^3 + 3x^2 - 4x + 1 = 0$
28. $5x^3 - x^2 + 5x - 1 = 0$
29. $\frac{1}{3}x^3 - \frac{1}{2}x^2 - \frac{1}{6}x + \frac{1}{6} = 0$ *Hint:* Clear the equation of fractions.
30. $\frac{2}{3}x^3 - \frac{1}{2}x^2 + \frac{2}{3}x - \frac{1}{2} = 0$

MISCELLANEOUS PROBLEMS

31. **The number $2 + i$ is one solution to $x^4 - 3x^3 + 2x^2 + x + 5 = 0$. Find the other solutions.**

32. Find the fourth degree polynomial with real coefficients and leading coefficient 1 that has 2, -4, and $2 - 3i$ as three of its zeros.

33. Find all solutions of $x^4 - 3x^3 - 20x^2 - 24x - 8 = 0$.

34. Find all solutions of $2x^4 - x^3 + x^2 - x - 1 = 0$.

35. Solve $x^5 + 6x^4 - 34x^3 + 56x^2 - 39x + 10 = 0$.

36. Solve $x^5 - 2x^4 + x - 2 = 0$. *Hint:* $x^4 + 1$ is easy to factor by adding and subtracting $2x^2$.

37. Show that a polynomial equation with real coefficients and of odd degree has at least one real solution.

38. A cubic equation with real coefficients has either 3 real solutions (not necessarily distinct), or 1 real solution and a pair of nonreal solutions. State all the possibilities for:
    (a) A fourth degree equation with real coefficients;
    (b) A fifth degree equation with real coefficients.

39. Show that if u is any complex number, then $u + \overline{u}$ and $u\overline{u}$ are real numbers.

40. Let u and v be complex numbers with $v \neq 0$. Show that $\overline{u/v} = \overline{u}/\overline{v}$. Use this to demonstrate that if $f(x)$ is a real rational function (a quotient of two polynomials with real coefficients) and if $f(a + bi) = c + di$, then $f(a - bi) = c - di$.

41. Write $x^4 + 3x^3 + 3x^2 - 3x - 4$ as a product of linear and quadratic factors with real coefficients as guaranteed by the real factors theorem.

42. Write $x^6 - 9x^5 + 38x^4 - 106x^3 + 181x^2 - 205x + 100$ as a product of linear and quadratic factors with real coefficients. *Hint:* $1 + 2i$ is a zero of multiplicity 2.

43. Write $x^8 - 1$ as a product of linear and quadratic factors with real coefficients. *Hint:* First factor as a difference of squares; then see the hint in Problem 36.

44. Let $x^n + a_{n-1}x^{n-1} + \cdots + a_1 x + a_0 = 0$ have integral coefficients.
    (a) Show that all real solutions are either integral or irrational.

(b) From (a), deduce that if m and n are positive integers and m is not a perfect nth power, then $\sqrt[n]{m}$ is irrational (in particular, $\sqrt[3]{3}$, $\sqrt[4]{17}$, and $\sqrt[5]{12}$ are irrational).

45. Find the exact value of $x = \sqrt[3]{\sqrt{5} - 2} - \sqrt[3]{\sqrt{5} + 2}$. *Hint:* Begin by showing that $x^3 + 3x + 4 = 0$.

46. TEASER Let (u, v, w) satisfy the following nonlinear system of equations.

$$u + v + w = 2$$
$$u^2 + v^2 + w^2 = 8$$
$$u^3 + v^3 + w^3 = 8$$

(a) Determine the cubic equation $x^3 + a_2 x^2 + a_1 x + a_0 = 0$ that has u, v, and w as its solutions. *Hint:* Problem 67 of Section 10–2 should be helpful.

(b) Solve this cubic equation thereby solving the system of equations.

Eighteen Months of Genius

The year and a half that began in January 1665 has been called the most fruitful period in the history of scientific thought. In that short time, Isaac Newton discovered the general binomial theorem, the differential and integral calculus, the theory of colors, and the laws of gravity.

"Nature and Nature's laws lay hid in night:
God said, let Newton be! and all was light."

Alexander Pope

Isaac Newton
1642–1727

10-4 The Method of Successive Approximations

Even with the theory developed so far, we are often unable to get started on solving an equation of high degree. Imagine being given a fifth degree equation whose true solutions are $\sqrt{2} + 5i$, $\sqrt{2} - 5i$, $\sqrt[3]{19}$, 1.597, and 3π. How would you ever find them? Nothing you have learned until now would be of much help.

There is a general method of solving problems known to all resourceful people. We call it "muddling through" or "trial and error." Given a cup of tea, we add sugar a bit at a time until it tastes just right. Given a stopper too large for a hole, we whittle it down until it fits. We change the solution a step at a time, continually improving the accuracy until we are satisfied. Mathematicians call it a **method of successive approximations.**

We explore two such methods in this section. The first is a graphical method; the second is Newton's algebraic method. Both are designed to find the real solutions of polynomial equations with real coefficients. Both require many computations. We suggest that you keep your pocket calculator handy.

METHOD OF SUCCESSIVE ENLARGEMENTS

Consider the equation

$$x^3 - 3x - 5 = 0$$

We begin by graphing

$$y = x^3 - 3x - 5$$

looking for the point (or points) where the graph crosses the *x*-axis (Figure 2). These points correspond to the real solutions.

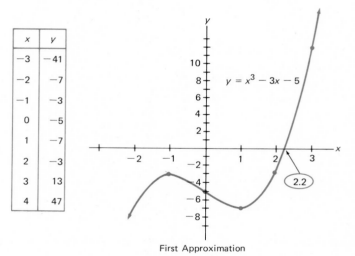

x	y
−3	−41
−2	−7
−1	−3
0	−5
1	−7
2	−3
3	13
4	47

First Approximation

Figure 2

Clearly there is only one real solution and it is between 2 and 3; a good first guess might be 2.2. Now we calculate *y* for values of *x* near 2.2 (for instance, 2.1, 2.2, and 2.3) until we find an interval of length .1 on which *y* changes sign. The interval is $2.2 \le x \le 2.3$. On this interval, we pretend the graph is a straight line. The point at which this line crosses the *x*-axis gives us our next approximation. It is about 2.28 (Figure 3).

Now we calculate *y* for values of *x* near 2.28 until we find an interval of length .01 on which *y* changes sign. This occurs on the interval $2.27 \le x \le 2.28$. Using a straight-line graph for this interval, we read our next approximation as 2.279 (Figure 4).

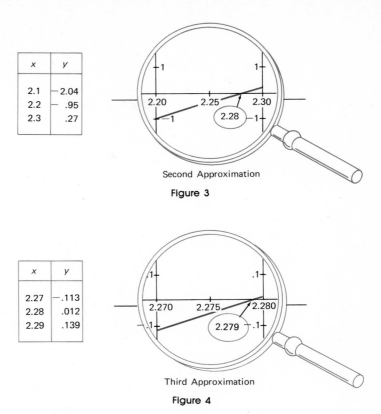

x	y
2.1	−2.04
2.2	−.95
2.3	.27

Second Approximation

Figure 3

x	y
2.27	−.113
2.28	.012
2.29	.139

Third Approximation

Figure 4

In effect, we are enlarging the graph (using a more and more powerful magnifying glass) at each stage, increasing the accuracy by one digit each time. We can continue this process as long as we have the patience to do the necessary calculations.

NEWTON'S METHOD

Let $P(x) = 0$ be a polynomial equation with real coefficients. Suppose that by some means (perhaps graphing), we discover that it has a real solution r which we guess to be about x_1. Then, as Figure 5 suggests, a better approximation to r is x_2, the point at which the tangent line to the curve at x_1 crosses the x-axis.

If the slope of the tangent line is m_1, then

$$m_1 = \frac{\text{rise}}{\text{run}} = \frac{P(x_1)}{x_1 - x_2}$$

Solving this for x_2 yields

$$x_2 = x_1 - \frac{P(x_1)}{m_1}$$

What has been done once can be repeated. A still better approximation to r would be x_3, obtained from x_2 as follows:

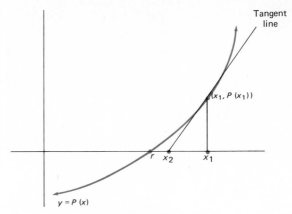

Figure 5

$$x_3 = x_2 - \frac{P(x_2)}{m_2}$$

where m_2 is the slope of the tangent line at $(x_2, P(x_2))$. In general, we find the $(k + 1)$st approximation from the kth by using Newton's formula

$$x_{k+1} = x_k - \frac{P(x_k)}{m_k}$$

where m_k is the slope of the tangent line at $(x_k, P(x_k))$.

The rub, of course, is that we do not know how to calculate m_k. That is precisely where Newton made his biggest contribution. He showed how to find the slope of the tangent line to any curve. This is done in every calculus course; we state one result without proof.

SLOPE THEOREM

If

$$P(x) = a_n x^n + a_{n-1} x^{n-1} + \cdots + a_2 x^2 + a_1 x + a_0$$

is a polynomial with real coefficients, then the slope of the tangent line to the graph of $y = P(x)$ at x is $P'(x)$, where $P'(x)$ is the polynomial

$$P'(x) = n a_n x^{n-1} + (n - 1) a_{n-1} x^{n-2} + \cdots + 2 a_2 x + a_1$$

For example, if

$$P(x) = 2x^4 + 4x^3 - 6x^2 + 2x + 15$$

then

$$P'(x) = 4 \cdot 2 \cdot x^3 + 3 \cdot 4 \cdot x^2 - 2 \cdot 6 \cdot x + 2$$
$$= 8x^3 + 12x^2 - 12 \cdot x + 2$$

In particular, the slope of the tangent line at $x = 2$ is $P'(2) = 8 \cdot 2^3 + 12 \cdot 2^2 - 12 \cdot 2 + 2 = 90$.

Taking the slope theorem for granted, we may write Newton's formula in the useful form

$$x_{k+1} = x_k - \frac{P(x_k)}{P'(x_k)}$$

USING NEWTON'S METHOD

Consider the equation

$$x^3 - 3x - 5 = 0$$

again. If we let

$$P(x) = x^3 - 3x - 5$$

then, by the slope theorem,

$$P'(x) = 3x^2 - 3$$

and Newton's formula becomes

$$x_{k+1} = x_k - \frac{x_k^3 - 3x_k - 5}{3x_k^2 - 3}$$

If we take $x_1 = 3$ as our initial guess, then

$$x_2 = x_1 - \frac{x_1^3 - 3x_1 - 5}{3x_1^2 - 3}$$

$$= 3 - \frac{3^3 - 3 \cdot 3 - 5}{3 \cdot 3^2 - 3}$$

$$\approx \boxed{2.5}$$

$$x_3 = x_2 - \frac{x_2^3 - 3x_2 - 5}{3x_2^2 - 3}$$

$$= 2.5 - \frac{(2.5)^3 - 3(2.5) - 5}{3(2.5)^2 - 3}$$

$$\approx \boxed{2.30}$$

$$x_4 = x_3 - \frac{x_3^3 - 3x_3 - 5}{3x_3^2 - 3}$$

$$= 2.30 - \frac{(2.30)^3 - 3(2.30) - 5}{3(2.30)^2 - 3}$$

$$\approx \boxed{2.279}$$

We can continue this repetitive process until we have the accuracy we desire.

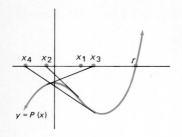

Figure 6

When you use Newton's method, it is important to make your initial guess reasonably good. Figure 6 shows how badly the method can lead you astray if you choose x_1 too far off the mark.

Problem Set 10-4

In Problems 1–6, use the slope theorem to find the slope polynomial $P'(x)$ if $P(x)$ is the given polynomial. Then find the slope of the tangent line at $x = 1$.

1. $2x^2 - 5x + 6$
2. $3x^2 + 4x - 9$
3. $2x^2 + x - 2$
4. $x^3 - 5x + 8$
5. $2x^5 + x^4 - 2x^3 + 8x - 4$
6. $x^6 - 3x^4 + 7x^3 - 4x^2 + 5x - 4$

Each of the equations in Problems 7–10 has exactly one real solution. By means of a graph, make an initial guess at the solution. Then use the method of successive enlargements to find a second and a third approximation.

7. $x^3 + 2x - 5 = 0$
8. $x^3 + x - 32 = 0$
9. $x^3 - 3x - 10 = 0$
10. $2x^3 - 6x - 15 = 0$

11. Use Newton's method to approximate the real solution of the equation $x^3 + 2x - 5 = 0$. Take as x_1 the initial guess you made in Problem 7, and then find x_2, x_3, and x_4.

12–14. Follow the instructions of Problem 11 for the equations in Problems 8–10.

EXAMPLE (Finding All Real Solutions by Newton's Method) Find all real solutions of the following equation by Newton's method.

$$P(x) = x^4 - 8x^3 + 22x^2 - 24x + 6 = 0$$

Solution. First we sketch the graph of $P(x)$ (Figure 7).

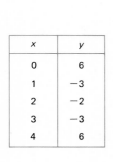

x	y
0	6
1	-3
2	-2
3	-3
4	6

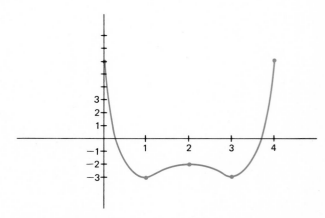

Figure 7

The graph crosses the x-axis at approximately .4 and 3.6. The slope polynomial is

$$P'(x) = 4x^3 - 24x^2 + 44x - 24$$

Take $x_1 = \boxed{4}$.

$$x_2 = .4 - \frac{P(.4)}{P'(.4)}$$

$$= .4 - \frac{(-.57)}{(-9.98)} \approx \boxed{.34}$$

$$x_3 = .34 - \frac{P(.34)}{P'(.34)}$$

$$= .34 - \frac{.0821}{(-4.657)} \approx \boxed{.347}$$

Take $x_1 = \boxed{3.6}$.

$$x_2 = 3.6 - \frac{P(3.6)}{P'(3.6)}$$

$$= 3.6 - \frac{(-.566)}{9.98} \approx \boxed{3.66}$$

$$x_3 = 3.66 - \frac{P(3.66)}{P'(3.66)}$$

$$= 3.66 - \frac{.0821}{11.66} \approx \boxed{3.653}$$

Each equation in Problems 15–18 has several real solutions. Draw a graph to get your first estimates. Then use Newton's method to find these solutions to three decimal places.

ⓒ 15. $x^4 + x^3 - 3x^2 + 4x - 28 = 0$
ⓒ 16. $x^4 + x^3 - 6x^2 - 7x - 7 = 0$
ⓒ 17. $x^3 - 3x + 1 = 0$
ⓒ 18. $x^3 - 12x + 1 = 0$

MISCELLANEOUS PROBLEMS

ⓒ 19. The line $y = x + 3$ intersects the curve $y = x^3 - 3x + 4$ at three points. Use Newton's method to find the x-coordinates of these points correct to 2 decimal places.

ⓒ 20. A spherical shell has a thickness of 1 centimeter. What is the outer radius r of the shell if the volume of the shell is equal to the volume of the space inside? First find an equation for r and then solve it by Newton's method.

ⓒ 21. The problem of the dog running around a marching column of soldiers (Problem 88 of Section 3-4) led to the equation $x^4 - 4x^3 - 2x^2 + 4x + 5 = 0$. Use Newton's method to find the solution of this equation that is near 4, correct to four decimal places. Multiply this answer by 50 to get the answer to the question asked in that earlier problem.

ⓒ 22. The dimensions of a rectangular box are 6, 8, and 10 feet. Suppose that the volume of the box is increased by 300 cubic feet by equal elongations of the three dimensions. Find this elongation correct to two decimal places.

ⓒ 23. What rate of interest compounded annually is implied in an offer to sell a house for $50,000 cash, or in annual installments of $20,000 each payable 1, 2, and 3 years from now? *Hint:* The amount of $50,000 with interest for 3 years should equal the sum of the first payment accumulated for 2 years, the second accumulated for 1 year, and the third payment. Hence, if i is the interest rate, $50,000(1 + i)^3 = 20,000(1 + i)^2 + 20,000(1 + i) + 20,000$. Dividing by 10,000 and writing x for $1 + i$ gives the equation $5x^3 = 2x^2 + 2x + 2$.

24. Find the rate of interest implied if a house is offered for sale at $80,000 cash or in 4 annual installments of $23,000, the first payable now. (See Problem 23.)

25. Find the equation of the tangent line to the graph of the equation $y = 3x^2$ at the point (2, 12). *Hint:* Remember that the line through (2, 12) with slope m has equation $y - 12 = m(x - 2)$. Find m by evaluating the slope polynomial at $x = 2$.

26. Find the equation of the tangent line to the curve
    (a) $y = x^2 + x$ at the point (2, 6);
    (b) $y = 2x^3 - 4x + 5$ at the point (−1, 7);
    (c) $y = \frac{1}{5}x^5$ at the point $(2, \frac{32}{5})$.

    See the hint for Problem 25.

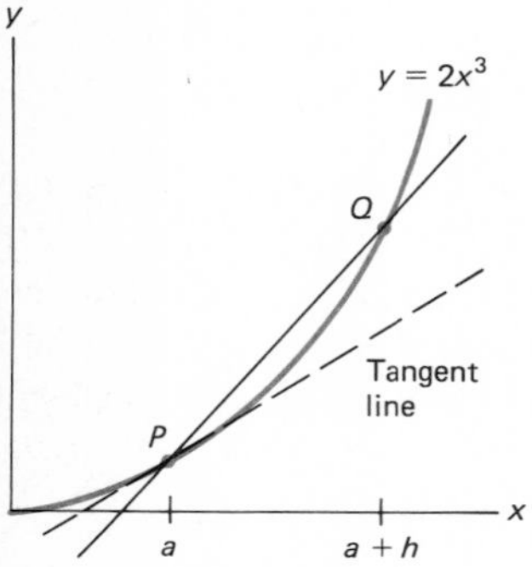

27. The points where the tangent line to a graph is horizontal are of great significance since among them are found the high and low points of the graph. Determine the x-coordinates of such points for the graphs of the following polynomials.
    (a) $P(x) = 2x^3 - 3x^2 - 36x + 10$ (b) $P(x) = 3x^4 - 8x^3 - 6x^2 + 9$

28. A box is to be made from a rectangular piece of sheet metal 24 inches long and 9 inches wide by cutting out identical squares of side x inches from each of the four corners and turning up the sides. Find the value of x that makes the volume of the box a maximum. What is this maximum volume? *Hint:* See Problem 27.

29. We stated the slope theorem without proof, but here we give a hint of its derivation (a subject treated at great length in calculus). Consider the graph of $y = 2x^3$, a part of which is shown in Figure 8. We are interested in finding the slope of the tangent line at the point P with x-coordinate a. Consider P to be a fixed point and let Q be a neighboring movable point with x-coordinate $a + h$. Note that as h tends to 0, the line PQ rotates toward the tangent line.
    (a) Show that the slope of the line PQ is m_h, where

$$m_h = \frac{2(a + h)^3 - 2a^3}{h}$$

    (b) Simplify the above expression for m_h as much as possible.
    (c) Find the limiting value of m_h as h tends to 0.
    (d) Compare the answer obtained in (c) with the value of the slope at $x = a$ given by the slope theorem.

30. TEASER Let $P(x)$ be a polynomial of degree at least 3. From the division algorithm of Section 10-1, we may write

$$P(x) = (x - a)Q_1(x) + R_1$$
$$Q_1(x) = (x - a)Q_2(x) + R_2$$

from which we conclude that

$$P(x) = (x - a)^2 Q_2(x) + R_2(x - a) + R_1$$

We know that $R_1 = P(a)$, but what is the meaning of R_2? Note that R_2 is the constant remainder we get after dividing $P(x)$ and then its quotient $Q_1(x)$ by $x - a$, a number found easily by a repeated synthetic division.
    (a) Carry out this double synthetic division on $P(x) = 2x^3$ to obtain $R_2 = 6a^2$.
    (b) Find R_2 for the polynomial $P(x) = x^4 - 2x^3 + x^2 - x + 1$.
    (c) Make a conjecture based on (a) and (b).
    (d) Give an argument to support your conjecture.

Figure 8

y = 2x³

Q

Tangent
line

P

a   a + h

Chapter Summary

The **division law for polynomials** asserts that if $P(x)$ and $D(x)$ are any given nonconstant polynomials, then there are unique polynomials $Q(x)$ and $R(x)$ such that

$$P(x) = D(x)Q(x) + R(x)$$

where $R(x)$ is either 0 or of lower degree than $D(x)$. In fact, we can find $Q(x)$ and $R(x)$ by the **division algorithm,** which is just a fancy name for ordinary long division. When $D(x)$ has the form $x - c$, $R(x)$ will have to be a constant R, since it is of lower degree than $D(x)$. The substitution $x = c$ then gives

$$P(c) = R$$

a result known as the **remainder theorem.** An immediate consequence is the **factor theorem,** which says that c is a zero of $P(x)$ if and only if $x - c$ is a factor of $P(x)$. Division of a polynomial by $x - c$ can be greatly simplified by use of **synthetic division.**

That every nonconstant polynomial has at least one zero is guaranteed by Gauss's **fundamental theorem of algebra.** But we can say much more than that. For any nonconstant polynomial

$$P(x) = a_n x^n + a_{n-1} x^{n-1} + \cdots + a_1 x + a_0$$

there are n numbers $c_1, c_2, \ldots, c_n$ (not necessarily all different) such that

$$P(x) = a_n(x - c_1)(x - c_2) \cdots (x - c_n)$$

We call the latter result the **complete factorization theorem.**

If the polynomial equation

$$P(x) = a_n x^n + a_{n-1} x^{n-1} + \cdots + a_1 x + a_0 = 0$$

has real coefficients, then its nonreal solutions (if any) must occur in conjugate pairs $a + bi$ and $a - bi$. If the coefficients are integers and if c/d is a rational solution in reduced form, then c divides a_0 and d divides a_n.

To find exact solutions to a polynomial equation may be very difficult; often we are more than happy to find good approximations. Two good methods for doing this are the **method of successive enlargements** and **Newton's method.** Both require plenty of calculating power.

Chapter Review Problem Set

1. Find the quotient and remainder if the first polynomial is divided by the second.
   (a) $2x^3 - x^2 + 4x - 5$; $x^2 + 2x - 3$
   (b) $x^4 - 8x^2 + 5$; $x^2 + 3x$

2. Use synthetic division to find the quotient and remainder if the first polynomial is divided by the second.
   (a) $x^3 - 2x^2 - 4x + 7; x - 2$
   (b) $2x^4 - 15x^2 + 4x - 3; x + 3$
   ⊡ (c) $x^3 + (3 - 3i)x^2 - (9 + 15i)x - 3 - 3i; x - 2 - 3i$
3. Without dividing, find the remainder if $2x^4 - 6x^3 + 17$ is divided by $x - 2$; if it is divided by $x + 2$.
⊡ 4. Find the zeros of the given polynomial and give their multiplicities.
   (a) $(x^2 - 1)^2(x^2 + 1)$
   (b) $x(x^2 - 2x + 4)(x + \pi)^3$

In Problems 5 and 6, use synthetic division to show that $x - c$ is a factor of $P(x)$. Then factor $P(x)$ completely into linear factors.

5. $P(x) = 2x^3 - x^2 - 18x + 9; c = 3$
6. $P(x) = x^3 + 4x^2 - 7x - 28; c = -4$

In Problems 7 and 8, find a polynomial $P(x)$ with integral coefficients that has the given zeros. Assume each zero to be simple unless otherwise indicated.

7. $3, -2, 4$ (multiplicity 2)
⊡ 8. $3 + \sqrt{7}, 3 - \sqrt{7}, 2 - i, 2 + i$
9. Show that 1 is a zero of multiplicity 2 of the polynomial $x^4 - 4x^3 - 3x^2 + 14x - 8$ and find the remaining zeros.
10. Find the polynomial $P(x) = a_3x^3 + a_2x^2 + a_1x + a_0$ which has zeros $\frac{1}{2}$, $-\frac{1}{3}$, and 4 and for which $P(2) = -42$.
11. Find the value of k so that $\sqrt{2}$ is a zero of $x^3 + 3x^2 - 2x + k$.
⊡ 12. Find a cubic polynomial with real coefficients that has $4 + 3i$ and -2 as two of its zeros.
⊡ 13. Solve the equation $x^4 - 4x^3 + 24x^2 + 20x - 145 = 0$, given that $2 + 5i$ is one of its solutions.
14. The equation $2x^3 - 15x^2 + 20x - 3 = 0$ has a rational solution. Find it and then find the other solutions.
15. The equation $x^3 - x^2 - x - 7 = 0$ has at least one real solution. Why? Show that it has no rational solution.
ⓒ 16. The equation $x^3 - 6x + 6 = 0$ has exactly one real solution. By means of a graph, make an initial guess at the solution. Then use the method of successive enlargements to find a second and a third approximation.
ⓒ 17. Use Newton's method to approximate the real solution of the equation in Problem 16. Take as x_1 the initial guess you made in Problem 16, and find x_2 and x_3.

Geometry may sometimes appear to take the lead over analysis but in fact precedes it only as a servant goes before the master to clear the path and light him on his way.

—*James Joseph Sylvester*

CHAPTER 11

Systems of Equations and Inequalities

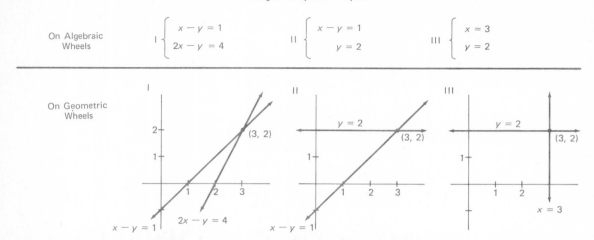

Moving from System to System

On Algebraic
Wheels

I $\begin{cases} x - y = 1 \\ 2x - y = 4 \end{cases}$ II $\begin{cases} x - y = 1 \\ y = 2 \end{cases}$ III $\begin{cases} x = 3 \\ y = 2 \end{cases}$

On Geometric
Wheels

11-1   Equivalent Systems of Equations

In Section 3-3, you learned how to solve a system of two equations in two unknowns, but you probably did not think of the process as one of replacing the given system by another having the same solutions. It is this point of view that we now want to explore.

In the display above, system I is replaced by system II, which is simpler; system II is in turn replaced by system III, which is simpler yet. To go from system I to system II, we eliminated x from the second equation; we then substituted $y = 2$ in the first equation to get system III. What happened geometrically is shown in the bottom half of our display. Notice that the three pairs of lines have the same point of intersection $(3, 2)$.

Because the notion of changing from one system of equations to another having the same solutions is so important, we make a formal definition. We say that two systems of equations are **equivalent** if they have the same solutions.

OPERATIONS THAT LEAD TO EQUIVALENT SYSTEMS

Now we face a big question. What operations can we perform on a system without changing its solutions?

Operation 1 We can interchange the position of two equations.

Operation 2 We can multiply an equation by a nonzero constant, that is, we can replace an equation by a nonzero multiple of itself.

Operation 3 We can add a multiple of one equation to another, that is, we can replace an equation by the sum of that equation and a multiple of another.

Operation 3 is the workhorse of the set. We show how it is used in the example of the opening display.

$$\text{I} \quad \begin{cases} x - y = 1 \\ 2x - y = 4 \end{cases}$$

If we add -2 times the first equation to the second, we obtain

$$\text{II} \quad \begin{cases} x - y = 1 \\ y = 2 \end{cases}$$

We then add the second equation to the first. This gives

$$\text{III} \quad \begin{cases} x = 3 \\ y = 2 \end{cases}$$

This is one way to write the solution. Alternatively, we say that the solution is the ordered pair $(3, 2)$.

THE THREE POSSIBILITIES FOR A LINEAR SYSTEM

We are mainly interested in linear systems—that is, systems of linear equations—and we shall restrict our discussion to the case where there is the same number of equations as unknowns. There are three possibilities for the set of solutions: The set may be empty, it may have just one point, or it may have infinitely many points. These three cases are illustrated in Figure 1.

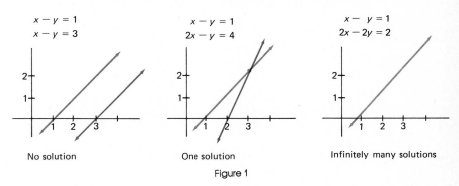

Figure 1

Someone is sure to object and ask if we cannot have a linear system with exactly two solutions or exactly three solutions. The answer is no. If a linear system has two solutions, it has infinitely many. This is obvious in the case of two equations in two unknowns, since two points determine a line, but it is also true for n equations in n unknowns.

LINEAR SYSTEMS IN MORE THAN TWO UNKNOWNS

When we consider large systems, it is a good idea to be very systematic about our method of attack. Our method is to reduce the system to **triangular form** and then use **back substitution.** Let us explain. Consider

$$x - 2y + z = -4$$
$$5y - 3z = 18$$
$$2z = -2$$

which is already in triangular form. (The name arises from the fact that the terms within the dotted triangle have zero coefficients.) This system is easy to solve. Solve the third equation first ($z = -1$). Then substitute that value in the second equation

$$5y - 3(-1) = 18$$

which gives $y = 3$. Finally substitute these two values in the first equation

$$x - 2(3) + (-1) = -4$$

which gives $x = 3$. Thus the solution of the system is $(3, 3, -1)$. This process, called **back substitution**, works on any linear system that is in triangular form. Just start at the bottom and work your way up.

If the system is not in triangular form initially, we try to operate on it until it is. Suppose we start with

$$2x - 4y + 2z = -8$$
$$2x + y - z = 10$$
$$3x - y + 2z = 4$$

Begin by multiplying the first equation by $\frac{1}{2}$ so that its leading coefficient is 1.

$$x - 2y + z = -4$$
$$2x + y - z = 10$$
$$3x - y + 2z = 4$$

Next add -2 times the first equation to the second equation. Also add -3 times the first equation to the third.

$$x - 2y + z = -4$$
$$5y - 3z = 18$$
$$5y - z = 16$$

Finally, add -1 times the second equation to the third.

$$x - 2y + z = -4$$
$$5y - 3z = 18$$
$$2z = -2$$

The system is now in triangular form and can be solved by back substitution. It is, in fact, the triangular system we discussed earlier. The solution is $(3, 3, -1)$.

In Examples A and B of the problem set, we show how this process works when there are infinitely many solutions and when there are no solutions.

Problem Set 11-1

Solve each of the following systems of equations.

1. $2x - 3y = 7$
   $y = -1$

2. $5x - 3y = -25$
   $y = 5$

3. $x = -2$
   $2x + 7y = 24$

4. $x = 5$
   $3x + 4y = 3$

5. $x - 3y = 7$
   $4x + y = 2$

6. $5x + 6y = 27$
   $x - y = 1$

7. $2x - y + 3z = -6$
   $2y - z = 2$
   $z = -2$

8. $x + 2y - z = -4$
   $3y + z = 2$
   $z = 5$

9. $3x - 2y + 5z = -10$
   $y - 4z = 8$
   $2y + z = 7$

10. $4x + 5y - 6z = 31$
    $y - 2z = 7$
    $5y + z = 2$

11. $x + 2y + z = 8$
    $2x - y + 3z = 15$
    $-x + 3y - 3z = -11$

12. $x + y + z = 5$
    $-4x + 2y - 3z = -9$
    $2x - 3y + 2z = 5$

13. $x - 2y + 3z = 0$
    $2x - 3y - 4z = 0$
    $x + y - 4z = 0$

14. $x + 4y - z = 0$
    $-x - 3y + 5z = 0$
    $3x + y - 2z = 0$

15. $x + y + z + w = 10$
    $y + 3z - w = 7$
    $x + y + 2z = 11$
    $x - 3y + w = -14$

16. $2x + y + z = 3$
    $y + z + w = 5$
    $4x + z + w = 0$
    $3y - z + 2w = 0$

EXAMPLE A (Infinitely Many Solutions)   Solve

$$x - 2y + 3z = 10$$
$$2x - 3y - z = 8$$
$$4x - 7y + 5z = 28$$

Solution.   Using Operation 3, we may eliminate x from the last two equations. First add -2 times the first equation to the second equation. Then add -4 times the first equation to the third equation. We obtain

$$x - 2y + 3z = 10$$
$$y - 7z = -12$$
$$y - 7z = -12$$

Next add -1 times the second equation to the third equation.

$$x - 2y + 3z = 10$$
$$y - 7z = -12$$
$$0 = 0$$

Finally, we solve the second equation for y in terms of z and substitute that result in the first equation.

$$y = 7z - 12$$

$$x = 2y - 3z + 10 = 2(7z - 12) - 3z + 10 = 11z - 14$$

Notice that there are infinitely many solutions; we can give any value we like to z, calculate the corresponding x and y values, and come up with a solution. Here is the format we use to list all the solutions to the system.

$$x = 11z - 14$$

$$y = 7z - 12$$

$$z \quad \text{arbitrary}$$

We could also say that the set of solutions consists of all ordered triples of the form $(11z - 14, 7z - 12, z)$. Thus if $z = 0$, we get the solution $(-14, -12, 0)$; if $z = 2$, we get $(8, 2, 2)$. Of course, it does not have to be z that is arbitrary; in our example, it could just as well be x or y. For example, if we had arranged things so y is treated as the arbitrary variable, we would have obtained

$$x = \frac{11}{7}y + \frac{34}{7}$$

$$z = \frac{1}{7}y + \frac{12}{7}$$

$$y \quad \text{arbitrary}$$

Note that the solution corresponding to $y = 2$ is $(8, 2, 2)$, which agrees with one found above.

Solve each of the following systems. Some, but not all, have infinitely many solutions.

17. $\begin{aligned} x - 4y + z &= 18 \\ 2x - 7y - 2z &= 4 \\ 3x - 11y - z &= 22 \end{aligned}$

18. $\begin{aligned} x + y - 3z &= 10 \\ 2x + 5y + z &= 18 \\ 5x + 8y - 8z &= 48 \end{aligned}$

19. $\begin{aligned} x - 2y + 3z &= -2 \\ 3x - 6y + 9z &= -6 \\ -2x + 4y - 6z &= 4 \end{aligned}$

20. $\begin{aligned} -4x + y - z &= 5 \\ 4x - y + z &= -5 \\ -24x + 6y - 6z &= 30 \end{aligned}$

21. $\begin{aligned} 2x - y + 4z &= 0 \\ 3x + 2y - z &= 0 \\ 9x - y + 11z &= 0 \end{aligned}$

22. $\begin{aligned} x + 3y - 2z &= 0 \\ 2x + y + z &= 0 \\ y - z &= 0 \end{aligned}$

EXAMPLE B (No Solution)   Solve

$$x - 2y + 3z = 10$$

$$2x - 3y - z = 8$$

$$5x - 9y + 8z = 20$$

Solution. Using Operation 3, we eliminate x from the last two equations to obtain

$$x - 2y + 3z = 10$$
$$y - 7z = -12$$
$$y - 7z = -30$$

It is already apparent that the last two equations cannot get along with each other. Let us continue anyway, putting the system in triangular form by adding -1 times the second equation to the third.

$$x - 2y + 3z = 10$$
$$y - 7z = -12$$
$$0 = -18$$

This system has no solution; we say it is **inconsistent.**

Solve the following systems or show that they are inconsistent.

23. $x - 4y + z = 18$
    $2x - 7y - 2z = 4$
    $3x - 11y - z = 10$

24. $x + y - 3z = 10$
    $2x + 5y + z = 18$
    $5x + 8y - 8z = 50$

25. $x + 3y - 2z = 10$
    $2x + y + z = 4$
    $5y - 5z = 16$

26. $x - 2y + 3z = -2$
    $3x - 6y + 9z = -6$
    $-2x + 4y - 6z = 0$

EXAMPLE C (Nonlinear Systems)   Solve the following system of equations.

$$x^2 + y^2 = 25$$
$$x^2 + y^2 - 2x - 4y = 5$$

Solution. We can use the same operations on this system as we did on linear systems. Adding (-1) times the first equation to the second, we get

$$x^2 + y^2 = 25$$
$$-2x - 4y = -20$$

We solve the second equation for x in terms of y, substitute in the first equation, and then solve the resulting quadratic equation in y.

$$x = 10 - 2y$$
$$(10 - 2y)^2 + y^2 = 25$$
$$5y^2 - 40y + 75 = 0$$
$$y^2 - 8y + 15 = 0$$
$$(y - 3)(y - 5) = 0$$

From this we get $y = 3$ or 5. Substituting these values in the equation $x = 10 - 2y$ yields two solutions to the original system, $(4, 3)$ and $(0, 5)$.

Solve each of the following systems.

27. $x + 2y = 10$
    $x^2 + y^2 - 10x = 0$

28. $x + y = 10$
    $x^2 + y^2 - 10x - 10y = 0$

29. $x^2 + y^2 - 4x + 6y = 12$
    $x^2 + y^2 + 10x + 4y = 96$

30. $x^2 + y^2 - 16y = 45$
    $x^2 + y^2 + 4x - 20y = 65$

31. $y = 4x^2 - 2$
    $y = x^2 + 1$

32. $x = 3y^2 - 5$
    $x = y^2 + 3$

MISCELLANEOUS PROBLEMS

In Problems 33–40, solve the given system or show that it is inconsistent.

33. $2x - 3y = 12$
    $x + 4y = -5$

34. $2x + 5y = -2$
    $y = -\frac{2}{5}x + 3$

35. $2x - 3y = 6$
    $x + 2y = -4$

36. $x - y + 3z = 1$
    $3x - 2y + 4z = 0$
    $4x + 2y - z = 3$

37. $x + y + z = -4$
    $x + y + z = 17$
    $x + y + z = -2$

38. $.43x - .79y + 4.24z = .67$
    $3.61y - 9.74z = 2$
    $y + 1.22z = 1.67$

39. $x^2 + y^2 = 4$
    $x + 2y = 2\sqrt{5}$

40. $x - \log y = 1$
    $\log y^x = 2$

41. If the system $x + 2y = 4$ and $ax + 3y = b$ has infinitely many solutions, what are a and b?

42. Helen claims that she has \$4.40 in nickels, dimes, and quarters, that she has four times as many dimes as quarters, and that she has 40 coins in all. Is this possible? If so, determine how many coins of each kind she has.

43. A three-digit number equals 19 times the sum of its digits. If the digits are reversed, the resulting number is greater than the given number by 297. The tens digit exceeds the units digit by 3. Find the number.

44. Find the equation of the parabola $y = ax^2 + bx + c$ that goes through $(-1, 6)$, $(1, 0)$, and $(2, 3)$.

45. Find the equation and radius of the circle that goes through $(0, 0)$, $(4, 0)$, and $(\frac{72}{25}, \frac{96}{25})$. *Hint:* Writing the equation in the form $(x - h)^2 + (y - k)^2 = r^2$ is not the best way to start. Is there another way to write the equation of a circle?

46. Determine a, b, and c so that

$$\frac{-12x + 6}{(x - 1)(x + 2)(x - 3)} = \frac{a}{x - 1} + \frac{b}{x + 2} + \frac{c}{x - 3}$$

47. Determine a, b, and c so that $(x - 1)^3$ is a factor of $x^4 + ax^2 + bx + c$.

48. Find the dimensions of a rectangle whose diagonal and perimeter measure 25 and 62 meters, respectively.

49. A certain rectangle has an area of 120 square inches. Increasing the width by 4 inches and decreasing the length by 3 inches increases the area by 24 square inches. Find the dimensions of the original rectangle.

50. TEASER   The ABC company reported the following statistics about its employees.
    (a) Average length of service for all employees: 15.9 years.

(b) Average length of service for male employees: 16.5 years.

(c) Average length of service for female employees: 14.1 years.

(d) Average hourly wage for all employees: $21.40.

(e) Average hourly wage for male employees: $22.50.

(f) Number of male employees: 300

For reasons not stated, the company did not report the number of female employees nor their average hourly wage, but you can figure them out. Do so.

Arthur Cayley, lawyer, painter, mountaineer, Cambridge professor, but most of all creative mathematician, made his biggest contributions in the field of algebra. To him we owe the idea of replacing a linear system by its matrix.

$$\begin{array}{rcl} 2x + 3y - z &=& 1 \\ x + 4y - z &=& 4 \\ 3x + y + 2z &=& 5 \end{array} \qquad \begin{bmatrix} 2 & 3 & -1 & 1 \\ 1 & 4 & -1 & 4 \\ 3 & 1 & 2 & 5 \end{bmatrix}$$

Arthur Cayley (1821–1895)

11-2   Matrix Methods

Contrary to what many people think, mathematicians do not enjoy long, involved calculations. What they do enjoy is looking for shortcuts, for labor-saving devices, and for elegant ways of doing things. Consider the problem of solving a system of linear equations, which as you know can become very complicated. Is there any way to simplify and systematize this process? There is. It is the method of matrices (plural of matrix).

A **matrix** is just a rectangular array of numbers. One example is shown in our opening panel. It has 3 rows and 4 columns and is referred to as a 3×4 matrix. We follow the standard practice of enclosing a matrix in brackets.

AN EXAMPLE WITH THREE EQUATIONS

Look at our opening display again. Notice how we obtained the matrix from the system of equations. We just suppressed all the unknowns, the plus signs, and the equal signs, and supplied some 1's. We call this matrix the **matrix of the system.** We are going to solve this system, keeping track of what happens to the matrix as we move from step to step.

$$\begin{aligned} 2x + 3y - z &= 1 \\ x + 4y - z &= 4 \\ 3x + y + 2z &= 5 \end{aligned} \qquad \begin{bmatrix} 2 & 3 & -1 & 1 \\ 1 & 4 & -1 & 4 \\ 3 & 1 & 2 & 5 \end{bmatrix}$$

Interchange the first and second equations.

$$\begin{aligned} x + 4y - z &= 4 \\ 2x + 3y - z &= 1 \\ 3x + y + 2z &= 5 \end{aligned} \qquad \begin{bmatrix} 1 & 4 & -1 & 4 \\ 2 & 3 & -1 & 1 \\ 3 & 1 & 2 & 5 \end{bmatrix}$$

Add -2 times the first equation to the second; then add -3 times the first equation to the third.

$$\begin{aligned} x + 4y - z &= 4 \\ -5y + z &= -7 \\ -11y + 5z &= -7 \end{aligned} \qquad \begin{bmatrix} 1 & 4 & -1 & 4 \\ 0 & -5 & 1 & -7 \\ 0 & -11 & 5 & -7 \end{bmatrix}$$

Multiply the second equation by $-\frac{1}{5}$.

$$\begin{aligned} x + 4y - z &= 4 \\ y - \tfrac{1}{5}z &= \tfrac{7}{5} \\ -11y + 5z &= -7 \end{aligned} \qquad \begin{bmatrix} 1 & 4 & -1 & 4 \\ 0 & 1 & -\frac{1}{5} & \frac{7}{5} \\ 0 & -11 & 5 & -7 \end{bmatrix}$$

Add 11 times the second equation to the third.

$$\begin{aligned} x + 4y - z &= 4 \\ y - \tfrac{1}{5}z &= \tfrac{7}{5} \\ \tfrac{14}{5}z &= \tfrac{42}{5} \end{aligned} \qquad \begin{bmatrix} 1 & 4 & -1 & 4 \\ 0 & 1 & -\frac{1}{5} & \frac{7}{5} \\ 0 & 0 & \frac{14}{5} & \frac{42}{5} \end{bmatrix}$$

Now the system is in triangular form and can be solved by backward substitution. The result is $z = 3$, $y = 2$, and $x = -1$; we say the solution is $(-1, 2, 3)$.

We make two points about what we have just done. First, the process is not unique. We happen to prefer having a leading coefficient of 1; that was the reason for our first step. One could have started by multiplying the first equation by $-\frac{1}{2}$ and adding to the second, then multiplying the first equation by $-\frac{3}{2}$ and adding to the third. Any process that ultimately puts the system in triangular form is fine.

The second and main point is this. It is unnecessary to carry along all the x's and y's. Why not work with just the numbers? Why not do all the operations on the matrix of the system? Well, why not?

EQUIVALENT MATRICES

Guided by our knowledge of systems of equations, we say that matrices **A** and **B** are **equivalent** if **B** can be obtained from **A** by applying the operations below (a finite number of times).

Operation 1 Interchanging two rows.
Operation 2 Multiplying a row by a nonzero number.
Operation 3 Replacing a row by the sum of that row and a multiple of another row.

When **A** and **B** are equivalent, we write **A** ~ **B**. If **A** ~ **B**, then **B** ~ **A**. If **A** ~ **B** and **B** ~ **C**, then **A** ~ **C**.

AN EXAMPLE WITH FOUR EQUATIONS

Consider

$$
\begin{aligned}
x + 3y + z \quad\quad &= 1 \\
2x + 7y + z - w &= -1 \\
3x - 2y \quad\quad + 4w &= 8 \\
-x + y - 3z - w &= -6
\end{aligned}
$$

To solve this system, we take its matrix and transform it to triangular form using the operations above. Here is one possible sequence of steps.

$$
\begin{bmatrix}
1 & 3 & 1 & 0 & 1 \\
2 & 7 & 1 & -1 & -1 \\
3 & -2 & 0 & 4 & 8 \\
-1 & 1 & -3 & -1 & -6
\end{bmatrix}
$$

Add -2 times the first row to the second; -3 times the first row to the third row; and 1 times the first row to the fourth row.

CAUTION

Be sure to note that the first row of the matrix was unchanged while we performed operation 3 on the other three rows.

$$
\begin{bmatrix}
1 & 3 & 1 & 0 & 1 \\
0 & 1 & -1 & -1 & -3 \\
0 & -11 & -3 & 4 & 5 \\
0 & 4 & -2 & -1 & -5
\end{bmatrix}
$$

Add 11 times the second row to the third and -4 times the second row to the fourth.

$$\begin{bmatrix} 1 & 3 & 1 & 0 & 1 \\ 0 & 1 & -1 & -1 & -3 \\ 0 & 0 & -14 & -7 & -28 \\ 0 & 0 & 2 & 3 & 7 \end{bmatrix}$$

Multiply the third row by $-\frac{1}{14}$.

$$\begin{bmatrix} 1 & 3 & 1 & 0 & 1 \\ 0 & 1 & -1 & -1 & -3 \\ 0 & 0 & 1 & \frac{1}{2} & 2 \\ 0 & 0 & 2 & 3 & 7 \end{bmatrix}$$

Add -2 times the third row to the fourth row.

$$\begin{bmatrix} 1 & 3 & 1 & 0 & 1 \\ 0 & 1 & -1 & -1 & -3 \\ 0 & 0 & 1 & \frac{1}{2} & 2 \\ 0 & 0 & 0 & 2 & 3 \end{bmatrix}$$

This last matrix represents the system

$$\begin{aligned} x + 3y + z & = 1 \\ y - z - w & = -3 \\ z + \tfrac{1}{2}w & = 2 \\ 2w & = 3 \end{aligned}$$

If we use back substitution, we get $w = \frac{3}{2}, z = \frac{5}{4}, y = -\frac{1}{4}$, and $x = \frac{1}{2}$. The solution is $(\frac{1}{2}, -\frac{1}{4}, \frac{5}{4}, \frac{3}{2})$.

THE CASES WITH MANY SOLUTIONS AND NO SOLUTION

A system of equations need not have a unique solution; it may have infinitely many solutions or none at all. We need to be able to analyze the latter two cases by our matrix method. Fortunately, this is easy to do. Consider Example A of Section 11-1 first. Here is how we handle it using matrices.

$$\begin{bmatrix} 1 & -2 & 3 & 10 \\ 2 & -3 & -1 & 8 \\ 4 & -7 & 5 & 28 \end{bmatrix} \sim \begin{bmatrix} 1 & -2 & 3 & 10 \\ 0 & 1 & -7 & -12 \\ 0 & 1 & -7 & -12 \end{bmatrix}$$

$$\sim \begin{bmatrix} 1 & -2 & 3 & 10 \\ 0 & 1 & -7 & -12 \\ 0 & 0 & 0 & 0 \end{bmatrix}$$

The appearance of a row of zeros tells us that we have infinitely many solutions. The set of solutions is obtained by considering the equations corresponding to the first two rows.

$$x - 2y + 3z = 10$$
$$y - 7z = -12$$

When we solve for y in the second equation and substitute in the first, we obtain

$$x = 11z - 14$$
$$y = 7z - 12$$
$$z \quad \text{arbitrary}$$

Next consider the inconsistent example treated in Section 11-1 (Example B). Here is what happens when the matrix method is applied to this example.

$$\begin{bmatrix} 1 & -2 & 3 & 10 \\ 2 & -3 & -1 & 8 \\ 5 & -9 & 8 & 20 \end{bmatrix} \sim \begin{bmatrix} 1 & -2 & 3 & 10 \\ 0 & 1 & -7 & -12 \\ 0 & 1 & -7 & -30 \end{bmatrix}$$

$$\sim \begin{bmatrix} 1 & -2 & 3 & 10 \\ 0 & 1 & -7 & -12 \\ 0 & 0 & 0 & -18 \end{bmatrix}$$

We are tipped off to the inconsistency of the system by the third row of the matrix. It corresponds to the equation

$$0x + 0y + 0z = -18$$

which has no solution. Consequently, the system as a whole has no solution.

We may summarize our discussion as follows. If the process of transforming the matrix of a system of n equations in n unknowns to triangular form leads to a row in which all elements but the last one are zero, then the system is inconsistent; that is, it has no solution. If the above does not occur and we are led to a matrix with one or more rows consisting entirely of zeros, then the system has infinitely many solutions.

Problem Set 11-2

Write the matrix of each system in Problems 1–8.

1. $2x - y = 4$
   $x - 3y = -2$

2. $x + 2y = 13$
   $11x - y = 0$

3. $x - 2y + z = 3$
   $2x + y = 5$
   $x + y + 3z = -4$

4. $x + 4z = 10$
   $2y - z = 0$
   $3x - y = 20$

5. $2x = 3y - 4$
   $3x + 2 = -y$

6. $x = 4y + 3$
   $y = -2x + 5$

7. $x = 5$
   $2y + x - z = 4$
   $3x - y + 13 = 5z$

8. $z = 2$
   $2x - z = -4$
   $x + 2y + 4z = -8$

Regard each matrix in Problems 9–18 as a matrix of a linear system of equations. Tell whether the system has a unique solution, infinitely many solutions, or no solution. You need not solve any of the systems.

9. $\begin{bmatrix} 1 & -2 & 3 \\ 0 & 1 & -4 \end{bmatrix}$

10. $\begin{bmatrix} 2 & 5 & 0 \\ 0 & -3 & 5 \end{bmatrix}$

11. $\begin{bmatrix} 1 & -3 & 5 \\ 2 & -6 & -10 \end{bmatrix}$

12. $\begin{bmatrix} 2 & 1 & -4 \\ -6 & -3 & 12 \end{bmatrix}$

13. $\begin{bmatrix} 1 & -2 & 4 & -2 \\ 0 & 3 & 1 & 4 \\ 0 & 0 & 1 & -3 \end{bmatrix}$

14. $\begin{bmatrix} 5 & 4 & 0 & -11 \\ 0 & 1 & -4 & 0 \\ 0 & 0 & 2 & -4 \end{bmatrix}$

15. $\begin{bmatrix} 2 & 1 & 5 & 4 \\ 0 & 3 & -2 & 10 \\ 0 & 3 & -2 & 10 \end{bmatrix}$

16. $\begin{bmatrix} 4 & 1 & -3 & 5 \\ 0 & 0 & 1 & -4 \\ 0 & 0 & 1 & -4 \end{bmatrix}$

17. $\begin{bmatrix} 3 & 2 & -1 & 0 \\ 0 & 1 & 0 & -4 \\ 0 & 1 & 0 & 5 \end{bmatrix}$

18. $\begin{bmatrix} -1 & 5 & 6 & -3 \\ 0 & 0 & 0 & 0 \\ 0 & 0 & 0 & 4 \end{bmatrix}$

In Problems 19–30, use matrices to solve each system or to show that it has no solution.

19. $x + 2y = 5$
    $2x - 5y = -8$

20. $2x + 4y = 16$
    $3x - y = 10$

21. $3x - 2y = 1$
    $-6x + 4y = -2$

22. $x + 3y = 12$
    $5x + 15y = 12$

23. $3x - 2y + 5z = -10$
    $y - 4z = 8$
    $2y + z = 7$

24. $4x + 5y + 2z = 25$
    $y - 2z = 7$
    $5y + z = 2$

25. $x + y - 3z = 10$
    $2x + 5y + z = 18$
    $5x + 8y - 8z = 48$

26. $x - 4y + z = 18$
    $2x - 7y - 2z = 4$
    $3x - 11y - z = 22$

27. $2x + 5y + 2z = 6$
$x + 2y - z = 3$
$3x - y + 2z = 9$

28. $x - 2y + 3z = -2$
$3x - 6y + 9z = -6$
$-2x + 4y - 6z = 0$

[c] 29. $x + 1.2y - 2.3z = 8.1$
$1.3x + .7y + .4z = 6.2$
$.5x + 1.2y + .5z = 3.2$

30. $3x + 2y = 4$
$3x - 4y + 6z = 16$
$3x - y + z = 6$

MISCELLANEOUS PROBLEMS

Regard each matrix in Problems 31–36 as the matrix of a linear system of equations. Without solving the system, tell whether it has a unique solution, infinitely many solutions, or no solution.

31. $\begin{bmatrix} 3 & -2 & 5 \\ 0 & 1 & -3 \end{bmatrix}$

32. $\begin{bmatrix} 2 & -1 & 5 \\ -4 & 2 & 8 \end{bmatrix}$

33. $\begin{bmatrix} 2 & -1 & 4 & 6 \\ 0 & 4 & -1 & 5 \\ 0 & 0 & 2 & 1 \end{bmatrix}$

34. $\begin{bmatrix} 3 & 3 & 0 & -4 \\ 0 & 1 & -3 & 2 \\ 0 & 0 & 0 & 0 \end{bmatrix}$

35. $\begin{bmatrix} 1 & 2 & 3 & 4 & 5 \\ 0 & 3 & 2 & 1 & 0 \\ 0 & 0 & 0 & 3 & -4 \\ 0 & 0 & 0 & -9 & 15 \end{bmatrix}$

36. $\begin{bmatrix} 0 & 0 & 0 & 2 & 3 \\ 0 & 0 & 3 & 4 & 5 \\ 0 & 4 & 5 & 6 & 7 \\ 5 & 6 & 7 & 8 & 9 \end{bmatrix}$

In Problems 37–40, use matrices to solve each system or to show that it has no solution.

37. $3x - 2y + 4z = 0$
$x - y + 3z = 1$
$4x + 2y - z = 3$

38. $-4x + y - z = 5$
$4x - y + z = -5$
$-24x + 6y - 6z = 10$

39. $2x + 4y - z = 8$
$4x + 9y + 3z = 42$
$8x + 17y + z = 58$

40. $2x + y + 2z + 3w = 2$
$y - 2z + 5w = 2$
$z + 3w = 4$
$2z + 7w = 4$

41. Find a, b, and c so that the parabola $y = ax^2 + bx + c$ passes through the points $(-2, -32)$, $(1, 4)$, and $(3, -12)$.

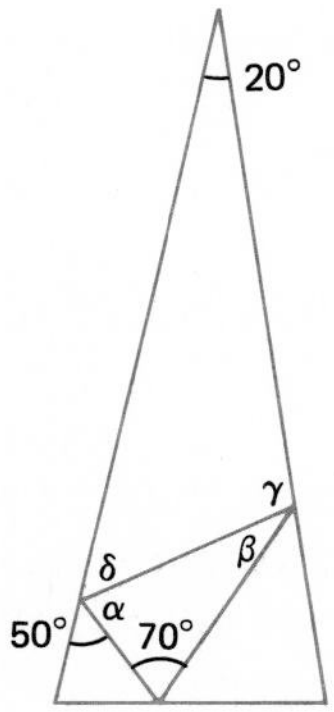

42. Find the equation and radius of a circle that passes through $(3, -3)$, $(8, 2)$, and $(6, 6)$.

43. Find angles α, β, γ, and δ in Figure 2 given that $\alpha - \beta + \gamma - \delta = 110°$.

44. A chemist plans to mix three different nitric acid solutions with concentrations of 25 percent, 40 percent, and 50 percent to form 100 liters of a 32 percent solution. If she insists on using twice as much of the 25 percent solution as the 40 percent solution, how many liters of each kind should she use?

45. The local garden store stocks three brands of phosphate-potash-nitrogen fertilizer with compositions indicated in the following table.

Figure 2

BRAND	PHOSPHATE	POTASH	NITROGEN
A	10%	30%	60%
B	20%	40%	40%
C	20%	30%	50%

Soil analysis shows that Wanda Wiseankle needs fertilizer for her garden that is 19 percent phosphate, 34 percent potash, and 47 percent nitrogen. Can she obtain the right mixture by mixing the three brands? If so, how many pounds of each should she mix together to get 100 pounds of the desired blend?

46. TEASER Tom, Dick, and Harry are good friends but have very different work habits. Together, they contracted to paint three identical houses. Tom and Dick painted the first house in $\frac{72}{5}$ hours; Tom and Harry painted the second house in 16 hours; Dick and Harry painted the third house in $\frac{144}{7}$ hours. How long would it have taken each boy to paint a house alone?

Cayley's Weapons

$$\begin{bmatrix} a & b \\ c & d \end{bmatrix} + \begin{bmatrix} A & B \\ C & D \end{bmatrix} = \begin{bmatrix} a+A & b+B \\ c+C & d+D \end{bmatrix}$$

$$\begin{bmatrix} a & b \\ c & d \end{bmatrix} \cdot \begin{bmatrix} A & B \\ C & D \end{bmatrix} = \begin{bmatrix} aA+bC & aB+bD \\ cA+dC & cB+dD \end{bmatrix}$$

"Cayley is forging the weapons for future generations of physicists."

P. G. Tait

11-3   The Algebra of Matrices

When Arthur Cayley introduced matrices, he had much more in mind than the application described in the previous section. There, matrices served as a device to simplify solving systems of equations. Cayley saw that these number boxes could be studied independently of equations, that they could be thought of as a new type of mathematical object. He realized that if he could give appropriate definitions of addition and multiplication, he would create a mathematical system that might stand with the real numbers and the complex numbers as a potential model for many applications. Cayley did all of this in a major paper in 1858. Some of his contemporaries saw little of significance in this new abstraction. But one of them, P. G. Tait, uttered the prophetic words quoted in the opening box. Tait was right. During the 1920's, Werner Heisenberg found that matrices were just the tool he needed to formulate his quantum mechanics. And by 1950, it was generally recognized that matrix theory provides the best model for many problems in economics and the social sciences.

To simplify our discussion, we initially consider only 2×2 matrices, that is, matrices with two rows and two columns. Examples are

$$\begin{bmatrix} -1 & 3 \\ 4 & 0 \end{bmatrix} \qquad \begin{bmatrix} \log .1 & \frac{6}{2} \\ \frac{12}{3} & \log 1 \end{bmatrix} \qquad \begin{bmatrix} a & b \\ c & d \end{bmatrix}$$

The first two of these matrices are said to be equal. In fact, two matrices are **equal** if and only if the entries in corresponding positions are equal. Be sure to distinguish the notion of equality (written $=$) from that of equivalence (written $\sim$) introduced in Section 11-2. For example,

$$\begin{bmatrix} 2 & 1 \\ -3 & 4 \end{bmatrix} \quad \text{and} \quad \begin{bmatrix} -3 & 4 \\ 2 & 1 \end{bmatrix}$$

are equivalent matrices; however, they are not equal.

ADDITION AND SUBTRACTION

Cayley's definition of addition is straightforward. To add two matrices, add the entries in corresponding positions. Thus

$$\begin{bmatrix} 1 & 3 \\ -1 & 4 \end{bmatrix} + \begin{bmatrix} 6 & -2 \\ 5 & 1 \end{bmatrix} = \begin{bmatrix} 1 + 6 & 3 + (-2) \\ -1 + 5 & 4 + 1 \end{bmatrix} = \begin{bmatrix} 7 & 1 \\ 4 & 5 \end{bmatrix}$$

and in general

$$\begin{bmatrix} a & b \\ c & d \end{bmatrix} + \begin{bmatrix} A & B \\ C & D \end{bmatrix} = \begin{bmatrix} a + A & b + B \\ c + C & d + D \end{bmatrix}$$

It is easy to check that the commutative and associative properties for addition are valid. Let $\mathbf{U}$, $\mathbf{V}$, and $\mathbf{W}$ be any three matrices.

1. **(Commutativity +)** $\mathbf{U} + \mathbf{V} = \mathbf{V} + \mathbf{U}$
2. **(Associativity +)** $\mathbf{U} + (\mathbf{V} + \mathbf{W}) = (\mathbf{U} + \mathbf{V}) + \mathbf{W}$

The matrix

$$\mathbf{O} = \begin{bmatrix} 0 & 0 \\ 0 & 0 \end{bmatrix}$$

behaves as the "zero" for matrices. And the additive inverse of the matrix

$$\mathbf{U} = \begin{bmatrix} a & b \\ c & d \end{bmatrix}$$

is given by

$$-\mathbf{U} = \begin{bmatrix} -a & -b \\ -c & -d \end{bmatrix}$$

We may summarize these statements as follows.

3. **(Neutral element +)**   There is a matrix **O** satisfying $\mathbf{O} + \mathbf{U} = \mathbf{U} + \mathbf{O} = \mathbf{U}$.

4. **(Additive inverses)**   For each matrix **U**, there is a matrix **−U** satisfying

$$\mathbf{U} + (-\mathbf{U}) = (-\mathbf{U}) + \mathbf{U} = \mathbf{O}$$

With the existence of an additive inverse settled, we can define subtraction by $\mathbf{U} - \mathbf{V} = \mathbf{U} + (-\mathbf{V})$. This amounts to subtracting the entries of **V** from the corresponding entries of **U**. Thus

$$\begin{bmatrix} 1 & 3 \\ -1 & 4 \end{bmatrix} - \begin{bmatrix} 6 & -2 \\ 5 & 1 \end{bmatrix} = \begin{bmatrix} -5 & 5 \\ -6 & 3 \end{bmatrix}$$

So far, all has been straightforward and nice. But with multiplication, Cayley hit a snag.

MULTIPLICATION

Cayley's definition of multiplication may seem odd at first glance. He was led to it by consideration of a special problem that we do not have time to describe. It is enough to say that Cayley's definition is the one that proves useful in modern applications (as you will see).

Here it is in symbols.

$$\begin{bmatrix} a & b \\ c & d \end{bmatrix} \cdot \begin{bmatrix} A & B \\ C & D \end{bmatrix} = \begin{bmatrix} aA + bC & aB + bD \\ cA + dC & cB + dD \end{bmatrix}$$

Stated in words, we multiply two matrices by multiplying the rows of the left matrix by the columns of the right matrix in pairwise entry fashion, adding the results. For example, the entry in the second row and first column of the product is obtained by multiplying the entries of the second row of the left matrix by the corresponding entries of the first column of the right matrix, adding the results. Until you get used to it, it may help to use your fingers as shown in the diagram below.

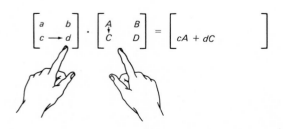

Here is an example worked out in detail.

$$\begin{bmatrix} 1 & 3 \\ -1 & 4 \end{bmatrix} \begin{bmatrix} 6 & -2 \\ 5 & 1 \end{bmatrix} = \begin{bmatrix} (1)(6) + (3)(5) & (1)(-2) + (3)(1) \\ (-1)(6) + (4)(5) & (-1)(-2) + (4)(1) \end{bmatrix}$$

$$= \begin{bmatrix} 21 & 1 \\ 14 & 6 \end{bmatrix}$$

Here is the same problem, but with the matrices multiplied in the opposite order.

$$\begin{bmatrix} 6 & -2 \\ 5 & 1 \end{bmatrix} \begin{bmatrix} 1 & 3 \\ -1 & 4 \end{bmatrix} = \begin{bmatrix} (6)(1) + (-2)(-1) & (6)(3) + (-2)(4) \\ (5)(1) + (1)(-1) & (5)(3) + (1)(4) \end{bmatrix}$$

$$= \begin{bmatrix} 8 & 10 \\ 4 & 19 \end{bmatrix}$$

Now you see the snag about which we warned you. The commutative property for multiplication fails. This is troublesome, but not fatal. We manage to get along in a world that is largely noncommutative (try removing your clothes and taking a shower in the opposite order). We just have to remember never to commute matrices under multiplication. Fortunately two other nice properties do hold.

5. (**Associativity** ·) $\mathbf{U} \cdot (\mathbf{V} \cdot \mathbf{W}) = (\mathbf{U} \cdot \mathbf{V}) \cdot \mathbf{W}$
6. (**Distributivity**) $\mathbf{U} \cdot (\mathbf{V} + \mathbf{W}) = \mathbf{U} \cdot \mathbf{V} + \mathbf{U} \cdot \mathbf{W}$
$\qquad\qquad\qquad (\mathbf{V} + \mathbf{W}) \cdot \mathbf{U} = \mathbf{V} \cdot \mathbf{U} + \mathbf{W} \cdot \mathbf{U}$

We have not said anything about multiplicative inverses; that comes in the next section. We do, however, want to mention a special operation called **scalar multiplication**—that is, multiplication of a matrix by a scalar (number). To multiply a matrix by a number, multiply each entry by that number. That is,

$$k \begin{bmatrix} a & b \\ c & d \end{bmatrix} = \begin{bmatrix} ka & kb \\ kc & kd \end{bmatrix}$$

Scalar multiplication satisfies the expected properties.

7. $k(\mathbf{U} + \mathbf{V}) = k\mathbf{U} + k\mathbf{V}$
8. $(k + m)\mathbf{U} = k\mathbf{U} + m\mathbf{U}$
9. $(km)\mathbf{U} = k(m\mathbf{U})$

LARGER MATRICES AND COMPATIBILITY

So far we have considered only 2×2 matrices. This is an unnecessary restriction; however, to perform operations on arbitrary matrices, we must make sure they are **compatible.** For addition, this simply means that the matrices must be of the same size. Thus

$$\begin{bmatrix} 1 & -1 & 3 \\ 4 & -5 & 2 \end{bmatrix} + \begin{bmatrix} 2 & 6 & 0 \\ -3 & 2 & 4 \end{bmatrix} = \begin{bmatrix} 3 & 5 & 3 \\ 1 & -3 & 6 \end{bmatrix}$$

but

$$\begin{bmatrix} 1 & -1 & 3 \\ 4 & -5 & 2 \end{bmatrix} + \begin{bmatrix} 2 & 6 \\ -3 & 2 \end{bmatrix}$$

makes no sense.

Two matrices are compatible for multiplication if the left matrix has the same number of columns as the right matrix has rows. For example,

$$\begin{bmatrix} 1 & -1 & 3 \\ 4 & -5 & 2 \end{bmatrix} \begin{bmatrix} 2 & 1 & 5 & 0 \\ -1 & 3 & 2 & 1 \\ 1 & -2 & 0 & 2 \end{bmatrix} = \begin{bmatrix} 6 & -8 & 3 & 5 \\ 15 & -15 & 10 & -1 \end{bmatrix}$$

The left matrix is 2×3, the right one is 3×4, and the result is 2×4. In general, we can multiply an $m \times n$ matrix by an $n \times p$ matrix, the result being an $m \times p$ matrix. All of the properties mentioned earlier are valid, provided we work with compatible matrices.

A BUSINESS APPLICATION

The ABC Company sells precut lumber for two types of summer cottages, standard and deluxe. The standard model requires 30,000 board feet of lumber and 100 worker-hours of cutting; the deluxe model takes 40,000 board feet of lumber and 110 worker-hours of cutting. This year, the ABC Company buys its lumber at $.20 per board foot and pays its laborers $9.00 per hour. Next year it expects these costs to be $.25 and $10.00, respectively. This information can be displayed in matrix form as follows.

	REQUIREMENTS A				UNIT COST B	
	Lumber	Labor			This year	Next year
Standard	30,000	100		Lumber	$.20	$.25
Deluxe	40,000	110		Labor	$9.00	$10.00

Now we ask whether the product matrix **AB** has economic significance. It does: It gives the total dollar cost of standard and deluxe cottages both for this year and next. You can see this from the following calculation.

$$\mathbf{AB} = \begin{bmatrix} (30{,}000)(.20) + (100)(9) & (30{,}000)(.25) + (100)(10) \\ (40{,}000)(.20) + (110)(9) & (40{,}000)(.25) + (110)(10) \end{bmatrix}$$

$$= \begin{bmatrix} \$6900 & \$8500 \\ \$8990 & \$11{,}100 \end{bmatrix} \begin{matrix} \text{Standard} \\ \text{Deluxe} \end{matrix}$$

$$\begin{matrix} \text{This year} & \text{Next year} \end{matrix}$$

Problem Set 11-3

Calculate **A** + **B**, **A** − **B**, *and* 3**A** *in Problems 1–4.*

1. $\mathbf{A} = \begin{bmatrix} 2 & -1 \\ 3 & 7 \end{bmatrix}$, $\mathbf{B} = \begin{bmatrix} 6 & 5 \\ -2 & 3 \end{bmatrix}$

2. $\mathbf{A} = \begin{bmatrix} -1 & 0 \\ 5 & 4 \end{bmatrix}$, $\mathbf{B} = \begin{bmatrix} 2 & -2 \\ 3 & 7 \end{bmatrix}$

3. $\mathbf{A} = \begin{bmatrix} 3 & -2 & 5 \\ 4 & 0 & -3 \end{bmatrix}$, $\mathbf{B} = \begin{bmatrix} 2 & 6 & -1 \\ 4 & 3 & -3 \end{bmatrix}$

4. $\mathbf{A} = \begin{bmatrix} 1 & 2 & 3 \\ 4 & 5 & 6 \\ 7 & 8 & 9 \end{bmatrix}$, $\mathbf{B} = \begin{bmatrix} -1 & -2 & -2 \\ -4 & -5 & -6 \\ -7 & -8 & -9 \end{bmatrix}$

Calculate **AB** *and* **BA** *if possible in Problems 5–12.*

5. $\mathbf{A} = \begin{bmatrix} 2 & -1 \\ 3 & 7 \end{bmatrix}$, $\mathbf{B} = \begin{bmatrix} 6 & 5 \\ -2 & 3 \end{bmatrix}$

6. $\mathbf{A} = \begin{bmatrix} -1 & 0 \\ 5 & 4 \end{bmatrix}$, $\mathbf{B} = \begin{bmatrix} 2 & -2 \\ 3 & 7 \end{bmatrix}$

7. $\mathbf{A} = \begin{bmatrix} 1 & -1 & 2 \\ 3 & 4 & -4 \\ 2 & 1 & 3 \end{bmatrix}$, $\mathbf{B} = \begin{bmatrix} 0 & 2 & -3 \\ 1 & 2 & 3 \\ -1 & -2 & 4 \end{bmatrix}$

8. $\mathbf{A} = \begin{bmatrix} -2 & 5 & 1 \\ 0 & -2 & 3 \\ 1 & 2 & -1 \end{bmatrix}$, $\mathbf{B} = \begin{bmatrix} -3 & 4 & 1 \\ 2 & 5 & 1 \\ 1 & 2 & 3 \end{bmatrix}$

9. $\mathbf{A} = \begin{bmatrix} 1 & -2 & 3 & 4 \\ 3 & 2 & -5 & 1 \end{bmatrix}$, $\mathbf{B} = \begin{bmatrix} 1 & 2 \\ 3 & 4 \end{bmatrix}$

10. $\mathbf{A} = \begin{bmatrix} -1 & 3 \\ 4 & 2 \\ 1 & 5 \end{bmatrix}$, $\mathbf{B} = \begin{bmatrix} -1 & 2 & 3 & 4 \\ 0 & -3 & 2 & 1 \end{bmatrix}$

11. $\mathbf{A} = \begin{bmatrix} 3 & 1 & -1 \\ 2 & 4 & 2 \\ -3 & 2 & -1 \end{bmatrix}$, $\mathbf{B} = \begin{bmatrix} 1 \\ 2 \\ 3 \end{bmatrix}$

12. $\mathbf{A} = \begin{bmatrix} 1 & 2 & -1 \end{bmatrix}$, $\mathbf{B} = \begin{bmatrix} 4 & 3 \\ 0 & 2 \\ -1 & 4 \end{bmatrix}$

13. Calculate $\mathbf{AB}$ and $\mathbf{BA}$ for

$$\mathbf{A} = \begin{bmatrix} 0 & 0 \\ 0 & 0 \end{bmatrix} \qquad \mathbf{B} = \begin{bmatrix} 2 & -1 \\ 3 & 4 \end{bmatrix}$$

14. State the general property illustrated by Problem 13.

15. Find $\mathbf{X}$ if

$$\begin{bmatrix} 2 & 1 & -3 \\ 1 & 5 & 0 \end{bmatrix} + \mathbf{X} = 2\begin{bmatrix} -1 & 4 & 3 \\ -2 & 0 & 4 \end{bmatrix}$$

16. Solve for $\mathbf{X}$.

$$-3\mathbf{X} + 2\begin{bmatrix} 1 & -2 \\ 5 & 6 \end{bmatrix} = -\begin{bmatrix} 5 & -14 \\ 8 & 15 \end{bmatrix}$$

17. Calculate $\mathbf{A}(\mathbf{B} + \mathbf{C})$ and $\mathbf{AB} + \mathbf{AC}$ for

$$\mathbf{A} = \begin{bmatrix} 2 & -1 \\ 3 & 4 \end{bmatrix} \qquad \mathbf{B} = \begin{bmatrix} 2 & 4 \\ 6 & 1 \end{bmatrix} \qquad \mathbf{C} = \begin{bmatrix} -1 & -2 \\ 3 & 6 \end{bmatrix}$$

What property does this illustrate?

18. Calculate $(\mathbf{A} + \mathbf{B})(\mathbf{A} - \mathbf{B})$ and $\mathbf{A}^2 - \mathbf{B}^2$ for

$$\mathbf{A} = \begin{bmatrix} 3 & -2 \\ 1 & 4 \end{bmatrix} \qquad \mathbf{B} = \begin{bmatrix} 6 & -3 \\ 2 & 5 \end{bmatrix}$$

Why are your answers different?

© 19. Find the entry in the third row and second column of the product

$$\begin{bmatrix} 1.39 & 4.13 & -2.78 \\ 4.72 & -3.69 & 5.41 \\ 8.09 & -6.73 & 5.03 \end{bmatrix} \begin{bmatrix} 5.45 & 6.31 \\ 7.24 & -5.32 \\ 6.06 & 1.34 \end{bmatrix}$$

© 20. Find the entry in the second row and first column of the product in Problem 19.

MISCELLANEOUS PROBLEMS

21. Compute $\mathbf{A} - 2\mathbf{B}$, $\mathbf{AB}$, and $\mathbf{A}^2$ for

$$\mathbf{A} = \begin{bmatrix} 4 & -1 & 3 \\ 2 & 5 & 3 \\ 6 & 2 & 1 \end{bmatrix} \qquad \mathbf{B} = \begin{bmatrix} 1 & -3 & 2 \\ 5 & 0 & 3 \\ -5 & 2 & 1 \end{bmatrix}$$

22. Let $\mathbf{A}$ and $\mathbf{B}$ be 3×4 matrices, $\mathbf{C}$ a 4×3 matrix, and $\mathbf{D}$ a 3×3 matrix. Which of the following do not make sense, that is, do not satisfy the compatibility conditions?

(a) $\mathbf{AB}$ (b) $\mathbf{AC}$ (c) $\mathbf{AC} - \mathbf{D}$

(d) $(\mathbf{A} - \mathbf{B})\mathbf{C}$ (e) $(\mathbf{AC})\mathbf{D}$ (f) $\mathbf{A}(\mathbf{CD})$

(g) $\mathbf{A}^2$ (h) $(\mathbf{CA})^2$ (i) $\mathbf{C}(\mathbf{A} + 2\mathbf{B})$

23. Calculate $\mathbf{AB}$ and $\mathbf{BA}$ for

$$\mathbf{A} = \begin{bmatrix} 1 & 2 & 3 & 4 \end{bmatrix} \qquad \mathbf{B} = \begin{bmatrix} 2 \\ 1 \\ -1 \\ -2 \end{bmatrix}$$

24. Show that if $\mathbf{AB}$ and $\mathbf{BA}$ both make sense, then $\mathbf{AB}$ and $\mathbf{BA}$ are both square matrices.

25. If $(\mathbf{A} + \mathbf{B})^2 = \mathbf{A}^2 + 2\mathbf{AB} + \mathbf{B}^2$, what conclusions can you draw about $\mathbf{A}$ and $\mathbf{B}$?

26. If the ith row of $\mathbf{A}$ consists of all zeros, what is true about the ith row of $\mathbf{AB}$ (assuming $\mathbf{AB}$ makes sense)?

27. Let

$$\mathbf{A} = \begin{bmatrix} 0 & a \\ 0 & 0 \end{bmatrix} \qquad \mathbf{B} = \begin{bmatrix} 0 & a & b \\ 0 & 0 & c \\ 0 & 0 & 0 \end{bmatrix}$$

Calculate $\mathbf{A}^2$ and $\mathbf{B}^3$ and then make a conjecture.

28. A matrix of the form

$$\mathbf{A} = \begin{bmatrix} 1 & a & b \\ 0 & 1 & c \\ 0 & 0 & 1 \end{bmatrix}$$

where a, b, and c are any real numbers is called a Heisenberg matrix. What is true about the product of two such matrices?

29. Let

$$\mathbf{A} = \begin{bmatrix} 3 & 0 & 0 \\ 0 & -4 & 0 \\ 0 & 0 & 5 \end{bmatrix}$$

If $\mathbf{B}$ is any 3×3 matrix, what does multiplication on the left by $\mathbf{A}$ do to $\mathbf{B}$? Multiplication on the right by $\mathbf{A}$?

30. Calculate $\mathbf{A}^2$ and $\mathbf{A}^3$ for the matrix $\mathbf{A}$ of Problem 29. State a general result about raising a diagonal matrix to a positive integral power.

31. Art, Bob, and Curt work for a company that makes Flukes, Gizmos, and Horks. They are paid for their labor on a piecework basis, receiving \$1 for each Fluke, \$2 for each Gizmo, and \$3 for each Hork. Below are matrices $\mathbf{U}$ and $\mathbf{V}$ representing their outputs on Monday and Tuesday. Matrix $\mathbf{X}$ is the wage/unit matrix.

	MONDAY'S OUTPUT U				TUESDAY'S OUTPUT V				WAGE/UNIT X
	F	G	H		F	G	H		
Art	4	3	2	Art	3	6	1	F	1
Bob	5	1	2	Bob	4	2	2	G	2
Curt	3	4	1	Curt	5	1	3	H	3

Compute the following matrices and decide what they represent.

(a) $\mathbf{UX}$ (b) $\mathbf{VX}$ (c) $\mathbf{U} + \mathbf{V}$ (d) $(\mathbf{U} + \mathbf{V})\mathbf{X}$

32. Four friends, A, B, C, and D, have unlisted telephone numbers. Whether or not one person knows another's number is indicated by the matrix $\mathbf{U}$ below, where 1 indicates knowing and 0 indicates not knowing. For example, the 1 in row 3 and column 1 means that C knows A's number.

$$
\mathbf{U} = \begin{array}{c} \\ A \\ B \\ C \\ D \end{array} \begin{array}{cccc} A & B & C & D \\ \left[\begin{array}{cccc} 1 & 0 & 1 & 0 \\ 0 & 1 & 1 & 0 \\ 1 & 0 & 1 & 1 \\ 0 & 1 & 0 & 1 \end{array}\right] \end{array}
$$

(a) Calculate $\mathbf{U}^2$.
(b) Interpret $\mathbf{U}^2$ in terms of the possibility of each person being able to get a telephone message to another.
(c) Can D get a message to A via one other person?
(d) Interpret $\mathbf{U}^3$.

33. Consider the set $\mathbf{C}$ of all 2×2 matrices of the form

$$
\begin{bmatrix} a & b \\ -b & a \end{bmatrix}
$$

where a and b are real numbers.
(a) Let

$$
\mathbf{U} = \begin{bmatrix} u_1 & u_2 \\ -u_2 & u_1 \end{bmatrix} \quad \text{and} \quad \mathbf{V} = \begin{bmatrix} v_1 & v_2 \\ -v_2 & v_1 \end{bmatrix}
$$

be two such matrices.
Calculate $\mathbf{U} + \mathbf{V}$ and $\mathbf{UV}$. Note that both $\mathbf{U} + \mathbf{V}$ and $\mathbf{UV}$ are in $\mathbf{C}$.
(b) Let $\mathbf{I} = \begin{bmatrix} 1 & 0 \\ 0 & 1 \end{bmatrix}$ and $\mathbf{J} = \begin{bmatrix} 0 & 1 \\ -1 & 0 \end{bmatrix}$. Calculate $\mathbf{I}^2$ and $\mathbf{J}^2$.
(c) Note that $\mathbf{U} = u_1\mathbf{I} + u_2\mathbf{J}$ and $\mathbf{V} = v_1\mathbf{I} + v_2\mathbf{J}$. Write $\mathbf{U} + \mathbf{V}$ and $\mathbf{UV}$ in terms of $\mathbf{I}$ and $\mathbf{J}$.
⊡ (d) What does all this have to do with the complex numbers?

34. TEASER Find the four square roots of the matrix

$$
\begin{bmatrix} 7 & 10 \\ 15 & 22 \end{bmatrix}
$$

"What was that?" inquired Alice. "Reeling and Writhing, of course, to begin with," the mock turtle replied, "and then the different branches of Arithmetic—Ambition, Distraction, Uglification, and Derision."

from Alice's Adventures in Wonderland

by Lewis Carroll

$$\begin{bmatrix} 2 & 3 \\ -4 & 1 \end{bmatrix}$$

$$\frac{}{\begin{bmatrix} 6 & 7 \\ 1 & 2 \end{bmatrix}} = \begin{bmatrix} ? & ? \\ ? & ? \end{bmatrix}$$

11-4   Multiplicative Inverses

Even for ordinary numbers, the notion of division seems more difficult than that of addition, subtraction, and multiplication. Certainly this is true for division of matrices. Look at the example displayed above. It could tempt more than a mock turtle to derision. However, Arthur Cayley saw no need to sneer. He noted that in the case of numbers,

$$\frac{U}{V} = U \cdot \frac{1}{V} = U \cdot V^{-1}$$

What is needed is a concept of "one" for matrices; then we need the concept of multiplicative inverse. The first is easy.

THE MULTIPLICATIVE IDENTITY FOR MATRICES

Let

$$\mathbf{I} = \begin{bmatrix} 1 & 0 \\ 0 & 1 \end{bmatrix}$$

Then for any 2×2 matrix $\mathbf{U}$,

$$\mathbf{UI} = \mathbf{U} = \mathbf{IU}$$

This can be checked by noting that

$$\begin{bmatrix} a & b \\ c & d \end{bmatrix}\begin{bmatrix} 1 & 0 \\ 0 & 1 \end{bmatrix} = \begin{bmatrix} a & b \\ c & d \end{bmatrix} = \begin{bmatrix} 1 & 0 \\ 0 & 1 \end{bmatrix}\begin{bmatrix} a & b \\ c & d \end{bmatrix}$$

The symbol $\mathbf{I}$ is chosen because it is often called the **multiplicative identity.** In accordance with Section 1-4, it is also called the neutral element for multiplication.

For 3×3 matrices, the multiplicative identity has the form

$$\begin{bmatrix} 1 & 0 & 0 \\ 0 & 1 & 0 \\ 0 & 0 & 1 \end{bmatrix}$$

You should be able to guess its form for 4×4 and higher order matrices.

INVERSES OF 2 × 2 MATRICES

Suppose we want to find the multiplicative inverse of

$$\mathbf{V} = \begin{bmatrix} 6 & 7 \\ 1 & 2 \end{bmatrix}$$

We are looking for a matrix

$$\mathbf{W} = \begin{bmatrix} a & b \\ c & d \end{bmatrix}$$

that satisfies $\mathbf{VW} = \mathbf{I}$ and $\mathbf{WV} = \mathbf{I}$. Taking $\mathbf{VW} = \mathbf{I}$ first, we want

$$\begin{bmatrix} 6 & 7 \\ 1 & 2 \end{bmatrix} \begin{bmatrix} a & b \\ c & d \end{bmatrix} = \begin{bmatrix} 1 & 0 \\ 0 & 1 \end{bmatrix}$$

which means

$$\begin{bmatrix} 6a + 7c & 6b + 7d \\ a + 2c & b + 2d \end{bmatrix} = \begin{bmatrix} 1 & 0 \\ 0 & 1 \end{bmatrix}$$

or

$$6a + 7c = 1 \qquad 6b + 7d = 0$$
$$a + 2c = 0 \qquad b + 2d = 1$$

When these four equations are solved for a, b, c, d, we have

$$\mathbf{W} = \begin{bmatrix} \frac{2}{5} & -\frac{7}{5} \\ -\frac{1}{5} & \frac{6}{5} \end{bmatrix}$$

as a tentative solution to our problem. We say tentative, because so far we know only that $\mathbf{VW} = \mathbf{I}$. Happily, $\mathbf{W}$ works on the other side of $\mathbf{V}$ too, as we can check. (In this exceptional case, we do have commutativity.)

$$\mathbf{WV} = \begin{bmatrix} \frac{2}{5} & -\frac{7}{5} \\ -\frac{1}{5} & \frac{6}{5} \end{bmatrix} \begin{bmatrix} 6 & 7 \\ 1 & 2 \end{bmatrix} = \begin{bmatrix} 1 & 0 \\ 0 & 1 \end{bmatrix}$$

Success! $\mathbf{W}$ is the inverse of $\mathbf{V}$; we denote it by the symbol $\mathbf{V}^{-1}$.

The process just described can be carried out for any specific 2 × 2 matrix, or better, it can be carried out for a general 2 × 2 matrix. But before we give the result, we make an important comment. There is no reason to think that every 2 × 2 matrix has a multiplicative inverse. Remember that the number 0 does not have such an inverse; neither does the matrix $\mathbf{O}$. But here is a mild surprise. Many other 2 × 2 matrices do not have inverses. The following theorem identifies in a very precise way those that do, and then gives a formula for their inverses.

THEOREM (MULTIPLICATIVE INVERSES)

The matrix

$$\mathbf{V} = \begin{bmatrix} a & b \\ c & d \end{bmatrix}$$

has a multiplicative inverse if and only if $D = ad - bc$ is nonzero. If $D \neq 0$, then

$$\mathbf{V}^{-1} = \begin{bmatrix} \dfrac{d}{D} & -\dfrac{b}{D} \\ -\dfrac{c}{D} & \dfrac{a}{D} \end{bmatrix}$$

Thus the number D determines whether a matrix has an inverse. This number, which we shall call a *determinant,* will be studied in detail in the next section. Each 2×2 matrix has such a number associated with it. Let us look at two examples.

$$\mathbf{X} = \begin{bmatrix} 2 & -3 \\ -4 & 6 \end{bmatrix} \qquad\qquad \mathbf{Y} = \begin{bmatrix} 5 & -3 \\ -4 & 3 \end{bmatrix}$$

$$D = (2)(6) - (-3)(-4) = 0 \qquad D = (5)(3) - (-3)(-4) = 3$$

$\mathbf{X}^{-1}$ does not exist $\qquad\qquad \mathbf{Y}^{-1} = \begin{bmatrix} \frac{3}{3} & \frac{3}{3} \\ \frac{4}{3} & \frac{5}{3} \end{bmatrix}$

INVERSES FOR HIGHER-ORDER MATRICES

There is a theorem like the one above for square matrices of any size, which Cayley found in 1858. It is complicated and, rather than try to state it, we are going to illustrate a process which yields the inverse of a matrix whenever it exists. Briefly described, it is this. Take any square matrix $\mathbf{V}$ and write the corresponding identity matrix $\mathbf{I}$ next to it on the right. By using the three row operations of Section 11-2, attempt to reduce $\mathbf{V}$ to the identity matrix while simultaneously performing the same operations on $\mathbf{I}$. If you can reduce $\mathbf{V}$ to $\mathbf{I}$, you will simultaneously turn $\mathbf{I}$ into $\mathbf{V}^{-1}$. If you cannot reduce $\mathbf{V}$ to $\mathbf{I}$, $\mathbf{V}$ has no inverse.

Here is an illustration for the 2×2 matrix $\mathbf{V}$ that we used earlier.

$$\left[\begin{array}{cc|cc} 6 & 7 & 1 & 0 \\ 1 & 2 & 0 & 1 \end{array}\right]$$

$$\sim \left[\begin{array}{cc|cc} 1 & 2 & 0 & 1 \\ 6 & 7 & 1 & 0 \end{array}\right] \qquad \text{(interchange rows)}$$

$$\sim \left[\begin{array}{cc|cc} 1 & 2 & 0 & 1 \\ 0 & -5 & 1 & -6 \end{array}\right] \qquad \text{(add -6 times row 1 to row 2)}$$

$$\sim \begin{bmatrix} 1 & 2 & 0 & 1 \\ 0 & 1 & -\frac{1}{5} & \frac{6}{5} \end{bmatrix} \qquad \text{(divide row 2 by } -5\text{)}$$

$$\sim \begin{bmatrix} 1 & 0 & \frac{2}{5} & -\frac{7}{5} \\ 0 & 1 & -\frac{1}{5} & \frac{6}{5} \end{bmatrix} \qquad \text{(add } -2 \text{ times row 2 to row 1)}$$

Notice that the matrix $\mathbf{V}^{-1}$ that we obtained earlier appears on the right. We illustrate the same process for a 3×3 matrix in Example A of the problem set.

AN APPLICATION

Consider the system of equations

$$2x + 6y + 6z = 8$$
$$2x + 7y + 6z = 10$$
$$2x + 7y + 7z = 9$$

If we introduce matrices

$$\mathbf{A} = \begin{bmatrix} 2 & 6 & 6 \\ 2 & 7 & 6 \\ 2 & 7 & 7 \end{bmatrix} \qquad \mathbf{X} = \begin{bmatrix} x \\ y \\ z \end{bmatrix} \qquad \mathbf{B} = \begin{bmatrix} 8 \\ 10 \\ 9 \end{bmatrix}$$

this system can be written in the form

$$\mathbf{AX} = \mathbf{B}$$

Now divide both sides by $\mathbf{A}$, by which we mean, of course, multiply both sides by $\mathbf{A}^{-1}$. We must be more precise. Multiply both sides on the left by $\mathbf{A}^{-1}$ (do not forget the lack of commutativity).

$$\mathbf{A}^{-1}\mathbf{AX} = \mathbf{A}^{-1}\mathbf{B}$$

$$\mathbf{IX} = \mathbf{A}^{-1}\mathbf{B}$$

$$\mathbf{X} = \mathbf{A}^{-1}\mathbf{B}$$

CAUTION

~~$\mathbf{AX} = \mathbf{B}$~~
~~$\mathbf{X} = \mathbf{BA}^{-1}$~~

$\mathbf{AX} = \mathbf{B}$
$\mathbf{X} = \mathbf{A}^{-1}\mathbf{B}$

In Example A on pages 461–462, $\mathbf{A}^{-1}$ is found to be

$$\mathbf{A}^{-1} = \begin{bmatrix} \frac{7}{2} & 0 & -3 \\ -1 & 1 & 0 \\ 0 & -1 & 1 \end{bmatrix}$$

Thus

$$\mathbf{X} = \begin{bmatrix} \frac{7}{2} & 0 & -3 \\ -1 & 1 & 0 \\ 0 & -1 & 1 \end{bmatrix} \begin{bmatrix} 8 \\ 10 \\ 9 \end{bmatrix} = \begin{bmatrix} 1 \\ 2 \\ -1 \end{bmatrix}$$

and therefore $(1, 2, -1)$ is the solution to our system.

This method of solution is particularly useful when many systems with the same coefficient matrix $\mathbf{A}$ are under consideration. Once we have $\mathbf{A}^{-1}$, we can obtain any solution simply by doing an easy matrix multiplication. If only one system is being studied, the method of Section 11-2 is best.

Problem Set 11-4

Find the multiplicative inverse of each matrix. Use matrix multiplication as a check.

1. $\begin{bmatrix} 2 & 3 \\ -1 & -1 \end{bmatrix}$

2. $\begin{bmatrix} 4 & 3 \\ 1 & 2 \end{bmatrix}$

3. $\begin{bmatrix} 6 & -14 \\ 0 & 2 \end{bmatrix}$

4. $\begin{bmatrix} 0 & 3 \\ 2 & 4 \end{bmatrix}$

5. $\begin{bmatrix} 1 & 0 \\ 0 & 1 \end{bmatrix}$

6. $\begin{bmatrix} 4 & 0 \\ 0 & 5 \end{bmatrix}$

7. $\begin{bmatrix} a & 0 \\ 0 & b \end{bmatrix}$

8. $\begin{bmatrix} 3 & 0 & 0 \\ 0 & 4 & 0 \\ 0 & 0 & 5 \end{bmatrix}$

EXAMPLE A (Inverses of Large Matrices)   Find the multiplicative inverse of

$$\begin{bmatrix} 2 & 6 & 6 \\ 2 & 7 & 6 \\ 2 & 7 & 7 \end{bmatrix}$$

Solution.   We use the reduction method described in the text.

$$\begin{bmatrix} 2 & 6 & 6 & | & 1 & 0 & 0 \\ 2 & 7 & 6 & | & 0 & 1 & 0 \\ 2 & 7 & 7 & | & 0 & 0 & 1 \end{bmatrix}$$

$\sim \begin{bmatrix} 1 & 3 & 3 & | & \frac{1}{2} & 0 & 0 \\ 2 & 7 & 6 & | & 0 & 1 & 0 \\ 2 & 7 & 7 & | & 0 & 0 & 1 \end{bmatrix}$ (divide row 1 by 2)

$\sim \begin{bmatrix} 1 & 3 & 3 & | & \frac{1}{2} & 0 & 0 \\ 0 & 1 & 0 & | & -1 & 1 & 0 \\ 0 & 1 & 1 & | & -1 & 0 & 1 \end{bmatrix}$ (add -2 times row 1 to row 2 and to row 3)

$\sim \begin{bmatrix} 1 & 3 & 3 & | & \frac{1}{2} & 0 & 0 \\ 0 & 1 & 0 & | & -1 & 1 & 0 \\ 0 & 0 & 1 & | & 0 & -1 & 1 \end{bmatrix}$ (add -1 times row 2 to row 3)

$\sim \begin{bmatrix} 1 & 0 & 3 & | & \frac{7}{2} & -3 & 0 \\ 0 & 1 & 0 & | & -1 & 1 & 0 \\ 0 & 0 & 1 & | & 0 & -1 & 1 \end{bmatrix}$ (add -3 times row 2 to row 1)

$\sim \begin{bmatrix} 1 & 0 & 0 & | & \frac{7}{2} & 0 & -3 \\ 0 & 1 & 0 & | & -1 & 1 & 0 \\ 0 & 0 & 1 & | & 0 & -1 & 1 \end{bmatrix}$ (add -3 times row 3 to row 1)

Thus the desired inverse is

$$\begin{bmatrix} \frac{7}{2} & 0 & -3 \\ -1 & 1 & 0 \\ 0 & -1 & 1 \end{bmatrix}$$

Use the method illustrated above to find the multiplicative inverse of each of the following.

9. $\begin{bmatrix} 1 & 3 \\ 2 & 4 \end{bmatrix}$

10. $\begin{bmatrix} 2 & 6 \\ 3 & 1 \end{bmatrix}$

11. $\begin{bmatrix} 1 & 1 & 1 \\ 1 & -1 & 2 \\ 3 & 2 & 0 \end{bmatrix}$

12. $\begin{bmatrix} 2 & 1 & 1 \\ 1 & 3 & 1 \\ -1 & 4 & 0 \end{bmatrix}$

13. $\begin{bmatrix} 3 & 1 & 2 \\ 4 & 1 & -6 \\ 1 & 0 & 1 \end{bmatrix}$

14. $\begin{bmatrix} 2 & 4 & 6 \\ 3 & 2 & -5 \\ 2 & 3 & 1 \end{bmatrix}$

15. $\begin{bmatrix} 1 & 2 & 1 & 1 \\ 0 & 2 & 3 & 2 \\ 0 & 0 & 1 & 3 \\ 0 & 0 & 0 & 4 \end{bmatrix}$

16. $\begin{bmatrix} 1 & 1 & 1 & 1 \\ 1 & 1 & 1 & -1 \\ 1 & 1 & -1 & 1 \\ 1 & -1 & 1 & 1 \end{bmatrix}$

Solve the following systems by making use of the inverses you found in Problems 11–14. Begin by writing the system in the matrix form **AX = B**.

17. $x + y + z = 2$
$x - y + 2z = -1$
$3x + 2y = 5$

18. $2x + y + z = 4$
$x + 3y + z = 5$
$-x + 4y = 0$

19. $3x + y + 2z = 3$
$4x + y - 6z = 2$
$x + z = 6$

20. $2x + 4y + 6z = 9$
$3x + 2y - 5z = 2$
$2x + 3y + z = 4$

EXAMPLE B (Matrices Without Inverses) Try to find the multiplicative inverse of

$$U = \begin{bmatrix} 1 & 4 & 2 \\ 0 & 2 & 4 \\ 0 & -3 & -6 \end{bmatrix}$$

Solution.

$$\left[\begin{array}{ccc|ccc} 1 & 4 & 2 & 1 & 0 & 0 \\ 0 & 2 & 4 & 0 & 1 & 0 \\ 0 & -3 & -6 & 0 & 0 & 1 \end{array}\right] \sim \left[\begin{array}{ccc|ccc} 1 & 4 & 2 & 1 & 0 & 0 \\ 0 & 1 & 2 & 0 & \frac{1}{2} & 0 \\ 0 & -3 & -6 & 0 & 0 & 1 \end{array}\right]$$

$$\sim \left[\begin{array}{ccc|ccc} 1 & 4 & 2 & 1 & 0 & 0 \\ 0 & 1 & 2 & 0 & \frac{1}{2} & 0 \\ 0 & 0 & 0 & 0 & \frac{3}{2} & 1 \end{array}\right]$$

Since we got a row of zeros in the left half above, we know we can never reduce it to the identity matrix $\mathbf{I}$. The matrix $\mathbf{U}$ does not have an inverse.

Show that neither of the following matrices has a multiplicative inverse.

21. $\begin{bmatrix} 1 & 3 & 4 \\ 2 & 1 & -1 \\ 4 & 7 & 7 \end{bmatrix}$
22. $\begin{bmatrix} 2 & -2 & 4 \\ 5 & 3 & 2 \\ 3 & 5 & -2 \end{bmatrix}$

MISCELLANEOUS PROBLEMS

In Problems 23–26, find the multiplicative inverse or indicate that it does not exist.

23. $\begin{bmatrix} 4 & -3 \\ 5 & -\frac{15}{4} \end{bmatrix}$
24. $\begin{bmatrix} 3 & -1 \\ 4 & 2 \end{bmatrix}$

25. $\begin{bmatrix} 1 & -2 & 1 \\ 3 & 0 & 2 \\ 1 & 2 & \frac{1}{2} \end{bmatrix}$
26. $\begin{bmatrix} -2 & 4 & 2 \\ 3 & 5 & 6 \\ 1 & 9 & 8 \end{bmatrix}$

27. Find the multiplicative inverse of

$$\begin{bmatrix} 2 & 0 & 0 \\ 0 & 3 & 0 \\ 0 & 0 & -4 \end{bmatrix}$$

28. Give a formula for $\mathbf{U}^{-1}$ if

$$\mathbf{U} = \begin{bmatrix} a & 0 & 0 \\ 0 & b & 0 \\ 0 & 0 & c \end{bmatrix}$$

When does the matrix $\mathbf{U}$ fail to have an inverse?

29. Use your result from Problem 25 to solve the system

$$x - 2y + z = a$$
$$3x \qquad + 2z = b$$
$$x + 2y + \tfrac{1}{2}z = c$$

30. Let $\mathbf{A}$ and $\mathbf{B}$ be 3×3 matrices with inverses $\mathbf{A}^{-1}$ and $\mathbf{B}^{-1}$. Show that $\mathbf{AB}$ has an inverse given by $\mathbf{B}^{-1}\mathbf{A}^{-1}$. *Hint:* The product in either order must be $\mathbf{I}$.

31. Show that

$$\begin{bmatrix} 1 & -1 \\ 3 & -3 \end{bmatrix}\begin{bmatrix} 2 & -4 \\ 2 & -4 \end{bmatrix} = \begin{bmatrix} 0 & 0 \\ 0 & 0 \end{bmatrix}$$

Thus $\mathbf{AB} = \mathbf{O}$ but neither $\mathbf{A}$ nor $\mathbf{B}$ is $\mathbf{O}$. This is another way in which matrices differ from ordinary numbers.

32. Suppose $\mathbf{AB} = \mathbf{O}$ and $\mathbf{A}$ has a multiplicative inverse. Show that $\mathbf{B} = \mathbf{O}$. See Problem 31.

33. Find the inverse of the matrix

$$\begin{bmatrix} 1 & 1 & 1 & 1 \\ 1 & 2 & 2 & 2 \\ 1 & 2 & 1 & 1 \\ 1 & 2 & 1 & 2 \end{bmatrix}$$

34. Consider the matrices

$$\mathbf{A} = \begin{bmatrix} 0 & 1 & 0 \\ 0 & 0 & 1 \\ 1 & 0 & 0 \end{bmatrix} \quad \text{and} \quad \mathbf{B} = \begin{bmatrix} 0 & 1 & 0 & 0 \\ 0 & 0 & 1 & 0 \\ 0 & 0 & 0 & 1 \\ 1 & 0 & 0 & 0 \end{bmatrix}$$

(a) Show that $\mathbf{A}^3 = \mathbf{I}$ and $\mathbf{B}^4 = \mathbf{I}$.
(b) What does (a) allow you to conclude about $\mathbf{A}^{-1}$ and $\mathbf{B}^{-1}$?
(c) Conjecture a generalization of (a)

35. Show that the inverse of a Heisenberg matrix (see Problem 28 of Section 11-3) is a Heisenberg matrix.

36. TEASER   The matrices

$$\mathbf{A} = \begin{bmatrix} 1 & \frac{1}{2} \\ \frac{1}{2} & \frac{1}{3} \end{bmatrix} \quad \text{and} \quad \mathbf{B} = \begin{bmatrix} 1 & \frac{1}{2} & \frac{1}{3} \\ \frac{1}{2} & \frac{1}{3} & \frac{1}{4} \\ \frac{1}{3} & \frac{1}{4} & \frac{1}{5} \end{bmatrix}$$

and their $n \times n$ generalizations are called Hilbert matrices; they play an important role in numerical analysis.

(a) Find $\mathbf{A}^{-1}$ and $\mathbf{B}^{-1}$.
(b) Let $\mathbf{C}$ be the column matrix with entries ($\frac{11}{6}$, $\frac{13}{12}$, $\frac{47}{60}$). Solve the equation $\mathbf{BX} = \mathbf{C}$ for $\mathbf{X}$.

Matrix	Determinant	Value of Determinant
$\begin{bmatrix} a & b \\ c & d \end{bmatrix}$	$\begin{vmatrix} a & b \\ c & d \end{vmatrix}$	$ad - bc$
$\begin{bmatrix} a_1 & b_1 & c_1 \\ a_2 & b_2 & c_2 \\ a_3 & b_3 & c_3 \end{bmatrix}$	$\begin{vmatrix} a_1 & b_1 & c_1 \\ a_2 & b_2 & c_2 \\ a_3 & b_3 & c_3 \end{vmatrix}$	$a_1 b_2 c_3 + a_2 b_3 c_1 + a_3 b_1 c_2$ $- a_1 b_3 c_2 - a_2 b_1 c_3 - a_3 b_2 c_1$

11-5   Second- and Third-Order Determinants

The notion of a determinant is usually attributed to the German mathematician Gottfried Wilhelm Leibniz (1646–1716), but it seems that Seki Kōwa of Japan had the idea somewhat earlier. It grew out of the study of systems of equations.

SECOND-ORDER DETERMINANTS

Consider the general system of two equations in two unknowns

$$ax + by = r$$
$$cx + dy = s$$

If we multiply the second equation by a and then add $-c$ times the first equation to it, we obtain the equivalent triangular system.

$$ax + \qquad by = r$$
$$(ad - bc)y = as - cr$$

If $ad - bc \neq 0$, we can solve this system by backward substitution.

$$x = \frac{rd - bs}{ad - bc}$$

$$y = \frac{as - rc}{ad - bc}$$

These formulas are hard to remember unless we associate special symbols with the numbers $ad - bc$, $rd - bs$, and $as - rc$. For the first of these, we propose

$$\begin{vmatrix} a & b \\ c & d \end{vmatrix} = ad - bc$$

The symbol on the left is called a **second-order determinant,** and we say that $ad - bc$ is its value. Thus

$$\begin{vmatrix} -2 & -1 \\ 5 & 6 \end{vmatrix} = (-2)(6) - (-1)(5) = -7$$

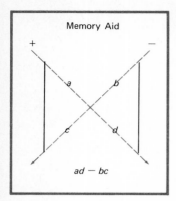

Memory Aid

$+$ $-$

a b

c d

$ad - bc$

Figure 3

Figure 3 may help you remember how to make the evaluation. With this new symbol, we can write the solution to

$$ax + by = r$$
$$cx + dy = s$$

as

$$x = \frac{rd - bs}{ad - bc} = \frac{\begin{vmatrix} r & b \\ s & d \end{vmatrix}}{\begin{vmatrix} a & b \\ c & d \end{vmatrix}}$$

$$y = \frac{as - rc}{ad - bc} = \frac{\begin{vmatrix} a & r \\ c & s \end{vmatrix}}{\begin{vmatrix} a & b \\ c & d \end{vmatrix}}$$

These results are easy to remember when we notice that the denominator is the determinant of the coefficient matrix, and that the numerator is the same except that the coefficients of the unknown we are seeking are replaced by the constants from the right side of the system.

Here is an example.

$$3x - 2y = 7$$
$$4x + 5y = 2$$

$$x = \frac{\begin{vmatrix} 7 & -2 \\ 2 & 5 \end{vmatrix}}{\begin{vmatrix} 3 & -2 \\ 4 & 5 \end{vmatrix}} = \frac{(7)(5) - (-2)(2)}{(3)(5) - (-2)(4)} = \frac{39}{23}$$

$$y = \frac{\begin{vmatrix} 3 & 7 \\ 4 & 2 \end{vmatrix}}{\begin{vmatrix} 3 & -2 \\ 4 & 5 \end{vmatrix}} = \frac{(3)(2) - (7)(4)}{23} = -\frac{22}{23}$$

The choice of the name *determinant* is appropriate, for the determinants of a system completely *determine* its character.

1. If $ad - bc \neq 0$, the system has a unique solution, the one given at the top of this page.
2. If $ad - bc = 0$, $as - rc = 0$, and $rd - bs = 0$, then a, b, and r are proportional to c, d, and s and the system has infinitely many solutions.

Here is an example.

$$3x - 2y = 7 \qquad \frac{3}{6} = \frac{-2}{-4} = \frac{7}{14}$$
$$6x - 4y = 14$$

3. If $ad - bc = 0$ and $as - rc \neq 0$ or $rd - bs \neq 0$, then a and b are proportional to c and d, but this proportionality does not extend to r and s; the system has no solution. This is illustrated by the following.

$$3x - 2y = 7 \qquad \frac{3}{6} = \frac{-2}{-4} \neq \frac{7}{10}$$
$$6x - 4y = 10$$

THIRD-ORDER DETERMINANTS

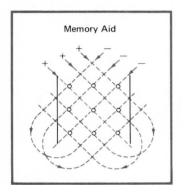

Memory Aid

Figure 4

When we consider the general system of three equations in three unknowns

$$a_1 x + b_1 y + c_1 z = d_1$$
$$a_2 x + b_2 y + c_2 z = d_2$$
$$a_3 x + b_3 y + c_3 z = d_3$$

things get more complicated, but the results are similar. The appropriate determinant symbol and its corresponding value are

$$\begin{vmatrix} a_1 & b_1 & c_1 \\ a_2 & b_2 & c_2 \\ a_3 & b_3 & c_3 \end{vmatrix} = a_1 b_2 c_3 + b_1 c_2 a_3 + c_1 b_3 a_2 - c_1 b_2 a_3 - b_1 a_2 c_3 - a_1 b_3 c_2$$

There are six terms in the sum on the right, three with a plus sign and three with a minus sign. The diagrams in Figures 4 and 5 will help you remember the products that enter each term. Just follow the arrows. Here is an example.

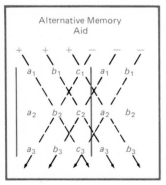

Alternative Memory Aid

Figure 5

$$\begin{vmatrix} 3 & 2 & 4 \\ 4 & -2 & 6 \\ 8 & 3 & 5 \end{vmatrix} = (3)(-2)(5) + (2)(6)(8) + (4)(3)(4)$$
$$-(4)(-2)(8) - (2)(4)(5) - (3)(3)(6)$$
$$= 84$$

CRAMER'S RULE

We saw that the solutions for x and y in a second-order system could be written as the quotients of two determinants. That fact generalizes to the third-order case. We present it without proof. Consider

$$a_1 x + b_1 y + c_1 z = d_1$$
$$a_2 x + b_2 y + c_2 z = d_2$$
$$a_3 x + b_3 y + c_3 z = d_3$$

If

$$D = \begin{vmatrix} a_1 & b_1 & c_1 \\ a_2 & b_2 & c_2 \\ a_3 & b_3 & c_3 \end{vmatrix} \neq 0$$

then the system above has a unique solution given by

$$x = \frac{1}{D} \begin{vmatrix} d_1 & b_1 & c_1 \\ d_2 & b_2 & c_2 \\ d_3 & b_3 & c_3 \end{vmatrix} \qquad y = \frac{1}{D} \begin{vmatrix} a_1 & d_1 & c_1 \\ a_2 & d_2 & c_2 \\ a_3 & d_3 & c_3 \end{vmatrix} \qquad z = \frac{1}{D} \begin{vmatrix} a_1 & b_1 & d_1 \\ a_2 & b_2 & d_2 \\ a_3 & b_3 & d_3 \end{vmatrix}$$

The pattern is the same as in the second-order situation. The denominator D is the determinant of the coefficient matrix. The numerator in each case is obtained from D by replacing the coefficients of the unknown by the constants from the right side of the system.

This method of solving a system of equations is named after one of its discoverers, Gabriel Cramer (1704–1752). Historically, it has been a popular method. However, note that even for a system of three equations in three unknowns, it requires the evaluation of four determinants. The method of matrices (Section 11-2) is considerably more efficient, both for hand and computer calculation. Consequently, Cramer's rule is now primarily of theoretical rather than practical interest.

PROPERTIES OF DETERMINANTS

We are interested in how the matrix operations considered in Section 11-2 affect the values of the corresponding determinants.

1. *Interchanging two rows changes the sign of the determinant; for example,*

$$\begin{vmatrix} a & b \\ c & d \end{vmatrix} = - \begin{vmatrix} c & d \\ a & b \end{vmatrix}$$

CAUTION

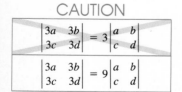

2. *Multiplying a row by a constant k multiplies the value of the determinant by k; for example,*

$$\begin{vmatrix} ka & kb \\ c & d \end{vmatrix} = k \begin{vmatrix} a & b \\ c & d \end{vmatrix}$$

3. *Adding a multiple of one row to another does not affect the value of the determinant; for example,*

$$\begin{vmatrix} a & b \\ c + ka & d + kb \end{vmatrix} = \begin{vmatrix} a & b \\ c & d \end{vmatrix}$$

We mention also the effect of a new operation.

4. *Interchanging the rows and columns (pairwise) does not affect the value of the determinant; for example,*

$$\begin{vmatrix} a & c \\ b & d \end{vmatrix} = \begin{vmatrix} a & b \\ c & d \end{vmatrix}$$

Though they are harder to prove in the third-order case, we emphasize that all four properties hold for both second- and third-order determinants. We offer only one proof, a proof of Property 3 in the second-order case.

$$\begin{vmatrix} a & b \\ c + ka & d + kb \end{vmatrix} = a(d + kb) - b(c + ka)$$

$$= ad + akb - bc - bka$$

$$= ad - bc$$

$$= \begin{vmatrix} a & b \\ c & d \end{vmatrix}$$

Property 3 can be a great aid in evaluating a determinant. Using it, we can transform a matrix to triangular form without changing the value of its determinant. But *the determinant of a triangular matrix is just the product of the elements on the main diagonal,* since this is the only nonzero term in the determinant formula. Here is an example.

$$\begin{vmatrix} 1 & 3 & 4 \\ -1 & -2 & 3 \\ 2 & -6 & 11 \end{vmatrix} = \begin{vmatrix} 1 & 3 & 4 \\ 0 & 1 & 7 \\ 0 & -12 & 3 \end{vmatrix} = \begin{vmatrix} 1 & 3 & 4 \\ 0 & 1 & 7 \\ 0 & 0 & 87 \end{vmatrix} = 87$$

Problem Set 11-5

Evaluate each of the determinants in Problems 1–8 by inspection.

1. $\begin{vmatrix} 4 & 0 \\ 0 & -2 \end{vmatrix}$

2. $\begin{vmatrix} 8 & 0 \\ 5 & 0 \end{vmatrix}$

3. $\begin{vmatrix} 11 & 4 \\ 0 & 2 \end{vmatrix}$

4. $\begin{vmatrix} 2 & -1 & 5 \\ 0 & 4 & 2 \\ 0 & 0 & -1 \end{vmatrix}$

5. $\begin{vmatrix} -1 & -7 & 9 \\ 0 & 5 & 4 \\ 0 & 0 & 10 \end{vmatrix}$

6. $\begin{vmatrix} 3 & -2 & 1 \\ 0 & 0 & 0 \\ 1 & 5 & -8 \end{vmatrix}$

7. $\begin{vmatrix} 3 & 0 & 8 \\ 10 & 0 & 2 \\ -1 & 0 & -9 \end{vmatrix}$

8. $\begin{vmatrix} 9 & 0 & 0 \\ 0 & 0 & -2 \\ 0 & 4 & 0 \end{vmatrix}$

9. If

$$\begin{vmatrix} a_1 & b_1 & c_1 \\ a_2 & b_2 & c_2 \\ a_3 & b_3 & c_3 \end{vmatrix} = 12$$

find the value of each of the following determinants.

(a) $\begin{vmatrix} a_1 & a_2 & a_3 \\ b_1 & b_2 & b_3 \\ c_1 & c_2 & c_3 \end{vmatrix}$

(b) $\begin{vmatrix} a_3 & b_3 & c_3 \\ a_2 & b_2 & c_2 \\ a_1 & b_1 & c_1 \end{vmatrix}$

(c) $\begin{vmatrix} a_1 & b_1 & c_1 \\ a_2 & b_2 & c_2 \\ 3a_3 & 3b_3 & 3c_3 \end{vmatrix}$

(d) $\begin{vmatrix} a_1 + 3a_3 & b_1 + 3b_3 & c_1 + 3c_3 \\ a_2 & b_2 & c_2 \\ a_3 & b_3 & c_3 \end{vmatrix}$

Evaluate each of the determinants in Problems 10–17.

10. $\begin{vmatrix} 3 & 2 \\ 5 & 6 \end{vmatrix}$

11. $\begin{vmatrix} 5 & 3 \\ 5 & -3 \end{vmatrix}$

12. $\begin{vmatrix} 3 & 0 & 0 \\ -2 & 5 & 4 \\ 1 & 2 & -9 \end{vmatrix}$

13. $\begin{vmatrix} 4 & 8 & -2 \\ 1 & -2 & 0 \\ 2 & 4 & 0 \end{vmatrix}$

14. $\begin{vmatrix} 3 & 2 & -4 \\ 1 & 0 & 5 \\ 4 & -2 & 3 \end{vmatrix}$

15. $\begin{vmatrix} 2 & 4 & 1 \\ 1 & 3 & 6 \\ 2 & 3 & -1 \end{vmatrix}$

□ 16. $\begin{vmatrix} 5.1 & -3.2 & 2.6 \\ 1.3 & 4.5 & 2.3 \\ 3.4 & -2.2 & 1.9 \end{vmatrix}$

□ 17. $\begin{vmatrix} 2.03 & 5.41 & -3.14 \\ 0 & 6.22 & 0 \\ -1.93 & 7.13 & 6.34 \end{vmatrix}$

Use Cramer's rule to solve the system of equations in Problems 18–21.

18. $2x - 3y = -11$
    $x + 2y = -2$

19. $5x + y = 7$
    $3x - 4y = 18$

20. $2x + 4y + z = 15$
    $x + 3y + 6z = 15$
    $2x + 3y - z = 11$

21. $5x - 3y + 2z = 18$
    $x + 4y + 2z = -4$
    $3x - 2y + z = 11$

MISCELLANEOUS PROBLEMS

22. Establish Property 2 for a general second-order determinant.

23. Evaluate each of the following determinants (the easy way).

(a) $\begin{vmatrix} 1 & 2 & 3 \\ 0 & 0 & 0 \\ 1.9 & 2.9 & 3.9 \end{vmatrix}$

(b) $\begin{vmatrix} 1 & 2 & 3 \\ 1.1 & 2.2 & 3.3 \\ 1.9 & 2.9 & 3.9 \end{vmatrix}$

(c) $\begin{vmatrix} 1.1 & 2.2 & 3.3 \\ 4.4 & 5.5 & 6.6 \\ 5.5 & 7.7 & 9.9 \end{vmatrix}$

(d) $\begin{vmatrix} 1 & 2 & 3 \\ 0 & 2 & 3 \\ 0 & 0 & 3 \end{vmatrix}$

(e) $\begin{vmatrix} 1 & 1 & 1 \\ 1 & 2 & 3 \\ 1 & 3 & 6 \end{vmatrix}$

(f) $\begin{vmatrix} 1 & 1 & 1 \\ 1 & 2 & 4 \\ 1 & 3 & 9 \end{vmatrix}$

24. Show that the value of a third-order determinant is 0 if either of the following are true.
    (a) Two rows are proportional.
    (b) The sum of two rows is the third row.

25. Consider the system of equations

$$kx + y = k^2$$
$$x + ky = 1$$

For what values of k does this system have (a) unique solution; (b) infinitely many solutions; (c) no solution?

26. If $\mathbf{C} = \mathbf{AB}$, where $\mathbf{A}$ and $\mathbf{B}$ are square matrices of the same size, then their determinants satisfy $|\mathbf{C}| = |\mathbf{A}||\mathbf{B}|$. Prove this fact for 2×2 matrices.

27. Suppose that $\mathbf{A}$ and $\mathbf{B}$ are 3×3 matrices with $|\mathbf{A}| = -2$. Use Problem 26 to evaluate (a) $|\mathbf{A}^5|$; (b) $|\mathbf{A}^{-1}|$; (c) $|\mathbf{B}^{-1}\mathbf{AB}|$; (d) $|3\mathbf{A}^3|$.

28. Show that

$$\begin{vmatrix} 1 & a & a^2 \\ 1 & b & b^2 \\ 1 & c & c^2 \end{vmatrix} = (a - b)(b - c)(c - a)$$

29. Let $(0, 0)$, (a, b), (c, d), and $(a + c, b + d)$ be the vertices of a parallelogram with a, b, c, and d being positive numbers. Show that the area of the parallelogram is the determinant

$$\begin{vmatrix} a & b \\ c & d \end{vmatrix}$$

30. Consider the following determinant equation.

$$\begin{vmatrix} x & y & 1 \\ a & b & 1 \\ c & d & 1 \end{vmatrix} = 0 \qquad \text{where } (a, b) \neq (c, d)$$

(a) Show that the above equation is the equation of a line in the xy-plane.
(b) How can you tell immediately that the points (a, b) and (c, d) are on this line?
(c) Write a determinant equation for the line that passes through the points $(5, -1)$ and $(4, 11)$.

31. Determine the polynomial $P(a, b, c)$ for which

$$\begin{vmatrix} a & b & c \\ b & c & a \\ c & a & b \end{vmatrix} = (a + b + c)P(a, b, c)$$

32. **TEASER**   If $\mathbf{A}$ is a square matrix, then $P(x) = |\mathbf{A} - x\mathbf{I}|$ is called the *characteristic polynomial* of the matrix. Let

$$\mathbf{A} = \begin{vmatrix} 1 & 1 & -1 \\ 0 & 0 & -1 \\ 1 & 1 & 4 \end{vmatrix}$$

(a) Write the polynomial $P(x)$ in the form $ax^3 + bx^2 + cx + d$.
(b) Solve the equation $P(x) = 0$. The solutions are called the *characteristic values* of $\mathbf{A}$.
(c) Show that $P(\mathbf{A}) = 0$, that is, $\mathbf{A}$ satisfies its own characteristic equation.

One of the most consistent workers on the theory of determinants over a period of 50 years was the Englishman, James Joseph Sylvester. Known as a poet, a wit, and a mathematician, he taught in England and in America. During his stay at Johns Hopkins University in Baltimore (1877–1883), he helped establish one of the first graduate programs in mathematics in America. Under his tutelage, mathematics began to flourish in the United States.

James Joseph Sylvester
1814–1897

11-6   Higher-Order Determinants

Having defined determinants for 2×2 and 3×3 matrices, we expect to do it for 4×4 matrices, 5×5 matrices, and so on. Our problem is to do it in such a way that Cramer's rule and the determinant properties of Section 11-5 still hold. This will take some work.

MINORS

We begin by introducing the standard notation for a general $n \times n$ matrix.

$$\begin{bmatrix} a_{11} & a_{12} & a_{13} & \cdots & a_{1n} \\ a_{21} & a_{22} & a_{23} & \cdots & a_{2n} \\ a_{31} & a_{32} & a_{33} & \cdots & a_{3n} \\ \cdot & \cdot & \cdot & & \cdot \\ \cdot & \cdot & \cdot & & \cdot \\ \cdot & \cdot & \cdot & & \cdot \\ a_{n1} & a_{n2} & a_{n3} & \cdots & a_{nn} \end{bmatrix}$$

Note the use of the double subscript on each entry: the first subscript gives the row in which a_{ij} is and the second gives the column. For example, a_{32} is the entry in the third row and second column.

Associated with each entry a_{ij} in an $n \times n$ matrix is a determinant M_{ij} of order $n - 1$ called the **minor** of a_{ij}. It is obtained by taking the determinant of the submatrix that results when we blot out the row and column in which a_{ij} stands. For example, the minor M_{13} of a_{13} in the 4×4 matrix

$$\begin{bmatrix} a_{11} & a_{12} & a_{13} & a_{14} \\ a_{21} & a_{22} & a_{23} & a_{24} \\ a_{31} & a_{32} & a_{33} & a_{34} \\ a_{41} & a_{42} & a_{43} & a_{44} \end{bmatrix}$$

is the third-order determinant.

$$\begin{vmatrix} a_{21} & a_{22} & a_{24} \\ a_{31} & a_{32} & a_{34} \\ a_{41} & a_{42} & a_{44} \end{vmatrix}$$

THE GENERAL nth-ORDER DETERMINANT

Here is the definition to which we have been leading.

$$\begin{vmatrix} a_{11} & a_{12} & \cdots & a_{1n} \\ a_{21} & a_{22} & \cdots & a_{2n} \\ \vdots & \vdots & & \vdots \\ a_{n1} & a_{n2} & \cdots & a_{nn} \end{vmatrix} = a_{11}M_{11} - a_{12}M_{12} + a_{13}M_{13} \cdots + (-1)^{n+1}a_{1n}M_{1n}$$

There are three important questions to answer regarding this definition.

Does this definition really define? Only if the minors M_{ij} can be evaluated. They are themselves determinants, but here is the key point: They are of order $n - 1$, one less than the order of the determinant we started with. They can, in turn, be expressed in terms of determinants of order $n - 2$, and so on, using the same definition. Thus, for example, a fifth-order determinant can be expressed in terms of fourth-order determinants, and these fourth-order determinants can be expressed in terms of third-order determinants. But we know how to evaluate third-order determinants from Section 11-5.

Is this definition consistent with the earlier definition when applied to third-order determinants? Yes, for if we apply it to a general third-order determinant, we get

$$\begin{vmatrix} a_1 & b_1 & c_1 \\ a_2 & b_2 & c_2 \\ a_3 & b_3 & c_3 \end{vmatrix} = a_1 \begin{vmatrix} b_2 & c_2 \\ b_3 & c_3 \end{vmatrix} - b_1 \begin{vmatrix} a_2 & c_2 \\ a_3 & c_3 \end{vmatrix} + c_1 \begin{vmatrix} a_2 & b_2 \\ a_3 & b_3 \end{vmatrix}$$

$$= a_1b_2c_3 - a_1c_2b_3 - b_1a_2c_3 + b_1c_2a_3 + c_1a_2b_3 - c_1b_2a_3$$

This is the same value we gave in Section 11-5.

Does this definition preserve Cramer's rule and the properties of Section 11-5? Yes, it does. We shall not prove this because the proofs are lengthy and difficult.

EXPANSION ACCORDING TO ANY ROW OR COLUMN

Our definition expressed the value of a determinant in terms of the entries and minors of the first row; we call it an expansion according to the first row. It is a remarkable fact that we can expand a determinant according to any row or column (and always get the same answer).

Before we can show what we mean, we must explain a sign convention. We associate a plus or minus sign with every position in a matrix. To the ij-position, we assign a plus sign if $i + j$ is even and a minus sign otherwise. Thus for a 4×4 matrix, we have this pattern of signs.

$$\begin{bmatrix} + & - & + & - \\ - & + & - & + \\ + & - & + & - \\ - & + & - & + \end{bmatrix}$$

There is always a $+$ in the upper left position and then the signs alternate.

With this understanding about signs, we may expand according to any row or column. For example, to evaluate a fourth-order determinant, we can expand according to the second column if we wish. We multiply each entry in that column by its minor, prefixing each product with a plus or minus sign according to the pattern above. Then we add the results.

$$\begin{vmatrix} a_{11} & a_{12} & a_{13} & a_{14} \\ a_{21} & a_{22} & a_{23} & a_{24} \\ a_{31} & a_{32} & a_{33} & a_{34} \\ a_{41} & a_{42} & a_{43} & a_{44} \end{vmatrix} = -a_{12}M_{12} + a_{22}M_{22} - a_{32}M_{32} + a_{42}M_{42}$$

EXAMPLE

To evaluate

$$\begin{vmatrix} 6 & 0 & 4 & -1 \\ 2 & 0 & -1 & 4 \\ -2 & 4 & -2 & 3 \\ 4 & 0 & 5 & -4 \end{vmatrix}$$

it is obviously best to expand according to the second column, since three of the four resulting terms are zero. The single nonzero term is just $(-1)(4)$ times the minor M_{32}—that is,

$$-4 \begin{vmatrix} 6 & 4 & -1 \\ 2 & -1 & 4 \\ 4 & 5 & -4 \end{vmatrix}$$

We could now evaluate this third-order determinant as in Section 11-5. But having seen the usefulness of zeros, let us take a different tack. It is easy to get two zeros in the first column. Simply add -3 times the second row to the first row and -2 times the second row to the third. We get

$$-4 \begin{vmatrix} 0 & 7 & -13 \\ 2 & -1 & 4 \\ 0 & 7 & -12 \end{vmatrix}$$

Finally, expand according to the first column.

$$(-4)(-1)(2) \begin{vmatrix} 7 & -13 \\ 7 & -12 \end{vmatrix} = 8(-84 + 91) = 56$$

The reason for the factor of -1 is that the entry 2 is in a minus position in the 3×3 pattern of signs.

Problem Set 11-6

Evaluate each of the determinants in Problems 1–6 according to a row or column of your choice. Make a good choice or suffer the consequences!

1. $\begin{vmatrix} 3 & -2 & 4 \\ 1 & 5 & 0 \\ 3 & 10 & 0 \end{vmatrix}$
2. $\begin{vmatrix} 4 & 0 & -6 \\ -2 & 3 & 5 \\ 1 & 0 & 8 \end{vmatrix}$

3. $\begin{vmatrix} 1 & 2 & 3 \\ 0 & 2 & 3 \\ 1 & 3 & 4 \end{vmatrix}$
4. $\begin{vmatrix} 2 & -1 & -1 \\ 3 & 4 & 2 \\ 0 & -1 & -1 \end{vmatrix}$

5. $\begin{vmatrix} 3 & 0 & 0 & 0 \\ -1 & 1 & 4 & 2 \\ 2 & 0 & 2 & -3 \\ -4 & 0 & 1 & 5 \end{vmatrix}$
6. $\begin{vmatrix} 0 & 5 & 0 & 0 \\ 1 & -3 & 0 & 2 \\ 4 & 1 & 2 & 8 \\ -3 & 2 & 0 & 5 \end{vmatrix}$

Evaluate each of the determinants in Problems 7–10 by first getting some zeros in a row or column and then expanding according to that row or column.

7. $\begin{vmatrix} 3 & 5 & -10 \\ 2 & 4 & 6 \\ -3 & -5 & 12 \end{vmatrix}$
8. $\begin{vmatrix} 2 & -1 & 2 \\ 4 & 3 & 4 \\ 7 & -5 & 10 \end{vmatrix}$

9. $\begin{vmatrix} 1 & -2 & 1 & 4 \\ -2 & 5 & -3 & 1 \\ 0 & 7 & -4 & 2 \\ 3 & -2 & 2 & 6 \end{vmatrix}$
10. $\begin{vmatrix} 1 & -2 & 0 & -4 \\ 3 & -4 & 3 & -10 \\ 2 & 1 & -2 & 1 \\ 4 & -5 & 1 & 4 \end{vmatrix}$

11. Solve the following system for x only.

$$x - 2y + z + 4w = 1$$
$$-2x + 5y - 3z + w = -2$$
$$7y - 4z + 2w = 3$$
$$3x - 2y + 2z + 6w = 6$$

(Make use of your answer to Problem 9.)

12. Solve the following system for z only.

$$x - 2y - 4w = -14$$
$$3x - 4y + 3z - 10w = -28$$
$$2x + y - 2z + w = 0$$
$$4x - 5y + z + 4w = 9$$

(Make use of your answer to Problem 10.)

Evaluate the determinants in Problems 13–18.

13. $\begin{vmatrix} 2 & -3 & 2 \\ 1 & 0 & -4 \\ -1 & 0 & 6 \end{vmatrix}$

14. $\begin{vmatrix} 3 & 1 & -5 \\ 2 & -2 & 7 \\ 1 & 0 & -1 \end{vmatrix}$

15. $\begin{vmatrix} 2 & -3 & 4 & 5 \\ 2 & -3 & 4 & 7 \\ 1 & 6 & 4 & 5 \\ 2 & 6 & 4 & -8 \end{vmatrix}$

16. $\begin{vmatrix} 2 & 2 & 3 & 7 \\ 1 & 2 & 3 & -2 \\ 4 & -3 & 9 & 6 \\ 1 & 2 & 3 & -1 \end{vmatrix}$

17. $\begin{vmatrix} 1 & 2 & -3 & 1 & 2 \\ -1 & 0 & 2 & 5 & -3 \\ 5 & 0 & 0 & -2 & 4 \\ 0 & 0 & 0 & 6 & 3 \\ 0 & 0 & 0 & 2 & -7 \end{vmatrix}$

18. $\begin{vmatrix} 1 & 2 & 3 & 4 & 5 \\ 2 & 1 & 1 & 1 & 1 \\ 3 & 1 & 1 & 1 & 1 \\ 4 & 1 & 1 & 1 & 1 \\ 5 & 1 & 1 & 1 & 1 \end{vmatrix}$

Hint: Subtract row 2 from row 3.

19. Evaluate the following determinant.

$$\begin{vmatrix} a & b & c & d \\ 0 & e & f & g \\ 0 & 0 & h & i \\ 0 & 0 & 0 & j \end{vmatrix}$$

Conjecture a general result about the determinant of a triangular matrix.

c 20. Use the result of Problem 19 to evaluate

$$\begin{vmatrix} 2.12 & 3.14 & -1.61 & 1.72 \\ 0 & -2.36 & 5.91 & 7.82 \\ 0 & 0 & 1.46 & 3.34 \\ 0 & 0 & 0 & 3.31 \end{vmatrix}$$

21. Evaluate by reducing to triangular form and using Problem 19.

$$\begin{vmatrix} 1 & 2 & 2.6 & 1.5 \\ 2.3 & 5.6 & -1.3 & 9.8 \\ 2.7 & 1.3 & 4.2 & -1.9 \\ 5.5 & 6.2 & 3.0 & 1.4 \end{vmatrix}$$

22. Show that

$$\begin{vmatrix} a_1 + d_1 & b_1 & c_1 \\ a_2 + d_2 & b_2 & c_2 \\ a_3 + d_3 & b_3 & c_3 \end{vmatrix} = \begin{vmatrix} a_1 & b_1 & c_1 \\ a_2 & b_2 & c_2 \\ a_3 & b_3 & c_3 \end{vmatrix} + \begin{vmatrix} d_1 & b_1 & c_1 \\ d_2 & b_2 & c_2 \\ d_3 & b_3 & c_3 \end{vmatrix}$$

MISCELLANEOUS PROBLEMS

23. Evaluate each of the following determinants.

(a) $\begin{vmatrix} 1 & 0 & 0 & 2 \\ 2.7 & 5 & 0 & 8.9 \\ 3.4 & 0 & 6 & 9.1 \\ 3 & 0 & 0 & 4 \end{vmatrix}$ (b) $\begin{vmatrix} a & 0 & 0 & b \\ ? & e & 0 & ? \\ ? & 0 & f & ? \\ c & 0 & 0 & d \end{vmatrix}$

24. Solve for x.

$$\begin{vmatrix} 2 - x & 0 & 0 & 5 \\ 2.7 & 3 - x & 0 & 8.9 \\ 3.4 & 0 & 4 - x & 9.1 \\ 5 & 0 & 0 & 2 - x \end{vmatrix} = 0$$

25. Evaluate the determinant below. *Hint:* Subtract the first row from each of the second and third rows.

$$\begin{vmatrix} n + 1 & n + 2 & n + 3 \\ n + 4 & n + 5 & n + 6 \\ n + 7 & n + 8 & n + 9 \end{vmatrix}$$

26. Show that if the entries in a determinant are all integers, then the value of the determinant is an integer. From this and Cramer's rule, draw a conclusion about the nature of the solution to a system of n linear equations in n unknowns if all the constants in the system are integers and the determinant of coefficients is nonzero.

27. Evaluate the given determinants, in which the entries come from Pascal's triangle (see page 537).

$$D_2 = \begin{vmatrix} 1 & 1 \\ 1 & 2 \end{vmatrix}$$

$$D_3 = \begin{vmatrix} 1 & 1 & 1 \\ 1 & 2 & 3 \\ 1 & 3 & 6 \end{vmatrix}$$

$$D_4 = \begin{vmatrix} 1 & 1 & 1 & 1 \\ 1 & 2 & 3 & 4 \\ 1 & 3 & 6 & 10 \\ 1 & 4 & 10 & 20 \end{vmatrix}$$

28. Based on the results of Problem 27, make a conjecture about the value of D_n, the nth-order determinant obtained from Pascal's triangle. Then support your conjecture by describing a systematic way of evaluating D_n.

29. Solve the given system. It should be easy after Problem 27.

$$
\begin{aligned}
x + y + z + w &= 0 \\
x + 2y + 3z + 4w &= 0 \\
x + 3y + 6z + 10w &= 0 \\
x + 4y + 10z + 20w &= 1
\end{aligned}
$$

30. Evaluate the determinants.

$$
E_1 = \begin{vmatrix} 2 & 1 \\ 1 & 2 \end{vmatrix}
\qquad
E_2 = \begin{vmatrix} 2 & 1 & 0 \\ 1 & 2 & 1 \\ 0 & 1 & 2 \end{vmatrix}
\qquad
E_3 = \begin{vmatrix} 2 & 1 & 0 & 0 \\ 1 & 2 & 1 & 0 \\ 0 & 1 & 2 & 1 \\ 0 & 0 & 1 & 2 \end{vmatrix}
$$

Now generalize by conjecturing the value of the nth-order determinant E_n that has 2's on the main diagonal, 1's adjacent to this diagonal on either side, and 0's elsewhere. Then prove your conjecture. *Hint:* Expand according to the first row to show that $E_n = 2E_{n-1} - E_{n-2}$ and from this argue that your conjecture must be correct.

31. Let $a_1, a_2, \ldots, a_n$ and $b_1, b_2, \ldots, b_n$ be two sequences of numbers.
    (a) Let $\mathbf{C}_n$ be the $n \times n$ matrix with entries $c_{ij} = a_i b_j$. Evaluate $|\mathbf{C}_n|$ for $n = 2$, 3, and 4 and make a conjecture.
    (b) Do the same if $c_{ij} = a_i - b_j$.

32. **TEASER** Generalize Problem 28 of Section 11-5 by evaluating the following determinant and writing your answer as the product of six linear factors.

$$
\begin{vmatrix}
1 & a & a^2 & a^3 \\
1 & b & b^2 & b^3 \\
1 & c & c^2 & c^3 \\
1 & d & d^2 & d^3
\end{vmatrix}
$$

Corn

Profit : $40 per acre

Labor : 2 hours per acre

Maximizing Profit

Farmer Brown has 480 acres of land on which he can grow either corn or wheat. He figures that he has 800 hours of labor available during the crucial summer season. Given the profit margins and labor requirements shown at the right, how many acres of each should he plant to maximize his profit? What is this maximum profit?

Wheat

Profit : $30 per acre

Labor : 1 hour per acre

11-7   Systems of Inequalities

At first glance you might think that Farmer Brown should put all of his land into corn. Unfortunately, however, that requires 960 hours of labor and he has only 800 available. Well, maybe he should plant 400 acres of corn, using his allocated 800 hours of labor on them, and let the remaining 80 acres lie idle. Or would it be wise to at least plant enough wheat so all his land is in use? This problem is complicated enough so that no one is likely to find the best solution without a lot of work. And would not a method be better than blind experimenting? That is our subject—a method for handling Farmer Brown's problem and others of the same type.

Like all individuals and businesses, Farmer Brown must operate within certain limitations; we call them **constraints.** Suppose he plants x acres of corn and y acres of wheat. His constraints can be translated into inequalities.

Land constraint:	$x + y \leq 480$
Labor constraint:	$2x + y \leq 800$
Nonnegativity constraints:	$x \geq 0 \qquad y \geq 0$

His task is to maximize the profit $P = 40x + 30y$ subject to these constraints. Before we can solve his problem, we will need to know more about inequalities.

THE GRAPH OF A LINEAR INEQUALITY

The best way to visualize an inequality is by means of its graph. Consider, for example,

$$2x + y \leq 6$$

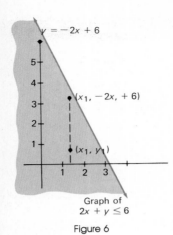

Graph of
$2x + y \leq 6$

Figure 6

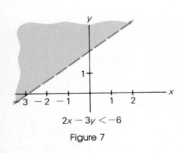

$2x - 3y < -6$

Figure 7

which can be rewritten as

$$y \leq -2x + 6$$

The complete graph consists of those points which satisfy $y = -2x + 6$ (a line), together with those that satisfy $y < -2x + 6$ (the points below the line). To see that this description is correct, note that for any abscissa x_1, the point $(x_1, -2x_1 + 6)$ is on the line $y = -2x + 6$. The point (x_1, y_1) is directly below that point if and only if $y_1 < -2x_1 + 6$ (Figure 6). Thus the graph of $y \leq -2x + 6$ is the **closed half-plane** that we have shaded on the diagram. We refer to it as *closed* because the edge $y = -2x + 6$ is included. Correspondingly, the graph of $y < -2x + 6$ is called an **open half-plane.**

The graph of any linear inequality in x and y is a half-plane, open or closed. To sketch the graph, first draw the corresponding edge. Then determine the correct half-plane by taking a sample point, not on the edge, and checking to see if it satisfies the inequality.

To illustrate this procedure, consider

$$2x - 3y < -6$$

Its graph does not include the line $2x - 3y = -6$, although that line is crucial in determining the graph. We therefore show it as a dotted line. Since the sample point $(0, 0)$ does not satisfy the inequality, we choose the half-plane on the opposite side of the line from it. The complete graph is the shaded open half-plane shown in Figure 7.

GRAPHING A SYSTEM OF LINEAR INEQUALITIES

The graph of a system of inequalities like Farmer Brown's constraints

$$x + y \leq 480$$

$$2x + y \leq 800$$

$$x \geq 0 \qquad y \geq 0$$

is simply the intersection of the graphs of the individual inequalities. We can construct the graph in stages as we do in Figure 8, though we are confident that you will quickly learn to do it in one operation.

The diagram on the right of Figure 8 is the one we want. All the points in the shaded region F satisfy the four inequalities simultaneously. The points $(0, 0)$, $(400, 0)$, $(320, 160)$, and $(0, 480)$ are called the **vertices** (or corner points) of F. Incidentally, the point $(320, 160)$ was obtained by solving the two equations $2x + y = 800$ and $x + y = 480$ simultaneously.

The region F has three important properties (Figure 9).

1. It is polygonal (its boundary consists of line segments).
2. It is convex (if points P and Q are in the region, then the line segment PQ lies entirely within the region).
3. It is bounded (it can be enclosed in a circle).

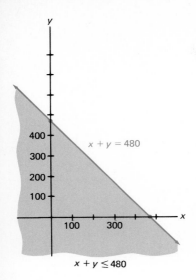

$x + y \leq 480$

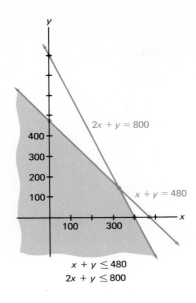

$x + y \leq 480$
$2x + y \leq 800$

Figure 8

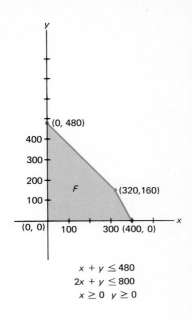

$x + y \leq 480$
$2x + y \leq 800$
$x \geq 0 \quad y \geq 0$

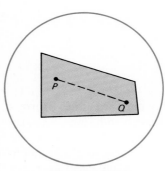

Polygonal, convex,
and bounded

Figure 9

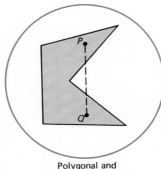

Polygonal and
bounded but
not convex

Figure 10

As a matter of fact, every region that arises as the solution set of a system of linear inequalities is polygonal and convex, though it need not be bounded. The shaded region in Figure 10 could not be the solution set for a system of linear inequalities because it is not convex.

LINEAR PROGRAMMING PROBLEMS

It is time that we solved Farmer Brown's problem.
    Maximize

$$P = 40x + 30y$$

subject to

$$\begin{cases} x + y \leq 480 \\ 2x + y \leq 800 \\ x \geq 0 \quad y \geq 0 \end{cases}$$

Any problem that asks us to find the maximum (or minimum) of a linear function subject to linear inequality constraints is called a **linear programming problem.** Here is a method for solving such problems.

1. Graph the solution set corresponding to the inequality constraints.
2. Find the coordinates of the vertices of the solution set.
3. Evaluate the linear function that you want to maximize (or minimize) at each of these vertices. The largest of these gives the maximum, while the smallest gives the minimum.

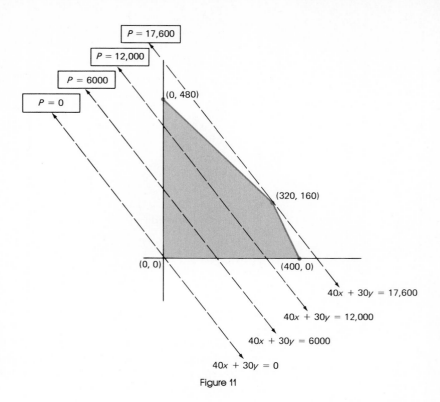

P = 17,600

P = 12,000

P = 6000

P = 0

(0, 480)

(320, 160)

(0, 0)

(400, 0)

$40x + 30y = 17{,}600$

$40x + 30y = 12{,}000$

$40x + 30y = 6000$

$40x + 30y = 0$

Figure 11

Vertex	$P = 40x + 30y$
(0, 0)	0
(0, 480)	14,400
(320, 160)	17,600
(400, 0)	16,000

Figure 12

Vertex	$P = 80x + 30y$
(0, 0)	0
(0, 480)	14,400
(320, 160)	30,400
(400, 0)	32,000

Figure 13

To see why this method works, consider the diagram for Farmer Brown's problem (Figure 11). The dotted lines are profit lines, each with slope $-\frac{4}{3}$; they are the graphs of $40x + 30y = P$ for various values of P. All the points on a dotted line give the same total profit. Imagine a profit line moving from left to right across the shaded region with the slope constant, as indicated in Figure 11 . During this motion, the profit is zero at (0, 0) and increases to its maximum of 17,600 at (320,160). It should be clear that such a moving line (no matter what its slope) will always enter the shaded set at a vertex and leave it at another vertex. The particular vertex depends upon the slope of the profit lines. In Farmer Brown's problem, the minimum profit of $0 occurs at (0, 0); the maximum profit of $17,600 occurs at (320, 160). In Figure 12, we show the total profit for each of the four vertices. Clearly Farmer Brown should plant 320 acres of corn and 160 acres of wheat.

Now suppose the price of corn goes up so that Farmer Brown can expect a profit of $80 per acre on corn but still only $30 per acre on wheat. Would this change his strategy? In Figure 13, we show his total profit $P = 80x + 30y$ at each of the four vertices. Evidently he should plant 400 acres of corn and no wheat to achieve maximum profit. Note that this means he should leave 80 acres of his land idle.

Finally, suppose that the profit per acre is $40 both for wheat and for corn. The table for this case (Figure 14) shows the same total profit at the vertices (0, 480) and (320, 160). This means that the moving profit line leaves

Vertex	$P = 40x + 40y$
(0, 0)	0
(0, 480)	19,200
(320, 160)	19,200
(400, 0)	16,000

Figure 14

the shaded region along the side determined by those two vertices, so that every point on that side gives a maximum profit. It is still true, however, that the maximum profit occurs at a vertex.

The situation with an unbounded constraint set is slightly more complicated. It is discussed in Example A. Here we simply point out that in the unbounded case, there may not be a maximum (or minimum), but if there is one, it will still occur at a vertex.

Problem Set 11-7

In Problems 1–6, graph the solution set of each inequality in the xy-plane.

1. $4x + y \le 8$ 2. $2x + 5y \le 20$ 3. $x \le 3$
4. $y \le -2$ 5. $4x - y \ge 8$ 6. $2x - 5y \ge -20$

In Problems 7–10, graph the solution set of the given system. On the graph, label the coordinates of the vertices.

7. $4x + y \le 8$
   $2x + 3y \le 14$
   $x \ge 0 \quad y \ge 0$

8. $2x + 5y \le 20$
   $4x + y \le 22$
   $x \ge 0 \quad y \ge 0$

9. $4x + y \le 8$
   $x - y \le -2$
   $x \ge 0$

10. $2x + 5y \le 20$
    $x - 2y \ge 1$
    $y \ge 0$

In Problems 11–14, find the maximum and minimum values of the given linear function P subject to the given inequalities.

11. $P = 2x + y$; the inequalities of Problem 7.
12. $P = 3x + 2y$; the inequalities of Problem 8.
13. $P = 2x - y$; the inequalities of Problem 9.
14. $P = 3x - 2y$; the inequalities of Problem 10.

EXAMPLE A (Unbounded Region) Find the maximum and minimum values of the function $3x + 4y$ subject to the constraints

$$\begin{cases} 3x + 2y \ge 13 \\ x + y \ge 5 \\ x \ge 1 \\ y \ge 0 \end{cases}$$

Solution. We proceed to graph the solution set of our system, noting that the region must lie above the lines $3x + 2y = 13$ and $x + y = 5$ and to the right of the line $x = 1$. It is shown in Figure 15. Notice that the region is unbounded and has (5, 0), (3, 2), and (1, 5) as its vertices.

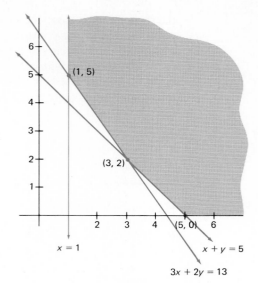

Figure 15

The point $(1, 4)$ at which the lines $x + y = 5$ and $x = 1$ intersect is not a vertex. It should be clear right away that $3x + 4y$ does not assume a maximum value in our region; its values can be made as large as we please by increasing x and y. To find the minimum value, we calculate $3x + 4y$ at the three vertices.

$$(5, 0): \quad 3 \cdot 5 + 4 \cdot 0 = 15$$

$$(3, 2): \quad 3 \cdot 3 + 4 \cdot 2 = 17$$

$$(1, 5): \quad 3 \cdot 1 + 4 \cdot 5 = 23$$

The minimum value of $3x + 4y$ is 15.

Solve each of the following problems.

15. Minimize $5x + 2y$ subject to

$$\begin{cases} x + y \geq 4 \\ \quad\; x \geq 2 \\ \quad\; y \geq 0 \end{cases}$$

16. Minimize $2x - y$ subject to

$$\begin{cases} x - 2y \geq 2 \\ \quad\;\;\; y \geq 2 \end{cases}$$

17. Minimize $2x + y$ subject to

$$\begin{cases} 4x + \;\; y \geq 7 \\ 2x + 3y \geq 6 \\ \quad\quad\;\; x \geq 1 \\ \quad\quad\;\; y \geq 0 \end{cases}$$

18. Minimize $3x + 2y$ subject to

$$\begin{cases} x - 2y \leq 2 \\ x - 2y \geq -2 \\ 3x - 2y \geq 10 \end{cases}$$

EXAMPLE B (Systems with Nonlinear Inequalities)   Graph the solution set of the following system of inequalities.

$$\begin{cases} y \geq 2x^2 \\ y \leq 2x + 4 \end{cases}$$

Solution. It helps to find the points at which the parabola $y = 2x^2$ intersects the line $y = 2x + 4$. Eliminating y between the two equations and then solving for x, we get

$$2x^2 = 2x + 4$$

$$x^2 - x - 2 = 0$$

$$(x - 2)(x + 1) = 0$$

$$x = 2 \qquad x = -1$$

The corresponding values of y are 8 and 2, respectively; so the points of intersection are $(-1, 2)$ and $(2, 8)$. Making use of these points, we draw the parabola and the line. Since the point $(0, 2)$ satisfies the inequality $y \geq 2x^2$, the desired region is above and including the parabola. The graph of the linear inequality $y \leq 2x + 4$ is to the right of and including the line. The shaded region in Figure 16 is the graph we want.

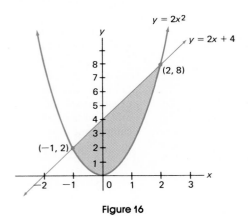

Figure 16

In Problems 19–22, graph the solution set of each system of inequalities.

19. $y \leq 4x - x^2$
    $y \leq x$
    $x \geq 0 \qquad y \geq 0$

20. $y \geq 2^x$
    $y \leq 8$
    $x \geq 0$

21. $y \leq \log_2 x$
    $x \leq 8$
    $y \geq 0$

22. $x^2 + y^2 \geq 9$
    $0 \leq x \leq 3$
    $0 \leq y \leq 3$

MISCELLANEOUS PROBLEMS

In Problems 23 and 24, graph the solution set and label the coordinates of the vertices.

23. $4x + y \leq 8$
    $x - y \geq -2$
    $x \geq 0 \qquad y \geq 0$

24. $2x + 5y \leq 20$
    $x - 2y \leq 1$
    $x \geq 0$

25. Find the maximum and minimum values of $P = 2x - y$ subject to the inequalities of Problem 23.

26. Find the maximum and minimum values of $P = 3x - 2y$ subject to the inequalities of Problem 24.

27. Find the maximum and minimum values of $x + y$ (if they exist) subject to the following inequalities.

$$x - y \geq -1$$
$$x - 2y \leq 5$$
$$3x + y \leq 10$$
$$x \geq 0 \qquad y \geq 0$$

28. Find the maximum and minimum values of $x - 2y$ (if they exist) subject to the inequalities of Problem 27.

29. A company makes a single product on two production lines, A and B. A labor force of 900 hours per week is available, and weekly running costs shall not exceed $1500. It takes 4 hours to produce one item on production line A and 3 hours on production line B. The cost per item is $5 on line A and $6 on line B. Find the largest number of items that can be produced in one week.

30. An oil refinery has a maximum production of 2000 barrels of oil per day. It produces two types of oil; type A, which is used for gasoline and type B, which is used for heating oil. There is a requirement that at least 300 barrels of type B be produced each day. If the profit is $3 a barrel for type A and $2 a barrel for type B, find the maximum profit per day.

31. A manufacturer of trailers wishes to determine how many camper units and how many house trailers she should produce in order to make the best possible use of her resources. She has 42 units of wood, 56 worker-weeks of labor and 16 units of aluminum. (Assume that all other needed resources are available and have no effect on her decision.) The amount of each resource needed to produce each camper and each trailer is given below.

	Wood	Worker-weeks	Aluminum
Per camper	3	7	3
Per trailer	6	7	1

If the manufacturer realizes a profit of $600 on a camper and $800 on a trailer, what should be her production in order to maximize her profit?

32. A shoemaker has a supply of 100 square feet of type A leather which is used for soles and 600 square feet of type B leather used for the rest of the shoe. The average shoe uses $\frac{1}{4}$ square feet of type A leather and 1 square foot of type B leather. The average boot uses $\frac{1}{4}$ square feet and 3 square feet of types A and B leather, respectively. If shoes sell at $40 a pair and boots at $60 a pair, find the maximum income.

33. Suppose that the minimum monthly requirements for one person are 60 units of carbohydrates, 40 units of protein, and 35 units of fat. Two foods A and B contain the following numbers of units of the three diet components per pound.

	Carbohydrates	Protein	Fat
A	5	3	5
B	2	2	1

If food A costs $3.00 a pound and food B costs $1.40 a pound, how many pounds of each should a person purchase per month to minimize the cost?

34. A grain farmer has 100 acres available for sowing oats and wheat. The seed oats costs $5 per acre and the seed wheat costs $8 per acre. The labor costs are $20 per acre for oats and $12 per acre for wheat. The farmer expects an income from oats of $220 per acre and from wheat of $250 per acre. How many acres of each crop should he sow to maximize his profit, if he does not wish to spend more than $620 for seed and $1800 for labor?

35. Sketch the polygon with vertices $(0, 3)$, $(4, 7)$, $(3, 0)$, and $(2, 4)$ taken in cyclic order. Then, find the maximum value of $|y - 2x| + y + x$ on this polygon.

36. **TEASER**   If $P = (a, b)$ is a point in the plane, then $tP = (ta, tb)$. Thus, if $0 \le t \le 1$, the set of points of the form $tP + (1 - t)Q$ is just the line segment PQ (see point-of-division formula, page 142). It follows that a set A is convex if whenever P and Q are in A, all points of the form $tp + (1 - t)Q, 0 \le t \le 1$ are also in A.

   (a) Let P, Q, and R be three fixed points in the plane and consider the set H of all points of the form $t_1 P + t_2 Q + t_3 R$, where the t's are nonnegative and $t_1 + t_2 + t_3 = 1$. Show that H is convex.

   (b) Let P, Q, R, and S be four fixed points in the plane and consider the set K of all points of the form $t_1 P + t_2 Q + t_3 R + t_4 S$ where the t's are non-negative and $t_1 + t_2 + t_3 + t_4 = 1$. Show that K is convex.

   (c) Describe the sets H and K geometrically.

Chapter Summary

Two systems of equations are **equivalent** if they have the same solutions. Three elementary operations (multiplying an equation by a nonzero constant, interchanging two equations, and adding a multiple of one equation to another) lead to equivalent systems. Use of these operations allows us to transform a system of linear equations to **triangular form** and then to solve the system by **back substitution** or to show it is **inconsistent.**

A **matrix** is a rectangular array of numbers. In solving a system of linear equations, it is efficient to work with just the **matrix of the system.** We solve the system by transforming its matrix to triangular form using the three operations mentioned above.

Addition and multiplication are defined for matrices with resulting algebraic rules. Matrices behave much like numbers, with the exception that the commutative law for multiplication fails. Even the notion of **multiplicative inverse** has meaning, though the process for finding such an inverse is lengthy.

Associated with every square matrix is a symbol called its **determinant.** For 2 × 2 and 3 × 3 matrices, the value of the determinant can be found by using certain arrow diagrams. For higher order cases, the value of a determinant is found by expanding it in terms of the elements of a row (or column) and their **minors** (determinants whose order is lower than the given determinant by 1). In doing this, it is helpful to know what happens to the determinant of a matrix when any of the three elementary operations are applied to it. **Cramer's rule** provides a direct way of solving a system of n equations in n unknowns using determinants.

The graph of a **linear inequality** in x and y is a **half-plane** (closed or open according as the inequality sign does or does not include the equal sign). The graph of a **system of linear inequalities** is the intersection of the half-planes corresponding to the separate inequalities. Such a graph is always **polygonal** and **convex** but may be **bounded** or **unbounded.** A **linear programming problem** asks us to find the maximum (or minimum) of a linear function (such as $2x + 5y$) subject to a system of linear inequalities called **constraints.** The maximum (or minimum) always occurs at a **vertex,** that is, at a corner point of the graph of the inequality constraints.

Chapter Review Problem Set

In Problems 1–5, solve each system or show that there is no solution.

1. $3x + y = 12$
   $2x - y = -2$

2. $x - 3y + 2z = 5$
   $2y - z = 1$
   $z = 3$

3. $2x - y + 3z = 10$
   $y - 2z = 4$
   $-2y + 4z = 8$

4. $x - 3y + 2z = -5$
   $4x + y - z = 4$
   $5x + 11y - 8z = 20$

5. $x + 2y + 3z + 4w = 20$
   $y - 4z + w = 6$
   $z + 2w = 4$
   $2z + 4w = 8$

6. Evaluate each expression for

$$\mathbf{A} = \begin{bmatrix} 1 & 1 & 1 \\ 3 & 1 & -1 \\ 2 & 2 & -1 \end{bmatrix} \quad \text{and} \quad \mathbf{B} = \begin{bmatrix} -3 & 6 & -4 \\ 2 & -3 & 5 \\ 1 & 9 & 2 \end{bmatrix}$$

   (a) $2\mathbf{A} + \mathbf{B}$ (b) $\mathbf{A} - 2\mathbf{B}$ (c) $\mathbf{AB}$ (d) $\mathbf{BA}$

7. Find $\mathbf{A}^{-1}$ for the matrix $\mathbf{A}$ of Problem 6.

8. Consider the system

$$x + y + z = 4$$
$$3x + y - z = -4$$
$$2x + 2y - z = -1$$

   Write this system in matrix form and then use the result of Problem 7 to solve it.

Evaluate the determinants in Problems 9–13.

9. $\begin{vmatrix} -2 & 5 \\ 2 & -6 \end{vmatrix}$

10. $\begin{vmatrix} 2 & 3 \\ -4 & -6 \end{vmatrix}$

11. $\begin{vmatrix} -2 & 1 & 4 \\ 0 & 5 & -1 \\ 4 & 0 & 3 \end{vmatrix}$ 12. $\begin{vmatrix} 1 & -2 & 3 \\ 4 & 1 & 5 \\ 7 & -5 & 14 \end{vmatrix}$

13. $\begin{vmatrix} 6 & 0 & 0 & 0 \\ -1 & 3 & 1 & 0 \\ 3 & 2 & 3 & 2 \\ 4 & 5 & 1 & -4 \end{vmatrix}$

14. Use Cramer's rule to solve the following system.

$$2x + y - z = -4$$
$$x - 3y - 2z = -1$$
$$3x + 2y + 3z = 11$$

In Problems 15 and 16, graph the solution set of the given system. On the graph, label the coordinates of the vertices.

15. $x + y \le 7$
    $3x + y \le 15$
    $x \ge 0 \qquad y \ge 0$

16. $x - 2y + 4 \ge 0$
    $x + y - 11 \ge 0$
    $x \ge 0 \qquad y \ge 0$

17. Find the maximum value of the function $P = x + 2y$ subject to the inequalities of Problem 15.

18. Find the minimum value of the function $P = x + 2y$ subject to the inequalities of Problem 16.

19. A certain company has 100 employees, some of whom get $4 an hour, others $5, and the rest $8. Half as many make $8 an hour as $5 an hour. If the total paid out in hourly wages is $544, find the number of employees who make $8 an hour.

20. A tailor has 110 yards of cotton material and 160 yards of woolen material. It takes $1\frac{1}{2}$ yards of cotton and 1 yard of wool to make a suit, while a dress requires 1 yard of cotton and 2 yards of wool. If a suit sells for $100 and a dress for $80, how many of each should the tailor make to maximize the total income?

Method consists entirely in properly ordering and arranging things to which we should pay attention.

—René Descartes

CHAPTER 12

Sequences and Counting Problems

(a) 1, 4, 7, 10, 13, 16, □, □, . . .

(c) 1, 4, 9, 16, 25, 36, □, □, . . .

(b) 2, 4, 6, 8, 10, 12, □, □, . . .

(d) 1, 4, 9, 16, 27, 40, □, □, . . .

(e) 1, 1, 2, 3, 5, 8, □, □, . . .

12-1 Number Sequences

Try filling in the boxes of our opening display. You will have little trouble with *a*, *b*, and *c*, but *d* and *e* may offer quite a challenge. We will give the answers we had in mind later. Right now, we merely point out that each sequence has a pattern; we used a definite rule in writing the first six terms of each of them.

The word *sequence* is commonly used in ordinary language. For example, your history teacher may talk about a sequence of events that led to World War II (for instance, the Versailles Treaty, world depression, Hitler's ascendancy, Munich Agreement). What characterizes this sequence is the notion of one event following another in a definite order. There is a first event, a second event, a third event, and so on. We might even give them labels.

E_1: Versailles Treaty

E_2: World depression

E_3: Hitler's ascendancy

E_4: Munich Agreement

We use a similar notation for number sequences. Thus

$$a_1, a_2, a_3, a_4, . . .$$

could denote sequence (a) of our opening display. Then

$$a_1 = 1$$
$$a_2 = 4$$
$$a_3 = 7$$
$$a_4 = 10$$
$$\vdots$$

Note that a_3 stands for the 3rd term; a_{10} would represent the tenth term. The subscript indicates the position of the term in the sequence. For the general term, that is, the nth term, we use the symbol a_n. The three dots indicate that the sequence continues indefinitely.

There is another way to describe a number sequence. A **number sequence** is a function whose domain is the set of positive integers. That means it is a rule that associates with each positive integer n a definite number a_n. In conformity with Chapter 5, we could use the notation $a(n)$, but tradition dictates that we use a_n instead. We usually specify functions by giving formulas; this is true of sequences also.

EXPLICIT FORMULAS

Rather than give the first few terms of a sequence and hope that our readers see the pattern intended (different people sometimes see different patterns in the first few terms of a sequence), it is better to give a formula. Take sequence (a) for example. The formula

$$a_n = 3n - 2$$

tells all there is to know about that sequence. In particular,

$$a_1 = 3 \cdot 1 - 2 = 1$$
$$a_2 = 3 \cdot 2 - 2 = 4$$
$$a_3 = 3 \cdot 3 - 2 = 7$$
$$a_{100} = 3 \cdot 100 - 2 = 298$$

How does one find the formula for a sequence? Look at sequence (b). Suppose we let b_n stand for the nth term. Then $b_1 = 2$, $b_2 = 4$, $b_3 = 6$, $b_4 = 8$, and so on. Our job is to relate the value of b_n to the subscript n. Clearly, it is just twice the subscript—that is,

$$b_n = 2n$$

Knowing this formula, we can calculate the value of any term. For example,

$$b_{10} = 2 \cdot 10 = 20$$
$$b_{281} = 2 \cdot 281 = 562$$

If we follow the same procedure for sequence (c), we have

$$c_1 = 1 \quad c_2 = 4 \quad c_3 = 9 \quad c_4 = 16$$

from which we infer the formula

$$c_n = n^2$$

Now look at sequence (d).

$$1, 4, 9, 16, 27, 40, \ldots$$

The fact that it starts out just like sequence (c) suggests that the pattern is subtle and incidentally warns us that we may have to look at many terms of a sequence before we can discover its rule of construction. Here, as in many sequences, it is a good idea to observe how each term relates to the previous one. Let us write the sequence again, indicating below it the numbers to be added as we progress from term to term.

$$1 \quad 4 \quad 9 \quad 16 \quad 27 \quad 40$$
$$3 \quad 5 \quad 7 \quad 11 \quad 13$$

You may recognize the second row of numbers as consecutive primes (starting with 3). Thus the next two terms in sequence (d) are

$$40 + 17 = 57$$
$$57 + 19 = 76$$

But observing a pattern does not necessarily mean we can write an explicit formula. Though many have tried, no one has found a simple formula for the nth prime and, thus, no one is likely to find a simple formula for sequence (d).

Sequence (e) is a famous one. It was introduced by Leonardo Fibonacci around 1200 A.D. in connection with rabbit reproduction (see Problem 31). If you are a keen observer, you have noticed that any term (after the second) is the sum of the preceding two. It was not until 1724 that mathematician Daniel Bernoulli found the explicit formula for this sequence. You will agree that it is complicated, but at least you can check it for $n = 1, 2, 3$.

$$e_n = \frac{1}{\sqrt{5}} \left[\left(\frac{1 + \sqrt{5}}{2} \right)^n - \left(\frac{1 - \sqrt{5}}{2} \right)^n \right]$$

If it took 500 years to discover this formula, you should not be surprised when we say that explicit formulas are often difficult to find (the problem set will give more evidence). There is another type of formula that is usually easier to discover.

RECURSION FORMULA

An explicit formula relates the value of a_n to its subscript n (for example, $a_n = 3n - 2$). Often the pattern we first observe relates a term to the preceding term (or terms). If so, we may be able to describe this pattern by a recursion formula. Look at sequence (a) again. To get a term from the preceding one, we always add 3, that is,

$$a_n = a_{n-1} + 3$$

Or look at sequence (b). There we add 2 each time.

$$b_n = b_{n-1} + 2$$

Sequence (e) is more interesting. There we add together the two previous terms, that is,

$$e_n = e_{n-1} + e_{n-2}$$

We summarize our knowledge about the five sequences in Table 4.

TABLE 4

Sequence	Explicit Formula	Recursion Formula
(a) 1, 4, 7, 10, 13, 16, . . .	$a_n = 3n - 2$	$a_n = a_{n-1} + 3$
(b) 2, 4, 6, 8, 10, 12, . . .	$b_n = 2n$	$b_n = b_{n-1} + 2$
(c) 1, 4, 9, 16, 25, 36, . . .	$c_n = n^2$	$c_n = c_{n-1} + 2n - 1$
(d) 1, 4, 9, 16, 27, 40, . . .	?	$d_n = d_{n-1} + n\text{th prime}$
(e) 1, 1, 2, 3, 5, 8, . . .	$e_n = \dfrac{1}{\sqrt{5}}\left[\left(\dfrac{1 + \sqrt{5}}{2}\right)^n - \left(\dfrac{1 - \sqrt{5}}{2}\right)^n\right]$	$e_n = e_{n-1} + e_{n-2}$

Recursion formulas are themselves not quite enough to determine a sequence. For example, the recursion formula

$$f_n = 3f_{n-1}$$

does not determine a sequence until we specify the first term. But with the additional information that $f_1 = 2$, we can find any term. Thus

$$f_1 = 2$$
$$f_2 = 3f_1 = 3 \cdot 2 = 6$$
$$f_3 = 3f_2 = 3 \cdot 6 = 18$$
$$f_4 = 3f_3 = 3 \cdot 18 = 54$$
$$\vdots$$

The disadvantage of a recursion formula is apparent. To find the 100th term, we must first calculate the 99 previous terms. But if it is hard work, it is at least possible. Programmable calculators are particularly adept at calculating sequences by means of recursion formulas.

Problem Set 12-1

1. Discover a pattern and use it to fill in the boxes.
   (a) 1, 3, 5, 7, □, □, . . .
   (b) 17, 14, 11, 8, □, □, . . .
   (c) 1, $\frac{1}{2}$, $\frac{1}{4}$, $\frac{1}{8}$, □, □, . . .
   (d) 1, 9, 25, 49, □, □, . . .

2. Fill in the boxes.
   (a) 1, 3, 9, 27, □, □, . . .
   (b) 2, 2.5, 3, 3.5, □, □, . . .
   (c) 1, 8, 27, 64, □, □, . . .
   (d) $\frac{1}{2}$, $\frac{2}{3}$, $\frac{3}{4}$, $\frac{4}{5}$, □, □, . . .

3. In each case an explicit formula is given. Find the indicated terms.
   (a) $a_n = 2n + 3$; $a_4 = $ □; $a_{20} = $ □

(b) $a_n = \dfrac{n}{n + 1}$; $a_5 = \square$; $a_9 = \square$

(c) $a_n = (2n - 1)^2$; $a_4 = \square$; $a_5 = \square$

(d) $a_n = (-3)^n$; $a_3 = \square$; $a_4 = \square$

4. Find the indicated terms.

(a) $a_n = 2n - 5$; $a_4 = \square$; $a_{20} = \square$

(b) $a_n = 1/n$; $a_5 = \square$; $a_{50} = \square$

(c) $a_n = (2n)^2$; $a_5 = \square$; $a_{10} = \square$

(d) $a_n = 4 - \frac{1}{2}n$; $a_5 = \square$; $a_{10} = \square$

5. Give an explicit formula for each sequence in Problem 1. Recall that you must relate the value of a term to its subscript (see the examples on page 493).

6. Give an explicit formula for each sequence in Problem 2.

7. In each case, an initial term and a recursion formula are given. Find a_5. *Hint:* First find a_2, a_3, and a_4, and then find a_5.

(a) $a_1 = 2$; $a_n = a_{n-1} + 3$

(b) $a_1 = 2$; $a_n = 3a_{n-1}$

(c) $a_1 = 8$; $a_n = \frac{1}{2}a_{n-1}$

(d) $a_1 = 1$; $a_n = a_{n-1} + 8(n - 1)$

8. Find a_4 for each of the following sequences.

(a) $a_1 = 2$; $a_n = 2a_{n-1} + 1$

(b) $a_1 = 2$; $a_n = a_{n-1} + 3$

(c) $a_1 = 1$; $a_n = a_{n-1} + 3n^2 - 3n + 1$

(d) $a_1 = 3$; $a_n = a_{n-1} + .5$

9. Try to find a recursion formula for each of the sequences in Problem 1.

10. Try to find a recursion formula for each of the sequences in Problem 2.

EXAMPLE (Sum Sequences) Corresponding to a sequence $a_1, a_2, a_3, \ldots$, we introduce another sequence A_n, called the *sum sequence*, by

$$A_n = a_1 + a_2 + a_3 + \cdots + a_n$$

Thus

$$A_1 = a_1$$

$$A_2 = a_1 + a_2$$

$$A_3 = a_1 + a_2 + a_3$$

$$\vdots$$

Find A_5 for the sequence given by $a_n = 3n - 2$.

Solution. We begin by finding the first five terms of sequence a_n.

$$a_1 = 1$$

$$a_2 = 4$$

$$a_3 = 7$$

$$a_4 = 10$$

$$a_5 = 13$$

Then

$$A_5 = a_1 + a_2 + a_3 + a_4 + a_5$$
$$= 1 + 4 + 7 + 10 + 13$$
$$= 35$$

In each of the following problems, find A_6.

11. $a_n = 2n + 1$ 12. $a_n = 2^n$
13. $a_n = (-2)^n$ 14. $a_n = n^2$
15. $a_n = n^2 - 2$ 16. $a_n = 3n - 4$
17. $a_1 = 4; a_n = a_{n-1} + 3$ 18. $a_1 = 1; a_2 = 1; a_n = a_{n-1} + 2a_{n-2}$

MISCELLANEOUS PROBLEMS

19. Find a_5 and a_{16} for each sequence.
    (a) $a_n = 5n + 2$ (b) $a_n = n(n - 1)$
20. Find a_5 in each case.
    (a) $a_1 = 9$, $a_n = \frac{2}{3}a_{n-1}$ (b) $a_1 = 6$, $a_2 = 2$, $a_n = (a_{n-1} + a_{n-2})/2$
21. Let $a_1 = 2$, $a_n = 3a_{n-1} - 1$; $b_n = n^2 - 3n$; $c_n = a_n - b_n$. Evaluate c_1, c_2, c_3, c_4, and c_5.
22. Find a pattern in each of the following sequences and use it to fill in the boxes.
    (a) 2, 6, 18, 54, □, □, . . .
    (b) 2, 6, 10, 14, □, □, . . .
    (c) 2, 4, 8, 14, □, □, . . .
    (d) 2, 4, 6, 10, 16, □, □, . . .
    (e) 2, 1, $\frac{1}{2}$, $\frac{1}{4}$, □, □, . . .
    (f) 2, 5, 10, 17, □, □, . . .
23. Find a recursion formula for each sequence in Problem 22.
24. Find (if you can) an explicit formula for each sequence in Problem 22.
25. Let a_n be the nth digit in the decimal expansion of $\frac{1}{7} = .1428 \ldots$. Thus, $a_1 = 1$, $a_2 = 4$, $a_3 = 2$, and so on. Find a pattern and use it to determine a_8, a_{27}, and a_{53}.
26. Suppose that January 1 occurs on Wednesday. Let a_n be the day of the week corresponding to the nth day of the year. Thus $a_1 =$ Wednesday, $a_2 =$ Thursday, and so on. Find a_{39}, a_{57}, and a_{84}.
27. Let $a_n = 2n - 1$ and $A_n = a_1 + a_2 + a_3 + \cdots + a_n$.
    (a) Calculate a_1, a_2, a_3, a_4, and a_5.
    (b) Calculate A_1, A_2, A_3, A_4, and A_5.
    (c) Find an explicit formula for A_n.
28. Follow the directions of Problem 27 for

$$a_n = \frac{1}{n(n + 1)} = \frac{1}{n} - \frac{1}{n + 1}$$

29. The Greeks were enchanted with sequences that arose in a geometric way (Figures 1 and 2).
    (a) Find an explicit formula for s_n.
    (b) Find a recursion formula for s_n.

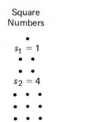

Square
Numbers

$s_1 = 1$

$s_2 = 4$

$s_3 = 9$

Figure 1

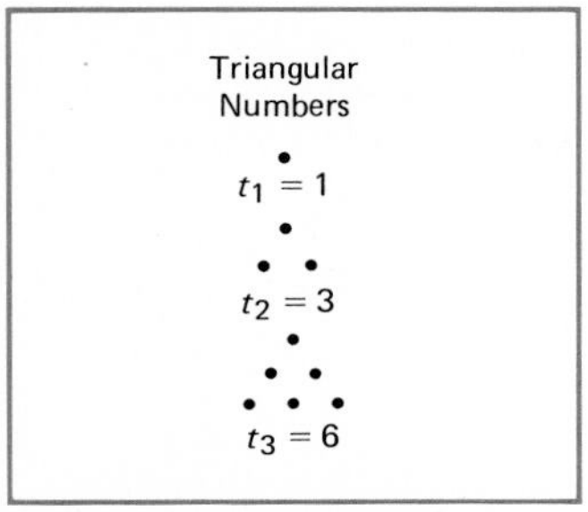

Triangular
Numbers

$t_1 = 1$

$t_2 = 3$

$t_3 = 6$

Figure 2

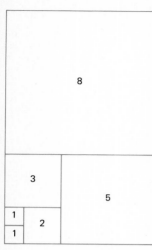

Figure 3

(c) Find a recursion formula for t_n.

(d) By considering a square of dots, $n + 1$ on a side, express s_{n+1} in terms of t_n.

(e) Use (d) to find an explicit formula for t_n.

30. The numbers 1, 5, 12, 22, . . . are called *pentagonal numbers*. See if you can figure out why and then guess at an explicit formula for p_n, the nth pentagonal number. Use diagrams.

31. Following Leonardo Fibonacci, suppose that a pair of rabbits consisting of a male and a female begin to bear young when they are 2 months old, and that thereafter they produce one male-female pair at 1-month intervals forever. Assume that each new male-female pair of rabbits has the same reproductive habits as its parents and that no rabbits die. Beginning with a single pair born on the first day of a month, how many rabbit pairs will there be on the last day of that month? the last day of the second month? Third month? Fourth month? Fifth month? nth month?

32. Let f_n denote the Fibonacci sequence determined by $f_1 = 1$, $f_2 = 1$, and $f_n = f_{n-1} + f_{n-2}$. Also let $F_n = f_1 + f_2 + \cdots + f_n$.

(a) Calculate F_1, F_2, F_3, F_4, F_5, and F_6.

(b) Based on (a), find a nice formula connecting F_n to f_{n+2}.

33. By considering adjoining geometric squares as in Figure 3, obtain a nice formula for $f_1^2 + f_2^2 + f_3^2 + \cdots + f_n^2$.

34. TEASER   Consider the matrix

$$\mathbf{Q} = \begin{bmatrix} 1 & \cdot & 1 \\ 1 & & 0 \end{bmatrix}$$

(a) Find $\mathbf{Q}^2$, $\mathbf{Q}^3$, and $\mathbf{Q}^4$.

(b) Express the matrix $\mathbf{Q}^n$ in terms of the Fibonacci numbers f_n.

(c) Use the fact that $|\mathbf{Q}^n| = |\mathbf{Q}|^n$ to obtain a nice formula for $f_{n+1}f_{n-1} - f_n^2$.

(d) From $\mathbf{Q}^{2n} = \mathbf{Q}^n\mathbf{Q}^n$, obtain a nice formula for f_{2n}.

(e) Use the result in (d) to show that the difference in the squares of alternate Fibonacci numbers is a Fibonacci number.

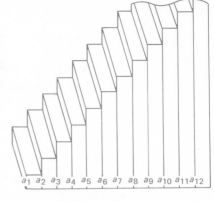

Jacob's Stairway

In the skyscraper that Jacob owns, there is a stairway going from ground level to the very top. The first step is 8 inches high. After that each step is 9 inches above the previous one. How high above the ground is the 800th step?

a_1 a_2 a_3 a_4 a_5 a_6 a_7 a_8 a_9 a_{10} a_{11} a_{12}

12-2 Arithmetic Sequences

We are going to answer the question above and others like it. Notice that if a_n denotes the height of the nth step, then

$$a_1 = 8$$
$$a_2 = 8 + 1(9) = 17$$
$$a_3 = 8 + 2(9) = 26$$
$$a_4 = 8 + 3(9) = 35$$
$$\vdots$$
$$a_{800} = 8 + 799(9) = 7199$$

The 800th step of Jacob's stairway is 7199 inches (almost 600 feet) above the ground.

FORMULAS

Now consider the following number sequences. When you see a pattern, fill in the boxes.

(a) 5, 9, 13, 17, $\square$, $\square$, . . .

(b) 2, 2.5, 3, 3.5, $\square$, $\square$, . . .

(c) 8, 5, 2, −1, $\square$, $\square$, . . .

What is it that these three sequences have in common? Simply this: In each case, you can get a term by adding a fixed number to the preceding term. In (a), you add 4 each time, in (b) you add 0.5, and in (c), you add −3. Such sequences are called **arithmetic sequences.** If we denote such a sequence by $a_1, a_2, a_3, \ldots$, it satisfies the recursion formula

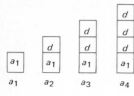

Figure 4

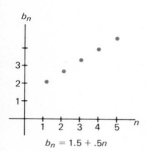

$b_n = 1.5 + .5n$

Figure 5

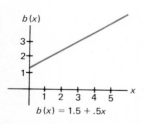

$b(x) = 1.5 + .5x$

Figure 6

$$a_n = a_{n-1} + d$$

where d is a fixed number called the **common difference.**

Can we also obtain an explicit formula? Yes. Figure 4 should help. Notice that the number of d's to be added to a_1 is one less than the subscript. This means that

$$a_n = a_1 + (n - 1)d$$

Now we can give explicit formulas for each of the sequences (a), (b), and (c) given earlier.

$$a_n = 5 + (n - 1)4 = 1 + 4n$$
$$b_n = 2 + (n - 1)(.5) = 1.5 + .5n$$
$$c_n = 8 + (n - 1)(-3) = 11 - 3n$$

ARITHMETIC SEQUENCES AND LINEAR FUNCTIONS

We have said that a sequence is a function whose domain is the set of positive integers. Functions are best visualized by drawing their graphs. Consider the sequence b_n discussed above; its explicit formula is

$$b_n = 1.5 + .5n$$

Its graph is shown in Figure 5.

Even a cursory look at this graph suggests that the points lie along a straight line. Consider now the function

$$b(x) = 1.5 + 0.5x = 0.5x + 1.5$$

where x is allowed to be any real number. This is a linear function, being of the form $mx + b$ (see Section 4-3). Its graph is a straight line (Figure 6), and its values at $x = 1, 2, 3, \ldots$ are equal to $b_1, b_2, b_3, \ldots$.

The relationship illustrated above between an arithmetic sequence and a linear function holds in general. An arithmetic sequence is just a linear function whose domain has been restricted to the positive integers.

SUMS OF ARITHMETIC SEQUENCES

There is an old story about Carl Gauss that aptly illustrates the next idea. We are not sure if the story is true, but if not, it should be.

When Gauss was about 10 years old, he was admitted to an arithmetic class. To keep the class busy, the teacher often assigned long addition problems. One day he asked his students to add the numbers from 1 to 100. Hardly had he made the assignment when young Gauss laid his slate on the teacher's desk with the answer 5050 written on it.

$$1 + 2 + \ldots + 49 + 50 + 51 + 52 + \ldots + 99 + 100$$

Here is how Gauss probably thought about the problem (Figure 7). Each of the indicated pairs has 101 as its sum and there are 50 such pairs. Thus the answer is $50(101) = 5050$. For a 10-year-old boy, that is good thinking.

Gauss's trick works perfectly well on any arithmetic sequence where we want to add an even number of terms. And there is a slight modification that works whether the number of terms to be added is even or odd.

Suppose $a_1, a_2, a_3, \ldots$ is an arithmetic sequence and let

$$A_n = a_1 + a_2 + a_3 + \cdots + a_{n-1} + a_n$$

Write this sum twice, once forwards and once backwards, and then add.

$$
\begin{array}{rcccccccc}
A_n = & a_1 & + & a_2 & + \cdots + & a_{n-1} & + & a_n \\
A_n = & a_n & + & a_{n-1} & + \cdots + & a_2 & + & a_1 \\
\hline
2A_n = & (a_1 + a_n) & + & (a_2 + a_{n-1}) & + \cdots + & (a_2 + a_{n-1}) & + & (a_1 + a_n)
\end{array}
$$

Each group on the right has the same sum, namely, $a_1 + a_n$. For example,

$$a_2 + a_{n-1} = a_1 + d + a_n - d = a_1 + a_n$$

There are n such groups and so

$$2A_n = n(a_1 + a_n)$$

$$\boxed{A_n = \frac{n}{2}(a_1 + a_n)}$$

We call this the **sum formula** for an arithmetic sequence. You can remember this sum as being n times the average term $(a_1 + a_n)/2$.

Here is how we apply this formula to Gauss's problem. We want the sum of 100 terms of the sequence $1, 2, 3, \ldots$, that is, we want A_{100}. Here $n = 100$, $a_1 = 1$, and $a_n = 100$. Therefore,

$$A_{100} = \frac{100}{2}(1 + 100) = 50(101) = 5050$$

For a second example, suppose we want to add the first 350 odd numbers, that is, the first 350 terms of the sequence $1, 3, 5, \ldots$. We can calculate the 350th odd number from the formula $a_n = a_1 + (n - 1)d$.

$$a_{350} = 1 + (349)2 = 699$$

Then we use the sum formula with $n = 350$.

$$A_{350} = 1 + 3 + 5 + \cdots + 699$$

$$= \frac{350}{2}(1 + 699) = 122{,}500$$

SIGMA NOTATION

There is a convenient shorthand that is frequently employed in connection with sums. The first letter of the word *sum* is *s*; the Greek letter for *S* is Σ (sigma). We use Σ in mathematics to stand for the operation of summation. In particular,

$$\sum_{i=1}^{n} a_i = a_1 + a_2 + a_3 + \cdots + a_n$$

The symbol $i = 1$ underneath the sigma tells where to start adding the terms a_i and the n at the top tells where to stop. Thus,

$$\sum_{i=1}^{4} a_i = a_1 + a_2 + a_3 + a_4$$

$$\sum_{i=3}^{7} b_i = b_3 + b_4 + b_5 + b_6 + b_7$$

$$\sum_{i=1}^{5} i^2 = 1^2 + 2^2 + 3^2 + 4^2 + 5^2$$

$$\sum_{i=1}^{30} 3i = 3 + 6 + 9 + \cdots + 90$$

If $a_1, a_2, a_3, \ldots$ is an *arithmetic sequence,* then the sum formula previously derived may be written

$$\sum_{i=1}^{n} a_i = \frac{n}{2}(a_1 + a_n)$$

Problem Set 12-2

1. Fill in the boxes.
   (a) 1, 4, 7, 10, $\square$, $\square$, . . .
   (b) 2, 2.3, 2.6, 2.9, $\square$, $\square$, . . .
   (c) 28, 24, 20, 16, $\square$, $\square$, . . .

2. Fill in the boxes.
   (a) 4, 6, 8, 10, $\square$, $\square$, . . .
   (b) 4, 4.2, 4.4, 4.6, $\square$, $\square$, . . .
   (c) 4, 3.8, 3.6, 3.4, $\square$, $\square$, . . .

3. Determine d and the 30th term of each sequence in Problem 1.

4. Determine d and the 101st term of each sequence in Problem 2.

5. Determine $A_{30} = a_1 + a_2 + \cdots + a_{30}$ for the sequence in part (a) of Problem 1. Similarly, find B_{30} and C_{30} for the sequences in parts (b) and (c).

6. Determine A_{100}, B_{100}, and C_{100} for the sequences of Problem 2.

7. If $a_1 = 5$ and $a_{40} = 24.5$ in an arithmetic sequence, determine d.

8. If $b_1 = 6$ and $b_{30} = -52$ in an arithmetic sequence, determine d.

Figure 8

9. Calculate each sum.
   (a) $2 + 4 + 6 + \cdots + 200$
   (b) $1 + 3 + 5 + \cdots + 199$
   (c) $3 + 6 + 9 + \cdots + 198$
   Hint: Before using the sum formula, you have to determine n. In part (a), n is 100 since we are adding the doubles of the integers from 1 to 100.

10. Calculate each sum.
    (a) $4 + 8 + 12 + \cdots + 100$
    (b) $10 + 15 + 20 + \cdots + 200$
    (c) $6 + 9 + 12 + \cdots + 72$

11. The bottom rung of a tapered ladder (Figure 8) is 30 centimeters long and the top rung is 15 centimeters long. If there are 17 rungs, how many centimeters of rung material are needed to make the ladder, assuming no waste?

12. A clock strikes once at 1:00, twice at 2:00, and so on. How many times does it strike between 10:30 A.M. on Monday and 10:30 P.M. on Tuesday?

13. If $3, a, b, c, d, 7, \ldots$ is an arithmetic sequence, find a, b, c, and d.

14. If $8, a, b, c, 5$ is an arithmetic sequence, find a, b, and c.

15. How many multiples of 9 are there between 200 and 300? Find their sum.

16. If Ronnie is paid \$10 on January 1, \$20 on January 2, \$30 on January 3, and so on, how much does he earn during January?

17. Calculate each sum.
    (a) $\displaystyle\sum_{i=2}^{6} i^2$ (b) $\displaystyle\sum_{i=1}^{4} \frac{2}{i}$
    (c) $\displaystyle\sum_{i=1}^{100} (3i + 2)$ (d) $\displaystyle\sum_{i=2}^{100} (2i - 3)$

18. Calculate each sum.
    (a) $\displaystyle\sum_{i=1}^{6} 2^i$ (b) $\displaystyle\sum_{i=1}^{5} (i^2 - 2i)$
    (c) $\displaystyle\sum_{i=1}^{101} (2i - 6)$ (d) $\displaystyle\sum_{i=3}^{102} (3i + 5)$

19. Write in sigma notation.
    (a) $b_3 + b_4 + \cdots + b_{20}$
    (b) $1^2 + 2^2 + \cdots + 19^2$
    (c) $1 + \dfrac{1}{2} + \dfrac{1}{3} + \cdots + \dfrac{1}{n}$

20. Write in sigma notation.
    (a) $a_6 + a_7 + a_8 + \cdots + a_{70}$
    (b) $2^3 + 3^3 + 4^3 + \cdots + 100^3$
    (c) $1 + 3 + 5 + 7 + \cdots + 99$

MISCELLANEOUS PROBLEMS

21. Find each of the following for the arithmetic sequence 20, 19.25, 18.5, 17.75, . . .
    (a) The common difference d.

(b) The 51st term.

(c) The sum of the first 51 terms.

22. Let a_n be an arithmetic sequence with $a_{19} = 42$ and $a_{39} = 54$. Find (a) d, (b) a_1, (c) a_{14}, and (d) k, given that $a_k = 64.2$.

23. Calculate the sum $6 + 6.8 + 7.6 + \cdots + 37.2 + 38$.

24. If $15, a, b, c, d, 24, \ldots$ is an arithmetic sequence, find $a, b, c,$ and d.

25. Let a_n be an arithmetic sequence and, as usual, let $A_n = a_1 + a_2 + \cdots + a_n$. If $A_5 = 50$ and $A_{20} = 650$, find A_{15}.

26. Find the sum of all multiples of 7 between 300 and 450.

27. Calculate each sum.

(a) $\displaystyle\sum_{i=1}^{100} (2i + 1)$

(b) $\displaystyle\sum_{i=1}^{100} \left(\frac{i}{i + 1} - \frac{i - 1}{i} \right)$

(c) $\displaystyle\sum_{i=1}^{100} \left(\frac{1}{i(i + 1)} - \frac{1}{(i + 1)(i + 2)} \right)$

Hint: Write out the first three or four terms to see a pattern.

28. Write in sigma notation.

(a) $c_3 + c_4 + c_5 + \cdots + c_{112}$

(b) $4^2 + 5^2 + 6^2 + \cdots + 104^2$

(c) $35 + 40 + 45 + \cdots + 185$

29. At a club meeting with 300 people present, everyone shook hands with every other person exactly once. How many handshakes were there? *Hint:* Person *A* shook hands with how many people, person *B* shook hands with how many people not already counted, and so on.

30. Mary learned 20 new French words on January 1, 24 new French words on January 2, 28 new French words on January 3, and so on, through January 31. By how much did she increase her French vocabulary during January?

31. A pile of logs has 70 logs in the bottom layer, 69 logs in the second layer, and so on to the top layer with 10 logs. How many logs are in the pile?

32. Calculate the following sum.

$$-1^2 + 2^2 - 3^2 + 4^2 - 5^2 + 6^2 - \cdots - 99^2 + 100^2$$

Hint: Group in a clever way.

33. Calculate the following sum.

$$\frac{1}{2} + \left(\frac{1}{3} + \frac{2}{3} \right) + \left(\frac{1}{4} + \frac{2}{4} + \frac{3}{4} \right) + \cdots + \left(\frac{1}{100} + \frac{2}{100} + \frac{3}{100} + \cdots + \frac{99}{100} \right)$$

34. Show that the sum of n consecutive positive integers plus n^2 is equal to the sum of the next n consecutive integers.

35. Approximately how long is the playing groove in a $33\frac{1}{3}$-rpm record that takes 18 minutes to play if the groove starts 6 inches from the center and ends 3 inches from the center? To approximate, assume that each revolution produces a groove that is circular.

36. **TEASER** Find the sum of all the digits in the integers from 1 to 999,999.

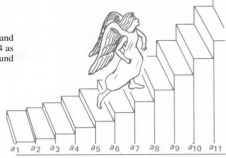

Jacob's Golden Staircase

In his dreams, Jacob saw a golden staircase with angels walking up and down. The first step was 8 inches high but after that each step was 5/4 as high above the ground as the previous one. How high above the ground was the 800th step?

a_1 a_2 a_3 a_4 a_5 a_6 a_7 a_8 a_9 a_{10} a_{11}

12-3   Geometric Sequences

The staircase of Jacob's dream is most certainly one for angels, not for people. The 800th step actually stands 3.4×10^{73} miles high. By way of comparison, it is 9.3×10^7 miles to the sun and 2.5×10^{13} miles to Alpha Centauri, our nearest star beyond the sun. You might say the golden staircase reaches to heaven.

To see how to calculate the height of the 800th step, notice the pattern of heights for the first few steps and then generalize.

$$a_1 = 8$$

$$a_2 = 8\left(\frac{5}{4}\right)$$

$$a_3 = 8\left(\frac{5}{4}\right)^2$$

$$a_4 = 8\left(\frac{5}{4}\right)^3$$

$$\vdots$$

$$a_{800} = 8\left(\frac{5}{4}\right)^{799}$$

With a pocket calculator, it is easy to calculate $8(\frac{5}{4})^{799}$ and then change this number of inches to miles; the result is the figure given above.

FORMULAS

In the sequence above, each term was $\frac{5}{4}$ times the preceding one. You should be able to find a similar pattern in each of the following sequences. When you do, fill in the boxes.

(a) 3, 6, 12, 24, □, □, . . .

(b) 12, 4, $\frac{4}{3}$, $\frac{4}{9}$, □, □, . . .

(c) .6, 6, 60, 600, □, □, . . .

The common feature of these three sequences is that in each case, you can get a term by multiplying the preceding term by a fixed number. In sequence (a), you multiply by 2; in (b), by $\frac{1}{3}$; and in (c), by 10. We call such sequences **geometric sequences.** Thus a geometric sequence $a_1, a_2, a_3, \ldots$ satisfies the recursion formula

$$a_n = r a_{n-1}$$

where r is a fixed number called the **common ratio.**

To obtain the corresponding explicit formula, note that

$$a_2 = r a_1$$
$$a_3 = r a_2 = r(r a_1) = r^2 a_1$$
$$a_4 = r a_3 = r(r^2 a_1) = r^3 a_1$$

In each case, the exponent on r is one less than the subscript on a. Thus

$$a_n = a_1 r^{n-1}$$

From this, we can get explicit formulas for each of the sequences (a), (b), and (c) above.

$$a_n = 3 \cdot 2^{n-1}$$

$$b_n = 12\left(\frac{1}{3}\right)^{n-1}$$

$$c_n = (.6)(10)^{n-1}$$

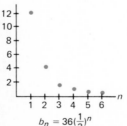

$b_n = 36(\frac{1}{3})^n$

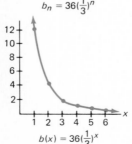

$b(x) = 36(\frac{1}{3})^x$

Figure 9

GEOMETRIC SEQUENCES AND EXPONENTIAL FUNCTIONS

Let us consider sequence (b) once more; its explicit formula is

$$b_n = 12\left(\frac{1}{3}\right)^{n-1} = 36\left(\frac{1}{3}\right)^n$$

We have graphed this sequence and also the exponential function

$$b(x) = 36\left(\frac{1}{3}\right)^x$$

in Figure 9. (See Section 6-2 for a discussion of exponential functions.) It should be clear that the sequence b_n is the function $b(x)$ with its domain restricted to the positive integers.

What we have observed in this example is true in general. A geometric sequence is simply an exponential function with its domain restricted to the positive integers.

SUMS OF GEOMETRIC SEQUENCES

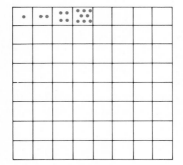

Figure 10

There is an old legend about geometric sequences and chessboards. When the king of Persia learned to play chess, he was so enchanted with the game that he determined to reward the inventor, a man named Sessa. Calling Sessa to the palace, the king promised to fulfill any request he might make. With an air of modesty, wily Sessa asked for one grain of wheat for the first square of the chessboard, two for the second, four for the third, and so on (Figure 10). The king was amused at such an odd request; nevertheless, he called a servant, told him to get a bag of wheat, and start counting. To the king's surprise, it soon became apparent that Sessa's request could never be fulfilled. The world's total production of wheat for a whole century would not be sufficient.

Sessa was really asking for

$$1 + 2 + 2^2 + 2^3 + \cdots + 2^{63}$$

grains of wheat, the sum of the first 64 terms of the geometric sequence 1, 2, 4, 8, ... We are going to develop a formula for this sum and for all others that arise from adding the terms of a geometric sequence.

Let $a_1, a_2, a_3, \ldots$ be a geometric sequence with ratio $r \neq 1$. As usual, let

$$A_n = a_1 + a_2 + a_3 + \cdots + a_n$$

which can be written

$$A_n = a_1 + a_1r + a_1r^2 + \cdots + a_1r^{n-1}$$

Now multiply A_n by r, subtract the result from A_n, and use a little algebra to solve for A_n. We obtain

$$
\begin{aligned}
A_n &= a_1 + a_1r + a_1r^2 + \cdots + a_1r^{n-1} \\
rA_n &= a_1r + a_1r^2 + \cdots + a_1r^{n-1} + a_1r^n \\
\hline
A_n - rA_n &= a_1 + 0 + 0 + \cdots + 0 - a_1r^n \\
A_n(1 - r) &= a_1(1 - r^n)
\end{aligned}
$$

$$A_n = \frac{a_1(1 - r^n)}{1 - r} \qquad r \neq 1$$

In sigma notation, this is

$$\sum_{i=1}^{n} a_i = \frac{a_1(1 - r^n)}{1 - r} \qquad r \neq 1$$

In the case where $r = 1$,

$$\sum_{i=1}^{n} a_i = a_1 + a_1 + \cdots + a_1 = na_1$$

Applying the sum formula to Sessa's problem (using $n = 64$, $a_1 = 1$, and $r = 2$), we get

$$A_{64} = \frac{1(1 - 2^{64})}{1 - 2} = 2^{64} - 1$$

Ignoring the -1 and using the approximation $2^{10} \approx 1000$ gives

$$A_{64} \approx 2^{64} \approx 2^4(1000)^6 = 1.6 \times 10^{19}$$

If you do this problem on a calculator, you will get $A_{64} \approx 1.845 \times 10^{19}$. Thus, if a bushel of wheat has one million grains, A_{64} grains would amount to more than 1.8×10^{13}, or 18 trillion, bushels. That exceeds the world's total production of wheat in 1 century.

THE SUM OF THE WHOLE SEQUENCE

Is it possible to add infinitely many numbers? Do the following sums make sense?

$$\frac{1}{2} + \frac{1}{4} + \frac{1}{8} + \frac{1}{16} + \cdots$$

$$1 + 3 + 9 + 27 + \cdots$$

Questions like this have intrigued great thinkers since Zeno first introduced his famous paradoxes of the infinite over 2000 years ago. We now show that we can make sense out of the first of these two sums but not the second.

Consider a string of length 1 kilometer. We may imagine cutting it into infinitely many pieces, as indicated in Figure 11.

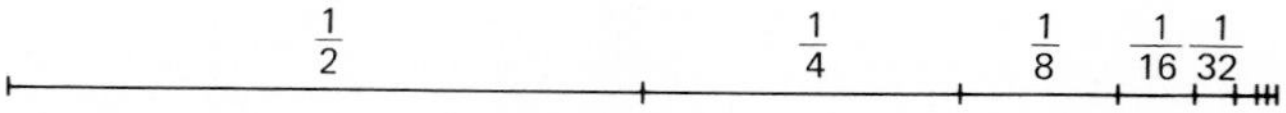

Figure 11

Since these pieces together make a string of length 1, it seems natural to say

$$\frac{1}{2} + \frac{1}{4} + \frac{1}{8} + \frac{1}{16} + \cdots = 1$$

Let us look at it another way. The sum of the first n terms of the geometric sequence $\frac{1}{2}, \frac{1}{4}, \frac{1}{8}, \frac{1}{16}, \ldots$ is given by

$$A_n = \frac{\frac{1}{2}\left[1 - \left(\frac{1}{2}\right)^n\right]}{1 - \frac{1}{2}} = 1 - \left(\frac{1}{2}\right)^n$$

As n gets larger and larger (tends to infinity), $(\frac{1}{2})^n$ gets smaller and smaller (approaches 0). Thus A_n tends to 1 as n tends to infinity. We therefore say that 1 is the sum of *all* the terms of this sequence.

Now consider any geometric sequence with ratio r satisfying $|r| < 1$. We claim that when n gets large, r^n approaches 0. (As evidence, try calculating $(0.99)^{100}$, $(0.99)^{1000}$, and $(0.99)^{10,000}$ on your calculator.) Thus, as n gets large,

$$A_n = \frac{a_1(1 - r^n)}{1 - r}$$

approaches $a_1/(1 - r)$. We write

$$\sum_{i=1}^{\infty} a_i = \frac{a_1}{1 - r} \qquad |r| < 1$$

For an important use of this formula in a familiar context, see the example in the problem set.

We emphasize that what we have just done is valid if $|r| < 1$. There is no way to make sense out of adding all the terms of a geometric sequence if $|r| \geq 1$.

Problem Set 12-3

1. Fill in the boxes.
   (a) $\frac{1}{2}$, 1, 2, 4, □, □, . . .
   (b) 8, 4, 2, 1, □, □, . . .
   (c) .3, .03, .003, .0003, □, □, . . .

2. Fill in the boxes.
   (a) 1, 3, 9, 27, □, □, . . .
   (b) 27, 9, 3, 1, □, □, . . .
   (c) .2, .02, .002, .0002, □, □, . . .

3. Determine r for each of the sequences in Problem 1 and write an explicit formula for the nth term.

4. Write a formula for the nth term of each sequence in Problem 2.

5. Evaluate the 30th term of each sequence in Problem 1.

6. Evaluate the 20th term of each sequence in Problem 2.

7. Use the sum formula to find the sum of the first five terms of each sequence in Problem 1.

8. Use the sum formula to find the sum of the first five terms of each sequence in Problem 2.

9. Find the sum of the first 30 terms of each sequence in Problem 1.

10. Find the sum of the first 20 terms of each sequence in Problem 2.

11. A certain culture of bacteria doubles every week. If there are 100 bacteria now, how many will there be after 10 full weeks?

12. A water lily grows so rapidly that each day it covers twice the area it covered the

day before. At the end of 20 days, it completely covers a pond. If we start with two lilies, how long will it take to cover the same pond?

13. Johnny is paid $1 on January 1, $2 on January 2, $4 on January 3, and so on. Approximately how much will he earn during January?

14. If you were offered 1¢ today, 2¢ tomorrow, 4¢ the third day, and so on for 20 days or a lump sum of $10,000, which would you choose? Show why.

15. Calculate

(a) $\displaystyle\sum_{i=1}^{\infty} \left(\frac{1}{3}\right)^i$;   (b) $\displaystyle\sum_{i=2}^{\infty} \left(\frac{2}{5}\right)^i$.

16. Calculate

(a) $\displaystyle\sum_{i=1}^{\infty} \left(\frac{2}{3}\right)^i$;   (b) $\displaystyle\sum_{i=3}^{\infty} \left(\frac{1}{6}\right)^i$.

17. A ball is dropped from a height of 10 feet. At each bounce, it rises to a height of $\frac{1}{2}$ the previous height. How far will it travel altogether (up and down) by the time it comes to rest? *Hint:* Think of the total distance as being the sum of the "down" distances $(10 + 5 + \frac{5}{2} + \cdots)$ and the "up" distances $(5 + \frac{5}{2} + \frac{5}{4} + \cdots)$.

18. Do Problem 17 assuming the ball rises to $\frac{2}{3}$ its previous height at each bounce.

EXAMPLE (Repeating Decimals) Show that $.333\overline{3}\ldots$ and $.2323\overline{23}\ldots$ are rational numbers by using the methods of this section.

Solution.

$$.333\overline{3} = \frac{3}{10} + \frac{3}{100} + \frac{3}{1000} + \cdots$$

Thus we must add all the terms of an infinite geometric sequence with ratio $\frac{1}{10}$. Using the formula $a_1/(1 - r)$, we get

$$\frac{\dfrac{3}{10}}{1 - \dfrac{1}{10}} = \frac{\dfrac{3}{10}}{\dfrac{9}{10}} = \frac{1}{3}$$

Similarly,

$$.2323\overline{23} = \frac{23}{100} + \frac{23}{10000} + \frac{23}{1000000} + \cdots$$

$$= \frac{\dfrac{23}{100}}{1 - \dfrac{1}{100}} = \frac{\dfrac{23}{100}}{\dfrac{99}{100}} = \frac{23}{99}$$

Use this method to express each of the following as the ratio of two integers.

19. $.11\overline{1}$

20. $.77\overline{7}$

21. $.2525\overline{25}$

22. $.99\overline{9}$

23. $1.234\overline{34}$

24. $.341\overline{41}$

MISCELLANEOUS PROBLEMS

25. If $a_n = 625(0.2)^{n-1}$, find a_1, a_2, a_3, a_4, and a_5.

26. Which of the following sequences are geometric, which are arithmetic, and which are neither?
    (a) 130, 65, 32.5, 16.25, . . .
    (b) $1, \frac{1}{2}, \frac{1}{3}, \frac{1}{4}, \ldots$
    (c) 100(1.05), 100(1.07), 100(1.09), 100(1.11), . . .
    (d) $100(1.05)$, $100(1.05)^2$, $100(1.05)^3$, $100(1.05)^4$, . . .
    (e) 1, 3, 6, 10, . . .
    (f) 3, −6, 12, −24, . . .

27. Write an explicit formula for each geometric or arithmetic sequence in Problem 26.

28. Use the formula for the sum of an infinite geometric sequence to express each of the following repeating decimals as a ratio of two integers.
    (a) $.499999 \ldots = .4\overline{9}$ (b) $.1234234234 \ldots = .1\overline{234}$

□ *Recall from Section 6-3 that if a sum of P dollars is invested today at a compound rate of i per conversion period, then the accumulated value after n periods is given by $P(1 + i)^n$. The sequence of accumulated values*

$$P(1 + i), P(1 + i)^2, P(1 + i)^3, P(1 + i)^4, \ldots$$

is geometric with ratio $1 + i$. In Problems 29–33, write a formula for the answer and then use a calculator to evaluate it.

29. If $1 is put in the bank at 8 percent interest compounded annually, it will be worth $(1.08)^n$ dollars after n years. How much will $100 be worth after 10 years? When will the amount first exceed $250?

30. If $1 is put in the bank at 8 percent interest compounded quarterly, it will be worth $(1.02)^n$ dollars after n quarters. How much will $100 be worth after 10 years (40 quarters)? When will the amount first exceed $250?

31. Suppose Karen puts $100 in the bank today and $100 at the beginning of each of the following 9 years. If this money earns interest at 8 percent compounded annually, what will it be worth at the end of 10 years?

32. José makes 40 deposits of $25 each in a bank at intervals of three months, making the first deposit today. If money earns interest at 8 percent compounded quarterly, what will it all be worth at the end of 10 years (40 quarters)?

33. Suppose the government pumps an extra billion dollars into the economy. Assume that each business and individual saves 25 percent of its income and spends the rest, so that of the initial one billion dollars, 75 percent is re-spent by individuals and businesses. Of that amount, 75 percent is spent, and so forth. What is the total increase in spending due to the government action? (This is called the *multiplier effect* in economics.)

34. Given an arbitrary triangle of perimeter 10, a second triangle is formed by joining the midpoints of the first, a third triangle is formed by joining the midpoints of the second, and so on forever. Find the total length of all line segments in the resulting configuration.

35. Find the area of the painted region in Figure 12 on the next page, which consists of an infinite sequence of 30°-60°-90° triangles.

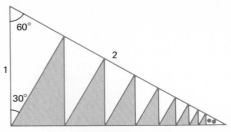

Figure 12

36. If the pattern in Figure 13 is continued indefinitely, what fraction of the area of the original square will be painted?

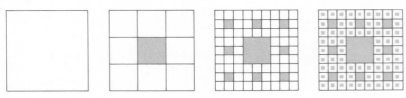

Figure 13

37. Expand in powers of 3; then evaluate P.

$$P = (1 + 3)(1 + 3^2)(1 + 3^4)(1 + 3^8)(1 + 3^{16})$$

38. By considering $S - rS$, find S.
    (a) $S = r + 2r^2 + 3r^3 + 4r^4 + \cdots,\ |r| < 1$.
    (b) $S = \frac{1}{3} + 2(\frac{1}{3})^2 + 3(\frac{1}{3})^3 + 4(\frac{1}{3})^4 + \cdots$

39. In a geometric sequence, the sum of the first 2 terms is 5 and the sum of the first 6 terms is 65. What is the sum of the first 4 terms. *Hint:* Call the first term a and the ratio r. Then $a + ar = 5$, so $a = 5/(1 + r)$. This allows you to write 65 in terms of r alone. Solve for r.

40. Imagine a huge maze with infinitely many adjoining cells, each having a square base 1 meter by 1 meter. The first cell has walls 1 meter high, the second cell has walls $\frac{1}{2}$ meter high, the third cell has walls $\frac{1}{4}$ meter high, and so on.
    (a) How much paint would it take to fill the maze with paint?
    (b) How much paint would it take to paint the floors of all the mazes?
    (c) Explain this apparent contradiction.

41. Starting 100 miles apart, Tom and Joel ride toward each other on their bicycles, Tom going at 8 miles per hour and Joel at 12 miles per hour. Tom's dog, Corky, starts with Tom running toward Joel at 25 miles per hour. When Corky meets Joel, he immediately turns tail and heads back to Tom. Reaching Tom, Corky again turns tail and heads toward Joel, and so on. How far did Corky run by the time Tom and Joel met? This can be answered using geometric sequences, but if you are clever, you will find a better way.

42. **TEASER** Sally walked 4 miles north, then 2 miles east, 1 mile south, $\frac{1}{2}$ mile west, $\frac{1}{4}$ mile north, and so on. If she continued this pattern indefinitely, how far from her initial point did she end?

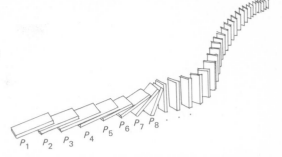

The Principle of Mathematical Induction
Let $P_1, P_2, P_3, \ldots$ be a sequence of statements with the following two properties:

1. P_1 is true.
2. The truth of P_k implies the truth of P_{k+1} ($P_k \Rightarrow P_{k+1}$).

Then the statement P_n is true for every positive integer n.

12-4   Mathematical Induction

The principle of mathematical induction deals with a sequence of statements. A **statement** is a sentence that is either true or false. In a sequence of statements, there is a statement corresponding to each positive integer. Here are four examples.

$$P_n: \quad \frac{1}{1 \cdot 2} + \frac{1}{2 \cdot 3} + \frac{1}{3 \cdot 4} + \cdots + \frac{1}{n(n+1)} = \frac{n}{n+1}$$

$Q_n: \quad n^2 - n + 41$ is a prime number

$R_n: \quad (a + b)^n = a^n + b^n$

$$S_n: \quad 1 + 2 + 3 + \cdots + n = \frac{n^2 + n - 6}{2}$$

To be sure we understand the notation, let us write each of these statements for the case $n = 3$.

$$P_3: \quad \frac{1}{1 \cdot 2} + \frac{1}{2 \cdot 3} + \frac{1}{3 \cdot 4} = \frac{3}{4}$$

$Q_3: \quad 3^2 - 3 + 41$ is a prime number

$R_3: \quad (a + b)^3 = a^3 + b^3$

$$S_3: \quad 1 + 2 + 3 = \frac{3^2 + 3 - 6}{2}$$

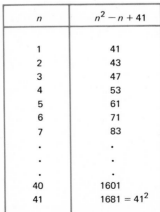

n	$n^2 - n + 41$
1	41
2	43
3	47
4	53
5	61
6	71
7	83
.	.
.	.
.	.
40	1601
41	$1681 = 41^2$

Figure 14

Of these, P_3 and Q_3 are true, while R_3 and S_3 are false; you should verify this fact. A careful study of these four sequences will indicate the wide range of behavior that sequences of statements can display.

While it certainly is not obvious, we claim that P_n is true for every positive integer n; we are going to prove it soon. Q_n is a well-known sequence. It was thought by some to be true for all n and, in fact, it is true for $n = 1, 2, 3, \ldots,$ 40 (see Figure 14). However, it fails for $n = 41$, a fact that allows us to make an important point. Establishing the truth of Q_n for a finite number of cases, no matter how many, does not prove its truth for *all* n. Sequences R_n and S_n are rather hopeless cases, since R_n is true only for $n = 1$ and S_n is never true.

PROOF BY MATHEMATICAL INDUCTION

How does one prove that something is true for all n? The tool uniquely designed for this purpose is the **principle of mathematical induction;** it was stated in our opening display. Let us use mathematical induction to show that

$$P_n: \quad \frac{1}{1 \cdot 2} + \frac{1}{2 \cdot 3} + \frac{1}{3 \cdot 4} + \cdots + \frac{1}{(n-1)n} + \frac{1}{n(n+1)} = \frac{n}{n+1}$$

is true for every positive integer n. There are two steps to the proof. We must show that

1. P_1 is true;
2. $P_k \Rightarrow P_{k+1}$; that is, the truth of P_k implies the truth of P_{k+1}.

The first step is easy. P_1 is just the statement

$$\frac{1}{1 \cdot 2} = \frac{1}{1 + 1}$$

which is clearly true.

To handle the second step ($P_k \Rightarrow P_{k+1}$), it is a good idea to write down the statements corresponding to P_k and P_{k+1} (at least on scratch paper). We get them by substituting k and $k + 1$ for n in the statement for P_n.

$$P_k: \quad \frac{1}{1 \cdot 2} + \frac{1}{2 \cdot 3} + \cdots + \frac{1}{(k-1)k} + \frac{1}{k(k+1)} = \frac{k}{k+1}$$

$$P_{k+1}: \quad \frac{1}{1 \cdot 2} + \frac{1}{2 \cdot 3} + \cdots + \frac{1}{k(k+1)} + \frac{1}{(k+1)(k+2)} = \frac{k+1}{k+2}$$

Notice that the left side of P_{k+1} is the same as that of P_k except for the addition of one more term, $1/(k+1)(k+2)$.

Suppose for the moment that P_k is true and consider how this assumption allows us to simplify the left side of P_{k+1}.

$$\left[\frac{1}{1 \cdot 2} + \frac{1}{2 \cdot 3} + \cdots + \frac{1}{k(k+1)} \right] + \frac{1}{(k+1)(k+2)}$$

$$= \frac{k}{k+1} + \frac{1}{(k+1)(k+2)}$$

$$= \frac{k(k+2) + 1}{(k+1)(k+2)}$$

$$= \frac{(k+1)^2}{(k+1)(k+2)}$$

$$= \frac{k+1}{k+2}$$

If you read this chain of equalities from top to bottom, you will see that we have established the truth of P_{k+1}, but under the *assumption that P_k is true*. That is, we have established that the truth of P_k implies the truth of P_{k+1}.

SOME COMMENTS ABOUT MATHEMATICAL INDUCTION

Students never have any trouble with the verification step (showing that P_1 is true). The inductive step (showing that $P_k \Rightarrow P_{k+1}$) is harder and more subtle. In that step, we do *not* prove that P_k or P_{k+1} is true, but rather that the truth of P_k implies the truth of P_{k+1}. For a vivid illustration of the difference, we point out that in the fourth example of our opening paragraph, the truth of S_k does imply the truth of S_{k+1} ($S_k \Rightarrow S_{k+1}$) and yet not a single statement in that sequence is true (see Problem 30). To put it another way, what $S_k \Rightarrow S_{k+1}$ means is that *if* S_k were true, then S_{k+1} would be true also. It is like saying that if spinach were ice cream, then kids would want two helpings at every meal.

Perhaps the dominoes in the opening display can help illuminate the idea. For all the dominoes to fall it is sufficient that

1. the first domino is pushed over;
2. if any domino falls (say the kth one), it pushes over the next one (the $(k + 1)$st one).

Figure 15 on the next page illustrates what happens to the dominoes in the four examples of our opening paragraph. Study them carefully.

ANOTHER EXAMPLE

Consider the statement

$$P_n: \quad 1^2 + 2^2 + 3^2 + \cdots + n^2 = \frac{n(n + 1)(2n + 1)}{6}$$

We are going to prove that P_n is true for all n by mathematical induction. For $n = 1$, k, and $k + 1$, the statements P_n are:

$$P_1: \quad 1^2 = \frac{1(2)(3)}{6}$$

$$P_k: \quad 1^2 + 2^2 + 3^2 + \cdots + k^2 = \frac{k(k + 1)(2k + 1)}{6}$$

$$P_{k+1}: \quad 1^2 + 2^2 + 3^2 + \cdots + k^2 + (k + 1)^2 = \frac{(k + 1)(k + 2)(2k + 3)}{6}$$

Clearly P_1 is true.

Why They Fall and Why They Don't

$P_n: \frac{1}{1 \cdot 2} + \frac{1}{2 \cdot 3} + \ldots + \frac{1}{n(n+1)} = \frac{n}{n+1}$ P_1 is true $P_k \Rightarrow P_k + 1$	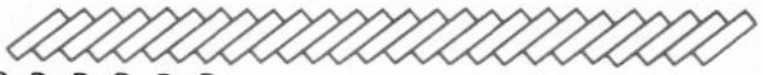 $P_1P_2P_3P_4P_5P_6 \cdots$ First domino is pushed over. Each falling domino pushes over the next one.
$Q_n: n^2 - n + 41$ is prime. $Q_1, Q_2, \ldots, Q_{40}$ are true. $Q_k \not\Rightarrow Q_k + 1$	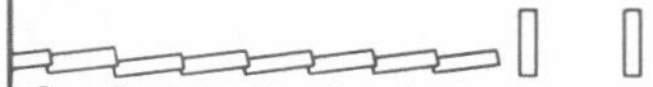 $Q_{35}\ Q_{36}\ Q_{37}\ Q_{38}\ Q_{39}\ Q_{40}\quad Q_{41}\quad Q_{42}$ First 40 dominoes are pushed over. 41st domino remains standing.
$R_n: (a+b)^n = a^n + b^n$ R_1 is true. $R_k \not\Rightarrow R_k + 1$	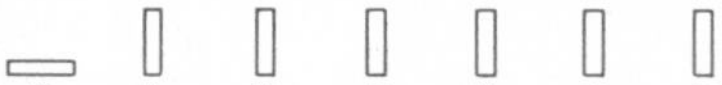 $R_1\quad R_2\quad R_3\quad R_4\quad R_5\quad R_6\quad R_7$ First domino is pushed over but dominoes are spaced too far apart to push each other over.
$S_n: 1 + 2 + 3 + \ldots + n = \frac{n^2 + n - 6}{2}$ S_1 is false. $S_k \Rightarrow S_k + 1$	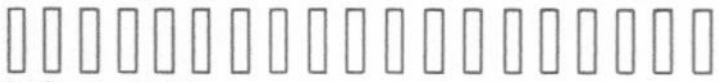 $S_1S_2S_3S_4S_5$ Spacing is just right but no one can push over the first domino.

Figure 15

Assuming that P_k is true, we can write the left side of P_{k+1} as shown in the following chain of equalities.

$$\begin{aligned} 1^2 + 2^2 + 3^2 + \cdots + k^2 + (k+1)^2 &= \frac{k(k+1)(2k+1)}{6} + (k+1)^2 \\ &= \frac{(k+1)[k(2k+1) + 6(k+1)]}{6} \\ &= \frac{(k+1)(2k^2 + 7k + 6)}{6} \\ &= \frac{(k+1)(k+2)(2k+3)}{6} \end{aligned}$$

Thus the truth of P_k does imply the truth of P_{k+1}. We conclude by mathematical induction that P_n is true for every positive integer n. Incidentally, the result just proved will be used in calculus.

Problem Set 12-4

In Problems 1–8, prove by mathematical induction that P_n is true for every positive integer n.

1. P_n: $1 + 2 + 3 + \cdots + n = \dfrac{n(n + 1)}{2}$

2. P_n: $1 + 3 + 5 + \cdots + (2n - 1) = n^2$

3. P_n: $3 + 7 + 11 + \cdots + (4n - 1) = n(2n + 1)$

4. P_n: $2 + 9 + 16 + \cdots + (7n - 5) = \dfrac{n(7n - 3)}{2}$

5. P_n: $1 \cdot 2 + 2 \cdot 3 + 3 \cdot 4 + \cdots + n(n + 1) = \frac{1}{3}n(n + 1)(n + 2)$

6. P_n: $\dfrac{1}{1 \cdot 3} + \dfrac{1}{3 \cdot 5} + \dfrac{1}{5 \cdot 7} + \cdots + \dfrac{1}{(2n - 1)(2n + 1)} = \dfrac{n}{2n + 1}$

7. P_n: $2 + 2^2 + 2^3 + \cdots + 2^n = 2(2^n - 1)$

8. P_n: $1^2 + 3^2 + 5^2 + \cdots + (2n - 1)^2 = \dfrac{n(2n - 1)(2n + 1)}{3}$

In Problems 9–18, tell what you can conclude from the information given about the sequence of statements. For example, if you are given that P_4 is true and that $P_k \Rightarrow P_{k+1}$ for any k, then you can conclude that P_n is true for every integer $n \geq 4$.

9. P_8 is true and $P_k \Rightarrow P_{k+1}$.

10. P_8 is not true and $P_k \Rightarrow P_{k+1}$.

11. P_1 is true but P_k does not imply P_{k+1}.

12. $P_1, P_2, \ldots , P_{1000}$ are all true.

13. P_1 is true and $P_k \Rightarrow P_{k+2}$.

14. P_{40} is true and $P_k \Rightarrow P_{k-1}$.

15. P_1 and P_2 are true; P_k and P_{k+1} together imply P_{k+2}.

16. P_1 and P_2 are true and $P_k \Rightarrow P_{k+2}$.

17. P_1 is true and $P_k \Rightarrow P_{4k}$.

18. P_1 is true, $P_k \Rightarrow P_{4k}$, and $P_k \Rightarrow P_{k-1}$.

EXAMPLE A (Mathematical Induction Applied to Inequalities)   Show that the following statement is true for every integer $n \geq 4$.

$$3^n > 2^n + 20$$

Solution.   Let P_n represent the given statement. You might check that P_1, P_2, and P_3 are false. However, that does not matter to us. What we need to do is to show that P_4 is true and that $P_k \Rightarrow P_{k+1}$ for any $k \geq 4$.

$$P_4:   3^4 > 2^4 + 20$$

$$P_k: 3^k > 2^k + 20$$

$$P_{k+1}:   3^{k+1} > 2^{k+1} + 20$$

Clearly P_4 is true (81 is greater than 36). Next we assume P_k to be true (for $k \geq 4$) and seek to show that this would force P_{k+1} to be true. Working with the left side of P_{k+1} and using the assumption that $3^k > 2^k + 20$, we get

$$3^{k+1} = 3 \cdot 3^k > 3(2^k + 20) > 2(2^k + 20) = 2^{k+1} + 40 > 2^{k+1} + 20$$

Therefore, P_{k+1} is true, provided P_k is true. We conclude that P_n is true for every integer $n \geq 4$.

In Problems 19–24, find the smallest positive integer n for which the given statement is true for it and all larger integers. Then prove that the statement is true for all integers greater than that smallest value.

19. $n + 5 < 2^n$
20. $3n \leq 3^n$
21. $\log_{10} n < n$ *Hint: $k + 1 < 10k$.*
22. $n^2 \leq 2^n$ *Hint: $k^2 + 2k + 1 = k(k + 2 + 1/k) < k(k + k)$.*
23. $(1 + x)^n \geq 1 + nx$, where $x \geq -1$
24. $|\sin nx| \leq |\sin x| \cdot n$ for all x

EXAMPLE B **(Mathematical Induction and Divisibility)** Prove that the statement

$$P_n: \quad x - y \text{ is a factor of } x^n - y^n$$

is true for every positive integer n.

Solution.   Trivially, $x - y$ is a factor of $x - y$ since $x - y = 1(x - y)$; so P_1 is true. Now suppose that P_k is true, that is, that

$$x - y \text{ is a factor of } x^k - y^k$$

This means that there is a polynomial $Q(x, y)$ such that

$$x^k - y^k = Q(x, y)(x - y)$$

Using this assumption, we may write

$$x^{k+1} - y^{k+1} = x^{k+1} - x^k y + x^k y - y^{k+1}$$
$$= x^k(x - y) + y(x^k - y^k)$$
$$= x^k(x - y) + yQ(x, y)(x - y)$$
$$= [x^k + yQ(x, y)](x - y)$$

Thus $x - y$ is a factor of $x^{k+1} - y^{k+1}$. We have shown that $P_k \Rightarrow P_{k+1}$ and that P_1 is true; we therefore conclude that P_n is true for all n.

Use mathematical induction to prove that each of the following is true for every positive integer n.

25. $x + y$ is a factor of $x^{2n} - y^{2n}$. *Hint: $x^{2k+2} - y^{2k+2} = x^{2k+2} - x^{2k}y^2 + x^{2k}y^2 - y^{2k+2}$*

26. $x + y$ is a factor of $x^{2n-1} + y^{2n-1}$.

27. $n^2 - n$ is even (that is, has 2 as a factor).

28. $n^3 - n$ is divisible by 6.

MISCELLANEOUS PROBLEMS

29. These four formulas can all be proved by mathematical induction. We proved (b) in the text; you prove the others.
    (a) $1 + 2 + 3 + \cdots + n = \frac{1}{2}n(n + 1)$
    (b) $1^2 + 2^2 + 3^2 + \cdots + n^2 = \frac{1}{6}n(n + 1)(2n + 1)$
    (c) $1^3 + 2^3 + 3^3 + \cdots + n^3 = \frac{1}{4}n^2(n + 1)^2$
    (d) $1^4 + 2^4 + 3^4 + \cdots + n^4 = \frac{1}{30}n(n + 1)(6n^3 + 9n^2 + n - 1)$
    From (a) and (c), another interesting formula follows, namely,

$$1^3 + 2^3 + 3^3 + \cdots + n^3 = (1 + 2 + 3 + \cdots + n)^2$$

30. Consider the statement

$$S_n: \quad 1 + 2 + 3 + \cdots + n = \frac{n^2 + n - 6}{2}$$

Show that
(a) $S_k \Rightarrow S_{k+1}$ for $k \geq 1$;
(b) S_n is not true for any positive integer n.

31. Use the results of Problem 29 to evaluate each of the following.

(a) $\displaystyle\sum_{k=1}^{100} (3k + 1)$ (b) $\displaystyle\sum_{k=1}^{10} (k^2 - 3k)$

(c) $\displaystyle\sum_{k=1}^{10} (k^3 + 3k^2 + 3k + 1)$ (d) $\displaystyle\sum_{k=1}^{n} (6k^2 + 2k)$

32. In a popular song titled "The Twelve Days of Christmas," my true love gave me 1 gift on the first day, $(2 + 1)$ gifts on the second day, $(3 + 2 + 1)$ gifts on the third day, and so on.
    (a) How many gifts did I get all together during the 12 days?
    (b) How many gifts would I get all together in a Christmas that had n days?

33. Prove that for $n \geq 2$,

$$\left(1 - \frac{1}{4}\right)\left(1 - \frac{1}{9}\right)\left(1 - \frac{1}{16}\right) \cdots \left(1 - \frac{1}{n^2}\right) = \frac{n + 1}{2n}$$

34. Prove that for $n \geq 1$,

$$\frac{1}{\sqrt{1}} + \frac{1}{\sqrt{2}} + \frac{1}{\sqrt{3}} + \cdots + \frac{1}{\sqrt{n}} < 2\sqrt{n}$$

35. Prove that for $n \geq 3$,

$$\frac{1}{n + 1} + \frac{1}{n + 2} + \frac{1}{n + 3} + \cdots + \frac{1}{2n} > \frac{3}{5}$$

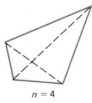

$n = 4$

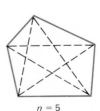

$n = 5$

Figure 16

36. Prove that the number of diagonals in an n-sided convex polygon is $n(n - 3)/2$ for $n \geq 3$. The diagrams in Figure 16 show the situation for $n = 4$ and $n = 5$.

37. Prove that the sum of the measures of the interior angles in an n-sided polygon

(without holes or self-intersections) is $(n - 2)180°$. What is the sum of the measures of the exterior angles of such a polygon?

38. Let $f_1 = 1$, $f_2 = 1$, and $f_{n+2} = f_{n+1} + f_n$ for $n \geq 1$. Call f_n the Fibonacci sequence (see Section 12-1) and let $F_n = f_1 + f_2 + f_3 + \cdots + f_n$. Prove by mathematical induction that $F_n = f_{n+2} - 1$ for all n.

39. For the Fibonacci sequence of Problem 38, prove that for $n \geq 1$,

$$f_1^2 + f_2^2 + f_3^2 + \cdots + f_n^2 = f_n f_{n+1}$$

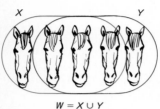

$W = X \cup Y$

Figure 17

40. What is wrong with the following argument?

Theorem. All horses in the world have the same color.

Proof. Let P_n be the statement: All the horses in any set of n horses are identically colored. Certainly P_1 is true. Suppose that P_k is true, that is, that all the horses in any set of k horses are identically colored. Let W be any set of $k + 1$ horses. Now we may think of W as the union of two overlapping sets X and Y, each with k horses. (The situation for $k = 4$ is shown in Figure 17.) By assumption, the horses in X are identically colored and the horses in Y are identically colored. Since X and Y overlap, all the horses in $X \cup Y$ must be identically colored. We conclude that P_n is true for all n. Thus the set of all horses in the world (some finite number) have the same color.

41. Let $a_0 = 0$, $a_1 = 1$, and $a_{n+2} = (a_{n+1} + a_n)/2$ for $n \geq 0$. Prove that for $n \geq 0$,

$$a_n = \frac{2}{3}\left[1 - \left(-\frac{1}{2}\right)^n\right]$$

Hint: In the inductive step, show that P_k and P_{k+1} together imply P_{k+2}.

42. Let f_n be the Fibonacci sequence of Problem 38. Use mathematical induction (as in the hint to Problem 41) to prove that

$$f_n = \frac{1}{\sqrt{5}}\left[\left(\frac{1 + \sqrt{5}}{2}\right)^n - \left(\frac{1 - \sqrt{5}}{2}\right)^n\right]$$

12-5 Counting Ordered Arrangements

The Senior Birdwatchers' Club, consisting of 4 women and 2 men, is about to hold its annual meeting. In addition to having a group picture taken, they plan to elect a president, a vice president, and a secretary. Here are some questions that they (and we) might consider.

1. In how many ways can they line up for their group picture?
2. In how many ways can they elect their three officers if there are no restrictions as to sex?
3. In how many ways can they elect their three officers if the president is required to be female and the vice-president, male?
4. In how many ways can they elect their three officers if the president is to be of one sex and the vice-president and secretary of the other?

In order to answer these questions and others of a similar nature, we need two counting principles. One involves multiplication; the other involves addition.

MULTIPLICATION PRINCIPLE IN COUNTING

Suppose that there are three roads a, b, and c leading from Clearwater to Longview and two roads m and n from Longview to Sun City. How many different routes can you choose from Clearwater to Sun City going through Longview? Figure 18 clarifies the situation. For each of the 3 choices from Clearwater to Longview, you have 2 choices from Longview to Sun City. Thus, you have $3 \cdot 2$ routes from Clearwater to Sun City. Here is the general principle.

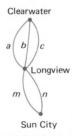

Clearwater

a b c

Longview

m n

Sun City

Possible Routes

am	*an*
bm	*bn*
cm	*cn*

Figure 18

MULTIPLICATION PRINCIPLE

*Suppose that an event H can occur in h ways and, after it has occurred, event K can occur in k ways. Then the number of ways in which both H **and** K can occur is hk.*

This principle extends in an obvious way to three or more events.

Consider now the Birdwatchers' third question. It involves three events.

> *P:* Elect a female president.
>
> *V:* Elect a male vice-president.
>
> *S:* Elect a secretary of either sex.

We understand that the election will take place in the order indicated and that no person can fill more than one position. Thus

P can occur in 4 ways   (there are 4 women);

V can occur in 2 ways   (there are 2 men);

S can occur in 4 ways   (after *P* and *V* occur, there are 4 people left from whom to select a secretary).

The entire selection process can be accomplished in $4 \cdot 2 \cdot 4 = 32$ ways.

PERMUTATIONS

Permutations
of *ART*

ART
ATR
RAT
RTA
TAR
TRA

Figure 19

To permute a set of objects is to rearrange them. Thus a **permutation** of a set of objects is an ordered arrangement of those objects (Figure 19). Take the set of letters in the word *FACTOR* as an example. Imagine that these 6 letters are printed on small cards so they can be arranged at will. Then we may form words like *COTARF*, *TRAFOC*, and *FRACTO*, none of which are in a dictionary but all of which are perfectly good words from our point of view. Let us call them code words. How many 6-letter code words can be made from the letters of the word *FACTOR;* that is, how many permutations of 6 objects are there?

Think of this as the problem of filling 6 slots.

We may fill the first slot in 6 ways. Having done that, we may fill the second slot in 5 ways, the third in 4 ways, and so on. By the multiplication principle, we can fill all six slots in

$$6 \cdot 5 \cdot 4 \cdot 3 \cdot 2 \cdot 1 = 720$$

ways.

Do you see that this is also the answer to the first question about the Birdwatchers, which asked in how many ways the 6 members could be arranged for a group picture? Let us identify each person by a letter; the letters of *FACTOR* will do just fine. Then to arrange the Birdwatchers is to make a 6-letter code word out of *FACTOR*. It can be done in 720 ways.

What if we want to make 3-letter code words from the letters of the word *FACTOR*, words like *ACT, COF*, and *TAC*? How many such words can be made? This is the problem of filling 3 slots with 6 letters available. We can fill the first slot in 6 ways, then the second in 5 ways, and then the third in 4 ways. Therefore we can make $6 \cdot 5 \cdot 4 = 120$ 3-letter code words from the word *FACTOR*.

The number 120 is also the answer to the second question about the Birdwatchers. If there are no restrictions as to sex, they can elect a president, vice-president, and secretary in $6 \cdot 5 \cdot 4 = 120$ ways (Figure 20).

Pres		VP		Sec
6	·	5	·	4

Figure 20

Consider the corresponding general problem. Suppose that from n distinguishable objects, we select r of them and arrange them in a row. The resulting arrangement is called a **permutation of n things taken r at a time.** The number of such permutations is denoted by the symbol $_nP_r$. Thus

$$_6P_3 = 6 \cdot 5 \cdot 4 = 120$$

$$_6P_6 = 6 \cdot 5 \cdot 4 \cdot 3 \cdot 2 \cdot 1 = 720$$

$$_8P_2 = 8 \cdot 7 = 56$$

and in general

$$_nP_r = n(n-1)(n-2) \cdots (n-r+2)(n-r+1)$$

Notice that $_nP_r$ is the product of r consecutive positive integers starting with n and going down. In particular, $_nP_n$ is the product of n positive integers starting with n and going all the way down to 1, that is,

$$_nP_n = n(n-1)(n-2) \cdots 3 \cdot 2 \cdot 1$$

The symbol $n!$ (read **n factorial**) is also used for this product. Thus

$$_5P_5 = 5! = 5 \cdot 4 \cdot 3 \cdot 2 \cdot 1 = 120$$

$$_4P_4 = 4! = 4 \cdot 3 \cdot 2 \cdot 1 = 24$$

ADDITION PRINCIPLE IN COUNTING

We still have not answered the fourth Birdwatchers' question. In how many ways can they elect their three officers if the president is to be of one sex and the vice president and secretary of the other? This means that the president should be female and the other two officers male, *or* the president should be male and the other two female. To answer a question like this we need another principle.

ADDITION PRINCIPLE

*Let H and K be disjoint events, that is, events that cannot happen simultaneously. If H can occur in h ways and K in k ways, then H **or** K can occur in h + k ways.*

This principle generalizes to three or more disjoint events.

Applying this principle to the question at hand, we define H and K as follows.

H: Elect a female president, male vice-president, and male secretary.

K: Elect a male president, female vice-president, and female secretary.

Clearly H and K are disjoint. From the multiplication principle,

H can occur in $4 \cdot 2 \cdot 1 = 8$ ways;
K can occur in $2 \cdot 4 \cdot 3 = 24$ ways.

Then by the addition principle, H or K can occur in $8 + 24 = 32$ ways.

Here is another question that requires the addition principle. Consider again the letters of *FACTOR*, which we supposed were printed on 6 cards. How many code words of any length can we make using these 6 letters? We immediately translate this into 6 disjoint events: make 6-letter words, or 5-letter words, or 4-letter words, or 3-letter words, or 2-letter words, or 1-letter words. We can do this in the following number of ways.

$$_6P_6 + {_6P_5} + {_6P_4} + {_6P_3} + {_6P_2} + {_6P_1}$$

$$= 6 \cdot 5 \cdot 4 \cdot 3 \cdot 2 \cdot 1 + 6 \cdot 5 \cdot 4 \cdot 3 \cdot 2 + 6 \cdot 5 \cdot 4 \cdot 3 + 6 \cdot 5 \cdot 4 + 6 \cdot 5 + 6$$

$$= \quad\quad 720 \quad\quad + \quad\quad 720 \quad + \quad 360 \quad + \quad 120 \quad + \quad 30 \quad + \quad 6$$

$$= 1956$$

Students sometimes find it hard to decide whether to multiply or to add in a counting problem. Notice that the words **and** and **or** are in boldface type in the statements of the multiplication principle and of the addition principle. They are the key words; **and** goes with multiplication; **or** goes with addition.

Problem Set 12-5

1. Calculate
   (a) $3!$;   (b) $(3!)(2!)$;   (c) $10!/8!$.
2. Calculate
   (a) $7!$;   (b) $7! + 5!$;   (c) $12!/9!$.
3. Calculate
   (a) $_5P_2$;   (b) $_9P_4$;   (c) $_{10}P_3$.
4. Calculate
   (a) $_4P_3$;   (b) $_8P_4$;   (c) $_{20}P_3$.
5. In how many ways can a president and a secretary be chosen from a group of 6 people?
6. Suppose that a club consists of 3 women and 2 men. In how many ways can a president and a secretary be chosen if

(a) the president is to be female and the secretary, male;

(b) the president is to be male and the secretary, female;

(c) the president and secretary are to be of opposite sex?

7. A box contains 12 cards numbered 1 through 12. Suppose one card is drawn from the box. Find the number of ways each of the following can occur.

(a) The number drawn is even.

(b) The number is greater than 9 or less than 3.

8. Suppose that two cards are drawn in succession from the box in Problem 7. Assume that the first card is not replaced before the second one is drawn. In how many ways can each of the following occur?

(a) Both numbers are even.

(b) The two numbers are both even or both odd.

(c) The first number is greater than 9 and the second one less than 3.

9. Do Problem 8 with the assumption that the first card is replaced before the second one is drawn.

10. In how many ways can a president, a vice-president, and a secretary be chosen from a group of 10 people?

11. How many 4-letter code words can be made from the letters of the word *EQUATION*? (Letters are not to be repeated.)

12. How many 3-letter code words can be made from the letters of the word *PROBLEM* if

(a) letters cannot be repeated;

(b) letters can be repeated?

13. Five roads connect Cheer City and Glumville. Starting at Cheer City, how many different ways can Smith drive to Glumville and return, that is, how many different round trips can he make? How many different round trips can he make if he wishes to return by a different road than he took to Glumville?

14. Filipe has 4 ties, 6 shirts, and 3 pairs of trousers. How many different outfits can he wear? Assume that he wears one of each kind of article.

15. Papa's Pizza Place offers 3 choices of salad, 20 kinds of pizza, and 4 different desserts. How many different 3-course meals can one order?

16. Minnesota license plate numbers consist of 3 letters followed by 3 digits (for example, AFF033). How many different plates could be issued? (You need not multiply out your answer.)

17. The letters of the word *CREAM* are printed on 5 cards. How many 3-, 4-, or 5-letter code words can be formed?

EXAMPLE A (Arrangements with Side Conditions)   Suppose that the letters of the word *COMPLEX* are printed on 7 cards. How many 3-letter code words can be formed from these letters if

(a) the first and last letters must be consonants (that is, C, M, P, L, or X);

(b) all vowels used (if any) must occur in the right-hand portion of a word (that is, a vowel cannot be followed by a consonant)?

Solution.

(a) Let c denote consonant, v vowel, and a any letter. We must fill the three slots on the next page.

c a c

We begin by filling the two restricted slots, which can be done in $5 \cdot 4 = 20$ ways. Then we fill the unrestricted slot using one of the 5 remaining letters. It can be done in 5 ways. There are $20 \cdot 5 = 100$ code words of the required type. The following diagram summarizes the procedure.

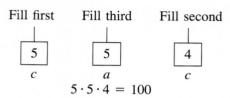

$5 \cdot 5 \cdot 4 = 100$

(b) We want to count words of the form cvv, ccv, or ccc. Note the use of the addition principle (as well as the multiplication principle) in the following solution.

$$5 \cdot 2 \cdot 1 + 5 \cdot 4 \cdot 2 + 5 \cdot 4 \cdot 3 = 10 + 40 + 60$$
$$= 110$$

18. Using the letters of the word *FACTOR* (without repetition), how many 4-letter code words can be formed
    (a) starting with R;
    (b) with vowels in the two middle positions;
    (c) with only consonants;
    (d) with vowels and consonants alternating;
    (e) with all the vowels (if any) in the left-hand portion of a word (that is, a vowel cannot be preceded by a consonant)?

19. Using the letters of the word *EQUATION* (without repetition), how many 4-letter code words can be formed
    (a) starting with T and ending with N;
    (b) starting and ending with a consonant;
    (c) with vowels only;
    (d) with three consonants;
    (e) with all the vowels (if any) in the right-hand portion of the word?

20. Three brothers and 3 sisters are lining up to be photographed. How many arrangements are there
    (a) altogether;
    (b) with brothers and sisters in alternating positions;
    (c) with the 3 sisters standing together?

21. A baseball team is to be formed from a squad of 12 people. Two teams made up of the same 9 people are different if at least some of the people are assigned different positions. In how many ways can a team be formed if
    (a) there are no restrictions;
    (b) only 2 of the people can pitch and these 2 cannot play any other position;

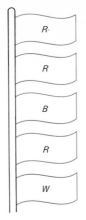

R

R

B

R

W

$R_1R_2BR_3W$
$R_1R_3BR_2W$
$R_2R_1BR_3W$
$R_2R_3BR_1W$ ⎬ RRBRW
$R_3R_1BR_2W$
$R_3R_2BR_1W$

Figure 21

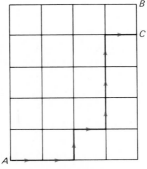

Figure 22

(c) only 2 of the people can pitch but they can also play any other position?

EXAMPLE B (Permutations with Some Indistinguishable Objects) Lucy has 3 identical red flags, 1 white flag, and 1 blue flag. How many different 5-flag signals could she display from the flagpole of her small boat?

Solution. If the 3 red flags were distinguishable, the answer would be $_5P_5 = 5! = 120$. Pretending they are distinguishable leads to counting an arrangement such as *RRBRW* six times, corresponding to the 3! ways of arranging the 3 red flags (Figure 21). For this reason, we must divide by 3!. Thus the number of signals Lucy can make is

$$\frac{5!}{3!} = \frac{5 \cdot 4 \cdot 3 \cdot 2 \cdot 1}{3 \cdot 2 \cdot 1} = 20$$

This result can be generalized. For example, given a set of n objects in which j are of one kind, k of a second kind, and m of a third kind, then the number of distinguishable permutations is

$$\frac{n!}{j! \, k! \, m!}$$

22. How many different signals consisting of 8 flags can be made using 4 white flags, 3 red flags, and 1 blue flag?

23. How many different signals consisting of 7 flags can be made using 3 white, 2 red, and 2 blue flags?

24. How many different 5-letter code words can be made from the 5 letters of the word *MIAMI*?

25. How many different 11-letter code words can be made from the 11 letters of the word *MISSISSIPPI*?

26. In how many different ways can a^4b^6 be written without using exponents? *Hint:* One way is *aaaabbbbbb*.

27. In how many different ways can a^3bc^6 be written without using exponents?

28. Consider the part of a city map shown in Figure 22. How many different shortest routes (no backtracking, no cutting across blocks) are there from A to C? Note that the route shown might be given the designation *EENENNNE*, with E denoting *East* and N denoting *North*.

29. How many different shortest routes are there from A to B in Problem 28?

MISCELLANEOUS PROBLEMS

30. Note that $_7P_3 = 7 \cdot 6 \cdot 5 = 7!/4!$. Write each of the following as a quotient of two factorials.
    (a) $_{10}P_5$ (b) $_{12}P_3$ (c) $_nP_r$

31. Simplify.
    (a) $\dfrac{11!}{8!}$ (b) $\dfrac{11!}{8! \, 3!}$ (c) $11! - 8!$

    (d) $\dfrac{8!}{2^6}$ (e) $\dfrac{(n + 1)! - n!}{n!}$ (f) $\dfrac{(n + 1)! + n!}{(n + 1)! - n!}$

32. Five chefs enter a pie-baking contest. In how many ways can a blue ribbon, a red ribbon, and a yellow ribbon be awarded for the three best pies?

33. Ten horses run in a race at Canterbury Downs.
    (a) How many different orders of finishing are there?
    (b) How many different possibilities are there for the first three places?

34. The Greek alphabet has 24 letters. How many different 3-letter fraternity names are possible if:
    (a) Repeated letters are allowed?
    (b) Repeated letters are not allowed?

35. The letters of the word *CYCLIC* are written on 6 cards.
    (a) How many 6-letter code words can be obtained?
    (b) How many of these have the three *C*s in consecutive positions?

36. I want to arrange my 5 history books, 4 math books, and 3 psychology books on a shelf. In how many ways can I do this if:
    (a) There are no restrictions as to arrangement;
    (b) I put the 5 history books on the left, the 4 math books in the middle, and the 3 psychology books on the right;
    (c) I insist only that books on the same subject be together?

37. Obtain a nice formula for

$$\frac{1}{2!} + \frac{2}{3!} + \frac{3}{4!} + \cdots + \frac{n}{(n+1)!}$$

Hint: Show first that

$$\frac{k}{(k+1)!} = \frac{1}{k!} - \frac{1}{(k+1)!}$$

38. Obtain a nice formula for $1 \cdot 1! + 2 \cdot 2! + 3 \cdot 3! + \cdots + n \cdot n!$ *Hint:* $k \cdot k! = (k+1)! - k!$.

39. A telephone number has 10 digits consisting of an area code (three digits, first is not 0 or 1, second is 0 or 1), an exchange (three digits, first is not 0 or 1, second is not 0 or 1), and a line number (four digits, not all are zeros). How many such 10 digit numbers are there?

40. The letters of *ENIGMA* are written on 6 cards. How many code words can be made?

41. How many different numbers are there between 0 and 60,000 that use only the digits 1, 2, 3, 4, or 5?

42. Consider making 6-digit numbers from the digits 1, 2, 3, 4, 5, and 6 without repetition.
    (a) How many such numbers are there?
    (b) Find the sum of these numbers.

43. In how many ways can 6 people be seated at a round table? (We consider two arrangements of people at a round table to be the same if everyone has the same people to the left and right in both arrangements.)

44. A husband and wife plan to invite 4 couples to dinner. The dinner table is rectangular. They decide on a seating arrangement in which the hostess will sit at the end nearest the kitchen, the host at the opposite end, and 4 guests on each side. Furthermore, no man shall sit next to another man, nor shall he sit next to his own wife. In how many ways can this be done?

45. Suppose that *n* teams enter a tournament in which a team is eliminated as soon

as it loses a game. Since $n \geq 2$ is arbitrary, a number of byes may be needed. How many games must be scheduled to determine a winner? *Hint:* There is a clumsy way to do this problem but there is also a very elegant way.

46. **TEASER** Here is an old problem. Suppose we start with an ordered arrangement $(a_1, a_2, \cdots, a_n)$ of n objects. Let d_n be the number of derangements of this arrangement. By a **derangement**, we mean a permutation that leaves no object fixed. For example, the derangements of ABC are BCA and CAB. Show each of the following.

(a) $d_1 = 0$, $d_2 = 1$, and $d_n = (n-1)d_{n-1} + (n-1)d_{n-2}$ *Hint:* To derange the n objects, we may either derange the first $n - 1$ objects and then exchange a_n with one of them or we may exchange a_n with a_j and then derange the remaining $n - 2$ objects.

(b) $d_n = n!\left[\dfrac{1}{2!} - \dfrac{1}{3!} + \dfrac{1}{4!} - \dfrac{1}{5!} + \cdots + (-1)^n\dfrac{1}{n!}\right]$

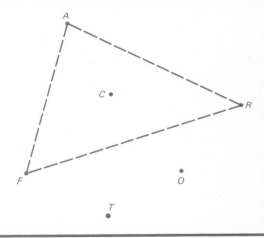

How Many Triangles?
Consider six points in the plane, no three on the same line. Label them F, A, C, T, O, R. How many triangles can be drawn using these points as vertices?

12-6   Counting Unordered Collections

For each choice of three points in the opening display, we can draw a triangle. But notice that the order in which we choose the three points does not matter. For example, *FAR*, *FRA*, *AFR*, *ARF*, *RAF*, and *RFA* all determine the same triangle, namely, the one that is shown by dotted lines in the picture. The question about triangles is very different from the question about 3-letter code words raised in the last section; yet there is a connection.

We learned that we can make

$$_6P_3 = 6 \cdot 5 \cdot 4 = 120$$

3-letter code words out of the letters of *FACTOR*. However, every triangle can be labeled with $3! = 6$ different code words. To find the number of triangles, we should therefore divide the number of code words by 3!. We conclude that the number of triangles that can be drawn is

$$\frac{_6P_3}{3!} = \frac{6 \cdot 5 \cdot 4}{3 \cdot 2 \cdot 1} = 20$$

COMBINATIONS

An unordered collection of objects is called a **combination** of those objects. If we select r objects from a set of n distinguishable objects, the resulting subset is called a **combination of n things taken r at a time.** The number of such combinations is denoted by $_nC_r$ or by $\binom{n}{r}$. Thus $_6C_3$ is the number of combinations of 6 things taken 3 at a time. We calculated it in connection with the triangle problem.

$$_6C_3 = \frac{_6P_3}{3!} = \frac{6 \cdot 5 \cdot 4}{3 \cdot 2 \cdot 1} = 20$$

More generally, if $1 \le r \le n$, the combination symbol $_nC_r$ is given by

$$_nC_r = \frac{_nP_r}{r!} = \frac{n(n-1)(n-2) \cdots (n-r+2)(n-r+1)}{r(r-1)(r-2) \cdots 2 \cdot 1}$$

A good way to remember this is that you want r factors in both numerator and denominator. In the numerator, you start with n and go down; in the denominator you start with r and go down. Incidentally, the answer must be an integer. That means the denominator has to divide the numerator. Here are some more examples.

$$_5C_2 = \frac{5 \cdot \overset{2}{\cancel{4}}}{\cancel{2} \cdot 1} = 10$$

$$_{12}C_4 = \frac{\cancel{12} \cdot 11 \cdot \overset{5}{\cancel{10}} \cdot 9}{\cancel{4} \cdot \cancel{3} \cdot \cancel{2} \cdot 1} = 495$$

$$_{10}C_8 = \frac{\overset{5}{\cancel{10}} \cdot 9 \cdot \cancel{8} \cdot \cancel{7} \cdot \cancel{6} \cdot \cancel{5} \cdot \cancel{4} \cdot \cancel{3}}{\cancel{8} \cdot \cancel{7} \cdot \cancel{6} \cdot \cancel{5} \cdot \cancel{4} \cdot \cancel{3} \cdot \cancel{2} \cdot 1} = 45$$

$$_{10}C_2 = \frac{\overset{5}{\cancel{10}} \cdot 9}{\cancel{2} \cdot 1} = 45$$

Notice that $_{10}C_8 = {_{10}C_2}$. This is not surprising, since in selecting a subset of 8 objects out of 10, you automatically select 2 to leave behind (you might call it selection by omission). The general fact that follows by the same reasoning is

$$\boxed{_nC_r = {_nC_{n-r}}}$$

We can express $_nC_r$ entirely in terms of factorials. Recall that

$$_nC_r = \frac{n(n-1)(n-2)\cdots(n-r+1)}{r!}$$

If we multiply both numerator and denominator by $(n-r)!$, we get

$$_nC_r = \frac{n!}{r!\,(n-r)!}$$

For this formula to hold when $r = 0$ and $r = n$, it is necessary to define $_nC_0 = 1$ and $0! = 1$. Even in mathematics, there is truth to that old proverb, "Necessity is the mother of invention."

COMBINATIONS VERSUS PERMUTATIONS

Whenever we are faced with the problem of counting the number of ways of selecting r objects from n objects, we are faced with this question. Is the notion of order significant? If the answer is yes, it is a permutation problem; if no, it is a combination problem.

Consider the Birdwatchers' Club of Section 12-5 again. Suppose the club members want to select a president, a vice-president, and a secretary. Is order significant? Yes. The selection can be done in

$$_6P_3 = 6 \cdot 5 \cdot 4 = 120$$

ways.

But suppose they decide simply to choose an executive committee consisting of 3 members. Is order relevant? No. A committee consisting of Filipe, Celia, and Amanda is the same as a committee consisting of Celia, Amanda, and Filipe. A 3-member committee can be chosen from 6 people in

$$_6C_3 = \frac{6 \cdot 5 \cdot 4}{3 \cdot 2 \cdot 1} = 20$$

ways.

The words *arrangement*, *lineup*, and *signal* all suggest order. The words *set*, *committee*, *group*, and *collection* do not.

Problem Set 12-6

1. Calculate each of the following.
   (a) $_{10}P_3$ (b) $_{10}C_3$
   (c) $_5P_5$ (d) $_5C_5$
   (e) $_6P_1$ (f) $_6C_1$

2. Calculate each of the following.
   (a) $_{12}P_2$ (b) $_{12}C_2$

(c) $_4P_4$ (d) $_4C_4$

(e) $_{10}P_1$ (f) $_{10}C_1$

3. Use the fact that $_nC_r = {}_nC_{n-r}$ to calculate each of the following.

   (a) $_{20}C_{17}$ (b) $_{100}C_{97}$

4. Calculate each of the following.

   (a) $_{41}C_{39}$ (b) $_{1000}C_{998}$

   Hint: See Problem 3.

5. In how many ways can a committee of 3 be selected from a class of 8 students?

6. In how many ways can a committee of 5 be selected from a class of 8 students?

7. A political science professor must select 4 students from her class of 12 students for a field trip to the state legislature. In how many ways can she do it?

8. The professor of Problem 7 was asked to rank the top 4 students in her class of 12. In how many ways could that be done?

9. A police chief needs to assign officers from the 10 available to control traffic at junctions A, B, and C. In how many ways can he do it?

10. If 12 horses are entered in a race, in how many ways can the first 3 places (win, place, show) be taken?

11. From a class of 6 members, in how many ways can a committee of any size be selected (including a committee of one)?

12. From a penny, a nickel, a dime, a quarter, and a half dollar, how many different sums can be made?

EXAMPLE A (More on Committees)   A committee of 4 is to be selected from a group of 3 seniors, 4 juniors, and 5 sophomores. In how many ways can it be done if

(a) there are no restrictions on the selection;

(b) the committee must have 2 sophomores, 1 junior, and 1 senior;

(c) the committee must have at least 3 sophomores;

(d) the committee must have at least 1 senior?

Solution.

(a) $_{12}C_4 = \dfrac{\cancel{12} \cdot 11 \cdot \overset{5}{\cancel{10}} \cdot 9}{\cancel{4} \cdot \cancel{3} \cdot \cancel{2} \cdot 1} = 495$

(b) Two sophomores can be chosen in $_5C_2$ ways, 1 junior in $_4C_1$ ways, and 1 senior in $_3C_1$ ways. By the multiplication principle of counting, the committee can be chosen in

$$_5C_2 \cdot {}_4C_1 \cdot {}_3C_1 = 10 \cdot 4 \cdot 3 = 120$$

ways. We used the multiplication principle because we choose 2 sophomores *and* 1 junior *and* 1 senior.

(c) At least 3 sophomores means 3 sophomores and 1 nonsophomore *or* 4 sophomores. The word *or* tells us to use the addition principle of counting. We get

$$_5C_3 \cdot {}_7C_1 + {}_5C_4 = 10 \cdot 7 + 5 = 75$$

(d) Let x be the number of selections with at least one senior and let y be the number of selections with no seniors. Then $x + y$ is the total number of selections, that is, $x + y = 495$ (see part (a)). We calculate y rather than x because it is easier.

$$y = {}_9C_4 = \frac{9 \cdot 8 \cdot 7 \cdot 6}{4 \cdot 3 \cdot 2 \cdot 1} = 126$$

$$x = 495 - 126 = 369$$

13. An investment club has a membership of 4 women and 6 men. A research committee of 3 is to be formed. In how many ways can this be done if
    (a) there are to be 2 women and 1 man on the committee;
    (b) there is to be at least 1 woman on the committee;
    (c) all 3 are to be of the same sex?

14. A senate committee of 4 is to be formed from a group consisting of 5 Republicans and 6 Democrats. In how many ways can this be done if
    (a) there are to be 2 Republicans and 2 Democrats on the committee;
    (b) there are to be no Republicans on the committee;
    (c) there is to be at most one Republican on the committee?

15. Suppose that a bag contains 4 black and 7 white balls. In how many ways can a group of 3 balls be drawn from the bag consisting of
    (a) 1 black and 2 white balls;
    (b) balls of just one color;
    (c) at least 1 black ball?
    Note: Assume the balls are distinguishable; for example, they may be numbered.

16. John is going on a vacation trip and wants to take 5 books with him from his personal library, which consists of 6 science books and 10 novels. In how many ways can he make his selection if he wants to take
    (a) 2 science books and 3 novels;
    (b) at least 1 science book;
    (c) 1 book of one kind and 4 books of the other kind?

EXAMPLE B (Bridge Card Problems)   A standard deck consists of 52 cards. There are 4 suits (spades, clubs, hearts, diamonds), each with 13 cards $(2, 3, 4, \ldots, 10,$ jack, queen, king, ace). A bridge hand consists of 13 cards.
(a) How many different possible bridge hands are there?
(b) How many of them have exactly 3 aces?
(c) How many of them have no aces?
(d) How many of them have cards from just 3 suits?

Solution.
(a) The order of the cards in a hand is irrelevant; it is a combination problem. We can select 13 cards out of 52 in ${}_{52}C_{13}$ ways, a number so large we will not bother to calculate it.
(b) The three aces can be selected in ${}_4C_3$ ways, the 10 remaining cards in ${}_{48}C_{10}$ ways. The answer (using the multiplication principle) is ${}_4C_3 \cdot {}_{48}C_{10}$.
(c) From 48 nonaces, we select 13 cards; the answer is ${}_{48}C_{13}$.
(d) We think of this as no clubs, or no spades, or no hearts, or no diamonds and use the addition principle.

$$_{39}C_{13} + {}_{39}C_{13} + {}_{39}C_{13} + {}_{39}C_{13} = 4 \cdot {}_{39}C_{13}$$

Problems 17–22 deal with bridge hands. Leave your answers in terms of combination symbols.

17. How many of the hands have only red cards? *Note:* Half of the cards are red.
18. How many of the hands have only honor cards (aces, kings, queens, and jacks)?
19. How many of the hands have one card of each kind (1 ace, 1 king, 1 queen, and so on)?
20. How many of the hands have exactly 2 kings?
21. How many of the hands have 2 or more kings?
22. How many of the hands have exactly 2 aces and 2 kings?

Problems 23–26 deal with poker hands, which consist of 5 cards.

23. How many different poker hands are possible?
24. How many of them have exactly 2 hearts and 2 diamonds?
25. How many have 2 pairs of different kinds (for example, 2 aces and 2 fives)?
26. How many are 5-card straights (for example, 7, 8, 9, 10, jack)? An ace may count either as the highest or the lowest card, that is, as 1 or 13.

MISCELLANEOUS PROBLEMS

27. A quarter, a dime, a nickel, and a penny are tossed. In how many ways can they fall?
28. From a committee of 11, a subcommittee of 4 is to be chosen. In how many ways can this be done?
29. From 5 representatives of labor, 4 representatives of business, and 3 representatives of the general public, how many different mediation committees can be formed with 2 people from each of the three groups?
30. In how many ways can a group of 12 people be split into three nonoverlapping committees of size 5, 4, and 3, respectively?
31. A class of 12 people will select a president, a secretary, a treasurer, and a program committee of 3 with no overlapping of positions. In how many ways can this be done?
32. A committee of 4 is to be formed from a group of 4 freshmen, 3 sophomores, 2 juniors, and 6 seniors. In how many ways can this be done if
    (a) each class must be represented;
    (b) freshmen are excluded;
    (c) the committee must have exactly two seniors;
    (d) the committee must have at least one senior?
33. A test consists of 10 true-false items.
    (a) How many different sets of answers are possible?
    (b) How many of these have exactly 4 right answers?
34. An ice cream parlor has 10 different flavors. How many different double-dip cones can be made if
    (a) the two dips must be of different flavors but the order of putting them on the cone does not matter;
    (b) the two dips must be different and order does matter;

(c) the two dips need not be different but order does matter;

(d) the two dips need not be different and order does not matter?

35. Mary has a penny, a nickel, a dime, a quarter, and a half dollar in her purse. How many different possible sums of money (consisting of at least one coin) could she give to her daughter Tosha?

36. In how many ways can 8 presents be split between John and Mary if
    (a) each is to get 4 presents;
    (b) John is to get 5 presents and Mary 3 presents;
    (c) there are no restrictions on how the presents are split?

37. Let 10 fixed points on a circle be given. How many convex polygons can be formed which have vertices chosen from among these points?

38. Calculate the sums in (a)–(c).
    (a) $_2C_0 + _2C_1 + _2C_2$
    (b) $_3C_0 + _3C_1 + _3C_2 + _3C_3$
    (c) $_4C_0 + _4C_1 + _4C_2 + _4C_3 + _4C_4$
    (d) Conjecture a formula for $_nC_0 + _nC_1 + _nC_2 + \cdots + _nC_n$.

39. Prove your conjecture in Problem 38 by considering two different methods of counting the ways of splitting n presents between two people. *Hint:* One method is to let each present choose the person it will go to.

40. Show that

$$_nC_0\,_nC_n + _nC_1\,_nC_{n-1} + _nC_2\,_nC_{n-2} + \cdots + _nC_n\,_nC_0 = _{2n}C_n$$

by counting the number of ways of drawing n balls from an urn that has n red and n black balls.

41. Find a nice formula for $\sum\limits_{j=0}^{n} (_nC_j)^2$. *Hint:* See Problem 40.

42. Use the factorial formula for $_nC_r$ to show that $_nC_{r-1} + _nC_r = _{n+1}C_r$.

43. Obtain a nice formula for

$$S = _{n+1}C_1 + _{n+2}C_2 + _{n+3}C_3 + \cdots + _{n+k}C_k$$

Hint: Add $_{n+1}C_0$ to S on the left; then use the result in Problem 42 repeatedly to collect two terms on the left.

44. Show that if $m \le n$, then $\sum\limits_{j=0}^{m} {_nC_j}\,_{n-j}C_{m-j} = {_nC_m}2^m$.

Figure 23

45. Consider an n-by-n checkerboard (a 5-by-5 checkerboard is shown in Figure 23). How many rectangles of all sizes are determined by this board?

46. TEASER   Obtain a formula for the number of squares of all sizes that are determined by the n-by-n checkerboard. Evaluate when $n = 10$.

$(x + y)(x + y)(x + y)(x + y) = xxxx + xxxy + xxyy + xyyy + yyyy$
$+ xxyx + xyxy + yxyy$
$+ xyxx + \boxed{xyyx} + yyxy$
$+ yxxx + yxxy + yyyx$
$+ yxyx$
$+ yyxx$

$(x + y)^4 = x^4 + 4x^3y + 6x^2y^2 + 4xy^3 + y^4$

$(x + y)^4 = {}_4C_0x^4y^0 + {}_4C_1x^3y^1 + {}_4C_2x^2y^2 + {}_4C_3x^1y^3 + {}_4C_4x^0y^4$

12-7 The Binomial Formula

In the opening box, we have shown how to expand $(x + y)^4$. Admittedly, it looks complicated; however, it leads to the remarkable formula at the bottom of the display. That formula generalizes to handle $(x + y)^n$, where n is any positive integer. It is worth a careful investigation.

To produce any given term in the expansion of $(x + y)^4$, each of the four factors $x + y$ contributes either an x or a y. There are $2 \cdot 2 \cdot 2 \cdot 2 = 16$ ways in which they can make this contribution, hence the 16 terms in the long expanded form. But many of these terms are alike; in fact, only five different types occur, namely, x^4, x^3y, x^2y^2, xy^3, and y^4. The number of times each occurs is ${}_4C_0$, ${}_4C_1$, ${}_4C_2$, ${}_4C_3$, and ${}_4C_4$, respectively. (Remember we defined ${}_4C_0$ to be 1.)

Why do the combination symbols arise in this expansion? For example, why is ${}_4C_2$ the coefficient of x^2y^2? If you follow the arrows in the opening display, you see how the term $xyyx$ comes about. It gets its two y's from the second and third $x + y$ factors (the x's then must come from the first and fourth $x + y$ factors). Thus, the number of terms of the form x^2y^2 is the number of ways of selecting two factors out of four from which to take y's (the x's must come from the remaining two factors). We can select two objects out of four in ${}_4C_2$ ways; hence the coefficient of x^2y^2 is ${}_4C_2$.

THE BINOMIAL FORMULA

What we have just done for $(x + y)^4$ can be carried out for $(x + y)^n$, where n is any positive integer. The result is called the **binomial formula.**

$$(x + y)^n = {}_nC_0x^ny^0 + {}_nC_1x^{n-1}y^1 + \cdots + {}_nC_{n-1}x^1y^{n-1} + {}_nC_nx^0y^n$$

$$= \sum_{k=0}^{n} {}_nC_kx^{n-k}y^k$$

Notice that the k in ${}_nC_k$ is the exponent on y and that the two exponents in each term sum to n.

Let us apply the binomial formula with $n = 6$.

$$(x + y)^6 = {}_6C_0x^6 + {}_6C_1x^5y + {}_6C_2x^4y^2 + \cdots + {}_6C_6y^6$$
$$= x^6 + 6x^5y + 15x^4y^2 + 20x^3y^3 + 15x^2y^4 + 6xy^5 + y^6$$

This same result applies to the expansion of $(2a - b^2)^6$. We simply think of $2a$ as x and $-b^2$ as y. Thus

$$[2a + (-b^2)]^6 = (2a)^6 + 6(2a)^5(-b^2) + 15(2a)^4(-b^2)^2 + 20(2a)^3(-b^2)^3$$
$$+ 15(2a)^2(-b^2)^4 + 6(2a)(-b^2)^5 + (-b^2)^6$$
$$= 64a^6 - 192a^5b^2 + 240a^4b^4 - 160a^3b^6$$
$$+ 60a^2b^8 - 12ab^{10} + b^{12}$$

THE BINOMIAL COEFFICIENTS

The combination symbols ${}_nC_k$ are often called **binomial coefficients,** for reasons that should be obvious. Their remarkable properties have been studied for hundreds of years. Let us see if we can discover some of them. We begin by expanding $(x + y)^k$ for increasing values of k, listing only the coefficients (Figure 24). The resulting triangle of numbers composed of the binomial coefficients is called **Pascal's triangle** after the gifted French philosopher and mathematician, Blaise Pascal (1622–1662). Notice its symmetry. If folded across a vertical center line, the numbers match. This corresponds to an algebraic fact you learned earlier.

$${}_nC_k = {}_nC_{n-k}$$

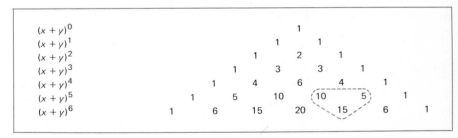

Figure 24

Next notice that any number in the body of the triangle is the sum of the two numbers closest to it in the line above the number. For example, $15 = 10 + 5$, as the dotted triangle

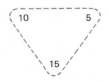

was meant to suggest. In symbols

$$_{n+1}C_k = \ _nC_{k-1} + \ _nC_k$$

a fact that can be proved rigorously by using the factorial formula for $_nC_k$ (see Problem 42 of Section 12-6).

Now add the numbers in each row of Pascal's triangle.

$$1 = 1 = 2^0$$
$$1 + 1 = 2 = 2^1$$
$$1 + 2 + 1 = 4 = 2^2$$
$$1 + 3 + 3 + 1 = 8 = 2^3$$

This suggests the formula

$$_nC_0 + \ _nC_1 + \ _nC_2 + \cdots + \ _nC_n = 2^n$$

Its truth can be demonstrated by substituting $x = 1$ and $y = 1$ in the binomial formula. It has an important interpretation. Take a set with n elements. This set has $_nC_n$ subsets of size n, $_nC_{n-1}$ subsets of size $n - 1$, and so on. The left side of the boxed formula is the total number of subsets (including the empty set) of a set with n elements; remarkably, it is just 2^n. For example, $\{a, b, c\}$ has $2^3 = 8$ subsets.

$$\{a, b, c\} \quad \{a, b\} \quad \{a, c\} \quad \{b, c\} \quad \{a\} \quad \{b\} \quad \{c\} \quad \phi$$

A related formula is

$$_nC_0 - \ _nC_1 + \ _nC_2 - \cdots + (-1)^n {}_nC_n = 0$$

which can be obtained by setting $x = 1$ and $y = -1$ in the binomial formula. When rewritten as

$$_nC_0 + \ _nC_2 + \ _nC_4 + \cdots = \ _nC_1 + \ _nC_3 + \ _nC_5 + \cdots$$

it says that the number of subsets with an even number of elements is equal to the number of subsets with an odd number of elements.

Problem Set 12-7

In Problems 1–6, expand and simplify.

1. $(x + y)^3$
2. $(x - y)^3$
3. $(x - 2y)^3$
4. $(3x + b)^3$
5. $(c^2 - 3d^3)^4$
6. $(xy - 2z^2)^4$

Write the first three terms of each expansion in Problems 7–10 in simplified form.

7. $(x + y)^{20}$ 8. $(x + y)^{30}$ 9. $\left(x + \dfrac{1}{x^5}\right)^{20}$ 10. $\left(xy^2 + \dfrac{1}{y}\right)^{14}$

11. Find the number of subsets of each of the following sets.
   (a) $\{a, b, c, d\}$ (b) $\{1, 2, 3, 4, 5\}$
   (c) $\{x_1, x_2, x_3, x_4, x_5, x_6\}$

EXAMPLE A (Finding a Specific Term of a Binomial Expansion)   Find the term in the expansion of $(2x + y^2)^{10}$ that involves y^{12}.

Solution.   This term will arise from raising y^2 to the 6th power. It is, therefore,

$$_{10}C_6(2x)^4(y^2)^6 = 210 \cdot 16x^4y^{12} = 3360x^4y^{12}$$

12. Find the term in the expansion of $(y^2 - z^3)^{10}$ that involves z^9.
13. Find the term in the expansion of $(3x - y^3)^{10}$ that involves y^{24}.
14. Find the term in the expansion of $(2a - b)^{12}$ that involves a^3.
15. Find the term in the expansion of $(x^2 - 2/x)^5$ that involves x^4.

EXAMPLE B (An Application to Compound Interest)   If $100 is invested at 12 percent compounded monthly, it will accumulate to $100(1.01)^{12}$ dollars by the end of one year. Use the binomial formula to approximate this amount.

Solution.
$$100(1.01)^{12} = 100(1 + .01)^{12}$$

$$= 100\left[1 + 12(.01) + \frac{12 \cdot 11}{2}(.01)^2 + \frac{12 \cdot 11 \cdot 10}{6}(.01)^3 + \cdots\right]$$

$$= 100[1 + .12 + .0066 + .00022 + \cdots]$$

$$\approx 100(1.12682) \approx 112.68$$

This answer of $112.68 is accurate to the nearest penny since the last nine terms of the expansion do not add up to as much as a penny.

In Problems 16–19 use the first three terms of a binomial expansion to find an approximate value of the given expression.

16. $20(1.02)^8$ 17. $100(1.002)^{20}$ 18. $500(1.005)^{20}$ 19. $200(1.04)^{10}$

20. Bacteria multiply in a certain medium so that by the end of k hours their number N is $N = 100(1.02)^k$. Approximate the number of bacteria after 20 hours.
21. Do Problem 20 assuming $N = 1000(1.01)^k$.

MISCELLANEOUS PROBLEMS

22. Find and simplify the first three terms of the expansion of $(a^2 + 4b^3)^{12}$.
23. Find and simplify the term in the expansion of $(x - 2z^3)^8$ that involves z^{15}.

24. Expand and simplify each of the following.

(a) $\left(2x^3 - \dfrac{1}{x}\right)^5$

(b) $\dfrac{(x + h)^4 - x^4}{h}$

25. In each of the following, find the term in the expanded and simplified form that does not involve h (a procedure very important in calculus).

(a) $\dfrac{(x + h)^n - x^n}{h}$

(b) $\dfrac{(x + h)^{10} + 2(x + h)^4 - x^{10} - 2x^4}{h}$

26. Given that i is the imaginary unit, calculate

(a) $(1 + i)^5$;

(b) $(1 - i)^6$.

27. Find the constant term in the expansion of $(3x^2 + 1/3x)^{12}$.

28. Without using a calculator, show that $(1.0003)^{10} > 1.003004$.

29. Without using a calculator, show that $(1.01)^{50} > 1.5$.

30. Without using a calculator, find $(.999)^{10}$ correct to six decimal places.

31. How many committees consisting of 3 or more members can be selected from a group of 12 people?

32. How many subsets with an odd number of elements does a set with 11 members have?

33. In the expansion of the trinomial $(x + y + z)^n$, the coefficient of $x^r y^s z^t$ where $r + s + t = n$ is $n!/r!s!t!$

(a) Expand $(x + y + z)^3$.

(b) Find the coefficient of the term $x^2 y^4 z$ in the expansion of $(2x + y + z)^7$.

34. Find the sum of all the coefficients in the expansion of the trinomial $(x + y + z)^n$.

35. Find a simple formula for $\sum\limits_{k=0}^{n} {}_nC_k \, 2^k$.

36. Find the sum of the coefficients in $(4x^3 - x)^6$ after it is expanded and simplified. *Hint:* This is a simple problem when looked at the right way.

37. Let $P(x)$ be the nth degree polynomial defined by

$$P(x) = 1 + x + \dfrac{x(x - 1)}{2!} + \dfrac{x(x - 1)(x - 2)}{3!}$$

$$+ \cdots + \dfrac{x(x - 1)(x - 2) \cdots (x - n + 1)}{n!}$$

Find a simple formula for each of the following.

(a) $P(k)$, $k = 0, 1, 2, \ldots, n$.

(b) $P(n + 1)$

(c) $P(n + 2)$

38. The integer 3 can be expressed as the sum of one or more positive integers in four ways, namely, as 3, 2 + 1, 1 + 2, and 1 + 1 + 1. In how many ways can the positive integer n be so expressed? It is not enough to give the answer; you must give an argument to show that your answer is correct.

Chapter Summary

A **number sequence** $a_1, a_2, a_3, \ldots$ is a function that associates with each positive integer n a number a_n. Such a sequence may be described by an **explicit formula** (for instance, $a_n = 2n + 1$), by a **recursion formula** (for instance,

$a_n = 3a_{n-1}$), or by giving enough terms so a pattern is evident (for instance, 1, 11, 21, 31, 41, . . .).

If any term in a sequence can be obtained by adding a fixed number d to the preceding term, we call it an **arithmetic sequence.** There are three key formulas associated with this type of sequence.

$$\text{Recursion formula:} \quad a_n = a_{n-1} + d$$

$$\text{Explicit formula:} \quad a_n = a_1 + (n - 1)d$$

$$\text{Sum formula:} \quad A_n = \frac{n}{2}(a_1 + a_n)$$

In the last formula, A_n represents

$$A_n = a_1 + a_2 + \cdots + a_n = \sum_{i=1}^{n} a_i$$

A **geometric sequence** is one in which any term results from multiplying the previous term by a fixed number r. The corresponding key formulas are

$$\text{Recursion formula:} \quad a_n = ra_{n-1}$$

$$\text{Explicit formula:} \quad a_n = a_1 r^{n-1}$$

$$\text{Sum formula:} \quad A_n = \frac{a_1(1 - r^n)}{1 - r}, \, r \neq 1$$

In the last formula, we may ask what happens as n grows larger and larger. If $|r| < 1$, the value of A_n gets closer and closer to $a_1/(1 - r)$, which we regard as the sum of *all* the terms of the sequence.

Often in mathematics, we wish to demonstrate that a whole **sequence of statements** P_n is true. For this, a powerful tool is the **principle of mathematical induction,** which asserts that if P_1 is true and if the truth of P_k implies the truth of P_{k+1}, then all the statements of the sequence are true.

Given enough time, anyone can count the number of elements in a set. But if the set consists of arrangements of objects (for example, the letters in a word), the work can be greatly simplified by using two principles, the **multiplication principle** and the **addition principle.** Of special interest are the number of **permutations** (ordered arrangements) of n things taken r at a time and the number of **combinations** (unordered collections) of n things taken r at a time. They can be calculated from the formulas

$$_nP_r = n(n - 1)(n - 2) \cdots (n - r + 1)$$

$$_nC_r = \frac{_nP_r}{r!} = \frac{n(n - 1)(n - 2) \cdots (n - r + 1)}{r(r - 1)(r - 2) \cdots 1}$$

One important use of the symbol $_nC_r$ is in the **binomial formula**

$$(x + y)^n = {_nC_0}x^n y^0 + {_nC_1}x^{n-1}y^1 + \cdots + {_nC_{n-1}}x^1 y^{n-1} + {_nC_n}x^0 y^n$$

Chapter Review Problem Set

Problems 1–6 refer to the sequences below.

(a) 2, 5, 8, 11, 14, . . .
(b) 2, 6, 18, 54, . . .
(c) 2, 1.5, 1, 0.5, 0, . . .
(d) 2, 4, 6, 10, 16, . . .
(e) 2, $\frac{2}{3}$, $\frac{2}{9}$, $\frac{2}{27}$, $\frac{2}{81}$, . . .

1. Which of these sequences are arithmetic? Which are geometric?
2. Give a recursion formula for each of the sequences (a)–(e).
3. Give an explicit formula for sequences (a) and (b).
4. Find the sum of the first 67 terms of sequence (a).
5. Write a formula for the sum of the first 100 terms of sequence (b).
6. Find the sum of *all* the terms of sequence (e).
7. If $a_n = 3a_{n-1} - a_{n-2}$, $a_1 = 1$, and $a_2 = 2$, find a_6.
8. If $a_n = n^2 - n$, find $A_5 = \sum_{n=1}^{5} a_n$.
9. Calculate $2 + 4 + 6 + 8 + \cdots + 1000$.
10. Write $.5\overline{55}$ as a ratio of two integers.
11. If \$100 is put in the bank today earning 8 percent interest compounded quarterly, write a formula for its value at the end of 12 years.
12. Show by mathematical induction that
    (a) $5 + 9 + \cdots + (4n + 1) = 2n^2 + 3n$;
    (b) $n! > 2^n$ when $n \geq 4$.
13. Suppose P_3 is true and $P_k \Rightarrow P_{k+3}$. What can we conclude about the sequence P_n?
14. John has 4 sport coats, 3 pairs of trousers, and 5 shirts. How many different outfits could he wear?
15. How many code words of all lengths can be made from the letters of the word *SNOW*, assuming letters cannot be repeated?
16. Evaluate.
    (a) $_{10}P_4$ (b) $_{10}C_4$ (c) $6!/4!$
    (d) $_{50}C_{48}$ (e) $8!/(2!\ 6!)$ (f) $10!/(2!\ 3!\ 5!)$
17. In how many ways can a class of 5 girls and 4 boys select
    (a) a president, vice president, and secretary;
    (b) a president, vice president, and secretary if the secretary must be a boy;
    (c) a social committee of 3 people;
    (d) a social committee of 3 people consisting of 2 girls and 1 boy?
18. The letters of the word *BARBARIAN* are written on 9 cards. How many 9-letter code words can be formed?
19. Consider 8 points in the plane, no 3 on the same line. How many triangles can be formed using these points as vertices?
20. How many subsets does a set of 8 elements have?
21. Find the first 4 terms in simplified form in the expansion of $(x + 2y)^{10}$.
22. Find the term involving a^3b^6 in the expansion of $(a - b^2)^6$.
23. Find $(1.002)^{20}$ accurate to 4 decimal places.

TABLE A. Natural Logarithms

	.00	.01	.02	.03	.04	.05	.06	.07	.08	.09
5.5	1.7047	1.7066	1.7084	1.7102	1.7120	1.7138	1.7156	1.7174	1.7192	1.7210
5.6	1.7228	1.7246	1.7263	1.7281	1.7299	1.7317	1.7334	1.7352	1.7370	1.7387
5.7	1.7405	1.7422	1.7440	1.7457	1.7475	1.7492	1.7509	1.7527	1.7544	1.7561
5.8	1.7579	1.7596	1.7613	1.7630	1.7647	1.7664	1.7682	1.7699	1.7716	1.7733
5.9	1.7750	1.7766	1.7783	1.7800	1.7817	1.7834	1.7851	1.7867	1.7884	1.7901
6.0	1.7918	1.7934	1.7951	1.7967	1.7984	1.8001	1.8017	1.8034	1.8050	1.8066
6.1	1.8083	1.8099	1.8116	1.8132	1.8148	1.8165	1.8181	1.8197	1.8213	1.8229
6.2	1.8245	1.8262	1.8278	1.8294	1.8310	1.8326	1.8342	1.8358	1.8374	1.8390
6.3	1.8406	1.8421	1.8437	1.8453	1.8469	1.8485	1.8500	1.8516	1.8532	1.8547
6.4	1.8563	1.8579	1.8594	1.8610	1.8625	1.8641	1.8656	1.8672	1.8687	1.8703
6.5	1.8718	1.8733	1.8749	1.8764	1.8779	1.8795	1.8810	1.8825	1.8840	1.8856
6.6	1.8871	1.8886	1.8901	1.8916	1.8931	1.8946	1.8961	1.8976	1.8991	1.9006
6.7	1.9021	1.9036	1.9051	1.9066	1.9081	1.9095	1.9110	1.9125	1.9140	1.9155
6.8	1.9169	1.9184	1.9199	1.9213	1.9228	1.9242	1.9257	1.9272	1.9286	1.9301
6.9	1.9315	1.9330	1.9344	1.9359	1.9373	1.9387	1.9402	1.9416	1.9430	1.9445
7.0	1.9459	1.9473	1.9488	1.9502	1.9516	1.9530	1.9544	1.9559	1.9573	1.9587
7.1	1.9601	1.9615	1.9629	1.9643	1.9657	1.9671	1.9685	1.9699	1.9713	1.9727
7.2	1.9741	1.9755	1.9769	1.9782	1.9796	1.9810	1.9824	1.9838	1.9851	1.9865
7.3	1.9879	1.9892	1.9906	1.9920	1.9933	1.9947	1.9961	1.9974	1.9988	2.0001
7.4	2.0015	2.0028	2.0042	2.0055	2.0069	2.0082	2.0096	2.0109	2.0122	2.0136
7.5	2.0149	2.0162	2.0176	2.0189	2.0202	2.0215	2.0229	2.0242	2.0255	2.0268
7.6	2.0282	2.0295	2.0308	2.0321	2.0334	2.0347	2.0360	2.0373	2.0386	2.0399
7.7	2.0412	2.0425	2.0438	2.0451	2.0464	2.0477	2.0490	2.0503	2.0516	2.0528
7.8	2.0541	2.0554	2.0567	2.0580	2.0592	2.0605	2.0618	2.0631	2.0643	2.0656
7.9	2.0669	2.0681	2.0694	2.0707	2.0719	2.0732	2.0744	2.0757	2.0769	2.0782
8.0	2.0794	2.0807	2.0819	2.0832	2.0844	2.0857	2.0869	2.0882	2.0894	2.0906
8.1	2.0919	2.0931	2.0943	2.0956	2.0968	2.0980	2.0992	2.1005	2.1017	2.1029
8.2	2.1041	2.1054	2.1066	2.1078	2.1090	2.1102	2.1114	2.1126	2.1138	2.1150
8.3	2.1163	2.1175	2.1187	2.1199	2.1211	2.1223	2.1235	2.1247	2.1258	2.1270
8.4	2.1282	2.1294	2.1306	2.1318	2.1330	2.1342	2.1353	2.1365	2.1377	2.1389
8.5	2.1401	2.1412	2.1424	2.1436	2.1448	2.1459	2.1471	2.1483	2.1494	2.1506
8.6	2.1518	2.1529	2.1541	2.1552	2.1564	2.1576	2.1587	2.1599	2.1610	2.1622
8.7	2.1633	2.1645	2.1656	2.1668	2.1679	2.1691	2.1702	2.1713	2.1725	2.1736
8.8	2.1748	2.1759	2.1770	2.1782	2.1793	2.1804	2.1815	2.1827	2.1838	2.1849
8.9	2.1861	2.1872	2.1883	2.1894	2.1905	2.1917	2.1928	2.1939	2.1950	2.1961
9.0	2.1972	2.1983	2.1994	2.2006	2.2017	2.2028	2.2039	2.2050	2.2061	2.2072
9.1	2.2083	2.2094	2.2105	2.2116	2.2127	2.2138	2.2148	2.2159	2.2170	2.2181
9.2	2.2192	2.2203	2.2214	2.2225	2.2235	2.2246	2.2257	2.2268	2.2279	2.2289
9.3	2.2300	2.2311	2.2322	2.2332	2.2343	2.2354	2.2364	2.2375	2.2386	2.2396
9.4	2.2407	2.2418	2.2428	2.2439	2.2450	2.2460	2.2471	2.2481	2.2492	2.2502
9.5	2.2513	2.2523	2.2534	2.2544	2.2555	2.2565	2.2576	2.2586	2.2597	2.2607
9.6	2.2618	2.2628	2.2638	2.2649	2.2659	2.2670	2.2680	2.2690	2.2701	2.2711
9.7	2.2721	2.2732	2.2742	2.2752	2.2762	2.2773	2.2783	2.2793	2.2803	2.2814
9.8	2.2824	2.2834	2.2844	2.2854	2.2865	2.2875	2.2885	2.2895	2.2905	2.2915
9.9	2.2925	2.2935	2.2946	2.2956	2.2966	2.2976	2.2986	2.2996	2.3006	2.3016

TABLE B. Common Logarithms

n	0	1	2	3	4	5	6	7	8	9
1.0	.0000	.0043	.0086	.0128	.0170	.0212	.0253	.0294	.0334	.0374
1.1	.0414	.0453	.0492	.0531	.0569	.0607	.0645	.0682	.0719	.0755
1.2	.0792	.0828	.0864	.0899	.0934	.0969	.1004	.1038	.1072	.1106
1.3	.1139	.1173	.1206	.1239	.1271	.1303	.1335	.1367	.1399	.1430
1.4	.1461	.1492	.1523	.1553	.1584	.1614	.1644	.1673	.1703	.1732
1.5	.1761	.1790	.1818	.1847	.1875	.1903	.1931	.1959	.1987	.2014
1.6	.2041	.2068	.2095	.2122	.2148	.2175	.2201	.2227	.2253	.2279
1.7	.2304	.2330	.2355	.2380	.2405	.2430	.2455	.2480	.2504	.2529
1.8	.2553	.2577	.2601	.2625	.2648	.2672	.2695	.2718	.2742	.2765
1.9	.2788	.2810	.2833	.2856	.2878	.2900	.2923	.2945	.2967	.2989
2.0	.3010	.3032	.3054	.3075	.3096	.3118	.3139	.3160	.3181	.3201
2.1	.3222	.3243	.3263	.3284	.3304	.3324	.3345	.3365	.3385	.3404
2.2	.3424	.3444	.3464	.3483	.3502	.3522	.3541	.3560	.3579	.3598
2.3	.3617	.3636	.3655	.3674	.3692	.3711	.3729	.3747	.3766	.3784
2.4	.3802	.3820	.3838	.3856	.3874	.3892	.3909	.3927	.3945	.3962
2.5	.3979	.3997	.4014	.4031	.4048	.4065	.4082	.4099	.4116	.4133
2.6	.4150	.4166	.4183	.4200	.4216	.4232	.4249	.4265	.4281	.4298
2.7	.4314	.4330	.4346	.4362	.4378	.4393	.4409	.4425	.4440	.4456
2.8	.4472	.4487	.4502	.4518	.4533	.4548	.4564	.4579	.4594	.4609
2.9	.4624	.4639	.4654	.4669	.4683	.4698	.4713	.4728	.4742	.4757
3.0	.4771	.4786	.4800	.4814	.4829	.4843	.4857	.4871	.4886	.4900
3.1	.4914	.4928	.4942	.4955	.4969	.4983	.4997	.5011	.5024	.5038
3.2	.5051	.5065	.5079	.5092	.5105	.5119	.5132	.5145	.5159	.5172
3.3	.5185	.5198	.5211	.5224	.5237	.5250	.5263	.5276	.5289	.5302
3.4	.5315	.5328	.5340	.5353	.5366	.5378	.5391	.5403	.5416	.5428
3.5	.5441	.5453	.5465	.5478	.5490	.5502	.5514	.5527	.5539	.5551
3.6	.5563	.5575	.5587	.5599	.5611	.5623	.5635	.5647	.5658	.5670
3.7	.5682	.5694	.5705	.5717	.5729	.5740	.5752	.5763	.5775	.5786
3.8	.5798	.5809	.5821	.5832	.5843	.5855	.5866	.5877	.5888	.5899
3.9	.5911	.5922	.5933	.5944	.5955	.5966	.5977	.5988	.5999	.6010
4.0	.6021	.6031	.6042	.6053	.6064	.6075	.6085	.6096	.6107	.6117
4.1	.6128	.6138	.6149	.6160	.6170	.6180	.6191	.6201	.6212	.6222
4.2	.6232	.6243	.6253	.6263	.6274	.6284	.6294	.6304	.6314	.6325
4.3	.6335	.6345	.6355	.6365	.6375	.6385	.6395	.6405	.6415	.6425
4.4	.6435	.6444	.6454	.6464	.6474	.6484	.6493	.6503	.6513	.6522
4.5	.6532	.6542	.6551	.6561	.6571	.6580	.6590	.6599	.6609	.6618
4.6	.6628	.6637	.6646	.6656	.6665	.6675	.6684	.6693	.6702	.6712
4.7	.6721	.6730	.6739	.6749	.6758	.6767	.6776	.6785	.6794	.6803
4.8	.6812	.6821	.6830	.6839	.6848	.6857	.6866	.6875	.6884	.6893
4.9	.6902	.6911	.6920	.6928	.6937	.6946	.6955	.6964	.6972	.6981
5.0	.6990	.6998	.7007	.7016	.7024	.7033	.7042	.7050	.7059	.7067
5.1	.7076	.7084	.7093	.7101	.7110	.7118	.7126	.7135	.7143	.7152
5.2	.7160	.7168	.7177	.7185	.7193	.7202	.7210	.7218	.7226	.7235
5.3	.7243	.7251	.7259	.7267	.7275	.7284	.7292	.7300	.7308	.7316
5.4	.7324	.7332	.7340	.7348	.7356	.7364	.7372	.7380	.7388	.7396

TABLE B. Common Logarithms

n	0	1	2	3	4	5	6	7	8	9
5.5	.7404	.7412	.7419	.7427	.7435	.7443	.7451	.7459	.7466	.7474
5.6	.7482	.7490	.7497	.7505	.7513	.7520	.7528	.7536	.7543	.7551
5.7	.7559	.7566	.7574	.7582	.7589	.7597	.7604	.7612	.7619	.7627
5.8	.7634	.7642	.7649	.7657	.7664	.7672	.7679	.7686	.7694	.7701
5.9	.7709	.7716	.7723	.7731	.7738	.7745	.7752	.7760	.7767	.7774
6.0	.7782	.7789	.7796	.7803	.7810	.7818	.7825	.7832	.7839	.7846
6.1	.7853	.7860	.7868	.7875	.7882	.7889	.7896	.7903	.7910	.7917
6.2	.7924	.7931	.7938	.7945	.7952	.7959	.7966	.7973	.7980	.7987
6.3	.7993	.8000	.8007	.8014	.8021	.8028	.8035	.8041	.8048	.8055
6.4	.8062	.8069	.8075	.8082	.8089	.8096	.8102	.8109	.8116	.8122
6.5	.8129	.8136	.8142	.8149	.8156	.8162	.8169	.8176	.8182	.8189
6.6	.8195	.8202	.8209	.8215	.8222	.8228	.8235	.8241	.8248	.8254
6.7	.8261	.8267	.8274	.8280	.8287	.8293	.8299	.8306	.8312	.8319
6.8	.8325	.8331	.8338	.8344	.8351	.8357	.8363	.8370	.8376	.8382
6.9	.8388	.8395	.8401	.8407	.8414	.8420	.8426	.8432	.8439	.8445
7.0	.8451	.8457	.8463	.8470	.8476	.8482	.8488	.8494	.8500	.8506
7.1	.8513	.8519	.8525	.8531	.8537	.8543	.8549	.8555	.8561	.8567
7.2	.8573	.8579	.8585	.8591	.8597	.8603	.8609	.8615	.8621	.8627
7.3	.8633	.8639	.8645	.8651	.8657	.8663	.8669	.8675	.8681	.8686
7.4	.8692	.8698	.8704	.8710	.8716	.8722	.8727	.8733	.8739	.8745
7.5	.8751	.8756	.8762	.8768	.8774	.8779	.8785	.8791	.8797	.8802
7.6	.8808	.8814	.8820	.8825	.8831	.8837	.8842	.8848	.8854	.8859
7.7	.8865	.8871	.8876	.8882	.8887	.8893	.8899	.8904	.8910	.8915
7.8	.8921	.8927	.8932	.8938	.8943	.8949	.8954	.8960	.8965	.8971
7.9	.8976	.8982	.8987	.8993	.8998	.9004	.9009	.9015	.9020	.9025
8.0	.9031	.9036	.9042	.9047	.9053	.9058	.9063	.9069	.9074	.9079
8.1	.9085	.9090	.9096	.9101	.9106	.9112	.9117	.9122	.9128	.9133
8.2	.9138	.9143	.9149	.9154	.9159	.9165	.9170	.9175	.9180	.9186
8.3	.9191	.9196	.9201	.9206	.9212	.9217	.9222	.9227	.9232	.9238
8.4	.9243	.9248	.9253	.9258	.9263	.9269	.9274	.9279	.9284	.9289
8.5	.9294	.9299	.9304	.9309	.9315	.9320	.9325	.9330	.9335	.9340
8.6	.9345	.9350	.9355	.9360	.9365	.9370	.9375	.9380	.9385	.9390
8.7	.9395	.9400	.9405	.9410	.9415	.9420	.9425	.9430	.9435	.9440
8.8	.9445	.9450	.9455	.9460	.9465	.9469	.9474	.9479	.9484	.9489
8.9	.9494	.9499	.9504	.9509	.9513	.9518	.9523	.9528	.9533	.9538
9.0	.9542	.9547	.9552	.9557	.9562	.9566	.9571	.9576	.9581	.9586
9.1	.9590	.9595	.9600	.9605	.9609	.9614	.9619	.9624	.9628	.9633
9.2	.9638	.9643	.9647	.9652	.9657	.9661	.9666	.9671	.9675	.9680
9.3	.9685	.9689	.9694	.9699	.9703	.9708	.9713	.9717	.9722	.9727
9.4	.9731	.9736	.9741	.9745	.9750	.9754	.9759	.9763	.9768	.9773
9.5	.9777	.9782	.9786	.9791	.9795	.9800	.9805	.9809	.9814	.9818
9.6	.9823	.9827	.9832	.9836	.9841	.9845	.9850	.9854	.9859	.9863
9.7	.9868	.9872	.9877	.9881	.9886	.9890	.9894	.9899	.9903	.9908
9.8	.9912	.9917	.9921	.9926	.9930	.9934	.9939	.9943	.9948	.9952
9.9	.9956	.9961	.9965	.9969	.9974	.9978	.9983	.9987	.9991	.9996

TABLE C. Trigonometric Functions (degrees)

Deg.	Sin	Tan	Cot	Cos		Deg.	Sin	Tan	Cot	Cos	
0.0	0.00000	0.00000	∞	1.0000	**90.0**	**6.0**	0.10453	0.10510	9.514	0.9945	**84.0**
.1	.00175	.00175	573.0	1.0000	89.9	.1	.10626	.10687	9.357	.9943	83.9
.2	.00349	.00349	286.5	1.0000	.8	.2	.10800	.10863	9.205	.9942	.8
.3	.00524	.00524	191.0	1.0000	.7	.3	.10973	.11040	9.058	.9940	.7
.4	.00698	.00698	143.24	1.0000	.6	.4	.11147	.11217	8.915	.9938	.6
.5	.00873	.00873	114.59	1.0000	.5	.5	.11320	.11394	8.777	.9936	.5
.6	.01047	.01047	95.49	0.9999	.4	.6	.11494	.11570	8.643	.9934	.4
.7	.01222	.01222	81.85	.9999	.3	.7	.11667	.11747	8.513	.9932	.3
.8	.01396	.01396	71.62	.9999	.2	.8	.11840	.11924	8.386	.9930	.8
.9	.01571	.01571	63.66	.9999	89.1	.9	.12014	.12101	8.264	.9928	83.1
1.0	0.01745	0.01746	57.29	0.9998	**89.0**	**7.0**	0.12187	0.12278	8.144	0.9925	**83.0**
.1	.01920	.01920	52.08	.9998	88.9	.1	.12360	.12456	8.028	.9923	82.9
.2	.02094	.02095	47.74	.9998	.8	.2	.12533	.12633	7.916	.9921	.8
.3	.02269	.02269	44.07	.9997	.7	.3	.12706	.12810	7.806	.9919	.7
.4	.02443	.02444	40.92	.9997	.6	.4	.12880	.12988	7.700	.9917	.6
.5	.02618	.02619	38.19	.9997	.5	.5	.13053	.13165	7.596	.9914	.5
.6	.02792	.02793	35.80	.9996	.4	.6	.13226	.13343	7.495	.9912	.4
.7	.02967	.02968	33.69	.9996	.3	.7	.13399	.13521	7.396	.9910	.3
.8	.03141	.03143	31.82	.9995	.2	.8	.13572	.13698	7.300	.9907	.2
.9	.03316	.03317	30.14	.9995	88.1	.9	.13744	.13876	7.207	.9905	82.1
2.0	0.03490	0.03492	28.64	0.9994	**88.0**	**8.0**	0.13917	0.14054	7.115	0.9903	**82.0**
.1	.03664	.03667	27.27	.9993	87.9	.1	.14090	.14232	7.026	.9900	81.9
.2	.03839	.03842	26.03	.9993	.8	.2	.14263	.14410	6.940	.9898	.8
.3	.04013	.04016	24.90	.9992	.7	.3	.14436	.14588	6.855	.9895	.7
.4	.04188	.04191	23.86	.9991	.6	.4	.14608	.14767	6.772	.9893	.6
.5	.04362	.04366	22.90	.9990	.5	.5	.14781	.14945	6.691	.9890	.5
.6	.04536	.04541	22.02	.9990	.4	.6	.14954	.15124	6.612	.9888	.4
.7	.04711	.04716	21.20	.9989	.3	.7	.15126	.15302	6.535	.9885	.3
.8	.04885	.04891	20.45	.9988	.2	.8	.15299	.15481	6.460	.9882	.2
.9	.05059	.05066	19.74	.9987	87.1	.9	.15471	.15660	6.386	.9880	81.1
3.0	0.05234	0.05241	19.081	0.9986	**87.0**	**9.0**	0.15643	0.15838	6.314	0.9877	**81.0**
.1	.05408	.05416	18.464	.9985	86.9	.1	.15816	.16017	6.243	.9874	80.9
.2	.05582	.05591	17.886	.9984	.8	.2	.15988	.16196	6.174	.9871	.8
.3	.05756	.05766	17.343	.9983	.7	.3	.16160	.16376	6.107	.9869	.7
.4	.05931	.05941	16.832	.9982	.6	.4	.16333	.16555	6.041	.9866	.6
.5	.06105	.06116	16.350	.9981	.5	.5	.16505	.16734	5.976	.9863	.5
.6	.06279	.06291	15.895	.9980	.4	.6	.16677	.16914	5.912	.9860	.4
.7	.06453	.06467	15.464	.9979	.3	.7	.16849	.17093	5.850	.9857	.3
.8	.06627	.06642	15.056	.9978	.2	.8	.17021	.17273	5.789	.9854	.2
.9	.06802	.06817	14.669	.9977	86.1	.9	.17193	.17453	5.730	.9851	80.1
4.0	0.06976	0.06993	14.301	0.9976	**86.0**	**10.0**	0.1736	0.1763	5.671	0.9848	**80.0**
.1	.07150	.07168	13.951	.9974	85.9	.1	.1754	.1781	5.614	.9845	79.9
.2	.07324	.07344	13.617	.9973	.8	.2	.1771	.1799	5.558	.9842	.8
.3	.07498	.07519	13.300	.9972	.7	.3	.1788	.1817	5.503	.9839	.7
.4	.07672	.07695	12.996	.9971	.6	.4	.1805	.1835	5.449	.9836	.6
.5	.07846	.07870	12.706	.9969	.5	.5	.1822	.1853	5.396	.9833	.5
.6	.08020	.08046	12.429	.9968	.4	.6	.1840	.1871	5.343	.9829	.4
.7	.08194	.08221	12.163	.9966	.3	.7	.1857	.1890	5.292	.9826	.3
.8	.08368	.08397	11.909	.9965	.2	.8	.1874	.1908	5.242	.9823	.2
.9	.08542	.08573	11.664	.9963	85.1	.9	.1891	.1926	5.193	.9820	79.1
5.0	0.08716	0.08749	11.430	0.9962	**85.0**	**11.0**	0.1908	0.1944	5.145	0.9816	**79.0**
.1	.08889	.08925	11.205	.9960	84.9	.1	.1925	.1962	5.079	.9813	78.9
.2	.09063	.09101	10.988	.9959	.8	.2	.1942	.1980	5.050	.9810	.8
.3	.09237	.09277	10.780	.9957	.7	.3	.1959	.1998	5.005	.9806	.7
.4	.09411	.09453	10.579	.9956	.6	.4	.1977	.2016	4.959	.9803	.6
.5	.09585	.09629	10.385	.9954	.5	.5	.1994	.2035	4.915	.9799	.5
.6	.09758	.09805	10.199	.9952	.4	.6	.2011	.2053	4.872	.9796	.4
.7	.09932	.09981	10.019	.9951	.3	.7	.2028	.2071	4.829	.9792	.3
.8	.10106	.10158	9.845	.9949	.2	.8	.2045	.2089	4.787	.9789	.2
.9	.10279	.10334	9.677	.9947	84.1	.9	.2062	.2107	4.745	.9785	78.1
6.0	0.10453	0.10510	9.514	0.9945	**84.0**	**12.0**	0.2079	0.2126	4.705	0.9781	**78.0**
	Cos	Cot	Tan	Sin	Deg.		Cos	Cot	Tan	Sin	Deg.

TABLE C. Trigonometric Functions (degrees)

Deg.	Sin	Tan	Cot	Cos		Deg.	Sin	Tan	Cot	Cos	
12.0	0.2079	0.2126	4.705	0.9781	**78.0**	**18.0**	0.3090	0.3249	3.078	0.9511	**72.0**
.1	.2096	.2144	4.665	.9778	77.9	.1	.3107	.3269	3.060	.9505	71.9
.2	.2113	.2162	4.625	.9774	.8	.2	.3123	.3288	3.042	.9500	.8
.3	.2130	.2180	4.586	.9770	.7	.3	.3140	.3307	3.024	.9494	.7
.4	.2147	.2199	4.548	.9767	.6	.4	.3156	.3327	3.006	.9489	.6
.5	.2164	.2217	4.511	.9763	.5	.5	.3173	.3346	2.989	.9483	.5
.6	.2181	.2235	4.474	.9759	.4	.6	.3190	.3365	2.971	.9478	.4
.7	.2198	.2254	4.437	.9755	.3	.7	.3206	.3385	2.954	.9472	.3
.8	.2215	.2272	4.402	.9751	.2	.8	.3223	.3404	2.937	.9466	.2
.9	.2233	.2290	4.366	.9748	77.1	.9	.3239	.3424	2.921	.9461	71.1
13.0	0.2250	0.2309	4.331	0.9744	**77.0**	**19.0**	0.3256	0.3443	2.904	0.9455	**71.0**
.1	.2267	.2327	4.297	.9740	76.9	.1	.3272	.3463	2.888	.9449	70.9
.2	.2284	.2345	4.264	.9736	.8	.2	.3289	.3482	2.872	.9444	.8
.3	.2300	.2364	4.230	.9732	.7	.3	.3305	.3502	2.856	.9438	.7
.4	.2317	.2382	4.198	.9728	.6	.4	.3322	.3522	2.840	.9432	.6
.5	.2334	.2401	4.165	.9724	.5	.5	.3338	.3541	2.824	.9426	.5
.6	.2351	.2419	4.134	.9720	.4	.6	.3355	.3561	2.808	.9421	.4
.7	.2368	.2438	4.102	.9715	.3	.7	.3371	.3581	2.793	.9415	.3
.8	.2385	.2456	4.071	.9711	.2	.8	.3387	.3600	2.778	.9409	.2
.9	.2402	.2475	4.041	.9707	76.1	.9	.3404	.3620	2.762	.9403	70.1
14.0	0.2419	0.2493	4.011	0.9703	**76.0**	**20.0**	0.3420	0.3640	2.747	0.9397	**70.0**
.1	.2436	.2512	3.981	.9699	75.9	.1	.3437	.3659	2.733	.9391	69.9
.2	.2453	.2530	3.952	.9694	.8	.2	.3453	.3679	2.718	.9385	.8
.3	.2470	.2549	3.923	.9690	.7	.3	.3469	.3699	2.703	.9379	.7
.4	.2487	.2568	3.895	.9686	.6	.4	.3486	.3719	2.689	.9373	.6
.5	.2504	.2586	3.867	.9681	.5	.5	.3502	.3739	2.675	.9367	.5
.6	.2521	.2605	3.839	.9677	.4	.6	.3518	.3759	2.660	.9361	.4
.7	.2538	.2623	3.812	.9673	.3	.7	.3535	.3779	2.646	.9354	.3
.8	.2554	.2642	3.785	.9668	.2	.8	.3551	.3799	2.633	.9348	.2
.9	.2571	.2661	3.758	.9664	75.1	.9	.3567	.3819	2.619	.9342	69.1
15.0	0.2588	0.2679	3.732	0.9659	**75.0**	**21.0**	0.3584	0.3839	2.605	0.9336	**69.0**
.1	.2605	.2698	3.706	.9655	74.9	.1	.3600	.3859	2.592	.9330	68.9
.2	.2622	.2717	3.681	.9650	.8	.2	.3616	.3879	2.578	.9323	.8
.3	.2639	.2736	3.655	.9646	.7	.3	.3633	.3899	2.565	.9317	.7
.4	.2656	.2754	3.630	.9641	.6	.4	.3649	.3919	2.552	.9311	.6
.5	.2672	.2773	3.606	.9636	.5	.5	.3665	.3939	2.539	.9304	.5
.6	.2689	.2792	3.582	.9632	.4	.6	.3681	.3959	2.526	.9298	.4
.7	.2706	.2811	3.558	.9627	.3	.7	.3697	.3979	2.513	.9291	.3
.8	.2723	.2830	3.534	.9622	.2	.8	.3714	.4000	2.500	.9285	.2
.9	.2740	.2849	3.511	.9617	74.1	.9	.3730	.4020	2.488	.9278	68.1
16.0	0.2756	0.2867	3.487	0.9613	**74.0**	**22.0**	0.3746	0.4040	2.475	0.9272	**68.0**
.1	.2773	.2886	3.465	.9608	73.9	.1	.3762	.4061	2.463	.9265	67.9
.2	.2790	.2905	3.442	.9603	.8	.2	.3778	.4081	2.450	.9259	.8
.3	.2807	.2924	3.420	.9598	.7	.3	.3795	.4101	2.438	.9252	.7
.4	.2823	.2943	3.398	.9593	.6	.4	.3811	.4122	2.426	.9245	.6
.5	.2840	.2962	3.376	.9588	.5	.5	.3827	.4142	2.414	.9239	.5
.6	.2857	.2981	3.354	.9583	.4	.6	.3843	.4163	2.402	.9232	.4
.7	.2874	.3000	3.333	.9578	.3	.7	.3859	.4183	2.391	.9225	.3
.8	.2890	.3019	3.312	.9573	.2	.8	.3875	.4204	2.379	.9219	.2
.9	.2907	.3038	3.291	.9568	73.1	.9	.3891	.4224	2.367	.9212	67.1
17.0	0.2924	0.3057	3.271	0.9563	**73.0**	**23.0**	0.3907	0.4245	2.356	0.9205	**67.0**
.1	.2940	.3076	3.251	.9558	72.9	.1	.3923	.4265	2.344	.9198	66.9
.2	.2957	.3096	3.230	.9553	.8	.2	.3939	.4286	2.333	.9191	.8
.3	.2974	.3115	3.211	.9548	.7	.3	.3955	.4307	2.322	.9184	.7
.4	.2990	.3134	3.191	.9542	.6	.4	.3971	.4327	2.311	.9178	.6
.5	.3007	.3153	3.172	.9537	.5	.5	.3987	.4348	2.300	.9171	.5
.6	.3024	.3172	3.152	.9532	.4	.6	.4003	.4369	2.289	.9164	.4
.7	.3040	.3191	3.133	.9527	.3	.7	.4019	.4390	2.278	.9157	.3
.8	.3057	.3211	3.115	.9521	.2	.8	.4035	.4411	2.267	.9150	.2
.9	.3074	.3230	3.096	.9516	72.1	.9	.4051	.4431	2.257	.9143	66.1
18.0	0.3090	0.3249	3.078	0.9511	**72.0**	**24.0**	0.4067	0.4452	2.246	0.9135	**66.0**
	Cos	Cot	Tan	Sin	Deg.		Cos	Cot	Tan	Sin	Deg.

TABLE C. Trigonometric Functions (degrees)

Deg.	Sin	Tan	Cot	Cos		Deg.	Sin	Tan	Cot	Cos	
24.0	0.4067	0.4452	2.246	0.9135	**66.0**	**30.0**	0.5000	0.5774	1.7321	0.8660	**60.0**
.1	.4083	.4473	2.236	.9128	65.9	.1	.5015	.5797	1.7251	.8652	59.9
.2	.4099	.4494	2.225	.9121	.8	.2	.5030	.5820	1.7182	.8643	.8
.3	.4115	.4515	2.215	.9114	.7	.3	.5045	.5844	1.7113	.8634	.7
.4	.4131	.4536	2.204	.9107	.6	.4	.5060	.5867	1.7045	.8625	.6
.5	.4147	.4557	2.194	.9100	.5	.5	.5075	.5890	1.6977	.8616	.5
.6	.4163	.4578	2.184	.9092	.4	.6	.5090	.5914	1.6909	.8607	.4
.7	.4179	.4599	2.174	.9085	.3	.7	.5105	.5938	1.6842	.8599	.3
.8	.4195	.4621	2.164	.9078	.2	.8	.5120	.5961	1.6775	.8590	.2
.9	.4210	.4642	2.154	.9070	65.1	.9	.5135	.5985	1.6709	.8581	59.1
25.0	0.4226	0.4663	2.145	0.9063	**65.0**	**31.0**	0.5150	0.6009	1.6643	0.8572	**59.0**
.1	.4242	.4684	2.135	.9056	64.9	.1	.5165	.6032	1.6577	.8563	58.9
.2	.4258	.4706	2.125	.9048	.8	.2	.5180	.6056	1.6512	.8554	.8
.3	.4274	.4727	2.116	.9041	.7	.3	.5195	.6080	1.6447	.8545	.7
.4	.4289	.4748	2.106	.9033	.6	.4	.5210	.6104	1.6383	.8536	.6
.5	.4305	.4770	2.097	.9026	.5	.5	.5225	.6128	1.6319	.8526	.5
.6	.4321	.4791	2.087	.9018	.4	.6	.5240	.6152	1.6255	.8517	.4
.7	.4337	.4813	2.078	.9011	.3	.7	.5255	.6176	1.6191	.8508	.3
.8	.4352	.4834	2.069	.9003	.2	.8	.5270	.6200	1.6128	.8499	.2
.9	.4368	.4856	2.059	.8996	64.1	.9	.5284	.6224	1.6066	.8490	58.1
26.0	0.4384	0.4887	2.050	0.8988	**64.0**	**32.0**	0.5299	0.6249	1.6003	0.8480	**58.0**
.1	.4399	.4899	2.041	.8980	63.9	.1	.5314	.6273	1.5941	.8471	57.9
.2	.4415	.4921	2.032	.8973	.8	.2	.5329	.6297	1.5880	.8462	.8
.3	.4431	.4942	2.023	.8965	.7	.3	.5344	.6322	1.5818	.8453	.7
.4	.4446	.4964	2.014	.8957	.6	.4	.5358	.6346	1.5757	.8443	.6
.5	.4462	.4986	2.006	.8949	.5	.5	.5373	.6371	1.5697	.8434	.5
.6	.4478	.5008	1.997	.8942	.4	.6	.5388	.6395	1.5637	.8425	.4
.7	.4493	.5029	1.988	.8934	.3	.7	.5402	.6420	1.5577	.8415	.3
.8	.4509	.5051	1.980	.8926	.2	.8	.5417	.6445	1.5517	.8406	.2
.9	.4524	.5073	1.971	.8918	63.1	.9	.5432	.6469	1.5458	.8396	57.1
27.0	0.4540	0.5095	1.963	0.8910	**63.0**	**33.0**	0.5446	0.6494	1.5399	0.8387	**57.0**
.1	.4555	.5117	1.954	.8902	62.9	.1	.5461	.6519	1.5340	.8377	56.9
.2	.4571	.5139	1.946	.8894	.8	.2	.5476	.6544	1.5282	.8368	.8
.3	.4586	.5161	1.937	.8886	.7	.3	.5490	.6569	1.5224	.8358	.7
.4	.4602	.5184	1.929	.8878	.6	.4	.5505	.6594	1.5166	.8348	.6
.5	.4617	.5206	1.921	.8870	.5	.5	.5519	.6619	1.5108	.8339	.5
.6	.4633	.5228	1.913	.8862	.4	.6	.5534	.6644	1.5051	.8329	.4
.7	.4648	.5250	1.905	.8854	.3	.7	.5548	.6669	1.4994	.8320	.3
.8	.4664	.5272	1.897	.8846	.2	.8	.5563	.6694	1.4938	.8310	.2
.9	.4679	.5295	1.889	.8838	62.1	.9	.5577	.6720	1.4882	.8300	56.1
28.0	0.4695	0.5317	1.881	0.8829	**62.0**	**34.0**	0.5592	0.6745	1.4826	0.8290	**56.0**
.1	.4710	.5340	1.873	.8821	61.9	.1	.5606	.6771	1.4770	.8281	55.9
.2	.4726	.5362	1.865	.8813	.8	.2	.5621	.6796	1.4715	.8271	.8
.3	.4741	.5384	1.857	.8805	.7	.3	.5635	.6822	1.4659	.8261	.7
.4	.4756	.5407	1.849	.8796	.6	.4	.5650	.6847	1.4605	.8251	.6
.5	.4772	.5430	1.842	.8788	.5	.5	.5664	.6873	1.4550	.8241	.5
.6	.4787	.5452	1.834	.8780	.4	.6	.5678	.6899	1.4496	.8231	.4
.7	.4802	.5475	1.827	.8771	.3	.7	.5693	.6924	1.4442	.8221	.3
.8	.4818	.5498	1.819	.8763	.2	.8	.5707	.6950	1.4388	.8211	.2
.9	.4833	.5520	1.811	.8755	61.1	.9	.5721	.6976	1.4335	.8202	55.1
29.0	0.4848	0.5543	1.804	0.8746	**61.0**	**35.0**	0.5736	0.7002	1.4281	0.8192	**55.0**
.1	.4863	.5566	1.797	.8738	60.9	.1	.5750	.7028	1.4229	.8181	54.9
.2	.4879	.5589	1.789	.8729	.8	.2	.5764	.7054	1.4176	.8171	.8
.3	.4894	.5612	1.782	.8721	.7	.3	.5779	.7080	1.4124	.8161	.7
.4	.4909	.5635	1.775	.8712	.6	.4	.5793	.7107	1.4071	.8151	.6
.5	.4924	.5658	1.767	.8704	.5	.5	.5807	.7133	1.4019	.8141	.5
.6	.4939	.5681	1.760	.8695	.4	.6	.5821	.7159	1.3968	.8131	.4
.7	.4955	.5704	1.753	.8686	.3	.7	.5835	.7186	1.3916	.8121	.3
.8	.4970	.5727	1.746	.8678	.2	.8	.5850	.7212	1.3865	.8111	.2
.9	.4985	.5750	1.739	.8669	60.1	.9	.5864	.7239	1.3814	.8100	54.1
30.0	0.5000	0.5774	1.732	0.8660	**60.0**	**36.0**	0.5878	0.7265	1.3764	0.8090	**54.0**
	Cos	Cot	Tan	Sin	Deg.		Cos	Cot	Tan	Sin	Deg.

TABLE C. Trigonometric Functions (degrees)

Deg.	Sin	Tan	Cot	Cos		Deg.	Sin	Tan	Cot	Cos	
36.0	0.5878	0.7265	1.3764	0.8090	**54.0**	**40.5**	0.6494	0.8541	1.1708	0.7604	**49.5**
.1	.5892	.7292	1.3713	.8080	53.9	.6	.6508	.8571	1.1667	.7593	.4
.2	.5906	.7319	1.3663	.8070	.8	.7	.6521	.8601	1.1626	.7581	.3
.3	.5920	.7346	1.3613	.8059	.7	.8	.6534	.8632	1.1585	.7570	.2
.4	.5934	.7373	1.3564	.8049	.6	.9	.6547	.8662	1.1544	.7559	49.1
.5	.5948	.7400	1.3514	.8039	.5	**41.0**	0.6561	0.8693	1.1504	0.7547	**49.0**
.6	.5962	.7427	1.3465	.8028	.4	.1	.6574	.8724	1.1463	.7536	48.9
.7	.5976	.7454	1.3416	.8018	.3	.2	.6587	.8754	1.1423	.7524	.8
.8	.5990	.7481	1.3367	.8007	.2	.3	.6600	.8785	1.1383	.7513	.7
.9	.6004	.7508	1.3319	.7997	53.1	.4	.6613	.8816	1.1343	.7501	.6
37.0	0.6018	0.7536	1.3270	0.7986	**53.0**	.5	.6626	.8847	1.1303	.7490	.5
.1	.6032	.7563	1.3222	.7976	52.9	.6	.6639	.8878	1.1263	.7478	.4
.2	.6046	.7590	1.3175	.7965	.8	.7	.6652	.8910	1.1224	.7466	.3
.3	.6060	.7618	1.3127	.7955	.7	.8	.6665	.8941	1.1184	.7455	.2
.4	.6074	.7646	1.3079	.7944	.6	.9	.6678	.8972	1.1145	.7443	48.1
.5	.6088	.7673	1.3032	.7934	.5	**42.0**	0.6691	0.9004	1.1106	0.7431	**48.0**
.6	.6101	.7701	1.2985	.7923	.4	.1	.6704	.9036	1.1067	.7420	47.9
.7	.6115	.7729	1.2938	.7912	.3	.2	.6717	.9067	1.1028	.7408	.8
.8	.6129	.7757	1.2892	.7902	.2	.3	.6730	.9099	1.0990	.7396	.7
.9	.6143	.7785	1.2846	.7891	52.1	.4	.6743	.9131	1.0951	.7385	.6
38.0	0.6157	0.7813	1.2799	0.7880	**52.0**	.5	.6756	.9163	1.0913	.7373	.5
.1	.6170	.7841	1.2753	.7869	51.9	.6	.6769	.9195	1.0875	.7361	.4
.2	.6184	.7869	1.2708	7859	.8	.7	.6782	.9228	1.0837	.7349	.3
.3	.6198	.7898	1.2662	.7848	.7	.8	.6794	.9260	1.0799	.7337	.2
.4	.6211	.7926	1.2617	.7837	.6	.9	.6807	.9293	1.0761	.7325	47.1
.5	.6225	.7954	1.2572	.7826	.5	**43.0**	0.6820	0.9325	1.0724	0.7314	**47.0**
.6	.6239	.7983	1.2527	.7815	.4	.1	.6833	.9358	1.0686	.7302	46.9
.7	.6252	.8012	1.2482	.7804	.3	.2	.6845	.9391	1.0649	.7290	.8
.8	.6266	.8040	1.2437	.7793	.2	.3	.6858	.9424	1.0612	.7278	.7
.9	.6280	.8069	1.2393	.7782	51.1	.4	.6871	.9457	1.0575	.7266	.6
39.0	0.6293	0.8098	1.2349	0.7771	**51.0**	.5	.6884	.9490	1.0538	.7254	.5
.1	.6307	.8127	1.2305	.7760	50.9	.6	.6896	.9523	1.0501	.7242	.4
.2	.6320	.8156	1.2261	.7749	.8	.7	.6909	.9556	1.0464	.7230	.3
.3	.6334	.8185	1.2218	.7738	.7	.8	.6921	.9590	1.0428	.7218	.2
.4	.6347	.8214	1.2174	.7727	.6	.9	.6934	.9623	1.0392	.7206	46.1
.5	.6361	.8243	1.2131	.7716	.5	**44.0**	0.6947	0.9657	1.0355	0.7193	**46.0**
.6	.6374	.8273	1.2088	.7705	.4	.1	.6959	.9691	1.0319	.7181	45.9
.7	.6388	.8302	1.2045	.7694	.3	.2	.6972	.9725	1.0283	.7169	.8
.8	.6401	.8332	1.2002	.7683	.2	.3	.6984	.9759	1.0247	.7157	.7
.9	.6414	.8361	1.1960	.7672	50.1	.4	.6997	.9793	1.0212	.7145	.6
40.0	0.6428	0.8391	1.1918	0.7660	**50.0**	.5	.7009	.9827	1.0176	.7133	.5
.1	.6441	.8421	1.1875	.7649	49.9	.6	.7022	.9861	1.0141	.7120	.4
.2	.6455	.8451	1.1833	.7638	.8	.7	.7034	.9896	1.0105	.7108	.3
.3	.6468	.8481	1.1792	.7627	.7	.8	.7046	.9930	1.0070	.7096	.2
.4	.6481	.8511	1.1750	.7615	.6	.9	.7059	.9965	1.0035	.7083	45.1
40.5	0.6494	0.8541	1.1708	0.7604	**49.5**	**45.0**	0.7071	1.0000	1.0000	0.7071	**45.0**
	Cos	Cot	Tan	Sin	Deg.		Cos	Cot	Tan	Sin	Deg.

TABLE D. Trigonometric Functions (radians)

Rad.	Sin	Tan	Cot	Cos	Rad.	Sin	Tan	Cot	Cos
.00	.00000	.00000	∞	1.00000	**.50**	.47943	.54630	1.8305	.87758
.01	.01000	.01000	99.997	0.99995	.51	.48818	.55936	1.7878	.87274
.02	.02000	.02000	49.993	.99980	.52	.49688	.57256	1.7465	.86782
.03	.03000	.03001	33.323	.99955	.53	.50553	.58592	1.7067	.86281
.04	.03999	.04002	24.987	.99920	.54	.51414	.59943	1.6683	.85771
.05	.04998	.05004	19.983	.99875	.55	.52269	.61311	1.6310	.85252
.06	.05996	.06007	16.647	.99820	.56	.53119	.62695	1.5950	.84726
.07	.06994	.07011	14.262	.99755	.57	.53963	.64097	1.5601	.84190
.08	.07991	.08017	12.473	.99680	.58	.54802	.65517	1.5263	.83646
.09	.08988	.09024	11.081	.99595	.59	.55636	.66956	1.4935	.83094
.10	.09983	.10033	9.9666	.99500	**.60**	.56464	.68414	1.4617	.82534
.11	.10978	.11045	9.0542	.99396	.61	.57287	.69892	1.4308	.81965
.12	.11971	.12058	8.2933	.99281	.62	.58104	.71391	1.4007	.81388
.13	.12963	.13074	7.6489	.99156	.63	.58914	.72911	1.3715	.80803
.14	.13954	.14092	7.0961	.99022	.64	.59720	.74454	1.3431	.80210
.15	.14944	.15114	6.6166	.98877	.65	.60519	.76020	1.3154	.79608
.16	.15932	.16138	6.1966	.98723	.66	.61312	.77610	1.2885	.78999
.17	.16918	.17166	5.8256	.98558	.67	.62099	.79225	1.2622	.78382
.18	.17903	.18197	5.4954	.98384	.68	.62879	.80866	1.2366	.77757
.19	.18886	.19232	5.1997	.98200	.69	.63654	.82534	1.2116	.77125
.20	.19867	.20271	4.9332	.98007	**.70**	.64422	.84229	1.1872	.76484
.21	.20846	.21314	4.6917	.97803	.71	.65183	.85953	1.1634	.75836
.22	.21823	.22362	4.4719	.97590	.72	.65938	.87707	1.1402	.75181
.23	.22798	.23414	4.2709	.97367	.73	.66687	.89492	1.1174	.74517
.24	.23770	.24472	4.0864	.97134	.74	.67429	.91309	1.0952	.73847
.25	.24740	.25534	3.9163	.96891	.75	.68164	.93160	1.0734	.73169
.26	.25708	.26602	3.7591	.96639	.76	.68892	.95045	1.0521	.72484
.27	.26673	.27676	3.6133	.96377	.77	.69614	.96967	1.0313	.71791
.28	.27636	.28755	3.4776	.96106	.78	.70328	.98926	1.0109	.71091
.29	.28595	.29841	3.3511	.95824	.79	.71035	1.0092	.99084	.70385
.30	.29552	.30934	3.2327	.95534	**.80**	.71736	1.0296	.97121	.69671
.31	.30506	.32033	3.1218	.95233	.81	.72429	1.0505	.95197	.68950
.32	.31457	.33139	3.0176	.94924	.82	.73115	1.0717	.93309	.68222
.33	.32404	.34252	2.9195	.94604	.83	.73793	1.0934	.91455	.67488
.34	.33349	.35374	2.8270	.94275	.84	.74464	1.1156	.89635	.66746
.35	.34290	.36503	2.7395	.93937	.85	.75128	1.1383	.87848	.65998
.36	.35227	.37640	2.6567	.93590	.86	.75784	1.1616	.86091	.65244
.37	.36162	.38786	2.5782	.93233	.87	.76433	1.1853	.84365	.64483
.38	.37092	.39941	2.5037	.92866	.88	.77074	1.2097	.82668	.63715
.39	.38019	.41105	2.4328	.92491	.89	.77707	1.2346	.80998	.62941
.40	.38942	.42279	2.3652	.92106	**.90**	.78333	1.2602	.79355	.62161
.41	.39861	.43463	2.3008	.91712	.91	.78950	1.2864	.77738	.61375
.42	.40776	.44657	2.2393	.91309	.92	.79560	1.3133	.76146	.60582
.43	.41687	.45862	2.1804	.90897	.93	.80162	1.3409	.74578	.59783
.44	.42594	.47078	2.1241	.90475	.94	.80756	1.3692	.73034	.58979
.45	.43497	.48306	2.0702	.90045	.95	.81342	1.3984	.71511	.58168
.46	.44395	.49545	2.0184	.89605	.96	.81919	1.4284	.70010	.57352
.47	.45289	.50797	1.9686	.89157	.97	.82489	1.4592	.68531	.56530
.48	.46178	.52061	1.9208	.88699	.98	.83050	1.4910	.67071	.55702
.49	.47063	.53339	1.8748	.88233	.99	.83603	1.5237	.65631	.54869
.50	.47943	.54630	1.8305	.87758	**1.00**	.84147	1.5574	.64209	.54030
Rad.	Sin	Tan	Cot	Cos	Rad.	Sin	Tan	Cot	Cos

TABLE D. Trigonometric Functions (radians)

Rad.	Sin.	Tan	Cot	Cos	Rad.	Sin	Tan	Cot	Cos
1.00	.84147	1.5574	.64209	.54030	**1.50**	.99749	14.101	.07091	.07074
1.01	.84683	1.5922	.62806	.53186	1.51	.99815	16.428	.06087	.06076
1.02	.85211	1.6281	.61420	.52337	1.52	.99871	19.670	.05084	.05077
1.03	.85730	1.6652	.60051	.51482	1.53	.99917	24.498	.04082	.04079
1.04	.86240	1.7036	.58699	.50622	1.54	.99953	32.461	.03081	.03079
1.05	.86742	1.7433	.57362	.49757	1.55	.99978	48.078	.02080	.02079
1.06	.87236	1.7844	.56040	.48887	1.56	.99994	92.621	.01080	.01080
1.07	.87720	1.8270	.54734	.48012	1.57	1.00000	1255.8	.00080	.00080
1.08	.88196	1.8712	.53441	.47133	1.58	.99996	− 108.65	− .00920	− .00920
1.09	.88663	1.9171	.52162	.46249	1.59	.99982	− 52.067	− .01921	− .01920
1.10	.89121	1.9648	.50897	.45360	**1.60**	.99957	− 34.233	− .02921	− .02920
1.11	.89570	2.0143	.49644	.44466	1.61	.99923	− 25.495	− .03922	− .03919
1.12	.90010	2.0660	.48404	.43568	1.62	.99879	− 20.307	− .04924	− .04918
1.13	.90441	2.1198	.47175	.42666	1.63	.99825	− 16.871	− .05927	− .05917
1.14	.90863	2.1759	.45959	.41759	1.64	.99761	− 14.427	− .06931	− .06915
1.15	.91276	2.2345	.44753	.40849	1.65	.99687	− 12.599	− .07937	− .07912
1.16	.91680	2.2958	.43558	.39934	1.66	.99602	− 11.181	− .08944	− .08909
1.17	.92075	2.3600	.42373	.39015	1.67	.99508	− 10.047	− .09953	− .09904
1.18	.92461	2.4273	.41199	.38092	1.68	.99404	− 9.1208	− .10964	− .10899
1.19	.92837	2.4979	.40034	.37166	1.69	.99290	− 8.3492	− .11977	− .11892
1.20	.93204	2.5722	.38878	.36236	**1.70**	.99166	− 7.6966	− .12993	− .12884
1.21	.93562	2.6503	.37731	.35302	1.71	.99033	− 7.1373	− .14011	− .13875
1.22	.93910	2.7328	.36593	.34365	1.72	.98889	− 6.6524	− .15032	− .14865
1.23	.94249	2.8198	.35463	.33424	1.73	.98735	− 6.2281	− .16056	− .15853
1.24	.94578	2.9119	.34341	.32480	1.74	.98572	− 5.8535	− .17084	− .16840
1.25	.94898	3.0096	.33227	.31532	1.75	.98399	− 5.5204	− .18115	− .17825
1.26	.95209	3.1133	.32121	.30582	1.76	.98215	− 5.2221	− .19149	− .18808
1.27	.95510	3.2236	.31021	.29628	1.77	.98022	− 4.9534	− .20188	− .19789
1.28	.95802	3.3413	.29928	.28672	1.78	.97820	− 4.7101	− .21231	− .20768
1.29	.96084	3.4672	.28842	.27712	1.79	.97607	− 4.4887	− .22278	− .21745
1.30	.96356	3.6021	.27762	.26750	**1.80**	.97385	− 4.2863	− .23330	− .22720
1.31	.96618	3.7471	.26687	.25785	1.81	.97153	− 4.1005	− .24387	− .23693
1.32	.96872	3.9033	.25619	.24818	1.82	.96911	− 3.9294	− .25449	− .24663
1.33	.97115	4.0723	.24556	.23848	1.83	.96659	− 3.7712	− .26517	− .25631
1.34	.97348	4.2556	.23498	.22875	1.84	.96398	− 3.6245	− .27590	− .26596
1.35	.97572	4.4552	.22446	.21901	1.85	.96128	− 3.4881	− .28669	− .27559
1.36	.97786	4.6734	.21398	.20924	1.86	.95847	− 3.3608	− .29755	− .28519
1.37	.97991	4.9131	.20354	.19945	1.87	.95557	− 2.2419	− .30846	− .29476
1.38	.98185	5.1774	.19315	.18964	1.88	.95258	− 3.1304	− .31945	− .30430
1.39	.98370	5.4707	.18279	.17981	1.89	.94949	− 3.0257	− 33.051	− .31381
1.40	.98545	5.7979	.17248	.16997	**1.90**	.94630	− 2.9271	− .34164	− .32329
1.41	.98710	6.1654	.16220	.16010	1.91	.94302	− 2.8341	− .35284	− .33274
1.42	.98865	6.5811	.15195	.15023	1.92	.93965	− 2.7463	− .36413	− .34215
1.43	.99010	7.0555	.14173	.14033	1.93	.93618	− 2.6632	− .37549	− .35153
1.44	.99146	7.6018	.13155	.13042	1.94	.93262	− 2.5843	− .38695	− .36087
1.45	.99271	8.2381	.12139	.12050	1.95	.92896	− 2.5095	− .39849	− .37018
1.46	.99387	8.9886	.11125	.11057	1.96	.92521	− 2.4383	− .41012	− .37945
1.47	.99492	9.8874	.10114	.10063	1.97	.92137	− 2.3705	− .42185	− .38868
1.48	.99588	10.983	.09105	.09067	1.98	.91744	− 2.3058	− .43368	− .39788
1.49	.99674	12.350	.08097	.08071	1.99	.91341	− 2.2441	− .44562	− .40703
1.50	.99749	14.101	.07091	.07074	**2.00**	.90930	− 2.1850	− .45766	− .41615
Rad.	Sin	Tan	Cot	Cos	Rad.	Sin	Tan	Cot	Cos

Answers to Odd-Numbered Problems

PROBLEM SET 1-1 (Page 5)

1. Let x be one number and y the other; $x + \frac{1}{3}y$.   **3.** Let x be one number and y the other; $2x/3y$.
5. Let x be the number; $0.10x + x$, or $1.10x$.   **7.** Let x be one side and y the other; $x^2 + y^2$.   **9.** xy **11.** y/x
13. $(30/x) + (30/y)$ **15.** x^2 **17.** $6x^2$ **19.** $4\pi(x/2)^2 = \pi x^2$ **21.** $10x^2$ **23.** $\frac{4}{3}\pi(x/2)^3 = \pi x^3/6$
25. $x^3 - 4\pi x$ **27.** $A = x^2 + \frac{1}{4}\pi x^2$; $P = 2x + \pi x$ **29.** Let x be the number; $x + \frac{1}{2}x = 45$; $x = 30$.
31. Let x be the smaller odd number; $x + x + 2 = 168$; $x = 83$.   **33.** $\frac{9}{2}x = 252$; $x = 56$ **35.** 10.5 meters
37. $800 - 200\pi \approx 171.68$ square feet.   **39.** 3 centimeters   **41.** \$3.24   **43.** $4\pi r^3$ **45.** 400 miles
47. 48,000 miles   **49.** 12π feet

PROBLEM SET 1-2 (Page 11)

1. 12   **3.** -31 **5.** -67 **7.** $-60 + 9x$ **9.** $14t$ **11.** $\frac{8}{9}$ **13.** $-\frac{3}{4}$ **15.** $(1 - 3x)/2$
17. $(-2x + 3)/2$ **19.** $\frac{21}{12} = \frac{7}{4}$ **21.** $\frac{19}{20}$ **23.** $\frac{23}{36}$ **25.** $\frac{106}{108} = \frac{53}{54}$ **27.** $\frac{1}{2}$ **29.** $\frac{3}{4}$ **31.** $\frac{5}{4}$ **33.** $\frac{3}{8}$
35. $\frac{54}{7}$ **37.** $\frac{17}{7}$ **39.** -17 **41.** $\frac{9}{17}$ **43.** $\frac{5}{9}$ **45.** $\frac{1}{18}$ **47.** $\frac{13}{24}$ **49.** $\frac{37}{7}$ **51.** $-\frac{1}{2}$ **53.** 1   **55.** $\frac{33}{5}$
57. $\frac{1}{19}$ **59.** 100   **61.** 64   **63.** 252 meters

PROBLEM SET 1-3 (Page 20)

1. $2 \cdot 5 \cdot 5 \cdot 5$ **3.** $2 \cdot 2 \cdot 2 \cdot 5 \cdot 5$ **5.** $2 \cdot 2 \cdot 3 \cdot 5 \cdot 5 \cdot 7$ **7.** $2 \cdot 2 \cdot 2 \cdot 5 \cdot 5 \cdot 5 = 1000$
9. $2 \cdot 2 \cdot 3 \cdot 5 \cdot 5 \cdot 5 \cdot 7 = 10,500$ **11.** $2 \cdot 2 \cdot 2 \cdot 3 \cdot 5 \cdot 5 \cdot 5 \cdot 7 = 21,000$ **13.** 97/1000   **15.** 289/10,500
17. $-1423/21,000$ **19.** $.\overline{6}$ **21.** $.625\overline{0}$ **23.** $.\overline{461538}$ **25.** 7/9   **27.** 235/999   **29.** 13/40
31. $318/990 = 53/165$ **33.** $\frac{1}{21}$ **35.** $\frac{50}{99}$ **37.** Suppose $\sqrt{3} = m/n$, where m and n are positive integers greater

than 1. Then $3n^2 = m^2$. Both n^2 and m^2 must have an even number of 3's as factors. This contradicts $3n^2 = m^2$, since $3n^2$ must have an odd number of 3's.   **39.** Let $r = a/b$ and $s = c/d$, where a, b, c, and d are integers, $b \neq 0$ and $d \neq 0$. Then $r + s = (ad + bc)/bd$, $rs = ac/bd$, both of which are rational.   **41.** If $\sqrt{2} + \frac{2}{3} = r$ is rational, then $\sqrt{2} = r + (-\frac{2}{3})$. This is impossible because $\sqrt{2}$ cannot be the sum of two rationals.   **43.** If $\frac{2}{3}\sqrt{2} = r$ is rational, then $\sqrt{2} = \frac{3}{2}r$. This is impossible because $\sqrt{2}$ cannot be the product of two rationals.   **45.** (d), (e), (f), and (i)

47.
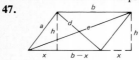
$a^2 = x^2 + h^2$, $e^2 = (b + x)^2 + h^2$, $d^2 = (b - x)^2 + h^2$,
$d^2 + e^2 = 2h^2 + (b + x)^2 + (b - x)^2 = 2(a^2 - x^2) + 2b^2 + 2x^2 = 2(a^2 + b^2)$

49. $d = \sqrt{(e^2 + f^2 + g^2)/2}$

PROBLEM SET 1-4 (Page 26)

1. 1431   **3.** 1700   **5.** 61   **7.** 3   **9.** 2   **11.** Associative and commutative properties of multiplication.
13. Commutative property of addition.
15. Associative property of addition; additive inverse; zero is neutral element for addition.   **17.** Distributive property.
19. Distributive property.   **21.** True; $a - (b - c) = a + (-1)[b + (-c)] = a + (-1)b + (-1)(-c) = a - b + c$.
23. False; $1 \div (1 + 1) = \frac{1}{2}$, but $(1 \div 1) + (1 \div 1) = 2$.   **25.** True; $ab(a^{-1} + b^{-1}) = aba^{-1} + abb^{-1} = b + a$.
27. False; $(1 + 2)(1^{-1} + 2^{-1}) = 3(1 + \frac{1}{2}) = 3(\frac{3}{2}) = \frac{9}{2}$.   **29.** False; $(1 + 2)(1 + 2) = 3 \cdot 3 = 9$, but $1^2 + 2^2 = 5$.
31. False; $1 \div (2 \div 3) = 1/\frac{2}{3} = \frac{3}{2}$ but $(1 \div 2) \div 3 = \frac{1}{2}/3 = \frac{1}{6}$.   **33.** (a) 64; 18; 512   (b) No; for example, $4 \# 3 = 64$ but $3 \# 4 = 81$ (c) No; for example, $2 \# (3 \# 2) = 512$ but $(2 \# 3) \# 2 = 64$ **35.** $(ab)(b^{-1}a^{-1}) = a(bb^{-1})a^{-1} = aa^{-1} = 1$; so $(ab)^{-1} = b^{-1}a^{-1}$. Since $a^{-1} \cdot a = 1$, it follows that a is the multiplicative inverse of a^{-1}; that is, $a = (a^{-1})^{-1}$. Since $\left(\frac{a}{b}\right) \cdot \left(\frac{b}{a}\right) = (ab^{-1})(ba^{-1}) = a(b^{-1}b)a^{-1} = aa^{-1} = 1$, $\frac{b}{a} = \left(\frac{a}{b}\right)^{-1}$ **37.** Additive inverse; zero is neutral element for addition; distributive property; associative property of addition; additive inverse; zero is neutral element for addition.
39. $(-a)(-b) + -(ab) = (-a)(-b) + (-a)b = -a(-b + b) = -a \cdot 0 = 0$. Thus, $(-a)(-b)$ is the additive inverse of $-(ab)$; so $(-a)(-b) = ab$.

PROBLEM SET 1-5 (Page 33)

1. >   **3.** >   **5.** >   **7.** >   **9.** <   **11.** =   **13.** $-\frac{3}{2}\sqrt{2}$; -2; $-\frac{\pi}{2}$; $\frac{3}{4}$; $\sqrt{2}$; $\frac{43}{24}$

15.   number line from -8 to 0, open circle at -2

17.   number line from -5 to 3, filled circle at -2

19.   number line from -3 to 5, open circle at -1

21.   number line from -3 to 5, open circle at 0, filled circle at 2

23.   number line from -3 to 5, filled circles at 0 and 2

25. $-2 \le x \le 3$ **27.** $x \ge 2$ **29.** $-2 < x \le 3$

31. $-1 < x < 2$ **33.** $-4 \le x \le 4$ number line from -6 to 6, filled circles at -4 and 4

35. $1 < x < 5$ number line from 0 to 6, open circles at 1 and 5

37. $-4 \le x \le 2$ number line from -6 to 6, filled circles at -4 and 2

39. $x < 0$ or $x > 10$ number line from -8 to 16, open circles at 0 and 10

41. (a) $x \le 12$ (b) $-11 \le x < 3$ (c) $|x - y| \le 4$ (d) $|x - 7| \ge 3$ (e) $|x - 5| < |x - y|$
43. 1.414, $(\sqrt{2} + 1.414)/2$, $1.41\overline{4}$, $\sqrt{2}$, $1.\overline{414}$, $1.41\overline{4}$ **45.** (a) $\frac{6}{25}$ (b) $\frac{4}{17}$ (c) 17.1/85   (d) They are equal.
47. $\frac{25}{3} < a < 12$ **49.** $\frac{1}{6} < y < \frac{1}{2}$ **51.** $ab \ge 0$ **53.** Suppose $\sqrt{a^2 + b^2} > |a| + |b|$. Then $a^2 + b^2 > (|a| + |b|)^2 = a^2 + b^2 + 2|a||b|$. This is impossible. Thus, $\sqrt{a^2 + b^2} \le |a| + |b|$

PROBLEM SET 1-6 (Page 39)

1. $-2 + 8i$ **3.** $-4 - i$ **5.** $0 + 6i$ **7.** $-4 + 7i$ **9.** $11 + 4i$ **11.** $14 + 22i$ **13.** $16 + 30i$

15. $61 + 0i$ **17.** $\frac{3}{2} + \frac{7}{2}i$ **19.** $2 - 5i$ **21.** $\frac{11}{2} + \frac{3}{2}i$ **23.** $\frac{2}{5} + \frac{1}{5}i$ **25.** $\sqrt{3}/4 - i/4$ **27.** $\frac{1}{5} + \frac{8}{5}i$

29. $(\sqrt{3} - 1/4) + (-\sqrt{3}/4 - 1)i$ **31.** -1 **33.** $-i$ **35.** $-729i$ **37.** $-2 + 2i$

39. $i^4 = (i^2)(i^2) = (-1)(-1) = 1$. The four 4th roots of 1 are $1, -1, i$, and $-i$.

41. $(1 - i)^4 = (1 - i)^2(1 - i)^2 = (-2i)(-2i) = 4i^2 = -4$

43. (a) $-1 - i$ (b) $1 - 2i$ (c) $-8 + 6i$ (d) $-\frac{8}{29} + \frac{20}{29}i$ (e) $\frac{14}{25}$ (f) $2\sqrt{3} - i$

45. (a) $a = 2, b = -\frac{4}{5}$ (b) $a = 3, b = 2$

47. (a) 1 (b) 0 (c) 3   **49.** $\pm[(\sqrt{2}/2) + (\sqrt{2}/2)i]$

51. (a) $\overline{x + y} = \overline{a + c + (b + d)i} = a + c - (b + d)i = a - bi + (c - di) = \overline{x} + \overline{y}$

(b) $\overline{xy} = \overline{ac - bd + (bc + ad)i} = ac - bd - (bc + ad)i = (a - bi)(c - di) = \overline{x}\ \overline{y}$

(c) $\overline{x^{-1}} = \overline{[1/(a + bi)]} = \overline{[a/(a^2 + b^2)] - [b/(a^2 + b^2)]i} = a/(a^2 + b^2) + [b/(a^2 + b^2)]i = (a + bi)/(a - bi)(a + bi) = 1/\overline{x} = (\overline{x})^{-1}$

(d) $\overline{x/y} = \overline{xy^{-1}} = \overline{x}\overline{y^{-1}} = \overline{x}(\overline{y})^{-1} = \overline{x}/\overline{y}$

CHAPTER 1. REVIEW PROBLEM SET (Page 42)

1. $V = x^2y; S = x^2 + 4xy$ **2.** $100/(x + y) + 100/(x - y)$ **3.** (a) $\frac{15}{24} = \frac{5}{8}$ (b) $\frac{1}{2}$ (c) $\frac{1}{9}$ (d) $\frac{1}{5}$ (e) $\frac{25}{7}$ (f) $\frac{27}{10}$

4. $2 \cdot 2 \cdot 2 \cdot 5 \cdot 5 \cdot 5; 2 \cdot 2 \cdot 3 \cdot 3 \cdot 5$ **5.** $2 \cdot 2 \cdot 2 \cdot 3 \cdot 3 \cdot 5 \cdot 5 \cdot 5 = 9000$ **6.** $.\overline{384615}; 1.\overline{571428}$ **7.** $257/999, 122/99$

8. Yes; no.   **9.** $a(bc) = (ab)c; a + b = b + a$ **10.** $1.4; \sqrt{2}; 1.\overline{4}; \frac{29}{20}, \frac{13}{8}$

11. $-8 < x < 4$

$-1 \le x \le 2$

12. No. If $x < 0$, then $|-x| = -x$. For example, $|-(-2)| = -(-2)$.

13. (a) $-6 + 7i$ (b) $17 - 8i$ (c) $\frac{29}{13} + \frac{28}{13}i$ (d) $2 + 11i$ (e) $\frac{5}{34} + \frac{3}{34}i$

14. Suppose that $5 + \sqrt{5} = r$, where r is rational. Then $\sqrt{5} = r - 5$, which is rational, a contradiction. Handle $5\sqrt{5}$ similarly.

15. 320 square centimeters

PROBLEM SET 2-1 (Page 49)

1. $3^3 = 27$ **3.** $2^2 = 4$ **5.** 8   **7.** $1/5^2 = 1/25$ **9.** $-1/5^2 = -1/25$ **11.** $1/(-2)^5 = -1/32$

13. $-27/8$ **15.** $27/4$ **17.** $81/16$ **19.** $3/64$ **21.** $1/72$ **23.** $81x^4$

25. x^6y^{12} **27.** $16x^8y^4/w^{12}$ **29.** $27y^6/(x^3z^6)$ **31.** $25x^8$ **33.** $1/(16y^6)$ **35.** $a/(5b^2x^2)$ **37.** $2z^3/(x^6y^2)$

39. $-x^3z^2/(2y^7)$ **41.** a^4b^3 **43.** $d^{40}/(32b^{15})$ **45.** $a^3/(a + 1)$ **47.** 0   **49.** $8x^3/y^6$ **51.** $\frac{1}{5}$

53. $2/x^6$ **55.** $4/x^8y^4$ **57.** xy **59.** (a) 2^{-15} (b) 2^0 **61.** 2^{1000}

63. (a) 15¢, 31¢, 63¢ (b) $2^n - 1$ cents (c) February 7.

PROBLEM SET 2-2 (Page 57)

1. 3.41×10^8 **3.** 5.13×10^{-8} **5.** 1.245×10^{-10} **7.** 8.4×10^{-4} **9.** 7.2×10^{-4} **11.** 1.08×10^{10}

13. 4.132×10^4 **15.** 4×10^9 **17.** 9.144×10^2

Note: Answers to Problems 19–43 may vary depending on the calculator used.

19. 48.35   **21.** -2441.7393 **23.** 303.27778   **25.** 2.7721×10^{15} **27.** 1.286×10^{10} **29.** -13.138859

31. 1.7891883   **33.** 1.067068   **35.** .90569641   **37.** .00000081   **39.** About 1.28 seconds.

41. About 4.068×10^{16}. **43.** About 9.3×10^{32}. **45.** (a) 7.04×10^5 (b) 6.72×10^{-5}
47. 1511.5 square centimeters **49.** $\bar{x} = 146.17$; $s = 18.85$ **51.** 3.3923×10^9
53. Hans will hear it .1710655 seconds earlier. **55.** (a) 6.69×10^{23} (b) 4.10×10^{23}
57. (a) 1.179×10^{61} (b) x_n grows without bound. (c) x_n tends toward zero.

PROBLEM SET 2-3 (Page 65)

1. Polynomial of degree 2. **3.** Polynomial of degree 5. **5.** Polynomial of degree 0. **7.** Not a polynomial.
9. Not a polynomial. **11.** $-2x + 1$ **13.** $4x^2$ **15.** $-2x^2 + 5x + 1$ **17.** $6x - 15$ **19.** $-10x + 12$
21. $35x^2 - 55x + 19$ **23.** $t^2 + 16t + 55$ **25.** $x^2 - x - 90$ **27.** $2t^2 + 13t - 7$ **29.** $y^2 + 2y - 8$
31. $7.9794x^2 + 0.1155x - 4.4781$ **33.** $x^2 + 20x + 100$ **35.** $x^2 - 64$ **37.** $4t^2 - 20t + 25$
39. $4x^8 - 25x^2$ **41.** $(t + 2)^2 + 2(t + 2)t^3 + t^6 = t^6 + 2t^4 + 4t^3 + t^2 + 4t + 4$
43. $(t + 2)^2 - (t^3)^2 = -t^6 + t^2 + 4t + 4$ **45.** $5.29x^2 - 6.44x + 1.96$ **47.** $x^3 + 6x^2 + 12x + 8$
49. $8t^3 - 36t^2 + 54t - 27$ **51.** $8t^3 + 12t^4 + 6t^5 + t^6$ **53.** $t^6 + 6t^5 + 15t^4 + 20t^3 + 15t^2 + 6t + 1$
55. $x^2 - 6xy + 9y^2$ **57.** $9x^2 - 4y^2$ **59.** $12x^2 + 11xy - 5y^2$ **61.** $2x^4y^2 - x^2yz - z^2$
63. $t^2 + 2t + 1 - s^2$ **65.** $8t^3 - 36t^2s + 54ts^2 - 27s^3$ **67.** $4x^4 - 9y^2$ **69.** $2s^6 - 5s^3t - 12t^2$
71. $8u^3 - 12u^2v^2 + 6uv^4 - v^6$ **73.** $-4x$ **75.** $12s + 18$ **77.** $x^4 + 4x^3 + 4x^2 - 9$ **79.** $2x^3 + 5x^2 + x - 2$
81. $x^5 + x^4y - 4x^3y^2 - 4x^2y^3$ **83.** $x^3 - 8y^3$ **85.** $x^4 + x^2y^2 + y^4$ **87.** -1 **89.** $(m^2 + 1)^2$
91. $2\sqrt{5}$ **93.** 243.5

PROBLEM SET 2-4 (Page 72)

1. $x(x + 5)$ **3.** $(x + 6)(x - 1)$ **5.** $y^3(y - 6)$ **7.** $(y + 6)(y - 2)$ **9.** $(y + 4)^2$ **11.** $(2x - 3y)^2$
13. $(y - 8)(y + 8)$ **15.** $(1 - 5b)(1 + 5b)$ **17.** $(2z - 3)(2z + 1)$ **19.** $(5x + 2y)(4x - y)$
21. $(x + 3)(x^2 - 3x + 9)$ **23.** $(a - 2b)(a^2 + 2ab + 4b^2)$ **25.** $x^3(1 - y^3) = x^3(1 - y)(1 + y + y^2)$
27. Does not factor over integers. **29.** Does not factor over integers. **31.** $(y - \sqrt{5})(y + \sqrt{5})$
33. $(\sqrt{5}z - 2)(\sqrt{5}z + 2)$ **35.** $t^2(t - \sqrt{2})(t + \sqrt{2})$ **37.** $(y - \sqrt{3})^2$ **39.** Does not factor over real numbers.
41. $(x + 3i)(x - 3i)$ **43.** $(x^3 + 7)(x^3 + 2)$ **45.** $(2x - 1)(2x + 1)(x - 3)(x + 3)$ **47.** $(x + 4y + 3)^2$
49. $(x^2 - 3y^2)(x^2 + 2y^2)$ **51.** $(x - 2)(x^2 + 2x + 4)(x + 2)(x^2 - 2x + 4)$ **53.** $x^4(x - y)(x + y)(x^2 + y^2)$
55. $(x^2 + y^2)(x^4 - x^2y^2 + y^4)$ **57.** $(x^2 + 1)(x - 4)$ **59.** $(2x - 1 - y)(2x - 1 + y)$ **61.** $(3 - x)(x + y)$
63. $(x + 3y)(x + 3y + 2)$ **65.** $(x + y + 2)(x + y + 1)$ **67.** $(x^2 - 4x + 8)(x^2 + 4x + 8)$
69. $(x^2 - x + 1)(x^2 + x + 1)$ **71.** $(2 - 3m)(2 + 3m)$ **73.** $(3x - 1)(2x - 1)$ **75.** $5x(x - 2)(x + 2)$
77. $x(3x - 1)(2x - 1)$ **79.** $(2u^2 - 5)(u - 1)(u + 1)$ **81.** $(a + 2b - 7)(a + 2b + 4)$
83. $(a + 3b - 1)(a + 3b + 1)(a^2 + 6ab + 9b^2 + 1)$ **85.** $(x - 3y)(x - 3y + 4)$
87. $(3x^2 - 4y^2 - y)(3x^2 - 4y^2 + y)$ **89.** $(x^2 - y^2 - xy)(x^2 - y^2 + xy)$ **91.** $(x + 3)(x + 2)^2(x + 7)(x - 2)$
93. $(x^n + 2)(x^n + 1)$ **95.** (a) 94,000 (b) 7 (c) $\frac{15}{29}$
97. $n^3 = [n(n + 1)/2]^2 - [n(n - 1)/2]^2$. It is clear that both $n(n + 1)/2$ and $n(n - 1)/2$ are integers.

PROBLEM SET 2-5 (Page 78)

1. $1/(x - 6)$ **3.** $y/5$ **5.** $(x + 2)^2/(x - 2)$ **7.** $z(x + 2y)/(x + y)$ **9.** $(9x + 2)/(x - 2)(x + 2)$
11. $2(4x - 3)/(x - 2)(x + 2)$ **13.** $(2xy + 3x - 1)/x^2y^2$ **15.** $(x^2 + 10x - 3)/(x - 2)^2(x + 5)$
17. $(4 - x)/(2x - 1)$ **19.** $(6y^2 + 9y + 2)/(3y - 1)(3y + 1)$ **21.** $(3m^2 + m - 1)/3(m - 1)^2$
23. $5x/(2x - 1)(x + 1)$ **25.** $1/(x - 3)(x - 2)$ **27.** $y^2(x^3 - y^3)$ **29.** $x/(x - 4)$ **31.** $5(x + 1)/x(2x - 1)$
33. $x/(x - 2)$ **35.** $(x - a)(x^2 + a^3)/(x + 2a)(x^2 + ax + a^2)$ **37.** $(y - 1)/2y$
39. $-2/(2x + 2h + 3)(2x + 3)$ **41.** $(-2x - h)/(x + h)^2x^2$ **43.** $(y - 1)(y + 2)/(7y + 19)$
45. $(a^2 + b^2)/ab$ **47.** 1 **49.** $(y - 1)/y$ **51.** $x^2/(x - y)$ **53.** 2 **55.** $(x - 6)/(x + 4)(x + 6)$
57. $a^2 - 3$ **59.** $3x + 2$
61. $(x^3 + y^3)/[x^3 + (x - y)^3] = (x + y)(x^2 - xy + y^2)/[x + (x - y)][x^2 - x(x - y) + (x - y)^2] =$
$(x + y)(x^2 - xy + y^2)/[x + (x - y)](x^2 - xy + y^2) = (x + y)/[x + (x - y)]$

63. (a) $2 + \dfrac{1}{1 + \dfrac{1}{1 + \frac{1}{2}}}$ (b) $2 + \dfrac{1}{1 + \dfrac{1}{1, + \dfrac{1}{1 + \frac{1}{3}}}}$ (c) $-2 + \dfrac{1}{1 + \frac{1}{3}}$

CHAPTER 2. REVIEW PROBLEM SET (Page 84)

1. 16/9 **2.** 36/25 **3.** 36/49 **4.** $y^5/2x^3$ **5.** $x^6/27y^9$ **6.** $b^{11}/2a^8$ **7.** 1.382×10^6
8. 6.82×10^{-3} **9.** 5×10^7 **10.** Polynomial of degree 2. **11.** Not a polynomial. **12.** Not a polynomial.
13. Not a polynomial. **14.** $2x^2 + x + 6$ **15.** $x^3 - x^2 - x + 7$ **16.** $2x^2 + 7x - 15$ **17.** $-4x + 1$
18. $z^6 - 16$ **19.** $2x^4 + 5x^2w - 12w^2$ **20.** $x^3 + 8a^3$ **21.** $6y^3 - 19y^2 + 27y - 18$
22. $9t^4 - 6t^3 + 7t^2 - 2t + 1$ **23.** $a^2 - b^2c^2d^2$ **24.** $x^2(2x^2 - x + 11)$ **25.** $(y - 4)(y - 3)$
26. $(3z - 1)(2z + 1)$ **27.** $(7a - 5)(7a + 5)$ **29.** $(a - 3)(a^2 + 3a + 9)$ **30.** $(2ab + 1)(4a^2b^2 - 2ab + 1)$
31. $x^4(x - y)(x + y)(x^2 + y^2)$ **32.** $(x + y - z^2)(x + y + z^2)$ **33.** $(2c + d)(2c - d - 3)$
34. $(3x - \sqrt{11})(3x + \sqrt{11})$ **35.** $(x - 4i)(x + 4i)$ **36.** $(x^2 + 2x + 4)/2$ **37.** $-2/(x - 1)$ **38.** $2/x$
39. $(x - 2)/(x + 1)$ **40.** $(x + 1)/10$

PROBLEM SET 3-1 (Page 92)

1. Conditional equation. **3.** Identity. **5.** Conditional equation. **7.** Conditional equation. **9.** Identity.
11. 2 **13.** $\frac{2}{5}$ **15.** $\frac{9}{2}$ **17.** $2/\sqrt{3}$ **19.** 7.57 **21.** -8.71×10^1 **23.** -24 **25.** $-\frac{1}{4}$ **27.** $\frac{22}{5}$
29. 3 **31.** 6 **33.** No solution (2 is extraneous). **35.** $-\frac{3}{4}$ **37.** $-\frac{13}{2}$ **39.** -10 **41.** -21
43. $P = A/(1 + rt)$ **45.** $r = (nE - IR)/nI$ **47.** $h = (A - 2\pi r^2)/2\pi r$ **49.** $R_1 = RR_2/(R_2 - R)$ **51.** $\frac{26}{5}$
53. $\frac{5}{14}$ **55.** 2 **57.** $-\frac{1}{4}$ **59.** No solution. **61.** $(1 - 2a)/(2a^2 - 1)$ **63.** (a) 86 (b) -40 (c) 160
65. (a) $11\frac{1}{9}$ (b) $100p/(100 - p)$

PROBLEM SET 3-2 (Page 99)

1. 9 **3.** 12 **5.** 23 **7.** 31 centimeters **9.** 89 **11.** 7 **13.** 12 **15.** 12:40 A.M. **17.** 4:20 P.M.
19. 9 miles per hour **21.** 15 **23.** 6000 **25.** 142.86 liters **27.** $145 **29.** 9 feet **31.** 7.5 days
33. 4.6 feet from fulcrum **35.** 125 **37.** 44.44 **39.** $17\frac{3}{4}$ feet **41.** 73 years old **43.** 32 **45.** $1511.36
47. 84 years old **49.** (a) $\frac{5}{4}$ hours (b) $\frac{16}{75}$ hours

PROBLEM SET 3-3 (Page 106)

1. $x = -13; y = 13$ **3.** $u = \frac{3}{2}; v = -4$ **5.** $x = -1; y = 3$ **7.** $x = 4; y = 3$ **9.** $x = 6; y = -8$
11. $s = 1; t = 4$ **13.** $x = 3; y = -1$ **15.** $x = 4; y = 7$ **17.** $x = 9; y = -2$ **19.** $x = 16; y = -5$
21. $x = \frac{1}{2}; y = \frac{1}{3}$ **23.** $x = 4; y = 9$ **25.** $x = 1; y = 2; z = 0$ **27.** $x = \frac{4}{3}, y = -\frac{2}{3}; z = -\frac{5}{3}$ **29.** $x = -1;$
$y = -2; z = -3$ **31.** $r = 36, s = 54$ **33.** $7200 and $5800 **35.** $\frac{5}{3}$ and $-\frac{4}{3}$ **37.** $\frac{11}{19}$
39. 36,000 $10 tickets and 9000 $15 tickets. **41.** 35 hours
43. $57\frac{1}{7}$ pounds of the first kind and $42\frac{6}{7}$ pounds of the second kind. **45.** 8 coats and 100 dresses.
47. $x = -10z + 33; y = -7z + 26$

PROBLEM SET 3-4 (Page 115)

1. $5\sqrt{2}$ **3.** $\frac{1}{2}$ **5.** $\frac{3}{2}$ **7.** 22 **9.** $(5 + 6\sqrt{2})/5$ **11.** $(6 + i)/2$ **13.** ± 5 **15.** $-1; 7$
17. $-\frac{15}{2}; \frac{5}{2}$ **19.** $\pm 3i$ **21.** $0; 3$ **23.** ± 3 **25.** $\pm.12$ **27.** $-2; 5$ **29.** $-2; \frac{1}{3}$ **31.** $-\frac{4}{3}; \frac{7}{2}$

33. $-9; 1$ **35.** $-\frac{1}{2}; \frac{3}{2}$ **37.** $-2 \pm \sqrt{5}i$ **39.** $-6; -2$ **41.** $(-5 \pm \sqrt{13})/2$ **43.** $(3 \pm \sqrt{42})/3$
45. $(-5 \pm \sqrt{5})/2$ **47.** $(3 \pm \sqrt{13}i)/2$ **49.** $-.2714; 1.8422$ **51.** $-1.6537; .8302$ **53.** $y = 2 \pm 2x$
55. $y = -6x$ or $y = 0$ **57.** $y = -2x + 3$ or $y = -2x + 5$ **59.** $-\frac{1}{4}; \frac{5}{4}$ **61.** 1 **63.** $-1 \pm \sqrt{5}$
65. $(-1 \pm i)/2$ **67.** $\pm\sqrt{6}; \pm i$ **69.** $-1, 4, 2 \pm 2\sqrt{2}$ **71.** 2 **73.** $\frac{1}{4}; \frac{3}{2}$ **75.** $0; 4$
77. $x = \frac{5}{2}, y = 8; x = -4, y = -5$ **79.** $x = 2\sqrt{2}, y = \sqrt{2}; x = -2\sqrt{2}, y = -\sqrt{2}$ **81.** 3 by 10
83. 12 inches by 12 inches **85.** At approximately 1:49 P.M. **87.** $(1 + \sqrt{5})/2$

PROBLEM SET 3-5 (Page 123)

1. Conditional. **3.** Unconditional. **5.** Conditional. **7.** Conditional. **9.** Conditional. **11.** Unconditional.

13. $\{x : x < -6\}$

15. $\{x : x > -24\}$

17. $\{x : x < \frac{30}{7}\}$

19. $\{x : -5 \le x \le 2\}$

21. $\{x : x < -3 \text{ or } x > \frac{1}{2}\}$

23. $\{x : x \le 1 \text{ or } x \ge 4\}$

25. $\{x : \frac{1}{2} < x < 3\}$

27. $\{x : -\frac{5}{2} < x < -\frac{1}{2}\}$

29. $\{x : -1 \le x \le 0\}$

31. $\{x : -4 \le x \le 0 \text{ or } x \ge 3\}$

33. $\{x : x < 5 \text{ and } x \ne 2\}$

35. $\{x : -2 < x \le 5\}$

37. $\{x : -2 < x < 0 \text{ or } x > 5\}$

39. $\{x : -2 < x < 2 \text{ or } x > 3\}$

41. $|x - 3| < 3$ **43.** $|x - 3| \le 4$ **45.** $|x - 6.5| < 4.5$

47. $\{x : -\sqrt{7} < x < \sqrt{7}\}$

49. $\{x : x \le 2 - \sqrt{2} \text{ or } x \ge 2 + \sqrt{2}\}$

51. $\{x : x < -5.71 \text{ or } x > -.61\}$

53. 4 **55.** 100

57. $\{x : x < \frac{25}{2}\}$ **59.** $\{x : -3 < x < \frac{1}{2}\}$ **61.** $\{x : x < -1 \text{ or } -1 < x < 1 \text{ or } 4 < x < 8\}$ **63.** $\{x : -2 < x < 2\}$
65. $\{x : x \le \frac{1}{4} \text{ or } x \ge \frac{5}{4}\}$ **67.** $\{x : -1 < x < \frac{5}{2}\}$ **69.** $\{x : x > -\frac{1}{2}\}$ **71.** $\{x : x < -2 \text{ or } 0 < x < 2 \text{ or } x > 4\}$
73. (a) $\{k : k \le 4\}$ (b) $\{k : k \le -6 \text{ or } k \ge 6\}$ (c) $\{k : k \le 0 \text{ or } k \ge 4\}$ (d) $\{k : k = 0\}$ **75.** $\$25,000 < S < \$53,000$
77. (a) 144 feet (b) Between $2 - \sqrt{3}$ seconds and $2 + \sqrt{3}$ seconds. (c) After 5 seconds.
79. $c^n = c^2 c^{n-2} = (a^2 + b^2)c^{n-2} = a^2 c^{n-2} + b^2 c^{n-2} > a^2 a^{n-2} + b^2 b^{n-2} = a^n + b^n$

PROBLEM SET 3-6 (Page 129)

1. About 18,333 feet. **3.** 525 miles **5.** About 21.82 minutes. **7.** 4 hours and 48 minutes
9. About 4.15 hours after takeoff. **11.** 12 standard; 20 deluxe **13.** 60 miles per hour **15.** 500 meters

17. 20 feet　　**19.** About 8.17 feet.　　**21.** $\frac{5}{6}$ kiloliters　　**23.** About 3.29 milligrams.
25. .298 grams sodium chloride; .202 grams sodium bromide　　**27.** 36 shares
29. 14 pounds walnuts; 11 pounds cashews　　**31.** 23,500　　**33.** $785,714.29
35. 2340 undergraduate students; 864 graduate students　　**37.** 2 feet　　**39.** 5; 12; 13　　**41.** $a = 8, b = 5, c = 10$
43. About 13.34 feet.　　**45.** Two solutions: longer piece is 28 inches or $\frac{236}{7}$ inches.

CHAPTER 3. REVIEW PROBLEM SET　(Page 135)

1. (a) Identity.　(b) Conditional equation.　(c) Conditional equation.　(d) Identity.
2. (a) $-\frac{11}{12}$　(b) 7　(c) $\frac{5}{2}$　(d) No solution.　　**3.** $v_0 = (s - \frac{1}{2}at^2)/t$　　**4.** (a) $-24.62°C$　(b) Above $-40°F$.
5. (a) $x = 2; y = -1$　(b) $x = 1; y = -2$　　**6.** (a) $(2 + \sqrt{2})/4$　(b) $5 - 5\sqrt{3}$　(c) $(-1 + \sqrt{2})/3$　　**7.** ± 7
8. $-5; 2$　　**9.** -3　　**10.** $-3; 7$　　**11.** $-5; 4$　　**12.** 4; 5　　**13.** 0; 4　　**14.** $-1 \pm \sqrt{5}$
15. $(-1 \pm \sqrt{13})/6$　　**16.** $(-m \pm \sqrt{m^2 - 8n})/2$　　**17.** $y = 2x \pm 2$　　**18.** $y = (3 - 3x)/2$ or $y = (1 - 3x)/2$
19. $\{x : x \le \frac{12}{7}\}$　　**20.** $\{x : x \le -6 \text{ or } x \ge 1\}$　　**21.** $\{x : (-1 - \sqrt{13})/2 < x < (-1 + \sqrt{13})/2\}$
22. $\{x : x < -1 \text{ or } x > 4\}$　　**23.** $3500 in the bank; $6500 in the credit union.　　**24.** 50 miles per hour
25. 2 miles per hour

PROBLEM SET 4-1　(Page 141)

1. 5　　**3.** 4　　**5.** 1.7　　**7.** $2\pi - 5$　　**9.** 3; 7　　**11.** $-6; 2$　　**13.** $\frac{7}{4}; \frac{13}{4}$　　**15.** Rectangle.
17. Parallelogram.　　**19.** 5; $(\frac{7}{2}, 1)$　　**21.** $2\sqrt{2}; (3, 3)$　　**23.** 3; $(\sqrt{3}/2, \sqrt{6}/2)$　　**25.** 14.54; (3.974, 1.605)
27. (a) $\sqrt{10}$; $\sqrt{5}$; $\sqrt{10}$; $\sqrt{5}$　(b) Each midpoint has coordinates (5/2, 5).　(c) Opposite sides have the same length; the two diagonals bisect each other.　　**29.** (a) $(-2, 3); (4, 0)$　(b) $(2, 7); (8, -1)$
31. If the points are labeled A, B, and C, respectively, then $d(A, B) = \sqrt{20}$, $d(B, C) = \sqrt{20}$, and $d(A, C) = \sqrt{40}$. Then note that $(\sqrt{40})^2 = (\sqrt{20})^2 + (\sqrt{20})^2$.
33. (a) The three distances are 5, 10, and 15, and $15 = 5 + 10$.　(b) The distances are 5, 10, and 15.
35. $(9, 0)$　　**37.** $(17, 35)$　　**39.** $1.47 cheaper by truck.　　**41.** $x \approx 1.9641, y \approx 4.5981$

43.

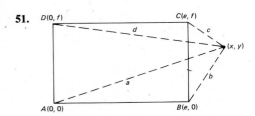

$$e^2 = \left(\frac{a}{2}\right)^2 + \left(\frac{b}{2}\right)^2, f^2 = \left(\frac{a}{2}\right)^2 + \left(\frac{b}{2}\right)^2, d^2 = \left(\frac{a}{2}\right)^2 + \left(\frac{b}{2}\right)^2$$

45. $x = 5, y = 7$　　**47.** $(1, 10), (6, 17), (11, 24), (16, 31)$
49. (a) $\frac{1}{3}(x_1 + x_2 + x_3), \frac{1}{3}(y_1 + y_2 + y_3)$　(b) They intersect at P.

51.

$$a^2 + c^2 = x^2 + y^2 + (x - e)^2 + (y - f)^2$$
$$b^2 + d^2 = (x - e)^2 + y^2 + x^2 + (y - f)^2$$
Thus $a^2 + c^2 = b^2 + d^2$.

PROBLEM SET 4-2 (Page 148)

1.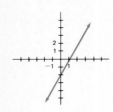

3.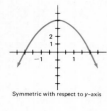
Symmetric with respect to y–axis

5.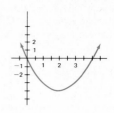

7.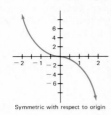
Symmetric with respect to origin

9.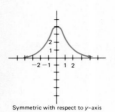
Symmetric with respect to y–axis

11.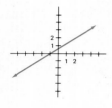

13.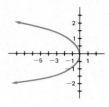

15.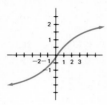

17. $x^2 + y^2 = 36$ **19.** $(x - 4)^2 + (y - 1)^2 = 25$ **21.** $(x + 2)^2 + (y - 1)^2 = 3$ **23.**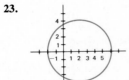

25. 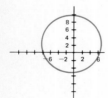   **27.** $(-1, 5); 1$ **29.** $(6, 0); 1$ **31.** $(-\frac{1}{2}, \frac{3}{2}); \frac{3}{2}$

33.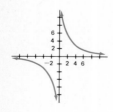

35.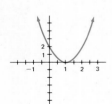

37.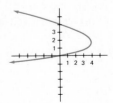

39. $(x - 5)^2 + (y + 7)^2 = 49$ **41.** $(x + 3)^2 + (y - 2)^2 = 50$
43. (a) Point: $(2, -3)$ (b) Circle: 2; $(\frac{1}{2}, \frac{-3}{2})$ (c) Empty set   (d) Circle: $\sqrt{5}$; $(0, \sqrt{3})$
45. (a) $(x - 4)^2 + (y - 4)^2 = 25$ (b) $(x - 4)^2 + (y - 4)^2 = \frac{25}{2}$
47. $6\sqrt{3} + 6\pi \approx 29.24$

1. $\frac{5}{2}$ **3.** $-\frac{2}{7}$ **5.** $-\frac{5}{3}$ **7.** 0.1920 **9.** $4x - y - 5 = 0$ **11.** $2x + y - 2 = 0$
13. $2x + y - 4 = 0$ **15.** $y - 5 = 0$ **17.** $5x - 2y - 4 = 0$ **19.** $5x + 3y - 15 = 0$
21. $1.56x + y - 5.35 = 0$ **23.** $x - 2 = 0$ **25.** 3; 5 **27.** $\frac{2}{3}$; $-\frac{4}{3}$ **29.** $-\frac{2}{3}$, 2 **31.** -4; 2
33. (a) $y + 3 = 2(x - 3)$ (b) $y + 3 = -\frac{1}{2}(x - 3)$ (c) $y + 3 = -\frac{2}{3}(x - 3)$ (d) $y + 3 = \frac{3}{2}(x - 3)$
(e) $y + 3 = -\frac{3}{4}(x - 3)$ (f) $x = 3$ (g) $y = -3$ **35.** $y + 4 = 2x$ **37.** $(-1, 2); y - 2 = \frac{3}{2}(x + 1)$
39. $(3, 1); y - 1 = -\frac{4}{3}(x - 3)$ **41.** $\frac{7}{5}$ **43.** $\frac{18}{13}$ **45.** $\frac{6}{5}$
47. (a) Parallel (b) Perpendicular (c) Neither (d) Perpendicular **49.** $3x - 2y - 7 = 0$ **51.** $x + 2y - 9 = 0$
53. Note that $(a, 0)$ and $(0, b)$ satisfy the equation $x/a + y/b = 1$, and use the fact that two points determine a line.
55. $2x + y - 8 = 0$ **57.** $P = 4x - 8500$; $-\$500$ (loss) **59.** $\frac{5}{13}$
61. Draw a picture. We may assume the triangle has vertices $(0, 0)$, $(a, 0)$, and (b, c). Midpoints are $(b/2, c/2)$ and
$((a + b)/2, c/2)$. The line joining midpoints has slope 0 as does the base.
63. $(10 \pm 4\sqrt{2}, 0)$ **65.** $x = (1 + \sqrt{7})/2, y = (1 - \sqrt{7})/2$

PROBLEM SET 4-4 (Page 163)

1.
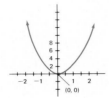

3.
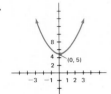

5.
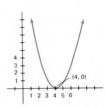

7.
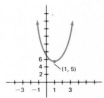

9.

11. $y = 2x^2 - 4x + 9$

13. $y = -\frac{1}{2}x^2 + 5x - 10$

15.
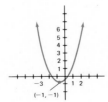

17.
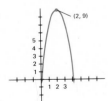

19.
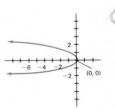

21.
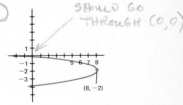

23.
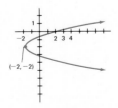

25. $(-3, 4); (0, 1)$ **27.** $(-1, 3); (2, -3)$ **29.** $(-.64, 2.25); (5.04, 10.75)$

31. $y = \frac{1}{12}x^2$ **33.** $y = -\frac{1}{8}(x + 2)^2$ **35.** $y = 2x^2 - 1$ **37.**

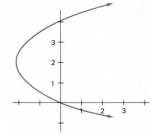

39. (a)

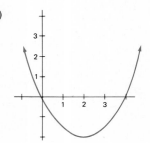

(b)

41. (a) $(-2, 14)$ and $(1, 5)$ (b) $(3, 3)$ (c) None   (d) $(-1, 10)$ and $(2, 7)$ **43.** $a = \frac{5}{2};\ (5, -\frac{45}{2})$ **45.** $\frac{45}{4}$
47. $P = (300 - 100x)(x - 2);\ \$2.50$ **49.** $(b - a)^3/8$

51.

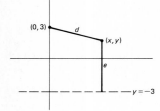

$$d^2 = e^2;\ x^2 + (y - 3)^2 = (y + 3)^2;\ x^2 = 12y$$

53. Let the coordinates of R be (x, y). $L = \overline{FR} + \overline{RG} = \sqrt{x^2 + (y - p)^2} + p - y = \sqrt{4py + (y - p)^2} + p - y = \sqrt{(y + p)^2} + p - y = y + p + p - y = 2p$

PROBLEM SET 4-5   (Page 172)

1.   Center: $(0,0)$;
Endpoints of major diameter: $(\pm 5, 0)$;
Endpoints of minor diameter: $(0, \pm 3)$.

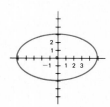

3. Center: (0, 0);
Endpoints of major diameter: (0, ±5);
Endpoints of minor diameter: (±3, 0).

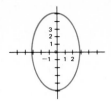

5. Center: (2, −1);
Endpoints of major diameter: (−3, −1), (7, −1);
Endpoints of minor diameter: (2, −4), (2, 2).

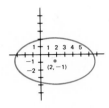

7. Center: (−3, 0);
Endpoints of major diameter: (−3, −4), (−3, 4);
Endpoints of minor diameter: (−6, 0), (0, 0).

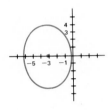

9. (0, 0); $x^2/36 + y^2/9 = 1$ **11.** (0, 0); $x^2/16 + y^2/36 = 1$ **13.** (2, 3); $(x - 2)^2/36 + (y - 3)^2/4 = 1$

15. Center: (0, 0);
Vertices: (±4, 0).

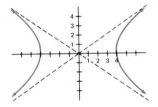

17. Center: (0, 0);
Vertices; (0, ±3).

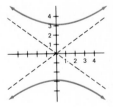

19. Center: (3, −2);
Vertices: (0, −2), (6, −2).

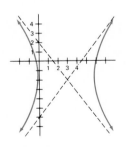

21. Center: (0, −3);
Vertices: (0, −5), (0, −1).

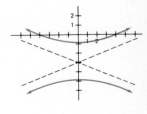

23. $x^2/16 - y^2/25 = 1$ **25.** $(x - 6)^2/4 - (y - 3)^2/4 = 1$ **27.** $(y - 9)^2/36 - (x - 4)^2/16 = 1$
29. $(x + 2)^2/16 + (y - 3)^2/9 = 1$; horizontal ellipse; center: (−2, 3); vertices: (−6, 3), (2, 3).
31. $(x - 2)^2/9 - (y + 1)^2/4 = 1$; horizontal hyperbola; center: (2, −1); vertices: (−1, −1), (5, −1).
33. $x^2/4 + (y - 2)^2/100 = 1$; vertical ellipse; center: (0, 2); vertices: (0, −8), (0, 12).
35. $(y + 2)^2/9 - (x - 4)^2/\frac{9}{4} = 1$; vertical hyperbola; center: (4, −2); vertices: (4, −5), (4, 1).

37. Ellipse.

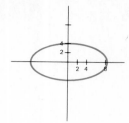

39. Hyperbola.

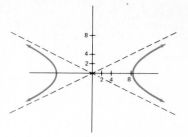

41. Ellipse.

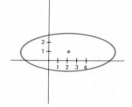

43. Hyperbola.

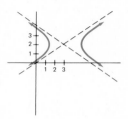

45. $(x - 3)^2/49 + (y - 2)^2/25 = 1$ **47.** $x^2/36 - y^2/16 = 1$ **49.** ± 3.7486 **51.** (a) $\pi\sqrt{77} \approx 27.5674$
(b) $\pi\sqrt{111} \approx 33.0987$ **53.** $9 - \sqrt{21} \approx 4.42$ centimeters   **55.** $y^2/441 + x^2/16 = 1$ **57.** $x^2/25 + y^2/9 = 1$
59. $x^2/9 - y^2/16 = 1$ **61.** $x^2/625 + y^2/400 = 1$

CHAPTER 4. REVIEW PROBLEM SET (Page 177)

1. (a) Line   (b) Circle   (c) Line   (d) Parabola   (e) Parabola   (f) Circle   (g) Ellipse   (h) Hyperbola
(i) Parabola   (j) Ellipse   (k) Hyperbola   (l) Hyperbola   **2.** (a)

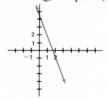

(b)

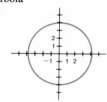

(c)

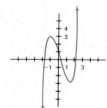

(d)

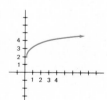

(e)

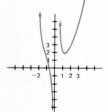

(f)

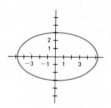

3. (a)

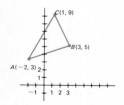

(b) $d(A, B) = \sqrt{29}$; $d(B, C) = 2\sqrt{5}$; $d(C, A) = 3\sqrt{5}$ (c) $\frac{2}{3}$; -2; 2 (d) $y - 3 = \frac{2}{3}(x + 2)$; $y - 5 = -2(x - 3)$;
$y - 3 = 2(x + 2)$ (e) $(\frac{1}{2}, 4)$, $(2, 7)$, $(-\frac{1}{2}, 6)$ (f) $y - 3 = -2(x + 2)$ (g) $y - 3 = \frac{1}{2}(x + 2)$ (h) $12/\sqrt{5} \approx 5.37$ (i) 12
4.   (a) $x - 4 = 0$ (b) $y + 1 = 0$ (c) $3x + 2y - 10 = 0$ (d) $5x - y - 15 = 0$ (e) $2x - 3y + 13 = 0$
(f) $4x - y - 1 = 0$ (g) $x + y - 10 = 0$ (h) $12x + 5y - 169 = 0$
5.   (a) $(0, 0)$; down.   (b) $(0, 0)$; to the right.   (c) $(-2, 1)$; up.   (d) $(\frac{1}{4}, -\frac{1}{2})$; to the left.   (e) $(2, -4)$; up.
(f) $(-17, -3)$; to the right.   **6.**   (a) $y = -\frac{1}{20}x^2$ (b) $x = -\frac{1}{72}y^2 + 3$ (c) $x = 2(y - 5)^2 - 3$ **7.** $(1, 2)$, $(4, 20)$
8.   (a) $(x + 1)^2 + (y - 2)^2 = 7$; circle with center $(-1, 2)$ and radius $\sqrt{7}$.   (b) $(x + 1)^2/10 + (y - 2)^2/20 = 1$;
vertical ellipse with center $(-1, 2)$.   (c) $(x + 1)^2/6 - (y + 2)^2/12 = 1$; horizontal hyperbola with center $(-1, -2)$.
(d) $(x + 1)^2 = \frac{1}{2}(y + 16)$; vertical parabola opening up with vertex $(-1, -16)$.
9.   (a) $x^2/9 + y^2/25 = 1$ (b) $(x - 2)^2/16 + y^2/4 = 1$ **10.** $(y - 8)^2/9 - (x + 2)^2/1 = 1$

PROBLEM SET 5-1   (Page 184)

1.   (a) 0 (b) -4 (c) $-15/4$ (d) -3.99 (e) -2 (f) $a^2 - 4$ (g) $(1 - 4x^2)/x^2$ (h) $x^2 + 2x - 3$
3.   (a) $\frac{1}{4}$ (b) $-\frac{1}{2}$ (c) 2 (d) -8 (e) Undefined.   (f) 100 (g) $x/(1 - 4x)$ (h) $1/(x^2 - 4)$ (i) $1/(h - 2)$ (j) $-1/(h + 2)$
5. All real numbers.   **7.** $\{x : x \neq \pm 2\}$ **9.** $\{x : x \neq -2 \text{ and } x \neq 3\}$ **11.** All real numbers.
13. $\{x : x \geq 2\}$ **15.** $\{x : x \geq 0 \text{ and } x \neq 25\}$ **17.** (a) 9 (b) 5 (c) 3; the positive integers.
19. $f(\#) = \boxed{(}\ \#\ \boxed{+}\ 2\ \boxed{)}\ \boxed{x^2}\ \boxed{=}$; $f(2.9) = 24.01$ **21.** $f(\#) = 3\ \boxed{\times}\ \boxed{(}\ \#\ \boxed{+}\ 2\ \boxed{)}\ \boxed{x^2}\ \boxed{-}\ 4\ \boxed{=}$; $f(2.9) = 68.03$
23. $f(\#) = \boxed{(}\ 3\ \boxed{\times}\ \#\ \boxed{+}\ 2\ \boxed{\div}\ \#\ \boxed{\sqrt{x}}\ \boxed{)}\ \boxed{y^x}\ 3\ \boxed{=}$; $f(2.9) = 962.80311$
25. $f(\#) = \boxed{(}\ \#\ \boxed{y^x}\ 5\ \boxed{-}\ 4\ \boxed{)}\ \boxed{\sqrt{x}}\ \boxed{\div}\ \boxed{(}\ 2\ \boxed{+}\ \#\ \boxed{1/x}\ \boxed{)}\ \boxed{=}$; $f(2.9) = 6.0479407$ **27.** $x + \sqrt{5}$
29. $2x^2 + 3x$ **31.** $\sqrt{3(x - 2)^2 + 9}$ **33.** 20 **35.** 5 **37.** 8 **39.** Undefined.   **41.** $18xy - 10x$
43. $3 - 5x$ **45.** $y = 4x$ **47.** $y = 1/x$ **49.** $I = 324s/d^2$ **51.** (a) $R = 2v^2/45$ (b) About $28{,}444$ feet.
53. (a) 3 (b) 3 (c) $-\frac{15}{4}$ (d) $2 - \sqrt{2}$ (e) $4 - 5\sqrt{2}$ (f) -199.9999 (g) $(1 - 2x^3)/x^2$ (h) $(a^6 - 2)/a^2$
(i) $(a^3 + 3a^2b + 3ab^2 + b^3 - 2)/(a + b)$ **55.** (a) $\{t : t \neq 0, -3\}$ (b) $\{t : t \leq -2 \text{ or } t \geq 2\}$ (c) $\{t : t \geq 0, t \neq \frac{1}{2}\}$
(d) $\{(s, t) : -3 \leq s \leq 3, t \neq \pm 1\}$ **57.** Domain: $\{x : 0 < x < \frac{1}{\pi}\}$; range: $\{y : 0 < y < \frac{1}{\pi}\}$.
59. (a) $\sqrt{3}x^2/36$ (b) $3\sqrt{3}x^2/2$ (c) $3\pi x^3/64$ (d) $(1300 + 240x)/x$
(e) $F(x) = \begin{cases} 180, & 0 \leq x \leq 100 \\ 180 + .22(x - 100), & x > 100 \end{cases}$ **61.** $S(x, y, z) = (5000/3)(xy^2/z)$; $333\frac{1}{3}$ pounds   **63.** $f(t) = 1 + t$

PROBLEM SET 5-2   (Page 193)

1.

3.

5.

7.

9.

11.

13.

15.

17.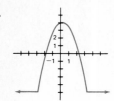

19. Even.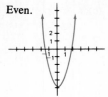

21. Neither even nor odd.

23. Odd.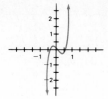

25. Even.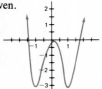

27.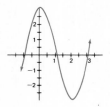

29.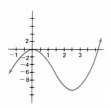

31.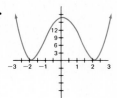

33. It must intersect each vertical line at most once.

35.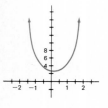

37.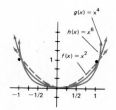

39.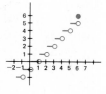

41.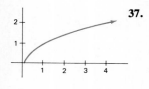

43.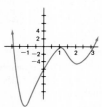

45.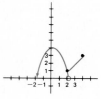

47. $C(x) = \begin{cases} 15 & \text{if } x < 1 \\ 15 + 10[x] & \text{if } x \geq 1 \end{cases}$

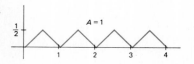

49. 792; 5022.3214; −86.55584

51.

PROBLEM SET 5-3 (Page 201)

1.
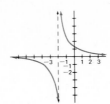

3.
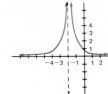

5.
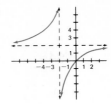

7.
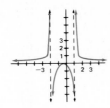

9.
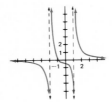

11.
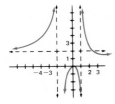

13.
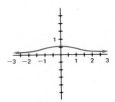

15.
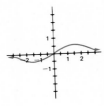

17.
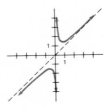

19.
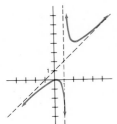

21.
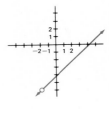

23.
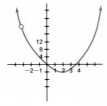

25.
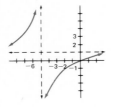

27.
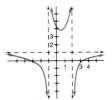

29.
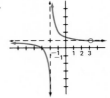

31.
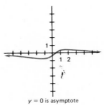
$y = 0$ is asymptote

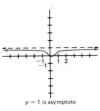
$y = 1$ is asymptote

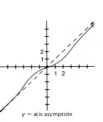
$y = x$ is asymptote

33.

n	Vertical	Horizontal	Oblique
1	$x = 1$	None	None
2	$x = -1, x = 1$	None	None
3	$x = 1$	None	$y = x$
4	$x = -1, x = 1$	$y = 1$	None
5	$x = 1$	$y = 0$	None
6	$x = -1, x = 1$	$y = 0$	None

35.

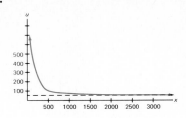

37. $f(x) = 2x + 3 + 2/(x - 3) = (2x^2 - 3x - 7)/(x - 3)$

PROBLEM SET 5-4 (Page 210)

1. (a) 3 (b) Undefined. (c) -1 (d) -10 (e) 2 (f) $\frac{1}{2}$ (g) 10 (h) $\frac{2}{5}$ (i) 3

3. $(f + g)(x) = x^2 + x - 2$, all real numbers; $(f - g)(x) = x^2 - x + 2$, all real numbers;
$(f \cdot g)(x) = x^3 - 2x^2$, all real numbers; $(f/g)(x) = x^2/(x - 2)$, $\{x : x \neq 2\}$.

5. $(f + g)(x) = x^2 + \sqrt{x}$, $\{x : x \geq 0\}$; $(f - g)(x) = x^2 - \sqrt{x}$, $\{x : x \geq 0\}$;
$(f \cdot g)(x) = x^2\sqrt{x}$, $\{x : x \geq 0\}$; $(f/g)(x) = x^2/\sqrt{x}$, $\{x : x > 0\}$.

7. $(f + g)(x) = (x^2 - x - 3)/(x - 2)(x - 3)$, $\{x : x \neq 2 \text{ and } x \neq 3\}$;
$(f - g)(x) = (-x^2 + 3x - 3)/(x - 2)(x - 3)$, $\{x : x \neq 2 \text{ and } x \neq 3\}$;
$(f \cdot g)(x) = x/(x - 2)(x - 3)$, $\{x : x \neq 2 \text{ and } x \neq 3\}$; $(f/g)(x) = (x - 3)/x(x - 2)$, $\{x : x \neq 0, x \neq 2, x \neq 3\}$.

9. $(g \circ f)(x) = x^2 - 2$, all real numbers; $(f \circ g)(x) = (x - 2)^2$, all real numbers.

11. $(g \circ f)(x) = (3x + 1)/x$, $\{x : x \neq 0\}$; $(f \circ g)(x) = 1/(x + 3)$, $\{x : x \neq -3\}$

13. $(g \circ f)(x) = x - 4$, $\{x : x \geq 2\}$; $(f \circ g)(x) = \sqrt{x^2 - 4}$, $\{x : |x| \geq 2\}$

15. $(g \circ f)(x) = x$, all real numbers; $(f \circ g)(x) = x$, all real numbers. **17.** $g(x) = x^3; f(x) = x + 4$

19. $g(x) = \sqrt{x}; f(x) = x + 2$ **21.** $g(x) = 1/x^3; f(x) = 2x + 5$ **23.** $g(x) = |x|; f(x) = x^3 - 4$

25. 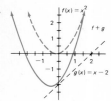 **27.** 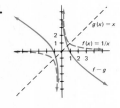 **29.** 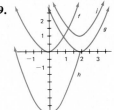 **31.**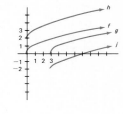

33. (a) $x^3 + 2x + 3$ (b) $x^3 - 2x - 3$ (c) $2x^4 + 3x^3$ (d) $(2x + 3)/x^3$ (e) $2x^3 + 3$ (f) $(2x + 3)^3$ (g) $4x + 9$ (h) x^{27}

35. $1/|x^2 - 4|$; $\{x : x \neq -2 \text{ and } x \neq 2\}$ **37.** (a) $2x + h$ (b) 2 (c) $-1/x(x + h)$ (d) $-2/(x - 2)(x + h - 2)$

39. (a) $x; |x|$ (b) $x; x$ (c) $x^6; x^6$ (d) $1/x^6; 1/x^6$ **41.** $\frac{1}{2}$ **43.**

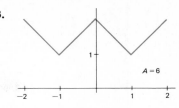

45. (a) Odd. (b) Odd. (c) Even. (d) Odd. (e) Neither. (f) Odd. (g) Even. (h) Even. (i) Even.

47. (a)

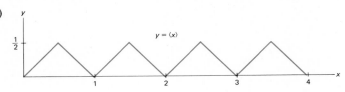

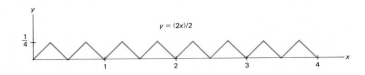

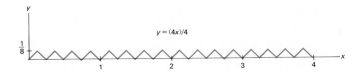

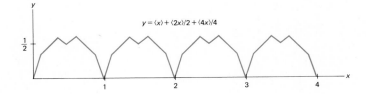

(b) $1; \frac{1}{2}; \frac{1}{4}; \frac{7}{4}$

PROBLEM SET 5-5 (Page 218)

1. (a) i, ii; iii; iv; vii; viii (b) i; ii; viii; (c) i; ii; viii **3.** (a) 1 (b) $-\frac{1}{3}$ (c) $\frac{16}{3}$ **5.** $f^{-1}(x) = (\frac{1}{5})x$

7. $f^{-1}(x) = (\frac{1}{2})(x + 7)$ **9.** $f^{-1}(x) = (x - 2)^2$ **11.** $f^{-1}(x) = 3x/(x - 1)$ **13.** $f^{-1}(x) = 2 + \sqrt[3]{(x - 2)}$

15.

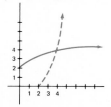

17. (a)

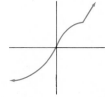

(b)

(c)

19. $(f\circ g)(x) = 3\left(\dfrac{2x}{3-x}\right) \bigg/ \left(\dfrac{2x}{3-x} + 2\right) = \dfrac{6x}{3-x} \bigg/ \dfrac{6}{3-x} = x;\ (g\circ f)(x) = 2\left(\dfrac{3x}{x+2}\right) \bigg/ \left(3 - \dfrac{3x}{x+2}\right) = \dfrac{6x}{x+2} \bigg/ \dfrac{6}{x+2} = x$

21. $\{x : x \geq 1\};\ f^{-1}(x) = 1 + \sqrt{x}$ **23.** $\{x : x \geq -1\};\ f^{-1}(x) = -1 + \sqrt{x+4}$

25. $\{x : x \geq -3\};\ f^{-1}(x) = -3 + \sqrt{x+2}$

27. $\{x : x \geq -2\};\ f^{-1}(x) = x - 2$ **29.** $\{x : x \geq 1\};\ f^{-1}(x) = 1 + \sqrt{2x/(1+x)}$

31.

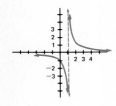

33. $f^{-1}(x) = (x+1)/x$

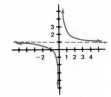

(a) $\frac{1}{2}$ (b) 3   (c) -1 (d) 0   (e) $\frac{4}{3}$ (f) $\frac{1}{2}$

35.

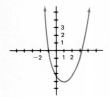

$\{x : x \geq 1\};\ f^{-1}(x) = 1 + \sqrt{x+4}$ **37.** $f^{-1}(x) = 1 + \sqrt{3/(2-x)}$

39. The graph must be symmetric about the line $y = x$. This means the xy-equation determining f is unchanged if x and y are interchanged.

41. (a) $f^{-1}(x) = (b - dx)/(cx - a)$ (b) $(f\circ f^{-1})(x)$ should be x. But $f(f^{-1}(x)) = \left(a\dfrac{b-dx}{cx-a} + b\right) \bigg/ \left(c\dfrac{b-dx}{cx-a} + d\right) = (bc - ad)x/(bc - ad) = x$, provided $bc - ad \neq 0$.   (c) $a = -d$

CHAPTER 5. REVIEW PROBLEM SET   (Page 222)

1. (a) 15   (b) 4   (c) Undefined.   (d) 0   (e) $-\frac{3}{4}$ (f) $\frac{2}{15}$ **2.** $\{x : x \geq -1 \text{ and } x \neq 1\}$ **3.** $y = \frac{1}{8}x^3$ **4.** 12

5. (a)

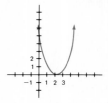

(b)

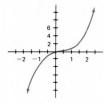

(c)

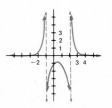

(d)

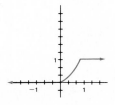

6.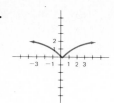

7.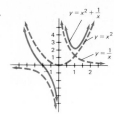

8. (a) $x^3 + 6x^2 + 12x + 9$ (b) $x^3 + 3$ (c) $x^9 + 3x^6 + 3x^3 + 2$ (d) $x + 4$ (e) $\sqrt[3]{x - 1}$ (f) $x - 2$ (g) $x + h + 2$
(h) 1 (i) $27x^3 + 1$ **9.** It is moved 2 units to the right and up 3 units.
10. (a) Even. (b) Odd, one-to-one. (c) Even. (d) One-to-one.

11. 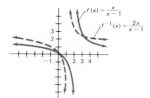 **12.** $\{x : x \geq -2\}$

PROBLEM SET 6-1 (Page 229)

1. 3 **3.** 2 **5.** 7 **7.** $\frac{3}{2}$ **9.** 25 **11.** 9 **13.** 2 **15.** $\frac{1}{100}$ **17.** $\sqrt{2}/2$ **19.** $\sqrt{5}$
21. $3xy\sqrt[3]{2xy^2}$ **23.** $(x + 2)y\sqrt[4]{y^3}$ **25.** $x\sqrt{1 + y^2}$ **27.** $x\sqrt[3]{x^3 - 9y}$ **29.** $(xz^2/y^2)\sqrt[3]{x}$
31. $2(\sqrt{x} - 3)/(x - 9)$ **33.** $2\sqrt{x + 3}/(x + 3)$ **35.** $\sqrt[4]{2x}/2x$ **37.** $(2y/x)\sqrt[3]{x^2}$ **39.** $\sqrt{2}$ **41.** 26
43. $\frac{9}{2}$ **45.** $-\frac{32}{15}$ **47.** 0 **49.** 4 **51.** $2[(\sqrt{x} - \sqrt{x + h})/\sqrt{x}\sqrt{x + h}]$ **53.** $(x + 7)/\sqrt{x + 6}$
55. $x/(2\sqrt[3]{x} + 2)$ **57.** $-9/(x^2\sqrt{x^2 + 9})$ **59.** (a) $2ab^2$ (b) $9b\sqrt{b}$ (c) $3\sqrt{3}$ (d) $5a^2b^3\sqrt{10}$ (e) $-2y^2/x$
(f) $y^2/4x^2$ (g) $(2a^2 + 3a)\sqrt{2a}$ (h) $4\sqrt[4]{2} - 5\sqrt{2} + 2\sqrt[6]{2}$ (i) $a\sqrt[4]{1 + b^4}$ (j) $\sqrt[3]{49b^2}/(7bc)$ (k) $2(\sqrt{a} + b)/(a - b^2)$
(l) $a + \frac{1}{a} = (a^2 + 1)/a$ **61.** (a) 13 (b) 0; 2 (c) 6 (d) No solution. (e) 9 (f) -8; 64

63. They are reflections of each other in the line $y = x$.

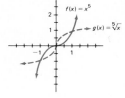

65. $\overline{AC} = \frac{15}{4}$

67. (a) Both sides are positive. $[(\sqrt{6} + \sqrt{2})/2]^2 = (6 + 2\sqrt{12} + 2)/4 = 2 + \sqrt{3} = (\sqrt{2} + \sqrt{3})^2$.
(b) Both sides are positive. $(\sqrt{2 + \sqrt{3}} + \sqrt{2 - \sqrt{3}})^2 = 2 + \sqrt{3} + 2\sqrt{1} + 2 - \sqrt{3} = 6$.
(c) $(\sqrt{3} - \sqrt{2})^3 = 3\sqrt{3} - 3 \cdot 3\sqrt{2} + 3\sqrt{3} \cdot 2 - 2\sqrt{2} = 9\sqrt{3} - 11\sqrt{2} = (\sqrt[3]{9\sqrt{3} - 11\sqrt{2}})^3$.

PROBLEM SET 6-2 (Page 236)

1. $7^{1/3}$ **3.** $7^{2/3}$ **5.** $7^{-1/3}$ **7.** $7^{-2/3}$ **9.** $7^{4/3}$ **11.** $x^{2/3}$ **13.** $x^{5/2}$ **15.** $(x + y)^{3/2}$
17. $(x^2 + y^2)^{1/2}$ **19.** $\sqrt[3]{16}$ **21.** $1/\sqrt{8^3} = \sqrt{2}/32$ **23.** $\sqrt[4]{x^4 + y^4}$ **25.** $y\sqrt[3]{x^4y}$ **27.** $\sqrt{\sqrt{x} + \sqrt{y}}$

29. 5 **31.** 4 **33.** $\frac{1}{27}$ **35.** .04 **37.** .000125 **39.** $\frac{1}{5}$ **41.** $\frac{1}{16}$ **43.** $\frac{1}{16}$ **45.** $-6a^2$ **47.** $8/x^4$
49. x^4 **51.** $4y^2/x^4$ **53.** y^9/x^{30} **55.** $(2y^3 - 1)/y$ **57.** $x + y + 2\sqrt{xy}$ **59.** $(7x + 2)/3(x + 2)^{1/5}$
61. $(1 - x^2)/(x^2 + 1)^{2/3}$ **63.** $\sqrt[6]{32}$ **65.** $\sqrt[12]{8x^2}$ **67.** $\sqrt{x}$ **69.** 2.53151 **71.** 4.6364
73. 1.70777 **75.** .0050463

77. 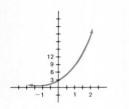 **79.** 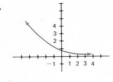 **81.**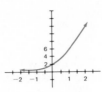

83. (a) $b^{3/5}$ (b) $x^{1/2}$ (c) $(a + b)^{2/3}$
85. (a) 72 (b) $3 \cdot 2^{1/6}$ (c) $a^{17/12}$ (d) $a^{5/6}$ (e) $a^3 + 2 + \dfrac{1}{a^3}$ (f) $\dfrac{a}{1 + a^2}$ (g) $a^{1/2}b^{1/12}$ (h) $a^8/b^{14/3}$ (i) 8
(j) $4a^{3/2}b^{9/4}$ (k) $3^{3/2}$ (l) $a - b$
87. (a) $-\frac{1}{2}$ (b) $-1; 2$ (c) All reals. (d) $-4; 3$ (e) 1; 8 (f) 2

89.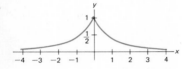

91. The graph of $y = f(x) = a^x$ has the x-axis as a horizontal asymptote. Also f is not a constant function. The graph of a nonconstant polynomial does not have a horizontal asymptote.

PROBLEM SET 6-3 (Page 242)

1. (a) Decays. (b) Grows. (c) Grows. (d) Decays. **3.** (a) 4.66095714 (b) 17.00006441
(c) 4801.02063 (d) 9750.87832 **5.** 1480 **7.** (a) 5.384 billion (b) 6.562 billion (c) 23.772 billion
9. (a) \$185.09 (b) \$247.60 **11.** (a) \$76,035.83 (b) \$325,678.40 **13.** $p(1 + r/100)^n$ **15.** \$7401.22
17. \$7102.09 **19.** \$7305.57 **21.** $\$1000(1 + .08/12)^{120} = \$2219.64$ **23.** 8100; 5400; 3600; 2400; 1600
25. 800 **27.** (a) 8680; 22,497 (b) About 44 years. **29.** (a) \$146.93 (b) \$148.59 (c) \$148.98 (d) \$149.18
31. (a) About 9 years. (b) About 11 years. **33.** (a) $k \approx .0005917$ (b) 10.76 milligrams **35.** 2270 years
37. \$320,057,300

PROBLEM SET 6-4 (Page 251)

1. $\log_4 64 = 3$ **3.** $\log_{27} 3 = \frac{1}{3}$ **5.** $\log_4 1 = 0$ **7.** $\log_{125}(1/25) = -\frac{2}{3}$ **9.** $\log_{10} a = \sqrt{3}$
11. $\log_{10} \sqrt{3} = a$ **13.** $5^4 = 625$ **15.** $4^{3/2} = 8$ **17.** $10^{-2} = .01$ **19.** $c^1 = c$ **21.** $c^y = Q$ **23.** 2
25. -1 **27.** 1/3 **29.** -4 **31.** 0 **33.** $\frac{4}{3}$ **35.** 2 **37.** $\frac{1}{27}$ **39.** -2.9 **41.** 49 **43.** .778
45. 1.204 **47.** $-.602$ **49.** 1.380 **51.** $-.051$ **53.** .699 **55.** 1.5314789 **57.** $.08990511 - 2$
59. 3.9878003 **61.** $\log_{10}[(x + 1)^3(4x + 7)]$ **63.** $\log_2[8x(x + 2)^3/(x + 8)^2]$ **65.** $\log_6(\sqrt{x}\sqrt[3]{x^3 + 3})$
67. 47 **69.** $-\frac{11}{4}$ **71.** $\frac{16}{7}$ **73.** 5; 2 is extraneous. **75.** 7 **77.** 4.0111687 **79.** 2.0446727
81. (a) 2 (b) $\frac{1}{2}$ (c) 125 (d) 32 (e) 10 (f) 4 **83.** $\frac{8}{9}$
85. (a) 13 (b) -5 (c) No solution. (d) 20 (e) 3 (f) 16 **87.** (a) $y = x/(x - 1)$ (b) $y = \frac{1}{2}(a^x + a^{-x})$
89. $x = 10^c$, $\log_2 x = c \log_2 10$, $\log_2 x = (\log_{10} x)(\log_2 10)$

91. By Problem 90, $\log_a a = \log_b a \cdot \log_a b$, or $1 = \log_b a \cdot \log_a b$. Therefore, $\log_a b = \dfrac{1}{\log_b a}$.

93.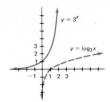

95.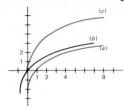

PROBLEM SET 6-5 (Page 259)

1. 1 **3.** 0 **5.** $\frac{1}{2}$ **7.** -3 **9.** 3.5 **11.** $-.2$ **13.** -7.5 **15.** 4.787 **17.** 6.537 **19.** .182
21. 9.1 **23.** .9 **25.** 90 **27.** 1.4609379 **29.** -2.0635682 **31.** -1.8411881 **33.** 50833303
35. 11.818915 **37.** 61.146135 **39.** 8.3311375 **41.** .8824969 **43.** 915.98 **45.** About 2.73.
47. About -0.737. **49.** About 6.84. **51.** Approximately 6.12 years. **53.** Approximately 4.71 years.

55. (a) 5^{10} (b) 9^{10} (c) 10^{20} (d) 10^{1000} **57.** 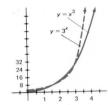 **59.** $y = ba^x$; $a \approx 1.5$, $b \approx 64$

61. $y = bx^a$; $a \approx 4$, $b \approx 12$ **63.** (a) 4.2 (b) 4 (c) $\frac{1}{2}$ **65.** (a) 7 (b) 12.25 (c) $-\frac{1}{2}$ (d) 0 (e) 125 (f) $\frac{1}{3}$
67. (a) $-.349$ (b) $-.823$ (c) $.633, -3.633$ (d) 2.166 (e) 4.560; .219 (f) $e^e \approx 15.154$ (g) $e \approx 2.718$
(h) $e \approx 2.718$ (i) $\pm\sqrt{e+5} \approx \pm 2.778$ **69.** $(\ln 2)/3 \approx .231$ years **71.** $(\ln 2)/240 \approx .00289$

73.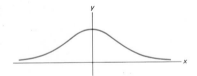

75. $e^{\pi/e - 1} > 1 + \pi/e - 1$, $e^{\pi/e}/e > \pi/e$, $e^{\pi/e} > \pi$, $e^\pi > \pi^e$
77. (a) $100(1 + .01)^{120} \approx \$330.04$ (b) $100(1 + .12/365)^{3650} \approx \$331.95$ (c) $100(1 + .12/(365)(24))^{(3650)(24)} \approx \$332.01$
(d) $100e^{(.12)(10)} \approx \$332.01$

PROBLEM SET 6-6 (Page 267)

1. 4 **3.** -2 **5.** $\frac{11}{2}$ **7.** -15 **9.** 10,000 **11.** .01 **13.** $10^{3/2}$ **15.** $10^{-3/4}$ **17.** .6355
19. 2.1987 **21.** $.5172 - 2$ **23.** 5.7505 **25.** 8.9652 **27.** 32.8 **29.** .0101 **31.** 3.98×10^8
33. 166 **35.** .838 **37.** .7191 **39.** 3.8593 **41.** $.0913 - 3$ **43.** 7.075 **45.** 8184 **47.** .03985
49. (a) $\frac{5}{4}$ (b) $-\frac{4}{3}$ (c) $\frac{5}{6}$ (d) -3 **51.** (a) 2.6926 (b) $.6726 - 2$ (c) 856.4 (d) .001861
53. (a) .0035703 (b) .0000845 **55.** $.2932 - 3$ **57.** .2123 **59.** 16 **61.** About 972.5 miles.

PROBLEM SET 6-7 (Page 272)

1. 128 **3.** .0959 **5.** .0208 **7.** 7.12×10^7 **9.** 3.50 **11.** .983 **13.** 6.05 **15.** 4762
17. 6.143 **19.** 3.530×10^{-6} **21.** 8.90 **23.** 18.2 **25.** 5.19 **27.** $-.5984$ **29.** 2.24
31. (a) .3495 (b) 100.7 (c) .8274 **33.** (a) $\frac{2}{9}$ (b) 2 (c) 1 (d) 3 **35.** About 4.395 hours from now.
37. (a) About 6.17 billion. (b) About the year 2018.

CHAPTER 6. REVIEW PROBLEM SET (Page 275)

1. (a) $-2y^2\sqrt[3]{z}/z^5$ (b) $2xy^2\sqrt[4]{2x}$ (c) $2\sqrt[6]{5}$ (d) $2(\sqrt{x} + \sqrt{y})/(x - y)$ (e) $5\sqrt{2 + x^2}$ (f) $6\sqrt{2}$ **2.** (a) 12 (b) 4
3. (a) $125a^3$ (b) $1/a^{1/2}$ (c) $1/5^{7/4}$ (d) $3y^{13/6}/x^6$ (e) $x - 2x^{1/2}y^{1/2} + y$ (f) $2^{7/6}$

4. 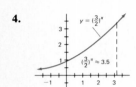 **5.** $(\frac{1}{2})^{81} \approx 4.14 \times 10^{-25}$ **6.** 16 million **7.** $220.80 **8.** (a) 3 (b) $\frac{1}{8}$ (c) 7
(d) 1 (e) $\frac{3}{2}$ (f) 5 (g) 10 (h) 1.14 **9.** $\log_4[(3x + 1)^2(x - 1)/\sqrt{x}]$
10. (a) $\frac{3}{2}$ (b) $\frac{4}{3}$ (c) $\frac{1}{2}$ (d) -2.773 **11.** (a) 1.680 (b) 9.3 (c) 3.517 (d) .9
12. 1.807 **13.** 13.9 years **14.** .1204 **15.** 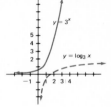 **16.** 3999

PROBLEM SET 7-1 (Page 281)

1. .6600 **3.** .6534 **5.** 3.133 **7.** 12.5° **9.** 66.6° **11.** 69.3° **13.** $16.97 \approx 17$ **15.** $41.34 \approx 41$
17. $66.60 \approx 67$ **19.** $\beta = 48°; a = 23.42 \approx 23; b = 26.01 \approx 26$ **21.** $\alpha = 33.8°; a = 50.8; b = 75.9$
23. $\beta = 50.6°; b = 146; c = 189$ **25.** $c = 15; \alpha = 36.9°; \beta = 53.1°$ **27.** $b = 30; \alpha = 53.1°; \beta = 36.9°$
29. $\alpha = 26.7°; \beta = 63.3°; b = 29.0$ **31.** $\alpha = 32.9°; \beta = 57.1°; c = 17.5$ **33.** 14.6° **35.** 7.0°
37. 31.2 feet **39.** (a) .25862 (b) 6.6568 (c) 17.445 **41.** 725 feet **43.** 65.369 **45.** 364 feet
47. 448 meters **49.** 41.77° **51.** $P = 24; A = 24\sqrt{3}$ **53.** $\sqrt{3}$

PROBLEM SET 7-2 (Page 288)

1. $2\pi/3$ **3.** $4\pi/3$ **5.** $7\pi/6$ **7.** $7\pi/4$ **9.** 3π **11.** $-7\pi/3$ **13.** $8\pi/9$ **15.** $\frac{1}{9}$ **17.** 240°
19. $-120°$ **21.** 540° **23.** 259.0° **25.** 18.2° **27.** (a) 2 (b) 3.14 **29.** (a) 3 centimeters (b) 5.5 inches
31. II **33.** III **35.** II **37.** IV **39.** $16\pi/3 \approx 16.76$ feet per second.
41. $320\pi \approx 1005.3$ inches per minute. **43.** (a) -8π (b) 5π (c) $\frac{1}{3}$
45. (a) 25.5 centimeters (b) .0327 centimeters (c) 37.83 centimeters
47. $9600\pi \approx 30,159$ centimeters **49.** $330\pi \approx 1037$ miles per hour. **51.** 8.6×10^5 miles
53. $\frac{33\pi}{20} \approx 5.184$ hours **55.** $264\pi \approx 829$ miles **57.** $7\pi \approx 21.99$ square inches. **59.** 130

PROBLEM SET 7-3 (Page 295)

1. $(\sqrt{3}/2, \frac{1}{2})$ **3.** $(-\sqrt{2}/2, \sqrt{2}/2)$ **5.** $(-\sqrt{2}/2, -\sqrt{2}/2)$ **7.** $(-\sqrt{3}/2, \frac{1}{2})$ **9.** $-\sqrt{2}/2$
11. $\sqrt{2}/2$ **13.** $-\sqrt{2}/2$ **15.** $-\frac{1}{2}$ **17.** 1 **19.** 0 **21.** $-\sqrt{3}/2$ **23.** $\frac{1}{2}$ **25.** $-\sqrt{2}/2$

27. $\frac{1}{2}$ **29.** $-\frac{1}{2}$ **31.** $-\sqrt{3}/2$ **33.** $-.95557; -.29476$ **35.** (a) $(1/\sqrt{5}, 2/\sqrt{5})$ (b) $2/\sqrt{5}, 1/\sqrt{5}$
37. (a) $\sin(\pi + t) = -y = -\sin t$ (b) $\cos(\pi + t) = -x = -\cos t$
39. (a) Negative. (b) Negative. (c) Positive. (d) Positive. (e) Positive. (f) Negative.
41. (a) $\pm\sqrt{3}/2$ (b) $7\pi/6; 11\pi/6$
43. (a) $\pi/4; 5\pi/4$ (b) $\pi/6 < t < \pi/3, 2\pi/3 < t < 5\pi/6$ (c) $0 \le t \le \pi/3; 2\pi/3 \le t \le 4\pi/3;$
$5\pi/3 \le t < 2\pi$ (d) $0 \le t < \pi/4; 3\pi/4 < t < 5\pi/4; 7\pi/4 < t < 2\pi$ **45.** (a) $\frac{3}{5}$ (b) $-\frac{7}{25}$
47. (a) $\frac{3}{5}$ (b) $\frac{4}{5}$ (c) $\frac{4}{5}$ (d) $\frac{4}{5}$ (e) $\frac{3}{5}$ (f) $-\frac{4}{5}$
49. (a) Period 1 (b) Period $\frac{1}{3}$ (c) Not periodic (d) Period 1   **51.** 0

PROBLEM SET 7-4   (Page 301)

1. (a) $-\frac{4}{3}$ (b) $-\frac{3}{4}$ (c) $-\frac{5}{3}$ (d) $\frac{5}{4}$ **3.** $-\sqrt{5}/2; \frac{3}{5}\sqrt{5}$ **5.** $\sqrt{3}/3$ **7.** $2\sqrt{3}/3$ **9.** 1
11. $2\sqrt{3}/3$ **13.** $-\sqrt{3}/2$ **15.** $\sqrt{3}$ **17.** 0   **19.** $-\sqrt{3}/3$ **21.** -2
23. (a) $\pi/2; 3\pi/2; 5\pi/2; 7\pi/2$ (b) $\pi/2; 3\pi/2; 5\pi/2; 7\pi/2$ (c) $0; \pi; 2\pi; 3\pi; 4\pi$ (d) $0; \pi; 2\pi; 3\pi; 4\pi$
25. $-12/13; -12/5; 13/5$ **27.** $-2\sqrt{5}/5; 2; -\sqrt{5}$ **29.** $\frac{3}{5}; \frac{5}{4}$ **31.** $-\frac{12}{13}; -\frac{12}{5}$ **33.** $(\frac{5}{13}, -\frac{12}{13})$
35. $111.8°$ **37.** (a) $-2\sqrt{3}/3$ (b) $\sqrt{3}$ (c) $\sqrt{2}$ (d) -1 (e) -2 (f) 1
39. (a) $\frac{24}{25}$ (b) $\frac{-7}{25}$ (c) $\frac{-24}{7}$ (d) $\frac{25}{24}$ (e) $\frac{-24}{7}$ (f) $\frac{-25}{7}$ **41.** (a) $3\pi/4; 7\pi/4$ (b) $\pi/4; 7\pi/4$ (c) $\pi/2; 3\pi/2$
43. (a) 1 (b) $\sin\theta - 1/\cos\theta$ (c) $1 + 2\sin\theta\cos\theta$ (d) $1/\sin\theta$ (e) $\cos\theta + \sin\theta$ (f) $-(\sin^2\theta + 1)/(\cos^2\theta)$
45. (a) $\tan(t + \pi) = \sin(t + \pi)/\cos(t + \pi) = (-\sin t)/(-\cos t) = \tan t$
(b) $\cot(t + \pi) = \cos(t + \pi)/\sin(t + \pi) = (-\cos t)/(-\sin t) = \cot t$
(c) $\sec(t + \pi) = 1/\cos(t + \pi) = 1/(-\cos t) = -1/(\cos t) = -\sec t$
(d) $\csc(t + \pi) = 1/\sin(t + \pi) = 1/(-\sin t) = -1/(\sin t) = -\csc t$
47. $\frac{119}{169}$ **49.** -10 **51.** (a) -1.3764 (b) $.3153$ **53.** 428.98 centimeters

PROBLEM SET 7-5   (Page 306)

1. $.98185$ **3.** $.7337$ **5.** $.93309$ **7.** $.9291$ **9.** 1.30 **11.** $.40$ **13.** 1.10 **15.** 1.06 **17.** 1.12
19. $.50$ **21.** $3\pi/8$ **23.** $\pi/3$ **25.** $.24$ **27.** $.24$ **29.** $\pi/2$ **31.** $.15023$ **33.** 5.4707
35. $.84147$ **37.** -1.2885 **39.** $-.82534$ **41.** $1.25; 1.89$ **43.** $1.65; 4.63$ **45.** $1.37; 4.51$
47. $1.84; 4.98$ **49.** $40.4°$ **51.** $11.3°$ **53.** $80.2°$ **55.** $.4051$ **57.** $-.1962$ **59.** $.4051$ **61.** $.15126$
63. $.9657$ **65.** $21.3°; 158.7°$ **67.** $26.3°; 206.3°$ **69.** $155.3°; 204.7°$
71. (a) $.79608$ (b) $-.79560$ (c) -1.5574 (d) $-.7513$ (e) -1.2349 (f) $-.9877$
73. (a) $.9999997$ (b) $.744399$ (c) 1.2338651
75. (a) $.679996; 2.461597$ (b) $1.222007; 5.061178$ (c) $1.878966; 5.020558$
77. (a) ϕ (b) $90° - \phi$ (c) $90° - \phi$ **79.** $-.514496$ **81.** $\phi \approx 126.9°$

PROBLEM SET 7-6   (Page 314)

1.

t	0	$\frac{\pi}{6}$	$\frac{\pi}{4}$	$\frac{\pi}{3}$	$\frac{\pi}{2}$	$\frac{3\pi}{4}$	π	$\frac{5\pi}{4}$	$\frac{3\pi}{2}$	$\frac{7\pi}{4}$	2π
$\cos t$	1	$\frac{\sqrt{3}}{2}$	$\frac{\sqrt{2}}{2}$	$\frac{1}{2}$	0	$-\frac{\sqrt{2}}{2}$	-1	$-\frac{\sqrt{2}}{2}$	0	$\frac{\sqrt{2}}{2}$	1

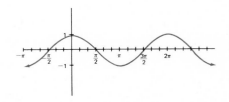

3.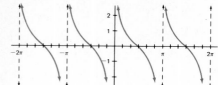

5. $\sec(t + 2\pi) = 1/\cos(t + 2\pi) = 1/\cos t = \sec t$

7. Domain: $\{t : t \neq \pi/2 + k\pi, k \text{ any integer}\}$; range: $\{y : |y| \geq 1\}$.　**9.** π; 2π　**11.** $\cot(-t) = -\cot t$

13. 3; 2π 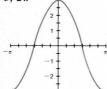　**15.** 1; 2π 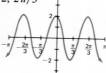　**17.** 1; $\pi/2$ 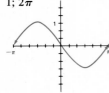　**19.** 2; 4π

21. 2; $2\pi/3$ 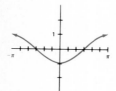　**23.** 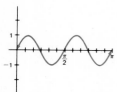　**25.** 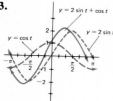　**27.**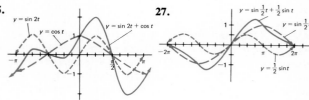

29. 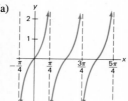　**31.** 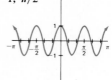　**33.** 1, 2π; 1, $\pi/2$

35. (a)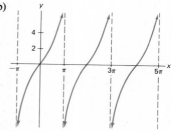

Period: $\pi/2$

(b)

Period: 2π

37.

39.

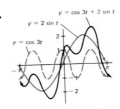

41.

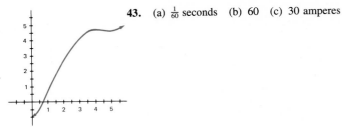

43. (a) $\frac{1}{60}$ seconds   (b) 60   (c) 30 amperes

45. (a) $1/\pi, 1/2\pi, 1/3\pi, 1/4\pi, \ldots$ (b) $1, -1, 1, -1, \ldots$ (c)

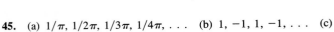

CHAPTER 7. REVIEW PROBLEM SET   (Page 318)

1. (a) $\beta = 42.9°; a = 27.0; b = 25.1$ (b) $\alpha = 46.7°; \beta = 43.3°; b = 393$ **2.** 5.01 feet   **3.** .576; 405°
4. 18,850 centimeters   **5.** (a) $-\frac{1}{2}$ (b) $\sqrt{3}/2$ (c) 1 (d) $\frac{1}{2}$ **6.** (a) .7771 (b) $-.6157$ (c) $-.5635$ (d) .5258
7. (a) $-\sin t$ (b) $\sin t$ (c) $-\sin t$ (d) $\sin t$ **8.** (a) $\{t : 0 \le t < \pi/2 \text{ or } 3\pi/2 < t \le 2\pi\}$
(b) $\{t : 0 \le t < \pi/4 \text{ or } 3\pi/4 < t < 5\pi/4 \text{ or } 7\pi/4 < t \le 2\pi\}$ **9.** (a) $\frac{5}{12}$ (b) $-\frac{13}{5}$ **10.** $-2/\sqrt{21}$

11.

12.

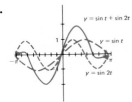

13. $\{y : -1 \le y \le 1\}; \{y : y \le -1 \text{ or } y \ge 1\}$

PROBLEM SET 8-1   (Page 324)

1. (a) $1 - \sin^2 t$ (b) $\sin t$ (c) $\sin^2 t$ (d) $(1 - \sin^2 t)/\sin^2 t$ **3.** (a) $1/\tan^2 t$ (b) $1 + \tan^2 t$ (c) $\tan t$ (d) 3
5. $\cos t \sec t = \cos t(1/\cos t) = 1$ **7.** $\tan x \cot x = \tan x(1/\tan x) = 1$ **9.** $\cos y \csc y = \cos y(1/\sin y) = \cot y$
11. $\cot \theta \sin \theta = (\cos \theta/\sin\theta) \sin \theta = \cos \theta$ **13.** $\tan u/\sin u = (\sin u/\cos u)(1/\sin u) = 1/\cos u$
15. $(1 + \sin z)(1 - \sin z) = 1 - \sin^2 z = \cos^2 z = 1/\sec^2 z$
17. $(1 - \sin^2 x)(1 + \tan^2 x) = \cos^2 x \sec^2 x = \cos^2 x(1/\cos^2 x) = 1$
19. $\sec t - \sin t \tan t = 1/(\cos t) - (\sin^2 t)/(\cos t) = (1 - \sin^2 t)/(\cos t) = (\cos^2 t)/(\cos t) = \cos t$
21. $(\sec^2 t - 1)/(\sec^2 t) = 1 - 1/(\sec^2 t) = 1 - \cos^2 t = \sin^2 t$
23. $\cos t(\tan t + \cot t) = \sin t + (\cos^2 t)/(\sin t) = (\sin^2 t + \cos^2 t)/(\sin t) = \csc t$

25. $\sin t = (1 - \cos^2 t)^{1/2}$; $\tan t = (1 - \cos^2 t)^{1/2}/\cos t$; $\cot t = \cos t/(1 - \cos^2 t)^{1/2}$; $\sec t = 1/\cos t$; $\csc t = 1/(1 - \cos^2 t)^{1/2}$

27. $\cos t = -3/5$; $\tan t = -4/3$; $\cot t = -3/4$; $\sec t = -5/3$; $\csc t = 5/4$

29. $\dfrac{\sec t - 1}{\tan t} \cdot \dfrac{\sec t + 1}{\sec t + 1} = \dfrac{\sec^2 t - 1}{\tan t(\sec t + 1)} = \dfrac{\tan^2 t}{\tan t(\sec t + 1)} = \dfrac{\tan t}{\sec t + 1}$

31. $\dfrac{\tan^2 x}{\sec x + 1} = \dfrac{\sec^2 x - 1}{\sec x + 1} = \dfrac{(\sec x - 1)(\sec x + 1)}{\sec x + 1} = \sec x - 1 = \dfrac{1 - \cos x}{\cos x}$

33. $\dfrac{\sin t + \cos t}{\tan^2 t - 1} \cdot \dfrac{\cos^2 t}{\cos^2 t} = \dfrac{(\sin t + \cos t) \cos^2 t}{\sin^2 t - \cos^2 t} = \dfrac{\cos^2 t}{\sin t - \cos t}$

35. (a) $2/\sin^2 x$ (b) $(2 + 2 \tan^2 x)/\tan^2 x$

37. $(1 + \tan^2 t)(\cos t + \sin t) = \sec^2 t \cos t + \sec^2 t \sin t = \sec t + \sec t \tan t = \sec t(1 + \tan t)$

39. $2 \sec^2 y - 1 = \dfrac{2}{\cos^2 y} - 1 = \dfrac{2 - \cos^2 y}{\cos^2 y} = \dfrac{1 + \sin^2 y}{\cos^2 y}$

41. $\dfrac{\sin z}{\sin z + \tan z} = \dfrac{\sin z}{\sin z + \sin z/\cos z} = \dfrac{1}{1 + 1/\cos z} = \dfrac{\cos z}{\cos z + 1}$

43. $(\csc t + \cot t)^2 = \left(\dfrac{1}{\sin t} + \dfrac{\cos t}{\sin t}\right)^2 = \dfrac{(1 + \cos t)^2}{1 - \cos^2 t} = \dfrac{1 + \cos t}{1 - \cos t}$

45. $\dfrac{1 + \tan x}{1 - \tan x} = \dfrac{1 + \sin x/\cos x}{1 - \sin x/\cos x} = \dfrac{\cos x + \sin x}{\cos x - \sin x}$

47. $(\sec t + \tan t)(\csc t - 1) = \left(\dfrac{1 + \sin t}{\cos t}\right)\left(\dfrac{1 - \sin t}{\sin t}\right) = \dfrac{\cos^2 t}{\cos t \sin t} = \cot t$

49. $\dfrac{\cos^3 t + \sin^3 t}{\cos t + \sin t} = \cos^2 t - \cos t \sin t + \sin^2 t = 1 - \sin t \cos t$

51. $\left(\dfrac{1 - \cos \theta}{\sin \theta}\right)^2 = \dfrac{(1 - \cos \theta)^2}{1 - \cos^2 \theta} = \dfrac{1 - \cos \theta}{1 + \cos \theta}$

53. $(\csc t - \cot t)^4(\csc t + \cot t)^4 = (\csc^2 t - \cot^2 t)^4 = 1^4 = 1$

55. $\sin^6 u + \cos^6 u = (1 - \cos^2 u)^3 + \cos^6 u = 1 - 3 \cos^2 u + 3 \cos^4 u$
$$= 1 - 3 \cos^2 u(1 - \cos^2 u) = 1 - 3 \cos^2 u \sin^2 u$$

57. $\cot 3x = \dfrac{1}{\tan 3x} = \dfrac{1 - 3 \tan^2 x}{3 \tan x - \tan^3 x}\left(\dfrac{\cot^3 x}{\cot^3 x}\right) = \dfrac{\cot^3 x - 3 \cot x}{3 \cot^2 x - 1} = \dfrac{3 \cot x - \cot^3 x}{1 - 3 \cot^2 x}$

PROBLEM SET 8-2   (Page 330)

1. (a) $(\sqrt{2} + 1)/2 \approx 1.21$ (b) $(\sqrt{2}\sqrt{3} + \sqrt{2})/4 \approx .97$
3. (a) $(\sqrt{2} - \sqrt{3})/2 \approx -.16$ (b) $(\sqrt{2}\sqrt{3} + \sqrt{2})/4 \approx .97$
5. $\sin(t + \pi) = \sin t \cos \pi + \cos t \sin \pi = -\sin t$
7. $\sin(t + 3\pi/2) = \sin t \cos(3\pi/2) + \cos t \sin(3\pi/2) = -\cos t$
9. $\sin(t - \pi/2) = \sin t \cos(\pi/2) - \cos t \sin(\pi/2) = -\cos t$
11. $\cos(t + \pi/3) = \cos t \cos(\pi/3) - \sin t \sin(\pi/3) = (1/2) \cos t - (\sqrt{3}/2) \sin t$
13. $\cos 2$ **15.** $\sin \pi$ **17.** $\cos 60°$ **19.** $\sin \alpha$ **21.** $\frac{56}{65}$; $-\frac{33}{65}$; in quadrant II.
23. $-(1 + 3\sqrt{3})/(2\sqrt{10}) \approx -.9797$; $(-3 + \sqrt{3})/(2\sqrt{10}) \approx -.2005$; quadrant III.
25. $\tan(s - t) = \tan(s + (-t)) = (\tan s + \tan(-t))/(1 - \tan s \tan(-t)) = (\tan s - \tan t)/(1 + \tan s \tan t)$
27. $\tan(t + \pi/4) = (\tan t + \tan \pi/4)/(1 - \tan t \tan \pi/4) = (1 + \tan t)/(1 - \tan t)$
29. (a) $-(\cos t + \sqrt{3} \sin t)/2$ (b) $(\sqrt{3} \cos t + \sin t)/2$

31. (a) $\sqrt{5}/3$ (b) $-2\sqrt{2}/3$ (c) $(4\sqrt{2} - \sqrt{5})/9$ (d) $(2\sqrt{10} - 2)/9$ (e) $(-\frac{2}{3})(\sqrt{2} + \sqrt{5})$ (f) $4\sqrt{2}/9$
33. (a) $\sqrt{3}/2$ (b) $-\sqrt{3}/2$ (c) $\sin 1 \approx .84147$

35. (a) $\sin(x+y)\sin(x-y) = (\sin x \cos y + \cos x \sin y)(\sin x \cos y - \cos x \sin y)$
$= \sin^2 x \cos^2 y - \cos^2 x \sin^2 y = \sin^2 x(1 - \sin^2 y) - \cos^2 x \sin^2 y$
$= \sin^2 x - \sin^2 y(\sin^2 x + \cos^2 x) = \sin^2 x - \sin^2 y$

(b) $\dfrac{\sin(x+y)}{\cos(x-y)} = \dfrac{\sin x \cos y + \cos x \sin y}{\cos x \cos y + \sin x \sin y} = \dfrac{\dfrac{\sin x \cos y}{\cos x \cos y} + \dfrac{\cos x \sin y}{\cos x \cos y}}{\dfrac{\cos x \cos y}{\cos x \cos y} + \dfrac{\sin x \sin y}{\cos x \cos y}} = \dfrac{\tan x + \tan y}{1 + \tan x \tan y}$

(c) $\dfrac{\cos 5t}{\sin t} - \dfrac{\sin 5t}{\cos t} = \dfrac{\cos 5t \cos t - \sin 5t \sin t}{\sin t \cos t} = \dfrac{\cos 6t}{\sin t \cos t}$

37.

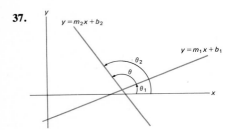

Since $\theta = \theta_2 - \theta_1$,

$$\tan \theta = \frac{\tan \theta_2 - \tan \theta_1}{1 + \tan \theta_2 \tan \theta_1} = \frac{m_2 - m_1}{1 + m_1 m_2}$$

39. (a) $\frac{1}{2}[\cos(s+t) + \cos(s-t)] = \frac{1}{2}[\cos s \cos t - \sin s \sin t + \cos s \cos t + \sin s \sin t] = \cos s \cos t$
(b) $-\frac{1}{2}[\cos s \cos t - \sin s \sin t - \cos s \cos t - \sin s \sin t] = \sin s \sin t$
(c) $\frac{1}{2}[\sin s \cos t + \cos s \sin t + \sin s \cos t - \cos s \sin t] = \sin s \cos t$
(d) $\frac{1}{2}[\sin s \cos t + \cos s \sin t - \sin s \cos t + \cos s \sin t] = \cos s \sin t$

41. (a) $(1 - \sqrt{3})/4$ (b) $-\sqrt{2}/2$ (c) $\dfrac{1 + \sqrt{2} + \sqrt{3} + \sqrt{6}}{2}$

43. $\tan(\alpha + \beta) = (\tan \alpha + \tan \beta)/(1 - \tan \alpha \tan \beta) = (\frac{1}{3} + \frac{1}{2})/(1 - \frac{1}{3} \cdot \frac{1}{2}) = 1 = \tan \gamma$. Thus $\alpha + \beta = \gamma$.

PROBLEM SET 8-3 (Page 337)

1. (a) $\frac{24}{25}$ (b) $\frac{7}{25}$ (c) $3\sqrt{10}/10$ (d) $\sqrt{10}/10$ **3.** $\sin 10t$ **5.** $\cos 3t$ **7.** $\cos(y/2)$
9. $\cos 1.2t$ **11.** $-\cos(\pi/4)$ **13.** $\cos^2(x/2)$ **15.** $\sin^2 2\theta$
17. (a) $\sin(\pi/8) = \sqrt{(1 - \cos \pi/4)/2} \approx .3827$ (b) $\cos 112.5° = -\sqrt{(1 + \cos 225°)/2} \approx -.3827$
19. $\tan 2t = \tan(t + t) = (\tan t + \tan t)/(1 - \tan^2 t) = 2 \tan t/(1 - \tan^2 t)$

21. $\tan \dfrac{t}{2} = \dfrac{\sin t/2}{\cos t/2} = \dfrac{\pm\sqrt{(1 - \cos t)/2}}{\pm\sqrt{(1 + \cos t)/2}} = \pm\sqrt{\dfrac{1 - \cos t}{1 + \cos t}}$

23. $\cos 3t = \cos(2t + t) = \cos 2t \cos t - \sin 2t \sin t = (2 \cos^2 t - 1) \cos t - 2 \sin^2 t \cos t$
$= (2 \cos^2 t - 1) \cos t - 2(1 - \cos^2 t) \cos t = 4 \cos^3 t - 3 \cos t$
25. $\csc 2t + \cot 2t = (1 + \cos 2t)/(\sin 2t) = (2 \cos^2 t)/(2 \sin t \cos t) = \cot t$
27. $\sin \theta/(1 - \cos \theta) = 2 \sin(\theta/2) \cos(\theta/2)/2 \sin^2(\theta/2) = \cot(\theta/2)$
29. $2 \tan \alpha/(1 + \tan^2 \alpha) = 2 \tan \alpha/\sec^2 \alpha = 2(\sin \alpha/\cos \alpha)\cos^2 \alpha = 2 \sin \alpha \cos \alpha = \sin 2\alpha$
31. $\sin 4\theta = 2 \sin 2\theta \cos 2\theta = 2(2 \sin \theta \cos \theta)(2 \cos^2 \theta - 1) = 4 \sin \theta(2 \cos^3 \theta - \cos \theta)$
33. (a) $\sin x$ (b) $\cos 6t$ (c) $-\cos(y/2)$ (d) $-\sin^2 2t$ (e) $\tan^2 2t$ (f) $\tan 3y$
35. (a) $120/169$ (b) $-2\sqrt{13}/13$ (c) $-\frac{3}{2}$ **37.** $\cos^4 z - \sin^4 z = (\cos^2 z + \sin^2 z)(\cos^2 z - \sin^2 z) = 1 \cdot \cos 2z$
39. $1 + (1 - \cos 8t)/(1 + \cos 8t) = 1 + \tan^2 4t = \sec^2 4t$

41. $\tan\frac{\theta}{2} - \sin\theta = (\sin\theta)/(1 + \cos\theta) - \sin\theta = (\sin\theta - \sin\theta - \sin\theta\cos\theta)/(1 + \cos\theta) = (-\sin\theta\cos\theta)/(1 + \cos\theta) = -(\sin\theta)/(\sec\theta + 1)$

43. $(3\cos t - \sin t)(\cos t + 3\sin t) = 3\cos^2 t - 3\sin^2 t + 8\sin t\cos t = 3\cos 2t + 4\sin 2t$

45. $2(\cos 3x\cos x + \sin 3x\sin x)^2 = 2\cos^2 2x = 1 + \cos 4x$

47. $\tan 3t = \tan(2t + t) = \dfrac{\tan 2t + \tan t}{1 - \tan 2t\tan t} = \dfrac{\dfrac{2\tan t}{1 - \tan^2 t} + \tan t}{1 - \dfrac{2\tan^2 t}{1 - \tan^2 t}} = \dfrac{3\tan t - \tan^3 t}{1 - 3\tan^2 t}$

49. $\sin^4 u + \cos^4 u = (\sin^2 u + \cos^2 u)^2 - 2\sin^2 u\cos^2 u = 1 - \frac{1}{2}\sin^2 2u = 1 - \frac{1}{2}\cdot(1 - \cos 4u)/2 = \frac{3}{4} + \frac{1}{4}\cos 4u$

51. $\cos^2 x + \cos^2 2x + \cos^2 3x = \dfrac{1 + \cos 2x}{2} + \cos^2 2x + \dfrac{1 + \cos 6x}{2} = 1 + \frac{1}{2}(\cos 2x + \cos 6x) + \cos^2 2x = $
$1 + \cos 4x\cos 2x + \cos^2 2x = 1 + \cos 2x(\cos 4x + \cos 2x) = 1 + \cos 2x(2\cos 3x\cos x) = 1 + 2\cos x\cos 2x\cos 3x$

53. Since $\alpha + \beta + \gamma = 180°$, $2\gamma = 360° - 2\alpha - 2\beta$. Thus $\sin 2\alpha + \sin 2\beta + \sin 2\gamma = \sin 2\alpha + \sin 2\beta - \sin(2\alpha + 2\beta) = \sin 2\alpha + \sin 2\beta - \sin 2\alpha\cos 2\beta - \cos 2\alpha\sin 2\beta = \sin 2\alpha(1 - \cos 2\beta) + \sin 2\beta(1 - \cos 2\alpha) = 2\sin\alpha\cos\alpha(2\sin^2\beta) + 2\sin\beta\cos\beta(2\sin^2\alpha) = 4\sin\alpha\sin\beta(\cos\alpha\sin\beta + \sin\alpha\cos\beta) = 4\sin\alpha\sin\beta\sin(\alpha + \beta) = 4\sin\alpha\sin\beta\sin\gamma$.

55. $(\frac{7}{9}, 4\sqrt{2}/9)$

PROBLEM SET 8-4 (Page 345)

1. $\pi/3$ **3.** $\pi/4$ **5.** 0 **7.** $\pi/3$ **9.** $2\pi/3$ **11.** $\pi/4$ **13.** .2200 **15.** $-.2200$ **17.** .2037
19. (a) .7938 (b) 1.9545 **21.** .3486; 2.7930 **23.** 1.2803; 4.4219 **25.** $\frac{2}{3}$ **27.** 10 **29.** $\pi/3$
31. $\pi/4$ **33.** $\frac{3}{5}$ **35.** $2/\sqrt{5}$ **37.** $\frac{1}{3}$ **39.** $2\pi/3$ **41.** .9666 **43.** .4508 **45.** 2.2913 **47.** $\frac{24}{25}$
49. $\frac{7}{25}$ **51.** $\frac{56}{65}$ **53.** $(6 + \sqrt{35})/12 \approx .993$ **55.** $\tan(\sin^{-1} x) = \sin(\sin^{-1} x)/\cos(\sin^{-1} x) = x/\sqrt{1 - x^2}$
57. $\tan(2\tan^{-1} x) = 2\tan(\tan^{-1} x)/[1 - \tan^2(\tan^{-1} x)] = 2x/(1 - x^2)$ **59.** $\cos(2\sec^{-1} x) = \cos[2\cos^{-1}(1/x)] = 2/x^2 - 1$
61. (a) $-\pi/3$ (b) $-\pi/3$ (c) $2\pi/3$ **63.** (a) 43 (b) $\frac{12}{13}$ (c) $7\sqrt{2}/10$ (d) $(4 - 6\sqrt{2})/15$
65. (a) $\pm\sqrt{7}/4$ (b) $\pm.9$ (c) $\frac{11}{6}$ (d) 1; 2 **67.** (a) $\pi/2$ (b) $\pi/4$ (c) $-\pi/4$ (d) $-\pi/2$ (e) π (f) $\pi/4$
69. (a) $\sin^{-1}(x/5)$ (b) $\tan^{-1}(x/3)$ (c) $\sin^{-1}(3/x)$ (d) $\tan^{-1}(3/x) - \tan^{-1}(1/x)$
71. (a) .6435011 (b) $-.3046927$ (c) .6435011 (d) 2.6905658 **73.** Show that the tangent of both sides is 120/119.
75. (a) $\theta = \tan^{-1}(6/b) - \tan^{-1}(2/b)$ (b) 22.83° (c) $2\sqrt{3}$

PROBLEM SET 8-5 (Page 353)

1. $\{0, \pi\}$ **3.** $\{3\pi/2\}$ **5.** No solution. **7.** $\{5\pi/6, 7\pi/6\}$ **9.** $\{\pi/4, 3\pi/4, 5\pi/4, 7\pi/4\}$
11. $\{\pi/4, 2\pi/3, 3\pi/4, 4\pi/3\}$ **13.** $\{0, \pi, 3\pi/2\}$ **15.** $\{0, \pi/3, \pi, 4\pi/3\}$ **17.** $\{\pi/3, \pi, 5\pi/3\}$
19. $\{.3649, 1.2059, 3.5065, 4.3475\}$ **21.** $\{0, \pi/2\}$ **23.** $\{\pi/6, \pi/2\}$ **25.** $\{0\}$
27. $\{\pi/6 + 2k\pi, 5\pi/6 + 2k\pi: k \text{ is an integer}\}$ **29.** $\{k\pi: k \text{ is an integer}\}$ **31.** $\{\pi/6 + k\pi, 5\pi/6 + k\pi: k \text{ is an integer}\}$
33. $\{0, \pi/2, \pi, 3\pi/2\}$ **35.** $\{\pi/8, 5\pi/8, 9\pi/8, 13\pi/8\}$ **37.** $\{3\pi/8, 7\pi/8, 11\pi/8, 15\pi/8\}$
39. $\{0, \pi/6, 5\pi/6, \pi\}$ **41.** $\{.9553, 2.1863, 4.0969, 5.3279\}$ **43.** $\{\pi/4, 5\pi/4\}$ **45.** $\{0, \pi/6, 5\pi/6, \pi, 7\pi/6, 11\pi/6\}$
47. $\{.3076, 2.8340\}$ **49.** $\{2\pi/3, 4\pi/3\}$ **51.** $\{2\pi/3, 5\pi/6, 5\pi/3, 11\pi/6\}$ **53.** $\{3\pi/2, 5.6397\}$
55. $\{\pi/4, 3\pi/4, 5\pi/4, 7\pi/4\}$ **57.** (a) 15 inches (b) $\tan\theta = \frac{2}{3}$ (c) 33.7° **59.** (a) 26.6° (b) 10.3°
61. $\{k\pi/3, 2\pi/3 + 2k\pi, 4\pi/3 + 2k\pi: k \text{ is an integer}\}$ **63.** $\{\pi/6, \pi/3, 2\pi/3, 5\pi/6\}$

CHAPTER 8. REVIEW PROBLEM SET (Page 356)

1. (a) $\cot\theta\cos\theta = \cos^2\theta/\sin\theta = (1 - \sin^2\theta)/\sin\theta = 1/\sin\theta - \sin\theta = \csc\theta - \sin\theta$
(b) $(\cos x\tan^2 x)(\sec x - 1)/(\sec x + 1)(\sec x - 1) = \cos x\tan^2 x(\sec x - 1)/\tan^2 x = 1 - \cos x$
2. (a) $-\sin^3 x/(1 - \sin^2 x)$ (b) $1 - \sin x$ **3.** (a) $\cos 45° = \sqrt{2}/2$ (b) $\sin 90° = 1$ (c) $\cos 45° = \sqrt{2}/2$

4. (a) $24/25 = .96$ (b) $3/\sqrt{10} \approx .95$ **5.** (a) $\sin 2t \cos t - \cos 2t \sin t = \sin(2t - t) = \sin t$
(b) $\sec 2t + \tan 2t = (1 + \sin 2t)/\cos 2t = (\cos t + \sin t)^2/(\cos^2 t - \sin^2 t) = (\cos t + \sin t)/(\cos t - \sin t)$
(c) $\cos(\alpha + \beta)/\cos \alpha \cos \beta = \cos \alpha \cos \beta/\cos \alpha \cos \beta - \sin \alpha \sin \beta/\cos \alpha \cos \beta = 1 - \tan \alpha \tan \beta = \tan \alpha(\cot \alpha - \tan \beta)$
6. (a) $\{5\pi/6, 7\pi/6\}$ (b) $\{0, \pi, 7\pi/6, 11\pi/6\}$ (c) $\{0\}$ (d) $\{\pi/2, \pi\}$ (e) $\{\pi/6, 5\pi/6, 3\pi/2\}$
7. $-\pi/2 \le t \le \pi/2; 0 \le t \le \pi; -\pi/2 < t < \pi/2$ **8.** (a) $-\pi/3$ (b) $5\pi/6$ (c) $-\pi/3$ (d) 6 (e) π (f) $\sqrt{5}/3$
(g) $-.02$ (h) $120/169$ **9.** See the graph in the text on page 343. **10.** -1.57
11. $\tan(\arctan \frac{1}{2} + \arctan \frac{1}{3}) = (\frac{1}{2} + \frac{1}{3})/(1 - \frac{1}{6}) = 1$. Since $\arctan \frac{1}{2}$ and $\arctan \frac{1}{3}$ are between 0 and $\pi/2$, their sum cannot be in quadrant III and so must equal $\pi/4$.

PROBLEM SET 9-1 (Page 363)

1. $\gamma = 55.5°; b \approx 20.9; c \approx 17.4$ **3.** $\beta = 56°; a = c \approx 53$ **5.** $\beta \approx 42°; \gamma \approx 23°; c \approx 20$
7. $\beta \approx 18°; \gamma \approx 132°; c \approx 12$ **9.** Two triangles: $\beta_1 \approx 53°, \gamma_1 \approx 97°, c_1 \approx 9.9; \beta_2 \approx 127°, \gamma_2 \approx 23°, c_2 \approx 3.9$
11. 93.7 meters **13.** 44.7° **15.** 192.8 **17.** 265.3 **19.** 78.4° **21.** 694.6 square feet
23. 1769 feet **25.** 40 **27.** $6 r^2 \sin \phi(\cos \phi + \sqrt{3} \sin \phi)$

PROBLEM SET 9-2 (Page 368)

1. $a \approx 12.5; \beta \approx 76°; \gamma \approx 44°$ **3.** $c \approx 15.6; \alpha \approx 26°; \beta \approx 34°$ **5.** $\alpha \approx 44.4°; \beta \approx 57.1°; \gamma \approx 78.5°$
7. $\alpha \approx 30.6°; \beta \approx 52.9°; \gamma \approx 96.5°$ **9.** 98.8 meters **11.** 24 miles **13.** 106°
15. $s = 6, A = \sqrt{6 \cdot 3 \cdot 2 \cdot 1} = 6$ **17.** 18.63 **19.** 41.68° **21.** 42.60 miles
23. $\cos^{-1}(3/4) \approx 41.41°; \frac{1}{2}(\sqrt{3} + 3\sqrt{7})$ **27.** $\sqrt{15}$

PROBLEM SET 9-3 (Page 374)

1. (a)

(b)

(c)

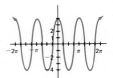

(d)

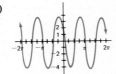

3. (a) $\pi; 4; 0$

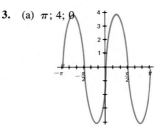

(b) $2\pi; 3, -\pi/8$

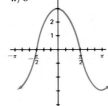

(c) $\pi/2; 1; -\pi/32$

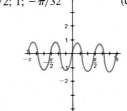

(d) $2\pi/3; 3; \pi/6$

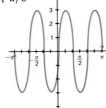

5. 4π; 2; $\pi/8$

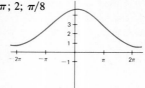

7.

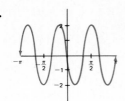

9.

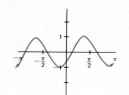

11.

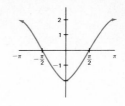

13. $(5 \cos 4t, 5 \sin 4t)$ **15.** $5 \cos 4t, -8 + 5 \sin 4t$

17. (a) $2\pi/5$; 1; 0   (b) 4π; $\frac{3}{2}$; 0   (c) $\pi/2$; 2; $\pi/4$ (d) $2\pi/3$; 4; $-\pi/4$ **19.** 4 feet; after 1.5 seconds

21. $\sin t + \sqrt{25 - \cos^2 t}$ **23.** 156; 55   **25.** (a) 0 (b) 0 (c) 53.53   **27.**

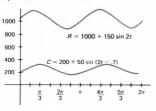

29. (a) $2\sqrt{2} \sin 2t - 2\sqrt{2} \cos 2t$ (b) $-\frac{3}{2}\sqrt{3} \sin 3t + \frac{3}{2} \cos 3t$

33. (a) $5 \sin[2t + \tan^{-1}(4/3)]$ (b) $2\sqrt{3} \sin(4t + 11\pi/6)$ **35.** $\sqrt{2}$; $-\sqrt{2}$

PROBLEM SET 9-4   (Page 381)

1–11.

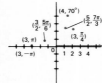

13. $(2\sqrt{2}, 2\sqrt{2})$ **15.** $(-3, 0)$ **17.** $(-5, -5\sqrt{3})$ **19.** $(\sqrt{2}, -\sqrt{2})$

21. $(4, 0)$ **23.** $(2, \pi)$ **25.** $(2\sqrt{2}, \pi/4)$ **27.** $(2\sqrt{2}, 3\pi/4)$ **29.** $(2, -\pi/3)$ **31.** $(2\sqrt{3}, 11\pi/6)$

33.

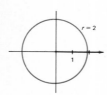

35.

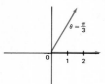

37.

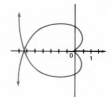

39.

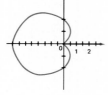

41.

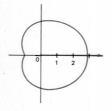

43. $r = 2$ **45.** $r = \tan \theta \sec \theta$ **47.** $y = 2x$ **49.** $(x^2 + y^2)^{3/2} = x^2 - y^2$

51.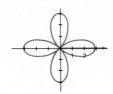

53.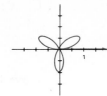

55.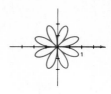

57. (a) $3y - 2x = 5$; a line   (b) $(x - 2)^2 + (y + 3)^2 = 13$; a circle

59.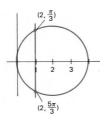

61.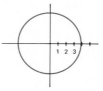

63.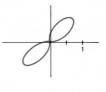

65. (a)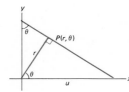

$(2, \pi/3), (2, 5\pi/3)$

(b)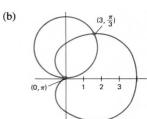

$(0, \pi), (3, \pi/3)$

67. Use the law of cosines; $4\sqrt{5}$.   **69.** $\frac{1}{2}(\beta - \alpha)(b - a)(b + a)$

71.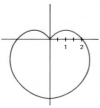

Since $\cos \theta = r/u$ and $\sin \theta = u/4$, it follows
that $r = u \cos \theta = 4 \sin \theta \cos \theta = 2 \sin 2\theta$.
Now compare with Example B.

PROBLEM SET 9-5 (Page 389)

1–11. 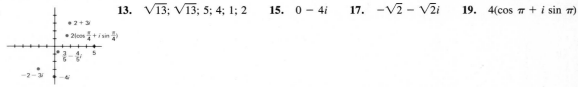   **13.** $\sqrt{13}; \sqrt{13}; 5; 4; 1; 2$ **15.** $0 - 4i$ **17.** $-\sqrt{2} - \sqrt{2}i$ **19.** $4(\cos \pi + i \sin \pi)$

21. $5(\cos 270° + i \sin 270°)$ **23.** $2\sqrt{2}(\cos 315° + i \sin 315°)$ **25.** $4(\cos \pi/6 + i \sin \pi/6)$
27. $6.403(\cos .6747 + i \sin .6747)$ **29.** $6(\cos 210° + i \sin 210°)$ **31.** $\frac{3}{2}(\cos 125° + i \sin 125°)$
33. $\frac{2}{3}(\cos 70° + i \sin 70°)$ **35.** $2(\cos 305° + i \sin 305°)$ **37.** $16 + 0i$ **39.** $-2 + 2\sqrt{3}i$

41. $16(\cos 60° + i \sin 60°)$ **43.** $1(\cos 120° + i \sin 120°)$ **45.** $16(\cos 270° + i \sin 270°)$

47.

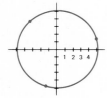

(a) $|-5 + 12i| = 13$ (b) $|-4i| = 4$ (c) $|5(\cos 60° + i \sin 60°)| = 5$

49. (a) $12(\cos 0° + i \sin 0°)$ (b) $2(\cos 135° + i \sin 135°)$ (c) $3(\cos 270° + i \sin 270°)$
(d) $4(\cos 300° + i \sin 300°)$ (e) $8(\cos 30° + i \sin 30°)$ (f) $2(\cos 315° + i \sin 315°)$
51. (a) $12(\cos 160° + i \sin 160°)$ (b) $3(\cos 40° + i \sin 40°)$ (c) $\cos 45° + i \sin 45°$
53. (a) $12 + 5i, -12 + 5i$ (b) $\pm(4\sqrt{2} + 4\sqrt{2}i)$
55. (a) $r^3(\cos 3\theta + i \sin 3\theta)$ (b) $r[\cos(-\theta) + i \sin(-\theta)]$ (c) $r^2(\cos 0 + i \sin 0)$ (d) $\frac{1}{r}[\cos(-\theta) + i \sin(-\theta)]$
(e) $r^{-2}[\cos(-2\theta) + i \sin(-2\theta)]$ (f) $r[\cos(\theta + \pi) + i \sin(\theta + \pi)]$
57. (a) The distance between U and V in the complex plane.   (b) The angle from the positive x-axis to the line joining U and V.
59. (a) -2^8 (b) 0

PROBLEM SET 9-6   (Page 396)

1. $8[\cos(3\pi/4) + i \sin(3\pi/4)]$ **3.** $125(\cos 66° + i \sin 66°)$ **5.** $16(\cos 0° + i \sin 0°)$ **7.** $1 + 0i$
9. $-16\sqrt{3} + 16i$
11. $5(\cos 15° + i \sin 15°)$;
$5(\cos 135° + i \sin 135°)$;
$5(\cos 255° + i \sin 255°)$

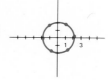

13. $2[\cos(\pi/12) + i \sin(\pi/12)]$; $2[\cos(5\pi/12) + i \sin(5\pi/12)]$;
$2[\cos(9\pi/12) + i \sin(9\pi/12)]$; $2[\cos(13\pi/12) + i \sin(13\pi/12)]$;
$2[\cos(17\pi/12) + i \sin(17\pi/12)]$; $2[\cos(21\pi/12) + i \sin(21\pi/12)]$

15. $\sqrt{2}(\cos 28° + i \sin 28°)$; $\sqrt{2}(\cos 118° + i \sin 118°)$;
$\sqrt{2}(\cos 208° + i \sin 208°)$; $\sqrt{2}(\cos 298° + i \sin 298°)$

17. ±2; $\pm2i$

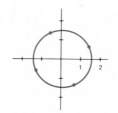

19. $\pm(\sqrt{2} + \sqrt{2}i)$ **21.** $\pm(\sqrt{2} + \sqrt{6}i)$

23. $\pm 1; \pm i$

25. $\cos(k \cdot 36°) + i \sin(k \cdot 36°)$, $k = 0, 1, \ldots, 9$

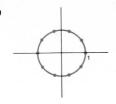

27. (a) $81(\cos 80° + i \sin 80°)$ (b) $90.09(\cos 7.7 + i \sin 7.7)$ (c) $8(\cos 240° + i \sin 240°)$
(d) $16[\cos(5\pi/3) + i \sin(5\pi/3)]$

29. $2(\cos 51° + i \sin 51°)$, $2(\cos 123° + i \sin 123°)$, $2(\cos 195° + i \sin 195°)$, $2(\cos 267° + i \sin 267°)$,
$2(\cos 339° + i \sin 339°)$

31. 1, $\sqrt{2}/2 + (\sqrt{2}/2)i$, i, $-\sqrt{2}/2 + (\sqrt{2}/2)i$, -1, $-\sqrt{2}/2 - (\sqrt{2}/2)i$, $-i$, $\sqrt{2}/2 - (\sqrt{2}/2)i$. Sum $= 0$. Product $= -1$.

33. $\sqrt[5]{2} (\cos 27° + i \sin 27°) \approx 1.0235 + .5215i$

35. Method 1: Use the formula $\cos(k\pi/3) + i \sin(k\pi/3)$, $k = 0, 1, 2, 3, 4, 5$.
Method 2: Write $x^6 - 1 = (x - 1)(x^2 + x + 1)(x + 1)(x^2 - x + 1) = 0$.
Both methods give the answers ± 1, $(-1 \pm \sqrt{3}i)/2$, $(1 \pm \sqrt{3}i)/2$.

37. $\pm i$, $\sqrt{2}/2 \pm (\sqrt{2}/2)i$, $-\sqrt{2}/2 \pm (\sqrt{2}/2)i$ **39.** (a) $\sqrt{6}/2 + (\sqrt{2}/2)i$ (b) $\sqrt{2}/2 + (\sqrt{6}/2)i$ **41.** -2^n

CHAPTER 9. REVIEW PROBLEM SET   (Page 399)

1. (a) $\beta = 39.1°$; $a \approx 228$; $c \approx 139$ (b) $\alpha \approx 4.0°$; $\beta \approx 154.5°$; $\gamma \approx 21.5°$ (c) $c \approx 13.2$; $\alpha \approx 37.3°$; $\beta \approx 107.7°$
(d) $\gamma \approx 30.7°$; $\alpha \approx 100.7°$; $a \approx 75.5$ **2.** $x \approx 37.1$; $A \approx 281$
3. (a) π; 1; 0 (b) $\pi/2$; 3; 0 (c) $2\pi/3$; 2; $\pi/6$ (d) 4π; 2; -2π

4. (a)
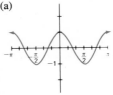
(b)
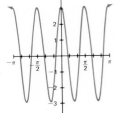
(c)
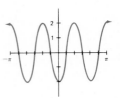
(d)
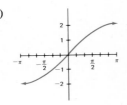

5. $(4 \cos(3\pi t/4 + \pi), 4 \sin(3\pi t/4 + \pi))$ **6.** (a) $2\pi/5$ seconds   (b) At $x = -3$ feet.   (c) When $t = \pi/5$ seconds.

7.
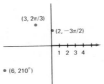

8. (a) $(-3/2, 3\sqrt{3}/2)$ (b) $(0, 2)$ (c) $(-3\sqrt{3}, -3)$

9. (a) $(5, 0)$ (b) $(4, 7\pi/4)$ (c) $(4, 5\pi/6)$

10. (a) 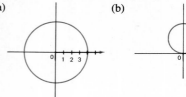 (b) (c)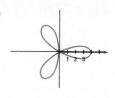

11. $r^2 = 4 \sec \theta \csc \theta$; $(x^2 + y^2)^{3/2} = 2xy$ **12.** 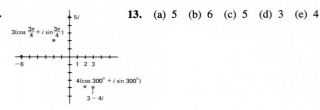 **13.** (a) 5 (b) 6 (c) 5 (d) 3 (e) 4

14. $-2\sqrt{3} + 2i$

15. (a) $3[\cos (\pi/2) + i \sin (\pi/2)]$ (b) $6(\cos \pi + i \sin \pi)$ (c) $\sqrt{2}[\cos (5\pi/4) + i \sin (5\pi/4)]$
(d) $4[\cos (11\pi/6) + i \sin(11\pi/6)]$

16. (a) $32(\cos 145° + i \sin 145°)$ (b) $2(\cos 65° + i \sin 65°)$ (c) $512(\cos 315° + i \sin 315°)$ (d) $4096(\cos 330° + i \sin 330°)$

17. $2(\cos 20° + i \sin 20°)$; $2(\cos 80° + i \sin 80°)$; $2(\cos 140° + i \sin 140°)$; $2(\cos 200° + i \sin 200°)$; $2(\cos 260° + i \sin 260°)$; $2(\cos 320° + i \sin 320°)$

18. $\cos 0° + i \sin 0°$; $\cos 72° + i \sin 72°$; $\cos 144° + i \sin 144°$; $\cos 216° + i \sin 216°$; $\cos 288° + i \sin 288°$

PROBLEM SET 10-1 (Page 405)

1. $x + 1$; 0 **3.** $3x - 1$; 10 **5.** $2x^2 + x + 3$; 0 **7.** $x^2 + 4$; 0 **9.** $x + 2 + 5/x^2$
11. $x - 1 + (-x + 3)/(x^2 + x - 2)$ **13.** $2 + (3 - 4x)/(x^2 + 1)$ **15.** $2x^2 + x + 2$; -2 **17.** $3x^2 + 8x + 10$; 0
19. $x^3 + 3x^2 + 7x + 21$; 62 **21.** $x^2 + x - 4$; 6 **23.** $2x^3 + 4x + 5$; $\frac{3}{2}$ **25.** $x^2 - 3ix - 4 - 6i$; 4
27. $x^3 + 2ix^2 - 4x - 8i$; -1 **29.** $2x - 1 + (-x^2 - 2x - 1)/(x^3 + 1) = 2x - 1 + (-x - 1)/(x^2 - x + 1)$
31. (a) $2x + 3$; $-11x + 9$ (b) $2x + 16$; -14 (c) $x^2 - 8x + 16$; $x^2 + x + 1$ (d) $x^4 + 6x^2 + 2x + 9$; $4x - 1$
33. (a) $x^4 + 3x^3 + 6x^2 + 12x + 8$ (b) $x^4 - 2x^3 + 4x^2 - 8x + 16$ (c) $x^3 - 2x^2 + 4x + 4$ (d) $x^2 + 1$
35. (a) -40 (b) 13 (c) 2 **37.** $a = 1$, $b = -9$, $c = 11$

PROBLEM SET 10-2 (Page 411)

1. -2 **3.** $-\frac{9}{4}$ **5.** -6 **7.** 14 **9.** 1, -2, and 3, each of multiplicity 1.
11. $\frac{1}{2}$ (multiplicity 1); 2 (multiplicity 2); 0 (multiplicity 3). **13.** $1 + 2i$ and $-\frac{2}{3}$, each of multiplicity 1.
15. $P(1) = 0$ **17.** $P(3) = 0$ **19.** $(5 \pm \sqrt{7}i)/4$ **21.** 3; 2 (multiplicity 2). **23.** $(x - 2)(x - 3)$
25. $(x - 1)(x + 1)(x - 2)(x + 2)$ **27.** $(x - 6)(x - 2)(x + 5)$ **29.** $(x + 1)(x + 1 - \sqrt{13})(x + 1 + \sqrt{13})$
31. $x^3 + x^2 - 10x + 8$ **33.** $12x^2 + 4x - 5$ **35.** $x^3 - 2x^2 - 5x + 10$ **37.** $4x^5 + 20x^4 + 25x^3 - 10x^2 - 20x + 8$
39. $x^4 - 3x^2 - 4$ **41.** $x^4 - x^3 - 9x^2 + 79x - 130$ **43.** $-3, -2; (x - 1)^3(x + 3)(x + 2)$ **45.** $\pm i$
47. $(x - 3)(x + 3)(x - 2i)$ **49.** $(x + 1)^2(x - 1 - i)$ **51.** (a) 2 (b) -2 (c) 0
53. (a) ± 2, each of multiplicity 3. (b) 1 and 2, each of multiplicity 2.
(c) $-1 \pm \sqrt{5}$, each of multiplicity 3; -2, of multiplicity 4.
55. (a) $(2x - 1)(3x + 1)(2x + 1)$ (b) $(2x - 1)(x - \sqrt{2})(x + \sqrt{2})$ (c) $(2x - 1)(x + i)(x - i)$

57.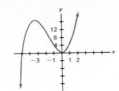

59. 3, 2.35   **61.** $16x^4 - 32x^3 + 24x^2 - 8x + 1$

63. An nth degree polynomial has at most n zeros. Therefore, $P(x)$ must be the zero polynomial (without degree). From this it follows that all coefficients are zero.

65. We are given that $c^6 - 5c^5 + 3c^4 + 7c^3 + 3c^2 - 5c + 1 = 0$. Note that $c \neq 0$ and so we may divide by c^6 to obtain $1 - 5(1/c) + 3(1/c)^2 + 7(1/c)^3 + 3(1/c)^4 - 5(1/c)^5 + (1/c)^6 = 0$, which is the desired conclusion.

67. $a_n x^n + a_{n-1}x^{n-1} + \cdots + a_1 x + a_0 = a_n(x - c_1)(x - c_2) \cdots (x - c_n) =$
$a_n[x^n - (c_1 + c_2 + \cdots + c_n)x^{n-1} + (c_1c_2 + c_1c_3 + \cdots + c_{n-1}c_n)x^{n-2} + \cdots + (-1)^n c_1 c_2 \cdots c_n] =$
$a_n x^n - a_n(c_1 + c_2 + \cdots + c_n)x^{n-1} + a_n(c_1c_2 + c_1c_3 + \cdots + c_{n-1}c_n)x^{n-2} + \cdots + (-1)^n a_n c_1 c_2 \cdots c_n$.
Thus, $a_{n-1} = -a_n(c_1 + c_2 + \cdots + c_n)$, $a_{n-2} = a_n(c_1c_2 + \cdots + c_{n-1}c_n)$, $a_0 = (-1)^n a_n c_1 c_2 \cdots c_n$.

PROBLEM SET 10-3   (Page 420)

1. $2 - 3i$ **3.** $-4i$ **5.** $4 + \sqrt{6}$ **7.** $(2 + 3i)^8$ **9.** $2(1 - 2i)^3 - 3(1 - 2i)^2 + 5$ **11.** $5 + i$
13. $3 + 2i; 5 - 4i$ **15.** $-i; \frac{1}{2}$ **17.** $1 - 3i; -2; -1$ **19.** $x^2 - 4x + 29$ **21.** $x^3 + 3x^2 + 4x + 12$
23. $x^5 - 2x^4 + 18x^3 - 36x^2 + 81x - 162$ **25.** $-1; 1; 3$ **27.** $\frac{1}{2}; -1 \pm \sqrt{2}$ **29.** $\frac{1}{2}; (1 \pm \sqrt{5})/2$
31. $2 - i, (-1 \pm \sqrt{3}i)/2$ **33.** $-1, -2, 3 \pm \sqrt{13}$ **35.** $1, 1, 1, 1, -10$
37. This follows from the fact that nonreal solutions occur in pairs.
39. If $u = a + bi$, $\overline{u} = a - bi$, then $u + \overline{u} = 2a$, $u\overline{u} = a^2 + b^2$, both of which are real.
41. $(x - 1)(x + 1)(x^2 + 3x + 4)$ **43.** $(x - 1)(x + 1)(x^2 + 1)(x^2 - \sqrt{2}x + 1)(x^2 + \sqrt{2}x + 1)$
45. Let $r = \sqrt{5} - 2$ and $s = \sqrt{5} + 2$. Then $x = r^{1/3} - s^{1/3}$. When we cube x and simplify, we obtain
$x^3 = -4 - 3(rs)^{1/3} = -4 - 3x$. Thus, $x^3 + 3x + 4 = 0$, that is, $(x + 1)(x^2 - x + 4) = 0$. Since x is real, x must be -1.

PROBLEM SET 10-4   (Page 428)

1. $4x - 5; -1$ **3.** $4x + 1; 5$ **5.** $10x^4 + 4x^3 - 6x^2 + 8; 16$

7.

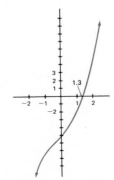 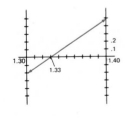 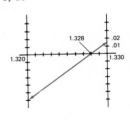

9.

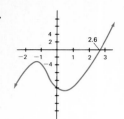

11. $x_1 = 1.3; x_2 = 1.33; x_3 = 1.328; x_4 = 1.3283$ **13.** $x_1 = 2.6; x_2 = 2.61; x_3 = 2.613; x_4 = 2.6129$

15.

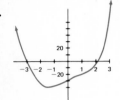

First estimate	Solution to thousandths
-3.2	-3.193
2.2	2.193

17.

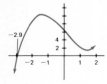

First estimate	Solution to thousandths
-1.9	-1.879
$.4$	$.347$
1.5	1.532

19. $-2.11; .25; 1.86$ **21.** $4.1811; 209$ **23.** 9.701 percent   **25.** $y - 12 = 12(x - 2)$
27. (a) $-2, 3$ (b) $0, 1 \pm \sqrt{2}$
29. (a) $m_h = [f(a + h) - f(a)]/h = [2(a + h)^3 - 2a^3]/h$ (b) $6a^2 + 6ah + 2h^2$ (c) $6a^2$ (d) They are identical.

CHAPTER 10. REVIEW PROBLEM SET   (Page 431)

1. (a) $2x - 5; 20x - 20$ (b) $x^2 - 3x + 1; -3x + 5$
2. (a) $x^2 - 4; -1$ (b) $2x^3 - 6x^2 + 3x - 5; 12$ (c) $x^2 + 5x + 1; -1$ **3.** $1; 97$
4. (a) -1 (two); 1 (two); i (one); $-i$ (one)   (b) 0 (one); $1 + \sqrt{3}i$ (one); $1 - \sqrt{3}i$ (one); $-\pi$ (three)
5. $2(x - 3)(x - \frac{1}{2})(x + 3)$ **6.** $(x + 4)(x - \sqrt{7})(x + \sqrt{7})$ **7.** $x^4 - 9x^3 + 18x^2 + 32x - 96$
8. $x^4 - 10x^3 + 31x^2 - 38x + 10$ **9.** Remaining zeros: $-2, 4$.   **10.** $6x^3 - 25x^2 + 3x + 4$ **11.** -6
12. $x^3 - 6x^2 + 9x + 50$ **13.** $2 \pm 5i; \pm\sqrt{5}$ **14.** $\frac{3}{2}; 3 \pm 2\sqrt{2}$
15. A cubic equation has 3 solutions (counting multiplicities) and the nonreal solutions for an equation with real coefficients occur in conjugate pairs. The only possible rational solutions are ± 1 and ± 7, but none of them work.
16.

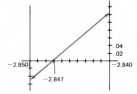

17. $x_1 = -2.9; x_2 = -2.85; x_3 = -2.847$

PROBLEM SET 11-1 (Page 437)

1. $(2, -1)$ **3.** $(-2, 4)$ **5.** $(1, -2)$ **7.** $(0, 0, -2)$ **9.** $(1, 4, -1)$ **11.** $(2, 1, 4)$ **13.** $(0, 0, 0)$
15. $(5, 6, 0, -1)$ **17.** $(15z - 110, 4z - 32, z)$ **19.** $(2y - 3z - 2, y, z)$ **21.** $(-z, 2z, z)$ **23.** Inconsistent.
25. $(-z + \frac{2}{5}, z + \frac{16}{5}, z)$ **27.** $(2, 4); (10, 0)$ **29.** $(5, -7); (6, 0)$ **31.** $(-1, 2); (1, 2)$ **33.** $(3, -2)$
35. $(0, -2)$ **37.** $(-6, -12, 24)$ **39.** $(2\sqrt{5}/5, 4\sqrt{5}/5)$ **41.** $a = \frac{3}{2}, b = 6$ **43.** 285
45. $x^2 + y^2 - 4x - 3y = 0, r = \frac{5}{2}$ **47.** $a = -6, b = 8, c = -3$ **49.** 15 inches by 8 inches

PROBLEM SET 11-2 (Page 446)

1. $\begin{bmatrix} 2 & -1 & 4 \\ 1 & -3 & -2 \end{bmatrix}$ **3.** $\begin{bmatrix} 1 & -2 & 1 & 3 \\ 2 & 1 & 0 & 5 \\ 1 & 1 & 3 & -4 \end{bmatrix}$ **5.** $\begin{bmatrix} 2 & -3 & -4 \\ 3 & 1 & -2 \end{bmatrix}$ **7.** $\begin{bmatrix} 1 & 0 & 0 & 5 \\ 1 & 2 & -1 & 4 \\ 3 & -1 & -5 & -13 \end{bmatrix}$

9. Unique solution. **11.** No solution. **13.** Unique solution. **15.** Infinitely many solutions.
17. No solution. **19.** $(1, 2)$ **21.** $(x, \frac{3}{2}x - \frac{1}{2})$ **23.** $(1, 4, -1)$ **25.** $(\frac{16}{3}z + \frac{32}{3}, -\frac{7}{3}z - \frac{2}{3}, z)$
27. $(3, 0, 0)$ **29.** $(4.36, 1.26, -.97)$ **31.** Unique solution. **33.** Unique solution.
35. No solution. **37.** $(0, 2, 1)$ **39.** $(\frac{21}{2}z - 48, -5z + 26, z)$ **41.** $a = -4, b = 8, c = 0$
43. $\alpha = 80°, \beta = 30°, \gamma = 110°, \delta = 50°$ **45.** A: 10 pounds, B: 40 pounds, C: 50 pounds.

PROBLEM SET 11-3 (Page 453)

1. $\begin{bmatrix} 8 & 4 \\ 1 & 10 \end{bmatrix}; \begin{bmatrix} -4 & -6 \\ 5 & 4 \end{bmatrix}; \begin{bmatrix} 6 & -3 \\ 9 & 21 \end{bmatrix}$ **3.** $\begin{bmatrix} 5 & 4 & 4 \\ 8 & 3 & -6 \end{bmatrix}; \begin{bmatrix} 1 & -8 & 6 \\ 0 & -3 & 0 \end{bmatrix}; \begin{bmatrix} 9 & -6 & 15 \\ 12 & 0 & -9 \end{bmatrix}$

5. $\begin{bmatrix} 14 & 7 \\ 4 & 36 \end{bmatrix}; \begin{bmatrix} 27 & 29 \\ 5 & 23 \end{bmatrix}$ **7.** $\begin{bmatrix} -3 & -4 & 2 \\ 8 & 22 & -13 \\ -2 & 0 & 9 \end{bmatrix}; \begin{bmatrix} 0 & 5 & -17 \\ 13 & 10 & 3 \\ 1 & -3 & 18 \end{bmatrix}$

9. **AB** not possible; $\mathbf{BA} = \begin{bmatrix} 7 & 2 & -7 & 6 \\ 15 & 2 & -11 & 16 \end{bmatrix}$

11. $\mathbf{AB} = \begin{bmatrix} 2 \\ 16 \\ -2 \end{bmatrix}$; **BA** not possible. **13.** $\mathbf{AB} = \mathbf{BA} = \begin{bmatrix} 0 & 0 \\ 0 & 0 \end{bmatrix}$ **15.** $\begin{bmatrix} -4 & 7 & 9 \\ -5 & -5 & 8 \end{bmatrix}$

17. $\mathbf{A(B + C)} = \mathbf{AB + AC} = \begin{bmatrix} -7 & -3 \\ 39 & 34 \end{bmatrix}$; the distributive property. **19.** 93.5917

21. $\begin{bmatrix} 2 & 5 & -1 \\ -8 & 5 & -3 \\ 16 & -2 & -1 \end{bmatrix}; \begin{bmatrix} -16 & -6 & 8 \\ 12 & 0 & 22 \\ 11 & -16 & 19 \end{bmatrix}; \begin{bmatrix} 32 & -3 & 12 \\ 36 & 29 & 24 \\ 34 & 6 & 25 \end{bmatrix}$

23. $\mathbf{AB} = -7$; $\mathbf{BA} = \begin{bmatrix} 2 & 4 & 6 & 8 \\ 1 & 2 & 3 & 4 \\ -1 & -2 & -3 & -4 \\ -2 & -4 & -6 & -8 \end{bmatrix}$

25. **A** and **B** are square, of the same size, and $\mathbf{AB} = \mathbf{BA}$.
27. $\mathbf{A}^2 = \mathbf{0}, \mathbf{B}^3 = \mathbf{0}$. The nth power of a strictly upper triangular $n \times n$ matrix is the zero matrix.
29. Multiplication of **B** on the left by **A** multiplies the three rows by 3, -4, and 5, respectively. Similarly, multiplication on the right by **A** multiplies the columns of **B** by 3, -4, and 5, respectively.

31. (a) $\begin{bmatrix} 16 \\ 13 \\ 14 \end{bmatrix}$ $\rightarrow$ Art's wages on Monday.   (b) $\begin{bmatrix} 18 \\ 14 \\ 16 \end{bmatrix}$ Each man's corresponding wages on Tuesday.
$\rightarrow$ Bob's wages on Monday.
$\rightarrow$ Curt's wages on Monday.

(c) $\begin{bmatrix} 7 & 9 & 3 \\ 9 & 3 & 4 \\ 8 & 5 & 4 \end{bmatrix}$ The combined output for Monday and Tuesday.   (d) $\begin{bmatrix} 34 \\ 27 \\ 30 \end{bmatrix}$ Each man's combined wages for the two days.

33. (a) $\mathbf{U} + \mathbf{V} = \begin{bmatrix} u_1 + v_1 & u_2 + v_2 \\ -(u_2 + v_2) & u_1 + v_1 \end{bmatrix}$, $\mathbf{UV} = \begin{bmatrix} u_1v_1 - u_2v_2 & u_1v_2 + u_2v_1 \\ -(u_1v_2 + u_2v_1) & u_1v_1 - u_2v_2 \end{bmatrix}$

(b) $\mathbf{I}^2 = \mathbf{I}$, $\mathbf{J}^2 = -\mathbf{I}$ (c) $\mathbf{U} + \mathbf{V} = (u_1 + v_1)\mathbf{I} + (u_2 + v_2)\mathbf{J}$, $\mathbf{UV} = (u_1v_1 - u_2v_2)\mathbf{I} + (u_1v_2 + u_2v_1)\mathbf{J}$

(d) This system of matrices behaves just like the complex number system provided we identify $\mathbf{I}$ with 1 and $\mathbf{J}$ with i.

PROBLEM SET 11-4   (Page 461)

1. $\begin{bmatrix} -1 & -3 \\ 1 & 2 \end{bmatrix}$ **3.** $\begin{bmatrix} \frac{1}{6} & \frac{7}{6} \\ 0 & \frac{1}{2} \end{bmatrix}$ **5.** $\begin{bmatrix} 1 & 0 \\ 0 & 1 \end{bmatrix}$ **7.** $\begin{bmatrix} 1/a & 0 \\ 0 & 1/b \end{bmatrix}$ **9.** $\begin{bmatrix} -2 & \frac{3}{2} \\ 1 & -\frac{1}{2} \end{bmatrix}$

11. $\begin{bmatrix} -\frac{4}{7} & \frac{2}{7} & \frac{3}{7} \\ \frac{6}{7} & -\frac{3}{7} & -\frac{1}{7} \\ \frac{5}{7} & \frac{1}{7} & -\frac{2}{7} \end{bmatrix}$ **13.** $\begin{bmatrix} -\frac{1}{9} & \frac{1}{9} & \frac{8}{9} \\ \frac{10}{9} & -\frac{1}{9} & -\frac{26}{9} \\ \frac{1}{9} & -\frac{1}{9} & \frac{1}{9} \end{bmatrix}$ **15.** $\begin{bmatrix} 1 & -1 & 2 & -\frac{5}{4} \\ 0 & \frac{1}{2} & -\frac{3}{2} & \frac{7}{8} \\ 0 & 0 & 1 & -\frac{3}{4} \\ 0 & 0 & 0 & \frac{1}{4} \end{bmatrix}$ **17.** $(\frac{5}{7}, \frac{10}{7}, -\frac{1}{7})$

19. $(\frac{47}{9}, -\frac{128}{9}, \frac{7}{9})$ **23.**   Inverse does not exist.

25. $\begin{bmatrix} -4 & 3 & -4 \\ \frac{1}{2} & -\frac{1}{2} & 1 \\ 6 & -4 & 6 \end{bmatrix}$ **27.** $\begin{bmatrix} \frac{1}{2} & 0 & 0 \\ 0 & \frac{1}{3} & 0 \\ 0 & 0 & -\frac{1}{4} \end{bmatrix}$

29. $(-4a + 3b - 4c, \frac{1}{2}a - \frac{1}{2}b + c, 6a - 4b + 6c)$

33. $\begin{bmatrix} 2 & -1 & 0 & 0 \\ -1 & 0 & 1 & 0 \\ 0 & 1 & 0 & -1 \\ 0 & 0 & -1 & 1 \end{bmatrix}$ **35.** $\begin{bmatrix} 1 & a & b \\ 0 & 1 & c \\ 0 & 0 & 1 \end{bmatrix}^{-1} = \begin{bmatrix} 1 & -a & ac - b \\ 0 & 1 & -c \\ 0 & 0 & 1 \end{bmatrix}$

PROBLEM SET 11-5   (Page 469)

1. -8 **3.** 22   **5.** -50 **7.** 0   **9.** (a) 12 (b) -12 (c) 36 (d) 12   **11.** -30 **13.** -16
15. 7   **17.** 42.3582   **19.** $(2, -3)$ **21.** $(2, -2, 1)$ **23.** (a) 0 (b) 0 (c) 0 (d) 6 (e) 1 (f) 2
25. (a) $k \neq \pm 1$ (b) $k = 1$ (c) $k = -1$ **27.** (a) -32 (b) $-\frac{1}{2}$ (c) -2 (d) -216

29.

$A = (a + c)(b + d) - 2bc - 2(\frac{1}{2}ab) - 2(\frac{1}{2}cd)$
$= ab + ad + bc + cd - 2bc - ab - cd$
$= ad - bc = \begin{vmatrix} a & b \\ c & d \end{vmatrix}$

31. $P(a, b, c) = ab + ac + bc - a^2 - b^2 - c^2$

PROBLEM SET 11-6 (Page 475)

1. −20   **3.** −1   **5.** 39   **7.** 4   **9.** 57   **11.** $x = 2$ **13.** 6   **15.** −72   **17.** −960
19. *aehj*   **21.** 156.8659   **23.** (a) −60   (b) $ef(ad − bc)$ **25.** 0   **27.** $D_2 = D_3 = D_4 = 1$
29. $x = −1, y = 3, z = −3, w = 1$
31. (a) $|C_2| = |C_3| = |C_4| = 0$; $|C_n| = 0$ for $n \geq 2$ (b) $|C_2| = (a_2 − a_1)(b_2 − b_1)$; $|C_n| = 0$ for $n \geq 3$

PROBLEM SET 11-7 (Page 483)

1.

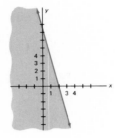

3.

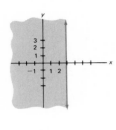

5.

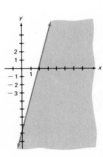

7.

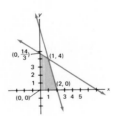

9.

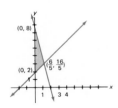

11. Maximum value: 6; minimum value: 0.   **13.** Maximum value: $−\frac{4}{5}$; minimum value: −8.
15. Minimum value of 14 at (2, 2).   **17.** Minimum value of 4 at $(\frac{3}{2}, 1)$.

19.

21.

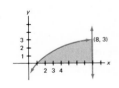

23.

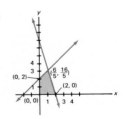

25. Maximum value of 4 at (2, 0); minimum value of −2 at (0, 2).
27. Maximum value of $\frac{11}{2}$ at $(\frac{9}{4}, \frac{13}{4})$; minimum value of 0 at (0, 0).   **29.** 266   **31.** 2 camper units and 6 house trailers.
33. 10 pounds of type *A* and 5 pounds of type *B*.

35.

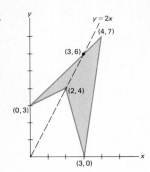

On A, $\left| y - 2x \right| + y + x = 2y - x$, which achieves its maximum value of 9 at $(3, 6)$.
On B, $\left| y - 2x \right| + y + x = 3x$, which achieves its maximum value of 12 at $(4, 7)$.

Therefore, the maximum value on the entire polygon is 12.

CHAPTER 11. REVIEW PROBLEM SET (Page 488)

1. $(2, 6)$ **2.** $(5, 2, 3)$ **3.** No solution.   **4.** No solution.   **5.** $(20w - 36, -9w + 22, -2w + 4, w)$

6. (a) $\begin{bmatrix} -1 & 8 & -2 \\ 8 & -1 & 3 \\ 5 & 13 & 0 \end{bmatrix}$ (b) $\begin{bmatrix} 7 & -11 & 9 \\ -1 & 7 & -11 \\ 0 & -16 & -5 \end{bmatrix}$ (c) $\begin{bmatrix} 0 & 12 & 3 \\ -8 & 6 & -9 \\ -3 & -3 & 0 \end{bmatrix}$ (d) $\begin{bmatrix} 7 & -5 & -5 \\ 3 & 9 & 0 \\ 32 & 14 & -10 \end{bmatrix}$

7. $\begin{bmatrix} \frac{1}{6} & \frac{1}{2} & -\frac{1}{3} \\ \frac{1}{6} & -\frac{1}{2} & \frac{2}{3} \\ \frac{2}{3} & 0 & -\frac{1}{3} \end{bmatrix}$ **8.** $(-1, 2, 3)$ **9.** 2   **10.** 0   **11.** -114 **12.** 0   **13.** -144 **14.** $(1, -2, 4)$

15.

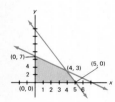

16.

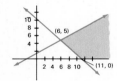

17. 14   **18.** 11   **19.** 24   **20.** 30 suits; 65 dresses

PROBLEM SET 12-1 (Page 495)

1. (a) 9; 11 (b) 5; 2 (c) $\frac{1}{16}$; $\frac{1}{32}$ (d) 81; 121   **3.** (a) 11; 43 (b) $\frac{5}{6}$; $\frac{9}{10}$ (c) 49; 81 (d) -27; 81
5. (a) $a_n = 2n - 1$ (b) $b_n = 20 - 3n$ (c) $c_n = (\frac{1}{2})^{n-1}$ (d) $d_n = (2n - 1)^2$ **7.** (a) 14 (b) 162 (c) $\frac{1}{2}$ (d) 81
9. (a) $a_n = a_{n-1} + 2$ (b) $b_n = b_{n-1} - 3$ (c) $c_n = c_{n-1}/2$ (d) $d_n = d_{n-1} + 8(n - 1)$ **11.** 48   **13.** 42
15. 79   **17.** 69   **19.** (a) 27; 82 (b) 20; 240   **21.** 4, 7, 14, 37, 112
23. $a_n = 3a_{n-1}$; $b_n = b_{n-1} + 4$; $c_n = c_{n-1} + 2n - 2$; $d_n = d_{n-1} + d_{n-2}$; $e_n = \frac{1}{2}e_{n-1}$; $f_n = f_{n-1} + 2n - 1$
25. 4, 2, 5   **27.** (a) 1, 3, 5, 7, 9 (b) 1, 4, 9, 16, 25 (c) $A_n = n^2$
29. (a) $s_n = n^2$ (b) $s_n = s_{n-1} + 2n - 1$ (c) $t_n = t_{n-1} + n$ (d) $s_{n+1} = 2t_n + n + 1$ (e) $t_n = n(n + 1)/2$
31. 1, 1, 2, 3, 5; f_n **33.** $f_n f_{n+1}$

PROBLEM SET 12-2 (Page 502)

1. (a) 13; 16 (b) 3.2; 3.5 (c) 12; 8   **3.** (a) 3; 88 (b) .3; 10.7 (c) -4; -88 **5.** (a) 1335 (b) 190.5
(c) -900 **7.** .5   **9.** (a) 10,100 (b) 10,000 (c) 6,633   **11.** 382.5   **13.** 3.8; 4.6; 5.4; 6.2   **15.** 11; 2,772

17. (a) 90 (b) $\frac{25}{6}$ (c) 15,350 (d) 9,801   **19.** (a) $\sum_{i=3}^{20} b_i$ (b) $\sum_{i=1}^{19} i^2$ (c) $\sum_{i=1}^{n} 1/i$ **21.** (a) $-.75$
(b) -17.5 (c) 63.75   **23.** 902   **25.** 375   **27.** (a) 10,200 (b) 100/101 (c) 2575/5151   **29.** 44,850
31. 2440   **33.** 2475   **35.** $5400\pi \approx 16,965$ inches

PROBLEM SET 12-3   (Page 509)

1. (a) 8; 16 (b) $\frac{1}{2}; \frac{1}{4}$ (c) .00003; .000003   **3.** (a) 2; 2^{n-2} (b) $\frac{1}{2}$; $8(\frac{1}{2})^{n-1} = 1/2^{n-4}$ (c) .1; $.3(.1)^{n-1} = 3 \times 10^{-n}$
5. (a) $2^{28} \approx 2.68 \times 10^8$ (b) $(\frac{1}{2})^{26} \approx 1.49 \times 10^{-8}$ (c) 3×10^{-30} **7.** (a) $\frac{31}{2}$ (b) $\frac{31}{2}$ (c) .33333
9. (a) $\frac{1}{2}(2^{30} - 1) \approx 5.3687 \times 10^8$ (b) $16(1 - 1/2^{30}) = 16 - (\frac{1}{2})^{26}$ (c) $.3[1 - (.1)^{30}]/.9 = \frac{1}{3}(1 - (.1)^{30})$
11. $100(2)^{10} = 102,400$ **13.** $\$(2^{31} - 1)$, which is over \$2 billion.   **15.** (a) $\frac{1}{2}$ (b) $\frac{4}{15}$
17. 30 feet   **19.** $\frac{1}{9}$ **21.** 25/99   **23.** 611/495   **25.** 625; 125; 25; 5; 1
27. $a_n = 130(\frac{1}{2})^{n-1}$, $c_n = 103 + 2n$, $d_n = 100(1.05)^{n-1}$, $f_n = 3(-2)^{n-1}$ **29.** \$215.89   **31.** \$1564.55
33. \$3 billion   **35.** $3\sqrt{3}/14$ **37.** $(3^{32} - 1)/2 \approx 9.2651 \times 10^{14}$ **39.** $r^2 = 3$. The sum of the first 4 terms is 20.
41. 125 miles

PROBLEM SET 12-4   (Page 517)

Note: In the text, several proofs by mathematical induction are given in complete detail. To save space, we show only the key step here, namely, that P_{k+1} is true if P_k is true.
1. $(1 + 2 + \cdots + k) + (k + 1) = k(k + 1)/2 + k + 1 = [k(k + 1) + 2(k + 1)]/2 = (k + 1)(k + 2)/2$
3. $(3 + 7 + \cdots + (4k - 1)) + (4k + 3) = k(2k + 1) + (4k + 3) = 2k^2 + 5k + 3 = (k + 1)(2k + 3)$
5. $(1 \cdot 2 + 2 \cdot 3 + \cdots + k(k + 1)) + (k + 1)(k + 2) = \frac{1}{3}k(k + 1)(k + 2) + (k + 1)(k + 2) = \frac{1}{3}(k + 1)(k + 2)(k + 3)$
7. $(2 + 2^2 + \cdots + 2^k) + 2^{k+1} = 2(2^k - 1) + 2^{k+1} = 2^{k+1} - 2 + 2^{k+1} = 2(2^{k+1} - 1)$ **9.** P_n is true for $n \geq 8$.
11. P_1 is true.   **13.** P_n is true whenever n is odd.   **15.** P_n is true for every positive integer n.
17. P_n is true whenever n is a positive integer power of 4.
19. $n = 4$. If $k + 5 < 2^k$, then $k + 6 < 2^k + 1 < 2^k + 2^k = 2^{k+1}$.
21. $n = 1$. Since $k + 1 < 10k$, $\log(k + 1) < 1 + \log k < 1 + k$.
23. $n = 1$. Multiply both sides of $(1 + x)^k \geq 1 + kx$ by $(1 + x)$: $(1 + x)^{k+1} \geq (1 + x)(1 + kx) =$
$1 + (k + 1)x + kx^2 > 1 + (k + 1)x.$
25. $x^{2k+2} - y^{2k+2} = x^{2k}(x^2 - y^2) + (x^{2k} - y^{2k})y^2$. Now $(x - y)$ is a factor of both $x^2 - y^2$ and $x^{2k} - y^{2k}$, the latter
by assumption.
27. $(k + 1)^2 - (k + 1) = k^2 + 2k + 1 - k - 1 = (k^2 - k) + 2k$. Now 2 divides $k^2 - k$ by assumption and clearly divides $2k$.
29. (a) $(1 + 2 + 3 + \cdots + k) + (k + 1) = \frac{1}{2}k(k + 1) + k + 1 = (k + 1)(\frac{1}{2}k + 1) = \frac{1}{2}(k + 1)(k + 2)$
(c) $(1^3 + 2^3 + \cdots + k^3) + (k + 1)^3 = \frac{1}{4}k^2(k + 1)^2 + (k + 1)^3 = [(k + 1)^2/4][k^2 + 4(k + 1)] = \frac{1}{4}(k + 1)^2(k + 2)^2$
(d) $(1^4 + 2^4 + \cdots + k^4) + (k + 1)^4 = \frac{1}{30}k(k + 1)(6k^3 + 9k^2 + k - 1) + (k + 1)^4 =$
$[(k + 1)/30](6k^4 + 9k^3 + k^2 - k + 30k^3 + 90k^2 + 90k + 30) = [(k + 1)/30](6k^4 + 39k^3 + 91k^2 + 89k + 30) =$
$\frac{1}{30}(k + 1)(k + 2)(6k^3 + 27k^2 + 37k + 15) = \frac{1}{30}(k + 1)(k + 2)[6(k + 1)^3 + 9(k + 1)^2 + (k + 1) - 1]$
31. (a) 15,250 (b) 220 (c) 4355 (d) $2n(n + 1)^2$
33. $(1 - \frac{1}{4})(1 - \frac{1}{9}) \cdots (1 - 1/k^2)(1 - 1/(k + 1)^2) = [(k + 1)/2k][1 - 1/(k + 1)^2] =$
$[(k + 1)/2k][(k^2 + 2k)/(k + 1)^2] = (k + 2)/2(k + 1)$
35. Let $S_k = 1/(k + 1) + 1/(k + 2) + \cdots + 1/2k$ and assume $S_k > \frac{3}{5}$. Then
$S_{k+1} = 1/(k + 2) + 1/(k + 3) + \cdots + 1/2k + 1/(2k + 1) + 1/(2k + 2) =$
$S_k + 1/(2k + 1) + 1/(2k + 2) - 1/(k + 1) > S_k + 2/(2k + 2) - 1/(k + 1) = S_k > \frac{3}{5}$
37. The statement is true when $n = 3$ since it asserts that the angles of a triangle have a sum of 180°. Now any $(k + 1)$-sided convex polygon can be dissected into a k-sided polygon and a triangle. Its angles add up to $(k - 2)180° + 180° = (k - 1)180°$.
39. $f_1^2 + f_2^2 + \cdots + f_k^2 + f_{k+1}^2 = f_k f_{k+1} + f_{k+1}^2 = f_{k+1}(f_k + f_{k+1}) = f_{k+1}f_{k+2}$
41. Assume the equality holds for a_k and a_{k+1}. Then $a_{k+2} = (a_k + a_{k+1})/2 = \frac{2}{3}[(1 - (-\frac{1}{2})^k + 1 - (-\frac{1}{2})^{k+1})/2] =$
$\frac{2}{3}[1 - \frac{1}{2}(-\frac{1}{2})^k - \frac{1}{2}(-\frac{1}{2})^{k+1}] = \frac{2}{3}[1 - (-\frac{1}{2})^{k+2}].$

PROBLEM SET 12-5 (Page 524)

1. (a) 6 (b) 12 (c) 90 **3.** (a) 20 (b) 3024 (c) 720 **5.** 30 **7.** (a) 6 (b) 5 **9.** (a) 36 (b) 72 (c) 6
11. 1680 **13.** 25; 20 **15.** 240 **17.** 300 **19.** (a) 30 (b) 180 (c) 120 (d) 120 (e) 450
21. (a) $12 \cdot 11 \cdot 10 \cdot 9 \cdot 8 \cdot 7 \cdot 6 \cdot 5 \cdot 4$ (b) $2 \cdot 10 \cdot 9 \cdot 8 \cdot 7 \cdot 6 \cdot 5 \cdot 4 \cdot 3$ (c) $2 \cdot 11 \cdot 10 \cdot 9 \cdot 8 \cdot 7 \cdot 6 \cdot 5 \cdot 4$ **23.** 210 **25.** 34,650
27. 840 **29.** 126 **31.** (a) 990 (b) 165 (c) (8!)(989) (d) 630 (e) n (f) $(n + 2)/n$
33. (a) $_{10}P_{10} = 10!$ (b) 720 **35.** (a) 120 (b) 24 **37.** $1 - 1/(n + 1)!$ **39.** $8 \cdot 2 \cdot 10 \cdot 8 \cdot 8 \cdot 10(10^4 - 1)$
41. 3905 **43.** $5! = 120$ **45.** The number of games = the number of losers = $n - 1$.

PROBLEM SET 12-6 (Page 531)

1. (a) 720 (b) 120 (c) 120 (d) 1 (e) 6 (f) 6 **3.** (a) 1140 (b) 161,700 **5.** 56 **7.** 495 **9.** 720
11. 63 **13.** (a) 36 (b) 100 (c) 24 **15.** (a) 84 (b) 39 (c) 130 **17.** $_{26}C_{13}$ **19.** 4^{13}
21. $_4C_2 \cdot {}_{48}C_{11} + {}_4C_3 \cdot {}_{48}C_{10} + {}_4C_4 \cdot {}_{48}C_9$ **23.** $_{52}C_5$ **25.** $_{13}C_2 \cdot {}_4C_2 \cdot {}_4C_2 \cdot 44$ **27.** 16
29. $_5C_2 \cdot {}_4C_2 \cdot {}_3C_2 = 180$ **31.** $12 \cdot 11 \cdot 10 \cdot {}_9C_3$ **33.** (a) 1024 (b) 210 **35.** 31
37. $_{10}C_3 + {}_{10}C_4 + \cdots + {}_{10}C_{10} = 2^{10} - {}_{10}C_2 - {}_{10}C_1 - {}_{10}C_0 = 968$
39. One way: $_nC_0 + {}_nC_1 + {}_nC_2 + \cdots + {}_nC_n$. The other way: $2 \cdot 2 \cdot 2 \cdots 2 = 2^n$.
41. $_nC_j \cdot {}_nC_j = {}_nC_j \cdot {}_nC_{n-j}$. So the answer is the same as for Problem 40, namely, $_{2n}C_n$
43. $_{n+1}C_0 + S = \underbrace{{}_{n+1}C_0 + {}_{n+1}C_1} + {}_{n+2}C_2 + {}_{n+3}C_3 + \cdots {}_{n+k}C_k$

$= \underbrace{{}_{n+2}C_1 + {}_{n+2}C_2} + {}_{n+3}C_3 + \cdots + {}_{n+k}C_k$

$= {}_{n+3}C_2 + {}_{n+3}C_3 + \cdots + {}_{n+k}C_k$

$= \cdots = {}_{n+k}C_{k-1} + {}_{n+k}C_k = {}_{n+k+1}C_k$

45. The two horizontal lines and the two vertical lines determining the vertices of a rectangle can be chosen in $_{n+1}C_2$ times $_{n+1}C_2$ ways, giving us $[n(n + 1)/2]^2$ as the answer.

PROBLEM SET 12-7 (Page 538)

1. $x^3 + 3x^2y + 3xy^2 + y^3$ **3.** $x^3 - 6x^2y + 12xy^2 - 8y^3$ **5.** $c^8 - 12c^6d^3 + 54c^4d^6 - 108c^2d^9 + 81d^{12}$
7. $x^{20} + 20x^{19}y + 190x^{18}y^2$ **9.** $x^{20} + 20x^{14} + 190x^8$ **11.** (a) 16 (b) 32 (c) 64 **13.** $405x^2y^{24}$ **15.** $40x^4$
17. 104.076 **19.** 294.4 **21.** 1219 (using 3 terms) **23.** $-1792x^3z^{15}$ **25.** (a) nx^{n-1} (b) $10x^9 + 8x^3$
27. 55/9 **29.** $(1.01)^{50} > 1 + {}_{50}C_1(.01) = 1 + .5$ **31.** $2^{12} - 1 - {}_{12}C_1 - {}_{12}C_2 = 4017$
33. (a) $x^3 + y^3 + z^3 + 3x^2y + 3xy^2 + 3x^2z + 3xz^2 + 3y^2z + 3yz^2 + 6xyz$ (b) 420

35. $\sum_{k=0}^{n} {}_nC_k 2^k = {}_nC_0 + {}_nC_1 \cdot 2 + {}_nC_2 \cdot 2^2 + \cdots + {}_nC_n \cdot 2^n = (1 + 2)^n = 3^n$

37. (a) 2^k (b) $2^{n+1} - 1$ (c) $2^{n+2} - n - 3$

CHAPTER 12. REVIEW PROBLEM SET (Page 542)

1. (a) and (c) are arithmetic; (b) and (e) are geometric.
2. (a) $a_n = a_{n-1} + 3$ (b) $b_n = 3b_{n-1}$ (c) $c_n = c_{n-1} - .5$ (d) $d_n = d_{n-1} + d_{n-2}$ (e) $e_n = e_{n-1}/3$
3. (a) $a_n = 2 + (n - 1)3 = 3n - 1$ (b) $b_n = 2 \cdot 3^{n-1}$ **4.** 6767 **5.** $3^{100} - 1$ **6.** 3 **7.** 89
8. 40 **9.** 250,500 **10.** $\frac{5}{9}$ **11.** $100(1.02)^{48}$ **13.** P_n is true for n a multiple of 3. **14.** 60 **15.** 64
16. (a) 5040 (b) 210 (c) 30 (d) 1225 (e) 28 (f) 2520 **17.** (a) 504 (b) 224 (c) 84 (d) 40 **18.** 15,120
19. 56 **20.** 256 **21.** $x^{10} + 20x^9y + 180x^8y^2 + 960x^7y^3$ **22.** $-20a^3b^6$ **23.** 1.0408

Index
of Teaser Problems

Index
of Names and Subjects

Compound interest, 241, 246 (*table*)
Condorcet, Marquis de, 359
Conic sections, 171
Conjugate, 38, 416
Conjugate pair theorem, 418
Constraints, 479
Continued fractions, 83
Continuous compounding, 255, 263
Convex region, 480, 487
Coordinate, 139
Coordinate axes, 139
Cosecant, 298
Cosine, 278, 291
Cotangent, 298
Cramer, Gabriel, 468
Cramer's rule, 467
Cube root, 19
Curve fitting, 261

DaVinci, Leonardo, 401
Davis, Harry, 52
Davis, Philip J., 1
Decimals, 17
   nonrepeating, 19
   repeating, 19, 20, 21, 510
   terminating, 18
   unending, 18
Decomposing functions, 208, 210
Degree measurement, 285
Degree of a polynomial, 61
DeMoivre, Abraham, 392
DeMoivre's theorem, 392
Depreciation, 245
Derangements, 529
Descartes, René, 36, 95, 119, 138,
   491
Determinant, 459, 465, 472
Determinant properties, 468
Diameter, 167
   major, 168
   minor, 168
Difference quotient, 81, 212
Directrix, 166
Dirichlet, P. G. L., 180
Discriminant, 114
Distance formula, 140, 328
Distance from point to line, 157
Distance to nearest integer function,
   196, 212, 298
Distributive property, 25, 451
Division:
   complex numbers, 38, 387

fractions, 13
functions, 206
polynomials, 402
rational expressions, 78
real numbers, 25
by zero, 26
Division algorithm, 402
Division law for polynomials, 403
Domain, 180, 182
Double-angle formulas, 335
Doubling time, 240, 261
Dudeney, H. E., 133

e, 255, 263
Einstein, Albert, 2
Ellipse, 167
Equality, 88
   complex numbers, 36
   matrices, 449
Equations, 88
   conditional, 88
   exponential, 255, 260, 273
   linear, 89, 90, 154
   logarithmic, 252
   polynomial, 416
   quadratic, 110
   with radicals, 93, 229
   system of, 104, 434
   trigonometric, 349
Equivalent matrices, 443
Equivalent systems, 434
Euclid, 15, 88, 151
Euler, Leonhard, 109, 133
Eves, Howard, 110
Expansion by minors, 474
Explicit formula for sequences, 493
Exponent, 44
   integral, 44
   negative, 46, 49
   rational, 232
   real, 233
   zero, 46
Exponential decay, 242
Exponential equation, 255, 260, 273
Exponential function, 234, 506
Exponential growth, 239
Extraneous solution, 91, 230

Factor, 68
Factor completely, 68, 409

Factorial, 523
Factoring identities, 333
Factor over, 72
Factor theorem, 408
Fermat, Pierre de, 138
Fibonacci, Leonardo, 494
Fibonacci sequence, 494, 498, 520
Focus, 166, 175
Foil method, 64
Fractions:
   continued, 83
   number, 12
   polynomial, 76
   signs of, 79
Frequency, 377
Functional notation, 181
Functions, 180
   absolute value, 191
   additive, 188
   composite, 208
   decreasing, 235
   distance to nearest integer, 196,
      212, 298
   even, 193
   exponential, 234, 506
   greatest integer, 196, 212, 298
   increasing, 235
   inverse, 213, 215, 250
   inverse trigonometric, 340
   logarithmic, 249
   odd, 193
   one-to-one, 214
   periodic, 294
   polynomial, 188
   power, 258
   rational, 197
   translations of, 209
   trigonometric, 291, 298
   of two variables, 183
Fundamental theorem of algebra,
   410
Fundamental theorem of arithmetic,
   16

Galileo, 277, 321
Gauss, Carl, 407, 500, 543
Geometric sequences, 505
Graphs:
   equations, 144
   functions, 188, 310
   inequalities, 31, 479

GRAPHS OF TRIGONOMETRIC FUNCTIONS

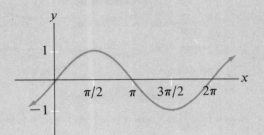

$y = \sin x$

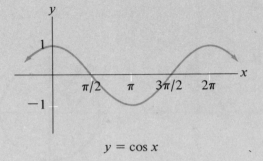

$y = \cos x$

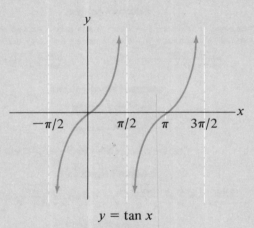

$y = \tan x$

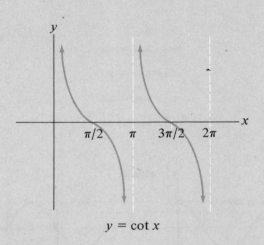

$y = \cot x$

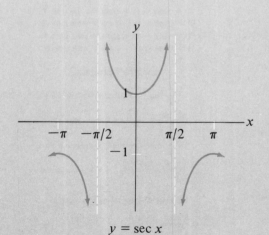

$y = \sec x$

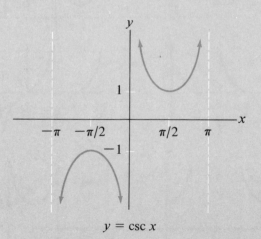

$y = \csc x$

Stormcrow Castle

Stormcrow Castle

AMANDA GRANGE

ROBERT HALE · LONDON

© Amanda Grange 2007
First published in Great Britain 2007

ISBN 978-0-7090-8201-9

Robert Hale Limited
Clerkenwell House
Clerkenwell Green
London EC1R 0HT

2 4 6 8 10 9 7 5 3 1

Typeset in 11¾/16pt Revival Roman
by Derek Doyle & Associates, Shaw Heath
Printed and bound in Great Britain
by Biddles Limited, King's Lynn

CHAPTER ONE

Helena Carlisle rested her valise on the dry stone wall and peered into the gathering gloom. The March daylight was fading and she was beginning to feel uneasy. The carrier had told her it was only two miles to Stormcrow Castle but she had already walked three miles across the moor. She strained her eyes but there was no sign of the castle, nor was there any sign of a house at which she could ask for directions. Looking over her shoulder, she wondered if she should retrace her steps, but it was a long way back to the nearest town and she decided to continue on her way. She picked up her valise and walked along the rutted road, bending her head against the icy wind and praying it would not snow.

A sound disturbed the silence and, looking back, she saw a speck in the distance. As it drew closer she could see that it was a coach, racing towards her. Four black horses were pulling it, and it was swaying from side to side. She stepped aside to let it pass, but, as it drew level with her, the horses were reined in and the coach rolled to an abrupt halt. The door was flung open and a man's voice said, 'Get in.'

She was about to back away when she caught sight of the gentleman inside. She heard her aunt's voice in her memory:

"Like a portrait, he is, with his gaunt face and his long, pointed chin. He should have been living in 1619, not 1819. Lord Torkrow his name is, but no one calls him that hereabouts. They all call him Stormcrow."

'Don't dawdle, you're late as it is,' he snapped.

Late? she thought uneasily. *But I didn't tell anyone I was coming.*

'Well?' he demanded.

She hesitated, but she had to reach the castle and, as she was already weary, she lifted the hem of her cloak and climbed inside. He slammed the door shut and rapped on the floor with his cane, then the coach pulled away, quickly building up speed and racing on again.

As she settled herself opposite him, Helena regarded Lord Torkrow covertly. He had no pretensions to being handsome. His face was thin and sharp, and his eyes looked as though they held secrets.

'Well?' he asked suddenly. 'Does my visage please you?'

Helena realized she had been staring.

'It is. . . .'

He raised one eyebrow in silent challenge.

'. . . striking,' she finished.

His lips curled. 'A good choice of word. It means precisely nothing. But let it pass.' He regarded her appraisingly. 'You're very young to be a housekeeper.'

'A housekeeper?' she asked, startled.

'I have been waiting two weeks for the registry office to send me a replacement. They have been very lax. If they do not do better in future, I shall use Jensen's office instead.'

She felt a cold chill, and pulled her cloak about her. Why did he think she was a housekeeper? Her aunt was the housekeeper at Stormcrow Castle.

'The registry office did tell you the vacancy was for a house-keeper? Or did they describe the post as a chatelaine?' he asked with a grimace. 'If you are imagining the castle to be a fashionable establishment, you will be sorely disappointed. There are no fine rooms; no army of servants; no touring visitors calling at the doors and begging to be shown round.'

'But I thought Mrs Carlisle was your housekeeper?' she said cautiously, wondering if she had mistaken him, and he was not Lord Torkrow after all.

'They have sent me a half-wit!' he muttered under his breath, then out loud he said, 'Mrs Carlisle left my service. She went to nurse her sister and will not be returning. That is why I need a replacement.'

Helena felt disoriented. Her aunt could not possibly have left to nurse her sister, for she did not have a sister.

'You do have experience as a housekeeper?' he asked.

'Yes,' said Helena, recalling the six months she had spent working for Mr and Mrs Hamilton, before they had left the country and removed to Wales.

'That is something, at least. You should have been here yesterday.'

She remained silent, considering what to say. If she revealed who she was then she would learn nothing more, for he had already told her that her aunt had left his service, and he would stop the carriage and expect her to return to town. But if she went to the castle in the guise of the new housekeeper, then she would have an opportunity to speak to the other servants and perhaps learn more about her aunt's sudden departure.

'No matter,' he said, 'you are here now.'

He leant back against the red squabs and his cloak spread across them like a cloud blotting out the sun.

The coach lurched as it turned off the main road and on to

a narrow track. She glanced out of the window, but there was nothing to be seen; nothing but the endless expanse of moor, gradually losing all colour in the fading light. At last the coach began to slow its pace. Up ahead, she dimly perceived the outline of a high stone wall, and then they were plunged into darkness as the coach passed beneath an archway. She felt her hands grow clammy. One heartbeat . . . two . . . three . . . then the darkness lifted and they emerged into a courtyard. A gravel road encircled a patch of lawn which might once have been fine, but which had now grown wild. Coarse grass had embedded itself there, as the moor had encroached on the civilized world.

'We're almost there,' he said.

The coach finally rolled to a halt. The coachman opened the door and lowered the step. Lord Torkrow climbed out. Helena followed him, and looked up at the forbidding walls of Stormcrow Castle.

It was a long, low building with a central square turret and two wings arranged symmetrically on either side. The door was arched, and above it there was a sickly yellow light, shining from a rose window. Crenellations ran along the top of the roof, thrusting their way into the sky like broken teeth. Helena felt a frisson of anxiety. There was an atmosphere surrounding the castle. Isolated and exposed, it seemed malevolent, and she shivered, reluctant to go inside.

'Cold?' he asked.

'A little,' she said, trying to speak bravely.

'It's colder inside.'

With these words he led her up to the door. Without waiting for a servant to open it he seized the iron ring and turned it, then pushed the heavy oak door inwards. He disappeared into the gloom and she followed him, finding herself in a cavernous hall with tapestries decorating the walls, and a huge fireplace, which was large enough to swallow her whole. The floor was bare, and was made of massive stone flags, discoloured with centuries of use.

'You will have plenty to do,' he said, taking a candelabra from a table next to the fireplace and removing one of the candles, putting it into a separate holder before handing it to her. 'The castle has been neglected for some time.'

Helena looked at the thick dust on the table and wondered how long it was since her aunt had left.

'Follow me,' he said.

His cloak stirred as it was caught by a draught, and it billowed around him as he set off at a brisk pace across the hall. The candles were small haloes of light in the gloom, revealing the dim outlines of suits of armour, plated and riveted in the semblance of men, gleaming dully in the fading light. Their silver was darkened with age and fantastic beasts were embossed on their breast plates, whilst above them hung weapons and shields. They had a sullen look about them, as though they resented the fact they were no longer used, like savage animals that had been caged.

Helena turned her head away from the warlike sight and looked straight ahead, but it was little better, for the cavernous space was ominous and she felt suddenly very small. Above her, the ceiling was too high to be seen. Lord Torkrow expected her to match his pace and she had to run to keep up with him.

At the far end of the hall they came to a massive stone staircase. The steps were wide and shallow, worn to a hollow in the centre with the passage of countless feet, and they led upwards into the darkness.

Lord Torkrow began to climb, and Helena followed. Her legs

felt like lead long before she reached the top, for the steps were numerous, and she had already walked a long way that day. She paused to rest at the top, but a curt 'Don't dawdle', set her hurrying after the earl again.

He turned left and led her along an ice-cold corridor, and then stopped abruptly at a door that blocked their way. It was forbidding, made of blackened oak and studded with iron.

'Your room is through the door and at the end of the corridor,' he said. 'You will wait upon me in the library at six o'clock, when we will discuss your previous experience, and I will instruct you in your duties, after which you may return to your own room and rest. Tomorrow you will start work in earnest.'

Helena opened her mouth to reply, but before she could say anything, he turned on his heel and disappeared into the shadows.

A drop of hot wax fell on to her hand, returning her thoughts to her own situation, and she was glad she was wearing gloves, for if her hands had been bare it would have burnt her.

It was a very irregular household, she thought, as she opened the door. There had been no servant to open the door, no footmen waiting in the hall, and no maid to show her to her room. Even more irregularly, his lordship had shown her the way himself, and seemed to be intent on giving her her instructions. There was no lady of the house, then. Feeling the chill from the old stone, she was not surprised. What lady would want to bury herself in a dank castle on the moors, with a dark and brooding man for a husband? Earl or no earl, he was the sort of man to strike terror into the heart, rather than any softer emotion.

She went through the door, knowing at once she was in the servants' part of the house, for there were no tapestries hanging on the wall. She was in a narrow passage with windows to her left, looking out on to the side of the castle, whilst to her right was a row of doors. At the end of the corridor was a final oak door which, gathering her courage, she opened. It was heavy, and it creaked as it moved, making her shiver. As she went in, ghostly shapes loomed out of the darkness, and, through the window she saw the moor looking bleak and dour. She had never seen such darkness before. In her rented room in Manchester there had always been a candle in a neighbouring window, or a glow from a nearby inn, or a flambeau on the street below. But here there was nothing; nothing but impenetrable blackness, unalleviated by a star or a sliver of moon.

Feeling suddenly afraid, she dropped her valise and quickly pulled the heavy curtains across the window, then hurriedly lit every candle in the room. As the flames sprang to life, the ghostly shapes resolved themselves into pieces of furniture that sat, squat and heavy, in the darkly panelled chamber. There was a four-poster bed with dark-red curtains, a large oak cupboard, a carved washstand, a cheval glass and, over by the empty grate, a table and chair.

She went over to the table and put her candlestick down. Was this where Aunt Hester had written her letters? she wondered. The surface was scored and pock-marked; it looked very old.

Overcome with a sudden loneliness, she took paper, ink, sand and a quill from her valise and sat down at the desk. Pulling off her gloves, she dipped the quill in the ink, and began to write.

My dear Caroline,

I have arrived at Stormcrow Castle, but something unsettling has happened. I have discovered that my aunt is

11

no longer here, and, even worse, Lord Torkrow has mistaken me for the new housekeeper. I cannot think where Aunt Hester has gone. Lord Torkrow says she left to look after a sick sister, but she does not have a sister. Why did she lie to him? And where is she?

I, too, have lied, for I have allowed him to think I am the housekeeper he was expecting. I am not easy about it. It does not sit well with my conscience, but I wanted to find out more about Aunt Hester's strange departure, and I could think of no other way. I hope to question the other servants, and, having done so, I will return to Manchester.

I will probably not post this letter. There do not seem to be many servants in the castle, and I might be able to speak to them all tomorrow, returning to Manchester before you could receive it, but I wanted to write because it makes me feel you are near, and I need to feel I have a friend. The castle is cold and dark, and it is taking all of my courage not to be afraid.

But enough of me. I hope you had good luck with Mrs Ling, and that you are now her new companion. You certainly deserve the position, but positions, alas, do not always go to those who deserve them. What a trial it is for us both, to be constantly having to seek work!

I know what you would say, that I should accept Mr Gradwell, but I am not sure I want to marry a man I do not love. That is why I need Aunt Hester's counsel.

I am worried, Caroline. She is all I have in the world, except you, my dear friend. Where can she have gone?

The faint sound of chimes from a far-away clock reached her ears. It was five o'clock. She had an hour before she had to see Lord Torkrow. Her stomach began to growl, reminding her that she had not eaten since that morning, and she resolved to find the kitchen and ask for something to eat.

I can write no more at present, but I hope to be with you before long.
Your affectionate friend,
Helena

She sanded the letter, then folded it and put it in the pocket of her gown, glad to have it with her, for it reminded her that Caroline was not too far away. Then, removing her pelisse and bonnet she picked up her candle and went out into the corridor.

The cook is Mrs Beal, she reminded herself, as she went in search of the stairs down to the kitchen. The cold from the stone floor bit into her feet, even through the soles of her shoes, and icy draughts lifted the hem of her gown. She walked briskly, feeling some welcome warmth creep into her body with the exercise, and was relieved when she saw the top of a back staircase. She went down the stairs, finding them narrower than those in the hall for, being used by servants, they had no need to be imposing.

She had never been in such an old building before, and the size of it was daunting. Down, down went the steps, and the walls were shrouded in shadows. Her footsteps had an eerie sound in the vastness, and she had to tell herself that the *tap, tap* following her was nothing but an echo of her own footsteps. Even so, twice she glanced over her shoulder, convinced that someone was following her. The second time, she thought she saw the hem of a gown pulling back into the shadows, but when she turned round and lifted her candle high, there was no one there.

13

Unnerved by the incident, she ran down the rest of the steps, but at the bottom she was forced to stop, because she was not sure which way to turn. She peered ahead into the gloom. In the distance, to her left, she saw what appeared to be the top of another flight of steps. She went over to them and descended once more, lifting her skirt in one hand and treading carefully, for the stone was smooth and slippery. She emerged in another corridor, and the smell of damp that had pervaded the stairwell was replaced by the smell of baking coming from a door in front of her. The warm, inviting scent lifted her spirits as she opened the door.

The kitchen was clean and well cared for. The table was scrubbed, the floor was gleaming, and copper pots and pans glowed red in the firelight.

Mrs Beal knows her business, Helena thought. Her eyes ran over a large woman of ample girth, who was standing at the kitchen table. She was wearing a clean dress protected by a floury apron; her sleeves were rolled up to her elbows, and she was busy kneading some pastry.

'Well,' said Mrs Beal, looking up, 'so you're here at last! I've set the kettle over the fire. I knew you'd be cold.'

'How did you know I'd arrived?' asked Helena.

'Effie saw you,' she said, glancing at the scullery maid who was peeling potatoes in the corner.

She spoke cheerfully, and Helena felt that here was someone who might be able and willing to help her discover what had become of her aunt.

'You'll be wanting something to eat,' went on Mrs Beal, knocking her hands together to remove the flour before wiping them on her apron. 'Leave those, Effie, and set the cups out on the table,' she said.

Effie did as she was instructed, and the cook said, 'I'm Mrs

14

Beal. I'm pleased you're here. We've been without a house-keeper for far too long. A place like this quickly gets disordered when there's no one to see to it. The pie'll be out of the oven in a few minutes and it'll do you good. You've had a long journey, I suppose?'

'Yes, I've been travelling all day.'

'And you'll have walked from the stage. It's a fair step, especially in the winter, with the wind whipping across the moor and the ground hard underfoot. You're lucky it didn't snow.'

Helena shivered, and Mrs Beale looked at her critically.

'You're even colder than I thought,' she said. 'Never mind tea, you'd better have some mulled wine. There's nothing like a mug of mulled wine to put new heart into you.'

She took a pitcher from the dresser and put it on the table, where the scents of cinnamon, cloves and nutmeg soon mingled with the scent of the wine. Taking the poker from its place by the fire, she plunged it into the wine and then poured the steaming drink into a mug. Helena took it gratefully, cupping her hands round it and feeling it warming her fingers. She took a sip, and felt the aromatic drink beginning to revive her.

As she began to relax, she wondered if she should take Mrs Beal into her confidence, and reveal that she was Mrs Carlisle's niece, but then she decided against it, for Mrs Beal might feel obliged to tell Lord Torkrow.

'It seems a strange household,' said Helena, as she watched Mrs Beal work. 'Lord Torkrow took me up in his carriage and then, when we arrived at the castle, he opened the door himself. He led me upstairs and told me where to find my room, and he means to instruct me in my duties. Has it always been this way?'

'There's usually a footman to open the door, but today's his

15

afternoon off. We used to have a couple of maids, but they left soon after Mrs Carlisle had gone. They didn't like to be upstairs without a housekeeper.'

'Oh? Why not?' asked Helena.

'There's things said about his lordship in the village. Stuff and nonsense, it is, all of it, but girls will be girls, and if they're not hearing noises, they're seeing things out of the corner of their eyes. Their fathers didn't like it, either, having their daughters here without a housekeeper. Always thinking something's going on, are people in a village.'

From Mrs Beal's demeanour, it was clear that she did not think there was anything going on, and remembering Lord Torkrow's cold manner, Helena could not imagine it, either.

'Of course, it was different in the old days, when his lordship's father was alive. Then the castle was full of servants: footmen, maids, valets, page boys, kitchen maids, hall boys. . . .' She looked around the table as if seeing it surrounded by servants. 'Jolly it was, at mealtimes. It's much quieter now.'

'Did his lordship dismiss the servants?' asked Helena.

'Ah, well,' said Mrs Beal, suddenly less forthcoming. 'Things change.' She got up and went over to the oven. 'I'm ready for a bit of something myself,' she said, as she took out the pie.

Helena looked at it longingly. The crust was a golden brown, and it smelled savoury.

Mrs Beal was about to sit down, when she appeared to remember herself and went on, 'But perhaps you'd prefer to eat in the housekeeper's room?'

Helena looked round. With its cheery fire and its air of wholesomeness, the kitchen was an inviting place. Besides, she hoped to learn something of use.

'No, I would far rather eat here with you.'

'It's nice to have a bit of company,' said Mrs Beal comfortably.

'Did the last housekeeper eat with you?' asked Helena, reminding herself that she was not meant to have known Mrs Carlisle, and that she must speak of her aunt only in the most general terms.

'Sometimes. She liked to take her breakfast in the kitchen with me,' said Mrs Beal, cutting the pie and putting a generous slice on to Helena's plate. Steam rose from it, and gravy ran round the plate, whilst large chunks of beef fell out of the pastry casing, with pieces of carrots and turnips.

'It must have been difficult for you since Mrs Carlisle left,' she said, sitting opposite Mrs Beal and taking up her fork.

'I won't deny it,' said the cook. 'I've had to do all the ordering and planning myself. Not that I didn't do a lot when Mrs Carlisle was here, but we shared it, and it was always useful to have someone to ask about the menus.'

'She left in a hurry, I understand,' said Helena, as she put a mouthful of the pie into her mouth. The pastry was light and feathery, and the meat was tender, for Mrs Beal was a very good cook.

'Yes, poor lady. It was her sister. She was taken ill. What could Mrs Carlisle do but go and look after her? One night she was drinking chocolate by the fire with me, the next morning she'd left the castle.'

'She left overnight?' asked Helena, putting her fork down in surprise.

'It was on account of the letter that came,' said Mrs Beal as she, too, ate her meal.

'A letter came late at night?' queried Helena.

Mrs Beal looked surprised. 'That does seem odd, now you mention it. It must have come earlier in the day, of course, but

17

likely she didn't have time to read it. There's always a lot of work in the castle, and she was kept busy.'

'It must have been a comfort for her to be able to talk to you about it,' said Helena.

Mrs Beal shook her head. 'She never mentioned it to me. I would have comforted her if I could have done, but I never saw her. She left before daybreak. It was his lordship who told me about it.'

Helena found the story more and more disturbing.

'She must have had a long walk over the moor. It can't have been pleasant for her in the dark. I hope she didn't miss her way,' she said, hoping to lead Mrs Beal to say more.

'His lordship ordered the carriage for her. He sent her to Draycot, so she could pick up the stagecoach from there.'

'That was very good of him.'

'There's things said about him in the village,' said Mrs Beal, between mouthfuls of pie, 'and of course there was . . . yes, well, least said soonest mended . . . but I've never had anything but kindness from him. There's many a master would have washed their hands of a housekeeper, once she'd decided to leave.'

Helena did not like the sound of *yes, well, least said soonest mended* but for the present she was more interested in her aunt.

'Did she have far to travel?' she asked, trying to sound as though hers was a casual interest.

'I don't rightly know. He didn't say. "If I'd known, I'd have packed her up a hamper", I said. "I could have put her up some bread and cheese, and a piece or two of chicken, and some of my apple pie." I'd made one that morning, and it would have helped her on her way,' she told Helena. 'But the poor lady went off with nothing. I've often wondered about her, and how she's getting on.'

'She has not written to you to let you know that she is all right, and to tell you how her sister is?'

'She won't have time for writing, any more than I have time for reading. Although she did write letters now and again.'

'Yes?' asked Helena, her pulse quickening.

'Yes, to her niece. "She's all I have in the world", she used to say to me. A nice girl, by all accounts.'

'That was a strange thing to say, if she also had a sister,' said Helena.

Mrs Beal looked surprised. 'So it was. She must have meant, all I have in the world beside a sister.'

Helena said nothing. It was becoming clear to her that, although Mrs Beal was very friendly, she did not have an enquiring mind. Thoughts of where the housekeeper had gone and what she was doing had not troubled her. She simply accepted what she had been told.

'And she did not tell you she was leaving before she went?' asked Helena. 'How very strange.'

'Folks do strange things when they're upset,' said Mrs Beal sagely. 'My sister once took the cat with her when her daughter was knocked down by a carriage. She meant to put a cushion in the basket, but she took Pussy Willow instead.'

Helena ate her pie and finished her wine, feeling, first of all her limbs, and then her fingers and toes grow warm, and the conversation turned to more practical matters. Mrs Beal told her about the castle, and gave her instructions on how to find the main rooms. As they talked of the housekeeper's room, Helena learnt that, although Mrs Carlisle had taken breakfast and dinner every day with Mrs Beal in the kitchen, she had had her lunch served in the housekeeper's room.

'I think I, too, will take lunch in the housekeeper's room,' said Helena.

19

She would be sure of one hot meal on the morrow, before she had to face the moor again.

'If you want a dish of tea at any time, just ring the bell. Effie will answer it.'

The scullery maid looked up briefly at the mention of her name, and then went back to peeling the potatoes.

'You'll have a bit of apple dumpling?' asked Mrs Beal, when they had finished their conversation.

Helena readily accepted, and by the time the clock struck a quarter to the hour she was feeling almost cheerful.

'I ought to be going to the library. I have to see his lordship there at six o'clock, and it may take me some time to find it.'

'Effie can show you the way.' Mrs Beal turned round, but Effie was no longer there. 'Never here when wanted,' said Mrs Beal. 'She must've gone to mend the fires. But you'll soon find the library. Just go up to the hall as I said, and it's on your left.'

Taking up a candle, Helena ventured out into the cold corridor once more, but as soon as the kitchen door closed behind her, some of her confidence began to leave her. She felt the cold bite into her, and she was glad of her shawl. After the light of many candles and the glow of the fire, the corridor seemed darker and colder than ever. She hurried along, tripping once on an uneven flagstone, and afterwards not knowing whether to watch her feet, or look at the way ahead. She had an urge to do neither, but instead to look over her shoulder, for she felt sure that someone was following her, but every time she turned round, there was no one there.

It is just my imagination, she told herself, I must not succumb to fancy. But the shadows danced beyond the light of her candle flame, and seemed to mock her with their shifting presence, assuming monstrous shapes before diminishing as she passed.

She came at last to the end of the corridor and went up the steps, and was soon crossing the hall. She counted the doors to her left, then stopped outside the library door as the clock chimed the hour. She smoothed her hair, arranged the folds of her skirt, took a deep breath, and knocked on the door.

CHAPTER TWO

There was a moment's silence and then the earl's voice called, 'Come in.'

Helena opened the door and found herself in a large room, its walls decorated with hangings and its stone floor covered with a rug. Two candelabras on the mantelpiece and another one on a large desk in the middle of the room did their best to provide light, but the walls were lost in darkness, save for glints of gold coming from the shadows that hinted the room was lined with books. There was a leaping fire in the grate, and in front of it stood the earl, holding a letter in his hand. He looked up as she entered.

'When I ask to see you in future, I expect you to arrive before the last chime has been struck. I will not tolerate tardiness,' he said.

Helena said nothing, not knowing what to reply.

'Well, come in,' he said.

She closed the door and stepped forward.

'So. Mrs Elizabeth Reynolds,' he said, looking down at the letter. 'You have three years' experience of housekeeping, two with the Right Honourable Mrs Keily, and one with the Revd James Plumley. Mr Keily was in business, I see.'

An expression of fastidious distaste crossed his face as he

said it, and she was forcibly reminded of the fact that he was an aristocrat. He had never had to earn his living, never known the fear of having nowhere to live, nowhere to go. She imagined a long line of ancestors stretching out behind him, reaching back through the centuries, governing the land and living in the castle. How long had he and his family lived there, she wondered, maintaining tradition, keeping the peace, ruling the neighbourhood? A hundred years? Two hundred years? Or even longer?

He went on, recalling her thoughts.

'You came by your position with Mr Plumley through the registry office, I see,' he said, referring to the letter. 'You wrote to the office again when Mr Plumley married, because his wife chose to manage the house herself, and the office recommended you for the job at the castle.'

'Yes, my lord.'

'Well, you will find it very different here. A castle is not a tradesman's house or a vicarage, even one as lost as this.'

'No, my lord.'

'Now, to your duties. You will make sure the dining-room and the other inhabited rooms are kept clean and warm. You will make sure there is a fire in the library at all times, and you will tend to it yourself. You will not allow any other member of staff to enter the library, except Miss Parkins.'

'Miss Parkins?'

'You will meet her later. She has my full confidence. You will take responsibility for everything that goes on inside the castle and you will make sure that I am not troubled with household matters, except for an appointed hour once a week, when you will report to me. As this is your first week, you will report to me tomorrow afternoon at four o'clock, when I will answer any questions that might have arisen once you have had

23

an opportunity to familiarize yourself with the castle.'

'Very good, my lord.'

'You know enough to begin with.' He waved his hand and said: 'That will be all for today.'

It was a dismissal. Helena inclined her head, then left the room.

As she closed the door behind her, she let out a sigh of relief, for she had passed the test, and been accepted as Elizabeth Reynolds.

She crossed the hall and mounted the stairs, returning to her room. I can do no more this evening, she thought, but tomorrow I must question the footman.

As she opened the door, she felt a welcome heat, and realized that a fire had been lit in her absence. She was about to cross to the fireplace when she saw, with a start, that she was not alone. A figure was standing in the corner, its dark eyes like blots of ink in a parchment face. It was dressed in a grey woollen gown, and its hair was drawn back into a severe chignon. It was holding a taper, and there was a sepulchral look about it. She wondered if one of the other servants had played a trick on her by putting a mannequin in her room to frighten her . . . until it suddenly moved, and Helena realized with a creeping sensation that it was not a mannequin, but a woman of flesh and blood.

The woman ignored Helena and used the taper to light the candles. The gesture seemed territorial, as though she was saying to Helena, *This is my room, and you do not belong here.*

'I thought you would like a fire,' said the woman at last, putting down the taper. Her words were welcoming, but her voice was hollow and it sent a chill down Helena's back. 'The day is very cold.'

'Yes, it is,' said Helena, having a sudden urge to flee.

'You have had a long journey?' the woman asked.

'I have been travelling all day.'

'You are very young to be a housekeeper,' said the woman.

Helena hesitated. Until she knew to whom she was speaking, she did not know what tone to take. Was this a member of the family? A distant relative of Lord Torkrow, perhaps? Or another servant?

'Perhaps,' she said cautiously. 'And you are?'

'Miss Parkins.'

So this was Miss Parkins, the servant his lordship trusted absolutely. They made a fine pair, she thought. They were both intimidating in different ways.

'His lordship mentioned you,' said Helena. 'I am not sure what your position is?'

'My *position*,' said Miss Parkins, allowing the word to pass her lips as though it was a spider, 'is unusual. I came here many years ago as a lady's maid. Now, I help his lordship in whatever capacity he desires.'

Helena gave a tentative smile, wondering if it was possible to make a friend of the woman, for if she had been at the castle for a long time she must have known her aunt.

'You must have been a great help to the previous housekeeper,' she said.

The woman said nothing. There was something unsettling about her, beyond her appearance, and as Helena looked at her she began to realize that the face had no expression. When the woman spoke, she did not smile, or frown, or look surprised. She seemed, as she had seemed at first sight, like a waxwork dummy. 'It must have been difficult for you when she left. She left very quickly,' said Helena, persevering.

'She did.'

'Her sister was ill, I understand? It must have been inconvenient for his lordship to have her leave so suddenly. Was her sister very ill? Could she not have given him some warning?'

'Servants these days care for no one but themselves,' said Miss Parkins in a toneless voice.

Helena felt a retort spring to her lips, but she was prevented from uttering it by a flicker of interest in Miss Parkins's eyes. She sensed a strong and malevolent personality at work behind the maid's immobile face, and realized she could not afford to make any mistakes, so she stifled her retort, and said, 'It was very good of you to light a fire for me; it's cheerful to see the flames. It was very cold on the moor, and the castle is not much warmer.'

'The scullery maid will do it for you in future,' said Miss Parkins.

'Then you do not usually. . . ?' asked Helena, trailing off as she saw a gleam of sour amusement in Miss Parkins's eye.

'Light fires for the housekeeper? No, I do not.'

'It was good of you to do it on this occasion.' As Miss Parkins showed no signs of leaving, Helena said, in a friendly manner, 'If you do not mind, I am very tired. I have had a long journey, and if I am to be capable of fulfilling my duties tomorrow, I must get some rest.'

'As you wish,' said Miss Parkins.

But although she had accepted her dismissal, Helena was under no illusions: it was Miss Parkins who held the power, Miss Parkins who had agreed to leave, and not Helena who had dismissed her.

As Miss Parkins left the room, she left a chill in the air and Helena crossed to the door, on an impulse pulling a chair against it.

She went over to the fire and held her hands out to feel the heat. Aunt Hester had never mentioned Miss Parkins, and yet she must have known her.

Helena stared into the fire, as though she would be able to see her aunt in the flames. Aunt Hester, she thought, why did you leave? Where did you go? Why did you not write to tell me you were leaving? And why did you lie to Lord Torkrow, telling him you needed to tend your sick sister, when I am your only relative?

Simon, Lord Torkrow, stood by the window in the library, looking out over the courtyard. It was too dark to see anything but the silhouette of the outer wall in the distance, with a patch of grey where the archway cut through it, and beyond, the deep dark of the moor.

He turned round as the door opened and Miss Parkins entered the room. She had not knocked, or waited for permission to enter. She stood before him in a respectful attitude, but her face was devoid of all emotion. Her dark eyes looked out from her white face, their large pupils seeming to swallow the light, but as she looked at him, he wondered what was going on behind them. Her black hair was pulled back into a bun, and he thought, with a passing fascination, that in all the years he had known her, he had never seen her hair loose. He did not know how old she was. She had a quality of stillness about her that made her seem scarcely alive. That had not always been the case. There had been a time when she had been vital.

'There is a woman here,' she said.

'Yes, I know. She is the new housekeeper.'

'Is she?'

'What do you mean?' he asked sharply.

'She is very young for such a position.'

'It is not easy to get servants these days, particularly in such a remote corner.'

27

'She is not wearing a wedding band,' said Miss Parkins.

'It is possible she calls herself Mrs for reasons of employment, or that she has had to sell her wedding ring,' he said.

Inwardly, however, he berated himself, for he had not noticed her lack of a ring, and had been too ready to accept her as the person she claimed to be.

'Perhaps.'

'I will call in at the registry office the next time I am in York,' he said. 'Someone there will have met Elizabeth Reynolds and I can find out what she looks like, and see if her description matches this young woman.'

'Have you questioned her about her previous employment?'

'Yes.'

'A pity. If you had been on your guard, you might have laid a trap for her.'

'What's done is done, but we must be careful. Watch her, Parkins. See where she goes, and what she does. Make yourself her shadow. Find out if she knows how to keep house. Because if she is not who she claims to be, then we must be prepared.'

'And if she discovers what has happened here?'

'She must be stopped.'

She looked at him unwaveringly.

'Very good, *my lord*.'

There was an almost imperceptible note of scorn in her voice when she said *my lord*, and it did not escape him.

You don't think I should be the earl, he thought. You think the title should belong to another.

'Very well, Parkins. You are dismissed.'

She did not blink. She did not speak. But when he addressed her as the servant she was, he could feel the venom coming from her.

She unfolded her hands and moved to the door, going

through it in a gliding action, and leaving the room on noiseless feet.

He knew what she felt about him. He knew that she blamed him, that she had always blamed him.

Perhaps she was right.

Helena unpacked her few belongings, hanging her two woollen gowns in the wardrobe, and putting her chemise and petticoat in the top drawer, together with her handkerchiefs and her woollen stockings. Her shoes she put next to the bed. Then she took the hot brick from its place by the fire and put it between the sheets.

It was not the first night she had been expecting. She had been hoping for a warm welcome from her aunt, and after their reunion she had been intending to tell her of Mr Gradwell's proposal, and to hear her aunt's advice.

Caroline had been in no doubt. 'Marry him, Helena,' she had said. 'He's a kind man, a gentleman. He'll take care of you. You'll have servants of your own, instead of having to be a servant. You'll never have to sleep in an attic again.'

But Helena was still uncertain. She wanted a home of her own, yes, and it would be good to be no longer at someone else's beck and call, but she was not sure she could face a future with Mr Gradwell. He had kissed her once, and although the experience had not been unpleasant, she had hoped for something more.

Where heart, and soul, and sense, in concert move . . .
Each kiss a heart-quake. . . .

Each kiss a heart-quake, she thought with longing. There had been no quaking of her heart when Mr Gradwell kissed her. But was Byron's poetry a true vision of love? Or was it simply a romantic dream?

What would it really be like, to be married? she wondered, as she brushed her hair; to live with a man every day, to share a home with him, and to be with him every day of her life?

Aunt Hester knew. Aunt Hester had been married to Uncle George, and could tell her what to expect, as well as helping her to decide whether or not she could be happy in a marriage to Mr Gradwell. But Aunt Hester had disappeared.

She undressed in front of the fire, stepping out of her gown and stripping off her underwear before lifting her nightgown over her head. As she did so, she caught sight of her hand and she froze. She was not wearing a wedding ring. She should have thought of it sooner, but it was too late to do anything about it now. Besides, Lord Torkrow seemed to have accepted her. He knew as well as she did that many women had become destitute after losing their husbands at Waterloo, and had been forced to sell their jewellery in order to stay alive.

She climbed into bed. The hot brick had warmed the sheets, and she pushed it further down the bed, resting her toes on it and basking in its heat. She blew out her candle then, worn out from her day, she fell asleep.

It seemed hardly any time before she awoke to the sound of scratching on her door. At first she did not know where she was. The bed felt strange, and the red hangings confused her, but then it came back to her, and she remembered that she was in the castle. Fumbling on the table next to her bed she found the tinderbox and lit her candle then, throwing a wrapper round her shoulders, she removed the chair she had set in front of the door before calling, 'Come in.'

The door opened and Effie stood there. She wore a shapeless dress, over which was a large, grubby apron. In one hand she carried a jug of water from which steam was rising, and in the other she carried a bucket of coal.

'Good morning,' said Helena.

The girl made a nervous noise that could have been 'Morning', and then hurried across the room lumpishly, without grace. As Helena watched her, she thought of her aunt's letters, and as she recalled that Aunt Hester had taken a motherly interest in the girl, she hoped she might learn something from her.

Effie went over to the washstand and deposited the jug of water there clumsily, spilling the water.

'Oh, mum, I'm sorry, mum, I'm sorry,' said Effie, mopping up the water nervously with her apron.

'That's all right. You did not mean to do it,' said Helena.

'No, mum.'

The girl left the water half mopped and crossed to the grate, putting the bucket of coal down with a clatter that made Helena start, and then knelt down in front of the fire. Her skirt rode up to reveal a few inches of leg, and Helena saw that she had holes in her woollen stockings, which had been badly darned.

Effie picked up the poker, setting the other fire irons jangling, and began to rake the coals, which had turned to ash as the fire had burnt down overnight. The poker made a scraping noise across the iron grate, and there was a soft, shifting sound as the ash fell through into the box beneath.

'It's an early start for you,' said Helena, trying to put the girl at ease.

Effie dropped the poker with a clatter.

'Sorry, mum, I didn't mean to do it, I didn't mean to,' she said, grabbing at the poker.

31

'It's all right,' said Helena, wondering how many more times she was going to have to soothe the girl. You knew my aunt, she longed to say, but instead she went on, 'It must be confusing for you to have a new housekeeper in the castle. Perhaps you did not expect to find me still in bed. I am usually awake early, but I had a tiring day yesterday. Mrs Carlisle, my predecessor, was an early riser, I suppose?' she enquired casually.

'Yes, mum. Always up early she was. "There's no use lying abed when there's work to be done", she used to say.'

'Quite right, too. There is plenty to do in the castle. You must be busy all day long.'

'Yes, mum. There's fires to be lit and there's that many steps, it's 'ard work.'

'Mrs Carlisle must have been sorry to leave the castle. She took a pride in her job, I believe.'

'Very particular, Mrs Carlisle was. The flowers 'ad to be fresh in summer. Very particular about 'er flowers, was Mrs Carlisle. I mustn't move anything on 'er desk, and I mustn't go through the drawers.'

'Did you used to go through the drawers?' Helena asked in surprise.

Effie dropped the poker.

'I were only looking for some string,' she said, but she seemed nervous, and Helena wondered if she was speaking the truth. 'My stockings were falling down. Mrs Carlisle said I needed garters, she showed me 'ow to make 'em.'

'Of course,' said Helena. 'Did you find anything interesting when you were looking for the string?' she asked nonchalantly.

'Very particular about her pens, she was. Mended 'em 'erself. Didn't want no one touching her pens,' said Effie obliquely, picking up the poker and hanging it back on its stand, then she took a piece of newspaper from the top of the bucket of coals and crumpling it vigorously before laying it in the grate.

Helena's eyes were drawn to the girl's hands. They were large and strong and, as they picked up another piece of newspaper and crushed it, Helena found herself wondering what else the girl's hands could crush.

Changing the subject, she said, 'It must have been a shock to you when Mrs Carlisle left so suddenly.'

'I didn't know she was going,' said Effie. 'She said nothing to me, just went. One day she was here and the next day she wasn't.'

'Do you know why she had to leave?' Helena asked.

Effie sat back on her heels and rolled up a sheet of newspaper, winding it round her hand and knotting it before laying it on top of the crumpled paper.

'Do you?' asked Helena patiently.

Effie glanced over her shoulder and seemed reluctant to speak.

'I believe her sister was ill?' Helena prompted her.

'That's what he said.'

Helena had the feeling she was concealing something.

'And did you believe him?'

'It's not my place, mum, if Master says it, then it must be true.'

'Ah, yes. Do you like him? The master?'

'I reckon.'

But the girl's open manner had disappeared, and once she had finished lighting the fire she wiped her hands on her black-streaked apron, then picking up the bucket she left the room.

Helena was left with much to think about. As she removed her nightgown and washed in the hot water, she thought that Effie had not told her everything she knew. But, if she stayed at the castle an extra day, there would be another morning, and another conversation whilst Effie lit the fire.

She dressed quickly, glad of her thick woollen gown and woollen stockings, brushed her hair and fastened it into a neat chignon, then, picking up her candle, she went down to the kitchen, following the route she had used on the previous day. As she went through the door into the servants' quarters, she once again had the unnerving feeling that she was being followed, but when she turned round, there was no one there.

She quickened her step, and was relieved to gain the sanctuary of the kitchen, where she found Mrs Beal baking bread. The smell of it filled the room and made Helena realize how hungry she was.

'Effie, set the kettle over the fire,' she said. Then, to Helena, she said, 'You'll have some rolls? They're freshly baked.'

Helena looked at in the newly baked rolls that were set on the dresser, laid out on a clean cloth. With their golden tops, they looked appetizing.

'Yes, please.'

Mrs Beal set jars of home-made jam and honey on a table in the corner of the kitchen, and put out cups, saucers and plates. She added a mound of freshly churned butter to the table, and a jar of frothing milk. Soon a bowl of sugar and a pot of tea joined the rest.

'I'm ready for a bit of something myself,' said Mrs Beal.

'I see you have finished the fires,' said Helena to Effie, hoping to reassure the girl, so that the next time they met, she would be agreeable to talking.

Mrs Beal answered for her.

'Yes, she does the fires in the mornings, but his lordship doesn't want anyone in the library except the housekeeper and Miss Parkins, so she left a bucket of coal outside, as she always does. His lordship's told you you're to keep the library clean yourself?'

'Yes, he has. Miss Parkins does not see to it, then?'

'Miss Parkins doesn't see to a lot, from what I can see.'

'I am not quite sure what Miss Parkins's position is in the castle,' said Helena, gently probing, as Mrs Beal poured out the tea.

'That makes two of us,' said Mrs Beal. 'I wouldn't have much to do with her, if I were you. She comes down here from time to time, but I won't have anyone interfering in my kitchen. She looks at you sometimes . . . well, I've said enough.'

As Helena ate her rolls and drank her tea, the conversation turned to the idleness of dairy maids and the impossibility of running the kitchen adequately without any kitchen maids.

'In the old days, there were seven people working in the kitchen: Mrs Barnstaple the cook, me as her assistant, three kitchen maids and two scullery maids. Mind, we had a castle full of people to feed. His lordship and Master Richard . . .' She tailed away, then finished, 'We'll not see those days back again.'

Helena tried to encourage her to say more, being sure there had been something important left unsaid, but Mrs Beal would not be drawn.

'Thank you for breakfast,' said Helena, when she had finished her meal. 'And now, I had better see to his lordship's fire.'

Taking up her candle, she left the kitchen, and then the servants' quarters, behind her, and emerged into the hall. A faint grey light could be seen coming through the windows. Outside, the sun was rising and it would soon be daylight.

She found the bucket of coal outside the library. Picking it up, she went in, but she was taken aback to see Lord Torkrow sitting behind the desk, looking at some papers. She had not expected him to rise so early, and she wished she had knocked.

He looked up as she stood there in the doorway. As she felt his eyes run over her, she was conscious of a sudden unease, and again she wondered if he had been fooled by her deception, or if he knew that she was not who she claimed to be. She told herself there was no way he would know, but even so she felt anxious, for there was something about the way he looked at her. . . .

'I'm sorry,' she said. 'I thought the room was empty.'

'You may see to the fire,' he replied.

She walked across the room, conscious of his eyes on her, and then poured coal on to the small flames.

'Tell me, how do you find the castle, now that you have had an opportunity to see it in daylight?' he said.

She was surprised by his question, for it was not the sort of thing that most earls would ask their servants. She replied, 'I find it . . . interesting.'

'You do not find it too remote?' he queried.

'No, my lord.'

'That is surprising. Most people are disinclined to work in such an isolated spot. It preys on their nerves. The loneliness becomes oppressive. After a time, they find themselves imagining things.'

There seemed to be something behind his words. Was he warning her about something, or was he trying to find out if she had heard anything unusual?

'I have no difficulty in working here,' she said.

'Perhaps you are used to the moors?'

'I have never seen them before. I think they are beautiful,' she added.

'You think so? I used to think so, once.' His voice dropped and his eyes fell to the desk. He was not seeing the desk, thought Helena, he was seeing something far away, and she wondered why he no longer liked his surroundings. He roused himself. 'If you are not used to the moors, then you perhaps grew up in gentler climes?'

'Yes, my lord. I grew up by the sea, in Cornwall,' she told him.

His eyes narrowed. 'You do not speak with a Cornish accent,' he remarked.

'I left Cornwall many years ago, when I was fifteen.'

'Ah, I see. Then why did you leave?'

'My father died, and my mother took me to live in Manchester. . . .' She trailed away, suddenly conscious of the fact that Mrs Reynolds might have mentioned her abode. She felt herself colouring and hoped he would not notice, or that he would put her sudden flush down to the heat of the fire, but instead she was disturbed to see him turning questioning eyes towards her, as if to say, *Now what were you about to tell me?* She began to think that his questions were more than a passing curiosity in a new servant; they were designed to find out if she was really Mrs Reynolds.

'And do you like Manchester?' he asked.

'It is my home,' she said, 'but no, I do not like it.'

'I am surprised. You are young. I thought you would enjoy the liveliness of a city. It must seem very quiet here by comparison.'

'It seems peaceful,' she said. 'I like the quiet.'

'And when did you go into service?' he asked, returning to his earlier theme.

She was about to say, 'A year ago,' when she realized that Mrs Reynolds had been in service for far longer.

'Well?' he asked.

'I cannot recall exactly.'

There was a silence. Then he said, 'In your letter, you stated that you had been a housekeeper for three years.' He shot her a sudden glance, and said, 'You do not have the look of a servant.'

She felt her heart beating more quickly.

'My father was a gentleman,' she said, 'and I was raised to be a lady. But he fell on hard times and our circumstances changed, so I had no choice but to earn a living.'

He said nothing, and she wondered what was going through his mind. Unwillingly, Helena found herself remembering some of the things her aunt had said about the man in front of her. *Afraid of him in the village, they are. The stories they tell! It's always the same in these remote places, but I've seen nothing amiss. He's not an easy master, but I'll say this for him, he's fair.*

She only hoped her aunt had been right.

At last he said, 'And now you are keeping house in a castle. Not many people wish to work in such a large establishment, especially with so few servants. What is your opinion of the castle, now that you have seen it?'

She looked round the large room.

'I think it has been neglected, but it is a beautiful building, and with hard work, I think it will be possible to bring it back to life.'

'You are an optimist, I see. Hard work will go some way to making it brighter, but hard work has its limits and will not remove the draughts.'

'Large fires and carefully placed screens can do much to limit their effect,' she said thoughtfully, wondering how best the disadvantages of such an old building could be overcome.

'There is no money to waste on large fires at Stormcrow.'

He appeared to become lost in his thoughts, and she said no more.

He roused himself.

'Very well. The fire will do now. You may return to your duties.'

'Very good, my lord.'

Helena left the room, relieved that she had escaped unscathed, and made her way to the housekeeper's room, where she hoped she might find a letter or a diary entry, perhaps, that would tell her something about her aunt's decision to leave the castle, and her intended destination.

CHAPTER THREE

As she opened the door, she breathed in the scent of lavender, and it awakened memories in her. She remembered how, as a little girl, she had helped her aunt to pick flowers and herbs, and how her aunt had showed her how to plait lavender. Her aunt had always had a plait of it attached to her belt. She remembered summer holidays when her father had been alive, and, in her mind's eye, she saw her mother and Aunt Hester cutting flowers and herbs, whilst she and her father sat on a rug beneath the chestnut tree with their books. She could remember the pulled thread on the rug, and she could feel its softness beneath her fingers.

She went in, thinking how lucky Aunt Hester was to have such an attractive room to work in. It was newly decorated in cheerful colours, with flowered wall hangings echoing the gold damask of the sofa. There were vases on the console tables, and although they were empty, they were still decorative.

Diamond-paned windows looked out on to the side of the castle. It was a bleak prospect at present, but under a summer sky it would be attractive. Her own position as a housekeeper had not been so grand, and her room had been a dingy room at the back of the house, with a window looking on to a brick wall. She had almost been glad to leave it when the Hamiltons had moved to Wales – almost, but not quite, as she had needed the position, and without it she had been reduced to sharing a room with Caroline.

She went over to the hearth, where there was a fire burning in the grate, noticing that the shelves had been dusted and the furniture polished. She set down the coal bucket and then let her eyes wander over the chintz sofa set beneath the window and the matching chair that was placed by the fire. There were two console tables, one by the chair and one by the sofa, and in the middle of the room was a desk. She went over to the desk, which had a number of pigeon holes down the side and across the top. On the desk was a large book, an inkstand, and a shaker of sand. In front of it was an inlaid chair.

Helena sat down and opened the book. It was in the form of a diary, but it held nothing useful, simply details of the work that needed to be done around the castle. The notes stopped just over three weeks before.

She turned her attention to the pigeon holes, but they revealed nothing more than sample menus, letters to and from tradesmen, and other household items. Then she opened the first drawer.

What did Effie find when she looked for some string? Helena wondered.

But a search of the drawers found nothing more than some household documents, some paper, a quill pen and a large bunch of keys, which she took out and fastened to her belt. There was nothing more of interest.

So what did Effie see? she asked herself. Why did it upset her? And where had it gone? Unless Effie had seen nothing, and had simply been nervous because she had let slip that she had been guilty of going through the housekeeper's desk.

As Helena looked round the room she began to think that

this must be the case. The chintz upholstery and the placid ticking of the clock were reassuring. Their very ordinariness reminded her that her aunt had been an ordinary woman, and that there must be an ordinary reason for her disappearance.

Maybe she did, indeed, have a sister, thought Helena, or perhaps, a half sister she had never mentioned. Maybe there was a reason for Aunt Hester not mentioning her. Perhaps they had been estranged.

Perhaps Aunt Hester wrote to me, she thought, but perhaps the letter was lost in the post, or perhaps my aunt gave it to the footman to post, and he forgot about it.

The more she thought about it, the more likely it seemed. There were very few servants in the castle, and with no one to keep them to their tasks, something like posting a letter could easily be overlooked.

She saw a row of bells on the wall by the fireplace, and rang the one labelled 'Footman'. Soon afterwards the footman entered the room. He was wearing livery, but some of the braid was missing from his coat, and the buttons were dull. His person reflected the same carelessness: his hair had been combed, but a tuft stuck up at the back, and his nails were dirty.

He stood in front of her with a strange expression. It was part insolence and part insinuation, and there was a sly look on his face. He rubbed his hands together in an unpleasant manner and looked at her from the corner of his eye, as though he was sizing her up.

She wondered if he had looked that way at her aunt, or if he was simply doing it to her because of her youth.

'Your name?' she asked him, injecting a note of authority into her voice: if he thought he could patronize her, he would soon learn his mistake. She had dealt with difficult footmen before, and would most probably have to do so again.

'Dawkins, missus,' he said.

'Dawkins. I have summoned you here to ask you how Mrs Carlisle went about sending her letters. I will have my own letters to send, and I need to know the routine at the castle. Do I leave them on my desk when they are ready to go?'

'I don't come in the housekeeper's room, not unless I'm sent for,' he said.

There was something self-consciously virtuous about his reply, and Helena found herself thinking that he probably did enter the housekeeper's room uninvited, though what he could want there she could not imagine, unless it was to snoop through the desk, in order to see if there was anything of use to him.

'What am I to do with my letters, then?' she asked.

'You have to leave them in the hall. There's a pewter bowl on a table under the window at the far end, in between two suits of armour. His lordship franks them, then I takes them to the village.'

'I see. And when do they go? Every week? Every day?'

'Whenever his lordship sees fit,' he said.

'And what happens to the letters until then? Do they remain in the bowl?'

'Nowhere else for them to go,' he said with an insolent grin.

Helena felt herself bridling.

'Did that suit Mrs Carlisle?' she asked.

'What d'you mean?'

'I mean, did she ever ask you to take one of her letters, even if his lordship had no letters to send? Perhaps she had something that needed to go urgently, and could not wait.'

'What kind of thing?' he asked craftily.

'I have no idea,' said Helena quellingly. 'Anything that might

43

need to be sent in a hurry, and might perhaps require a speedy answer.'

'No, missus, there were nothing like that.'

'Did she send letters often, then?'

A sly look crept into his eye, and Helena was sure he knew something she didn't. It seemed he could say more if he wanted to.

'No, missus. Once a week, as a general rule. She weren't a great letter writer.'

'I see.' She paused, to give him a chance to say more, but he remained silent. 'Very well, thank you, Dawkins.'

'Thank you, missus. Will that be all?'

'No. Not quite. I need to find out a little more about the castle, to help me with my duties. Tell me, what other servants work here? There is a butler, I suppose? And his lordship must have a valet.'

'There's no butler. The last one died, and his lordship never replaced him. And his lordship's valet left when . . . he don't have a valet any more. He likes to see to himself. There's not many as'll work in the castle. Servants are hard to come by.'

He puffed his chest out, and she realized that he was taunting her, daring her to interfere with him, and warning her that, if she did, he might decide not to work there either.

'Then what other servants are there in the castle besides you, Mrs Beal, Effie and Miss Parkins?'

'There ain't no more.'

'None? How did Mrs Carlisle keep the castle clean without any maids to help her?'

'It weren't always that way. There were two maids here when Mrs Carlisle worked here. Sally and Martha, they were. But they wouldn't stay in the castle.'

'Oh? Why not?' asked Helena, wondering if he would tell her more than Mrs Beal had done.

'It was the stories, missus. About his lordship.'

Helena felt her pulse quicken, but she gave no sign of it.

'What kind of stories?' she asked.

'People likes to talk in a village,' he said. 'There's always been things said about the Stormcrows.'

'A lot of nonsense, I expect,' said Helena encouragingly.

He gave another sly smile.

'You, at least, do not seem to believe them, or you would not still be working here,' she said, hoping to coax him into saying something further.

'Oh, I'm safe enough. Nothing'll happen to me. There's never anything happened to a man,' he said.

He was toying with her, trying to unsettle her.

'I'm glad to hear it. But surely there hasn't been anything happening to women, either?' she asked.

He said nothing.

'Why did the maids leave?' she prompted him.

'It were on account of Mrs Carlisle,' he said, his desire to talk overcoming his desire to have her in his power. 'Disappeared in the dead of night, she did, and Sally said she heard crying from the east wing, up in the attic, and the following day, Martha said she heard it, too. "It's a cat", I said to them, but they wouldn't listen. Gave in their notice and went home.'

Helena felt a shiver run up her spine.

'Did you find it?' she asked. 'The cat?'

'Didn't need to. The crying stopped, so it must have got out. But it's better not to go near the attics, all the same.'

'Oh? Why?' she asked.

'Rotting floorboards, missus. Dangerous, they are. Could

45

give way at any minute. Anyone who goes up there could go crashing right through and break their necks.'

He gave her an insolent look, and the thought flashed through her mind that she would not like to be alone with Dawkins in the attic.

She questioned him further about his fellow servants, but he had nothing to say, other than that Mrs Beal was a good cook and that Effie was a clumsy thing.

'And Miss Parkins?' asked Helena.

He hesitated, and she thought: He is afraid of Miss Parkins, too.

When he had told her all he could, she dismissed him.

Once he had left the room, she took her letter out of her pocket. She had not been going to send it, thinking that she would see Caroline soon, but she changed her mind. She wanted to see if a letter sent from the castle would arrive. If it never reached its destination, then it was possible that Aunt Hester had written to her, but that Aunt Hester's letter had never reached its destination, either.

Finding sealing wax in the drawer, she was about to apply it to her letter when she paused. If Dawkins read it – and having met him, she would not put it past him – she did not want him to discover that she was not Mrs Reynolds. She found paper and a quill, and she rewrote the letter. As she began to write, she was pleased with the pen's smoothness, and was reminded of Aunt Hester, who had prided herself on her quills. She had told Helena on more than one occasion that she could not hope to write a neat hand with an ill-mended pen, advice that had gone home, for Helena had always admired her aunt's hand-writing.

She thought for a few minutes, composing the letter carefully in her head, and then began to write.

My dearest Caroline,

I have arrived at the castle, and his lordship has given me the position as his new housekeeper. I have not found what I was looking for, but I have not despaired of finding it either, and mean to persevere. I am sure you will be pleased to know that I am well. You will not have time to write me more than a line or two, I don't suppose, but let me know if you are well, and if you hear anything of H, please let me know. You may send your reply to me here at the castle. Address it to:

Mrs Reynolds
Torkrow Castle
Seremoor
Yorkshire
Fondest regards,
Your dear friend

She scrawled an illegible signature at the bottom of the letter, then sanded it, and, when it had dried, she folded it and fastened it with sealing wax, and then she went out into the hall, and looked about her for the table.

Seen in full daylight, the hall was even larger than she had imagined, and just as austere. The light glinted on the silver armour and lit the stone with a cold light.

Her eye fell on the oak table, and she crossed to it and put her letter in the bowl. There were no further letters there, and she wondered how long it would be before it was sent.

She heard a clanking sound and started, but, turning round, she saw that it was only Effie, carrying a bucket of coal towards the housekeeper's room. As she watched her, Helena thought that, although the girl was young and nervous, if she was capable of going through the housekeeper's desk, she might also be capable of tampering with the mail. Perhaps she had interfered

47

with it innocently, dropping the bowl as she dusted beneath it, and seeing that a letter was damaged, perhaps she had taken it in order to escape a scolding. It was possible. She questioned the girl gently, but Effie maintained that she never touched the mail, so she let her go about her business.

Who else crossed the hall in the course of the day? she wondered, as she glanced at her letter, which lay defenceless in the bowl. Mrs Beal might venture into the hall occasionally, but Helena did not believe Mrs Beal would interfere with the post. And then there was Miss Parkins. Helena shivered as she thought of the waxen face and the long, cold hands. Miss Parkins would be capable of taking one of Aunt Hester's letters, but why?

There was no one else . . . except Martha and Sally. They had both been at the castle when her aunt had been there, and perhaps one of them had seen it, or taken it.

There was a sound of footsteps behind her, and his lordship came into view, followed by Dawkins, who was hurrying to keep up.

'Go to the stables. Tell them to ready my horse. I want it brought round to the front of the castle.'

'Yes, my lord, very good, my lord,' said Dawkins, bowing, before heading towards the door.

Summoning her courage, Helena spoke to the earl as he passed.

'Might I speak to you, my lord?' she asked.

He turned towards her, and she wondered what he was thinking. Nothing very pleasant, if his expression was any guide. His mouth was grim, and his deep-set eyes looked haggard.

'Well?' he demanded.

'It is about the maids, my lord. I understand there used to

48

be some working here. I do not believe I can keep the castle clean without help. There is a great deal of dusting and polishing to be done, to say nothing of the floors to be washed. Mrs Carlisle had some housemaids to help her, I understand.'

He looked at her as though weighing his words and then said, 'And so you would like me to appoint some?'

'I could take care of that, my lord, if I had permission to employ, perhaps, two girls.'

'Very well. You may walk in to the village on Friday. See to it, Mrs Reynolds, but don't disturb me with this matter again.'

'Very good, my lord.'

He strode past her, and went into the library.

Perhaps Martha and Sally could shed new light on her aunt's sudden departure, she thought . . . and perhaps they could tell her more about the crying in the attic.

For some reason the tale had disturbed her. It had only been the sound of a cat . . . and even if it had, by an chance, been a human being, it would not have been Aunt Hester. Helena could not remember Aunt Hester ever crying.

But a small voice asked her: what if it had been Aunt Hester? What if Aunt Hester had had some bad news, and had left the castle accordingly?

She found that she was walking towards the stairs, almost without her own volition, and she knew she would have no peace until she had been to the attic, to see if, perhaps, there might be any evidence that her aunt had been there, and now was a good time, for there was no chance of encountering Dawkins, who was on an errand for Lord Torkrow.

Lifting the hem of her skirt, she mounted the stairs, going up to the second floor and then looking for the steps that led to the attic. She found them at last, tucked away on a corner, a narrow spiral staircase, lit by arrow slits in the walls.

She went up as fast as she safely could, and finally reached the top. To her left was a row of windows, and from them she could see the moors stretching out before her, their undulating hills and hummocks a dull green against the grey sky. Set in their midst, the castle was isolated and cut off, and she was forcefully reminded of the fact that it was a long way back to town, and civilization. Anything could happen in the castle, and no one would ever know. . . .

She turned her attention back to the task in hand. She saw a long corridor on either side of her, from which various doors opened off. At the end of each corridor was a heavy oak door, the doors to the east and west wings, she supposed.

Dawkins had said the crying came from the east wing, and, glancing at the dim sun that shone weakly through a rent in the clouds to get her bearings, she chose the east door. She tried to open it, but it was locked.

She began to try the keys. One by one, she tried them all, but none of them fitted. She listened at the door, but could hear nothing, so she knocked on the door, and called out, but there was no reply.

There is no one there, she thought. The attic is disused. The crying was nothing more than a cat, and the animal escaped weeks ago.

But a need to get into the east wing and see for herself had taken hold of her, and she went into the large attic room that was nearest to the east wing, hoping that there might be a way through. It was a vast space, and draughts swirled around her. It was full of old pieces of furniture, a selection of childhood toys and assorted broken chairs, tables and household objects. The floorboards were bare. She went into the corners, but there was no sign of a door, or a way into the east wing, and reluctantly she had to admit defeat.

She went out on to the landing and a movement below caught her eye. Through the window she saw a solitary figure in the courtyard below: Lord Torkrow. Where was he going? she wondered.

As he headed towards his horse, he stopped suddenly, and she felt an unaccountable sense of alarm. She shrank back as he looked up, and his eyes raked the window. As her heart began to race, she wondered why she was so afraid. She had every right to be in the attic. But even so, she felt a sense of relief when she heard the horse's hoofs on the gravel and knew he was on his way.

CHAPTER FOUR

Helena was relieved to join Mrs Beal in the kitchen again for dinner. After an unsettling day, here, with cheerful company, the castle seemed less menacing, and Helena felt her confidence returning. She knew she needed to be careful of what she said, but she was not as frightened of making a slip in front of Mrs Beal as she had been in front of Lord Torkrow, because Mrs Beal would probably not notice. And if she did, she would probably forget it the moment a pie needed taking out of the oven.

'Yes, you'll need some maids,' said Mrs Beal, as the two sat down to a nourishing meal of chicken and potatoes, and Helena told her that she had spoken to the earl. 'There's no way you can run the castle without them. It's a long walk to the village across the moor, mind, so make sure you're wrapped up warm, and mind you wear stout shoes.'

'I will,' said Helena.

'Go and see the rector's wife, Mrs Willis. She's used to finding maids for the castle. The last two left, silly girls. Said they'd heard a ghost, or some such nonsense. But work's scarce hereabouts, and there'll be two more to take their place.'

'Do you think the same two could be persuaded to return?' asked Helena. 'They would know their business,' she explained, when Mrs Beal looked surprised.

Mrs Beal considered. 'Maybe. Their fathers will want them working, that's for sure. If it hadn't been for the fact there was no housekeeper at the castle, they'd have made the girls go back to work at once, ghosts or no. But there were those in the village who said it wasn't right for girls to be working at the castle with no one but his lordship here. I'm down in the kitchen all day long, and the villagers know it.'

'But they would consider Miss Parkins a suitable chaperon, surely?' asked Helena, voicing the concern that had been troubling her.

Mrs Beal pulled a face. 'There's not many that like Miss Parkins hereabouts. Why keep a lady's maid when there's no lady? That's what the gossips say.'

'They can't think. . . ?'

'Why, bless you no, there's none so crazed as that, but there are those who say she knows things about him, things that could harm him, and that's why he keeps her here. There are those who say he can't afford to turn her away.'

'Do you believe it?' Helena asked, putting down her cup.

'Not I. He's a good master. Some are forever finding fault: the food's too rich, the food's too plain, there's too much spent, there's too much waste . . . nothing but complaints with some people. But he never criticizes. I can make what I want, as long as the housekeeper agrees. It's a good place, and I mean to keep it.'

As they ate, Helena asked, 'Dawkins doesn't eat in the kitchen, then?'

'He has his meal earlier, at four o'clock, with Effie. It leaves him free to attend to his lordship when his lordship eats his meal.'

'Do you know where his lordship has gone?' asked Helena.

'He's gone to York, maybe, to see to business.'

'And do you know when he is likely to return?'

'He never says. It would be easier if he did. He'll expect a hot meal when he gets back. But there, it's not his place to think of my convenience, it's mine to think of his.'

They finished their meal, and Helena retired for the night. Her footsteps sounded ominously on the stone floor, pattering like a frightened animal scurrying for shelter, and she thought she detected the sound of footsteps following her. Her mind worked feverishly, trying to convince herself that any stray footfalls were merely echoes, but she quickened her step nonetheless. Then she stopped abruptly, trying to catch whoever was following her, but there was no extra footfall. The echo was nothing but her imagination, she told herself, and hurried on.

The flickering light of her candle cast strange shadows on the walls, and she jumped at the sound of a door creaking somewhere below. The castle seemed full of mysteries, and she longed for the safety of her room.

She began to run, hastening up the stairs and along the corridor . . . and then stopped. She quickly retreated into an open doorway as she saw Miss Parkins at the end of the corridor, standing just outside her room. The maid's hand was on the door knob.

Helena's thoughts began to race. Was Miss Parkins about to go into her room? Or had she already been inside?

She shrank back as she heard Miss Parkins coming towards her and, afraid of being discovered, she slipped into an empty room. She snuffed her candle, for a strange fear had gripped her, and it did not leave her until Miss Parkins had walked past.

She waited until she was sure Miss Parkins had gone before stepping out again. The corridor was dark, and she had to let her eyes adjust to the gloom before she could go on. She began to regret having snuffed her candle. Feeling the wall at her right with one hand, she continued down the corridor and fumbled with her door knob, then turned it and went in. The fire was glowing in the hearth, and she quickly lit her candle from the flames, then lit the other candles. She looked around, wondering if Miss Parkins had entered the room but she could see no signs of it. Nothing seemed to have been disturbed. But Helena was still not comfortable. If Miss Parkins had not entered the room, then Helena felt that she had been about to do so. She determined to lock her door every time she left her room in future. She wanted no more unwelcome visits.

Simon, Lord Torkrow, arrived in York and then made his way to the office that had sent him Mrs Reynolds. He went in.

'May I help you?' asked the man behind the desk.

'You supplied me with a housekeeper, a Mrs Elizabeth Reynolds. I would like to speak to the person who interviewed her and recommended her for the post.'

'If I might have a name?' enquired the young man.

'Lord Torkrow.'

'I will apprise Mr Wantage of your visit,' said the young man, bobbing into an inner office and returning a minute later to usher him in.

'Lord Torkrow, this is a pleasant surprise – an honour, an unexpected honour. I hope all is well at the castle? Mrs Reynolds suits, I trust?'

'Did you interview her?' asked Simon, taking the seat that was offered to him.

'No, that was my colleague, Mr Brunson.'

'I would like to speak with him.'

'I am afraid he is not here, he was taken ill on Monday with

55

a putrid sore throat, but if I may be of assistance?'

'You met Mrs Reynolds?'

'No, I did not. I read her references, however, and they appeared to be in order. We have recommended her for positions before, and she has always given satisfaction. I hope there is nothing wrong?'

'I would like to speak to Mr Brunson as soon as he is well enough. You will write to me, and let me know when he is fit to be seen.'

'Yes, my lord, of course, my lord.'

'Good.' He thought for a moment, and then said, 'Your boy in the outer office. He met Mrs Reynolds?'

'Alas no, my lord. He has only just joined us. His predecessor is sadly deceased.'

'I see. Very well. Inform me when I might speak to Mr Brunson.'

'Yes, my lord, very good, my lord.'

He took his leave, with Mr Wantage bowing him out of the office, then set about paying attention to business.

The following morning, Helena was awake early, and was already dressed when Effie entered the room. She wanted to question the girl, and find out what she had discovered in the drawer in the housekeeper's room.

'Good morning,' she said, when Effie entered the room.

Effie grunted a reply, and set about seeing to the fire.

'I wonder if you can help me,' she said. She considered asking a direct question, but suspected it would produce nothing but anxiety in the girl, as it had when she had questioned her about the mail, and so she decided to lead up to it in a roundabout manner.

'I would like to make an inventory of the drawers in the

housekeeper's desk – that means making a list of everything that's inside them,' she explained, as Effie turned and looked at her blankly. 'I need to know which of the things belong to the castle, and which belong to Mrs Carlisle. She might want to claim her belongings when her sister is feeling better, and I do not want to use them by mistake.'

'No, missus.'

'Can you remember what you saw there when you looked for some string?'

Effie turned back to the fire hurriedly, knocking the fire irons over in the process. They fell with a clatter, and Effie jumped, picking them up nervously and trying to hang them back in place, with hands that shook so much she had to make several attempts before succeeding.

'Can you remember what there was?' Helena prompted her.

Effie shook her head.

'Was there, perhaps, some writing paper?'

Effie jumped.

'There was a letter, perhaps?'

Effie's mouth clamped together, and her hands shook as she raked the grate.

'Do you remember anything at all?' Helena asked.

Effie shook her head, and concentrated vigorously on her task.

It was clear she was going to get nothing from the girl, at least for the moment, so she complimented her on her ability to lay a clean fire. Her praise went some way towards relaxing Effie, who picked up the empty bucket and hurried out of the room.

One avenue of exploration had led nowhere, but she hoped she might have better luck with another, and after taking breakfast with Mrs Beal in the kitchen, she went out to the stables, for she had remembered something overnight: Mrs Beal had mentioned in an earlier conversation that the coachman had taken her aunt to Draycot to catch the stage.

The stables were situated behind the castle, and the block was well tended. The noise of horses snuffling came from the stalls, and a glossy chestnut head looked out.

The black carriage, which Helena had ridden in on her journey across the moor, was standing in the stable yard, and the coachman was polishing the brass lamps.

'Good morning,' she said.

He looked up briefly and acknowledged her presence, before returning to his work.

'I wanted to thank you for driving me across the moor on my arrival here,' said Helena.

'His lordship's orders,' said the coachman.

'Quite so. It was good of him to take me up. It is not every earl who would make room in his carriage for his housekeeper.'

He grunted a reply and went on with his work.

'He seems to be a good master to work for,' she said.

He grunted again.

'He set you to drive my predecessor to the nearest town, so that she could catch the stage coach a few weeks ago, I understand.'

'Aye.'

'It was a very kind thing for him to do. Poor lady, having to leave in such a hurry.'

'Ah.'

'And so late at night. Was it not difficult for you to harness the horses?' she asked. He looked at her as though he thought she was a half wit. 'I'm afraid I know very little about horses. I have never learnt to ride, and I have ridden in a carriage only once or twice. But don't horses sleep, as we do? Did you have to wake them? Or had they not yet gone to bed?'

'Horses work here, same as everyone else,' he said.

'Even late at night?'

'Whenever his lordship commands.'

'It must be difficult driving across the moor in the dark,' she said. 'I am surprised Mrs Carlisle wanted to venture out in the middle of the night. Did she not think it would be better to wait until morning?'

'No.'

A horse snorted.

'Was there a stagecoach to take her on when you left her? I hope she did not have to wait in an isolated spot, all on her own.'

'I left her at the inn,' he said.

'What a distressing thing for her, to have to make such a long journey.'

She paused, hoping he would reply, but he was a taciturn individual, more used to dealing with horses than with people, and he said nothing, just continued with his work.

'Where do the stagecoaches go from here?' she asked.

'North. South,' he said.

'And west and east, I suppose,' she said in disappointment.

'Most ways,' he agreed

'That is very convenient.'

He did not reply and reluctantly she left the stable yard. She bent her footsteps towards the castle, but she was disinclined to go back inside. She feared she would be overcome by the oppressive atmosphere, and her lack of progress in discovering her aunt's whereabouts. Instead, she decided to take a walk. It was a bright morning. The air was fresh, and the sun was shining. There was even a little warmth in its rays.

She began to walk across the lawns that led to the outer

wall. To her right the drive led through the arch and out on to the moor. Directly in front of her was a set of stone steps leading up to the top of the wall.

As she took the steps, she wondered if the coachman had really taken her aunt to Draycot, or if he had simply said so on Lord Torkrow's orders.

In the fresh air, with the solid feel of the stone steps beneath her, it was easy to dismiss such suspicions.

She reached the top of the wall. It was windy, and she pulled her cloak tightly round her. She looked out across the moors. The landscape looked gentler than it had done the previous day. The colours were brighter, and the air softer. Far off, she saw a gleam of yellow. The cheerful colour stood out against the muted greens of the moor, and she saw that a few early daffodils were in flower, nestling in a sheltered hollow.

She descended the steps and went out of the gate, making her way towards the bright flowers, which were nodding their heads in the breeze. She picked a bunch and then carried them back to the castle. Taking them into the flower room, she arranged them in a vase, and then carried them back to the housekeeper's room.

On the way she passed the library, and thinking that the fire might need mending, she went in. She put the vase on the mantelpiece whilst she poured more coal on the dying flames, then allowed herself a few minutes to look at the books that lined the walls. She had read a great deal as a child, but after her father's death there had been little money for books and she had purchased only two the previous year. But here was a feast of literature. There were works by Shakespeare, Marlowe, Chaucer and many more, some in fine covers, and some in books that were falling apart with age. She took down a copy of *Le Morte D'Arthur* and lost track of time as she became absorbed. She was lost to the world, but the sound of the door opening shocked her back to reality. She turned round to see Lord Torkrow standing in the doorway.

'I have just been repairing the fire,' she said, hastily putting the book back on the shelf.

He glanced round the room, and his eyes fell on the vase of flowers. She was about to hurry over to the mantelpiece when, to her surprise, his face relaxed. It was warmer and more open than before, and she felt a rush of some strange feeling rise up within her. She had not realized he could look so appealing.

'There haven't been daffodils in here since. . . .' he said.

There was such a wistful tone in his voice that she held her breath, wondering what he would say next, but he never finished the sentence. Instead, his voice trailed away, and Helena dare not move. He was lost in thought, going back to some previous time, and the memory seemed to please him. But it was made up of pain as well as pleasure, she thought, because there was a twist to his mouth that cut her to the quick. She was surprised at the stab of pain that shot through her, because she had not been prepared for it, and for a moment she saw him not as an enigmatic and forbidding figure, but as a man of flesh and blood.

What had hurt him? she wondered. Why did the simple sight of daffodils bring him pain?

He roused himself, and turning towards her, he said, 'You have done well.' He noticed that she was standing by the book-case and said, 'You are interested in books?'

'Yes,' she said.

'Then you must use the library. You may choose something to read whenever you wish.'

For a moment there was a gleam of friendship illuminating the room. It warmed her, as the unexpected gleam of daffodils had warmed the moor. It relaxed something deep inside her, something that had long been frozen, but in this strange place and stranger situation, it started to come to life.

'Thank you,' she said.

'You were looking at this?' he asked, going over to the shelf and taking out *Le Morte D'Arthur*, which she had not pushed far enough back on the shelf.

'Yes.'

'Then take it. I think you will enjoy it.'

He handed her the volume.

'It must have taken generations to assemble a library like this,' she said, looking round at the laden shelves as she took it.

'Yes, it did.'

A sense of longing welled up inside her. She had no home, and, saving her aunt, no family. She did not know where she would be in a year, or even a month's time. She would have to go where the wind blew her. But he belonged to the castle. He was lucky. He had his place in the world by right. She sometimes wondered, in the dead of night, if she would ever find hers.

'It must be a wonderful feeling, to have a home, to belong,' she said.

He looked at her strangely and she realized that she had forgotten to whom she was speaking. The gleam of friendship he had shown her had lowered her defences and made her forget her position, so that she had spoken to him as an equal, but she quickly reminded herself that she and Lord Torkrow were not equals. They were master and servant, separated not only by rank but by deception and the disappearance of her aunt.

'I have work to do. . . .' she said.

She began to head towards the door, but, as she tried to pass him, he put out his arm, resting it on the desk so that he was blocking her path.

'You call it belonging,' he said. 'I call it being trapped.'

He looked down at her, and she felt herself being pulled into the strange aura that surrounded him, a magnetic strength that held her fast.

'The weight of the castle oppresses me,' he said, looking deep into her eyes as though seeking understanding. 'At night, the walls close in.'

'But it is your home,' she said, searching his eyes.

'It is not my home. It is my tomb.'

All light and warmth had gone from his voice, and she was once more afraid of him, but the fear was tempered with intrigue. She clenched and unclenched her hands, then said, 'But you can leave the castle if you want to.'

'Can I?' he said with bitterness.

'You were returning to it on the day I arrived, so you must have left,' she said, striving to remain calm, 'and you left again yesterday.'

'Briefly, yes. But the castle keeps drawing me back. It is not fond of letting its inhabitants go.' He looked deep into her eyes once more, and his words sounded like a warning. 'One way or another, it finds a way to keep them.'

He fell silent, and Helena stood there, unable to pass, but unwilling to disturb him. He had become lost in thought, and his eyes were fixed on the floor. After a minute he roused himself.

'You must tell me what you think of the book when you have read it,' he said, dropping his arm so that she could pass. 'We are not unlike the knights of old, you and I. We, too, have monsters to fight.'

Helena quietly left the room. She went upstairs, taking her

book into her bedroom. As she put it on the table, she wondered if her aunt had sat at that very table writing her letters, and wondered, if the table could talk, what tales it would tell?

Simon scarcely noticed the door closing. He was lost in his thoughts, seeing the past, when the castle had flourished. It had been full of noise and colour when his parents had been alive, until. . . .

Strange how the sight of the daffodils had taken him back to that time, their bright yellow and green reminding him that there were colours beyond the stone, oak and metal of the castle.

How soft they had seemed, how fragile, as she had been soft and fragile. . . .

But he must be on his guard, he must not lower his defences again.

It was evening. Having dined with Mrs Beal, Helena was sitting by the fire in her room. The curtains were drawn, shutting out the black night. She was leafing through *Le Morte D'Arthur*, looking at the illustrations, which were beautifully done. It was as she looked at a picture of a man with a candle that a thought struck her. In the absence of a key to the east wing of the attic, she might be able to find out if anyone was in there by going out into the grounds and seeing if there was a light in the window. It was still early, not yet seven o'clock, and she decided to go before it was too late.

She laid aside her book, uncurled herself from the chair and put on her cloak and heavy shoes. She tied her bonnet under her chin, pulled on her gloves then went down the stairs.

She slipped out of a side door, and walked across the lawn,

which was silvered by the moon. She walked away from the castle, so that she could see the windows clearly when she looked back. When she felt she had gone far enough she turned and looked up, but they were dark. There was not a glimmer of light anywhere. She had been hoping to see something, but there was nothing.

She was just about to go back inside when she caught sight of a lantern bobbing along in the distance. Her senses were immediately alert. Who would be going out with a lantern at this time of night? And where were they going? She hesitated. A part of her wanted to ignore it, but curiosity won over caution, and she began to follow cautiously.

The light disappeared briefly and Helena realized that whoever had been carrying it had gone through the archway in the outer wall. She followed quickly, taking care to stay well back so that she would not be seen.

She passed through the arch and saw the light again, in the distance. It looked unearthly, bobbing along, detached from the ground, a ball of glowing yellow in the darkness. She followed it, but soon she began to grow uneasy as she felt the gravel give way to coarse grass and found herself walking across the moor.

The wind whipped round her, stronger than it had been in the courtyard, pulling her cloak open and knifing her with freezing air. She pulled it around her, holding it closed with folded arms, and went on.

An owl hooted as it flew by her on silent wings, making her jump, and she turned and looked at the castle, nervously wondering if she should turn back. But if she did, she would learn nothing.

The grass beneath her feet was tufted with hillocks that made the going uneven, and once or twice she stumbled as her foot caught in a ditch. Then her shin hit against something hard and she found that she had reached a low wall. She felt along it with her hands until she found a gap and went through.

The light was now further away and she hurried forward, only to trip over a large stone. When she looked down, she dimly made out the shape of a headstone. It had fallen on to its side and lay, neglected, on the turf. She stepped back in alarm, and found the back of her legs were against another tomb. Icy fingers of fear crawled up her spine. She was in a graveyard.

She wished she was safely in her own room, in front of the fire, reading about knights and battles, but as her panic began to dissipate, she reminded herself that it was only seven o'clock, and that there was probably a down-to-earth explanation for the lantern, which she would soon discover.

The light had disappeared and she moved forward cautiously. Her footsteps halted. She could see a figure kneeling ahead, silhouetted against the lantern, which was on the ground. As she stood there uncertainly, the moon sailed out from behind a cloud, and in the cold light she saw that the figure was Lord Torkrow. To her shock, his shoulders were heaving, and she realized he was crying.

Her heart lurched at the desolate sound, and she found herself privy to a terrible grief. She was torn between a desire to leave and an impulse to go and comfort him, but caught between the two impulses she remained where she was.

She was frozen, lost in a timeless expanse, until at last his grief was spent. His cries subsided and he stood up, reclaiming his lantern. Helena shrank back against the gravestone and he passed by without seeing her, his lantern bobbing away from her in the dark.

When he had gained a sufficient lead she followed the light

back across the moors, back through the arch and back to the castle. She slipped round the side to the small door and let herself in, her fingers trembling as they lifted the latch.

What had she just witnessed? she asked herself. Was it grief for the loss of a loved one, or could there a more sinister explanation? Could it be that his tears had been produced by guilt?

As she slipped upstairs, she felt the atmosphere of the castle beginning to oppress her.

She took off her outdoor clothes, glad to be safe in her room. But as she sat down by the fire, Lord Torkrow's desolate cries echoed in her ears.

CHAPTER FIVE

Helena dreamt that she was outside, late at night, and flying across the moor. Above her was a gibbous moon, with torn clouds blowing across its face, and ahead of her was a blasted tree, its twigs spreading like fingers and its joints creaking as it was bent and twisted by the wind. She sped towards it, then passed through the branches and emerged untouched on the other side. Before her lay a graveyard, with tombs scattered across it like bones picked clean by the crows. Beside them was a man, wrapped in a cloak, with a lantern at his side. As she glided closer, she saw that his face was ghostly. Black shadows filled the hollows, and a sickly pallor marked the planes. He was shaking with grief, and his shoulders were heaving as racking sobs filled the air. She flew closer, around and behind him, until she was looking over his shoulder into the grave.

Then all of a sudden she realized his shoulders were shaking, not with grief, but with mirth and, as she looked past him, she saw, to her horror, that the body in the grave was that of her aunt. She turned and fled, moving rapidly away, carried on the wind, floating higher and higher as she approached the castle, rising up and up, until she was on a level with the attic, and she found herself looking through the windows. There was nothing to be seen, only the ghostly shapes of furniture cloaked in dust sheets, and a clock ticking, ticking by the wall. And then a dust sheet moved, and was thrown back, and her aunt's corpse rose from a chair.

Helena awoke with a shock. She was covered in cold sweat and was trembling all over. It was icy in the room. She shivered, and her breath formed clouds in front of her. With numb fingers she reached out for her wrapper and threw it round her shoulders, then climbed out of bed on shaking legs. She went over to the fire, which had all but gone out. She raked the ashes, encouraging a small spark, and fed it with small pieces of paper. She piled on twigs, and when they had caught light she put on a few pieces of coal. Still shivering, she returned to her bed . . . but she stopped as she approached it, for there was something under the covers. Her skin began to crawl. She saw the covers rise and fall. Someone was under there!

Someone, or some thing.

She reached out and twitched back the cover, and her aunt sat up in the bed, two weeks dead and laughing—

She sat up with a start.

Am I really awake this time? Helena wondered, her heart hammering in her chest. Or am I still dreaming? She looked around the room, fearing another nightmare vision, but everything was peaceful. The fire was burning low in the grate, casting a mellow glow over the furniture. All was as it should be. Her pulse began to slow, and her breathing became less shallow. She reached for her wrapper, still not convinced that she was awake. Warily, she threw it round her shoulders and slipped out of bed. She went over to the fire and knelt down beside it, warming her hands and taking comfort from the glowing coals. She lit a candle, then sat on the hearth, loath to go back to bed. She glanced towards it, but there was no strange shape under the covers. The blanket was still thrown back, revealing the white sheets beneath.

She heard the clock strike in the hall. It would soon be time for her to rise. She was glad of it. She had no desire to go back to bed. She waited only for Effie to bring her hot water and relight the fire before slipping out of her nightgown and, once washed, putting on her dress. Having completed her toilette, she left the room. The stone corridor was unwelcoming. Her candle seemed feeble, a puny attempt to light the space. Walls and ceiling waited in the shadows. The castle seemed a living thing. Old, monstrous, biding its time, before it claimed another victim.

She tried to banish such thoughts, but they would not leave her. She quickened her steps and the patter of her feet was matched by the patter of her heart.

Quicker and quicker, down the stairs, through the hall, into the kitchen, where sanity was restored. Candles filled the space with light. The hearty fire added its glow. Mrs Beal was brewing a pot of tea, a beacon of homeliness in the brooding atmosphere of the castle.

'It's colder this morning,' said Mrs Beal cheerily, as she put the finishing touches to the table, adding a pot of honey to the rolls and butter that were already there. 'There was frost on the inside of the window when I came downstairs. It's still there, look, even now.'

'Yes,' said Helena, relieved to be talking about something so ordinary after her disturbed night.

She blew out her candle and put it on the dresser.

'There now, we're ready,' said Mrs Beal, looking at her handiwork with pleasure.

Helena sat on one side of the table, and Mrs Beal sat opposite her.

'Tell me, Mrs Beal, do you have a key to the east wing of the attic?' Helena asked, for she knew she must unravel its secrets soon, or say goodbye to sleep. 'I would like to air it, but the door is locked and there isn't a key on my ring.'

'No, I don't have any keys for upstairs. Now if you find you're missing a key to the wine cellar or the dairy, I can help you there. I've the keys for all the rooms below stairs.'

'No, thank you, it's only the attic key I need. Do you know if Mrs Carlisle kept any spare keys anywhere?'

Effie dropped a handful of cutlery, which clattered against the flags.

Mrs Beal shook her head and tut-tutted as the girl picked up the kitchen utensils.

'Sorry, Mrs Beal,' gasped Effie.

'Just you make sure you clean everything properly,' said Mrs Beal.

'Yes, Mrs Beal.'

'Now what were we talking of?' asked Mrs Beal.

'The key to the attic. I wanted to know if Mrs Carlisle had a spare set.'

'Not that I know of. She was a very organized lady, though, and I'm surprised there's a key missing. are you sure it's not on the ring?'

'Quite sure.'

'It's possible she never had one. Some of the rooms are never used. They probably haven't been opened since her lady-ship was alive.'

'Her ladyship? Did his lordship have a wife?' asked Helena, thinking that here was the answer to the mystery of him crying over a grave.

'Lor' bless you, no, his lordship's never been married,' said Mrs Beal. 'I meant her old ladyship, his mother. Ah, a wonder-ful woman she was. A great lady. Always had a kind word for

71

everyone. "That was a very good stew you served us up yesterday, Mrs Beal", or "I want to thank you for all your hard work,
Mrs Beal. The banquet was a great success".'

'So his lordship never married,' mused Helena.

'He never needed to, not with his brother—'

She stopped suddenly.

'I didn't know he had a brother,' said Helena.

'Oh, yes. But you don't want to hear about all that,' said Mrs
Beal. She took the kettle from the fire and made the tea.

'On the contrary, I'm interested in the family,' said Helena,
and waited for the cook to go on.

Mrs Beal looked to be weakening, but there was another
clatter as Effie dropped a pan and Mrs Beal's attention was
distracted.

'What are you doing?' asked Mrs Beal, going over to the
young girl.

'Sorry, Mrs Beal,' gasped Effie.

'That pan's given years of good service, and if it's properly
looked after it'll give years more,' said Mrs Beal admonishingly.

'Yes, Mrs Beal,' said Effie, picking up the pan and putting it
back into the sink.

Mrs Beal returned to the table, grumbling about the difficulty of finding good help in such an isolated spot. Helena
tried to induce her to talk about his lordship's family again, but
Mrs Beal had evidently decided that discretion was called for,
and Helena could not persuade her to say more.

Instead, Mrs Beal recounted the troubles of her position,
talking about the likelihood of the fishmonger retiring, and the
scandalous way the dairymaids paid attention to the farm
hands instead of keeping their minds on churning butter.

As she talked, Helena listened with only half an ear as she
wondered about Lord Torkrow's brother. He must be a

younger brother, otherwise he would have inherited the title. She wondered if he was still at school: that would explain why he was not living at Stormcrow Castle.

But she still found it odd that Mrs Beal did not want to talk about him, unless he was the black sheep of the family.

Such a fantastic idea was ridiculous, she told herself, looking round the cosy warmth of the kitchen. She resolutely banished it from her mind. But once she left the kitchen's safety and comfort, her dreams returned to haunt her. In the cold stone corridors they did not seem so far-fetched. The castle was dark and mysterious. It was also very old. It must have seen some terrible things. And so must the graveyard.

She knew she would have no peace until she had visited it again. She decided to go past it on her way to the village, in order to see Mrs Willis about appointing some maids, and read the inscription on the tombstone. If it bore the inscription H Carlisle . . . she could think no further. First, she must find out who it belonged to, then she could worry.

There were some tasks that she needed to complete first, but she resolved to go after lunch.

True to her resolve, she set out as the clock on the stables chimed one. She paused at the threshold. A light rain was falling. Lifting her hood, she made sure it covered her head, tucking in a stray wisp of hair, then she set out across the courtyard. From the direction of the stables she could hear the muffled sound of horses snorting, but there was no other sound in the stillness.

It was cold and wet underfoot, and she was glad of her stout boots. The drizzle was dispiriting. The clammy air made her face damp, and her cloak was soon beaded with water. She went under the arch and then across the moor until she reached the low wall she had struggled over the night before.

She saw the gap and went through, trying to remember the direction she had taken. She had walked forward until she had tripped over a fallen headstone . . . she saw it . . . and then she had moved forward again.

She walked more slowly, hoping to find the exact spot, but the rows of graves all looked the same. She stopped when she thought she had reached the right place and examined the tombs and headstones. There was nothing remarkable about them. Some were simple and some were ornate. Some told of long lives, and some of short. John Taylor, Bella Watson and Henry Carter had all lived for more than ninety years. Richard and Lucinda Pargeter had died before they were twenty-two. But she could see nothing that would account for Lord Torkrow's grief, nor could she see anything bearing the name Carlisle.

She walked slowly through the graveyard, looking for any signs of a recent burial, but she could not see any disturbed earth. The graves were all at least a year old, and most of them were much older.

She began to feel more easy, knowing that her aunt had not been consigned to the ground. She felt ridiculous for having pictured Lord Torkrow as a murderer who had visited the grave of his victim, overwhelmed by remorse, when instead he was simply a taciturn man, who was at that very moment probably doing nothing more alarming than taking luncheon and dealing with his business for the day. As for the grave he had been crying over, it was a private matter, and she should not meddle in it.

Leaving the graveyard behind, she continued on her way to the village. She walked briskly, enjoying the warmth her movement brought her. She had need of it, for the drizzle had intensified and she bent her head against it. She only hoped it would not rain until she reached her destination.

She was not to be so fortunate. Before long it was raining steadily. The rain came down more and more heavily, and she was just wondering whether she had better turn back when she heard an 'Urgh!' and, raising her head, she saw a woman hurrying along the road towards her. The woman was wrapped in a cloak and wore a bonnet on her head. She looked up and their eyes met. They smiled briefly, two strangers acknowledging the dreadful weather, and Helena was emboldened to ask, 'How far is it to the village?'

'It's a tidy step, and there's more rain to come,' said the woman, looking at the darkening sky. She hesitated, and then added, 'If you would like to take shelter, my cottage is not far away.' She glanced to her left, where a track ran off from the road.

Helena hesitated, but her cloak was sodden and if she remained out of doors she would soon be soaked to the skin.

'Thank you,' she said.

The two women fell into step and turned off the road. The track was rutted, and they trod carefully, trying to keep out of the mud and puddles. Before long they came to the cottage. It was a sturdy building of stone, with small windows set deeply into the thick walls. There was a wall around the garden, and a gate was set into it. The garden contained a few hardy shrubs which were looking as bedraggled as Helena felt, and she was glad when they reached the door. The woman opened it and they stepped inside.

The hall was small, but it was clean and well cared for. The woman removed her pelisse, bonnet and gloves and hung them on a peg, then turned to Helena.

'I am Mary Debbet,' she said, laughing as she pushed her wet hair out of her eyes, 'and I am very pleased to make your acquaintance.'

'I am . . . Elizabeth Reynolds,' said Helena, wishing that she did not have to lie. 'I am the new housekeeper at the castle.'

She thought Mary might withdraw, and tell her she would be welcome to sit in the kitchen until the rain stopped, but instead Mary said, 'Here, let me help you off with your wet things.'

She helped Helena remove her sodden cloak, and hung it up to dry, then led the way to a door on their left. She paused with her hand on the door knob.

'You will meet my brother in the sitting-room,' Mary went on. 'Please, do not be offended if he does not get up. He was wounded at Waterloo, and his nerves have not recovered. The doctor prescribed complete rest, and that is why we have taken a cottage on the moors.'

'I understand,' said Helena.

Mary opened the door and they went into the sitting-room. It had rough walls painted in shades of cream, and oak beams supported the ceiling. The small window was latticed, and the window ledges were very deep. To the right was a large fire, and in front of it sat a gentleman of perhaps five and thirty years.

'We have a visitor,' said Mary.

He looked up, but did not stand.

'Mrs Reynolds is the new housekeeper at the castle,' she said.

'I am pleased to meet you,' he said. His speech was slurred, but his words were polite and sounded genuine. He held out his hand, but it trembled and he dropped it again.

'Please, do sit down,' said Mary to Helena. She rang the bell, and a neat maid appeared. 'Tea, Jane, please. We are cold and wet and need something to cheer us.'

Jane bobbed a curtsey and withdrew.

'I hope your business was not urgent,' said Mary, glancing out of the window, where the rain poured down. 'I think you will be with us for some time.'

'No, luckily it can wait. It is very good of you to take me in.'

'On the moors, we help each other,' said Mary. 'We have to. It is very different from living in a town. Out here, it is possible to freeze to death when the snow falls, and whilst I don't believe it's possible to drown, it is certainly very unpleasant when the heavens open.'

The maid returned with a tray, and Mary poured the tea. She was a beautiful young woman. Her dark hair was sparkling with raindrops, which clung to it like diamonds. It was drawn back from her face in a tight chignon, but the severity of the style only served to enhance the beauty of her face. She had a creamy complexion and dark eyes. Her cheekbones were high, and her nose and mouth were well shaped. Her figure was good, and her well-cut gown suited her. She must have had many suitors in town, and Helena found it admirable that she had chosen to accompany her brother to an isolated spot for the good of his health, rather than indulge in the frivolities that must have been her lot in a more civilized location.

'Did you have some shopping to do in the village?' Mary asked.

'No,' said Helena, sipping her tea. 'I was going to see Mrs Willis, to ask her if she could help me to find some maids.'

'Ah.'

'There were girls working at the castle until recently, but they left, and I am finding it difficult to manage without them.'

'Yes, I'm sure you are. The castle is very large to manage alone. But it isn't surprising the girls left. There is a lot of superstition hereabouts. They were frightened of the strange

77

noises in the attic, and instead of attributing them to the creaking of old wood and the sighing of the wind, they attributed them to ghosts and ghouls.'

'You know about that?' asked Helena in surprise.

'There is not much to talk about in a small village,' said Mary with a smile. 'We all know everything. Tell me, have you heard any chains rattling or children crying? They are apparently everyday sounds at the castle.'

'No,' said Helena, smiling, too, at Mary's humorous tone of voice.

'You do not seem very happy, however,' said Mary, her expression becoming more serious. 'I don't suppose the castle is really haunted?'

'No, of course not,' said Helena quickly.

'But there is something troubling you,' said Mary thoughtfully.

Helena put down her cup.

'There are noises,' Helena admitted. 'But they are just the noises typical of an old building. It is taking me some time to get used to it, however. I have never lived in a castle before.'

'It must be exciting,' said Mary.

'I believe it would be, if I did not have so much work to do' – and if I was not so worried about my aunt, she added to herself.

'Yes, it must be difficult to keep clean. Old buildings always are. I hope you find the servants you need – though they may not be much use. Mrs Carlisle had a hard time making them work, I believe. They were more interested in gossiping, or so she told me.'

'Did you know her?' asked Helena in surprise.

'Oh, yes, we both did. We were very fond of Mrs Carlisle. She was an intelligent and interesting woman. We made her acquaintance at church, and she was good enough to visit us when she had an afternoon off. We do not have much company, and her visits were a treat for us, until. . . .'

'Until?'

'Until they stopped.' Mary picked up the teapot. 'Would you like another cup of tea?'

'Yes, please.'

Mary poured the tea, but did not continue.

'It must have been disappointing for you,' said Helena, prompting her.

'Yes, indeed. I asked Lord Torkrow if she was ill, thinking this must be why she had not called, and he said—'

'Yes?'

'He said she had gone to care for a sick sister and would not be returning.' Mary hesitated, and Helena had the feeling she was going to say that she did not believe in the story of the sick sister. She was clearly troubled, but did not seem to know how to begin. Then her expression changed, and Helena guessed that her sense of propriety had won out over her need to talk about her anxieties. Instead of expressing any fears, she simply said, 'It is a pity.'

'It is indeed,' said Helena.

There was a pause and then, in a slightly artificial tone of voice, Mary said, 'I am sorry not to have seen her one last time. We have a poetry book of hers, which she kindly lent to my brother.'

'I didn't know my—' Helena stopped herself saying *aunt* just in time, 'predecessor liked poetry.'

'Oh, yes, she was a very cultured woman. Do you have a forwarding address?' she asked casually.

How clever, thought Helena, admiring Mary's subterfuge.

'No, I'm afraid I don't.'

79

'No matter. How are you finding it working with Lord Torkrow? I believe he is a difficult man,' she said, changing the subject.

'I have seen very little of him. As I am new, though, everything seems strange.'

The conversation moved on, but Helena was sure that Mary's question had been a ruse. She had no doubt that Mary would have liked to write to her aunt, to reassure herself that everything was all right. So Mary, too, was anxious. Helena kept waiting for her to return to the subject, but she did not raise it again. Mary, she suspected, was in a similar position to her own: she did not know whom to trust.

They spoke of the moors, of the weather, and of Mr Debbet's health, and an hour passed very pleasantly. The rain began to abate at last.

'Thank you for your hospitality, but I think I must be going,' said Helena, as a weak gleam of sunshine found its way into the room.

'Of course,' said Mary. 'You will not walk to the village now, I hope? The light is fading, and it will soon be dark.'

'I must,' said Helena regretfully.

'Tom, our man, will be driving to the village tomorrow in the trap. I can have him take a note to Mrs Willis if you care to write one. I don't like to think of you walking on the moors so late in the day. The road is not well marked, and you might become lost on your return.'

Helena thanked her gratefully and accepted her kind offer. Mary gave her paper and pen, and Helena composed a note to Mrs Willis, asking her to send any willing girls to the castle. She particularly asked her to try and secure the return of the girls who had worked there before. She gave the note to Mary, and then the two of them went out into the hall.

'It was lucky for me I met you this afternoon,' said Helena.

'On the contrary, it was lucky for me. We see very few people. My brother, I know, enjoyed your company. He says very little but his spirits improve with diversion. I hope you will call on us again. You are welcome at any time.'

Helena thanked her, then having donned her outdoor things she took her leave. As she retraced her steps to the castle, she felt heartened to have met Mary. She felt, at least, that she had a friend in the neighbourhood, someone she could turn to if she had need. She was convinced that Mary was worried about her aunt's sudden disappearance, and decided that the next time they met she would broach the subject. If all went well, she might be able to take Mary into her confidence. Perhaps if she learnt nothing from Sally and Martha, she and Mary could think of what to do next.

She returned to the castle feeling tired, but happier than she had been for days.

Her happiness faded as she reached the courtyard however, for there, looking down at her from an upstairs window, was Miss Parkins. Seen in the distance, Miss Parkins looked like a statue, and even in the dim light, Helena felt sure the maid was watching her. She could feel the maid's malignancy spreading out to cover her.

What is she doing at the castle? thought Helena. Does she really have a hold over Lord Torkrow? Does she know something to the detriment of his brother? Is that why he allows her to remain?

Helena went on.

I might have an ally in Mary, she thought, but I have an enemy in Miss Parkins.

As she crossed the courtyard, she saw Lord Torkrow was just emerging from the front door. He was swathed in his black

81

coat, which flapped around his ankles. She found him a conundrum. He treated her with hostility, yet he had shown her sudden gleams of friendship; he frequented graveyards at night, but once there, he was overcome with grief; he inspired fear in his neighbours, but respect in his servants.

As she looked at him, she thought to herself, Enemy or ally, which is he?

CHAPTER SIX

Helena was just about to go in the side door when she heard the sound of wheels on gravel and, turning her head, she saw that a carriage was arriving at the castle. By the check in Lord Torkrow's step, and by the fact that he was wearing his cloak, she guessed the visitors had not been expected. The carriage rolled to a halt. The coachman jumped down from the box and opened the door, and a beautifully slippered foot set itself on the step. A moment later, a young woman robed in an emerald cloak with a trim of swansdown emerged. She had flame-red hair, which was elaborately coiffured, and which was topped by a hat with a large plume. She was followed out of the carriage by an older woman, who had the same brilliant hair, and who was dressed in an equally fashionable, if more matronly style, with an amber pelisse and turban.

Helena slipped in the side door and went upstairs to remove her cloak. As she did so, she passed Miss Parkins on the landing. Miss Parkins was looking down at the party below.

'He should have married her,' said Miss Parkins suddenly.

Helena did not know if the maid was speaking to her, and so she did not reply.

'*His* parents wished it,' said Miss Parkins, with a trace of bitterness. '*Her* parents wished it. It was a good match for both of them. If he had married her, he would have been on honeymoon when. . . .'

The sound of tinkling laughter came up from below, as the guests entered the hall. Miss Parkins seemed to recollect herself and she turned to Helena.

'You will have to hurry. You will be wanted downstairs.'

'Will you not be helping?' asked Helena.

Miss Parkins's gaze rested on Helena, making her squirm inwardly.

'His lordship and I do not see eye to eye on the subject of Miss Fairdean. He will not require my presence.'

'And Dawkins?'

'Dawkins has gone on an errand for his lordship.'

'Very well,' said Helena.

Miss Parkins moved away, leaving a chill behind her, making Helena shiver.

Helena returned to her room, removing her cloak and tidying her hair. As she did so, she regarded herself in the cheval glass. She had a well-shaped face with fine eyes and was passably pretty, as pretty, perhaps, as Miss Fairdean, but there any similarity between them ended. Miss Fairdean's hair had been arranged in the most becoming coiffure, artistically arranged with small curls framing her face, whereas Helena's chignon was scraped back from her face, with no curls to soften the style. Miss Fairdean's cloak had been made of velvet, and had shimmered like an emerald in the dim light, whereas Helena's cloak was made of grey wool. Her dress, too was made of dark grey wool, and she found herself wondering what Miss Fairdean's gown would be like.

Just for a moment she longed for beautiful clothes. She had never worn silk or satin; never possessed anything made out of velvet or lace; and never had a colour more interesting than dark blue. Miss Fairdean's hair and fashionable clothes had brightened up the afternoon like a beacon, whereas her own appearance was as dreary as the weather.

Fortunately, she had no time to linger. She could do nothing about her dull appearance, and besides, she had work to do. She went downstairs. The bell was ringing in the drawing-room as she reached the hall, and she answered it promptly. She went in to see Miss Fairdean reclining elegantly on a *chaise-longue*, her well-cut morning gown showing off her Rubenesque curves. She reminded Helena of a painting she had once seen in London, voluptuous and enticing, like a Venus come down to earth. Her mother, who was still a handsome woman, sat beside her.

Helena's eyes turned to Lord Torkrow, who was standing on the other side of the fireplace. His body was blocking the fire-light, and cast a black shadow across the gold damask-covered chair.

To her surprise, he was not looking at Miss Fairdean, he was looking at her. His eyes were fixed on her for fully a minute, as though committing her to memory. So long did he look, that Miss Fairdean and her mother looked, first at him, and then at Helena. They exchanged glances, and Miss Fairdean gave one exquisite shrug of her shoulder.

He roused himself.

'Mrs Reynolds,' he said. 'Some refreshments for my guests.'

'Very good, my lord.'

'You have managed to find a new housekeeper, I see,' said Miss Fairdean, as Helena walked towards the door.

'I have.'

'At least this one isn't as ugly as the last one,' said Miss Fairdean. 'The last one was a dreadful woman. She had a sour face, as though she'd been drinking vinegar. It must have been a torment for you to have to look at her. Where the offices find such frights is beyond me.'

Helena felt her teeth clench and, happening to glance in the mirror hanging by the door, she noticed that Lord Torkrow was watching her with a curious expression on his face. She quickly smoothed her expression and went out of the room, but Miss Fairdean's words would not leave her. A sour face? thought Helena angrily. Her aunt was a beautiful woman. She was lined with age and hard work, it was true, but beautiful nonetheless.

She hurried down to the kitchen.

'Tea, please, and cakes, Mrs Beal,' she said, when she entered the kitchen. 'We have visitors.'

Mrs Beal looked at her and then said, 'Miss Fairdean?'

Helena was surprised. 'How did you guess?'

'She's made you angry, and there's only one person round here that can anger a body so soon. What was she saying?'

'She made a remark about Mrs Carlisle,' said Helena bitterly.

'Ah. She's a spoilt young woman,' said Mrs Beal, as she set the kettle over the fire. 'His lordship's parents wanted him to marry her, but he was having none of it. They couldn't understand it. But fair by name is not fair by nature, and I reckon his lordship can tell the difference between the two.'

Mrs Beal set cups and saucers on the tray, followed by the tea pot, milk jug and sugar bowl. Helena carried them upstairs to the drawing-room. When she went in, a lively discussion was taking place.

'Oh, do say you'll let it go ahead,' said Miss Fairdean in an enticing voice. 'The spring won't be the same without a costume ball. It is such a feature of the castle. It is not such a very great amount of work, and besides, half of it must already be done.' She leant towards Lord Torkrow. 'Do say it will go ahead.'

Lord Torkrow turned to Helena.

'Miss Fairdean would like me to host a costume ball at the castle,' he said. 'Traditionally, we have one here each spring. I decided to cancel it this year when my housekeeper left, but perhaps it is not necessary. What do you think of the idea?'

'It is not my place to say,' said Helena, surprised that he had asked her.

'Really, how very eccentric, asking the housekeeper,' said Mrs Fairdean with a startled, and not altogether pleased, expression.

'It will be Mrs Reynolds's place to do the work,' he said. 'Why shouldn't she have a say?'

'My dear Lord Torkrow, she is paid to do it, as she is paid to do whatever your heart desires.'

'Whatever my heart desires? If she can do that, then she is cheap at twice the price,' he said.

Miss Fairdean looked confused, unable to understand his speech, and his words darkened the air.

'It is for you to decide, my lord,' Helena said.

'Is it? I think not. Not before I know something more about you. Have you ever organized a costume ball before?' He turned to his guests. 'Mrs Reynolds comes to me with three years' experience of housekeeping, but she has never been in such a large establishment. It takes a certain kind of woman to make a success of such a venture.'

Helena did not know why he was behaving so strangely; whether he wanted to flurry her into saying something that would reveal she was not a housekeeper, or whether he wanted to discomfit his guests. His manner to them was polite, but there were hidden barbs beneath the surface, and she suspected he did not like them. Mrs Fairdean had looked uncomfortable at first, but now ignored his strange manner, as did her daughter.

'No, my lord, I have not,' Helena replied.

'What does that signify?' asked Miss Fairdean impatiently. 'She can learn. Please, Simon, let us have one,' she went on in a wheedling voice. 'I have thought of my costume already.'

'I suppose it is very beautiful?' he asked her.

Helena was shocked to hear that he spoke with barely concealed contempt, but Miss Fairdean did not seem to notice.

'It is,' she said coquettishly.

'Then we must not disappoint you. Mrs Reynolds, you will continue with the arrangements for the costume ball. It will be held at the start of next month. You will engage any extra staff you need to help you. Miss Fairdean will delight us all with a beautiful costume, and I. . . .'

'Yes?' said Mrs Fairdean encouragingly.

'I will come as a crow.'

Miss Fairdean looked startled, but then she carried on as though he had not said anything.

'We must move quickly, Mama. That sluggard of a seamstress must be made to work harder. She is always dragging her heels, and making some excuse or other. She is idle, like all of her kind. We will make her see she must work for her money. We will go to London tomorrow and chivvy her. There are gloves to buy, jewels to be set. . . .'

Helena poured the tea whilst they continued to talk about the ball, roundly abusing the seamstresses, wig makers, milliners and shopkeepers who would provide them with everything they needed. Lord Torkrow said nothing, but the Fairdeans did not seem to notice. Helena, having poured the tea, returned to the kitchen.

'I've just learnt we're to arrange a costume ball for the start of next month,' she said

'Ah, so he's going ahead with it, is he?' asked Mrs Beal. 'I thought the Fairdeans wouldn't want him to cancel it, but I'm surprised he gave in to them so easily. He's never liked that sort of thing.'

'When are we to hold it?' asked Helena.

'On the third,' said Mrs Beal. 'And a lot of work it will be. Did Mrs Willis say she would find out some maids?'

'I didn't speak to her,' said Helena. 'It was raining too heavily and I had to turn back. But I managed to send a message to her.' She didn't mention Mary. She felt instinctively that the fewer people who knew about Mary the better. She felt safer for having a place to run to, should she need it. 'What has been arranged so far?' she asked.

'The invitations have all been written, and the guests have all very likely had their costumes made. The ball's held every year, it's a big event hereabouts, and everyone looks forward to it.'

'The food will not have been ordered?'

'No. That's something that will have to be done, and done soon. We'll need a sight of meat and vegetables, and eggs, we must have plenty of eggs – there'll be puddings to make, and custards and meringues. Cream, too,' she said. 'Ah, well, the shopkeepers are expecting it, that's one thing in our favour, they'll see to it we have everything we need. A chance for them to make some money, it is, and that's always welcome.'

'Who sees to the wine?' asked Helena.

Mrs Beal shook her head.

'Dawkins,' she said. That one word conveyed her dissatisfaction, and Helena guessed that he drank the wine he was meant to guard.

'He has the key to the wine cellar?' asked Helena.

'One of them. I keep the other one. I look in every week, to make sure that not too much has gone missing.'

'I'm surprised his lordship does not want a butler.'

'His lordship's lost heart, since. . . . Ah, well, it was a long time ago, and he never bothered to replace the butler when he left. "Dawkins can manage", he said.'

Her tone plainly said that Dawkins could not manage, but that she could do nothing about the situation.

'Now, about the desserts. . . .'

They fell to discussing the arrangements, until the bell rang again.

'They'll have finished with the tea tray,' said Mrs Beal.

Helena returned to the drawing-room, and to her surprise she found that Lord Torkrow's visitors had gone. Only the used tea cups and the hollows in the furniture showed they had ever been there.

'Mrs Reynolds. Come in.'

The fire had burnt down low, and its flames created odd patches of light across his body, throwing one shoulder and one side of his face into relief. His forehead, chin and cheek were lit brightly, and a gleam of gold was awakened in his eye. He turned his face to hers, and she wondered why she had never noticed how fine his cheeks were. They were like the rocks outside, sharp-angled, but with the stone made smooth by the constant onslaught of the elements.

'You have been speaking to Mrs Beal about the ball?' he asked.

'Yes, my lord.'

'Good. She has been here for many years, and knows what is required. Your predecessor had already done much of the work. You will find her notes in the housekeeper's room, no doubt. You have spoken to Mrs Willis about finding some more maids?'

'No, my lord. I was driven back by the weather. But I managed to send her a note, asking her to help me find two girls.'

'You will need more than two maids if the ball is to go ahead. You had better go and see her tomorrow, and tell her of the change of plan.'

'Yes, my lord.'

He stood there, saying nothing more, and Helena was conscious of a disturbing atmosphere in the room. It was as though he was keeping himself on a tight rein, and she felt that if he let the reins go, the power released would change her life for ever.

He considered her intently, and then he surprised her by saying, 'You were in the graveyard last night.'

Her heart jumped at the unexpected shift in the conversation. She wondered if he had seen her, or if someone else had told him.

'It's a strange place for a young woman to be after dark,' he continued. 'What were you doing there?'

'I went out for a breath of air,' she answered. 'I did not know where I was going. I walked across the courtyard and then on to the moor.'

'And just stumbled across the graveyard?'

She hesitated, wondering what to say. It would be easier to let him think she had found it by accident, but she wanted to say something, something that would help him, for she knew that he had been in pain. And he was still in pain now. She could see it etched across his face, in the lines around his mouth and by the haunted look in his eyes.

She heard herself saying, 'I was drawn to it by the sound of someone crying.' He went pale, but gave no other sign that he had been the person crying by the grave. She went on, 'I wanted to comfort them. It is a desolate thing, to cry

91

alone, in the dark.'

His eyes locked on to hers and she felt something pass between them. Won't he tell me? she wondered, without even knowing what it was she wanted him to say, only that he had secrets, and burdens, and she felt she could help him carry them if he would only let her.

With the question, she no longer felt like a housekeeper talking to her employer, she felt like a woman talking to a man. Even so, she was unprepared for his reaction to her words. He suddenly grasped her hand and, saying, 'Come with me', he pulled her along behind him, out of the room, up the broad, shallow stairs, so quickly that she had to run to keep up with him; along the corridor and into the portrait gallery. Then he let her go.

She looked about her. A long line of Torkrows hung on the wall. These were the men who had built the castle. They were also the men who had given rise to the tales in the village; superstitious nonsense most likely, arising from nothing more than the family living in a castle, and coming and going at will. Or so she tried to reassure herself.

The portraits began many centuries before. There were maidens in wimples and men in doublet and hose. There were cavaliers in silk and satin, and ladies in velvet and lace. There were men in tailcoats and women in panniered gowns; family portraits and wedding portraits; old men and little children. She traced the progression of family features, from the first Lord Torkrow to the man beside her.

There were several recent paintings of him. The first showed his family: his father and mother with their three children, two boys and a girl. He and his brother looked to be about the same age, whilst the girl appeared to be three or four years younger. His brother was like their mother, fair-haired and blue-eyed, and looked like a cherub, whilst he and his sister were dark-haired. His eyes looked out at her and she was shaken by the change in them. The eyes in the portrait were not haunted and secretive, as they were now, they were clear and happy.

Her gaze moved on until she stood in front of a portrait of the three children, fully grown, and dressed in the fashions of a few years previously. It was of the fair-haired son's wedding day. Helena remembered what Mrs Beal had said, that Lord Torkrow had no need to marry because of his brother. She must have meant that, as he had an heir in his brother, and as his brother looked set to carry on the family line, Lord Torkrow had no need to marry if he did not want to. Helena looked at the portrait of the bride, who stood next to his brother. She was a beautiful young woman with soft fair hair, and she seemed happy.

What had happened to the brother? she wondered. Where was he now? Not at school, that much was clear. So where was he? And where was his wife?

'Do you know what they call my family in the village?' he asked.

'Yes,' she acknowledged. 'They call you Stormcrow.'

She turned towards him and she was preternaturally aware of him. Though not handsome, his face was striking, and she found her eyes tracing the lines of his forehead, nose and mouth. It was not prone to laughter as it had been in his portrait, and she wondered if it would ever be again.

'Do you know why they call us that?' he asked.

She shook her head.

'Do you know what a stormcrow is?'

'No.'

'A stormcrow is a bird of ill omen,' he said. 'It brings bad news.' He led her over to the first portrait. It was of a thin, sinewy man in middle age, with bright amber eyes.

'This is the first earl. He brought the news of the Yorkist defeat at the Battle of Bosworth back to his father. As you can see, he was a man with a thin face and bright eyes. As he rode across the moors to break the news, a storm followed him. A crow flying before the storm alighted on his shoulder, and they rode in through the gate together. When it was known what news he brought, an old man, playing on our name of Torkrow, quipped, *Here they are, two stormcrows.*'

They moved on.

'That is the second earl,' he said.

He stood behind her. He lifted his hand as they looked at the portrait, and for a moment, she thought he was going to rest it on her shoulder. She felt an awareness ripple through her in anticipation of his touch, but instead he gestured at the painting, and the lack of his touch left her feeling strangely empty.

'Richard brought his father the news that his mother was dead, thrown by her palfrey,' he continued. ' "My son, you are a true stormcrow", his father said.'

Helena looked up at the face of Richard, who was dressed fashionably for his era, in a slashed doublet and breeches. He looked carefree.

'He had not earnt his nickname when this portrait was painted,' she said.

'No. He had no idea what was about to happen. He was still happy, then.'

He moved to the next portrait. The third Earl was standing with his hands on his hips and with his legs wide apart, looking solid and secure. He was wearing a doublet that accentuated the width of his shoulders, with wide sleeves that billowed outwards, before being confined at his wrist.

'He looks as though nothing can topple him, doesn't he?' asked Lord Torkrow, standing behind her. He was so close that she could feel his body heat, and she had a disturbing urge to lean backwards and feel his warmth envelop her, but she resisted the strange impulse.

'What happened to him?' asked Helena.

'There was a fire, and the family house in York was razed to the ground. He brought the news to his mother, an old woman of ninety-nine, who had been making plans to celebrate her one hundredth birthday. The news caused his mother's death, three hours before she would have achieved her ambition. He was ostracized for giving his mother the news, instead of letting her hear it through other means, after her birthday.'

He went on, telling her the story of each Stormcrow, and of how each one had earned his name, until at last they stood before his own family portrait.

'And you?' asked Helena. 'How did you earn the name?'

He said nothing, and a profound silence engulfed them. Helena turned to look at him, and she saw that his face had gone white. His eyes, in contrast, were dark and hollow, and the rings around them were black. He was staring at the portrait, and she knew he was far away, back in the past. His hands had dropped to his side, and she saw that they were clenched into fists. He opened his mouth, and she thought he was going to speak, but then he turned and strode out of the gallery, leaving her alone.

She looked again at the portrait of the boy he had been, a happy, carefree child. But now he was a man sunk in mystery, and darkness wrapped itself around him like a shroud.

*

95

Why did I do that? Simon asked himself as he descended the massive staircase and went into the library, wondering why he had tried to make her understand.

He tried to settle to estate business, but he could not concentrate. He heard Helena's light step as she descended the stairs and went into the housekeeper's room. He picked up his quill, then threw it down and went out of the library, climbing the stairs two at a time, returning to the gallery and pacing to the end, then pressing the embossing on the wall and waiting impatiently for the secret door to open. It swung inwards ponderously, and he went inside.

The room was small and panelled. A window looked out on to the moor. An empty grate held blackened ashes. Above the fireplace hung a portrait. It was of a young woman, his brother's bride. She was looking radiantly beautiful. She wore her fair hair loose, hanging round her shoulders in soft curls. Her muslin gown, with its high waist, revealed a slight figure with gentle curves. Her lips were pink, and her eyes were blue. She was standing in a garden, and the dew was on the roses.

He stood, lost in thought, until a sound disturbed him. Miss Parkins had entered the hidden room. She was the last person he wanted to see, especially here, now.

'Did you wish to speak to me?' he asked her coldly.

'I understand you are to go ahead with the ball, my lord.'

'Yes, I am.'

'Do you think it wise? A masked ball can hide many secrets.'

'I have made my decision. The ball will go ahead.'

'Very good, my lord,' she said, with a trace of insolence.

She walked over to him and stood beside him, looking at the portrait.

'She was very beautiful,' said Miss Parkins.

'Yes, she was.' He could not keep the wistfulness out of his voice.

'Your brother chose well. He loved her dearly. Until you killed her.'

CHAPTER SEVEN

The following morning, Helena began to organize the castle in earnest. Unsettled by everything that had happened, she was glad to take refuge in physical labour. The library, drawing-room and dining-room were well cared for, so she decided to rescue a further room from its state of neglect. If there was to be a ball, then the castle must be brought back to life again. All thoughts of leaving quickly had left her, for she did not intend to go before she had had a chance to speak to Sally and Martha.

She chose a small sitting-room that overlooked the front of the castle, and she began by removing the dust sheets, taking them off and folding them carefully so as not to disturb the dust that had settled on them. She was surprised to see that the furniture was of good quality, and delicate. Gold chairs in elegant styles were upholstered with red brocade, a padded sofa was covered in a matching brocade, and, as she removed the dust sheets from the floor, she discovered a flowered carpet. It had been a lady's room, then, she thought, as she looked about her.

She rang the bell, and whilst she waited for it to be answered, she began to dust the mantelpiece and other surfaces, revealing the beauty of the wood beneath.

The door opened and Effie entered hesitantly.

'It's all right, Effie, come in. I am preparing this room for use. I want you to light a fire here, and then I would like you to bring a bucket of water and wash the windows. Make sure Mrs Beal does not need you first.'

'Yes, mum.'

Effie departed, but returned soon afterwards.

As they worked, Helena asked the girl about her family. Reluctantly at first, Effie began to speak, saying that she had been orphaned and that a cousin had found her work at the castle. Once or twice, Helena led the conversation round to Mrs Carlisle, but Effie became nervous when she did so, and she talked of other things. Gradually, though, she began to win the girl's trust, and thought that, before many more days had passed, she might induce Effie to confide in her. That the girl knew something she was convinced, though whether it was important remained to be seen.

By late afternoon, the room was looking cheerful. Helena had wound the ormolu clock, which was ticking on the mantelpiece, and polished the gilded mirrors. Effie had washed the windows, both inside and out, and they sparkled where they caught the light. The fire was crackling merrily in the grate.

'It's a pity there is no one to use it,' said Helena to Effie, pushing a stray strand of hair out of her eyes with the back of her hand.

'Yes, mum.'

'Whose room was it? Do you know?'

'It was 'ers,' said Effie, not very helpfully.

'Was it used by Lord Torkrow's mother?'

Effie did not reply.

'Or his sister-in-law?'

Effie nodded.

'She liked it 'ere.'

'Does she live here now?' asked Helena.

Effie dropped the poker with a clatter, and was clearly frightened.

'Where is she, Effie?' asked Helena. 'Is she in the castle? Or on holiday?'

'No, mum. She's dead.'

'Dead?'

'Yes, mum.'

'Then the crying in the attic—' *is not her*, Helena was about to say, when Effie interrupted her.

'Yes, mum, it's 'er. Dawkins says she walks,' said Effie.

'Nonsense,' said Helena reassuringly. 'The dead don't walk, Effie. There was nothing in the attic but a cat. Together we have made a very good job of this,' she went on more cheerfully. 'The room looks bright and welcoming.'

'P'raps she'll stop crying now, mum. P'raps she'll walk in 'ere, not in the attic.'

As the thought clearly cheered her, Helena did not gainsay it.

'Now, you must return to the kitchen. I'm sure Mrs Beal will be wanting you. I will finish here.'

Effie picked up her bucket and left the room.

As Helena put a few finishing touches to the room, she wondered what had happened to Lord Torkrow's sister-in-law, thinking: How did she die? How long ago was it?

And where is she buried?

Helena joined Mrs Beal for dinner that evening, and as Mrs Beal dished out the mutton stew, she said, 'I will be going to see Mrs Willis tomorrow afternoon about finding some more maids for the castle. How many do you think I will need?'

'Take as many as you can find,' said Mrs Beal. 'There's plenty of work to be done.'

'I have made a start on the downstairs rooms already, opening up a sitting-room overlooking the front of the castle. Effie tells me it used to belong to his lordship's sister-in-law. I saw her portrait in the gallery. She was very beautiful.'

'Yes, she was, poor lady.'

'It was a tragedy when she died.'

'Master Richard went mad with grief,' said Mrs Beal with a sigh, then, recollecting herself, added, 'Least said, soonest mended, I always say. You're going to see Mrs Willis this afternoon?'

'Yes.'

'You'd better ask her to help you find some footmen, too. There's going to be a lot of work fetching and carrying beforehand, and we'll need someone to carry the drinks on the day.'

'It's all rather daunting,' said Helena. 'Did Mrs Carlisle find it so?'

'Bless you no, she'd arranged a dozen balls for his lordship.'

'If only I had her sister's address, I could write to her and ask her for her advice.'

'You wouldn't want to bother her, not with her sister being so ill,' said Mrs Beal, 'and besides, you've no need to worry. It will all come right in the end.'

As Helena set out for the village after luncheon she was glad to leave the castle behind. She felt herself being drawn deeper and deeper into its tangled world, but Lord Torkrow and his family had nothing to do with her. She had come to the castle for one reason and one reason only: she wanted to find her aunt.

The day was fine, with a weak sun shining out of a blue sky, and she was pleased to see that there was no threat of rain. It was miles to the village, across the moors, so she set off at a brisk pace. Hardy sheep were grazing, and she was glad of their bleating, which broke the silence and made the walk less lonely.

As she approached the turning to Mary's cottage, she decided to take it and pay Mary a call. She longed for someone to confide in, someone outside the castle, who was immune to its strange atmosphere and past. Perhaps Mary could shed some insight on to her aunt's disappearance.

The rough track was dry, unlike the last time she had visited, when the rain had turned it to mud, and was much easier to walk on. She soon found herself outside the cottage, and knocked on the door. It was opened by the maid, but Helena quickly learnt that neither Mary nor her brother were at home, and that the maid did not know when they would be back.

Helena swallowed her disappointment, thanked the maid, asked for Mr and Miss Debbet to be told that she had called, and carried on to the village. As she approached, she passed a small cottage, and then a few more, scattered haphazardly across the harsh landscape. She passed an old woman, dressed in black, as she entered the village, and a serious-looking little boy who was carrying a large basket into a cottage.

Helena greeted them with a 'Good afternoon', but they did not reply, instead favouring her with suspicious looks.

The village was larger than she had expected, and better favoured. It was sheltered from the prevailing wind by being built in a hollow of the moors, and it consisted of a collection of cottages and houses, with an inn at one end and a church at the other. Next to the church was a large, square stone building which Helena took to be the rectory. It was set back from the road, and separated from it by a low stone wall. There was a white-painted gate which creaked as Helena opened it, and a stone path snaked between barren borders to the door.

Helena lifted the brass knocker, which fell with a satisfying clunk, and a minute later the door was opened by an elderly maid.

'I am Mrs Reynolds, the new housekeeper at the castle,' said Helena. 'I'd like to see Mrs Willis.'

The maid bade her enter, then left her in the hall. It was well cared for, and Helena took pleasure in seeing a house she did not have the responsibility of cleaning. The living was perhaps not wealthy, but it seemed to keep the rector and his wife in some comfort. The hall was painted a muted green, and there was a rug on the polished floorboards, whilst a staircase led upwards on the left.

The maid returned to say, 'Mrs Willis says, "Please come in." '

'Thank you,' said Helena, as the maid helped her off with her cloak.

She went into the drawing-room. Whilst the hall had been plain, here there were pretensions of gentility. There was gold wallpaper on the walls, a brocade sofa, and an inlaid console table beneath the window. On it was a vase of fine porcelain, matched by two others of similar design on the mantelpiece. The candlesticks were of silver, and there was a good painting hanging above it. A square piano was set against the far wall, and a brocade-covered stool was set in front of it. A fire was burning in the grate, and the fire irons gleamed in the light of the flames.

Mrs Willis stood up. Her dress was simple yet well cut, and to Helena's surprise, it was made of silk.

'Mrs Reynolds, how very nice to meet you,' she said in a cultured voice. 'My husband and I heard there was to be a new housekeeper at the castle. It is not before time. I dread to think how his lordship has managed without one. Won't you sit down?'

Helena thanked her and took a seat.

'I have come to ask for your help,' said Helena. 'I will need some maids to assist me in the castle and, as you know the neighbourhood and the people I thought you might be able to help me find some suitable girls.'

'Ah, yes, I received your note.'

'When I wrote it, I needed only two girls, but as I now need more help, footmen as well as maids, I thought it better to come and see you in person. I understand that the two girls who worked at the castle under Mrs Carlisle left in a hurry. It is a great pity. It would have been much easier for me if they had remained,' she said.

Mrs Willis's face expressed her exasperation.

'The people round here are very insular,' she said. 'They have their prejudices and their superstitions. They mutter and whisper about Lord Torkrow, poor man, as they would mutter and whisper about anyone who lived in a castle. And the stories they tell about the castle itself! You would think it was unsafe to spend half an hour within its walls, the way some of them talk!'

Helena was reassured by Mrs Willis's disgusted manner: she, at least, did not appear to think ill of Lord Torkrow.

'I suppose it is understandable,' said Helena. 'The girls heard crying in the attic. The footman believed it to be a cat, but the girls were convinced that something dreadful had happened.'

'Exactly! That is just the sort of story I'm talking about. As if anything dreadful *would* happen.'

'It was sparked by the housekeeper's disappearance, I believe,' said Helena. 'I suppose an incident like that was bound to cause gossip. A servant does not usually leave without giving notice.'

'There was nothing suspicious about it. The poor woman left for a very ordinary reason, to tend her sick sister.'

'Did she not leave in the middle of the night? Or is that just another tale?'

'No, that is true, and of course, that fuelled the talk, but again there was a sensible reason for it. There is a stagecoach to London early in the morning. I imagine she wanted to make an early start.'

'Ah, I didn't know her sister lived in London,' said Helena.

'I don't know that she does' said Mrs Willis. 'That is the stagecoach's ultimate destination, but it stops a number of times on the way. Of course, Mrs Carlisle could also have gone north, in which case she would have caught the stagecoach to Edinburgh, which passes a little later.'

'Have you heard from her?' asked Helena, with more hope than confidence.

'No. I did not know her very well. We saw each other at church; a very sensible woman, well spoken and an asset to the congregation. I helped her to find staff for the castle, but other than that I did not speak to her. I am only sorry I did not find her some girls with more common sense.'

'Do you think they would return to the castle now that I am there?' asked Helena. 'It would be a great help to have girls who know their way about.'

'Perhaps. There is little work round here. I will try and persuade them to come and see you, and if not then I will try to find you some other girls. You said that you needed more than two?'

'Yes. His lordship means to go ahead with the costume ball, so I will need as much help as possible.'

'It is to go ahead? Oh, I'm so pleased,' she said, with a spark of excitement in her eye. 'My husband and I have already chosen our costumes. We are to go as King Henry VIII and Katharine of Aragon. I have the costumes left over from another ball,' she explained. 'My husband has put on weight since then, and I have spent the last few weeks letting his costume out. I am so glad my work will not go to waste.'

'Are the balls generally large? I haven't had time to look at the guest list yet,' said Helena.

'Oh, yes, everyone from the surrounding neighbourhood is invited. They all look forward to the ball. It is a big event, in fact it is the biggest event we have in this village. The castle is something to be seen when it is *en fête*. The light pours out of the windows, and then there is the music! The orchestra is always excellent. And the food! You don't need me to tell you that Mrs Beal is an excellent cook, and on these occasions she always surpasses herself. Carriages roll up in front of the castle by the dozen, and everyone wears the most wonderful costumes. There is a great deal of imagination brought into play, and although there are always a few duplications, the local gentry for the most part try and find a more unusual character to portray.'

'Miss Fairdean and her mother have already ordered their costumes. They were at the castle yesterday,' Helena explained.

'Yes, the Fairdeans always make a special effort where the castle is concerned. They will be having their costumes made in London, I expect, complete with wigs and jewels. They will be portraying royalty, I've no doubt. Last year, Miss Fairdean dressed as Elizabeth I. With her red hair, she looked the part. Her mother must have spent a fortune on her dress. It was encrusted with pearls. I suppose she thought it was worth it.

There was some talk that his lordship would marry Miss Fairdean – I believe his mother, as well as hers, wished it – but nothing has come of it. Miss Fairdean is not well liked in the neighbourhood,' she went on. 'She is very rude to her servants, and indeed to most of her neighbours. She seems to think she is above them. She said to me—' She stopped herself, as if remembering to whom she was speaking. 'There was no call for it.'

Helena waited, hoping she would say more, but Mrs Willis was silent. Then, with the appearance of a woman turning her thoughts into new channels by an act of will, she continued.

'We will be seeing you at church, I hope? His lordship never comes, but Mrs Carlisle used to attend regularly, as long as the weather was fine enough for her to walk. She was a great supporter of the church. It was a pity she was all alone in the world, with no one to miss her when she was gone.'

Helena felt a shock at the unexpected words. *No one to miss her.*

'She had a niece, I believe?' she said quickly. 'Mrs Beal said Mrs Carlisle wrote to her niece regularly.'

'I didn't know that,' said Mrs Willis slowly.

'And then, of course, she had a sister,' said Helena.

'Oh, yes, her sister,' said Mrs Willis dismissively.

Helena was disquieted. There was something decidedly odd about Mrs Willis's manner.

The conversation moved on, but as Mrs Willis spoke about other parishioners, Helena watched her covertly. Strange stories came back to her, stories of people who disappeared mysteriously in remote places, innocent-seeming locals who were not what they appeared. . . .

The chiming of the clock broke her thoughts, and she returned to her senses. Mrs Willis was now talking about the village girls in the most matter-of-fact way, and the idea of her being mixed up in a strange disappearance seemed ridiculous.

A few minutes later, the rector, Mr Willis, entered the room. He was a stout, kindly-looking man with white whiskers, and the idea of him being mixed up in anything untoward seemed even more ridiculous than his wife's involvement.

'This is Mrs Reynolds,' said Mrs Willis, performing the introductions. After a few minutes of polite conversation, she said, 'I will do what I can for you in the village, and I will send any willing workers to see you at the castle.'

Helena thanked her, then, having taken her leave of them, she reclaimed her outdoor clothes and set out.

The day had turned colder, but it was dry, and within the hour Helena found herself once more approaching the castle. She was pleased that she had made arrangements to acquire more staff, but disappointed that she had learnt nothing of use about her aunt.

She had almost reached the outer wall when a gleam of sunshine breaking through the clouds made her look up and she let out a startled cry as she saw there was someone on the battlements. From such a distance she could not be sure if it was a man or a woman, but she meant to find out. She hurried inside and went swiftly up to the attic, thinking that she did not remember a staircase to the battlements. When she had searched the attic in the west wing she realized there wasn't one.

She wondered if someone had gone through from the east wing, or. . . . She looked up, then went through the attic again, looking at the ceiling, and there, sure enough, in the corner of one room, was a small door. Tied to a handle in the middle of it was a piece of rope, and beneath it was a chair. She was about to stand on it and go through when she thought better of it, for she had no idea who was on the battlements or what they were doing there. She was just wondering what to do when she heard footsteps above her and hid herself behind a screen. Through the gap around the hinges she could still see the room. A minute later there came a creaking sound as the small door opened and a leg appeared, waving round as it tried to find the chair. Another followed, and then a pair of breeches, and then . . . Dawkins. He closed the door above him, then climbed down from the chair and put it against the wall before leaving the attic room. He was swaying as he walked, and Helena guessed what he had been doing, but she wanted to make sure. Waiting for his footsteps to die away, she replaced the chair, opened the door, and with some difficulty she climbed through.

She found herself on the battlements, with the wind whipping at her cloak and trying to pull her hair from its pins. Beneath her was the moor, grey and green in the dull light. Far off, she could see the village, with its collection of cottages and the church. She looked all round, wondering if there was any other human habitation nearby, but there was nothing except a few isolated cottages, Mary's amongst them.

Turning her attention back to the battlements, she searched them, and soon found a large cache of bottles, cushioned by sodden blankets and resting in the lee of the wall. There were perhaps a hundred bottles of wine and port, and half of them were empty. He must have taken them when the butler left, and before Mrs Beal started checking the cellars. No wonder he tried to warn people away from the attics: he did not want anyone noticing his comings and goings, or deciding to take a turn on the battlements and discovering his secret store. And if anyone heard his footsteps, why, he could blame them on a ghost.

Had it also been Dawkins crying in the attic? she wondered. She would have to try and find out.

She took one last look at the view, which was splendid from such a high vantage point, and would be even better in summer under a blue sky, and then climbed back into the attic. She grasped the piece of rope and pulled the door shut behind her, then replaced the chair and went down to her room. Once there, she took off her cloak and stout shoes, peeling off her gloves before removing her bonnet.

She was going down to the housekeeper's room when, passing the gallery, she had an urge to look at the portrait of Lord Torkrow's sister-in-law again. She went in, and had almost reached the end of the gallery when she noticed something odd. There was an open door at the end where no door should be. Curious, she went forward, and then stopped suddenly, as she saw that Lord Torkrow was in a hidden room, looking at a portrait of a beautiful girl: his sister-in-law.

Helena shrank back, then hurried from the gallery. There had been something in his face when he looked at his sister-in-law's portrait that had cut her to the heart.

She was about to go into the housekeeper's room when she changed her mind. She was tired after her exertions, and she went down to the kitchen. Mrs Beal was there, busy baking cakes.

'Well, so you're back, and cold, I'll warrant,' said Mrs Beal. 'Effie, set the kettle over the fire. How did you get on with Mrs Willis?' she asked.

'Very well. She has promised to try and find me some help, and will send any likely workers to the castle.'

'That's one job done, then,' said Mrs Beal.

The tea was made, and Mrs Beal poured it.

'I think I'll join you,' she said. 'I've some biscuits just come out of the oven. You'll have one with your tea?'

Helena thanked her, and was glad of something to eat and drink.

They fell to talking about the arrangements for the ball. Some of the suppliers had expressed doubts about being able to produce such large quantities of food, and Mrs Beal talked of alternatives whilst Helena gave her opinion.

'And now, I had better tend to my own work,' she said, as she finished her tea. 'Then, after dinner, I need to sort through the linen and make sure there are enough sheets for those guests who are staying overnight. I am hoping they are clean and dry.'

'Mrs Carlisle always took care of that. Clean, dry and smelling of lavender, they'll be.'

Helena felt a pang as she thought of Aunt Hester, and she found she could almost smell the lavender.

'Then I had better count them and make sure we have enough.'

Helena had just reassured herself that there would be enough clean linen for the overnight guests at the ball, and was about to retire for the night, when she was startled to find Effie waiting for her in the corridor.

'Yes, Effie, what is it?'

'Please, mum, it's about the key to the attic,' said Effie, twisting her apron in her big, clumsy hands.

'Yes, Effie?'

'I knows where I thinks it is, mum.'

Helena's pulse quickened.

'Mrs Carlisle, she kept some spare keys in the scullery, missus. I saw 'er with them once. She used to go in and out of the attic, quiet-like.'

'Quiet-like, you say?' asked Helena, wondering if her aunt

could have suspected Dawkins of taking wine from the cellar, and if she had perhaps followed him.

'Yes, missus. I saw her when I was doing the fires.'

'But you don't do the fires in the attic.'

'I was doing them in the bedrooms, mum, and I 'eard a noise. Manners – he was one of the footmen, missus, we used to 'ave ever so many footmen – he said to me, "It's a ghost", and he dared me to go 'ave a look.'

'And do you mean to say you did it?' asked Helena, looking at Effie with surprise.

'No, mum. But later, when I saw Mrs Carlisle going up there, I thought, I'll go after 'er and see if there's a ghost, and if there is, she won't let it 'urt me, and if there isn't, I don't need to be frightened of what Manners says to me no more.'

'And did she go into the east wing?' asked Helena. 'Did she go into the locked attic?'

'Yes, missus. That's where the noises were coming from.'

'And was it a ghost?' asked Helena, hardly daring to breathe.

'Don't know, missus. There were something in there, I 'eard it, but I don't know what it was. Mrs Carlisle, she went in, and then about ten minutes later she come out again.'

'Was there anyone with her? Dawkins, perhaps?'

'No, mum, she were by 'erself.'

'Did she seem agitated?' asked Helena.

'Don't know, mum.'

'Did she seem happy?'

'Don't know, mum.'

'Did you see her face?' asked Helena.

'No. I runned down the stairs so she wouldn't see me.'

'Very well, thank you, Effie.' She added casually, 'Is Mrs Beal in the kitchen?'

'No, mum, she's gone to bed.'

'No matter, I will speak to her tomorrow. I have a few spare minutes, I think I will come down and look for the key now,' said Helena.

'Yes, mum.'

As she went down to the scullery, Helena's thoughts were racing. So her aunt had been into the east wing, and she had discovered something there. Was it Dawkins? But he had climbed out on to the battlements from the west wing. What else could it have been?

Could it have been Lord Torkrow's brother? There was something about his brother, she was sure, something no one was telling her. She thought of Mrs Beal saying he had been driven mad with grief. What if he had literally been driven mad, and his family had confined him in the attic?

She thought of Miss Parkins. What if Miss Parkins was looking after his lordship's brother? Or perhaps her aunt had been the one who was looking after him. Perhaps his mad brother killed her. Or perhaps her aunt threatened to tell someone about him, because he had killed or injured someone else.

She had time for no more thoughts. Going into the scullery, she asked Effie to show her where the key was kept. She was determined to solve the mystery of the attic once and for all. Effie took her to a drawer at the back of the scullery. Helena opened it . . . and it was empty.

Helena stood staring at the empty drawer with disbelief.

'It were there, mum. I saw it,' said Effie.

'Yes, I'm sure you did, Effie,' said Helena soothingly.

But the key had nonetheless gone. Who had taken it? thought Helena. And why?

CHAPTER EIGHT

The following morning brought a letter to the castle from Caroline. It came as a welcome relief to Helena to know that she was not entirely cut off from the outside world. The atmosphere in the castle was oppressive, but Caroline's letter brought the noise and bustle of Manchester back to her. She could see Caroline, in her mind's eye, sitting at the cramped table beneath the window, with its view of the noisy street and its glimpse of the canal. She could see Caroline lifting her head, as she always did, and then resting it on her hand as she watched the bakers walking past with trays on their heads and ragged children playing, and dogs scavenging for food. There would be a restlessness about her, for Caroline was always restless inside. And when Caroline had finished the letter she would have thrown her cloak over her shoulders in a swirling movement, picked up her basket and gone out, threading her way purposefully between the street merchants and other shoppers, stopping to talk to neighbours, and sending the letter, before looking longingly in the windows of the milliners on her way home.

Helena examined the seal and was relieved to see that it had not been tampered with, so it seemed that the mail went from and came to the castle undisturbed. If Aunt Hester had written to her, then it seemed unlikely the letter had been interfered with.

She broke the seal and began to read.

My dear friend.

Good. So Caroline had guessed something was wrong, and was writing in a guarded style.

I was very pleased to get your letter. What a pity you have heard nothing of H. I have had no news, either. I hope all is well and that we will soon hear something.

I have some news of my own. I secured the position with Mrs Ling and I am writing to you from her home in Chester. She is not too demanding and she treats me with respect, which is the most I can hope for.

You, however, deserve more.

I have seen our friend G several times and I hope you will see him before long, too. I have not given him your direction, but if you wish to write to him, I'm sure a letter would be most welcome.

I will await your next letter with interest.

She included Mrs Ling's address, and signed the letter Caroline.

As Helena folded it and put it in her pocket, she found her thoughts returning to Mr Gradwell. Life with him would be safe. He would help her when needed, indeed, he would help her now if he knew of her troubles, though there was little help he could give. Yet she had no desire to hurry home and confide in him. Quite the reverse, she was glad of some time away from him, for it enabled her to think more clearly.

She tried to imagine what life would be like with him. She would be the mistress of her own home, with a maid and a cook to serve her. She would have new clothes to wear, and a carriage to ride in, and she would be able to spend her time visiting and shopping and inviting friends to supper, instead of working all day long. She would have the companionship of Mr Gradwell, and there would be trips to the theatre and to the museums, and in the summer there would be picnics and outings to the seaside, but although it seemed very inviting, her heart sank at the thought of it. Perhaps she was just tired. She would not think about it for the moment. There would be time enough to think about it when she had found her aunt.

She began to draw up a plan for cleaning the castle, in the hope that Mrs Willis would find her some willing helpers, and was rewarded for her hope by the arrival at the castle of seven girls and six men, shortly before ten o'clock. On asking them their names, she was pleased to learn that Sally and Martha, the two girls who had worked at the castle before, were among them.

She went down to the kitchen to speak to them.

'There is plenty of work to be done,' she told them. 'Can you start today?'

They had all come prepared to stay, and Helena set them to work. Whilst two of the young men began polishing the silver under the direction of Dawkins, the other four took down one of the more recent tapestries and carried it outside and Helena set three of the girls to work beating it with brooms. The men then moved on to fetching buckets of water so that the rest of the girls could wash the floor. Helena dropped some sprigs of dried lavender into the water, to perfume it, then set the girls to work.

Fortunately the day was fine, and she joined the girls who were working outside. It was pleasant to be out of doors, and though the air was cold, beating the tapestry was heavy work and it soon warmed them.

'I think you have you worked at the castle before?' she asked

Sally and Martha.

'That's right, missus.'

They were perhaps seventeen or eighteen years of age, and although they seemed ready enough to work, the glances they kept throwing at the younger footmen whenever they walked by suggested they would not be reliable if left alone.

'What were your duties?'

'We kept the rooms clean, missus. We dusted 'em and polished the grates and kept the fire-irons shiny. We swept the floors and made the beds.'

'Then I would like you to do the same now you have returned. Did you air the rooms before?'

'Yes, missus, some of 'em. The ones that 'ad someone in 'em.'

'Good, then you can continue to do so. Did you air the attics?'

'No, missus, we daren't go near the attics.'

'Why not?' asked Helena.

'There was noises,' said Sally.

'At night,' said Martha, with wide eyes.

'Made my blood curdle, they did,' said Sally. 'All that screeching and wailing.'

'You said it was crying,' put in one of the other girls, as she hit the tapestry with a broom.

'Screeching,' said Sally emphatically, 'and wailing.'

The story grew in the telling, and Helena was not surprised when the girl asserted that she had heard chains clanking behind the door. However, Helena believed there had been something in the girl's story.

'When did you hear it?' she asked.

'It were just before Mrs Carlisle disappeared. A week before, mebbe.'

'And you stayed in the castle a whole week with such noises?' asked Helena.

'They weren't so bad after that. Just sobbing now and then.'

'Ghost must've got a sore throat,' said one of the footmen cheekily, as he walked past on his way to the well.

'I'd like to see you spend a night there, for all your talk,' retorted the girl.

Helena was not sure what the girl had heard, and she knew she couldn't rely on what she said, but nevertheless she was sure Sally had heard something.

'Perhaps it was a cat,' suggested Helena.

'That's what Dawkins said, but it weren't no cat,' said Sally definitely.

'Was he with you when you heard it?'

'Right next to me, 'e was.'

So, the sound had not been made by Dawkins, at any rate.

In an attempt to find out more about Lord Torkrow's family, Helena tried to induce the maids to talk of them, but they answered her questions briefly and would not be drawn. Whether it was deliberate, or whether they were simply more interested in their own affairs, Helena did not know.

It was a pity, because something was tugging at her memory, and she thought it might be important, if only she could remember what it was.

At last she returned to the housekeeper's room to finish her plans. There was a lot of hard work to be done before the castle was ready for the ball.

The fire burned low, and Effie arrived with a bucket of coal to mend it.

Helena was about to ask her again what she had seen in the housekeeper's desk, when she had a better idea. Going over to the window, she toyed with the curtains, then said: 'Bring me some string from the drawer, would you please, Effie?'

Effie hesitated.

'The top drawer,' Helena prompted her.

The girl reluctantly went over to the desk, wiping her hands on her apron. She opened the drawer, and stood looking at something inside. She appeared to wrestle with herself, then blurted out incoherently: 'If someone knew something and someone 'ad said something but someone thought it wasn't what they said it was, what should they to do?'

'They should tell the housekeeper,' said Helena promptly.

'Very particular about 'er quills she was,' said Effie, looking at Helena nervously. 'Always used 'er own quills for letters.'

Yes, thought Helena, she did. She wished the girl would hurry up and tell her something she did not know.

'Always used 'er own quills, mum, but it's still 'ere.'

Helena's eyes widened as she realized what Effie was telling her. If Aunt Hester had left the castle, she would have taken her quill with her.

Lord Torkrow's ominous words came back to her: *The castle has a way of keeping people.*

Effie was looking at her with a frightened expression, and Helena quickly reassured her.

'Don't worry, Effie, Mrs Carlisle knew her sister's pens were well mended, I am sure.'

'Really, mum?'

'Yes, really.'

Effie's face shone with relief.

'I've been that worried, mum. It wasn't like 'er to go without saying goodbye. Always good to me were Mrs Carlisle.'

'I am sure she wanted to say goodbye, but did not want to wake you,' said Helena.

'Yes, mum,' said Effie, nodding.

119

'And now you had better go back to the kitchen. Mrs Beal will be wondering where you are.'

As soon as she had gone, Helena went out into the hall. Lord Torkrow had left the castle on horseback after luncheon, and Miss Parkins was also out of the castle, it being her day off. Their absence gave Helena an idea.

'Leave that,' she said to the footmen, who had rehung the tapestry and were preparing to take down the next one. 'I have something else for you to do.'

She took them upstairs, and then into the attic.

'The key to the attic has been lost,' she said. 'I would like you to break the door down.'

The footmen looked at each other uneasily.

'You're not afraid of ghosts, I hope?' she said with a smile.

'No, missus. But smashing a door . . . what will his lordship say?'

'His lordship has given me responsibility for the castle,' she said.

The footmen looked at each other, then shrugged and set their shoulders to the door. After much heaving they managed to break the lock. They stood back, and Helena went in, her heart racing. She expected to find a madman, or her aunt, or a body . . . but she found nothing. The attic was empty. She went through into the room leading off from it. Again there was nothing. The entire east wing was empty, save for an assortment of discarded furniture and a few odds and ends. She did not know whether to be relieved or disappointed.

She went through the attic rooms again, looking for any signs that someone had been there recently. There was less dust in the central room, and a few items of bedding that could have been used, but it told her nothing. Whatever secrets the castle was nursing, they were no longer to be found in the east wing.

'Thank you,' she said to the footmen.' You may return to your work downstairs.'

They departed, leaving her to wander through the rooms again. There must be something, some sign, she thought. . . . But she could find nothing.

As she looked round the bare room, it seemed ridiculous to remember that she had fancied it housing Lord Torkrow's mad brother. She was ashamed of herself for such a thought. He was probably away, abroad, perhaps, or attending to business in London. Or . . . something Mrs Beal had said came back to her. It had been nagging at her mind for some time, and now she remembered what it was.

She went down to the kitchen, asking Mrs Beal if she needed any of the maids to help her, before suggesting they take tea together.

Mrs Beal was agreeable, and they talked over the likelihood of the maids and footmen remaining at the castle.

When they had done, Helena asked casually: 'What is his lordship's name? His family name?'

'Pargeter,' said Mrs Beal. 'His lordship is Simon Pargeter. Why do you ask?'

'I was just curious,' said Helena.

But hers had been no idle curiosity. As soon as she had finished her tea, she put on her cloak and, slipping out of the side door, made her way to the graveyard. An icy wind was blowing across the moor, and she wrapped her cloak tightly round her. She crested the rise and then went through the gap in the low stone wall, where she found the grave she had been looking for. It was very simple and said, Richard Pargeter. *Master Richard*, Mrs Beal had called him. Lord Torkrow's brother. He wasn't in the attic, he was here in the graveyard. And next to him was his wife.

She heard a slight movement, and turning her head she saw Lord Torkrow sitting on his horse at the edge of the graveyard, watching her. So absorbed had she been that she had not heard his approach. Their eyes met; then he dismounted, tethered his horse to the dry stone wall and entered the graveyard.

'So. You discovered whose graves these are.'

'Yes. I'm sorry.'

He did not answer her immediately, and she did not break the silence, for he was lost in his own thoughts.

The sun went behind a cloud and the landscape darkened, the bright green of the grass fading to sage. The dry stone wall, which had been silvered by the sun, returned to its sombre dark hue. It was bitterly cold and, as the chill wind blew across the moor, Helena shivered. He did not seem to feel it, even when it blew his cloak open and whipped at the tails of his coat, for he stood there, motionless, making no move to fasten it.

At last he spoke.

'You asked me once how I earnt my name.'

She was very still, waiting for him to continue.

'I do not know why, but I have a mind to tell you.' He looked at the gravestone, as if he could see his brother's face there. 'It was on a dark night in the summer, when I had been to a neighbour's ball. It had been a tedious evening, the conversation had been shallow and the company bored me. I left early, and returned to the castle. There was an . . . incident. . . .' He became lost in his thoughts, then seemed to rouse himself with difficulty. 'I knew at once that I had found my curse, or that my curse had found me. We are all cursed, we storm-crows. We are all fated to carry terrible news. It was my fate to carry it to my brother. I had to tell him that the woman he loved, his bride of a year, my sister-in-law, was dead.'

The wind moaned, and rain began to fall.

'I will never forget his face. I should have known better than to leave him. He went up to the battlements, his favourite place in times of sorrow – it had been so since he was a boy. He could barely see or think, driven mad by his grief.'

The wind howled.

'He fell from the battlements. I earnt my name not once, but twice: for then I had to carry the news of his death to my mother.'

'It was not your fault,' she said.

'No?' To her consternation he cupped her face, looking deeply into her eyes. 'I failed my brother, and I failed my sister-in-law. I will not fail again.'

His tone was sombre, and his words were strange. She could not make sense of them, but she was finding it difficult to think clearly. Something about his touch confused her, blocking rational thought. Instead, she was a mass of feeling. She felt the wind; the wetness of the rain; the roughness of his skin against hers; the warmth of his breath on her face; and she began to tremble.

Each kiss a heart quake. . . .

Byron's words came back to her.

And then, to her frustration, he lowered his hand and let her go. She had an impulse to take his hand and return it to her face, and it was only with an effort of will that she was able to resist. But she could not turn away from him.

What had happened? she asked herself, as she looked into his eyes. Why had he touched her? Why had he stroked her face? He was a strange man; secretive and haunted; but also a man of strong feelings, and a man who could arouse strong feelings in return.

Aloud, she said, 'I should return to the castle.'

'We will return together.'

He untethered his horse and they began to walk, and without willing it to be so, Helena found her steps coinciding with his. She felt wrapped around by an energy that encompassed them both, and for the first time in her life she knew she was not alone.

They walked on in silence, and she was seized by a strange thought, that it could be a thousand years in the past, or a thousand years in the future, and she would never know, for the moor was unchanging, a primitive landscape outside of time and place. She would not be surprised to see an elf or a hobgoblin walking across her path, some figure from folk tale long forgotten by civilization but remembered here, in the wilds, on an isolated pocket of land.

They walked in through the archway and the spell was broken, for there before them lay the castle, and in the court-yard, the maids were busy working. A groom came from the direction of the stables to take Lord Torkrow's horse. He relin-quished the reins, and they went into the castle together. Once over the threshold, they heard the banter between the foot-men and the maids who were washing the floor.

She was about to retire to the housekeeper's room when he said, 'I need to speak to you about the arrangement of the rooms for the ball.'

For a moment she thought it was a ruse, because he was finding it as hard to part from her as she was finding it to part from him, but his manner had returned to normal, and she quickly dismissed the idea.

'You will attend me in five minutes,' he said.

'Very well, my lord.'

She had time only to divest herself of her outdoor clothes before she went into the library, where he was waiting for her. The fire was dancing, the large flames licking the inner walls of the fireplace and filling the room with their crackling.

'The dancing will be held in the ballroom,' he said. 'It will need to be cleaned and polished. I will not have it disgracing the castle. The supper will be laid out in the dining-room. My overnight guests will dine with me at four o'clock, which will give you time to clear the room and arrange it for the ball before my other guests arrive.'

She had not expected him to take such a personal interest in the ball, but as there was no mistress of the house, she realized that he had no choice. He seemed to take no pleasure in it, but to regard it as a duty to his neighbours and a tribute to his ancestors.

'The ball will not finish until about three o'clock in the morning, but my overnight guests will require breakfast. Their servants will collect it from the kitchen, probably some time after midday.'

She listened as he told her of the castle traditions, and she noted everything he said, but all the time she was thinking of his hand raised to her face and the feel of his skin on hers, and wondering what it would feel like if he kissed her.

'I will be going away for a few days, or possibly longer, but I will back before the ball,' he said at last.

'Very good, my lord,' she said, wondering where he was going.

He did not enlighten her, and she turned to go, but he said, 'I have not dismissed you.'

He sat down in a wing chair which was set on one side of the fire, and motioned her to sit in the other.

'I don't think I should sit,' she said.

'But I have chosen to do so, and as I have no intention of getting a stiff neck from looking up at you, you will oblige me,' he said.

125

She hesitated, then she smoothed her skirt beneath her and sat down on the edge of a chair.

'Do you need to take another book from the library, or are you still reading *Le Morte D'Arthur?*' he asked.

'I am still reading it, my lord. I have almost finished it – I read in the evenings when my work is done,' she added.

A ghost of a smile crossed his face. 'I was not about to castigate you for neglecting your duties. I am glad you have had a chance to begin. I am interested to know what you think of it.'

'I am enjoying it. It is very pleasant to be spirited out of this world and into another for a time.'

'This world does not suit you?' he asked.

'It has its trials,' she said cautiously.

His reply was ironic. 'So it does. Very well. You like it for transporting you to another world. You do not find the tales realistic, then?'

She was surprised by the question, for the stories of knights and ladies, kings and queens, wizards and magic were far removed from reality.

'No,' she said.

'I think, perhaps, I do.' His shoulders sank, and his eyes turned in. 'Love is at the heart of the stories. Love of power. Love of men. Love of women. It is strange the things that love can do to a man, the journeys on which it can take him, the things it can make him feel and do. It is not a gentle thing, but a wild animal, without reason or pity; it rends and tears, making a mockery of goodness, destroying people. Love is a terrible thing.'

He fell silent, but was roused by a knocking at the door.

'Come!' he called.

Miss Parkins entered the room. She looked at Helena with hostility, and Helena felt her skin crawl, for she felt certain that Miss Parkins was her enemy. The woman's eyes might be dead, but Helena could feel her malice as a living thing.

'A letter has arrived by messenger, my lord. It has come from York.'

There was a sudden change in him; so sudden that Helena was shocked. His eyes flicked to hers, and it was as though a shutter had come down between them, breaking the bond that had been forming since their meeting in the graveyard.

'Thank you, Miss Parkins.' He turned to Helena, and said coldly, 'You have your instructions, Mrs Reynolds. You know all you need to know for the ball.'

'Very good, my lord.'

Helena rose, but as she left the room, she did not miss the look of malevolent triumph on Miss Parkins's face.

Simon waited only for the door to close, and then he broke the seal on the letter and read it.

'Well?' asked Miss Parkins.

'Mr Brunson has recovered and has returned to work,' he said. 'He is at my disposal. I think I will not see him here, it will seem odd, and I do not want anyone alerted to our suspicions. I will go to York instead. I will be able to get a description of Mrs Reynolds, and find out if she is the woman we have in the castle or not. I will go first thing tomorrow. And what of you? Have you discovered anything?'

'Nothing. She sent a letter to a friend, but it divulged very little.'

'You read it?'

'Of course, but I sealed it again afterwards.'

'And there was nothing incriminating?'

He saw her mind working behind her eyes.

'There was one strange sentence. She said she had not found

what she was looking for, but did not despair of finding it.'

His expression darkened.

'It could mean she knows . . .' – he shook his head – 'or it could mean nothing more than a lost shawl. Has she had a reply?'

'Yes. I did not manage to read it before she saw it.'

'A pity. Never mind. We must continue to be vigilant.'

Miss Parkins's look was derisory.

She knows I lowered my guard, he thought. I should not have done it. But there is something about Mrs Reynolds . . . if she is Mrs Reynolds, he reminded himself.

'Very good, Miss Parkins. That will be all.'

'Very good, my lord.'

As the door closed behind her, he walked over to the fire. His brother's death had been a terrible thing.

And his sister-in-law's death had been worse.

CHAPTER NINE

Helena was woken by the noise of hail pelting against the window and pulled the covers up over her ears. The sound was dispiriting; even more so when she shook away the last vestiges of sleep and remembered that it was Sunday, and that she had been hoping to go to church, so that she could talk to the villagers about her aunt. If the weather did not improve, that would be impossible, and she would have to remain in the castle, where she was in danger of being tangled up in the dark mysteries that hung about it, and even worse, where she was in danger of becoming attracted to Lord Torkrow.

In danger of becoming attracted to him? she thought, scoffing at herself. She was attracted to him already. There was no use denying her feelings, for they had undergone a change since arriving at the castle, turning from apprehension to intrigue, and then to something more.

It was compassion, she told herself firmly, nothing more than a kindly sympathy for a man who had had to carry the news of two deaths in one night. But it was no good. She knew it was not compassion, it was something much deeper.

She threw back the covers and climbed out of bed. At least the cold and the damp took her thoughts away from her other, more disturbing, subjects.

She washed and dressed quickly, then went downstairs to the kitchen. Over breakfast, Mrs Beal remarked that she would need more help in the kitchen as the ball approached, and she and Helena arranged for all the maids to spend a spell in the kitchen, so that Mrs Beal could choose the most useful girl to help her as the time drew near.

'The men will all have to learn how to carry a tray,' Mrs Beal reminded her.

Helena nodded, but her thoughts were less placid than her expression suggested. She knew she could not stay at the castle forever, and if not for the ball she would be thinking of leaving already, for she had explored almost every avenue of information open to her.

'And make sure they know how to behave,' said Mrs Beal. 'No talking to his lordship's guests. It's a good thing he's going away, it will make it easier for us to get on. We're lucky. He always sends word to the kitchen. He's a good master that way, though he can't say for sure when he'll be back. I expect he'll be gone for a few days, anyway.'

Helena did not know whether to be relieved or disappointed. He affected her in curious ways. He was secretive and alarming, but he moved her, too, and that was something no one else had ever done.

She finished her breakfast and then, leaving the kitchen behind her, she decided to revisit the east wing of the attic, in case she had missed anything. She found the door as she had left it. Ignoring the broken lock she went in, examining the blankets and pieces of broken furniture, then looking for other tell-tale signs that someone had been there. She breathed on the windows, remembering the times she and her aunt had written messages to each other in the steam when she had been a child, and recalling that once the steam had gone, the

130

messages remained, to be revealed the next time the window misted over. But there was nothing.

She turned towards the door . . . and jumped, as she saw Miss Parkins standing there. She moves like a cat, thought Helena, unnerved by Miss Parkins's silent approach. Miss Parkins was standing with her hands folded in front of her and she looked at Helena with expressionless eyes.

'Miss Parkins,' said Helena, feeling that she must speak. 'Did you want to see me about something?'

'I wanted to inform you that the door into the attic has been broken, but I see you have already discovered it,' said Miss Parkins, watching her. 'Do you know what happened?'

'Yes,' said Helena, standing up to her. 'One of the maids thought she heard a cat in here, and I ordered the footmen to break the lock. I was concerned that the animal might be trapped.'

'That is strange. Dawkins believed he heard a cat in there several weeks ago.'

'It must have crawled back in.'

Helena could tell that Miss Parkins did not believe her, but did not say so. Instead, it was as though they were playing some deadly game.

'And did you find it?' asked Miss Parkins.

'No. I believe it must have got out again without our help.'

'So there was no need to break the lock.'

'I did not know that at the time.' There was an awkward silence, then Helena said, 'I must not keep you.'

'You are not keeping me,' said Miss Parkins.

'Then you must excuse me. I have work to do.'

Miss Parkins did not move out of the way. Instead she said, 'There was no mention of you destroying property in your previous positions.'

'I can see no point in discussing the matter further,' said Helena, unwilling to be drawn into a conversation about her supposed previous positions. 'If his lordship wishes to take the cost of repairing the lock from my wages, then I am sure he will do so.'

She did not ask Miss Parkins to leave again, she simply walked past her. She could feel Miss Parkins's eyes on her, but she resisted the urge to look back . . . until Miss Parkins spoke again, just as she reached the landing.

'I found a pile of handkerchiefs in the cupboard in your room,' she said. 'I expected to find them embroidered with your initials, ER, but instead they were embroidered with the initial C.'

Helena's throat constricted. She had forgotten to lock her door when she had gone downstairs for breakfast, and she was glad her back was to Miss Parkins, so that Miss Parkins would not see the consternation that swept across her face. Taking a moment to gather herself, she turned round and said, 'Do you make a habit of going into other people's rooms and going through their cupboards?'

'I feared it might be damp,' said Miss Parkins. 'I did not want your clothes to become mildewed.'

'It is not your job to see to damp cupboards. That is my preserve,' said Helena.

'I help his lordship in any way I can. Mrs Carlisle was also pleased to have my assistance. I am surprised you do not feel the same way. But you have not told me why your handkerchiefs bear the initial C.'

Helena had had time to think, and a simple solution to the problem had presented itself as her aunt's initial was also C.

'They are not my handkerchiefs. I found them in the cupboard. They must have belonged to the previous housekeeper.'

'How very singular,' said Miss Parkins, in a voice devoid of emotion.

'In what way?' asked Helena, suddenly apprehensive.

A malicious gleam of triumph entered Miss Parkins's eye.

'Mrs Carlisle did not occupy that room.'

Helena felt as though the floor had suddenly given way beneath her.

'I thought . . . that is, I assumed . . . that she had the bed chamber I now occupy.'

'No, she did not. So if she had had any handkerchiefs they would not have been in that room.'

'Perhaps she had just had them laundered,' said Helena, thinking quickly, 'and one of the maids returned them to the wrong room by mistake.'

'Strange that the maid should return them to a room with no fire, and no sheets on the bed,' said Miss Parkins.

Helena felt as though she was a mouse who had been caught by a particularly malignant cat.

'Perhaps she went into the room to air it, and then forgot to reclaim the handkerchiefs when she left,' Helena said.

'Mrs Carlisle was not a great believer in instructing the maids to air the rooms,' said Miss Parkins.

'Oh, yes—' said Helena, about to say that her aunt had been a great believer in fresh air. She recovered herself quickly. 'She must have been. The room smelt fresh, not as though it had been shut up. Any good housekeeper knows the value of opening the windows when the weather is fine.'

Miss Parkins stared at her, and Helena felt an urge to squirm. She was held, mesmerized, by Miss Parkins's strange eyes, and she found herself wondering if she had ever seen Miss Parkins blink.

'Do you have a forwarding address? Then I can send them

on to her,' said Helena, trying to turn the situation to her advantage.

'No. I have no address.'

'That is a pity. Then it seems I will have to keep them.'

'Perhaps you can use them yourself,' said Miss Parkins.

'I do not believe I would wish to do so,' said Helena, conscious of the fact that she had one such handkerchief tucked up her sleeve at that very moment. 'Now, Miss Parkins, I am very busy, and I will bid you good day.'

Helena turned and once more walked out on to the landing. As she did so, Miss Parkins said, 'The castle is a strange place. It has seen many strange things. No doubt, it will see many more.'

Helena did not look back, but felt uneasy as she went downstairs. Was it a threat? she wondered. Or was it a warning? Either way, she felt she must be on her guard.

Preparing the castle for the ball was hard work but, by and by, it began to take on a brighter air. The hall was clean and fresh, the downstairs rooms were dusted and the ballroom was ready. She and Mrs Beal were putting the finishing touches to the supper menu for the night of the ball.

'Of course, the balls aren't nearly as big as they were in her late ladyship's time. She had a lot of society friends and they came from all over: London, Edinburgh and Paris. Very good to the servants, she was, her ladyship. "The servants must have their fun as well", she used to say. The day after the ball the family would be up late, and they'd have a bit of a sandwich for lunch. Then we'd lay out a cold supper for them in the dining-room at six o'clock and they'd help themselves. We had our own ball then, and didn't have to touch a bit of housework till the following morning. . . . A Solomon Grundy for the centre of the table?' she asked, breaking off in mid sentence to suggest a dish for the ball.

'Yes, that will look impressive,' said Helena.

Mrs Beal wrote it down.

'There were ever so many of us,' she said, reminiscing again. 'Mr Vance the butler – a very stately gentleman he was – kept everyone in order, and there were the outside staff, too, stable boys, grooms – and didn't they just chase the maids! – and coachmen, all dressed up.'

'The servants dressed up, too?' asked Helena, surprised.

'Yes. It was our own costume ball. Many a maid's been a queen on ball night. There were pirates and monks – Mr Vance was once Julius Caesar. "Oh, Julius, seize her!" said one of the grooms, a cheeky young monkey, when Mr Vance was trying to pluck up the courage to dance with the housekeeper – I was just the kitchen maid then. Well, Mr Vance, he went bright red, but he did it all the same, seized her, that is, and the two of them whirled round the room. He married her in the end, and the two of them are living in Hull.

'We might get up some kind of dance on the night after the ball ourselves,' said Mrs Beal. 'Nothing so grand, but it might persuade the maids to stay if they think there's some fun to be had from time to time. I've still got the costumes, packed away in a tea chest.'

'I will speak to Lord Torkrow and see if he will allow it,' Helena said.

When they had agreed the final menu, she said, 'Where are the costumes? If his lordship approves of the ball, I will ask the maids to help me sort through them and see if they need wash-ing or mending. They can choose what they will wear, and it will help me to encourage them to work hard. Some of them are prone to stop and gossip the minute my back is turned.'

'They're in the last pantry,' said Mrs Beal. She took a key off a chain round her waist and handed it to Helena.

'I will go and make sure the moths haven't attacked them, before I mention it,' said Helena. 'Disappointed hopes will lead to less work, not more.'

Mrs Beal agreed.

'I've to go to the dairy, but you can give the key back to me when I return.'

Helena took the key and went through the first pantry and into a smaller one. After going through five similar rooms, each with its own purpose, she came to one that was empty apart from a large tea chest. She knelt down and opened the chest. There was a musty smell as she lifted the lid, but she saw that the costumes were in good condition. She took out a medieval gown made of dark red velvet. Beneath it was a gold mask. She picked it up and an idea came to her. If she wore it on the night of the ball, then she could pass unnoticed amongst Lord Torkrow's guests. She had been getting nowhere in her quest to find her aunt, but if she could talk to the neighbouring gentry she might learn something. She had been hoping she might overhear something of use at the ball, but, as a guest, she could ask questions.

One of them might have seen her aunt board the stagecoach if they had been returning home after a night of carousing, and she might learn, at least, if her aunt had gone north or south.

She closed the chest, resolving to return and take the medieval gown up to her room later, together with the hat, mask and shoes that went with it. If she mixed them in with a pile of freshly laundered sheets, then no one would see what she was doing.

As she stood up, she noticed that a small door led out of the pantry, opposite the door through which she had come. It was only two and a half feet high, and she was curious as to its purpose. She tried the handle, and the door opened. It led into a low passageway. It was dark inside. She looked around the pantry, finding a candle and tinder box in a drawer. Lighting the candle, she knelt down and peered into the tunnel. It smelled dank. The floor was made of hard-packed earth, and as she put out her hand to feel it, she discovered it was damp. She did not want to go through, but she could not ignore what she had found, so, pushing the candle in front of her, she began to crawl through the tunnel. She felt the damp seeping through her dress, and her knees were cold. She looked over her shoulder, wondering if she should close the door, for it would look strange if anyone entered the room, but she had a fear of being shut in. Besides, the light was a help, both to her eyes and her nerves.

She went on, shivering as the dank walls closed about her and watching the candle flame anxiously as it flickered and spurted in the gloom.

She had not gone far, however, when the roof began to rise, and before long she could stand up. She walked for some distance before she found her way blocked by another door. She tried to open it, but it was heavy, and it was not until she set her shoulder to it that she felt it give. She pushed with all her might and slowly it opened. She went through and found herself in a mausoleum. The desolate place made her shiver, and the candle flickered with the trembling of her hand. In the corners of the stone edifice were leaf skeletons, dry and brittle and decayed with age. Dust lay thickly on the floor, and spiders' webs hung from the stone ceiling. In the centre of the mausoleum was a tomb. Helena went forward and examined it. The stone figure of a man lay on the tomb with his feet on a lion. He was dressed in stone armour and a stone sword was at his side. He was, Helena guessed, one of Lord Torkrow's ancestors.

As she lifted the candle higher, her eye was caught by something odd in the far corner. She moved forward to see it better. It was a bare patch on the floor, where the dust had been swept clean. Someone had been there recently. Lovers, she wondered, seeking a place of solitude? Or someone else?

She brushed the floor with her hand, hoping to find some clue: a plaited piece of dried lavender, perhaps, or a brooch; but there was nothing. She stood up and went over to the main door. It opened inwards, and, with difficulty, she managed to pull it towards her. The daylight hurt her eyes. The sun had come out and, after the darkness of the tunnel and the mausoleum, it seemed dazzling. She blinked several times, then, loath to return to the dank tunnel, she blew out her candle, closed the door, and set off across the moor.

It was a beautiful day. The weather was unseasonably warm, and there was a strength in the sunshine that reminded her that spring was just around the corner. The sky above her was blue, and the breeze blew a few wispy white clouds across it. The grass beneath her feet was springy, and was enlivened with bright patches of heather. Red and purple tried to outdo each other with their vivid display.

On such a beautiful day the moor looked peaceful, not a brooding enemy, but a friend. Far off, the castle basked in the sunshine, its stone appearing mellow in the light. Even the crenellations seemed less threatening, reminding her of the gimping on the edges of Mrs Beal's pastry instead of broken teeth.

It seemed impossible to think that any evil had befallen her aunt. But if not, where could she be? Could she have gone away for a holiday? Or could she, perhaps, have been lured away from the castle with an offer of a higher salary? But then why would she lie to Lord Torkrow about it?

The more Helena thought about it, the less her aunt's disappearance made sense. But it had happened, and perhaps, at the ball, she would have a chance to find out something more.

As she approached the castle, she wondered if the maids would have taken advantage of her absence, but there was a flurry of activity when she went into the castle, and she was pleased to see that everyone was working.

'We're going to be short of chairs,' said Manners, one of the footmen, coming up to her.

Manners had been at the castle the last time a costume ball had been held, and he had been a great help with the preparations.

'What did you do last year?' asked Helena.

'Brung some down from the attic.'

'Then that is what we will do this year,' said Helena.

'Yes, missus.'

'Is there anything else we need from the attic?'

'There's a trestle table up there. It needs its leg fixing, but then it'll be good as new.'

'See to it, Manners, if you please.'

'Yes, missus.'

'I will leave you to organize it.'

He nodded, and went away, calling for two of the other men to help him. She left them to their task, and, seeing Effie going into the housekeeper's room to mend the fire, she went down to the kitchen again. Mrs Beal had not returned from the dairy and, going into the pantry, she put the candle back in its place and closed the door into the tunnel. She laid the medieval costume on top of the tea chest, then collected some sheets from the laundry. Concealing the costume in the pile of linen, she made her way up to her chamber.

She was about to put the costume in her wardrobe when she remembered that Miss Parkins had been through her things once before. She did not want to take the chance of the maid finding her costume if she should happen to forget to lock her door again. She thought for a few minutes, and then she went into one of the disused bedchambers at the end of the corridor. She took the precaution of placing a chair in front of the door and then she relaxed.

She looked about her. The room was almost bare. There was a dusty cheval glass and a wardrobe. There was also a bed, but it had no sheets on it and the mattress was tattered. The grate was empty, and from the look of it there had not been a fire there for a long time. The mantelpiece was chipped and the wallpaper was hanging from the walls.

As she had already apportioned the rooms for Lord Torkrow's overnight guests, she knew it would not be needed on the night of the ball, which was fortunate, as it was not a room anyone would wish to use. It would be perfect for her purposes.

She went over to the wardrobe and examined it. Outside, it was dusty, but inside, the shelves were clean. There was an old straw hat and a fan inside, and on the bottom shelf there was a blanket, but nothing else. She was about to put her costume inside when she was overcome with a strong urge to try it on. Quickly, she stripped off her woollen gown and petticoats, shivering in the cold air as she was left standing in her chemise and drawers.

She picked up the medieval dress and dropped it over her head. The velvet slipped over her skin, and she felt a sensuous pleasure at the feel of the fabric as it slid over her arms, down over her chest and then fell in folds around her legs and feet.

She had never worn a dress like this before, and she ran her hands over it, revelling in the feel of the velvet. The pile was deep, and she stroked it both ways, enjoying the contrast between the rough and the smooth sensations. It was so different from the thick woollen gowns she usually wore, and the feel of it against her skin was luxurious.

She turned round to pick up the wig, and stood transfixed as she saw herself in the cheval glass. Gone was her dumpy figure, padded out by layers of petticoats and a thick woollen gown. In its place was a willowiness she had not suspected. The simple lines of the dress followed the contours of her body. The rich red accentuated her smooth cream skin, and gave more colour to her lips.

She put on the wig, and she could hardly believe that the person in the mirror was her. The dark wig made her eyes seem deeper set, and the style changed the shape of her face from a heart to an oval. She put on the tall, pointed hat and the transformation was complete. No one who did not know her well would suspect who she was, and with a mask the disguise would be impenetrable.

She felt things were coming to a head. She would ask as many questions as she could at the ball, and she had a sense that some of them might be answered. At last, she would learn some clue to her aunt's whereabouts and strange disappearance.

She was loath to remove the costume, with its rich colours and its sensuous feel, but she did not want to be away from the other servants for too long, for she did not want anyone asking awkward questions about where she had been, so she changed quickly, then hid the costume under the blanket at the bottom of the wardrobe. She dressed in her own clothes then removed the chair from in front of the door. She listened, making sure no one was coming along the landing, then she went out, and was soon downstairs.

'Did you find everything you needed?' she asked Manners, as she saw him standing by a line of chairs.

'Yes, it was all there,' he said. 'I've had a look at the table, and I can mend the leg. I'll have it done by tomorrow.'

'Good.'

She went into the ballroom. The dust sheets had gone and the floor had been swept. The *bobèches* for the chandelier had been washed, and two of the maids were putting them back in place, so that they would catch the hot wax that fell from the candles. Chairs had been arranged down either side of the room, and the mirrors had been polished.

'This is looking very well,' she said to the maids. 'You've worked hard.'

It seemed that the preparations for the ball might be finished on time, after all.

Simon, Lord Torkrow, returned to the castle, weary from his journey, and weary, too, from the waste of his time. As he rode into the courtyard, he thought of his unsuccessful interview with Mr Brunson, whose description of Mrs Reynolds had been so vague as to be worthless. *A very pleasant widow, of medium build . . . medium height . . . brown hair . . . possibly twenty-five or thirty-five years of age. . . .*

He was reluctant to dismount as his horse came to a halt. He had been glad to be away from the castle, for he had been thinking more and more about his housekeeper, which was foolish, when he did not know who she was, and disastrous, when he recalled the pain of love. It was not an emotion to be welcomed; it was one to be fought.

A groom came out to meet him and he could delay no

longer. He dismounted and went inside. He saw Mrs Reynolds as he crossed the hall, and although he was loath to speak to her he knew he must, for there were some details he needed to arrange for the ball.

'Mrs Reynolds, a word, if you please.'

She joined him in the library, standing before him with hands folded, perfectly poised. Was she who she claimed to be? He could not believe any evil of her. And yet he could not be sure. She asked too many questions, and wanted to know too much.

'I want to speak to you about the final arrangements for the ball,' he said.

'Yes, my lord.'

'Tell me, will everything be ready on time?'

'Yes, my lord.'

'Good.' He had no desire to hold a ball, but now that he was doing so, he expected everything to go well. The castle had a long tradition to uphold, for the balls in his parents' time were spoken of far and wide, and he was determined that this year's ball should not be an exception.

'Here is the final list of overnight guests.'

He handed it to her, and she looked at it

'The Harcourts will be arriving early. They have a long way to come, and as Mrs Harcourt does not travel well, she prefers to arrive well in advance. I have made a note of those people who should have rooms together and those who should not under any circumstances be in the same wing. It is unfortunately necessary for me to invite my family, many of whom dislike each other, but I do not intend to compound the problem by housing them too close to each other.'

'I will make sure they are accommodated as you desire.'

'Is there anything else you need to help you?' he asked.

143

'No, thank you.' She hesitated, then said, 'There is just one thing, my lord.'

'Well?' he asked.

'Mrs Beal tells me that the staff used to be able to hold their own costume ball on the evening after the castle ball. If you are agreeable, I would like to revive the custom.'

He was thoughtful, as he recalled the custom.

'It will be a way of thanking the servants for their hard work,' said Helena.

'I remember . . . yes, very well, Mrs Reynolds, as long as they do not begin their celebrations until after my last guest has been attended to. See to it.'

'Thank you, my lord.'

Helena returned to the sitting-room and found the maids were lifting the rugs.

'Don't know why we 'ave to lift the rugs,' said Sally.

'The colours will be much brighter when you have beaten them,' said Helena.

'It's a powerful lot of work,' grumbled Martha.

'Yes, it is, but once it is done the room will look much more cared for.'

'All right for some folks who 'as balls to go to,' muttered Sally, under her breath. 'It's others who 'ave to do all the work.'

'But when all the work is done, and the ball is over, then we will have our own ball,' said Helena.

The girls looked up hopefully.

'His lordship has given me permission to revive the servants' costume ball. Once you have finished your work, you may go down to the pantry and choose a costume from the tea chest. But the rugs must be beaten first, and beaten well.'

'Yes, missus, it will be,' said the girls.

They set to work with renewed vigour. They took the rugs outside and hit them with all their might, sending clouds of dust spiralling into the early spring sunshine.

CHAPTER TEN

It was fortunate the previous weeks had been fine, thought Helena, as she oversaw the last minute preparations for the ball. It had allowed the maids to beat the carpets, wash the curtains, and air the rooms. She had been so busy that her worries about her aunt had been pushed to the back of her mind, to resurface in quiet moments. If she did not learn anything at the ball – she did not allow herself to think of it: she *would* learn something at the ball.

She turned her attention back to her work. There would be few flowers to decorate the castle, but there were plenty of other things to brighten the April gloom. Huge fires were lit in every room, and the reds and oranges of the flames cast a rosy glow over the walls and furniture. The carpets, newly beaten, were colourful, and she had brought more paintings down from the attic. Some had needed their frames mending and some had needed restringing, but all now adorned the walls. There were portraits, hunting scenes and beautiful views of the castle, painted in the summer, which brought the promise of blue skies and sunshine to the rooms.

She went down into the kitchen to see how Mrs Beal was getting along. The cook was directing the women and girls who had been hired to help her, and the kitchen was a mass of loaves, cakes and other tempting food. Joints of meat were roasting over the spit, and pies and pasties were being filled with sweet and savoury fillings. The room was warm, and full of all the varied scents of cooking: the smell of meat mixed with the smell of herbs and spices to create a heady brew.

Helena went into the pantry, where the cool marble surfaces displayed a collection of cheese, butter, milk, eggs and cream.

Satisfied that Mrs Beal had everything well in hand, she paused only to offer a few words of encouragement and then went to the housekeeper's room. She was about to go in when she saw Lord Torkrow crossing the hall. He was looking about him, taking an interest in everything that had been done.

'You have done well,' he said to her. 'I have never seen the castle looking better.'

'Thank you.'

'Now, to business. I will greet my guests as they arrive, and you will escort them to their rooms.'

'Very good, my lord.'

'I will be glad when this evening is over,' he said, looking around once more, and speaking as though he had forgotten she was there. 'If the company were more congenial and the chatter not so inconsequential, then perhaps. . . . But there is not one single person I wish to see and I have a horror of playing the charming host to people I despise.'

Helena's feelings were written across her face.

'You think I will not play the charming host?' he asked. 'Perhaps you are right. My charm has long since deserted me.'

He continued on his way. As Helena went into the housekeeper's room, she found herself wondering what he would say if he knew that there was going to be an unexpected guest at the ball, and that she would be the lady in medieval costume.

*

147

The first guests, Mr and Mrs Harcourt, arrived at midday. Lord Torkrow greeted them with civility if not warmth, and Helena conducted them to their room. They did not seem a happy couple, despite their evident wealth. Mr Harcourt was a man approaching forty years of age, with features that had once been handsome but were now growing slack with dissipation. His breath smelled of brandy, though the hour was early, and there was a restless look in his eye. His clothes were impeccably cut, but the collar and cuffs showed signs of fraying, indicating that he had seen better times. He wore no jewellery save a signet ring on one finger, and paid no attention to his wife, although without her he would not have been invited, for it was his wife who was a cousin of Lord Torkrow.

Mrs Harcourt herself had an ill-humoured look. She, too, was dressed in expensive clothes that had seen better days. She set about abusing her maid, an elderly, tired-looking woman, before declaring she had a headache and commanding Helena to send her an infusion of camomile at once.

'Of course,' said Helena.

As she left the room, Mr Harcourt cast a speculative glance in her direction.

'Don't send any of the girls up to Mr and Mrs Harcourt's room,' she warned Mrs Beal, as she entered the kitchen and set about making the tea.

'I wouldn't think of it,' said Mrs Beal with a snort. 'We've had Mr Harcourt here before. Mrs Carlisle had her work cut out for her, keeping him away from the maids.'

Helena pitied her aunt, knowing it could not have been easy. Her aunt would have been polite but strong and Mr Harcourt . . . her thoughts stopped. What would Mr Harcourt have done if her aunt had crossed him?

'When was Mr Harcourt last here?' she asked.

'Not for a long time, at least not to stay. He visits the castle from time to time on county business, but his lordship won't have him here overnight if it can be helped.'

Helena dropped a handful of camomile flowers in the pot and poured the water on to them.

'And when did he last come on business?' she asked.

'I don't know. I don't see him come and go. I'm down in my kitchen, and glad of it. Dawkins shows him in and out.'

Helena resolved to speak to Dawkins about it when she had a chance. If Mr Harcourt had been at the castle on the day of her aunt's disappearance, perhaps he had had something to do with it.

'Are there any more guests I need to be wary of?' asked Helena, as she set a cup and saucer on a tray.

'Stay away from Lady Jassry. She's a tongue as sharp as my kitchen knife, and Mrs Yorke will likely accuse you of stealing her jewels if you go in her room. But the others are mostly well behaved.'

Helena gave the tray to the oldest village woman, a stoutly made matron of ample girth, and told her to carry it upstairs.

She was kept busy throughout the afternoon as carriages rolled up at the door, spilling out their guests. They were elegant and well dressed, and were accompanied by valets and maids, who hastened to do everything to ensure the comfort of their masters and mistresses, whilst managing to banter between themselves.

Helena was kept busy showing guests to their rooms and making sure that the servants knew where they were to eat and sleep. Many had been to the castle before, but for some it was their first time, and twice Helena came across tearful maids who were lost in the castle's corridors.

Mrs Beal was like three women, seeing to the roasting and

boiling of meats, overseeing the preparation of mountains of vegetables, putting the finishing touches to the pies and puddings that had already been made, and were in the pantry, ready for their grand entrance at the end of dinner. She chivvied the maids who were making the tea and checked each tray before it left the kitchen, making sure there was a good selection of cakes and biscuits to go to each room.

The musicians arrived. They knew where to go, having played at the castle before, and they established themselves in the minstrels' gallery, tuning their instruments before trying out a variety of tunes.

The holiday atmosphere was infectious, and for the first time Helena saw the castle as it must have been when Lord Torkrow's parents were still alive. Every downstairs room was open, and every bedroom in the west wing. The dust sheets had gone and the fires had been lit. The candelabras were set in front of polished mirrors that reflected the dancing light.

At last, the overnight guests had all arrived, and were safely in their rooms. Helena retired to the kitchen, where she and Mrs Beal had a sandwich and a slice of pie before turning their attention to preparing dinner for Lord Torkrow's guests.

'I'll be glad when dinner's over,' said Mrs Beal. 'Then we can get on with laying out supper in the dining-room.'

Four o'clock arrived, and dinner was served to Lord Torkrow and his guests. Mrs Beal waited impatiently for them to finish, and as soon as they had left the dining-room and retired to their bedchambers to prepare for the ball, she began to organize the maids and footmen, who quickly cleared dinner and then started carrying the less delicate supper foods upstairs.

At seven o'clock, Helena paused for breath, looking over the groaning tables. The white damask cloths could barely be seen beneath silver platters, candelabras and crystal bowls on tall stems containing pyramids of fruit.

At ten minutes to eight, Lord Torkrow appeared in the hall. He was dressed in a black tailcoat with black pantaloons.

'Is everything ready?' he asked Helena, as she hurried through the hall.

'Yes, my lord.'

The sound of a carriage crunching on the gravel outside could be heard. Helena glanced to the door. Dawkins was there, dressed in his best livery, ready to open it. Helena had taken the precaution of locking the door to the west wing of the attic so that he could not get on to the battlements, and if he had noticed, he had not said so. Indeed, how could he mention it, without revealing that he had a reason for wanting to go there? The ruse had kept him sober and, as a loud knock came at the door, he opened it with aplomb.

It was the Fairdeans who had arrived. Helena showed them to the ladies' withdrawing-room, where their maid helped them to remove their cloaks. Miss Fairdean looked exquisite in a daring costume proclaiming her to be a wood nymph. The diaphanous material of her gown was skilfully woven in different shades of blue and green which changed with the light, giving a magical impression. She preened herself in front of the mirror, ignoring Helena in the way she ignored the chairs and washstand, whilst her maid and her mother both flattered her.

Helena followed them out of the room, whereupon they sought out Lord Torkrow, congratulating him on the splendour of the castle – 'a magnificent sight'; his attire – 'so clever of you to resist the urge to dress up, I'm sure the rest of us must seem like children to you'; and his goodness in holding the ball – 'for it must be quite a burden to you, but all your friends do so enjoy it'.

Helena saw his look of contempt, and thought that Miss

Fairdean should say less if she wanted to attract him more.

Helena was kept busy as more and more guests arrived. Footmen hurried past her, carrying trays of wine, the musicians played lively airs, and the ballroom began to fill with dancers.

When all the guests had arrived, Helena allowed herself a few quiet minutes in the housekeeper's room, glad of the forethought that had led her to place a tray there so that she could refresh herself before proceeding with her plan.

It was still not too late to abandon the idea. If her masquerade was discovered, she could find herself in danger. But if not, she could learn something useful.

She had a small glass of ratafia and several biscuits. Her energy renewed, she was about to go upstairs to change when the door opened and Mr Harcourt entered. He was dressed as Don Juan, with a short black cloak over black trousers and white shirt, and on his head he wore a black hat. Strings dangled from it, falling loosely beneath his chin. His eyes were covered with a black mask, but Helena recognized him by the ring on his finger and the smell of brandy on his breath.

'I've been waiting for this moment all evening,' he said. 'I thought I'd never get you alone, and then I saw you slipping in here. All you seem to do is work. That doesn't seem fair, when everyone else is enjoying the ball.'

'It is my job,' said Helena.

'It doesn't have to be,' he said, sitting down on a corner of the desk, with one foot touching the floor and one foot left dangling. He leant forwards in a familiar way. 'There are better ways to earn a living. Easier, too. No more lighting fires and cleaning grates.'

'I am a housekeeper. I do not light fires or clean grates,' she told him coolly. 'Now, if you will excuse me, I am just on my way to the ballroom.'

He stood up and blocked her path.

'It's a pity to see such a young woman wearing such an old gown,' he said, stroking her shoulder. 'And a pretty woman, too. Such beautiful hair . . . so rich and thick . . . It could be properly dressed, if you had a little money to play with.'

'Is this the way you always behave with housekeepers?' she demanded, as she shrugged away from him. 'Did you insult Mrs Carlisle in this way too?'

'Mrs Carlisle?'

'His lordship's previous housekeeper.'

He laughed.

'I'd have had to be desperate to do that. The women was old and sour, and fit for nothing but drudgery. But you—'

'Enough,' said Helena.

She attempted to side step past him, but he stepped to the side as well.

'Why so hasty? I have a proposition to put to you, one that would be worth you listening to. You don't have to go around in old rags, you could have something new and pretty to wear.' He plucked at the sleeve of her thick woollen gown. 'Not something coarse like this, but something made of silk, or satin. Something bright, like a butterfly. You could have jewels at your throat and a bracelet on your wrist.'

He stroked her arm as he said it. She shuddered, and pulled it away.

'I am satisfied with what I have,' she said quellingly.

'Oh, no, not *satisfied*. You don't know what it is to be *satisfied*,' he said suggestively. 'But I can teach you. You'd be a good student, I'll be bound.'

'I must go,' said Helena. 'I am needed to give instructions to the maids. I will be missed.'

'Not for a few minutes you won't, and a few minutes is all

it takes for you to earn a golden guinea.'

He took one out of his pocket and held it up in front of her.

She was enraged. It was bad enough that he should think the sight of gold would dazzle her and it was a hundred times worse that he should think one guinea should suffice to buy her. The final insult was that he would think she would earn money in that way in the first place.

'Get out of my way,' she said, all politeness gone.

'So you're a woman of passion,' he said, bending forward to whisper in her ear. His breath was hot and wet and made her shudder. 'I like that. But I can teach you how to channel that passion in other, more exciting ways, and I can teach you how to earn money from it as well. There's a guinea for you now, and another one when you come to my room tonight. Or would you rather have a lesson here?'

She pushed past him and ran into the hall, losing herself quickly in the crowd of guests. She glanced in the mirror hanging on the wall and saw no sign of him having followed her, so she put the incident behind her and went upstairs. She made her way to the empty bedchamber and then closed and locked the door behind her. She was suddenly nervous. If she went through with her plan – if she dressed up and went to the ball – then she would probably lose her position if she was discovered. But if she did not, then she might never find out what had happened to her aunt.

She slipped off her dress, feeling the cold bite of the air as her limbs were exposed. She almost wished she had ordered a fire lit in the room, but no, it was better this way, for with its abandoned air, the room had not attracted any unwelcome guests; young ladies retreating from the noise, or couples keeping secret assignations.

Quickly she took off her petticoats and donned the red

velvet dress. As it whispered down over skin, she felt herself taking on a new persona, and as she put on the wig and fastened the mask over her eyes she thought to herself that it was a disguise within a disguise: the medieval lady was a disguise for Mrs Reynolds, and Elizabeth Reynolds was a disguise for Helena Carlisle. She put on a pair of shoes with red high heels that had been in the tea chest with the dress, and then put on the hat.

Even Aunt Hester would not know me now, she thought.

She opened the door and the sound of music became louder. Voices rose up from below, chattering and laughing. She went along the corridor and down the stairs, keeping to the shadows so as not to draw attention to herself. As she descended, she cast her eyes over all the people. Kings and queens, monks and fairies, knights and ladies all mingled together, filling the sombre castle with their brilliance. Masks covered their faces, some no larger than was necessary to cover their eyes, some obscuring their entire heads.

Everywhere she looked she saw illusion, as people pretended to be something they were not. And then she saw Lord Torkrow, black against the dazzling background. He disdained pretence, and proclaimed to the world who he was. His face was unmasked. He was something real and solid in the sea of disguise. His strong features were shown off by the candlelight, and it created light and shade in patterns across his face. It was like his character, she thought, a perplexing mix of light and shade.

'Champagne?' came a respectful voice at her side.

It was Dawkins. She stiffened, afraid he might recognize her, but his face was impassive. She took a glass and he moved on. She felt her confidence grow. She moved through the room with ease, as though it was her right to be there, sipping her champagne. But her new-found security vanished when, going into the ballroom some five minutes later, she saw Lord Torkrow walking towards her. He was not looking at her, though, and after briefly faltering she continued, but as she drew level with him, his eyes flicked to hers, and she knew a moment of panic. Her pulse escalated still further when he stopped and looked at her curiously, as if trying to remember where he had seen her before. Then he said, 'May I have the pleasure of this dance?'

She searched her mind for an excuse, but before she could think of one he had taken her hand and led her on to the floor.

There was a stir of interest around them. The opening chords of the dance sounded, and Helena swept a curtsy. The dress made the action extravagant, and she was beginning to find the evening stimulating. She had never been to a ball before, and the sounds and scents were intoxicating.

Opposite her, Lord Torkrow bowed. She seemed to be seeing him more clearly than usual, as though the stimulation of her other senses had stimulated her sight as well. The deep-set eyes, the high cheekbones, the pointed chin all drew her eye. He was not handsome, and yet she found his features strangely compelling, and her gaze roved over his face, taking it in.

They began to dance. As they walked towards each other, Helena felt a shiver of anticipation as their hands touched each other in a star. They separated, and she found herself looking forward to the next contact. They repeated the measure, and came together again.

'I don't believe I know you,' said Lord Torkrow, looking at her curiously.

Helena did not reply.

'Are you a friend of Miss Cartwright's?'

She smiled and shook her head.

'Then a cousin of Mr Kerson's?'

Again she shook her head.

The dance parted them, and she was relieved to be away from him. So far, she had answered all his questions with a nod or a shake of the head, but there would be other questions, more difficult to answer.

When they came together again, he asked, 'Will you be staying in the neighbourhood long?'

She shook her head.

'Do you never speak?' he asked, his voice intrigued.

She shook her head again.

'*Can* you speak?'

She felt a vibration in the air, as though the very effort to hold him away from her was setting up a resonance in the atmosphere. He seemed to feel it too. The music faded into the background and the chatter died away. She forgot the other dancers existed. She was aware of only the two of them, and everything else was a blur. She saw that his eyes were not brown as she had thought, but were flecked with different colours, gold and green, and his lashes were long and thick. She was aware of his scent, deep and masculine, and she shivered every time he took her hand, for there were no gloves with her costume, nor did he wear them, and the feel of his skin on hers was exhilarating. The music came to an end, but she was scarcely aware of it, and remained standing opposite him, connected to him by an invisible thread.

It was only when a dowager bumped into her that she was recalled to her surroundings, and felt as though she had awakened from a dream. The chatter returned, and the people, and she was once more in the ballroom with all the guests.

She remembered why she had entered into the masquerade,

reminding herself that, once she had found out what had happened to her aunt, she would have to leave the castle, and she was conscious of a strange reluctance to discover the truth.

'This is a wonderful ball,' came a voice she recognized, and she found that Miss Fairdean had joined them. Miss Fairdean had placed herself between Helena and Lord Torkrow, and, recalled to her senses, Helena took the opportunity to slip away. She went into the supper room, feeling disquieted, and still feeling the after effects of the dance as, in the early mornings, she remembered the lingering traces of her dreams.

She shook her head, in an effort to shake it away.

She found herself standing next to a gentleman dressed in a brightly-coloured harlequin's costume. It was made from diamonds of red, yellow and blue cloth, and it had a matching mask, with a red diamond over his left eye and a yellow diamond over his right.

'Good evening, Harlequin,' said a woman dressed in a gown of white feathers.

'Evening, Mistress Swan,' he said. 'Or should I say Mrs Cranfield?'

She fluttered her fan and giggled.

Then his eyes drifted to Helena. 'And who are you, m' beauty?' he asked, as he turned to look at her.

Despite his juicy relish in calling her 'm'beauty', Helena sensed no harm in him and replied laughingly, 'I am not allowed to tell.'

'Ah! I've caught you out, Miss Garson,' he said, as he took a bite of a chicken leg.

'Not Miss Garson,' said another gentleman close by. 'Miss Garson's dressed as the Queen of Sheba. I saw her earlier.'

'Not Miss Garson?' said the Harlequin.

'No, sir, and please, don't guess any further,' said Helena.

158

'All will be revealed at midnight, eh?' he said.

'It will,' she agreed. 'Until then, we must enjoy ourselves. The castle is looking splendid. When his lordship's housekeeper disappeared, I feared his lordship would not go ahead with the ball.'

'That'd have been a pity. I wouldn't be here, talking to you, now, would I?'

'Do you know what happened to her?' asked Helena.

'Who? Miss Garson?' he asked, with another chicken leg halfway to his mouth.

'No. His lordship's housekeeper.'

'His lordship's housekeeper?' he said with a roar of laughter. 'How should I know? I don't keep a watch on his servants, m'pretty!'

'I was hoping to employ her,' said a woman standing next to him, who was dressed as a milkmaid. Her hat was askew and her dress was bunched up at one side. 'If she can keep a castle clean, she can manage my manor. You cannot imagine how hard it is to find good servants. My last housekeeper left after a week, saying the moors preyed on her nerves. I said to her: "They prey on all our nerves, but we don't give in to it. We stiffen our backbones." But it was no good. She didn't listen. Said she wanted to go back to Nottingham and left the following morning. No staying power, that's what's wrong with servants these days.'

'Do you know where his lordship's housekeeper went?' asked Helena.

'I wish I did. I can't ask his lordship, I don't want him to think I'm the sort of woman who goes about taking other people's servants. But if I find out, I will offer her double her present wage to come to me.'

'That's the spirit!' said Harlequin.

'Sir Hugh Greer? Is that you?' asked the milkmaid, peering at him.

'Give us a kiss and I'll tell you!' he said.

'It is you!' she said. 'I thought I recognized the voice. Such behaviour from a justice of the peace.'

'Not a justice tonight,' he said jovially. 'I'm Harlequin, and Harlequin doesn't deal with trouble, he makes it!'

He lunged good-naturedly at the milkmaid, who picked up her skirts and, with a whoop of laughter, ran off into the crowd, pursued by Sir Hugh.

Helena was about to move on when she heard a voice at her shoulder and froze as she realized Lord Torkrow had found her.

'If I didn't know better, I would think you were running away from me,' he said.

She could not avoid speaking to him for ever, but she hoped that the noise of the room would disguise her voice. Even so, she deliberately pitched it lower than usual.

'Perhaps I was,' she said.

'Are you afraid of me?' he asked, his eyes looking into her face as though he could see through the mask and discover her identity.

'No,' she replied.

'Then you are unusual,' he said. 'Everyone else here is.'

'Miss Fairdean doesn't seem to be afraid of you,' she remarked.

The mask had given her courage, and she found she could say things to him that would have been unthinkable in her housekeeper's clothes.

He raised his eyebrows, but replied, 'You are wrong. She is. But she is also avaricious and she fancies herself as the mistress of a castle, so she hides her fear deep. If she ever found herself alone with me, she would repent her bargain. Whereas you. . . .'

160

'. . . have my reputation to protect, and would never be alone with a gentleman,' she said.

'No?' he asked. His eyes glittered, and she felt her own widen in response.

'No,' she said, though her breathing became shallow.

'Perhaps you are wise. Temptation is a terrible thing. But you're not eating,' he said, abruptly changing the conversation.

He took a plate and began to put some of the choicest food on it.

'I am not hungry,' she said.

'You must have something. I am your host. I insist. Try a sugared almond. They are very good.'

He picked one up and held it up to her lips. She took it into her mouth, tasting the sweet sugar and the nutty flavour, and alongside it she tasted the saltiness of his skin. She had an almost overwhelming urge to taste more, but she jerked her head away before she could give in to temptation.

'I am looking forward to midnight,' he said softly.

Helena thought with relief, I will not be here at midnight. When the rest of the guests unmask, I will be safely in the kitchen, dressed in my housekeeper's clothes.

'My lord, at last!' came a voice at their side. 'I have been looking for you everywhere. I am sure you remember that I promised to introduce you to my niece when last we met. Talia, make your curtsy to his lordship. She is staying with us for a while, your lordship, and we are very glad to have her with us. Such a good girl! Such pretty manners. Now, now, child, don't blush.'

The poor young girl had gone scarlet, and was looking at Lord Torkrow with a mixture of fear and awe.

He replied politely. 'Miss Winson. It is good of you to come. I hope you are enjoying your first costume ball.'

161

As the girl mumbled a reply, Helena slipped away and went into the ballroom. A young man dressed as a knight asked for her hand and led her to safety out on to the floor. She began to talk of the splendour of the castle and mentioned Mrs Carlisle, but her partner could shed no light on Mrs Carlisle's disappearance. He was far more interested in trying to discover Helena's identity. Helena parried his questions easily, but she did not learn anything of use.

She went out into the hall when the dance was over, hoping that she would learn more from his lordship's female guests. They might have heard something from their own servants, or have made enquiries if they wanted to hire Mrs Carlisle themselves.

'Lord Torkrow will never marry her,' she heard a young woman saying. The young woman had a clumsy build, and was dressed unbecomingly as Joan of Arc. 'She's been setting her cap at his lordship for the last three years, but he's never so much as looked at her. I cannot think why she wants to attract him. He makes me shiver. There's something in his eyes – he's a cold man.'

No, thought Helena, remembering the flash in his eye as he had fed her the almond. He's far from cold.

'He wasn't so cold with his sister-in-law,' said another woman who was dressed as Maid Marion.

'Sh,' said the Amazon next to her.

'Why?' asked Maid Marion belligerently. 'I'm only saying what everyone knows.'

'I don't know it,' said a young woman dressed as a Greek goddess.

'Better not say anything more,' said Nell Gwyn.

'I want to know,' said the goddess. 'Was he in love with his sister-in-law? Is that what you mean? I never heard that.'

Helena recalled the expression she had seen on his face when she had seen him in the secret room, looking at his sister-in-law's portrait.

'Why do you think she was running out to meet him that night – the night she died – when he came home from a neighbouring ball? She couldn't go to it with him, it would have made a scandal, but everyone knew they were in love with each other. She couldn't wait to see him and she went to him the moment he returned to the castle. They were lovers. Everyone knows it.'

'No!' said the goddess.

'Everyone knows no such thing,' retorted Nell Gwyn. 'It was a rumour, and nothing more. Some people have nothing better to do than to gossip about their neighbours.'

'He fell in love with her when his brother brought her to the castle just before their wedding,' went on Maid Marion, ignoring the interruption. 'She came with her family. It was her father who'd arranged the match. They stayed for a week and at the end of the week she was married. But when she said "I do" to one brother, in her heart she was saying, "I do" to the other.'

'Scandal and nonsense,' said Nell Gwyn, her oranges dancing with her emotion.

'I heard that she was besotted with him, but that he would not look at her,' piped up a buxom Viking. 'She set her cap at him, but he ignored her, so she married the other brother to spite him.'

'And I heard that he was madly in love with her, but that she was in love with her husband,' said an Italian contessa.

'Everyone—' said the fairy, before stopping and looking at Helena.

All the women turned to look at her, finally realizing there

163

was an outsider amongst them.

'Can you tell me where the ladies' withdrawing-room is?' asked Helena.

'I don't know, I'm sure,' said the goddess.

'The balls used to be so well arranged when his lordship's old housekeeper was here, but tonight I can find nothing I want. It is a pity she left in such a hurry. I wonder what became of her,' said Helena.

'Tempted away by higher wages,' said the Viking promptly. 'She went to Lady Abbinghale in London.'

'I heard it was the Honourable Mrs Ingle,' said Nell Gwyn, her interest caught.

'No, it was Lady Abbinghale. She steals everyone's servants. She stole Lord Camring's chef. Paid the man double, and left Lord Camring with no one to cook for him when he entertained the Prince. So then what does Lord Camring do but steal his chef back again at treble his original wages. We're slaves to our servants, and anyone who says otherwise doesn't know what they're talking about.'

'The withdrawing-room is at the end of the corridor, on the right,' said a young woman who had previously said nothing, and who was dressed as Lady Macbeth.

Helena was disappointed in the answer, for now she had no excuse to remain, but she thanked Lady Macbeth and moved away. She went into the withdrawing-room in case anyone was watching her, and adjusted her hat, settling it more firmly on her head. It was very tall, and it had a tendency to slip to one side. As she secured it with a pin, she noticed that the woman next to her was dressed as Katharine of Aragon, and she remembered Mrs Willis saying that that would be her costume. More, she remembered Mrs Willis's strange manner when she had visited her, and found herself wondering about the rector's wife.

When Mrs Willis left the room, Helena followed her discreetly, and saw Mrs Willis going up the stairs as silently as a shadow. She reached the top, and caught a glimpse of Mrs Willis's hem going along the corridor until she reached a room at the end. She stopped and looked round furtively, and Helena shrank back against the wall. Appearing satisfied that no one was following her, Mrs Willis slipped into the room.

Helena followed, wondering what she would find. She reached the door and turned the handle slowly, hoping it would not creak. There was a slight noise as the door started to swing open and immediately she stopped, inching it open further when there was no commotion from within. She eventually opened it enough to see into the room, and what she saw shocked her. Mrs Willis was locked in a passionate embrace with a young Poseidon, a man who was clearly not her husband.

She hastily left the room, closing the door softly behind her. Mrs Willis was not all she seemed to be. If she was concealing a lover, could she possibly be concealing other things as well?

Helena's thoughts were whirling and she felt in need of some time to think. She was passing the long gallery and slipped inside. It was far away from the bustle of the ballroom, and she welcomed its coolness. The dim light was soothing. Here there were no candles and no mirrors, only the soft moonlight coming in through the windows. It was coloured by the stained glass, making red and blue patterns on the floor.

She began to pace the length of the gallery, walking in and out of the pools of coloured light as she thought over everything she had seen and heard. She had not gone more than halfway when she started, for there was a figure at the end. In the eerie light she could see no more than his silhouette, but she knew who he was at once, by a stirring of the air and a lift of something inside her. It was Lord Torkrow. She started to back away, but it was too late! He had heard her.

'We meet again,' he said, moving forward, his skin dappled red and then blue by the light. He looked down into her eyes. 'I wonder, was it by accident or design?'

'Forgive me, my lord, it was an accident,' she said. 'I did not know anyone was in here. I wanted to get away for a while. I did not mean to disturb you.'

'No matter,' he said. 'I was ready to rejoin my guests.'

'Then I must not prevent you,' she said, although she felt a powerful force emanating from him, and found it hard to turn away.

'I have changed my mind,' he said. 'It is time for the unmasking, and I am intrigued. Who are you?'

'I cannot tell you yet,' she prevaricated. 'It is still five minutes to midnight. I will unmask in the ballroom at the appointed hour.'

'Will you? Or will you disappear like a will-o'-the-wisp, never to be seen again?'

'Of course not,' she said. 'The idea is absurd.'

But, standing the in the long gallery, it did not seem absurd. The supernatural seemed to be all around them, from the dappled light to the strange atmosphere.

'I am not so sure,' he said. 'I am beginning to think you are a creature from folk tale who will evaporate as midnight strikes, leaving me bereft, and I have a mind to discover your identity now, before it is too late.'

'That would spoil the game entirely,' she said, turning to go.

He caught her by the arm and said, 'It is near enough the appointed time, and I will not be denied.'

So saying, he pulled off her mask. Its strings caught on her wig, and the mask, wig and hat came off together. She felt a surge of alarm and she had a desire to run away, but he was still holding her arm, and flight was impossible.

Her only hope lay in the dim light, but it was dashed as she saw recognition dawn in his eyes. For a long time, he just looked at her. And then he said again, 'Who are you?'

Helena's pulse jumped at the question. So he knew she wasn't Mrs Reynolds! Or perhaps he did not know for certain; perhaps he just had doubts; in which case, she must not confirm them.

'I know I should not have done it, but I could not resist. I heard the music and I was overcome with a longing to dance, and so I slipped upstairs and put on the costume I had been intending to wear for the servants' ball,' she said.

'Then if you want to dance, you must dance.'

He slipped his hand round her waist and before she knew what was happening, they were waltzing, whirling in and out of the shadows whilst the light played strange tricks all around them. Was he a man or a monster? she wondered, as the faint strains of music drifted up from the minstrels' gallery, like the wail of an unearthly creature howling in the dark.

'Well? Was it worth it?' he asked her, as they reached the end of the gallery.

'I cannot answer that,' she said, looking up at him and trying to read his thoughts.

'Why not?'

'Because I do not know yet what the consequences will be.'

'So, disguising yourself does not trouble you unless there are consequences?'

'That is not what I said.'

'But it is what you meant. Was it worth the deception, to get what you wanted? Did the end justify the means?'

167

She felt that he was not talking to her about her disguise, but about something much more sinister, and she began to be frightened. She tried to pull away from him, but he held her fast.

'Just what *would* you do, if you felt there would not be any consequences? You did not hesitate to impersonate Elizabeth Reynolds. What else would you not hesitate to do?'

'I don't understand you,' she said, feeling a rising tide of panic.

'Would you lie . . . steal . . . kill?'

His fingers tightened round her wrist like a vise.

'Let me go.'

With a strength born of desperation, she wrenched herself free, but he stood in front of her and would not let her pass.

'Who are you?' he demanded menacingly.

'I am Elizabeth Reynolds.'

'No, you are not Elizabeth Reynolds. She never arrived. A messenger arrived from York earlier this evening, saying that Mrs Reynolds had written to apologize for not taking up her position, because she had been ill, and was still not well enough to work. And so I ask you again, who are you? And what are you doing at Stormcrow Castle?'

For a brief moment she thought of telling him the truth, but it was too dangerous. If he had done away with her aunt, and if he knew she had come looking for her, then he would do away with her, too.

'A friend,' she said. 'I'm a friend of Mrs Reynolds. She told me she would not be able to take up the position as she was not well, but she did not want to acquire a reputation for being unreliable with the registry office. I was looking for a position at the time, and so we agreed that I would take her place.'

He looked at her searchingly, and then his face twisted.

168

'You are lying,' he said roughly. 'You will leave the castle first thing in the morning. The carriage will be at the door at seven o'clock. It will take you to the stagecoach. And to make sure you go, I will put you on the stage myself. You will leave this neighbourhood, and you will not return. If you do, I will know how to deal with you.'

His eyes were hard, and in the candlelight they glittered like obsidian. He loomed over her, and she wondered what he would be capable of if he was crossed. But she would never find out, because she had no intention of remaining. She had learnt all she could at the castle.

'Very well,' she said. She thought of the coming journey, and realized that she had no money. 'What of my wages?' she asked.

'Your wages?' he returned incredulously.

'I have worked for you faithfully, and my wages are owing,' she said defiantly.

He looked as though he was about to make a cutting retort, but then thought better of it.

'I will have them waiting for you in the morning,' he said. 'Make sure you are in the courtyard at five minutes to seven.'

'I will be there.'

And with that she picked up her mask, hat and wig, then swept past him, out of the gallery. Once she was out of sight she gave in to an urge to flee, and she ran back to her room, closing and locking the door behind her.

Only then did she let out a deep breath. She was safe at last. She went over to the fire and knelt in front of it, wrapping her arms around herself. As she did so, she began to shiver with reaction to the frightening encounter. She had not known what he would say or do, and at one point she had been afraid that she might not even escape with her life. Thoughts whirled round her head – graveyards and ballrooms, castles and crypts – all was jumble and confusion.

The fire was hot, with flames leaping in the grate, but it did little to warm her. She was cold through and through. She glanced at the bed, and wondered if there was a hot brick in it. She went over to it and discovered, to her relief, that there was. She undressed and slipped her nightdress over her head, then climbed between the sheets, but although she lay down and closed her eyes, Lord Torkrow aroused such conflicting emotions in her that she could not sleep.

At last she got out of bed and, throwing her shawl round her shoulders, she went over to the fire. Sitting beside it, she looked into the flames.

There was one chance more for her to learn something about her aunt. If she went to Mary and told her the truth, then perhaps Mary could tell her something.

The more she thought of it, the more the idea appealed to her. She would leave early, before the carriage was ready, for it would be better by far to be well away from the castle by the time he started looking for her. With Mary she would feel safe.

She went back to bed and at last she fell asleep, but vivid dreams gave her no rest. She was running through the castle, holding up the skirt of her medieval gown as she ran along the corridors, looking for something she could not find, her task made more difficult by a swirling mist. The mist parted, and she saw a door. She seemed to be moving in slow motion as she opened it, to reveal a large room with a four-poster bed, hung with red curtains. A man and woman were embracing passionately by the bed. As Helena watched, the woman opened her eyes and turned towards her, smiling as the man kissed her throat. And then the woman's face changed, becoming her own, and as the man spun round, Helena saw it was Lord Torkrow.

Shocked, she closed the door and ran on down the corridor, but it was hung with cobwebs. She brushed them aside, but they became thicker and thicker as she went along, until she was flailing wildly in an effort to keep them away from her. They were in her hair and her mouth, and they were beginning to suffocate her. She fought them wildly . . . and woke up to find that she was wrestling with the sheets. She was panting with the exertion, and she lay still, until she heard a noise and realized what had woken her: it was Effie, scratching on the door.

She rose, bleary eyed and feeling unrefreshed, and let the scullery maid into the room. As Effie saw to the fire, she washed and dressed. She put on her warmest clothes and her stout shoes, then she went down to the kitchen. It was empty apart from Effie, who had finished seeing to the fires and who was busy washing dishes.

'Where is Mrs Beal?' asked Helena.

'She's seeing to the clearing up,' said Effie.

Helena felt sorry to be leaving Mrs Beal to so much work. If things had been otherwise she would have overseen the servants as they returned the spare furniture to the attic and instructed them as they cleared the rooms, but she could not linger.

She helped herself to some rolls and chocolate, then sat by the fire to break her fast.

When she had done, she went upstairs and packed her few possessions. She checked the drawers and wardrobe to make sure that nothing had been forgotten, and looked under the bed, then closed her valise and set it by the door.

She glanced at the clock on the mantelpiece. It was nearly half past six.

Throwing her cloak over her shoulders, she put on her bonnet and pulled on her gloves then, picking up her valise

171

went swiftly down the back stairs. She had hoped to see Mrs Beal before she left, but time was moving on and she did not want to risk looking for her in case she bumped into Lord Torkrow.

She opened the side door carefully and looked out. There was no one about. She went out, closing the door behind her.

She hurried across the courtyard, looking over her shoulder as she did so to make sure she was not being followed. She had a dread of seeing Lord Torkrow or Miss Parkins standing at one of the windows, watching her, and she scanned them nervously, but, to her relief, there was no one there.

Her gaze reached the gallery window . . . and her heart almost stopped, for she was suddenly reminded of the fact that the castle was symmetrical. Every room had its counterpart.

So the galleries must be symmetrical, and the hidden room in the portrait gallery must have its counterpart in the long gallery.

There was another secret room.

CHAPTER ELEVEN

The enormity of the revelation froze her for a moment then, turning on her heel, she ran back to the castle. In the side door she went, up the stairs, along the corridor and into the long gallery. She walked along its length, her footsteps sounding loud to her ears, despite her attempts to walk quietly, and it was with relief that she reached the end of the gallery. She dropped her valise, and then began to feel the wainscoting, running both hands across it. There must be a way of opening it, and she guessed it must have something to do with the embossing. She pressed the flowers and turned the grapes and, as she did so, she called out softly, 'Aunt Hester! Aunt Hester! It's me, Helena!'

But there was no reply.

She pushed the centre of a small flower, and it gave. She heard a click, and then a door in the panelling swung open. She took a deep breath and went in.

She found herself in a small room. There was a window to the west, but the grey light of morning did little to illuminate the chamber. The air was stale, and she wrinkled her nose against it. She stood motionless whilst her eyes adjusted to the dim light and then went forward. As she did so, she saw that the room was empty, except for some blankets on the floor in the corner.

173

There were no pictures on the walls, and the floor was bare.

She went over to the blankets, which had been arranged to make a bed. She crouched down next to them and turned them over, then she sat back, shocked, as she saw that, in between the folds was a piece of plaited lavender. She picked it up with trembling fingers. So her aunt had been here!

She shook the blankets, hoping to find another clue, and something fell out. It was a wooden soldier. She picked it up and examined it. It had been painted but the paint was coming off. It was evidently a much-loved and much-used child's toy. But what had a child been doing in this room, and what had Aunt Hester been doing with him?

Could the child have been playing here, and could Aunt Hester have been looking after him? But why would anyone make a child play in a cramped, gloomy apartment? And what child could it be? Lord Torkrow had never married.

But his brother had. . . .

A sliver of fear crawled down her back. Every dark thing she had ever heard about Lord Torkrow and every unsettling thing she had experienced since entering the castle, returned to haunt her. Had he been responsible for her aunt's disappearance, and perhaps worse besides?

What had her aunt been doing in the secret room? Tending to the child? Or protecting him? Because if Lord Torkrow's brother had been the older of the two, and if he had had a son, then the boy was the true heir of Stormcrow Castle. . . .

Helena left the room, closing the door behind her. There was a click, and then it merged into the wall.

She abandoned her plan to leave the neighbourhood, for she knew she could not ignore what she had found. She feared that a terrible crime had been perpetrated at the castle, but who to tell?

Her mind went back to the costume ball, and the man dressed as Harlequin: Sir Hugh Greer, the local Justice of the Peace.

Helena made up her mind to visit him and lay the facts before him: that her aunt had gone missing, and that she had found evidence of her aunt and a child having been kept in a secret room in the castle.

He would know what to do.

She did not know where to find him, so first she must go to Mary's cottage, for Mary would know, and she might even lend Helena the trap to take her there.

She picked up her valise and went down the stairs, moving cautiously. It was nearing seven o'clock. She could hear the sound of the carriage being brought round. The crunching of the gravel under the wheels was like the sound of bones, and a new fear assailed her. She had delayed so long that, if she set off on foot, she feared she would soon be caught because Lord Torkrow would overtake her in the carriage.

A quick glance out of the front door showed her that he had not yet appeared, and hurrying through the hall, she reached the carriage before he came in sight. Its black body seemed ominous, and she was afraid of climbing inside, to be swallowed by the red interior, but she mastered her dread as Eldridge climbed down from the box.

'His lordship has been delayed,' she said. 'You are to take me to Miss Debbet's cottage, where I am to deliver a message. You will then proceed to the stage post alone and await his lordship's instructions.'

'That's not what 'e said to me,' said Eldridge, his dour face glowering suspiciously. ' 'E said I was to go to the stage, but 'e didn't say nothing about no cottage.'

'He has changed his mind. If you don't believe me, then you

must go and ask him yourself. He is in his study. But make haste! He has commanded me to deliver his message without delay.'

She climbed into the carriage. Eldridge looked towards the door, then at Helena's impassive face, and gave a brief nod before folding the step and shutting her in. He mounted his box, and then they were away. She breathed a sigh of relief to think that one problem, at least, had been overcome.

The carriage seemed to crawl away from the castle, and she sat forward on her seat, willing it to go faster. At any minute she expected Lord Torkrow to emerge from the castle, shouting, 'Stop!'

She was so fearful that she could not help looking back, but everything was quiet. The carriage rolled on slowly, through the arch, and then it began to pick up speed as it emerged on to the road.

She looked forward again, but her eyes did not see the moor as it rolled past. Instead they turned inwards, and she was consumed by her thoughts. What had really happened at the castle? Had Lord Torkrow tried to murder his nephew, and had Aunt Hester hidden the boy in an effort to protect him? If so, where had she gone? Had she taken the child with her? And was she alive, or were they both. . . ? She did not want to finish the thought.

The carriage turned off the main road and she recalled her thoughts from their dark paths. Ahead of her, she could see Mary's cottage. Never had a sight been more welcome. She opened the door as the carriage rolled to a halt and jumped out. Eldridge looked surprised at her behaviour, but said nothing, merely closing the door behind her.

'Go on to the stage post and await his lordship's further instructions,' she said.

At least, if Sir Hugh had to force his way into the castle, he would find one less man blocking his way.

Eldridge looked dubious, but he nodded his head, and the carriage rolled away. As soon as he had gone, Helena went up the path and knocked at the door. It was early, but she hoped Mary would be awake.

She need not have worried. The door was opened by the maid, and she was shown in, to find Mary sitting in the parlour.

'Mrs Reynolds,' said Mary, standing up in surprise.

'I am sorry to disturb you at such an hour, but I am in dire need of help,' said Helena without preamble, afraid that at any moment there could be a knock on the door and that Lord Torkrow could walk in.

'Whatever has happened?' asked Mary in concern. 'Has there been an accident? Is someone hurt? Are you ill?'

'Please, have the trap readied. I will explain everything when it is brought round.'

Mary looked surprised, but she hesitated for only a moment, and then she gave the maid instructions to see that the trap was to be brought round to the front door.

'And tell Tom to make haste,' she said, as the maid left the room.

'Thank you,' said Helena gratefully.

'I do not know where you need to go in such a hurry, but won't you have something to eat whilst we wait?' said Mary. 'I was just having breakfast, and you cannot set off until the horse has been harnessed. Some food will help sustain you on the journey ahead, wherever you are going.'

Helena accepted gratefully. She had already breakfasted, but it seemed a long time ago. Mary poured her a cup of chocolate and handed it to her with a piece of seed cake. Helena ate gratefully, then accepted a second cup of chocolate, but she left it half finished as she heard the trap outside. She leapt up.

'I must go at once.'

Mary rose calmly and followed her into the hall, putting on her cloak.

'I do not know what has happened, but I think you need a friend,' said Mary. 'I cannot let you go off by yourself. I am coming with you.'

Helena felt a rush of relief. With the groom and Mary beside her, she would feel much safer if Lord Torkrow should happen to ride after the carriage and come across her on the way.

'I would be glad of your company,' said Helena.

'Then come, let us be off.'

Together they went outside.

'Now, where to?' asked Mary, as she followed Helena into the trap.

'Sir Hugh Greer's house,' said Helena. 'I need to see a justice of the peace.'

'But isn't Lord Torkrow the nearest justice?' asked Mary with a frown.

'Yes, he is, but I cannot speak to him. It is about him I have to lay a complaint.'

Mary looked surprised, and she seemed about to protest, but then she simply instructed the groom to drive to Sir Hugh's house.

'Now, don't you think you had better tell me what this is all about?' she asked, once they were safely on their way.

Helena gave a deep sigh.

'There is so much I have to tell you.' She could maintain the deception no longer, and she was relieved to be able to tell Mary the truth. 'First of all, you must know that I am not Mrs Reynolds,' she said, and then her story came pouring out in a rush.

Mary listened silently, and when the recital was over, she said, 'So you think Lord Torkrow has hidden his brother's son, or done something worse, so that he can rob the boy of his inheritance, and keep the title and the castle for himself?'

'I am afraid it is possible, yes,' she said.

'It seems incredible,' said Mary musingly. 'And yet you found a hidden room, with your aunt's plaited lavender and the child's toy, and your aunt is definitely missing.'

'Yes,' Helena agreed. 'You were concerned as well, weren't you?' she asked. 'You were worried about my aunt? Your story about needing to return a book to her was a ruse, so that you could find out her forwarding address?'

Mary nodded. 'It was. I thought, if you had an address for her, then I could find her. How did you guess?'

'I knew the book could not belong to my aunt. She had never had much time for reading, and as far as I know she has never owned a book. Besides, she does not like poetry.'

'Ah. I see. It was a poor story, but it was the best I could think of at the time. You do not blame me for the ruse?'

'Not at all. I am grateful to you for it, and for trying to find her. I hope she is all right, but with every passing day and still no word. . . .' said Helena anxiously.

'Perhaps word has reached your lodgings?'

'No, my friend collects the mail and would have sent me news.'

'Even so, things might not be as bad as you fear. Perhaps your aunt managed to escape with the child. If she needed to retreat to a place of safety, where would she go?'

'I cannot think of anywhere,' said Helena, as she turned her mind to this new possibility.

'Does she have any relatives she could turn to?'

'No, only me, and she did not come to me.'

'But she must have taken him somewhere,' said Mary thoughtfully. 'Can you not think of anywhere?'

'No. Unless . . . Mrs Beal mentioned that the old butler, Vance, went to live in Hull, when he and his wife retired. My aunt had worked with Vance before, and it was he who had recommended her for the position at the castle—'

'Then that's where she must have gone. Never fear, you will find her yet. Do you know exactly whereabouts in Hull the butler lives?'

'No. I never thought to ask.'

'Why should you? But it is of no importance. We will go there and seek them out. They cannot be hard to find. Someone will know of them by name, or of a woman and a child who are newly arrived in the town. You do not object to my plan? If you wish, we can continue on our way and consult Sir Hugh, but it is not certain that he will be at home, or that he will believe us. And even if he does, he might not like to move against a neighbour, particularly not one of Lord Torkrow's standing. It seems to me that we would be better finding your aunt and the boy ourselves.'

Helena agreed. For the first time in many weeks she had hope. If only Mary was right, then she might be seeing Aunt Hester before the day was out.

They had travelled some miles across the moor, and were approaching The Dog and Cart. Mary suggested they should change the horse before going on.

'We will ask for a hamper to take with us, too. We might be delayed on the journey, and it could take us hours to find the right address. It will be quicker if we eat on the road, rather than wasting time looking for an inn once we reach Hull.'

Helena agreed, and when they pulled up in the yard, Mary suggested that Helena go inside to order the food, whilst she made sure the horse was changed for a satisfactory animal.

Helena climbed down and went into the inn. It was a small but respectable establishment, and as she entered, the innkeeper came forward to greet her. She told him what she wanted, and he showed her into a private parlour until the provisions should be ready.

The parlour had a table and two settles, but Helena was too restless to sit down. She paced the room, anxious to be on her way again.

The innkeeper seemed to be taking a long time with the provisions. She went out into the corridor to find him, but as she did so, she was horrified to see Lord Torkrow walking in at the inn's main door at the far end of the corridor.

She shrank back, wondering what he was doing there. Had he followed her, or was his presence there a coincidence? Perhaps he had decided to ride to the stage post when he had discovered the carriage was missing, and perhaps he had stopped at the inn to find if the carriage had passed. Once he had learnt what he wanted to know, she hoped he would be on his way again, but until then she would have to stay out of sight.

She ran back along the corridor and slipped back into the parlour. She listened intently, every nerve straining, but she heard nothing and began to relax. And then she heard foot-steps coming down the corridor. They made an ominous click-ing noise as they crossed the flagged floor, and stopped outside the door. But was it the innkeeper, or was it Lord Torkrow?

She saw the door knob turning, and, suddenly panicking, she leant against the door, but it heaved, and in a moment it was flung open. She was thrown back against the wall, but by good fortune she was hidden by the door. She saw Lord Torkrow stride into the room like a dark creature of the night, intent on finding his prey. Helena shrank back.

He looked round, and for a moment she thought he would not see her, but then his eyes alighted on her and he closed the door, revealing her.

'So,' he said menacingly. 'This is where you are. Then it is as I had suspected. You took a post in the castle under false pretences. Well, your master will be disappointed. You will not be able to tell him anything.'

'My master?' she asked in confusion, wondering what he was talking about.

'Or did you not see him? Has he remained in the shadows? Is it only Maria you have dealt with? Then you are fortunate. And I suppose it is possible, for she could pay you as well as he.'

Helena was perplexed.

'I don't know what you're talking about,' she said.

'No?'

He took her arm and pulled her over to the window. She saw the trap, complete with a fresh horse. Mary's coachman was climbing up on to the box, and Mary herself was already in the trap. They were ready to leave. She must go to them!

Pulling free of him, she ran for the door and wrenched it open. But then a sound from the yard gave her pause, a clattering on the cobbles, and turning her head she saw that the trap had set off without her. She ran to the window and in a lightning quick move, threw it open, and shouted, 'Mary!'

Mary turned her head and saw her, but then she turned away and the trap continued across the yard.

Helena was dumbfounded. Why had Mary deserted her? Had Lord Torkrow paid her to leave? Impossible! Mary would not give in to intimidation or bribery, she was sure. But the trap continued, and turned into the road.

Lord Torkrow joined her at the window and said savagely, 'Damn you! So she was here after all. And where she is, he will not be far behind. Where is she going? What have you told her? You found the secret room, but what more have you learnt? Answer me! What have you told her? Where is she going?'

Helena did not know why Mary had left her, but she knew one thing: she could not answer his question if she wanted her aunt to be safe.

'I will tell you nothing,' she said, rounding on him.

'You will tell me everything I want to know, or it will be the worse for you,' he said threateningly.

'Never,' she said, between gritted teeth.

'I don't know what you think to gain by protecting them. They have already paid you – or perhaps not,' he added appraisingly. 'Perhaps that is why you are protecting them. Perhaps you are afraid they will go back on their part of the bargain if you give them away.'

'I don't know what you're talking about,' she said.

'Tell me, *have* they paid you?' he asked roughly.

'Have who paid me?'

'Maria and Morton'

'I don't know anyone by the name of Maria, but if you mean Mary, why should she pay me? I do not work for her.'

'Then you work for Morton.'

'I know no one by the name of Morton. The only man I have ever seen her with is her brother, and surely even you would not blacken the character of a poor, sick man who was wounded at the Battle of Waterloo in defence of his country?' she returned scathingly.

He regarded her closely, then said: 'You are a very good actress, or they have lied to you.'

'They have never told me anything, other than that they

were living in the country for the good of Mr Debbet's health. You are either evil or mad.'

He searched her face.

'And you are either a hapless pawn or a willing accomplice, but it makes no difference,' he said. 'Where were you going with Maria?'

'I don't know,' she said.

'You do, and you will tell me,' he said brutally.

'Do you really believe I would deliver a woman and a child up to you, so that you can finish what you have begun? Do your worst. You will never have them,' she flared.

He looked at her curiously, then asked her the question he had asked her in the gallery. 'Who are you?' he said.

'Your nemesis,' she returned.

'Your name,' he demanded.

She flung it at him defiantly, glad to be rid of the pretence. 'My name is Carlisle,' she said. 'I am Mrs Carlisle's niece.'

He look shocked, then said, as if to himself, 'So that is it. The handkerchief – C. Carlisle.'

She saw understanding dawn on his face, and she wondered what he was going to do, now that he knew the truth. Would he imprison her as he had imprisoned her aunt?

Her question was soon answered, for saying 'There's no time to waste', he took her hand and pulled her along in his wake as he strode out of the parlour. 'You are coming with me,' he said.

She resisted, but he was too strong for her. She looked about her for help, but the corridor was empty. He pulled her towards the entrance, through which she could see his carriage. So! He had caught up with it, and learned of her ruse.

She dug her heels into the gap between the flags, knowing that if she climbed into the carriage she would be at his mercy. It gave her the resistance she needed to bring him to a halt.

'There is no escape,' he said, tightening his grip. 'You are coming with me, and you will tell me everything I need to know on the way.'

'So that you can find my aunt, and kill your nephew, as you killed your brother?' she demanded.

'*What?*' he said, rounding on her.

In his surprise he dropped her arm, and she ran, but she had only gone a few feet when he caught her again.

'You will never find them,' she said, turning on him. 'My aunt has hidden the boy, and she will look after him until Mary rescues them.'

He suddenly dropped her arm, and to her shock, his hand cupped her face. He looked deeply into her eyes.

'I don't know what Maria has told you,' he said, 'but I am not your enemy.'

Again, his words confused her.

'Who is Maria, and why do you keep talking about her?' she asked.

'She is the woman you arrived with at the inn. She is dangerous. I don't know how she has imposed on you, but I must know where she is going, because if she finds your aunt before we do, then she is likely to kill her.'

He was so convincing that she faltered.

'I don't know what to believe,' she said uncertainly.

Lord Torkrow was the villain, wasn't he? But somehow his words had the ring of truth, and he was no longer behaving like a villain. He was not threatening her. He had dropped her hand and although he could have dragged her to the carriage and thrown her in, he had not done so.

'Your aunt is in danger, and so is my nephew,' he said. 'You have to decide who to trust; I cannot take the decision for you. You must trust Maria, or you must trust me.'

She felt very still. She remembered everything she had heard about him, every whisper, every rumour, and then she thought of her own experiences at the castle. He had been hostile, but she had never seen him do anything amiss: he had never mistreated the servants, and he had not filled the castle with debauched friends. Even his hostility could be explained if Mary was dangerous, as he claimed, and if he had thought she was in league with Mary.

But Mary could not be dangerous . . . could she? Helena thought of everything she knew about Mary, and was surprised to realize it was so little, and that even that little came from Mary herself. She remembered that Mary had tried to discover her aunt's whereabouts by pretending she had a poetry book belonging to her and asking for a forwarding address; and then again by saying that her aunt must have run away with the boy, and asking if Helena knew where they might have gone. And Mary had abandoned her at the inn as soon as she had learnt what she wanted to know.

Helena looked at the man in front of her, and thought of everything that had passed between them since she had entered the castle. He was dark and dangerous and she was half afraid of him, but she realized that her fear had never been for her safety. She was not, nor had she ever been, afraid he would hurt her, no matter how he had behaved; she was afraid because he awakened feelings in her that she could neither control nor ignore.

'Well?' he said. 'What is it to be? Do you trust Maria, or do you trust me?'

'I trust you,' she said.

He gave a ghost of a smile, then said: 'We must go.'

'My aunt is in—'

'Hull. I know. All I need to know now is if that is where

Maria has gone.'

'Yes, it is.'

They went out into the corridor. The landlord was bustling towards them with a hamper. Lord Torkrow paid him a guinea, and gave him another, saying, 'Tell me, which is the best way to Hull?'

'Why, bless me, you're the second person to ask me that this morning. The young lady was just making enquiries. You follow the road here until you reach the main road. . . .'

Helena grew more and more impatient as the man went on. She could not understand why Lord Torkrow listened patiently to the innkeeper's directions. Surely he knew the way?

'Thank you,' said Lord Torkrow.

The landlord went about his business, and they went out into the courtyard.

'Why did you ask for directions?' she said.

'Because I hoped that Maria might have done the same, and then I could discover which way the landlord had sent her. He will have told her the main route, as he told me, but it is not the quickest. We will beat her yet.'

His carriage was waiting for them. Fresh horses had been harnessed, and were champing at the bit. Eldridge opened the door and let down the step. Lord Torkrow stood back, letting Helena go in, and then followed her inside. The step was folded, the door closed, and as soon as Eldridge had climbed on to his box they were on their way.

Helena remembered the last time she had been in a carriage with him. How long ago it seemed! She had been afraid of him, and worried about her aunt. She was still worried. Where had Aunt Hester gone? And why?

'Why did my aunt leave the castle?' she said. 'Did you send her away with your brother's child? And why?'

187

'Yes, I sent her away, but she was glad to go. She wanted to protect George as much as I did, and she knew it was the best way.'

'But why was he in danger? His parents are dead, so surely you are the boy's guardian?'

'My nephew is not the son of my brother. My brother died childless. The boy is my sister's son.'

'Your sister?'

'Yes. My sister. Anna. You saw her portrait.'

Helena cast her mind back to the portrait gallery and remembered the family portrait hanging there. It had shown Lord Torkrow and his brother as children, and there had been a little girl standing next to them.

'She was a beautiful girl,' he said with a far-away look on his face. 'She had soft, dark hair and a mouth that was made to smile. She was my father's favourite child, and my brother and I did not mind, because she was a favourite with us as well. She was the youngest, and we all took pleasure in looking after her.'

He turned to her. 'I owe you an explanation. I did not know you were Mrs Carlisle's niece, and so I said nothing about your aunt when you came to the castle. But you must be worried.'

'Yes, I am.'

'She is safe, in Hull, with my sister. My sister has not had a happy life,' he explained. 'When she was eighteen years old, my parents took her to London for a Season. She met a man there, John Morton, and she quickly became besotted with him. My father was displeased. Morton was older than her by some fifteen years. My father's warnings fell on deaf ears, however, because Morton had a way with women, and my sister was entranced.

'My father hoped it would come to nothing, that Morton would tire of her and turn to someone nearer his own age, but instead, Morton asked for her hand in marriage. My father was reluctant to agree, but Anna was in love, and so he gave his consent.

'At first all went well. Anna was happy, living with her husband in Norfolk. But then things started to change. She seemed quiet and withdrawn. I was anxious, and one day I paid her a surprise visit. I found her with a bruised face, and I told her she must leave her husband and come back with me to the castle, where my father, my brother – for he was still alive at the time – and I would protect her. She refused. I thought of carrying her bodily out of the manor house, but she told me she was with child. Her husband had been delighted with the news, she said. He had stopped hitting her, and had told her that he had only done it because he had been frustrated at the lack of an heir.

'I called on Anna many times over the next few years. My brother and sister-in-law also called on her, often unexpect-edly, to make sure she was all right. She never had any bruises, and she was radiantly happy. She loved her little boy, and her husband doted on the child, too.

'I was reassured. Besides, I had other things to think of. My sister-in-law and my brother both died. Anna came to the funeral, but she never visited the castle after that, and I never went to the manor. I was lost in my own labyrinth of darkness, and had no thought to spare for anyone else.

'So things went on. And then came a day, a few months ago, when, on a wild night, my sister came to the castle. She was almost collapsing with exhaustion when I took her in, and she had been badly beaten. The mark of a whip was on her back – but I will say no more. Her cur of a husband had taken to hitting her again, and she had borne it silently, because she had known that if she left him he would never let her see her son again. But then he had threatened little George. She waited until he became unconscious through drink and, persuading a groom to help her, she took her son and set out for the castle.

'She had a little money, which she had kept hidden from her husband, and she used it on a ticket for the stagecoach, before walking the last part of the journey.

'Even when she reached the castle she did not feel safe. She was terrified that her husband might follow her, and so I took her to the attic, and I called on Mrs Carlisle to take care of her and the boy.'

'So that is why the maid heard crying in the attic,' said Helena. 'It was George.'

'Yes. The rumours spread. It was easy to dismiss the sound as ghosts amongst the villagers, but I knew that if Morton came near, he would know well enough what it meant. I moved Anna and the boy to the secret room, but even so, Anna did not want to stay at the castle – she lived in hourly dread of him finding her – but she was by this time too ill to leave. She had caught a fever, travelling through the cold and the snow. And so I asked your aunt if she would take little George to Hull, where Mr and Mrs Vance would care for him. She agreed readily. She felt sorry for my sister and nephew. She was to stay with George until my sister was well enough to join her in Hull, and then my sister would go to Italy, where she has a godmother. Anna would be able to raise her son in safety, somewhere her husband would never find her.'

'But I still don't understand about Mary – Maria,' said Helena.

'She is Morton's mistress, and would do anything for him. I followed you this morning – I left the castle in time to see the carriage disappearing into the distance, so I saddled a horse and rode after it. I lost it for a time, and rode aimlessly across the moor, but then I caught sight of it heading towards the stage post. When I arrived there, Eldridge told me he had taken you to Mary's cottage. I rode across the moor and saw you setting out in the trap, and I was finally able to catch up with you at the inn. I saw Maria clearly as I relinquished my horse to one of the ostlers, and I recognized her at once. I assumed you must be in league with her.'

'And so you followed me into the inn.'

'Yes. I wanted to know who you were and what you had told her. I also wanted to know where Morton was. I knew he would not be far from Maria.'

'I have met him, I think,' said Helena. 'He is tall, with dark hair and a slack face?'

'Yes. That is the result of the drink. But how did you meet him?'

'He and Mary – Maria – took a cottage near the castle. I met Maria when I was out walking one day. It came on to rain and she invited me in to shelter.'

'That was a fortunate meeting for her. Otherwise, if she had not come across you accidentally, she would have had to approach you at church.'

'She introduced me to Morton, but she said he was her brother. She tried to find out where my aunt had gone by saying Aunt Hester had left a book in the cottage. She asked for a forwarding address.'

'Ah, that was clever.'

'I knew the talk of the book was a ruse, but at the time I thought—'

'Yes?'

'I thought that you had done away with my aunt, and I thought Mary was worried about my aunt's disappearance, too.'

191

'And now? Are you convinced I have not done your aunt any harm?'

'Yes, I believe I am.'

He smiled. She had never seen such a smile on his face. It was like a sliver of sunshine on a squally day.

'May I know your name? Your true name?' he asked.

'It is Helena.'

'I am pleased to meet you, Helena Carlisle,' he said. 'Even though you have caused me a great deal of unease over the last few weeks,' he added with a wry smile.

'Did you suspect I was an impostor from the beginning?' she asked curiously.

'No, not then. I thought there was something you were hiding, but when I saw you had no wedding ring I thought you were lying about being married. You seemed very young, and I thought you had lied about your age in order to secure a position.'

'I was dismayed when you looked at my finger,' she said. 'I began to realize it would be difficult to keep up the pretence. When did you begin to suspect I might be in league with your brother-in-law?'

'It is difficult to know. From the beginning, I felt that something was not right, but I did not know what. You were more outspoken than my previous housekeepers, and you asked more questions. Then, too, you seemed very interested in the attic. It could just have been a natural desire to explore the castle, or indeed it could have been fear: you could have heard the rumours about ghosts and wanted to set your mind at rest. But when Miss Parkins found the handkerchiefs in your room embroidered with the initials C we began to be more suspicious, although we still could not be certain. Neither of us realized the truth, that the C stood for Carlisle. We both thought it stood for a Christian name: Catherine, Caroline ... we thought of every name, and tried to remember if Morton had known any women whose names had begun with C. We could not recall one, but he had many women and we did not know them all.'

'I was alarmed when I knew Miss Parkins had been in my room, and it was worse when I realized she had found the handkerchiefs,' said Helena. 'She made me feel like a mouse being watched by a cat. I was frightened of her.'

'And you were wise to be so. She is a formidable adversary. She came to the castle with my mother, when my mother was a bride, and she was devoted to her, as she was devoted to my mother's children. If you had meant harm to my sister, Miss Parkins would have stopped at nothing to protect her.'

The carriage turned to the right and ahead of them there was softer countryside. They were leaving the moors behind. In the distance, a town could be seen. Smoke was rising from the chimneys. To the people who lived there, it was an ordinary day. They were paying visits, shopping, visiting the circulating library, going riding, attending to business ... but to Helena, it was a day of hope and revelation.

'Why did you not turn me out of the castle?' she asked, as the carriage finally turned into the main road and bowled along between houses and neatly kept gardens.

'Because if you were innocent, I did not want to deprive you of your livelihood, and if you were a pawn of Morton's I wanted you close by, so that I could watch you. And so I bided my time.'

'Until I showed my hand by taking your carriage?'

'Yes.'

'And so now what do we do?'

193

'We go to Hull. We hope to arrive before Maria and Morton. We tell Anna she can delay no longer, and we put her and her son on a ship bound for Italy.'

CHAPTER TWELVE

They fell into a companionable silence. Now that she was no longer afraid of him, Helena felt herself relax in his company. There was a softer side to him, one she had only glimpsed, but one she would like to know better.

But that was unlikely, she told herself. She would have no reason to remain at the castle once she had found her aunt. She would have to leave, and, unless she visited her aunt in the future, she would never see him again. She felt her spirits sink. He had become important to her, and she could not bear to think of their parting. She looked out of the window in an attempt to distract her thoughts.

They were now travelling along busier roads, which were better kept than those on the moors. There were fewer ruts, and the potholes were not so frequent. Helena watched the scenery change, going from countryside to town, and finally to coast. As the coach crested a hill, she saw the sea sparkling blue and placid beneath her. There was the cry of sea gulls, and the smell of salt was in the air. She licked her lips, and found that she could taste it.

'We are almost there,' he said.

Helena felt her interest quicken. She would soon see Aunt Hester again! And she would see Lord Torkrow's sister. She wondered what Anna would be like. She tried to remember the portrait, but she could remember very little, and besides, Anna had been much younger there. She was dark, that much Helena remembered, but little more.

The carriage rolled to a halt outside a small cottage. The coachman opened the door and let down the step. Helena climbed out. She was immediately hit by the wind, which tried to whip the bonnet from her head, and she held on to it, to prevent it blowing away. Her cloak was whipped around her ankles, and the sound of the wind battered her ears.

'It is often blustery on the coast,' he said, following her.

They went up to the door. As he lifted the knocker he looked all round, and Helena, too, was vigilant, knowing that, at any moment, Maria and Morton could appear.

The curtain moved a little and Helena caught sight of a face at the window. It was an elderly woman with an anxious look. As soon as the woman saw Lord Torkrow, however, her look of anxiety faded and was replaced with a smile. A minute later the door opened.

'My lord,' she said. 'We did not expect to see you.' Then her smile faded. 'Is something wrong?'

'I think we had better talk inside,' he said.

He stood back so that Helena could precede him. The woman gave her a curious look, but no more, and Helena found herself in a cosy hall. There was a staircase rising to her left, and ahead of her was a door leading to the back of the house. Beyond the staircase was another door, and it was through this that Helena was shown. She found herself in a whitewashed room, which was furnished with a blue sofa, blue curtains and a blue rug. Standing by the sofa, her knitting abandoned, was Aunt Hester.

'Aunt Hester!' said Helena, going forward and taking her hands.

'My dear Helena!' cried her aunt in astonishment. 'What are you doing here?'

'I wanted to see you, and I arrived at the castle to find you had gone.'

'My poor Helena. You went all that way for nothing. It was good of Lord Torkrow to bring you here,' she said. She looked searchingly at Helena. 'I can see from your face there is a lot you have to tell me, but now is not the time. I fear you bring bad news with you?'

'We do,' said Lord Torkrow. 'Morton is on his way here. Where is Anna?'

Aunt Hester glanced at a door behind her, and Lord Torkrow went through. Helena followed him into a small room at the back of the house. It was brightly furnished, and there was a cheery fire in the grate. A woman and a little boy were playing on the rug in front of the fire, where the boy was lining up a row of toy soldiers. The woman looked up.

'Simon!' she exclaimed, jumping up in delight and embracing him. 'I am so glad to see you.'

As Helena witnessed the warmth between brother and sister, she thought that she had never known he could be so affectionate, and that to be loved by such a man would be something indeed.

Anna's face became anxious. 'My husband has found me?'

'Yes, my dear, he has.'

'I know it would happen,' said Anna. 'I am only grateful it took him so long. At least I have had time to regain my health.'

She looked at Helena.

Quickly, Lord Torkrow introduced Helena and explained her presence, then said, 'There is no time to lose.'

Anna nodded, her expression grave.

'We must go. Once we are out of the country we will be safe from him. I have a bag packed already. I have had it packed ever since I arrived. Come, George, it is time for us to go on our journey.'

'To stay with Godmama?' asked the little boy.

'Yes, to stay with Godmama in Italy.'

'I'm going on a ship,' the boy said to Helena. 'A big ship. It's going to take me over the sea to a hot country where there are lots of flowers and we're going to live by the sea. We can go out, not like here. Here we have to stay indoors.'

A shadow crossed his face.

'But when we are in Italy, we will not have to hide any more,' Anna told him.

She took him out of the room to put on his outdoor clothes.

'You know what you do when he gets here?' said Lord Torkrow to the Vances.

'Yes, my lord, that we do. We'll send him to the cemetery so that he can see the grave.'

Helena looked at Lord Torkrow questioningly.

'I knew he would never let Anna go, so I had my stonemason make a headstone for her and my nephew,' he said. 'One night, we placed it in an out of the way corner in the graveyard. When her husband arrives, the Vances will tell him that she came here with the boy, but that she died soon afterwards and the boy followed her to the grave.'

'Will he believe it?'

'I think so. He beat them both badly, and it was snowing when they escaped. It is not unlikely that they would have caught a fever on their flight and, already weakened by the beating, have succumbed. It is, at all events, worth a try.'

Anna and George returned, ready for their journey. George

was prattling happily about Italy as they all went out to the carriage.

Helena saw that Aunt Hester was dressed for a journey.

'I am going with Anna, as her chaperon,' she said, 'but once I've seen her safely to her godmother's house, I'll return. Will you wait for me at the castle? Then I can tell you everything.'

'Of course,' said Helena.

She sat next to her aunt in the carriage. Anna sat opposite them, with Simon next to her, and little George on her knee.

'There's a ship sailing to Italy this evening,' said Anna. 'I've taken notice of their comings and goings, and I always know when the next ship will sail. I would have left soon anyway. I have regained my health, and I was only waiting to regain my strength completely before leaving.'

'But you are strong enough for the journey?' asked Simon in concern.

'Yes. I have Mrs Carlisle with me. We will manage.'

The cry of the gulls became louder as they approached the sea. The wind stiffened and the carriage swayed from side to side. The streets became busier. Carts were heading for the dock, laden with sacks and barrels. Women in rough skirts and thick woollen shawls jostled seamen, who swore and cursed and spat. There was a smell of fish, overlaid with the pervading smell of salt, and there was a clamour of creaking rigging which mixed in with the clatter of wheels on cobblestones and the sound of sailors' cries.

As they neared the water, their progress became much slower as Eldridge picked his way between carts and carriages, avoiding urchins and stray dogs, until he finally stopped by the shipping office.

The door opened and the step was let down. Simon climbed out first, looking round with alertness in case Maria had caught up with them. Anna followed, with George, then Helena stepped out with Aunt Hester.

A cry of 'Anna!' rent the air, and Helena looked round in horror, but it was not Maria. It was an elderly woman, who was waving to a young girl, another, different, Anna.

Helena breathed a sigh of relief, and they went into the office.

The arrangements were soon made. Anna's ship was to sail with the evening tide. As they emerged once more on to the dock, they saw the ship not far off. Looking around all the time, they crossed the dock and reached the vessel.

As she set foot on the gangplank, Helena felt it sway in the breeze and she clutched the rope, provoking laughter from a sailor nearby. Regaining her dignity, she ran up the remainder of the gangplank and was relieved to be on the ship.

Interested in all she heard and saw, she accompanied the small party to Anna's cabin. It was surprisingly well appointed, and George ran round it in delight.

'Will you stay with me until I sail?' Anna asked Simon.

'Yes. There are some things I must see to, first, but I will not leave the ship.'

She glanced at Helena, then nodded and gave her attention to her son.

'Will you join me?' said Simon to Helena.

'Of course,' she said.

They did not go on deck; they would be too obvious should Maria arrive; but they stood outside the cabin, talking.

'I want you to return to the castle,' he said.

'But would it not be easier for me to stay with you? The ship sails in a few hours, and we could return together.'

'I do not want you here. If Morton finds us, things could get ugly. It is bad enough that I cannot protect Anna from such a scene, if he arrives. I will not have you exposed to it. Take the carriage. Return to the castle. I will hire an equipage once the ship sets sail and join you later. And then we will have much to discuss.'

Her dismissal, and her wages, she thought, with a sinking feeling. Her time at the castle was drawing to its end.

'Very good, my lord.'

Helena returned to the cabin to take her leave. She embraced Aunt Hester, wished Anna a safe journey, and then she left the ship. As she set foot on the gangplank she looked around the dock, but there was still no sign of Maria. Bracing herself for the swaying underfoot, she succeeded in reaching the dock without difficulty, and then she went over to the carriage.

'We are to return to the castle,' she said. 'His lordship will follow.'

Eldridge nodded, then Helena climbed into the carriage and it pulled away, the horses' hoofs clattering on the cobblestones as they left the harbour.

Simon watched the carriage as it pulled away, and found himself wishing he was going with it. He wanted to see Anna safely on her way, but he wanted to be with Helena, too. He could finally acknowledge his feelings, now that he knew who she was, and he found they were even deeper than he had suspected. But to acknowledge them and to welcome them were two different things.

He returned to the cabin. Mrs Carlisle was playing with George, and Anna was watching them. She turned her head as he entered and said: 'Has she gone?'

'Yes. I have sent her back to the castle.'

'I feel in need of a breath of air,' said Anna.

'You cannot go on deck, it is not safe.'

201

'In the corridor, then. You will come with me?'

He agreed, and together they went into the corridor.

'You like her,' she said, when they were alone.

'Yes. I do,' he said.

He had never been able to keep anything from Anna, even when they had been children, for she had always known what he was thinking, and it was a relief to say the words out loud.

'I am glad,' she said with a sigh. 'You have suffered too much, Simon. I think you should marry her, and be happy at last.'

He shook his head.

'Love does not bring happiness,' he said, his mood darkening.

'It brought happiness to Richard.'

'It brought him torture!' he returned.

'I don't understand you,' she said, puzzled.

'You did not see him as I did, Anna. It was I who had to carry him the news; it was I who had to tell him I'd killed his wife. It was I who destroyed his world.'

She touched his hand.

'It was an accident, Simon. You cannot blame yourself.'

'Can I not? If I had not returned to the castle when I did, she would still be alive.'

'You could not know she would run out to greet you.'

'For a moment . . . one moment . . . I thought she wanted to see me,' he said, as he remembered his elation at seeing her, and seeing the smile on her lips. 'But she thought it was Richard, returning from the Doyles. It was dark, the horse was startled. . . .'

'You could not help it,' she said gently. 'No blame attached to you.'

'I remember it all so clearly. It is etched on my memory. I

can still see her running up to my horse and being knocked aside. I can see her falling, I can remember how I felt as I leapt from the saddle and tried to catch her, but it all happened so quickly, and before I knew what was happening she was lying on the ground with a trickle of blood wetting her hair. If there had not been a stone just there, where she had fallen, she might still be alive, but it was jagged and she hit her head . . . I can still remember my anguish when I knew she was dead.'

'You loved her,' said Anna quietly.

'No,' he said. 'I simply thought I did. I picked her up and carried her inside, fancying my feelings the grief of love, but when I told Richard . . . when he understood she was dead . . . I saw the pit of hell open up in his eyes. I had never, until that moment, known what love was, but I knew it then, and it terrified me. My feelings had been but a pale reflection. I decided at that moment that I would never fall in love. I never wanted to open myself up to such pain.'

'We cannot choose where or when we will love,' she said softly.

'I choose, and I have chosen.'

'Then I pity you,' she said sadly, 'for if I found love, I would not let it pass me by.'

'I cannot love her,' he said, wrestling with himself.

'You *will* not. That is a different thing. Don't let Richard's grief destroy you,' she said, stroking his cheek. 'You have been like a ghost for long enough. It is time for you to live again.'

But he only shook his head.

'If I love her, one day I will lose her. I cannot bear that pain.'

CHAPTER THIRTEEN

Helena leant back against the squabs. She felt suddenly tired, as all the excitement of the day caught up with her: waking early, finding the secret room, begging Mary for help, finding out Mary's true nature, discovering that Simon was not a monster, meeting Anna and her child, and finding her aunt.

She felt the tension that had been gripping her for the last few weeks fade away. Her muscles relaxed, and she felt at peace in a way she had not for a long time. Aunt Hester was not missing, or dead, she was safe. Helena recalled her aunt's face, cheerful and healthy, and she smiled.

The carriage left the sea behind. The cry of the gulls faded, and the tang of salt grew less marked until it disappeared altogether. The view outside the window changed from blue to green, and Helena found her thoughts moving forwards again. In a few hours she would be back at the castle, and then she would resume her masquerade as the housekeeper. She thought of revealing the truth to Mrs Beal, but it would involve divulging secrets which were not hers to tell. She would have to play her part for a little while longer.

Would Simon let her stay until her aunt returned from Italy? she wondered. Aunt Hester had seemed to think so. And yet he might ask her to leave at once. She could not bear the thought of it. He had become a part of her life, and although he had often unsettled her as well as intrigued her, she could not bear the thought of being without him.

The carriage stopped to change horses. Helena alighted, and ordered a cold collation, for she was hungry after her exertions, then she was once more on her way.

It was not long before the carriage turned off the main thoroughfare and began to cross the moors. She was nearing the castle. She looked out of the window, tracing the landmarks of her first journey: the twisted tree, the dry stone wall . . . and then the castle came in sight. It looked less threatening than it had done when she had first arrived. Then, it had been unknown. Now, it was the place where she had lived for many weeks, and although it had held terrors, it had held pleasures, too. She remembered the warm kitchen, the beauty of the ballroom, the music, waltzing with Simon in the gallery. . . .

Simon! She had thought of him by his Christian name, but she had no right to call him that, not even in the privacy of her own thoughts.

Had he been in love with his sister-in-law? she wondered, or had that just been a rumour? Had he any intention of marrying Miss Fairdean, or one of the other young women at the ball? Or would he remain alone?

The carriage passed under the arch and came to a halt outside the castle. Eldridge opened the door and let down the step, and Helena climbed out. As she entered the hall, she saw signs of the ball everywhere and with a shock, she realized that it had been held only the day before. So much had happened that it seemed like a week ago. From the corridor to her left she heard the sound of tables being moved and glasses clinking. The servants were busy clearing away everything they had had to leave the night before.

Helena went to her room and removed her outdoor things, then went downstairs. She followed the sound of clinking and went into the dining-room, where a scene of chaos met her eyes. The footmen were chaffing the maids, who were making a half-hearted attempt to pile glasses on trays.

'There will be no ball this evening until the work has been finished,' she said briskly, becoming the housekeeper once again. 'Manners, take the spare chairs to the attic,' she said. 'Dawkins, take the crockery and glasses down to the kitchen, and be quick about it. All the plates and glasses have to be washed. Martha, you will have to help Effie.' Martha pulled a face. 'It is no use looking like that, this should have been done hours ago.'

'We couldn't find you. . . .' began Martha.

'That is no reason not to get on with your work. I have had other things to see to,' said Helena. 'Now, quickly. This must all be done before Lord Torkrow returns. Have any of the guests been downstairs yet?'

'No, not a one. Never get up early after a ball, that's what their servants say. Up and down they've been, with trays, though.'

'Good, it will make it easier for us to finish our work if they remain in their chambers. Once everything is finished, we can change for our ball.'

Talk of their own costume ball brightened the servants, and they became busy, clearing away the remnants of the previous night's festivities. Spare furniture was returned to the attic, the tables were cleared, cloths were removed and spirited away, the floor was swept, odd wigs, gloves and shoes were put safely aside, and the room slowly began to return to normal.

The clock struck five. The overnight guests were now downstairs, and Helena explained to them that Lord Torkrow had been called away on urgent business, but that she had laid out a cold collation for them in the dining-room.

Some of them expressed their intention of leaving after the meal, whilst others intended to stay. They were subdued, however, and many of them were nursing sore heads and stomachs. After eating, those who did not leave the castle retired to their rooms.

Once more, the servants were free to devote their attentions to finishing clearing the ballroom, supper-room, hall and minstrels' gallery, which were littered with debris from the party, and it was seven o'clock before everything was restored to a semblance of its former state.

The remaining few guests were served supper on trays in their rooms, and then the day's work was done.

'Reckon we deserve our ball,' said Martha.

'You certainly do,' said Helena. 'A fiddler will be here in half an hour. I suggest you all go and put on your costumes, then assemble in the servants' hall.'

As the servants dispersed, Helena went upstairs to change. She did not put on her costume, having no taste for revelry, but simply changed into a clean dress. As she did so, she could not help remembering the previous night, when she had danced with Simon. As she thought of it, she remembered the feel of his fingers on hers, and the weight of his hand on her waist.

She pushed such thoughts aside and brushed her hair, then wrapped it into a neat chignon.

There was a scratching at the door, and Effie entered with a bucket of coal. She was sniffing, and when she put the bucket down, she wiped her nose with the back of her hand.

'Leave that! Go and change, or you will be late for the ball,' said Helena. 'And make sure you wash first,' she said, eyeing Effie's hands dubiously.

'Can't go. Not invited,' mumbled Effie.

'What do you mean, you're not invited? Of course you are!' said Helena. 'All the servants are invited.'

'Dawkins says it's not for scullery maids.'

'What nonsense. And besides, you are not going as a scullery maid. What costume have you chosen?'

'Haven't got one.'

'Did you not look in the chest?'

'All the lasses' costumes've gone.'

Helena felt exasperated at the girl's lack of initiative, but nevertheless she spoke kindly.

'Then we had better look in the attic. There are chests of clothes up there. We are sure to find something to fit you. I think you should go as Cinderella. You spend your days among the cinders.'

Her humour did not make Effie smile. The girl looked more woebegone than ever.

'I couldn't, mum,' said Effie. 'Mrs Beal'd give me what for if I went through the things in the attic.'

'Not if you are with me,' said Helena.

Then, taking Effie firmly by the hand – the one the girl had not used for wiping her nose – she led her up to the attic and together they looked through an old trunk.

'Now, what do you think of this?' she asked, as she held out a panniered gown.

'I couldn't wear nothing like that. That's for a lady,' said Effie.

'Tonight, you are a lady,' said Helena. She picked up the dress, and led the girl downstairs again. 'Now, you just have a wash.'

Effie needed a great deal of encouragement, but in the end she stripped down to her chemise. Helena breathed a sigh of relief when she saw it was clean. Mrs Beal evidently took a motherly interest in the girl. Then Effie washed at Helena's washstand, before putting on the gown. Helena helped her to fasten it, before turning Effie round so that she could see herself in the cheval glass.

'You look beautiful,' said Helena.

Effie looked at herself in amazement.

'Like a real lady,' she said, plucking at the dress in wonder.

'Here,' said Helena, handing her a wig. 'You will be Lady. . . .' she trailed away, then asked, 'What is your favourite name?'

'Charlotte,' said Effie promptly.

'Then tonight you will be Lady Charlotte. If anyone asks, you must give that as your name.'

Effie scratched her head, knocking her wig and giving it a lopsided appearance. Helena straightened it again and said, 'No scratching.'

'No, mum.'

'Now, go downstairs, Lady Charlotte, and enjoy yourself.'

Effie walked out of the chamber, picking up the bucket of coal as she passed.

'Not tonight, Effie. I will see to the fires,' said Helena.

She took the bucket of coal and put it back on the hearth.

She waited for Effie to reach the end of the corridor, and then followed her, to make sure the other servants did not tease her, but she need not have worried. There was already a mood of jollity in the servants' hall, and Effie was swept into a dance by a young man dressed as a pirate. Who he was, Helena did not know. One of the footmen, probably. There seemed to be an awful lot of them, and she guessed that some of the visiting servants and possibly some of the villagers had sneaked into the ball under cover of a costume. A few extra tankards of ale would be drunk, and a few extra sandwiches eaten, but no harm would be done. In fact, some good might come out of it, because the villagers might lose some of their superstitious fear of Simon and the castle.

The fiddler was scraping his bow across his fiddle, and stamping his foot to provide a drum. Helena found herself caught round the waist, and was soon whirling round with the other dancers, not stopping until her partner released her to get a tankard of ale.

'Well, this is fun and no mistake,' said Mrs Beal, joining her at the side of the room. Mrs Beal was dressed, in rather unlikely fashion, as a nymph. 'This dress is very tight,' she complained, 'but it was the only costume left.' She surveyed the dancers. 'Who's the young girl in the panniered gown?'

'That's Effie,' said Helena.

'Why, bless my soul, the girl's light on her feet. I never would have thought it. Just look at her!'

Effie was lifted from the floor by her partner, then dropped carefully back to the ground, where she twirled lightly around before swapping partners.

'There'll be some surprised faces when she takes her mask off,' said Mrs Beal. 'Is everyone here?'

'Eldridge is not coming.'

'That doesn't surprise me. He's never been one for fun and games.' She looked round again, then said, 'I'd best encourage everyone to eat something. There's a deal of ale being drunk. We don't want sore heads in the morning.'

Helena left the noise of the dance, and went upstairs. She looked into the main rooms and made sure the fires were ablaze, for Lord Torkrow would be cold when he returned home.

The long-case clock in the library struck ten o'clock.

He could be here soon, she thought. The ship sailed at seven o'clock.

She hoped that all had gone well, and that Anna and George had set sail, escaping England and Morton for ever.

She went over to the window and pulled back the curtains. The moon was high, and cast a silvery glow into the world below. The stars were out, and it was a magical scene. The moors, which had once seemed threatening, now seemed serene.

The room overlooked the front of the castle, and she strained her eyes, hoping to see a speck in the distance, something moving, which could be a carriage, but she could see nothing. She sat on the window seat, with her knees pulled up in front of her and her arms wrapped round them. She was at last rewarded. In the distance, she caught sight of movement. Simon was coming!

He must not find her in his library. She had no right to wait for him. She must return to the ball.

She left the window seat and pulled the curtains, then went out into the hall. She was halfway across when one of the footmen, dressed as a cavalier emerged from the direction of the kitchen. His costume gave him a swagger, and it was not surprising, thought Helena. His bucket boots, with their deep turned-over tops, were made for swaggering, and so was the extravagant costume, with its doublet and breeches in bright blue satin, and its falling collar of white lace. He wore a short cloak, which he had thrown back over one of his shoulders in a careless attitude. Over his face he wore a black silk mask, and on his head was a wide-brimmed hat with an extravagant plume.

'Are you taking the air?' she asked him.

He raised his hand, and she saw that he was holding a pistol.

'You have picked the wrong person to rob,' she said, entering into the spirit of the masquerade. 'I have no money.'

He said nothing, and something about his stance made her falter. The hairs on the back of her neck rose, and she had the unsettling thought that the pistol could be loaded.

'You thought you were very clever,' he said.

She felt a chill as she heard his voice. She had heard it before, in Mary's cottage. The man before her was Morton. But what was he doing here? How did he get in?

She had no time to wonder. He was walking towards her.

She began to back away from him. There was a bell on the wall behind her. In the confusion of the evening its summons might be ignored, but it might be answered, if only she could reach it.

'What do you want? How did you get here?' she asked, hoping to distract him so that he would not see what she was doing.

'You know what I want. I want my wife and son.'

'Your wife and son are dead, killed by you and your cruelty. You can see their gravestone in Hull. They died shortly after they arrived there.'

'I've already seen the gravestone. I've dug up the ground beneath it, too, and seen there was nothing there.'

'What?' gasped Helena, horrified that he would do such a thing.

'There were no bodies. My wife and son are not dead. He's hidden them somewhere, and you are going to tell me where.'

'I don't know,' said Helena.

'You were in Hull today.'

'I was abandoned before I got there,' she said. 'Your sister – or should I say, mistress – left me at the inn.'

'But you found your way there with Simon. Don't bother to

212

deny it. I saw his carriage leaving Hull with you inside. At the time I was more interested in finding my wife and child. But after following the false scent laid by my dear brother-in-law, then returning to the Vances to punish them for telling me lies, only to find they had fled in the meantime, I decided to visit the castle and find out what I want to know.'

'How did you get in?'

'Simon hasn't changed his habits. The kitchen door was unlocked, and the servants' costume ball was about to begin. A mask, a wig, and a cavalier's clothes let me pass unnoticed. You've been elusive. I thought I would find you downstairs. But it seems you have lost your taste for dancing. You are waiting for Simon, I suppose. And I – I am waiting for an answer to my question. Where are my wife and child?'

'You will never find them,' she said.

He took off his mask and his face was hard.

'Believe me, I will.' He levelled his pistol. 'Now where are they?'

'You would do better to leave them alone. They do not want you,' she returned.

'They do not have a choice in the matter. They are my property. They belong to me.'

To her surprise, she heard Simon's voice cut through the air, and she felt a wave of relief as she realized that he had arrived.

'My housekeeper does not know where they are. I sent her home so that she would not be involved. I have put Anna and the boy beyond your reach, somewhere you will never find them. Now drop the pistol.'

Morton turned and levelled the pistol at Simon instead.

'Not until you tell me where they are.'

'That I will never do,' said Simon, walking forward.

Morton cocked the pistol.

213

'What good will it do to shoot me?' asked Simon. 'If you kill me, you will never find out what you want to know.'

'I will never find out if you live. I know you, Simon. You'd rather die than see your precious sister resume her duties as my wife, so I might as well kill you,' said Morton.

'Then do it,' said Simon.

'*No!*' cried Helena.

Morton turned the pistol back to her, but he spoke to Simon.

'You might play with your life, but I'll wager you won't play with anyone else's.'

'You're bluffing,' said Simon. 'You have only one shot in that pistol. If you shoot her, I will be on you, and I will make sure you pay for your crime.'

'You've gambled with me often enough to know that I don't bluff,' said Morton. 'Besides' – he pulled a second pistol from his cloak – 'I have two pistols. One for her, and one for you. So tell me, Simon, are you willing to let her die? If you are, say nothing. If not, then tell me where Anna is. I'll let her go, and you will have saved your housekeeper's life. You might save Anna's life as well, if you can reach her before me. No one has to die here tonight.'

'Don't listen to him,' said Helena.

Simon's eyes turned to her, and she saw something in them she had not expected: she saw fear.

'I cannot let him kill you,' he said.

And then there came the sound of another pistol cocking behind her, and all three of them stopped in surprise. Helena caught sight of the mirror on the wall. She was standing in the centre of the hall. Morton was levelling a pistol at her. Between them and the front door was Simon. But standing behind Morton, almost invisible in the shadows, was Miss Parkins, and she was holding a pistol to his head. 'You will never have my lady's child,' she said, and her voice was as dead as a sepulchre.

Morton recovered his composure.

'If you pull the trigger, you will kill me, but not before I kill her,' he said.

'Do you think I care about her? A servant?' said Miss Parkins. 'I care about one thing, and one thing alone: the oath I swore to my lady. I promised her I would care for her children. It was I who nursed Miss Anna as she lay feverishly in the castle, brought low by your whippings, and when I saw what you had done to her, I swore that one day I would have the whipping of you.'

She stood there like an avenging demon, and Morton faltered. Helena saw it, and without thinking she knocked the pistol out of his raised hand. He lifted his other hand, but Simon was upon him and wresting the second pistol from him, sending it hurtling to the floor.

'Damn you!' said Morton, as Simon held him fast.

'There will be no bloodshed here,' said Simon to Miss Parkins. 'Give me your weapon.'

She did not respond.

'Anna is safe. Now give me the pistol.'

Slowly Miss Parkins handed it to him, and he put it in his pocket.

Helena breathed again. She was about to pick up the two dropped pistols when a voice came from the shadows.

'Let him go.'

Morton looked round and saw Maria standing there. He wrenched himself free of Simon and ran over to her. He was about to take the pistol she was holding out to him when there was a loud crack! and Helena turned in astonishment to see that Miss Parkins was wielding a whip. It flicked around

215

Maria's wrist . . . Maria, shocked, jerked her hand in an attempt to get it free . . . the pistol went off . . . everyone froze with shock . . . and then Morton's hands rose to his chest as a look of surprise spread across his face.

When he removed his hands, they were covered in blood.

'No!' cried Maria, as he began to fall.

She caught him, and his weight dragged her to her knees.

Helena looked on in horror as Morton's blood seeped across the flagstones.

'Don't leave me!' said Maria.

'Never thought . . . you . . . would be the one to kill me,' he said to her in surprise.

Then his eyes closed, and Maria began to cry.

Helena stood rooted to the spot. It had all happened so quickly that she was still having difficulty in taking it in. It was only a few minutes since she had been sitting in the library, looking forward to Simon's return, and now here she was in the hall, with Morton dead at her feet. Simon was rooted, too. But Miss Parkins was fully in command of herself.

She had the whip all ready, thought Helena, recalling Miss Parkins's words: "When I saw what you had done to her, I swore that one day I would have the whipping of you." *She must have seen Morton arrive, and come to the hall prepared to carry out her threat.*

'Something must be done,' said Miss Parkins.

Simon shook himself, as though clearing his head.

'She must be charged with murder,' said Miss Parkins, looking balefully at Maria.

Maria did not even look up, but went on weeping.

'No,' said Simon, taking the whip from Miss Parkins and coiling it round his hand. 'I will not have Anna's name tainted with scandal, and so the true circumstances of the evening must never come out. Anna has a chance now to come home and to live in England, where she can raise George in peace and safety, and where, when he is older, he can claim his inheritance. I will not have his future ruined by this night's work.'

'What do you mean to do?' asked Helena, looking at him.

'I don't know. I have not decided yet. Say that Morton's death was an accident, perhaps.'

'She will never let your sister live in peace,' said Miss Parkins, her eyes still on Maria.

Helena looked at Maria and saw that she had become quieter. Her sobbing had all but ceased, and now she sat quietly on the floor, looking at the man in her arms.

'I think she will,' said Helena. 'She has too much to lose if she tells the truth.'

Simon nodded.

'Maria,' he said.

Maria turned red-rimmed eyes on him.

'If you give me your word you will never return to Stormcrow Castle and that you will never harm any member of my family, then I will see to it that you go free.'

She nodded dully.

'You give me your word?'

'Yes,' she said.

'Very well,' said Simon. He thought. 'Then we will say this: that you were my house guest; that you expressed an interest in learning to shoot; that Morton said he would teach you; that your aim went wide, and that you shot him by mistake. Do you understand?'

'I do.'

'Good. Then I will send Eldridge for the undertaker. I will send for Sir Hugh Greer, too. I do not want anyone to suspect that anything is amiss.' He turned to Miss Parkins. 'I need you to get Maria out of the hall—'

'You can put her in the housekeeper's room,' said Helena.

Simon agreed. 'And Morton's body, too. We don't want any of the servants coming upstairs and stumbling across it.'

Miss Parkins inclined her head.

He turned to Helena, and saying, 'Wait for me in the library. I would like to speak to you when I return,' he set off for the stables.

Helena walked across the hall, her body feeling heavy. A reaction was starting to set in, and she felt cold. She went into the library, where the fire cast a mellow glow over the furniture and the clock ticked contentedly on the mantelpiece. To her surprise, she saw that only a quarter of an hour had passed since she had been sitting there last, waiting for Simon's return.

She went over to the fire and knelt down in front of it, feeling glad of its warmth. She thought over everything that had happened, until the jumble of images at last began to resolve themselves into an orderly pattern, and she felt her lethargy leave her.

It was some time before Simon joined her. As he entered the library she could see there were lines of strain on his face. She stood up, lifting her hand to soothe them away, but then dropped it again, for she knew she must not touch him.

'I have sent for Sir Hugh,' he said.

'How is Maria?' she asked.

'Still quiet.'

'Do you trust her?' asked Helena.

'No. I think it possible that, once she recovers from Morton's death, she will want revenge, so I propose to have her watched, to make sure she can do no more harm.'

'What will happen now?' asked Helena.

'We will hold the funeral as soon as possible. Morton has no family, so I propose to bury him here. He is my brother-in-law, and it will not seem too strange that I should do so. The funeral will be a quiet affair. I doubt if many people will come, for my neighbours did not know him, and I do not intend to noise it abroad: the sooner it is dealt with, the better.'

'I understand.'

There came the sound of voices from the hall, and the sound of footsteps: Sir Hugh Greer had arrived.

'I must leave you,' said Simon, stepping back.

The door was thrown open and Sir Hugh strode into the room, blowing into his hands.

'Now then, Pargeter, what's all this? There's been an accident, I understand.'

'Yes.'

'It's a good thing your man found me on the road or he wouldn't have got hold of me until tomorrow. I'm due at the Bancrofts' in an hour, so you'll have to be quick. What's happened?'

'Unfortunately, my brother-in-law has been shot. He was showing one of my guests, a lady, how to fire a pistol. She had never held one before, and although she did her best to follow his instructions, her aim went wide.'

'Dead?' asked Sir Hugh succinctly.

'Dead.'

'Never put a gun into the hands of a woman,' said Sir Hugh, shaking his head. 'They mean well, bless 'em, but it's asking for trouble.'

As the two men talked, Helena resumed her role as the housekeeper and quietly withdrew.

CHAPTER FOURTEEN

It was a cold, dreary day when Morton was buried. As Simon had foreseen, few people attended the funeral, and none of them accompanied him back to the castle afterwards, although he had been scrupulous about asking them. There had been a little gossip, but it had soon been overtaken as a subject of interest by news of Mrs Willis's expectation of a happy event.

'We all thought her husband was too old,' said Mrs Beal to Helena at breakfast a few days later, 'but there, she'll be delighted, poor thing. Always wanted children, she did. "It must be awful to be alone in the world," she said to me once. She was thinking of it even then. Two years married and not a sign of a child. But now . . . yes, it's a happy event.'

Helena thought of Mrs Willis and the young man at the ball, and then she thought of Mrs Willis's strange words when they had taken tea together – *It was a pity she was all alone in the world, with no one to miss her when she was gone* – and knew now that Mrs Willis had not been thinking her aunt was an easy target for wrongdoing, as she had suspected at the time, but had simply been thinking of her own situation.

Helena finished her breakfast and then went to the housekeeper's room to start on the day's work. As she went in, she saw *Le Morte D'Arthur* sitting on her desk, and she picked it up, meaning to return it to the library, for she had finished it. As she did so, she thought about the many forms love could take: the courtly love of her book, Maria's love for Morton, Simon's love for Anna, Anna's love for her son.

And then she thought about her own love: her love for her parents, her love for Caroline, her love for aunt . . . and her love for Simon. She could no longer hide it from herself; she was in love with him.

She was crossing the hall when she saw Miss Parkins coming down the stairs. Miss Parkins was dressed in her outdoor clothes, with a long grey cloak covering her bony body, and in her hand was a valise.

'Are you going out?' asked Helena in surprise.

Miss Parkins turned calm eyes on her, and Helena was surprised at the change in them. They looked human at last. Her face had smoothed, as though she had been holding herself rigid for a long time and had finally allowed herself to relax.

'My time here is done,' she said.

'You don't mean you're leaving?' asked Helena in surprise.

'I have done what I promised. I have looked after my lady's children. Her oldest son I could not save; he was dead before I made my vow. But her remaining children will now be happy. Her daughter is rid of a monstrous brute, and her younger son . . . I blamed him for a time, but now all is forgiven. I have forgiven him, and he has forgiven himself.' Miss Parkins walked towards the door, then turned and said, 'I wish you well.'

There was a flicker of a smile in her eyes and Helena saw in her a completely different person; not a terrifying, unnatural mannequin, but a devoted woman who had loved her mistress and who had loved her mistress's children.

It seemed strange to think she had been so frightened of

Miss Parkins when she had arrived at the castle, for now she knew that, although Miss Parkins had been alarming, she had been dangerous only to those who had threatened the Pargeters, and had been dangerous to Helena only whilst she had thought that Helena was a threat.

'And I you,' said Helena. 'Where will you go?'

'To my sister. She lives in Dorset. It is where we grew up. I am looking forward to going home.'

Simon stood on the landing, watching Miss Parkins through the window as she climbed into the carriage and set out on her journey. She had been a part of his life ever since he could remember. She had given him a sense of security in his child-hood, for she had always been there, always the same . . . until the day his sister-in-law had died.

He remembered how Miss Parkins had blamed him; not for the death of his sister-in-law, nor even the death of his brother, but for the way Richard's death had killed his mother.

But now Miss Parkins had forgiven him. And she was right, he thought, as he remembered the words he had overheard, he had forgiven himself. It was as though a great burden had been lifted from him, and now that it was gone, he could look to the future again. A future with Helena.

He began to walk downstairs. He had been determined never to fall in love, because love led to loss, and loss led to pain. But something had happened to him when Morton had turned the pistol on Helena. He had known in that moment that it was impossible to avoid love, because love had found him anyway. But he had known something else, too: that, terri-ble though it would be to lose Helena, it would be better than never having loved her, because the joy and the pleasure of loving her had been worth any pain.

And now he wanted to tell her so.

Helena returned *Le Morte D'Arthur* to its place on the book-shelves and was about to leave the library when Simon walked in. He stopped and looked at her with such intensity that her hands clenched and unclenched themselves. He seemed about to speak, but then he closed his mouth and walked over to her until he was standing in front of her, so close that the front of his coat was touching the front of her dress. She could feel the warmth of his breath on her cheeks and she felt as though something momentous was about to happen.

'Helena . . . there is so much I want to say to you. . . .' he began.

She turned up her face to his expectantly and saw the words die on his lips. His head came closer and her own tilted in response. And then he kissed her.

And her heart quaked.

'I suppose it is too much to hope that the villagers will stop calling you Stormcrow,' she said, as they walked outside in the garden some hours later. Though the day was dull, it was fine, and it felt good to be out of doors.

'It is. But it is not a bad name, and when our children are old enough, I will tell them so.'

She looked at him and he took her hands in his.

'Helena, I'm in love with you. Will you marry me?' he said.

'Yes, Simon, I will.'

He smiled, a natural smile, with no shadows in it, then he put his arm around her and they walked on.

'A stormcrow brings warning of a storm, it is true, but it also flies before the storm and, in the end, outraces it,' he said. 'We are all stormcrows, each one of us, for we all, at some time, bring bad news. But whilst I had to tell my brother that his wife was dead, and my mother that her son had died, I was also the bearer of good news, for I told my sister she was free. Our children will have their own storms and their own sanctuaries in the course of their lives, their own good news and bad.'

'And, if they are lucky, they will find their own loves, as we have,' said Helena.

She thought of the first time she had seen Simon. She had had no idea, when he had taken her up in his carriage, that she would fall in love with him. It had come upon her so slowly that she could not pinpoint the day or the time when it had happened, but it was now so much a part of her existence that she could no longer conceive of life without him.

'A long, long kiss, a kiss of youth, and love,
And beauty, all concentrating like rays
Into one focus, kindled from above . . .
Each kiss a heart quake. . . .

I never thought I would find it, that kind of love, nor my place in the world,' she said, 'but here it is, at Stormcrow Castle, with you.'